VOLUME FORTY NINE

ANNUAL REPORTS IN MEDICINAL CHEMISTRY

VOLUME FORTY NINE

ANNUAL REPORTS IN
MEDICINAL CHEMISTRY

Editor-in-Chief

MANOJ C. DESAI
Gilead Sciences, Inc.
Foster City, CA, USA

Section Editors

ROBICHAUD • DOW • WEINSTEIN • McALPINE •
WATKINS • LOWE • BERNSTEIN • BRONSON

AMSTERDAM • BOSTON • HEIDELBERG • LONDON
NEW YORK • OXFORD • PARIS • SAN DIEGO
SAN FRANCISCO • SINGAPORE • SYDNEY • TOKYO
Academic Press is an imprint of Elsevier

Academic Press is an imprint of Elsevier
32 Jamestown Road, London NW1 7BY, UK
525 B Street, Suite 1800, San Diego, CA 92101-4495, USA
225 Wyman Street, Waltham, MA 02451, USA
The Boulevard, Langford Lane, Kidlington, Oxford OX5 1GB, UK

First edition 2014

ISBN: 978-0-12-800167-7
ISSN: 0065-7743

For information on all Academic Press publications
visit our website at store.elsevier.com

CONTENTS

Section 4
Oncology

Section Editor: Shelli R. McAlpine, School of Chemistry, University of New South Wales, Sydney, Australia

Section 5
Infectious Diseases

Section Editor: William J. Watkins, Gilead Sciences, Inc., Foster City, California

Section 8
Case Histories and NCEs

Section Editor: Joanne Bronson, Bristol-Myers Squibb, Wallingford, Connecticut

CONTRIBUTORS

Michael A. Ackley	27
John J. Baldwin	3
Amelia Black	437
Michel Bouvier	285
Eli Breuer	269
Joanne Bronson	437
Matthew F. Brown	399
Sean P. Brown	77
David Burns	135
Allorie T. Caldwell	87
Monalisa Chatterji	209
Murali Dhar	437
Kirk Dinehart	383
Alexander Dömling	167
James J. Doherty	27
Bruce Ellsworth	437
Natalia Estrada-Ortiz	167
Heather Finlay	101
Mark E. Flanagan	399
Pontus Forsell	59
Stephen Frye	301
Paul Galatsis	43
Jean-Luc Girardet	151
William J. Greenlee	11
Peter D.J. Grootenhuis	383
Vincent Guerlavais	331
Sabine Hadida	383
Yong He	249
Jaclyn Henderson	43
Warren D. Hirst	43
John Hynes Jr.	117
Pravin S. Iyer	209
Ravi Jalluri	135
Sylvie E. Kandel	347
Kareem Khoury	167
Bethany L. Kormos	43
Jed N. Lampe	347
Jonathan Lawrence	249
Christopher Liu	249
Adam R. Looker	383
Gabriel Martinez-Botella	27
John P. Miller	317
Jeffrey N. Miner	151
Peter Mueller	383
Michael J. Munchhof	399
Satheesh K. Nair	117
Shridhar Narayanan	209
Constantinos Neochoritis	167
Pascal Neuville	285
Gunnar Nordvall	59
Gregory T. Notte	417
Gregory R. Ott	189
Reuven Reich	269
J. Robert Merritt	437
Jonathan R.A. Roffey	189
Francesco G. Salituro	27
Tomi K. Sawyer	331
Stephan Schann	285

PREFACE

The *Annual Reports in Medicinal Chemistry* is dedicated to furthering genuine interest in learning, chronicling, and sharing information about the discovery of compounds and the methods that lead to new therapeutic advances. ARMC's tradition is to provide disease-based reviews and highlight emerging technologies of interest to medicinal chemists.

A distinguishing feature of *ARMC* is its knowledgeable section editors in the field who evaluate invited reviews for scientific rigor. The current volume contains 27 chapters in 8 sections. The first five sections have a therapeutic focus. Their topics include CNS diseases (edited by Albert J. Robichaud); cardiovascular and metabolic diseases (edited by Robert L. Dow); inflammatory, pulmonary, and gastrointestinal diseases (edited by David S. Weinstein); oncology (edited by Shelli R. McAlpine); and infectious diseases (edited by William J. Watkins). Sections VI and VII review important topics in biology (edited by John A. Lowe) and new technologies for drug optimization (edited by Peter R. Bernstein). The last section deals with case histories and drugs approved by the FDA in the previous year (edited by Joanne Bronson).

The format for Volume 49 follows our previous issue. We continue with the personal essays, which express the personal stories and scientific careers of "drug hunters"; the volume opens with essays by John Baldwin and William Greenlee Jr. In all, there are 14 therapeutic focused chapters as well as 6 chapters that review important topics in biology and new technologies for drug optimization. The last section includes case histories for the three recently approved drugs, dapagliflozin, ivacaftor, and tofacitinib.

This would not have been possible without our panel of section editors, to whom I am indeed very grateful. I would like to thank the authors of this volume for their hard work, patience, dedication, and scholarship in the lengthy process of writing, editing, and making last-minute edits and revisions to their contributions. Further, I extend my sincere thanks to the following reviewers who have provided independent edits to the manuscripts: George Chang, Margaret Chu-Moyer, Kevin Currie, Dario Doller, Jennifer Davoren, Carolyn Dzierba, Bruce Ellsworth, Gary Flynn, Allen Hopper, Doug Johnson, Jeff Kropff, Mary Mader, Amanda Mikels-Vigdal, Bernard Murray, Anandan Palani, Greg Roth, Joachim Rudolf, Subas Sakya, Adam Schrier, Steven Spregel, Kirk Stevens, James Taylor, Lorin Thompson, and Jeff Zablocki.

Finally, I would like to thank John Primeau, who was the section editor for the infectious diseases section for Volumes 43 (2008) through 48 (2013), for his dedication to identifying the most relevant topics of interest to the medicinal chemists and his editorial expertise.

I hope the material provided in this volume will serve as a precious resource on important aspects of medicinal chemistry and, in this way, maintain the tradition of excellence that *ARMC* has brought to us for more than four decades. I am excited about our new initiatives; I look forward to hearing from you about them and welcome your suggestions for future content.

MANOJ C. DESAI, Ph.D.
Gilead Sciences, Inc.
Foster City, CA

Personal Essays

CHAPTER ONE

A Personal Essay: My Experiences in the Pharmaceutical Industry

John J. Baldwin
Gwynedd Valley, PA, USA

Content

I am a chemist, a medicinal chemist, born in Wilmington, Delaware, the cradle of the U.S. chemical industry, the home of DuPont and its post-monopoly offspring, Hercules and Atlas. Wilmington borders on three rivers, the Delaware, the Brandywine, and the Christiana. The fourth connecting side was dominated by the three research centers of DuPont and its spin-offs.

Like most budding scientists, I had the obligatory chemistry sets. My dream was to work someday in those beautiful research centers in the Wilmington suburbs, surrounded by parks and golf courses. (Of course there were the other dreams, maybe being a Bishop or riding on horseback into the sunset with imaginary cowboy friends, but that's another story.)

When choosing science as a career, it is important to develop a strong foundation in mathematics, languages, biology, chemistry, and physics. It is also important to immerse yourself in a motivating and competitive environment. A strong work ethic is critical. When I attended the University of Delaware as a chemistry major, I carried a full schedule each year while working 24 h a week in the analytical lab at Halby Chemical. My interest in biology was evident in my course selection: biology, physiology, and psychology. My thesis, under John Wriston, focused on the possible fate of one-carbon fragments produced during metabolism. At the University of Delaware, I felt that I had prepared myself to dig deeper into the biological sciences in graduate school. Work hard; study harder.

Annual Reports in Medicinal Chemistry, Volume 49
ISSN 0065-7743
http://dx.doi.org/10.1016/B978-0-12-800167-7.00001-8

For graduate school, I chose the University of Minnesota and was fortunate to receive a teaching fellowship. I majored in Organic Chemistry and minored in Biochemistry, which was structured within the Medical School. My mentor in synthetic chemistry was Lee Smith, a member of the National Academy of Science and an expert in quinone chemistry, especially as it related to vitamin E and coenzyme Q. My minor in Biochemistry required 2 years of heavy course work. My mentor here was Paul Boyer, the Nobel Prize winner. Working with both of these men was a great experience. My research in synthetic organic chemistry and the courses within the Medical School prepared me well for a future in drug discovery. Work hard; study harder.

After saying farewell to graduate school in 1960, I joined the Medicinal Chemistry Department of Merck and Company's West Point, Pennsylvania, laboratory. The company, then known as Merck Sharp & Dohme, had formed through a merger following the discovery at Sharp & Dohme of hydrochlorothiazide, a game changing medication for the treatment of hypertension and, in my thinking, the beginning of modern hypothesis-driven drug discovery.

The Merck laboratory in Rahway, New Jersey, focused on antibiotic chemistry, which had evolved from the penicillin program; steroids, which grew out of cortisone synthesis; and vitamins, stemming from their fine chemicals background. The West Point laboratory was devoted to cardiovascular and CNS diseases, antisecretory/antiulcer agents, atherosclerotic disease, and antiviral agents.

I became involved in many of the West Point programs, including loop diuretics, xanthine oxidase inhibitors, beta adrenergic blockers, vasodilators, dopamine agents for Parkinson's disease, antivirals, and carbonic anhydrase inhibitors for glaucoma. Learning pharmacology was never ending in the Merck environment. Work hard; study harder.

The areas that were especially attractive to me were the mechanistic approach to drug discovery and understanding the biochemistry of disease, using computationally intensive predictive methods and X-ray crystallography of ligand–protein complexes. From this work came Edecrin, Crixivan, Trusopt, Cosopt, and the antiulcer agent famotidine (Pepcid), which I identified in the patent literature and championed through the Merck system. The work, that led to the first topically available carbonic anhydrase inhibitor, dorzolamide (Trusopt), has been described in terms of its design, computational understanding, conformational analysis, and X-ray crystallographic details of the enzyme/ligand structure.[1] This integration of disciplines established a powerful approach to drug design that I repeated in the discovery of the HIV protease inhibitors at Merck and the renin inhibitors at Vitae.

A compound that created excitement within research but failed to reach the market was the topically penetrating, direct-acting dopamine agonist for Parkinson's disease. The compound was administered via patch and effectively controlled symptoms. However, as sometimes happens, the Research Division could not convince the Clinical/Marketing Organization to continue with clinical trials.

In 1993, Merck began to prepare for the twenty-first century and the predicted patent cliff which lay ahead. One step was to decrease staff through a retirement incentive plan. Cut-backs not only pose difficult decisions for management but impose difficult decisions on the people affected as well. Such actions by Merck and other companies marked the disappearance of the lifelong commitment of an employee to a single company and the belief that the commitment of the company to this contract would also be honored. A new era in the company/employee relationship had begun. This new reality became even more apparent over the past decade, and it must be considered by professionals and students alike as they make choices about employment. I adjusted to this apparent evolutionary change by deciding not only to take Merck's offer but to start a new company, Pharmacopeia, which was based on cutting-edge technology not being practiced in Big Pharma. Jack Chabala, also from Merck, and Larry Bock of Avalon Ventures joined me in launching the new company around encoded combinatorial chemistry and high-throughput screening. This technology played a key role in exploiting the genomic revolution and the target-specific approach to drug discovery.

Pharmacopeia perfected the synthesis of encoded solid-phase combinational libraries, which allowed the preparation and screening of large compound collections across numerous *in vitro* assays. These screening collections, built on a common theme, would be followed by smaller libraries focused on the bioactive lead.

Within a few years, the approach was adopted by all pharmaceutical companies, and the ability to generate large screening collections became a technology that was required within every drug discovery organization. The only advantage that Pharmacopeia had in selling what had become a commodity was price and its experience in drug discovery. One obvious solution to the price issue was to move compound production to a lower-cost environment. After first looking into a joint venture possibility in China and finding no interest at Pharmacopeia, I, along with one of our founding scientists, Ge Li, and three others decided to finance a new company, WuXi Pharma Tech. This Contract Research Organization grew rapidly and is the largest company in China servicing the research needs of the U.S. and European pharmaceutical industries.

With WuXi Pharma Tech successfully listed on the New York Stock Exchange, it was time to start something new. That new endeavor was Concurrent Pharma, later Vitae Pharma, which was based on computational methods from Eugene Shakhnovich's laboratory at Harvard. The method defines in detail a ligand–protein complex that was constructed *in silico*, fragment by fragment, along the surface of the target. The computational exercise was followed by chemical synthesis of the predicted ligand and appropriate *in vitro* testing. The predicted complex then could be verified by X-ray crystallography. We found a problem that was ideally suited to test the technology among the aspartic acid proteases, renin and beta-secretase. The approach produced novel, nonpeptidic, and bioavailable inhibitors of both enzymes. A similar approach was used to find inhibitors of 11-beta-hydroxysteroid dehydrogenase, a flexible enzyme that offered an additional challenge. This *de novo* design strategy holds great promise for the future, especially for the virtual company that lacks the internal architecture of Big Pharma.

During this evolutionary period of Vitae Pharma, the drug-discovery capability in China continued to develop such that the time seemed right for a new company not only capable of research support but also of drug discovery, development, toxicology, registration, and human clinical trials, each component present in a virtual organization. So, Bob Nelson of Arch Ventures and I decided to start a new company, Hua Medicine. This China-centric company will move its first drug candidate into human Phase I clinical trials this year, 2013.

Looking back to the start of the biotech era, it is clear that start-up companies continue to be a high-risk exercise. Whether the company is large or small, it still takes years and over a billion dollars to discover and develop a New Chemical Entity (NCE). This forces the small biotech firm to be in a constant search for financing and for a viable exit strategy.

Recent years have not been easy ones, especially for the major pharmaceutical companies, which faced a slowing rate of discovery, major products going "off patent" into the generic class, higher R&D costs, and longer approval times. To combat this, a range of new survival strategies has been developed and adopted,[2,3] including a series of mega-mergers with the resulting cut-back in employment. Since NCEs are the main factor in raising valuation, it is unlikely that the merger strategy will increase value to the pharmaceutical industry. Mergers like Merck/Schering-Plough, Pfizer/Wyeth, Bayer/Schering, and Sanofi/Aventis will, at best, produce only short-term stability. History teaches us that such mergers simply do not increase the productivity on which valuation is based. Such retrenchment stimulated an outsourcing trend that has accelerated over the past decade. The layoffs from mergers and outsourcing reached 130,000 between 2005 and 2008, with the total number now well over 300,000 people.[4,5] Few companies were spared: the top 10 pharma layoffs in 2011, in rank order, were Merck, Pfizer, Novartis, Abbott, Astra Zeneca, Teva, Sanofi, Johnson & Johnson, Eisai, and Bayer.[6] WuXi Pharma Tech, one of the first to recognize the growth in outsourcing and profit from it, went public on the New York Stock Exchange in 2000 and now has over 6000 employees. Along with the outsourcing of science, there has been growth in alliance among large pharmaceutical companies themselves as well as with academic institutions, all with the idea of sharing both cost and risk.[7–9] With the Western markets mature, Big Pharma has turned to the evolving economies as part of its growth strategy. However, emerging markets have shown resistance to patented medication and a greater interest in improving health through the availability of low-cost generic drugs.[10,11]

To overcome the generic competition, Big Pharma has turned to "branded generics." Even with cost only somewhat above the local generic equivalent, pricing remains a problem for Western companies.[12] The cost of patented Western drugs in these economies is being countered by price controls and forced licensing.[13] Considering both of these marketing issues, fast-growing evolving economies may not be the answer for the problems of today's pharmaceutical companies.

With the uptick in new FDA approvals in the past 2 years (2011–2012), it has been suggested that the worst may be over for the productivity decline we have seen in the pharmaceutical industry.[14] However, considering that programs for these approvals were initiated in the late 1990s to early 2000s, it is reasonable to predict that the decrease in productivity expected from the cut-backs will not be felt for at least another 10–12 years.

Things have changed in the pharmaceutical industry since I first walked the halls of Merck. These changes and the resulting strategies discussed here have had a negative impact for those who have worked as scientists on drug R&D and even those who hope to join the exciting adventure of drug discovery. The sky is red, but it is difficult to tell whether it is the sunrise or the sunset. Looking to the future, it is certain that there are many road blocks along the road to recovery. For synthetic organic chemists, the path ahead is not bright but should stabilize as the cost advantage of outsourcing to India and China decreases. In the USA, there will be fewer opportunities for chemists than there will be for biologically trained scientists as the revolution in biotechnology continues. Many of these opportunities will be in industry-sponsored academic laboratories, although the productivity of this industrial/academic strategy may decline over the next decade as Big Pharma increasingly depends on licensing new products from non-U.S. biotechs. Government laboratories, such as the National Institutes of Health and the National Cancer Institute, are unlikely to replace the industrial discovery machine that gave the world new drugs. Therefore, society will become increasingly dependent on generic medications as new drugs constitute a smaller percentage of total sales. With government budget pressures, generic drugs will become commodities with low profit margins. This may create the kind of shortages seen recently due to deteriorating production facilities and poor quality control, which leads to forced closings and recalls by government agencies.

The introduction of biosimilars will be slower than expected, and their price will not decrease as dramatically as in the case of small-molecule drugs. The high prices for niche drugs will come under increasing pressure, as demonstrated by the Sloan-Kettering Cancer Center's experience with the anticancer drug Zaltrap. Similar forced pricing and licensing decisions by India and other countries on expensive, patented drugs will be a growing problem for the Western pharmaceutical industry. Companies will begin to rethink whether it is worth pursuing low-volume drugs where high prices are needed to recover cost. Similarly, society must decide whether it is willing to pay high prices for drugs that extend the life of a very ill person for just a short period of time. These questions pose ethical issues that are difficult to solve.

For young scientists who received their first chemistry set under the Christmas tree, all these issues are only challenges for tomorrow. To play in the sandbox of drug discovery, you must be driven. Work hard; study harder.

REFERENCES

1. Greer, J.; Erickson, J. W.; Baldwin, J. J.; Varney, M. D. *J. Med. Chem.* **1994**, *37*, 1035–1054.
2. Baldwin, J. J. *Future Med. Chem.* **2011**, *3*(15), 1873–1876.

3. Edwards, J. Leaner Isn't Meaner for 9 Big Pharma Firms That Tried to Cut Their Way to Greatness. www.industry.bnet.com/Pharma/10008581/leaner-isnt-meaner-for-9-big-Pharma.
4. Johnson, L. Merck Plans More Job Cuts. *The Reporter*, Jul 30, 2011.
5. Novartis Cuts 1,400 US Jobs. *The Wall Street Journal*, Nov 30, 2011.
6. McBridem, R.; Hallmer, M. http://www.fiercepharma.com/special-reports/top-10-pharma-layoffs-2011.
7. DeArment, A. http://www.drugstorenews.com/article/astrazeneca-broad-institute-partner-antibiotic-antiviral-drugs.
8. PharmaLive. http://pharmalive.com/news/Print.cmf?articleid=877597.
9. Timmerman, L. http://www.xconomy.com/national/2011/06/20/pfizers-idea-to-fix-the-drug-development.
10. Silverman, E. http://www.pharmalot.com/2012/10/India-pushes-to-end-sale-of-branded-drugs/.
11. Staton, T. http://www.fiercepharma.com/node/91359/print.
12. Herper, M. http://www.forbes.com/sites/matthewherper/2012/07/12/The-global-drug-market.
13. Silverman, E. http://www.pharmabt.com/2012/10/australia-plans-to-preview.
14. Christel, M. D. Tide Turns for Innovation. *R&D Directions*, Jan/Feb 10–11, 2012.

CHAPTER TWO

Adventures in Medicinal Chemistry: A Career in Drug Discovery

William J. Greenlee
MedChem Discovery Consulting, LLC, Teaneck, New Jersey, USA

Contents

I was born into a family of chemists. My father, uncle, and grandfather were all chemists. As a result, my brother Mark and I were exposed to laboratories and the basic concepts of chemistry from an early age. We got used to having our father enliven family outings by throwing a lump of sodium into the nearest body of water and having complex chemistry discussions with our grandfather. I got my own start in the laboratory during high school, synthesizing acetylenes using reactions in liquid ammonia, running Grignard reactions, and monitoring large-scale distillations for my father's new catalog company "Chemical Samples Company." During that time, my intention was to study medicine, but I was drawn into the world of organic chemistry around me. I went on to major in chemistry at Ohio State University, doing undergraduate research with Paul Gassman. I completed my Ph.D. with Robert B. Woodward at Harvard University, and after postdoctoral work with Gilbert Stork at Columbia University, I joined Merck Research Laboratories in Rahway, New Jersey in 1977. It was not a surprise that Mark also got his Ph.D. in chemistry and has had a long and successful career as a medicinal chemist.

One thing that drew me to organic and (especially) medicinal chemistry was the concept of creating a new chemical compound that had never existed before. The idea that I could do that, and that the compound might be useful in treating disease was a powerful motivation. Later, when I was working at the bench, one of my favorite occurrences was spotting the first

Annual Reports in Medicinal Chemistry, Volume 49
ISSN 0065-7743
http://dx.doi.org/10.1016/B978-0-12-800167-7.00002-X

crystals growing in a solution of a new compound. Creating a new compound and watching how it forms crystals for the first time was a near-religious experience for me. Although I have been out of the laboratory for many years, I still dream about being back there doing experiments and experiencing the excitement of drug discovery.

I began my career at Merck in the New Lead Discovery group led by Art Patchett, who had also received his Ph.D. at Harvard University with R.B. Woodward. The goal of my first project at Merck was to synthesize halovinyl amino acids as irreversible inhibitors for DOPA decarboxylase (for Parkinson's disease (PD)) and alanine racemase (as potential antibacterial agents).[1] This type of inhibitor was referred to as a "suicide substrate," one that became reactive only after turnover by the enzyme target (also referred to as a "mechanism-based enzyme inactivator"). I was also able to work out a more general synthesis of β,γ-unsaturated amino acids using the Strecker reaction.[2] Drugs with an irreversible mode of action fell out of favor for many years, but irreversible inhibitors of protein kinases have become popular, and there are now many examples of marketed drugs that form covalent bonds with the target protein.

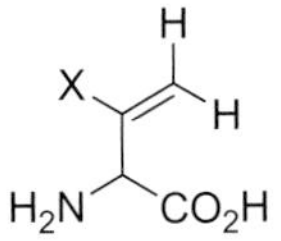

H
X H
H_2N CO_2H

X = –F
Fluorovinylglycine

X = –Cl
Chlorovinylglycine

By the time I arrived at Merck from Columbia, Art had already started a program to identify inhibitors of angiotensin-converting enzyme (ACE), to follow up on earlier discoveries made at Squibb (now Bristol-Myers Squibb). At the time, the antihypertensive drug Aldomet™ was a successful product for Merck, but would be off patent in a few years, and a drug that would treat hypertension with fewer side effects was a major goal. I was asked to join the effort and was privileged to be part of the team, along with Matt Wyvratt, Eugene Thorsett, and others, who discovered enalapril and lisinopril, both of which became important drugs for Merck.[3] For many of us, this was our first experience working with amino acids and small peptides, and doing reactions and workups in aqueous solution. The ACE inhibitor program was successful even though the X-ray crystal structure of ACE was not available. Amazingly, the details of the structure and the fact that ACE has two nonidentical active sites were not known until many years later.

After ACE inhibitors were moving ahead in development, Merck and other companies shifted their attention to renin inhibitors, which block an earlier step in the renin–angiotensin pathway, and were thought to have potential advantages. Merck had started working on renin even before the ACE inhibitor program began, but the initial screen using porcine renin had not yielded viable hits. Like most other companies, Merck pursued a peptidomimetic approach to renin inhibitors, and the team at the Merck West Point site, led by Dan Veber and Joshua Boger, used the X-ray crystal structure of a fungal aspartic protease (Rhizopus pepsin) as a model enzyme.

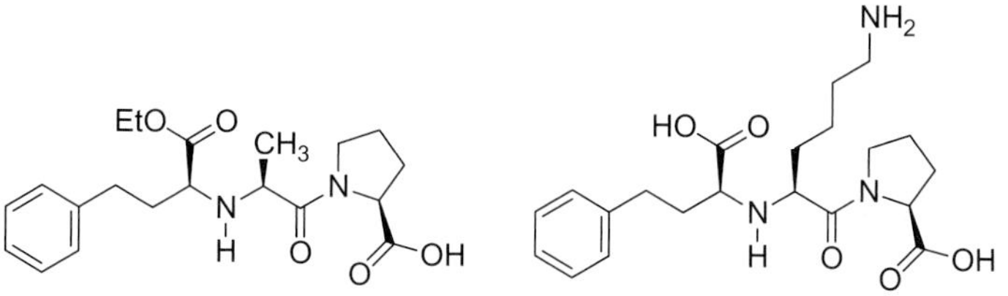

Enalapril (Vasotec™)

Lisinopril (Prinivil™)

Their elegant work led to a potent and selective hexapeptide inhibitor, but oral bioavailability was low, and the program was eventually put on hold. A few years later, we began a new effort in Rahway, now guided by a published X-ray structure of human renin. We made significant progress in reducing the peptide nature of our inhibitors and identifying potent macrocyclic inhibitors,[4] but we could not achieve the desired profile for an orally bioavailable inhibitor. At one point, there were over 20 companies with renin inhibitor programs, but none of the resulting candidates progressed beyond early clinical trials. It became clear that identifying peptidomimetics with drug-like properties would be a difficult challenge. It was over 20 years later (2007) when the first renin inhibitor, aliskiren, was approved for treatment of hypertension.[5]

It was therefore very exciting when DuPont Pharmaceuticals disclosed the first potent nonpeptide angiotensin II receptor antagonist and reported that it had efficacy in a rat model of hypertension. Rather than inhibiting the formation of angiotensin II, the receptor antagonist losartan blocks the action of the peptide at its receptor. With encouragement from senior management, the Merck renin inhibitor program was immediately abandoned, and all effort shifted to the new target. I soon found myself leading a large medicinal chemistry effort with an ambitious goal to identify a development candidate without delay. However, after a few months, Merck announced a new collaboration with DuPont Pharmaceuticals to develop their candidate losartan (Cozaar™). Our medicinal chemistry program and the ongoing program at DuPont Pharmaceuticals led by Ruth Wexler were merged into a combined effort to identify backup and second-generation antagonists. This work led to several backup candidates, including **1**, MK-996, and DuP532 that were derived from the losartan chemotype.[6,7] It had become clear that there are two receptors for angiotensin II, AT1 and AT2, and that losartan blocks only AT1 receptors. A potent AT2 antagonist had been disclosed by a group at Parke-Davis, but despite numerous studies, the signaling pathways and roles of this receptor were unclear (and still are). By modifying MK-996 and losartan, the team was able to identify potent dual AT1/AT2 antagonists such as **2** and XR510, but in the absence of a clear understanding of the AT2 receptor, none of these were taken forward into development.[8] In the meantime, Cozaar™ was launched as a monotherapy and in combination with hydrochlorothiazide (Hyzaar™), and both were very successful drugs for treatment of hypertension. Several other AT1 antagonists (also called angiotensin receptor blockers) based on the losartan lead were launched by competitors and also became successful drugs (valsartan, candesartan, telmisartan, irbesartan, eprosartan, azilsartan, olmesartan).

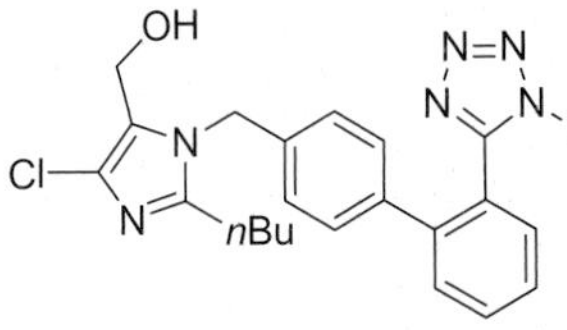

Losartan (Cozaar™)

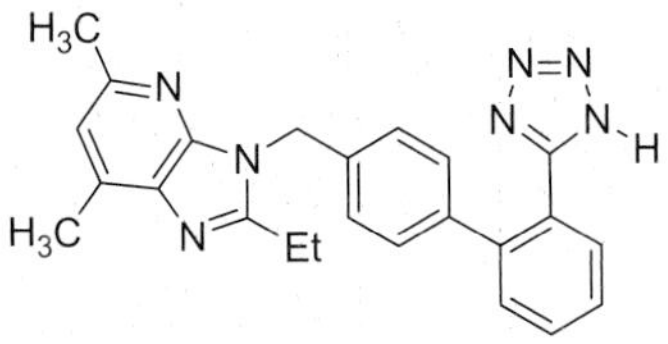

1

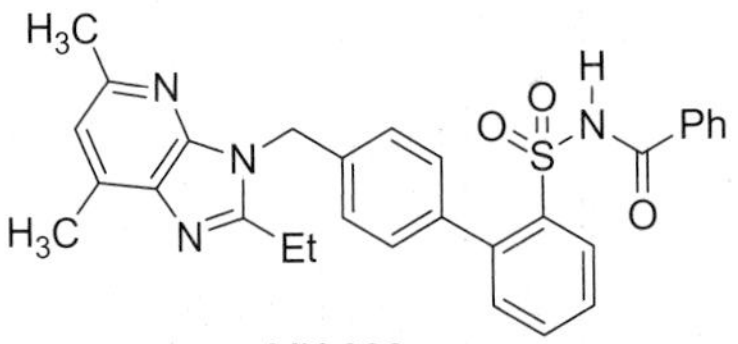

MK-996

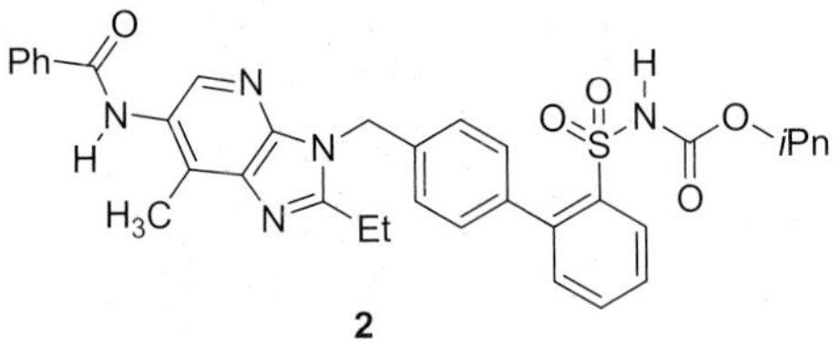

2

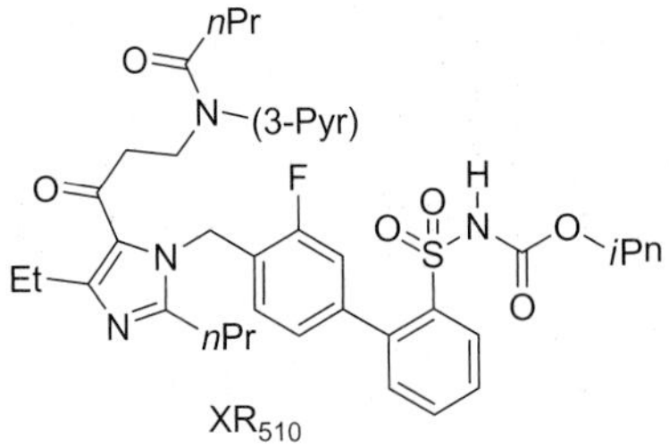

XR_{510}

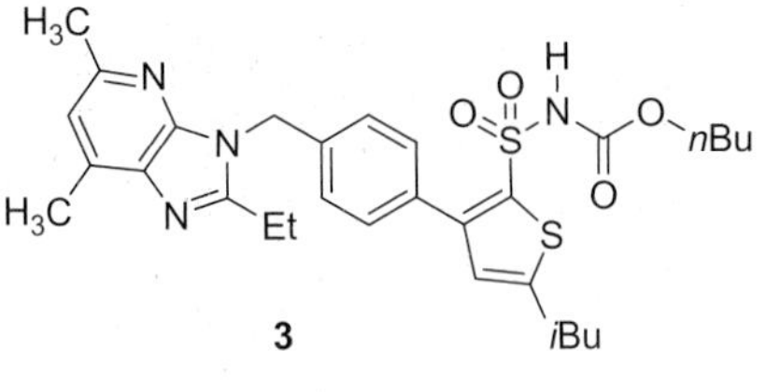

3

One day during our backup program, Pete Siegl, who directed the Merck pharmacology effort at West Point, called to tell me that one of our "antagonists" had raised blood pressure when dosed to rats. In fact, we had discovered the first nonpeptide AT1 agonist, **3**.[9] This compound was found to be a full agonist of the AT1 receptor, with a much longer duration of action than angiotensin II, which is rapidly degraded. Although nonpeptide agonists of peptide G-protein-coupled receptors are now common, at that time **3** was the only such compound identified outside of the opioid receptor field. We were astounded by how small a modification in structure was required to shift an antagonist to an agonist.[10] Recently, **3** became the starting point for design of AT2 selective agonists (without effects on blood pressure) that are being explored for potential utilities.[11]

Soon after we completed our angiotensin II program, the discovery of the vasoactive peptide endothelin was reported, and it was demonstrated to be the most potent vasoconstricting peptide to date. There was hope that blocking the endothelin receptors ET_A and ET_B, or preventing biosynthesis of endothelin by blocking endothelin-converting enzyme, would provide a novel (and possibly superior) approach to treating hypertension. Working from screening leads, we were able to identify potent dual ET_A/ET_B receptor antagonists, including **4**.[12] A few of our antagonists also had activity as angiotensin receptor antagonists, and remarkably we discovered compounds that blocked all four receptors (ET_A, ET_B, AT1, AT2) with nanomolar affinity.[13] While this work did not continue at Merck, others followed up on similar leads and were able to demonstrate reduction of blood pressure in hypertensive patients.[14] A number of selective endothelin receptor antagonists have been developed by others, but (similar to the situation with the angiotensin II receptors) the ET_A receptor appears to be linked to vasoconstriction, while the roles of the ET_B receptor are still unclear. Interestingly, both ET_A selective (ambrisentan) and dual ET_A/ET_B antagonists (bosentan, macitentan) are approved for treatment of pulmonary arterial hypertension.

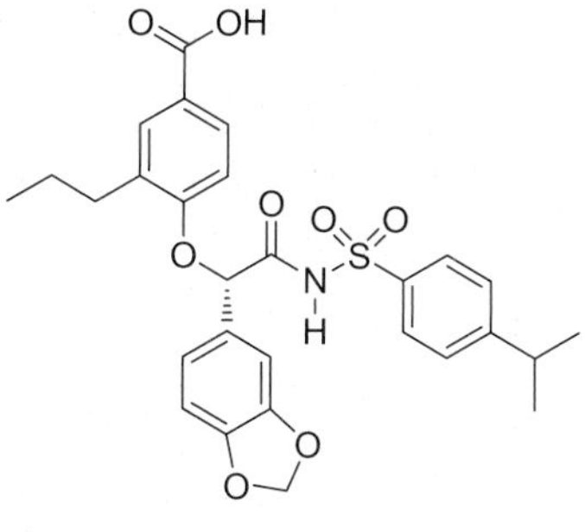

4

Art Patchett's interests in medicinal chemistry were very broad, and working with him provided me and others with broad training in medicinal chemistry. I also benefitted greatly from working with other managers and mentors at Merck, including Burt Christensen, Ralph Hirschmann, and Tom Salzmann. Although I was mostly involved in cardiovascular programs, I also had the opportunity to work on medicinal chemistry programs to discover antibacterial and antifungal agents and several projects in inflammation. Merck was an outstanding place to work, and I was privileged to be there during the "most admired" era of a world-class company.

After nearly two decades at Merck, I accepted a position at Schering-Plough in Kenilworth, NJ, to lead the Cardiovascular and CNS Medicinal Chemistry group. I was excited to join the Schering-Plough team and to work directly with Catherine Strader, who had also recently relocated from Merck. I had been fascinated by neuroscience for many years, and the opportunity to contribute in this area was particularly attractive. I quickly found myself involved in programs for Alzheimer's disease (muscarinic M2 receptor antagonists), PD (adenosine A_{2A} antagonists), and schizophrenia (dopamine D4 antagonists). Many of our CV/metabolic diseases projects (e.g., NPY5 antagonists, MCHR1 antagonists, MC4 agonists) also involved targets residing in the CNS, so most of the programs we took on required drug candidates with good brain penetration. This was a steep learning curve for me, since designing drugs for CNS penetration adds a new dimension of complexity to lead optimization. I was also gaining an appreciation for the liabilities of CNS side effects in otherwise excellent drug candidates.

One of our most challenging CNS programs was to identify a potent and selective adenosine A_{2A} receptor antagonist for the treatment of PD. This project was initiated and actively promoted by Ennio Ongini at Schering-Plough's Neuroscience Center in Milan, Italy. Ennio and his group had produced a very convincing set of data in rodent models of PD for the potent A_{2A} antagonist SCH 58261, which had been designed and synthesized by his collaborator Professor Pier Baraldi at the nearby University of Ferrara.[15] This antagonist, which was only modestly selective versus the A_1 receptor (10-fold), had extremely low aqueous solubility and had to be dosed in DMSO by intraperitoneal injection.

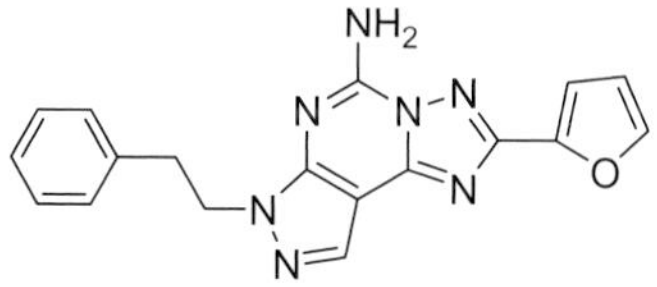

SCH 58261

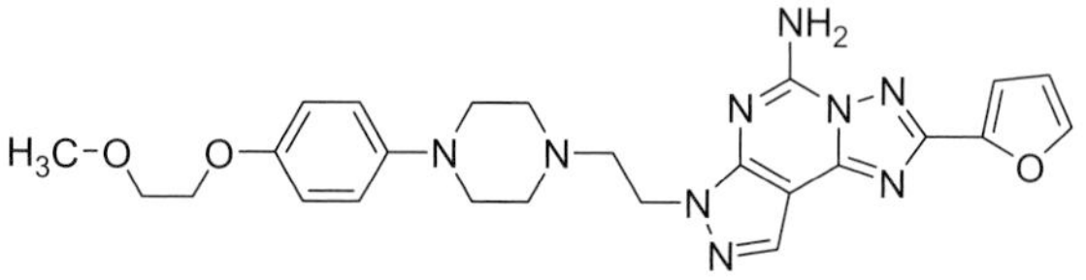

Preladenant

Although improving selectivity was straightforward, increasing solubility while maintaining robust activity in the *in vivo* models turned out to be a major challenge. Many attempts to reduce the complexity of the tricyclic core and replace the furan ring gave antagonists with high binding potency (1–5 nM), but led to loss of activity in the primary rat catalepsy model, despite high free drug levels in the brain. The adenosine antagonist program was not unique in this respect, and similar experiences in more recent programs have convinced me that we do not yet fully understand how to design effective CNS drugs. The CNS multiparameter optimization approach introduced recently by Pfizer[16] may be a step in the right direction. Of several thousand antagonists synthesized for the program, preladenant emerged as our development candidate, and it possessed the properties we believed to be essential for success.[17] Preladenant demonstrated robust activity in rodent and primate models of PD and had excellent pharmacokinetics and receptor occupancy in humans. Unfortunately, although it showed promising activity in Phase 2 studies in PD, preladenant did not demonstrate sufficient efficacy in Phase 3 studies to support continued development.

One of the first projects initiated soon after I joined Schering-Plough was a program to identify a potent thrombin receptor antagonist as a potential antiplatelet agent. Activation of the thrombin receptor, also known as protease-activated receptor-1 (PAR-1), activates platelets and promotes their aggregation as part of thrombus formation. Unlike most G-protein-coupled receptors that are activated by an external ligand, PAR-1 receptors are cleaved at their amino terminus by the enzyme thrombin, creating a new peptide sequence that functions as a "tethered ligand" to activate the receptor. We were extremely fortunate that one of the hits to emerge from high-throughput screening became an excellent lead for the program. The hit was an analog of the natural product Himbacine, which had been synthesized for another program at Schering-Plough. The chemistry team, headed by Sam Chackalamannil, brought two development candidates forward, but each of these ran into a toxicity issue. Fortunately, the team was able to address these issues, and the third candidate, SCH 530348 (vorapaxar), moved into clinical trials for prevention of arterial thrombosis.[18]

I was excited to have the opportunity to co-chair the Early Development Team for vorapaxar with Madhu Chintala, Head of Pharmacology for the preclinical program. After completion of the Phase 3 clinical program for vorapaxar (which enrolled over 41,000 patients), the NDA was filed in 2013[19] and was approved by the FDA in May of 2014. Vorapaxar (proposed trade name Zontivity™) is hoped to provide benefits in secondary prevention of heart attack and stroke.[20]

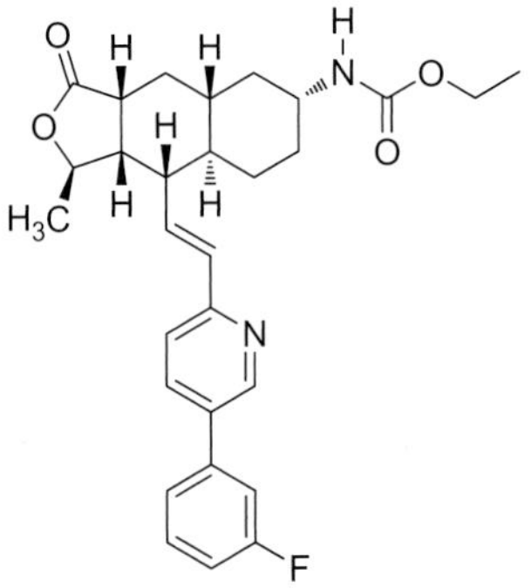

Vorapaxar
(Zontivity™)

Soon after the discovery of beta-amyloid-converting enzyme-1 (BACE-1) in 1999, Schering-Plough initiated a program to discover inhibitors of this enzyme which plays a key role in the biosynthesis pathway of beta-amyloid and formation of the extracellular plaques found in Alzheimer's disease. I was highly motivated to be a part of this effort, since Alzheimer's disease has affected my family, and I had seen the effects of this terrible disease. BACE-1 is an exceedingly challenging target, due to its location in organelles inside neurons in the brain. BACE-1 is anchored in the membrane, but its active site faces an aqueous environment that requires inhibitors with substantial aqueous solubility. At Schering-Plough, we screened our entire sample collection and drew upon external screening resources, but were unable to generate leads for our program. Efforts to work from nonpeptide renin inhibitors in the scientific and patent literature were also unsuccessful. Reluctantly, we began an effort to design a peptidomimetic BACE-1 inhibitor based on the amyloid precursor protein substrate. Although we were able to identify potent inhibitors with oral bioavailability using this approach, these inhibitors were P-glycoprotein (PGP) efflux substrates and showed exceedingly low brain penetration.[21]

Given these setbacks, it was very exciting when Dan Wyss and Yu-Sen Wang in our structural chemistry group presented us with fragment hits that they had identified using HSQC NMR screening. Follow up by X-ray crystallography provided several confirmed hits, including an isothiourea that bound in the BACE-1 active site, making multiple interactions with the two active site aspartic acids. The isothiourea was clearly an undesirable group to have in a drug molecule, and a key insight provided by Zhaoning (Johnny) Zhu was to replace this basic group with a five-membered heterocycle (iminohydantoin) in order to make similar interactions. A large team, led by Andy Stamford, worked to optimize the weak iminohydantoin lead, which later evolved into a six-membered ring series of iminopyrimidinones.[22,23] The BACE-1 program was supported by chemists at Pharmacopeia, and later by chemists at Albany Molecular Research Inc. As with most CNS programs, achieving potent binding affinity was only part of the challenge. Although most potent inhibitors in the series reduced beta-amyloid levels in plasma, reducing them in cerebrospinal fluid (CSF) and (especially) in cortex was much more difficult. We expected that a balance of lipophilicity, aqueous solubility, and low PGP efflux would be essential for good activity in the CNS. However, it also became clear that a high degree of inhibition of BACE-1 was required for CNS activity. Our failure to achieve good inhibition of BACE-1 in the cortex became a big concern, and it was fortunate that we identified an iminopyrimidinone (**5**) that showed good activity in both CSF and cortex with an ED_{50} of 6 mg/kg (po) in rats.[24] SCH 900931, later renamed MK-8931, was shown to reduce A-beta-40 levels by over 80% in a Phase 1 rising multiple dose study.[25] This inhibitor is now in Phase 3 clinical studies in Alzheimer's disease patients. Throughout the BACE-1 program, the medicinal chemistry team received strong support from our structural chemistry group led by Corey Strickland, and from other groups in discovery research. Merck has made a strong commitment to develop MK-8931 in both mild-to-moderate AD and in patients with mild cognitive impairment.[26] If successful, this inhibitor could make a major contribution to the treatment (and possibly prevention) of this terrible disease.

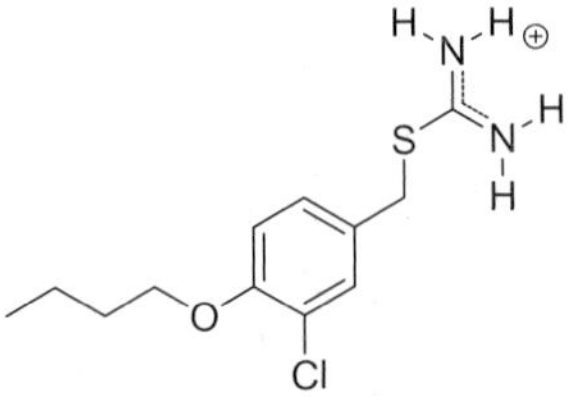

Fragment lead

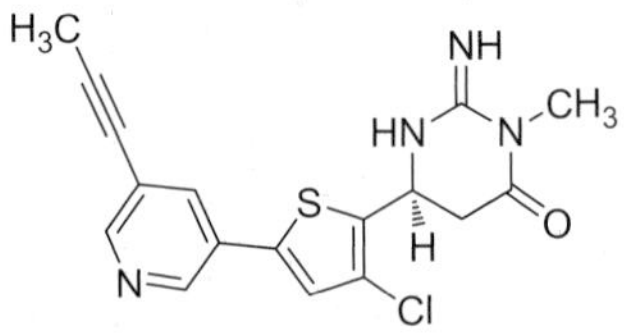

5

After having succeeded in identifying a potent BACE inhibitor for development, we wondered whether the same iminoheterocycle chemotype could be used for inhibition of the related aspartyl protease renin. We were able to convince senior management at Schering-Plough to give us 6 months to come up with a renin inhibitor with good oral bioavailability, a property that was lacking in the successful renin inhibitor aliskiren. Knowing the extensive literature on the roles of the renin–angiotensin system in the brain, we were also interested to see what advantages a brain-penetrant renin inhibitor might have. By mining Schering-Plough's BACE-1 inhibitor collection and carrying out rapid and creative lead optimization, Brian McKittrick and Tanweer Khan were able to discover potent and selective renin inhibitors such as inhibitor **6** (human renin $K_i = 0.6$ nM).[27] Having struggled with peptidomimetic renin inhibitors during 1980s, it was particularly gratifying to have this success over 25 years later. We wondered whether iminoheterocycles might prove to be a general scaffold for aspartyl protease inhibitors, and interestingly, a series of iminohydantoins have recently been reported which are potent inhibitors of the aspartyl protease plasmepsin, with potential for treatment of malaria.[28]

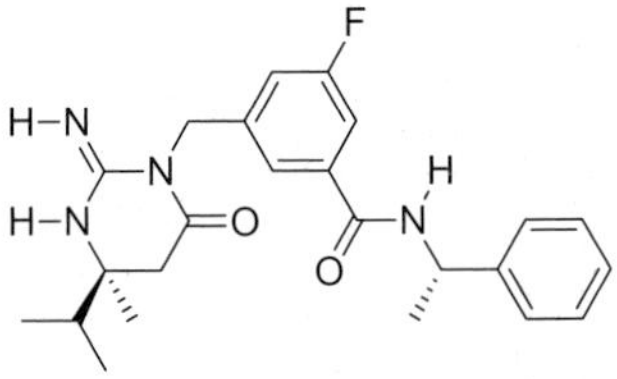

Renin inhibitor **6**

Throughout my career, I have been privileged to work with talented and dedicated scientists whose dream has been to contribute to the discovery of new medicines to reduce human suffering. Their commitment and persistence in the face of continual setbacks have been a source of inspiration. I am grateful for the support and encouragement that I have received from so many colleagues, and the opportunity to make a difference. There have been many changes since I joined Merck over 35 years ago. When I began as a new medicinal chemist there, I had every expectation that I would be at Merck my entire career, and in the end I moved only a few miles away to Schering-Plough, and later had the chance to rejoin Merck after the merger. No chemist hired today can have the expectation of a long, secure career with a single company. News of layoffs of medicinal chemists has become so frequent that it is easy to become cynical about the future of medicinal chemistry as a career path. In spite of the new uncertainty, there has never been a more exciting time to be involved in drug discovery, and I believe that new targets and approaches in discovery research offer a bright future for medicinal chemists.

What guidance can we offer to medicinal chemists in the industry now or contemplating a career there? Here are my words of advice: (1) Do your best, stay focused on drug discovery, and work hard to gain a deep and critical understanding of medicinal chemistry. Expand your reading beyond chemistry to acquire a basic knowledge of related disciplines such as pharmacology, drug metabolism, and toxicology. (2) Build your resume and try to get one publication, presentation, and patent (at least) from each project you contribute to, so that your contributions can be recognized. Do not put off writing the paper, even if you are busy with the next project. (3) Find ways to contribute more to your projects and to medicinal chemistry. Become an expert on an important topic and write a review. Look for opportunities to become involved in medicinal chemistry outside your company, including volunteering in support of the American Chemical Society. (4) Above all, maintain a sense of urgency in your work, since what you are doing is important to the lives of patients. Remember that you have limited time in your career to make a difference, and take advantage of your opportunities. I wish you success and happiness in your efforts.

ACKNOWLEDGMENTS

I would like to acknowledge the many outstanding colleagues who have contributed so much to my career in medicinal chemistry. I have mentioned just a few in the text, without the intention to overlook others. In addition, I would like to acknowledge Ashit Ganguly, Michael Czarniecki, Duane Burnett, Deen Tulshian, John Clader, Stuart McCombie, Cecil Pickett, Ismail Kola, Ann Weber, and Malcolm MacCoss.

REFERENCES

1. Thornberry, N. A.; Bull, H. G.; Taub, D.; Greenlee, W. J.; Patchett, A. A.; Cordes, E. H. *J. Am. Chem. Soc.* **1987**, *109*, 7543.
2. Greenlee, W. J. *J. Org. Chem.* **1984**, *49*, 2632.

3. Patchett, A. A.; Harris, E.; Tristram, E. W.; Wyvratt, M. J.; Wu, M. T.; Taub, D.; Peterson, E. R.; Ikeler, T. J.; ten Broeke, J.; Payne, L. G.; Ondeyka, D. L.; Thorsett, E. D.; Greenlee, W. J.; Lohr, N. S.; Hoffsommer, R. D.; Joshua, H.; Ruyle, W. V.; Rothrock, J. W.; Aster, S. D.; Maycock, A. L.; Robinson, F. M.; Hirschmann, R. *Nature* **1980**, *288*, 280.
4. Weber, A. E.; Halgren, T. A.; Doyle, J. J.; Lynch, R. J.; Siegl, P. K. S.; Parsons, W. H.; Greenlee, W. J.; Patchett, A. A. *J. Med. Chem.* **1991**, *34*, 2692.
5. Maibaum, J.; Stutz, S.; Göschke, R.; Rigollier, P.; Yamaguchi, Y.; Cumin, F.; Rahuel, J.; Baum, H.-P.; Cohen, N.-C.; Schnell, C. R.; Fuhrer, W.; Gruetter, M. G.; Schilling, W.; Wood, J. M. *J. Med. Chem.* **2007**, *50*, 4832.
6. Mantlo, N. B.; Chakravarty, P. K.; Ondeyka, D. L.; Siegl, P. K. S.; Chang, R. S.; Lotti, V. J.; Faust, K. A.; Chen, T.-B.; Schorn, T. W.; Sweet, C. S.; Emmert, S. E.; Patchett, A. A.; Greenlee, W. J. *J. Med. Chem.* **1991**, *34*, 2919.
7. Chakravarty, P. K.; Naylor, E. M.; Chen, A.; Chang, R. S. L.; Chen, T.-B.; Faust, K. A.; Lotti, V. J.; Kivlighn, S. D.; Gable, R. A.; Zingaro, G. J.; Schorn, T. W.; Schaffer, L. W.; Broten, T. P.; Siegl, P. K. S.; Patchett, A. A.; Greenlee, W. J. *J. Med. Chem.* **1994**, *37*, 4068.
8. Kivlighn, S. D.; Zingaro, G. J.; Gabel, R. A.; Broten, T. P.; Chang, R. S. L.; Ondeyka, D. L.; Mantlo, N. B.; Gibson, R. E.; Greenlee, W. J.; Siegl, P. K. *Eur. J. Pharmacol.* **1995**, *294*, 439.
9. Kivlighn, S. D.; Zingaro, G. J.; Rivero, R. A.; Huckle, W. R.; Lotti, V. J.; Chang, R. S. L.; Schorn, T. W.; Kevin, N.; Johnson, R. G., Jr.; Greenlee, W. J.; Siegl, P. K. S. *Am. J. Physiol.* **1995**, *268*, R820.
10. Perlman, S.; Costa-Neto, C. M.; Miyakawa, A. A.; Schambye, H. T.; Hjorth, S. A.; Paiva, A. C. M.; Rivero, R. A.; Greenlee, W. J.; Schwartz, T. W. *Mol. Pharmacol.* **1997**, *51*, 301.
11. Wan, Y.; Wallinder, C.; Plouffe, B.; Beaudry, H.; Mahalingam, A. K.; Wu, X.; Johansson, B.; Holm, M.; Botoros, M.; Karlen, A.; Pettersson, A.; Nyberg, F.; Fandriks, L.; Gallo-Payet, N.; Hallberg, A.; Alterman, M. *J. Med. Chem.* **2004**, *47*, 5995.
12. Williams, D. L., Jr.; Murphy, K. L.; Nolan, N. A.; O'Brien, J. A.; Pettibone, D. J.; Kivlighn, S. D.; Krause, S. M.; Lis, E. V., Jr.; Zingaro, G. J.; Gabel, R. A.; Clayton, F. C.; Siegl, P. K. S.; Zhang, K.; Naue, J.; Vyas, K.; Walsh, T. F.; Fitch, F. K.; Chakravarty, P. K.; Greenlee, W. J.; Clineschmidt, B. V. *J. Pharmacol. Exp. Ther.* **1995**, *275*, 1518.
13. Walsh, T. F.; Fitch, K. J.; Williams, D. L., Jr.; Murphy, K. L.; Nolan, N. A.; Pettibone, D. J.; Chang, R. S. L.; O'Malley, S. S.; Clineschmidt, B. V.; Veber, D. F.; Greenlee, W. J. *Bioorg. Med. Chem. Lett.* **1995**, *5*, 1155.
14. Murugesan, N.; Gu, Z.; Fadnis, L.; Tellew, J. E.; Baska, R. A. F.; Yang, Y.; Beyer, S. M.; Monshizadegan, H.; Dickinson, K. E.; Valentine, M. T.; Humphreys, W. G.; Lan, S.-J.; Ewing, W. R.; Carlson, K. E.; Kowala, M. C.; Zahler, R.; Macor, J. E. *J. Med. Chem.* **1995**, *48*, 171.
15. Baraldi, P. G.; Cacciari, B.; Spalluto, G.; Pineda de las Infantas y Villatoro, M. J.; Zocchi, C.; Dionisotti, S.; Ongini, E. *J. Med. Chem.* **1996**, *39*, 1164.
16. Wager, T. T.; Hou, X.; Verhoest, P. R.; Villalobos, A. *ACS Chem. Neurosci.* **2010**, *1*, 435.
17. Neustadt, B. R.; Hao, J.; Lindo, N.; Greenlee, W. J.; Stamford, A. W.; Ongini, E.; Hunter, J.; Monopoli, A.; Bertorelli, R.; Foster, C.; Arik, L.; Lachowicz, J.; Ng, K.; Feng, K.-I. *Bioorg. Med. Chem. Lett.* **2007**, *17*, 1376.
18. Chackalamannil, S.; Wang, Y.; Greenlee, W. J.; Hu, Z.; Xia, Y.; Ahn, H.-S.; Boykow, G.; Hsieh, Y.; Palamanda, J.; Agans-Fantuzzi, J.; Kurowski, S.; Graziano, M.; Chintala, M. *J. Med. Chem.* **2008**, *51*, 3061.
19. http://www.mercknewsroom.com/press-release/research-and-development-news/merck-announces-fda-acceptance-new-drug-application-vora.

20. Scirica, B. M.; Bonaca, M. P.; Braunwald, E.; De Ferrari, G. M.; Isaza, D.; Lewis, B. S.; Mehrhof, F.; Merlini, P. A.; Murphy, S. A.; Sabatine, M. S.; Tendera, M.; Van de Werf, F.; Wilcox, R.; Morrow, D. A.; TRA 2°P-TIMI 50 Steering Committee Investigators *Lancet* **2012**, *380*, 1317.
21. Cumming, J.; Babu, S.; Huang, Y.; Carrol, C.; Chen, X.; Favreau, L.; Greenlee, W.; Guo, T.; Kennedy, M.; Kuvelkar, R.; Le, T.; Li, G.; McHugh, N.; Orth, P.; Ozgur, L.; Parker, E.; Saionz, K.; Stamford, A.; Strickland, C.; Tadesse, D.; Voigt, J.; Zhang, L.; Zhang, Q. *Bioorg. Med. Chem. Lett.* **2010**, *20*, 2837.
22. Wang, Y.-S.; Strickland, C.; Voigt, J. H.; Kennedy, M. E.; Beyer, B. M.; Senior, M. M.; Smith, E. M.; Nechuta, T. L.; Madison, V. S.; Czarniecki, M.; McKittrick, B. A.; Stamford, A. W.; Parker, E. M.; Hunter, J. C.; Greenlee, W. J.; Wyss, D. F. *J. Med. Chem.* **2010**, *53*, 942.
23. Zhu, Z.; Sun, Z.-Y.; Ye, Y.; Voigt, J.; Strickland, C.; Smith, E. M.; Cumming, J.; Wang, L.; Wong, J.; Wang, Y.-S.; Wyss, D. F.; Chen, X.; Kuvelkar, R.; Kennedy, M. E.; Favreau, L.; Parker, E.; McKittrick, B. A.; Stamford, A.; Czarniecki, M.; Greenlee, W.; Hunter, J. C. *J. Med. Chem.* **2010**, *53*, 951.
24. Stamford, A. W.; Scott, J. D.; Li, S. W.; Babu, S.; Tadesse, D.; Hunter, R.; Wu, Y.; Misiaszek, J.; Cumming, J. N.; Gilbert, E. J.; Huang, C.; McKittrick, B. A.; Hong, L.; Guo, T.; Zhu, Z.; Strickland, C.; Orth, P.; Voigt, J. H.; Kennedy, M. E.; Chen, X.; Kuvelkar, R.; Hodgson, R.; Hyde, L. A.; Cox, K.; Favreau, L.; Parker, E. M.; Greenlee, W. J. *ACS Med. Chem. Lett.* **2012**, *3*, 897.
25. Merck Presents Findings from Phase 1b Study of Investigational BACE Inhibitor, MK-8931, in Patients with Alzheimer's Disease – http://www.mercknewsroom.com/printpdf/495.
26. Merck Advances Development Program for Investigational Alzheimer's Disease Therapy, MK-8931 – http://www.mercknewsroom.com/printpdf/573.
27. McKittrick, B. In: *American Chemical Society National Meeting, San Diego, CA, May 25, 2012, MEDI # 18* 2012.
28. Meyers, M. J.; Tortorella, M. D.; Xu, J.; Qin, L.; He, Z.; Lang, X.; Zeng, W.; Xu, W.; Qin, L.; Prinsen, M. J.; Sverdrup, F. M.; Eickhoff, C. S.; Griggs, D. W.; Oliva, J.; Ruminski, P. G.; Jon Jacobsen, E.; Campbell, M. A.; Wood, D. C.; Goldberg, D. E.; Liu, X.; Lu, Y.; Lu, X.; Tu, Z.-C.; Lu, X.; Ding, K.; Chen, X. *ACS Med. Chem. Lett.* **2014**, *5*, 89.

SECTION 1

Central Nervous System Diseases

Section Editor: Albert J. Robichaud
Sage Therapeutics, Inc. Cambridge, Massachusetts

CHAPTER THREE

Natural and Synthetic Neuroactive Steroid Modulators of $GABA_A$ and NMDA Receptors

Gabriel Martinez-Botella, Michael A. Ackley, Francesco G. Salituro, James J. Doherty
Sage Therapeutics, Cambridge, Massachusetts, USA

Contents

1. INTRODUCTION

Neuroactive steroids (NASs) are a family of steroid-based compounds of both natural and synthetic origin, which have been shown to impact central nervous system (CNS) function through allosteric modulation of the GABA(γ-aminobutyric acid)$_A$ receptor[1–4] and the *N*-methyl-D-aspartic acid (NMDA) class of glutamate receptors.[5] Respectively, these are two of the major inhibitory and excitatory neurotransmitter receptor families involved in synaptic transmission in the brain.

Research over recent years has demonstrated that these receptors can be either positively (PAM) or negatively (NAM) allosterically modulated by NAS compounds, an action mediated through binding to allosteric sites.[6] Evidence suggests that endogenous NASs (Fig. 3.1), such as allopregnanolone (**1**), are capable of modulating $GABA_A$ receptors at concentrations that are relevant to their endogenous levels.[4,9] Additionally, recent

Annual Reports in Medicinal Chemistry, Volume 49
ISSN 0065-7743
http://dx.doi.org/10.1016/B978-0-12-800167-7.00003-1

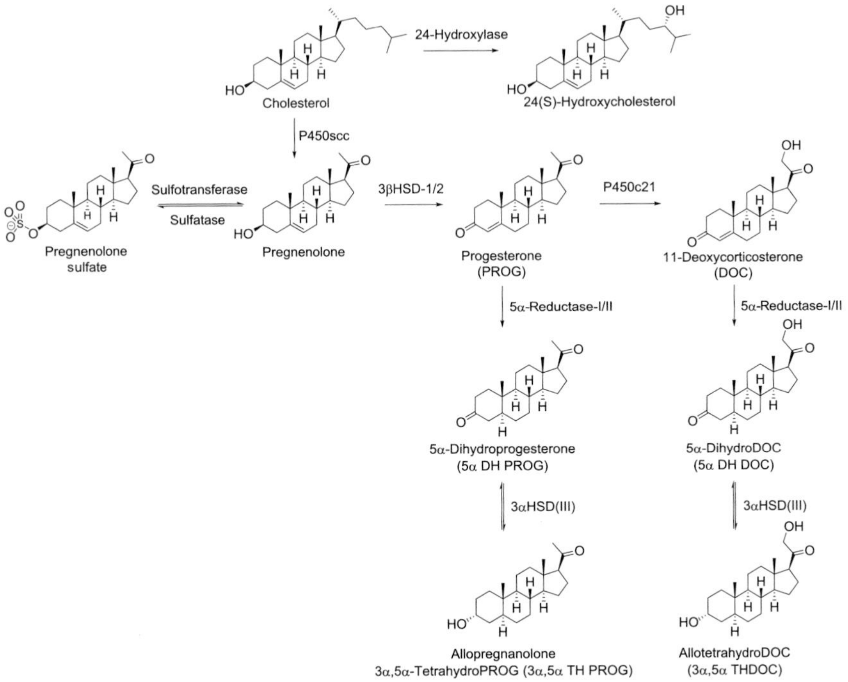

Figure 3.1 Biosynthesis of endogenous neuroactive steroids (NASs) of the $GABA_A$ and NMDA receptor.[7,8]

evidence has shown that cholesterol (**2**) metabolism in the brain is responsible for the generation of a specific, endogenous NAS modulator of NMDA receptors (24(S)hydroxycholesterol, **3**).[10]

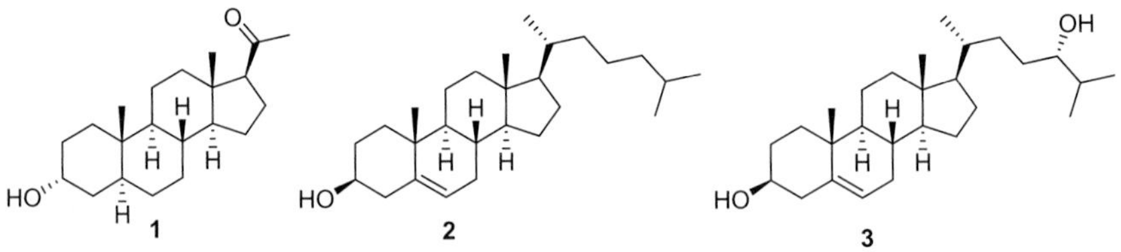

Due to their potential to modulate the overall excitatory and inhibitory tone in the nervous system, it is proposed that specific NAS modulators may be used to treat a broad range of CNS disorders, such as epilepsy, autism, schizophrenia, depression, and pain, in which the excitatory/inhibitory homeostasis is perturbed. The focus of this report is to briefly review endogenous and synthetic NASs that target $GABA_A$ and NMDA receptors.

2. NAS MODULATORS OF THE $GABA_A$ RECEPTOR

The $GABA_A$ receptor is a pentameric ion channel formed by assembling two alpha subunits (α1–α6), two beta subunits (β1–β3), and one additional subunit (γ1–γ3, δ, ε, π, or θ).[11] The subunit composition determines the biophysical and pharmacological characteristics of the channel as well as influencing its location at synaptic or extrasynaptic sites. For example, γ subunits have been associated with clustering the receptor at synaptic sites and thus mediating fast inhibitory synaptic transmission. On the other hand, receptors that contain the δ subunit have been associated with extrasynaptic receptors that mediate tonic inhibitory conductances.[12] Furthermore, certain subunit combinations are restricted to specific anatomic locations affording the opportunity to target individual neuronal circuits. Gating of the channel allows chloride ions to flow along their electrochemical gradient resulting in a modification of the resting membrane potential and change in the resistance of the cell membrane. Under the majority of conditions, this results in an inhibitory influence on the cell, reducing its ability to transduce excitatory inputs into action potentials. As the major inhibitory neurotransmitter in the nervous system, GABA acting via an array of $GABA_A$ receptors can influence a wide range of brain circuits that are central to a variety of behavioral states, such as anxiety levels, sleep, and memory. $GABA_A$ dysregulation also sits at the center of a range of neuropsychiatric diseases such as schizophrenia and mood disorders. Perhaps not surprisingly, given its critical role in the function of neuronal circuits, the $GABA_A$ receptor is the target for numerous clinically prescribed drugs such as benzodiazepines, barbiturates, and certain anesthetics.

NASs have the potential to differentiate from classical $GABA_A$ receptor modulators in several ways. First, NASs differentiate from benzodiazepines by targeting different populations of $GABA_A$ receptors. NASs have been suggested to putatively bind to three or four distinct sites on the $GABA_A$ receptor. At low concentrations, NASs bind to a site in the M3/M4 domains of the α subunit,[6] leading to allosteric modulation of the GABA-induced current. This modulation could be in either a positive (PAM) or a negative (NAM) direction. At high concentrations, direct gating of $GABA_A$ currents may be achieved in the absence of GABA via binding at the α/β interface near the GABA-binding site.[6] At least one or two alternate sites have also been suggested to be responsible for the inhibitory effects of sulfated NASs on GABA currents.[13] On the other hand, benzodiazepines bind to an allosteric site distinct from the GABA-binding site, at the interface of the α and γ subunits.[14–17] This limits their ability to potentiate synaptic GABA currents to receptor assemblies that contain a γ subunit.[17] Conversely, NASs potentiate the $GABA_A$ receptor by binding to residues within the obligatory α subunit and so modulate receptors in a manner which is agnostic to their subunit composition. In this way, and unlike benzodiazepines, NASs are capable of targeting extrasynaptic GABA receptors that include the δ subunit in addition to synaptic γ-containing receptors. As such, NASs may exhibit a therapeutic advantage over benzodiazepines, for example, in indications such as benzodiazepine-resistant seizures.[18]

Interestingly, data also suggest that NASs can exert profound effects on expression levels of $GABA_A$ receptors.[19–21] For example, in the hippocampus, brief exposure to low concentrations of NASs can enhance tonic currents while having little effects on phasic synaptic currents even after a prolonged wash.[20,22] These data suggest that NASs could influence the trafficking and surface expression of certain $GABA_A$ receptors. Consistent with this hypothesis, recent data suggest that NASs do indeed enhance trafficking of $GABA_A$ receptor subunits.[23] Such an effect would further differentiate NAS allosteric modulators of the $GABA_A$ receptor from benzodiazepines. Indeed, benzodiazepines typically display tolerance to their effects on repeated dosing.[18]

As NASs have complex modulatory actions via allosteric sites on the receptor complex, this has historically had implications for determining structure–activity relationships (SARs) using available pharmacological tools. The ability of NASs to modulate the binding of the picrotoxin site ligand, *t*-butylbicyclophosphorothionate (TBPS), has been used to identify compounds that bind to the $GABA_A$ receptor.[24] As this represents an allosteric interaction between the NAS- and TBPS-binding sites, there is not necessarily a linear relationship between functional potency of the NAS and inhibition of TBPS. Additional functional assays can be helpful to further characterize compounds that were promising based on the initial data. In fact, the gold standard for measuring functional activity of ligand-gated ion channels is using patch clamp electrophysiology and it is only with the recent advent of automated high-throughput patch clamp systems such as the QPatch systems (Sophion) and Ionworks Barracuda (Molecular Devices) that drug discovery has been able to screen in a high-throughput manner and to fully understand SARs.[25]

2.1. Endogenous NAS Modulators of the $GABA_A$ Receptor

Endogenous metabolites of progesterone (**4**) and deoxycorticosterone (**5**) are perhaps the most widely studied NASs that show activity at the $GABA_A$ receptor.[5] One essential specificity determinant that these metabolites have in common is a hydroxyl group in the α-C3 position, a small beta face functionality at C20, and either a 5α (*trans*) or 5β (*cis*) hydrogen (Table 3.1).[26] Metabolites such as allopregnanolone and pregnanolone (**6**) have been characterized as positive allosteric modulators of the $GABA_A$ receptor.[4,26]

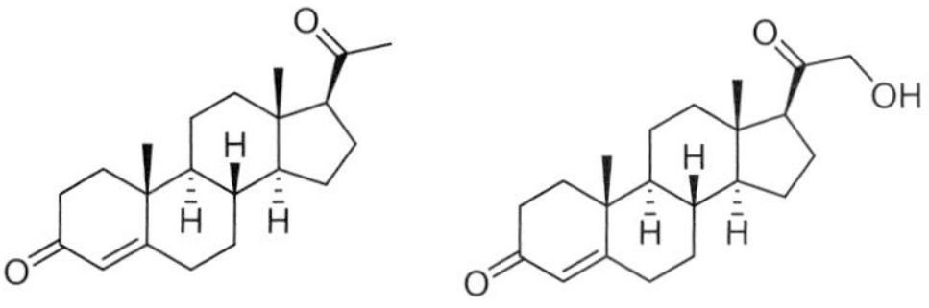

4

5

On the other hand, the sulfates of pregnenolone (PS, **7**) and dehydroepiandrosterone (**8**) have been characterized as $GABA_A$ negative allosteric modulators with μM EC_{50} values.[27]

Table 3.1 $GABA_A$ receptor endogenous NASs show a preference for α-C3 alcohol

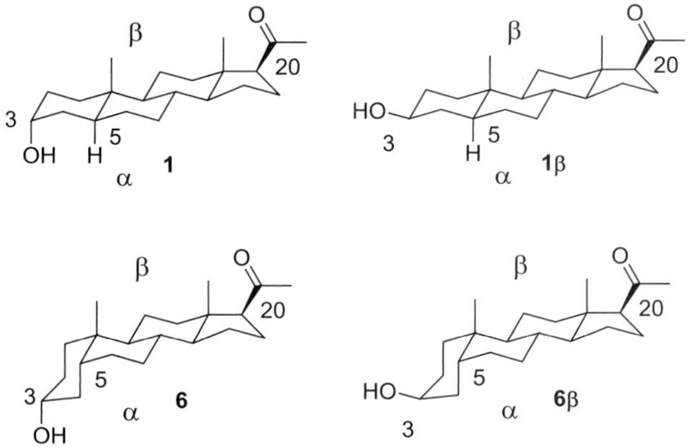

	TBPS IC_{50} (nM)	**E-Phys MEC (nM)**
(**1**) 3-α-OH/5-α	275	100
1β 3-β analog	Inactive	Inactive
(**6**) 3-α-OH/5-β	130	100
6β 3-β analog	Inactive	Inactive

Data shown are IC_{50} values in a TBPS-binding assay and minimal effective concentration (MEC) in a patch clamp electrophysiology assay.[26]

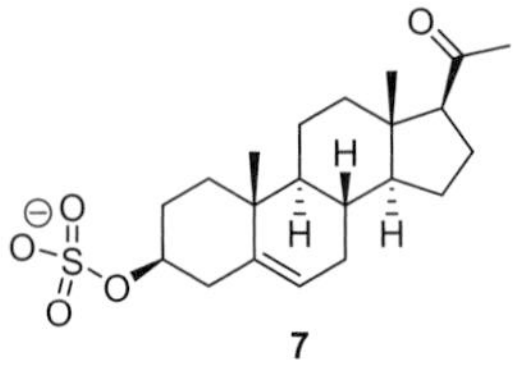

7

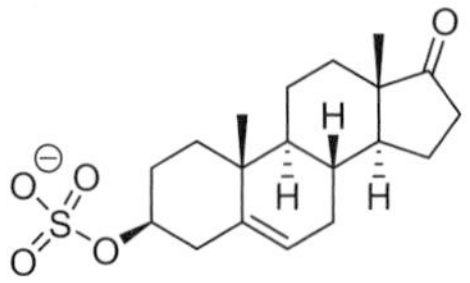

8

Endogenous compounds such as allopregnanolone ((3α,5α)-hydroxy-pregnane-20-one, **1**) and pregnanolone ((3α,5β)-hydroxypregnane-20-one, **6**) have emerged as agents for clinical development in a variety of indications. For example, Sage Therapeutics has evaluated this agent for the treatment of a patient with refractory status epilepticus in an open-label compassionate-use case study.[28] Furthermore, Sage Therapeutics and The University of California, Davis are conducting a Phase 2 clinical study for the treatment of traumatic brain injury (TBI) using an intravenous solution of all-opregnanolone (**1**).[29] In addition, pregnanolone (**6**) (eltanolone) has been evaluated as a potential anesthetic in humans.[30,31] Due to their low oral bio-availability, these endogenous NASs have been dosed as i.v. formulations.

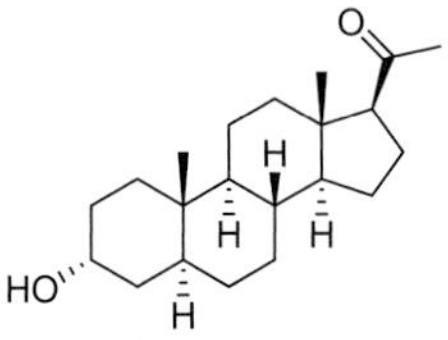

1

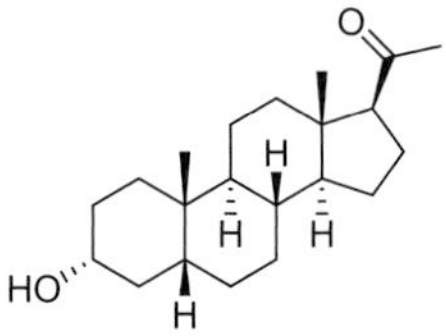

6

2.2. Synthetic NAS Modulators of the GABA$_A$ Receptor

2.2.1 Anesthetics

The ability of progesterone (**4**) and 5β-pregnane-3,20-dione (**9**) to induce a depressant effect on the CNS was reported in 1941 in rodents.[32] Since then, multiple groups have embarked in the search for novel synthetic analogs for use as anesthetics.[33] The main focus of this work was improving the intrinsic water solubility of these steroids through traditional approaches introducing solubilizing groups such as amines, succinates, or phosphates (**10**, **11**).[34,35] Although some of the compounds retain their anesthetic effects, they also introduce side effects such as thrombophlebitis or paresthesia.

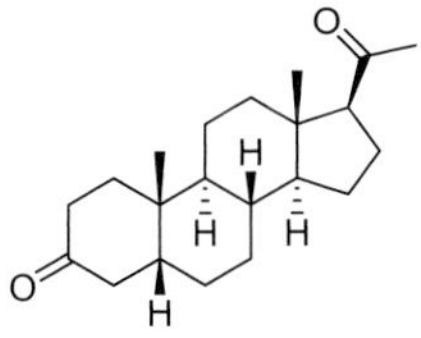

9 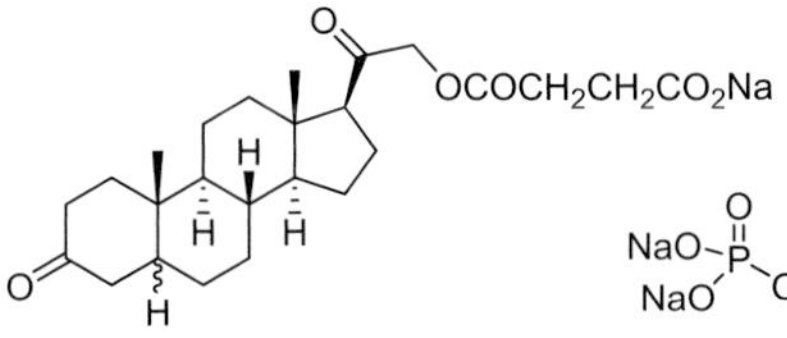 10 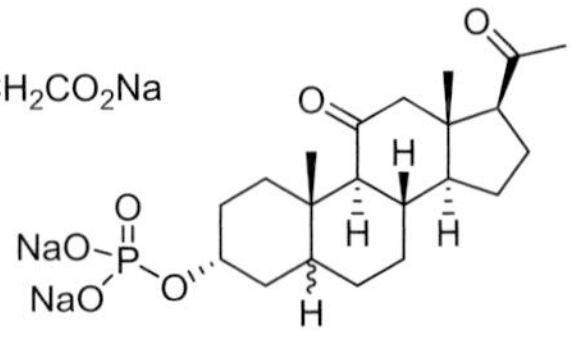 11

In 1971, researches at Glaxo utilized novel formulations such as Cremophor EL (a polyoxyethylated castor oil) to formulate poorly soluble compounds such as 3α-hydroxy-5α-pregnane-11,20-dione (alfaxolone, **12**).[36] They subsequently introduced a product which was a mixture of alfaxolone, and its 21-acetoxy derivative alfadolone acetate (**13**), in a 20% Cremophor EL water formulation. This product marketed as Althesin was withdrawn by Glaxo from human use in 1984 after 12 years on the market, due to anaphylactic reactions likely resulting from the Cremophor vehicle.[37] Independently, both Althesin and alfaxolone (**12**) are currently branded as Saffan and Alfaxan, respectively, for the use in veterinary medicine. Furthermore, an aqueous cyclodextrin formulation of alfaxalone, Phaxan, is currently being developed by Drawbridge Pharmaceuticals for general anesthesia in humans. Finally, althesin was tested as an agent to control seizures in status epilepticus patients,[38,39] but further development was not pursued. In 1984, alfaxolone (**12**) (TBPS $IC_{50}=538$ nM)[40] was reported to modulate the $GABA_A$ receptor providing the first evidence that the anesthetic effect of NASs can be linked to $GABA_A$ receptor pharmacology.[41] Several agents were developed after althesin, notably minaxolone (**14**)[42,43] (TBPS $IC_{50}=71$ nM)[40] and more recently Org21465 (**15**)[44] (TBPS $IC_{50}=504$ nM).[40] Both compounds were discontinued due to the observance of significant myoclonus (excitation and involuntary movements during anesthesia), among other side effects.[45–47] It is worth noting that most of the SARs unveiled by these efforts, except for the most recent Organon work,[40] were achieved in the absence of a $GABA_A$ assay to guide the design, the focus being on tracking water solubility and *in vivo* pharmacology by systematically attaching solubilizing group at C2, C6, C11, or C16 to the basic steroid scaffold. Finally, Washington University in St. Louis[48] and Sage Therapeutics[49] have recently reported novel compounds that may be developed as anesthetics, e.g., **16** (TBPS $IC_{50}=42$ nM)[48] and **17** (TBPS $IC_{50}=100–500$ nM).[49]

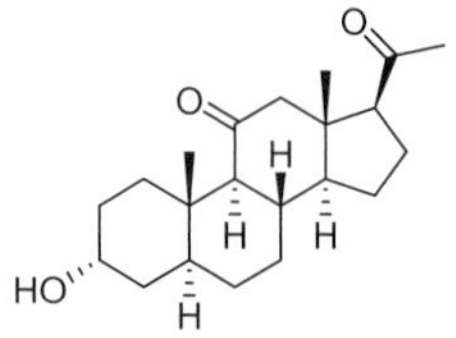

2.2.2 Compounds Suitable for Nonanesthetic Indications

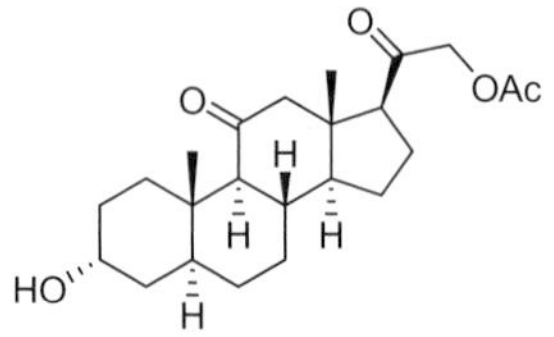

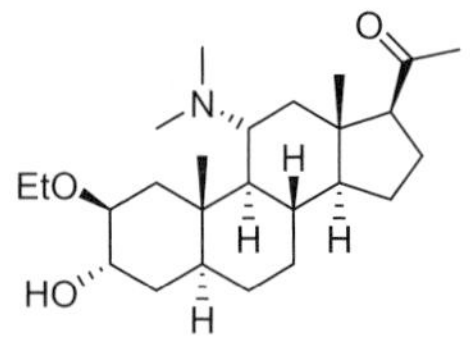

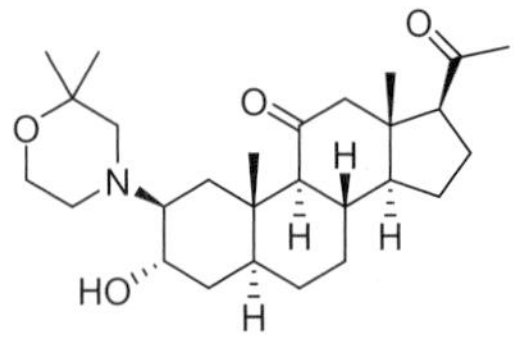

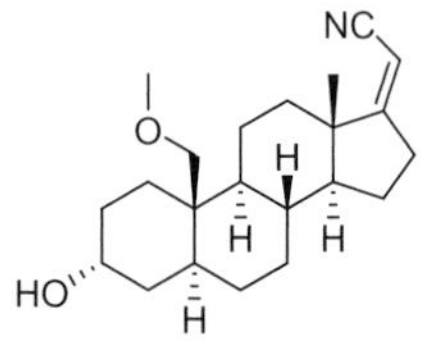

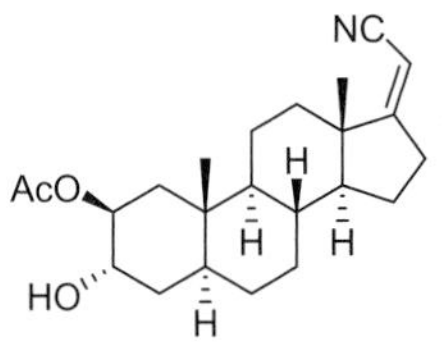

Over the years, as the linkage between NASs and $GABA_A$ receptors became apparent,[4] interest in developing orally bioavailable drugs to treat a range of CNS disorders became a primary target for drug discovery. The low oral bioavailability of allopregnanolone (**1**) has been hypothesized to arise mainly from high *in vivo* clearance resulting from rapid primary and secondary metabolism at the C3-hydroxyl oxidation to the ketone and glucuronidation of the alcohol, respectively.[33,50,51] In the 1990s, Cocensys[52–55] expanded on early work from Glaxo and explored the effects of substituents at the β-C3 with regard to both potency and DMPK[52,53]: for example, alkanes (**18**), alkenes, alkynes (**19**), alkoxy (**22**, **23**), and alkyl halides (**20**, **21**). In recent years, Marinus Pharmaceuticals has shown that ganaxolone (**18**),[50,56] the β-C3 methyl derivative of allopregnanolone, is safe[57] and active in adult patients with partial onset seizures[58,59] and in children with epilepsy.[60] Marinus is currently conducting a Phase 2 clinical trial in adults with drug-resistant partial onset seizures.[61] Marinus is also developing ganaxolone (**18**) as a treatment for children with fragile X syndrome and are conducting a Phase 2 clinical trial.[62] Ganaxolone (**18**) oral doses range from 1.2 to 1.8 g per day.[61,62] Expanding on the SAR at β-C3, the Cocensys team built large substituents, such as alkynylphenyls,[52] which retain activity against the $GABA_A$ receptor, e.g., Co152791 (**19**).[63]

Additional modifications were also explored, including replacing the C19 methyl with a hydrogen (CCD3693, **20**; Co26749, **21**)[64,65] and appending various heterocycles at C21 (Co134444, **22**; Co177843, **23**).[66,67] Several of these compounds have been tested in multiple *in vivo* models after oral administration, showing activity as anticonvulsants[52,66,68] and anxiolytics (Tables 3.2 and 3.3).[65,66] As in the case of allopregnanolone (**1**), minimal tolerance to effect after chronic dosing has been shown with this family of analogs, differentiating NASs from the established benzodiazepine drug

Table 3.2 *In vitro* (rat) and *in vivo* (mouse) data for compounds **18** and **19**

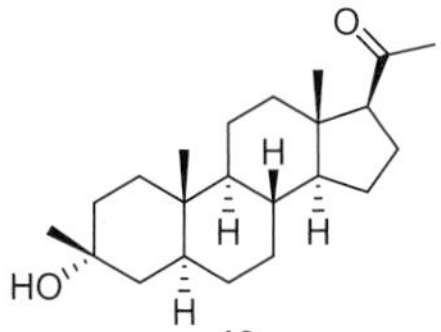

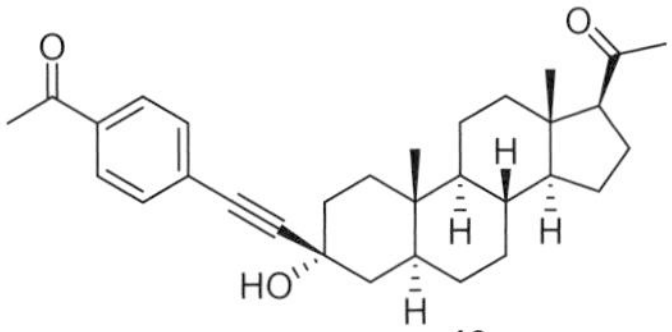

	TBPS IC_{50} (nM)	PTZ ED_{50} (mpk)	Rotarod TD_{50} (mpk)
18	80	4.3 (IP)	33.4 (IP)
19	4.7	1.1 (PO)	7.7 (PO)

PTZ efficacy against pentylenetetrazol-induced seizure.[50,53,63]

Table 3.3 Rat *in vitro* and *in vivo* data for compounds **20–23**

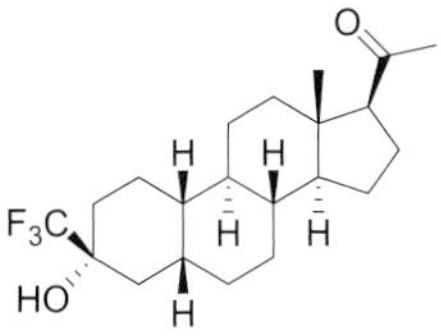

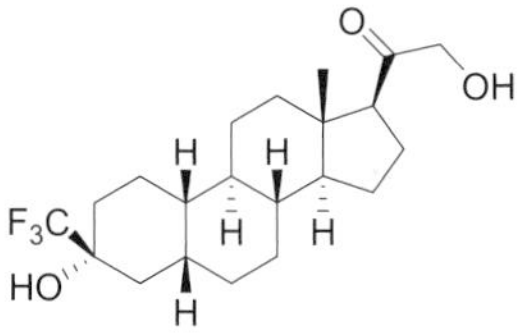

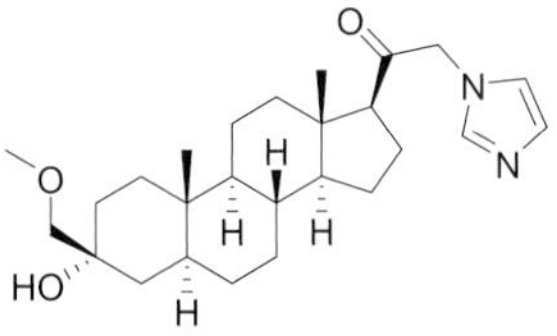

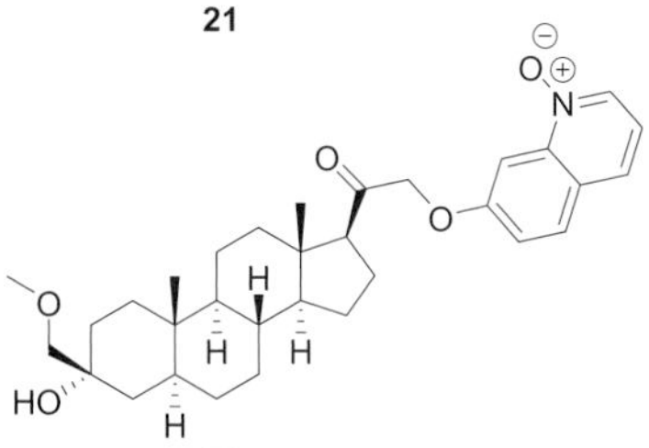

	TBPS IC_{50} (nM)	PTZ ED_{50} PO (mpk)	Rotarod TD_{50} PO (mpk)	Punished drinking MED PO (mpk)
20	76	30	26.1	–
21	230	–	25.3	1.6
22	–	23.6	39	3
23	–	–	29.1	–

PTZ efficacy against pentylenetetrazol-induced seizure.[64–67]

class.[18,69] In the past 2 years, Sage Therapeutics has expanded the arena of NAS analogs that act as allosteric modulators of the $GABA_A$ receptor and has identified several classes of novel orally bioavailable (40–70%) NAS with *in vitro* potencies <250 nM in a TBPS assay and robust *in vivo* activity against pentylenetetrazol (PTZ)-induced seizures (MED ≤1 mg/kg after i.p. dosing).[70,71]

3. NAS MODULATORS OF THE NMDA RECEPTOR

NMDA receptors are heterotetrameric ligand-gated ion channels that are involved in fast excitatory transmission and the synaptic plasticity involved in learning and memory. They are tetrameric receptors that are comprised of two obligatory GluN1 along with two GluN2(A–D) or GluN3(A,B) subunits. Subunit composition affects the biophysical and pharmacological properties of the channel such as glutamate affinity and permeability to specific cations.[72] In order to gate the channel, the receptor must bind two molecules of glutamate along with either glycine or D-serine as a co-agonist. The glutamate-binding site is situated on the GluN2 subunit while the glycine/D-serine-binding site is within the GluN1 subunit. Interestingly, the GluN3 subunit also binds glycine/D-serine and so heterodimers of GluN1/GluN3 lack a glutamate-binding site and this receptor ensemble is instead gated entirely by glycine. An interesting property of NMDA receptors is their sensitivity to open channel block by magnesium ions.[73] Like other pharmacological properties of the receptor, the sensitivity of this block is dependent on subunit composition. Interestingly, this block is voltage sensitive, so that under hyperpolarizing conditions, the channel will open but cannot flux cations due to the block by magnesium. As the cell depolarizes, the magnesium block is removed allowing cations such as calcium to enter the cell. This property makes the channel uniquely positioned to serve as a coincidence detector of neuronal activity and synaptic activation. Importantly, this role as a coincidence detector coupled with an array of second messenger systems allows the NMDA receptor to mediate neuroplastic changes such as long-term potentiation.[74] As such, the receptor has been intimately associated with the processes of learning and memory.

The NMDA receptor has historically been an intriguing target for drug development.[75] Activation of the NMDA receptor is of interest for enhancing cortical plasticity in diseases such as schizophrenia, where cognitive dysfunction is prevalent. Conversely, antagonists of the NMDA receptor have been of interest for their ability to alleviate chronic pain, as therapies for stroke, neuroprotection in neurodegenerative diseases such as Alzheimer's, and as rapid onset antidepressants.[76,77] Development of NMDA ligands that bind to the glutamate and/or glycine site has been hampered, however, by lack of clinical efficacy combined with a prevalence of undesirable side effects. Furthermore, direct agonism of the NMDA receptor can be associated with neurotoxic cell death. The glutamate-binding site is remarkably conserved across the GluN2 subunits and so it has proved difficult to develop subunit-selective ligands that target this site. Thus, interest has now turned to the development of ligands that bind to allosteric sites on the NMDA complex in the hope that they will be able to fine-tune receptor function and thus produce efficacious compounds with fewer undesirable adverse events.[78]

PS (**7**) is an endogenous NAS that acts as an allosteric modulator of NMDA receptor[79,80] as well as a $GABA_A$ receptor negative allosteric modulator.[27,81] Interestingly, PS (**7**) appears to show selectivity in its functional effects on NMDA receptor with different subunit compositions. PS (**7**) appears to act as a positive allosteric modulator of receptors containing GluN2A or GluN2B subunits, while acting as a negative modulator at receptors containing GluN2C or GluN2D subunits.[82] PS (**7**) also appears to display state-dependent activity whereby its affinity for the receptor complex is decreased following receptor activation.[83] Sage Therapeutics has recently reported that 24(S)hydroxycholesterol (**3**), a cholesterol metabolite, is also a potent endogenous allosteric modulator of NMDA receptors.[10] It is hypothesized that 24(S)hydroxycholesterol (**3**) binds to a site that is distinct from PS (**7**). The enzyme responsible for the production of this NAS is a highly conserved cytochrome P450 enzyme (cholesterol 24-hydroxylase, CYP46A1)[8] that is present within neuronal cell bodies and dendrites.[84] Interestingly, deficiency in this enzyme results in disrupted hippocampal LTP and memory acquisition in mice,[85,86] suggesting that endogenous 24(S)hydroxycholesterol (**3**) plays a critical role in normal hippocampal function.

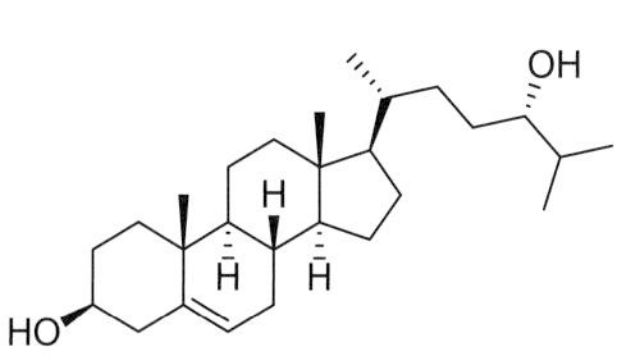

3 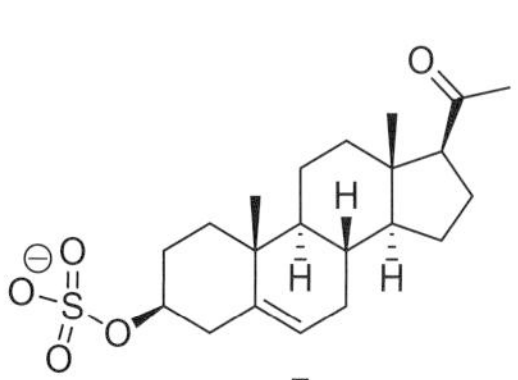 **7**

Parallel research has additionally identified two synthetic oxysterols reported to show potentiation of NMDA receptors, SGE-201 (**24**)[87] and

SGE-301 (**25**).[10] SGE-201 (**24**) potentiated NMDA currents with an EC_{50} of 110 nM.[10] *In vitro* and *in vivo* models have demonstrated that these synthetic sterols enhanced hippocampal LTP at sub-μM concentrations, as well as reversed a PCP-induced impairment of cognitive and social behavior with an MED <1 mg/kg using i.p. dosing.[10] These compounds represent the basis for further development in the area of NMDA receptor potentiators.[88] PAM modulation of NMDA receptor appears to favor the hydroxyl group in the β-C3. Whereas the opposite configuration, the hydroxyl group in the α-C3, appears to be required to modulate the $GABA_A$ receptor.

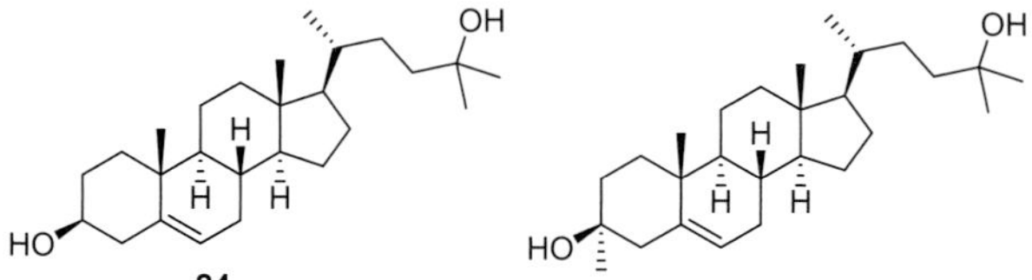

24 **25**

In addition to potentiation of NMDA, there are also promising data suggesting inhibitory effects on NMDA receptor by NAS structural subclasses.[82,89] As clinical trials with ketamine, an NMDA receptor blocker, has sparked significant promise in treatment of depression,[76,77] the search for novel orally bioavailable NAS NMDA receptor inhibitors has intensified and opened new and interesting opportunity for drug discovery in this difficult arena. For example, both sulfates of allopreganolone (**26**) and pregnanolone (**27**) appear to act as negative allosteric modulators of NMDA receptor with functional EC_{50} values in the mid-μM range.[82]

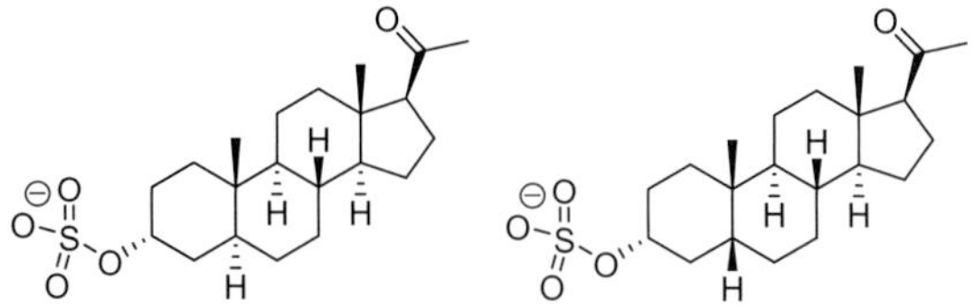

26 **27**

4. CONCLUSIONS

Since the early observation in the 1940s of the depressant effects of progesterone and the subsequent discovery four decades later that these effects were due to positive modulation of $GABA_A$ receptors, the field of NASs has grown and matured dramatically. New biological tools and starting points to interrogate biology and medicinal chemistry will further push this field beyond the pioneering work of Glaxo, Organon, and Cocensys. The recent discovery of 24(S)hydroxycholesterol (**3**) as an endogenous NAS has the potential to bring a novel NMDA receptor allosteric site as a viable target for drug development.

The ability to toggle a single class of compounds toward either positive or negative modulation amplifies the attractiveness of these compounds for drug discovery. Medicinal chemistry has revealed several selectivity determinants that distinguish between $GABA_A$ and NMDA receptors in a broad sense. For example, PAM modulation of $GABA_A$ receptor appears to require a hydroxyl group in the α-C3 configuration, while the NMDA receptor appears to favor the opposite configuration, the hydroxyl group in the β-C3.

Finally, coupling the understanding of emerging SAR with advances in DMPK strategies to optimize compounds for i.v., i.m., or oral administration will yield further advances in optimizing compounds toward viable clinical candidates. Finally, the efforts of both Marinus Pharmaceuticals, developing NAS treatments for fragile X patients, and Sage Therapeutics, developing NASs for treatment of status epilepticus[90] and TBI,[29] may yield the first approved NAS for clinical use by the end of this decade, almost 50 years after the approval of Althesin.

REFERENCES

1. Covey, D. F.; Evers, A. S.; Mennerick, S.; Zorumski, C. F.; Purdy, R. H. *Brain Res. Brain Res. Rev.* **2001**, *37*, 91–97.
2. Hamilton, N. M. *Curr. Top. Med. Chem.* **2002**, *2*, 887–902.
3. Veleiro, A. S.; Burton, G. *Curr. Med. Chem.* **2009**, *16*, 455–472.
4. Majewska, M. D.; Harrison, N. L.; Schwartz, R. D.; Barker, J. L.; Paul, S. M. *Science* **1986**, *232*, 1004–1007.
5. Zorumski, C. F.; Mennerick, S. *JAMA Psychiatry* **2013**, *70*, 659–660.
6. Hosie, A. M.; Wilkins, M. E.; da Silva, H. M.; Smart, T. G. *Nature* **2006**, *444*, 486–489.
7. Mellon, S. H.; Griffin, L. D. *Trends Endocrinol. Metab.* **2002**, *13*, 35–43.
8. Russell, D. W.; Halford, R. W.; Ramirez, D. M. O.; Shah, R.; Kotti, T. *Annu. Rev. Biochem.* **2009**, *78*, 1017–1040.
9. Genazzani, A. R.; Petraglia, F.; Bernardi, F.; Casarosa, E.; Salvestroni, C.; Tonetti, A.; Nappi, R. E.; Luisi, S.; Palumbo, M.; Purdy, R. H.; Luisi, M. *J. Clin. Endocrinol. Metab.* **1998**, *83*, 2099–2103.
10. Paul, S. M.; Doherty, J. J.; Robichaud, A. J.; Belfort, G. M.; Chow, B. Y.; Hammond, R. S.; Crawford, D. C.; Linsenbardt, A. J.; Shu, H.-J.; Izumi, Y.; Mennerick, S. J.; Zorumski, C. F. *J. Neurosci.* **2013**, *33*, 17290–17300.
11. Rudolph, U.; Möhler, H. *Curr. Opin. Pharmacol.* **2006**, *6*, 18–23.
12. Belelli, D.; Harrison, N. L.; Maguire, J.; Macdonald, R. L.; Walker, M. C.; Cope, D. W. *J. Neurosci.* **2009**, *29*, 12757–12763.
13. Hosie, A. M.; Wilkins, M. E.; Smart, T. G. *Pharmacol. Ther.* **2007**, *116*, 7–19.
14. Chang, Y.; Weiss, D. S. *Biophys. J.* **1999**, 77, 2542–2551.

15. Amin, J.; Brooks-Kayal, A.; Weiss, D. S. *Mol. Pharmacol.* **1997**, *51*, 833–841.
16. Wieland, H.; Lüddens, H.; Seeburg, P. *J. Biol. Chem.* **1992**, *267*(3), 1426–1429.
17. Wafford, K. A.; Macaulay, A. J.; Fradley, R.; O'Meara, G. F.; Reynolds, D. S.; Rosahl, T. W. *Biochem. Soc. Trans.* **2004**, *32*, 553–556.
18. Rogawski, M. A.; Loya, C. M.; Reddy, K.; Zolkowska, D.; Lossin, C. *Epilepsia* **2013**, *54*(Suppl. 6), 93–98.
19. Maguire, J.; Mody, I. *Neuron* **2008**, *59*, 207–213.
20. Maguire, J.; Mody, I. *J. Neurosci.* **2007**, *27*, 2155–2162.
21. Maguire, J. L.; Stell, B. M.; Rafizadeh, M.; Mody, I. *Nat. Neurosci.* **2005**, *8*, 797–804.
22. Stell, B. M.; Brickley, S. G.; Tang, C. Y.; Farrant, M.; Mody, I. *Proc. Natl. Acad. Sci. U.S.A.* **2003**, *100*, 14439–14444.
23. Abramian, A. M.; Comenencia-Ortiz, E.; Modgil, A.; Vien, T. N.; Nakamura, Y.; Moore, Y. E.; Maguire, J. L.; Terunuma, M.; Davies, P. A.; Moss, S. J. *Proc. Natl. Acad. Sci. U.S.A.* **2014**, *111*(19), 7132–7137.
24. Evers, A. S.; Chen, Z.; Manion, B. D.; Han, M.; Jiang, X.; Darbandi-tonkabon, R.; Kable, T.; Bracamontes, J.; Zorumski, C. F.; Mennerick, S.; Steinbach, J. H.; Covey, D. F. *J. Pharmacol. Exp. Ther.* **2010**, *333*, 404–413.
25. McManus, O. B. *Curr. Opin. Pharmacol.* **2014**, *15C*, 91–96.
26. Harrison, N. L.; Majewska, M. D.; Harrington, J. W.; Barker, J. L. *J. Pharmacol. Exp. Ther.* **1987**, *241*, 346–353.
27. Park-Chung, M.; Malayev, A.; Purdy, R. H.; Gibbs, T. T.; Farb, D. H. *Brain Res.* **1999**, *830*, 72–87.
28. Vaitkevicius, H.; Ng, M.; Moura, L.; Rosenthal, E.; Westover, M.; Rosand, J.; Rogawski, M.; Reddy, K.; Cole, A. *The 4th London—Innsbruck Colloquium on Status Epilepticus and Acute Seizures*, 2013, p. 124 (Abstract P29).
29. University of California. D. Clin. [Internet]. Bethesda Natl. Libr. Med. NCT0167382, 2012.
30. Tang, J.; Qi, J.; White, P. F.; Wang, B.; Wender, R. H. *Anesth. Analg.* **1997**, *85*, 801–807.
31. Carl, P.; Høgskilde, S.; Lang-Jensen, T.; Bach, V.; Jacobsen, J.; Sørensen, M. B.; Grälls, M.; Widlund, L. *Acta Anaesthesiol. Scand.* **1994**, *38*, 734–741.
32. Selye, H. *Exp. Biol. Med.* **1941**, *46*, 116–121.
33. Phillipps, G. H. *J. Steroid Biochem.* **1975**, *6*, 607–613.
34. Laubach, G.; P'an, S.; Rudel, H. *Science* **1955**, *122*, 78.
35. Robertson, J. In: *Recent Advances in Anaesthesia and Analgesia*; Hewer, C., Ed.; Churchill: London, 1963; pp 30–78.
36. Child, K. J.; Currie, J. P.; Dis, B.; Dodds, M. G.; Pearce, D. R.; Twissell, D. J. *Br. J. Anaesth.* **1971**, *43*, 2–13.
37. Tachon, P.; Descotes, J.; Laschi-Loquerie, A.; Guillot, J. P.; Evreux, J. C. *Br. J. Anaesth.* **1983**, *55*, 715–717.
38. Chin, L. S.; Havill, J. H.; Rothwell, R. P. *Anaesth. Intensive Care* **1979**, *7*, 50–52.
39. Munari, C.; Casaroli, D.; Matteuzzi, G.; Pacifico, L. *Epilepsia* **1979**, *20*, 475–483.
40. Anderson, A.; Boyd, A. C.; Byford, A.; Campbell, A. C.; Gemmell, D. K.; Hamilton, N. M.; Hill, D. R.; Hill-Venning, C.; Lambert, J. J.; Maidment, M. S.; May, V.; Marshall, R. J.; Peters, J. A.; Rees, D. C.; Stevenson, D.; Sundaram, H. *J. Med. Chem.* **1997**, *40*, 1668–1681.
41. Harrison, N. L.; Simmonds, M. A. *Brain Res.* **1984**, *323*, 287–292.
42. Phillips, G. H.; Ayres, B. E.; Bailey, E. J.; Ewan, G. B.; Looker, B. E.; May, P. J. *J. Steroid Biochem.* **1979**, *11*, 79–86.
43. Aveling, W.; Sear, J. W.; Fitch, W.; Chang, H.; Waters, A.; Cooper, G. M.; Simpson, P.; Savege, T. M.; Prys-Roberts, C.; Campbell, D. *Lancet* **1979**, *2*, 71–73.
44. Sear, J. W. *Br. J. Anaesth.* **1997**, *79*, 417–419.

45. Sneyd, J. R.; Wright, P. M.; Harris, D.; Taylor, P. a.; Vijn, P. C.; Cross, M.; Dale, H.; Voortman, G.; Boen, P. *Br. J. Anaesth.* **1997**, *79*, 433–439.
46. McNeill, H. G.; Clarke, R. S.; Dundee, J. W.; Briggs, L. P. *Anaesthesia* **1981**, *36*, 592–596.
47. Sear, J. W.; Prys-Roberts, C.; Dye, A. *Br. J. Anaesth.* **1983**, *55*, 603–609.
48. Covey, D. Neuroactive 19-Alkoxy-17(20)-Z-Vinylcyano-Substituted Steroids, Prodrugs Thereof, and Methods of Treatment Using Same. WO2014058736, 2014.
49. Salituro, F. G.; Robichaud, A. J. Neuroactive Steroids, Compositions, and Uses Thereof. WO2013188792, 2014.
50. Carter, R. B.; Wood, P. L.; Wieland, S.; Hawkinson, J. E.; Belelli, D.; Lambert, J. J.; White, H. S.; Wolf, H. H.; Mirsadeghi, S.; Tahir, S. H.; Bolger, M. B.; Lan, N. C.; Gee, K. W. *J. Pharmacol. Exp. Ther.* **1997**, *280*, 1284–1295.
51. He, X.-Y.; Wegiel, J.; Yang, S.-Y. *Brain Res.* **2005**, *1040*, 29–35.
52. Upasani, R. B.; Yang, K. C.; Acosta-Burruel, M.; Konkoy, C. S.; McLellan, J. A.; Woodward, R. M.; Lan, N. C.; Carter, R. B.; Hawkinson, J. E. *J. Med. Chem.* **1997**, *40*, 73–84.
53. Hogenkamp, D. J.; Tahir, S. H.; Hawkinson, J. E.; Upasani, R. B.; Alauddin, M.; Kimbrough, C. L.; Acosta-Burruel, M.; Whittemore, E. R.; Woodward, R. M.; Lan, N. C.; Gee, K. W.; Bolger, M. B. *J. Med. Chem.* **1997**, *40*, 61–72.
54. Hawkinson, J. E.; Kimbrough, C. L.; Belelli, D.; Lambert, J. J.; Purdy, R. H.; Lan, N. C. *Mol. Pharmacol.* **1994**, *46*, 977–985.
55. Hawkinson, J. E.; Drewe, J. A.; Kimbrough, C. L.; Chen, J. S.; Hogenkamp, D. J.; Lan, N. C.; Gee, K. W.; Shen, K. Z.; Whittemore, E. R.; Woodward, R. M. *Mol. Pharmacol.* **1996**, *49*, 897–906.
56. Reddy, D. S.; Woodward, R. *Drugs Future* **2004**, *29*, 227.
57. Monaghan, E. P.; Navalta, L. A.; Shum, L.; Ashbrook, D. W.; Lee, D. A. *Epilepsia* **1997**, *38*, 1026–1031.
58. Laxer, K.; Blum, D.; Abou-Khalil, B. W.; Morrell, M. J.; Lee, D. A.; Data, J. L.; Monaghan, E. P. *Epilepsia* **2000**, *41*, 1187–1194.
59. Kerrigan, J. F.; Shields, W. D.; Nelson, T. Y.; Bluestone, D. L.; Dodson, W. E.; Bourgeois, B. F.; Pellock, J. M.; Morton, L. D.; Monaghan, E. P. *Epilepsy Res.* **2000**, *42*, 133–139.
60. Pieribone, V. A.; Tsai, J.; Soufflet, C.; Rey, E.; Shaw, K.; Giller, E.; Dulac, O. *Epilepsia* **2007**, *48*, 1870–1874.
61. Marinus Pharmaceuticals. Clin. [Internet]. Bethesda Natl. Libr. Med. NCT0196320, 2013.
62. Pharmaceuticals. M. Clin. [Internet]. Bethesda Natl. Libr. Med. NCT0167382, 2012.
63. Hawkinson, J. E.; Acosta-Burruel, M.; Yang, K. C.; Hogenkamp, D. J.; Chen, J. S.; Lan, N. C.; rewe, J. A.; Whittemore, E. R.; Woodward, R. M.; Carter, R. B.; Upasani, R. B. *J. Pharmacol. Exp. Ther.* **1998**, *287*, 198–207.
64. Edgar, D. M.; Seidel, W. F.; Gee, K. W.; Lan, N. C.; Field, G.; Xia, H.; Hawkinson, J. E.; Wieland, S.; Carter, R. B.; Wood, P. L. *J. Pharmacol. Exp. Ther.* **1997**, *282*, 420–429.
65. Vanover, K. E.; Rosenzweig-Lipson, S.; Hawkinson, J. E.; Lan, N. C.; Belluzzi, J. D.; Stein, L.; Barrett, J. E.; Wood, P. L.; Carter, R. B. *J. Pharmacol. Exp. Ther.* **2000**, *295*, 337–345.
66. Vanover, K.; Hogenkamp, D.; Lan, N.; Gee, K.; Carter, R. *Psychopharmacology (Berl.)* **2001**, *155*, 285–291.
67. Vanover, K. E.; Edgar, D. M.; Seidel, W. F.; Hogenkamp, D. J.; Fick, D. B.; Lan, N. C.; Gee, K. W.; Carter, R. B. *J. Pharmacol. Exp. Ther.* **1999**, *291*, 1317–1323.
68. Gasior, M.; Ungard, J. T.; Beekman, M.; Carter, R. B.; Witkin, J. M. *Neuropharmacology* **2000**, *39*, 1184–1196.

69. Kokate, T. G.; Yamaguchi, S.; Pannell, L. K.; Rajamani, U.; Carroll, D. M.; Grossman, A. B.; Rogawski, M. A. *J. Pharmacol. Exp. Ther.* **1998**, *287*, 553–558.
70. Upasani, R. B.; Askew, B. C.; Harrison, B. L.; Distefano, P. S.; Salituro, F. G.; Robichaud, A. J. 3,3 Disubstituted 19-nor Pregnane Compounds, Compositions, and Uses Thereof. WO2013056181, 2013.
71. Loya, C.; Hammond, R.; Ackley, M.; Reddy, K.; Maciag, C.; Christian, E.; Martinez, G.; Hoffman, E.; Salituro, F.; Robichaud, A.; Doherty, J. In American Epilepsy Society; 2013; p. 3.335.
72. Paoletti, P.; Bellone, C.; Zhou, Q. *Nat. Rev. Neurosci.* **2013**, *14*, 383–400.
73. Nowak, L.; Bregestovski, P.; Ascher, P.; Herbet, A.; Prochiantz, A. *Nature* **1984**, *307*, 462–465.
74. Harris, E. W.; Ganong, A. H.; Cotman, C. W. *Brain Res.* **1984**, *323*, 132–137.
75. Collingridge, G. L.; Volianskis, A.; Bannister, N.; France, G.; Hanna, L.; Mercier, M.; Tidball, P.; Fang, G.; Irvine, M. W.; Costa, B. M.; Monaghan, D. T.; Bortolotto, Z. A.; Molnár, E.; Lodge, D.; Jane, D. E. *Neuropharmacology* **2013**, *64*, 13–26.
76. Zarate, C. A.; Singh, J. B.; Carlson, P. J.; Brutsche, N. E.; Ameli, R.; Luckenbaugh, D. A.; Charney, D. S.; Manji. H. K. *Arch. Gen. Psychiatry* **2006**, *63*(8), 856–864.
77. Berman, R. M.; Cappiello, A.; Anand, A.; Oren, D. A.; Heninger, G. R.; Charney, D. S.; Krystal, J. H. *Biol. Psychiatry* **2000**, *47*, 351–354.
78. Monaghan, D. T.; Irvine, M. W.; Costa, B. M.; Fang, G.; Jane, D. E. *Neurochem. Int.* **2012**, *61*, 581–592.
79. Wu, F. S.; Gibbs, T. T.; Farb, D. H. *Mol. Pharmacol.* **1991**, *40*, 333–336.
80. Bowlby, M. R. *Mol. Pharmacol.* **1993**, *43*, 813–819.
81. Majewska, M. D.; Schwartz, R. D. *Brain Res.* **1987**, *404*, 355–360.
82. Malayev, A.; Gibbs, T. T.; Farb, D. H. *Br. J. Pharmacol.* **2002**, *135*, 901–909.
83. Horak, M.; Vlcek, K.; Petrovic, M.; Chodounska, H.; Vyklicky, L. *J. Neurosci.* **2004**, *24*, 10318–10325.
84. Ramirez, D. M. O.; Andersson, S.; Russell, D. W. *J. Comp. Neurol.* **2008**, *507*, 1676–1693.
85. Kotti, T.; Head, D. D.; McKenna, C. E.; Russell, D. W. *Proc. Natl. Acad. Sci. U.S.A.* **2008**, *105*, 11394–11399.
86. Kotti, T. J.; Ramirez, D. M. O.; Pfeiffer, B. E.; Huber, K. M.; Russell, D. W. *Proc. Natl. Acad. Sci. U.S.A.* **2006**, *103*, 3869–3874.
87. Madau, P.; Clark, A.; Neale, S.; Smith, L.; Hamilton, N.; Thomson, F.; Connick, J.; Belelli, D.; Lambert, J. Neuroscience Meeting Planner. Program #613.2. Society for Neuroscience: Chicago, IL, 2009; 2009.
88. Upasani, R. B.; Harrison, B. L.; Askew, B. C.; Dodart, J. C.; Salituro, F. G.; Robichaud, A. J. Neuroactive Steroids, Compositions, and Uses Thereof. WO2013036835, 2013.
89. Borovska, J.; Vyklicky, V.; Stastna, E.; Kapras, V.; Slavikova, B.; Horak, M.; Chodounska, H.; Vyklicky, L. *Br. J. Pharmacol.* **2012**, *166*, 1069–1083.
90. Therapeutics. S. Clin. [Internet]. Bethesda Natl. Libr. Med. NCT0205273, 2014.

CHAPTER FOUR

Development of LRRK2 Kinase Inhibitors for Parkinson's Disease

Paul Galatsis, Jaclyn Henderson, Bethany L. Kormos, Warren D. Hirst
Worldwide Medicinal Chemistry and Neuroscience Research Unit, Pfizer Global R&D, Cambridge, Massachusetts, USA

Contents

1. INTRODUCTION

Parkinson's disease[1] (PD) is the most common movement disorder and the second most common neurodegenerative disorder after Alzheimer's disease (AD). The etiology of PD is complex but the most common phenotype is the loss of dopaminergic neurons of the substantia nigra leading to the clinical symptoms of bradykinesia, resting tremors, rigidity, and postural instability. The predominant form of the disease is idiopathic, but a genetic disposition to PD has been observed. The most common genetic cause of autosomal dominant PD is mutations in leucine-rich repeat kinase 2 (LRRK2) which accounts for 1% of sporadic and 4% of familial cases. There are over 45 known LRRK2 mutations, but only 6 of these (N1437H, R1441C, R1441G, Y1699C, G2019S, and I2020T) can be considered as definitively disease causing, on the basis of cosegregation with disease in families, and an absence in controls.[2–4] The most common is G2019S and occurs in the kinase domain of the protein, increasing its activity. Recent studies suggest the GTPase mutations (R1441C/G and Y1699C) cause a

Annual Reports in Medicinal Chemistry, Volume 49
ISSN 0065-7743
http://dx.doi.org/10.1016/B978-0-12-800167-7.00004-3

two-fold reduction in GTPase activity which resulted in an increase in kinase activity[5] that can also be measured by autophosphorylation.[6] Consequently, a great deal of effort has been directed toward identifying potent LRRK2 kinase inhibitors with a biopharmaceutical profile congruent with clinical evaluation.

2. LRRK2 BIOLOGY

LRRK2 was first reported in 2004[2,3] and its genetic link to PD was later confirmed by several genome-wide association studies.[4] Examination of LRRK2 mutant carriers identified the G2019S point mutation as pathophysiological. The LRRK2 gene encodes a 286 kDa protein comprising 2527 amino acids.[2,7] LRRK2 protein belongs to the Roco family of proteins defined by a tandem ROC (Ras of complex proteins)—COR (C-terminal of Roc) motif.[8,9] As with the other members of the Roco family, LRRK2 contains multiple domains (Fig. 4.1): the enzymatic ROC–COR and kinase domains, which are flanked by putative protein–protein interaction domains, including the N-terminal armadillo repeats, ankyrin repeats, leucine-rich repeats, and the C-terminal WD40 repeats.[10,11]

The kinase domain is characterized as a Ser/Thr kinase in the TKL group of kinases and its highest homology is to LRRK1 and the receptor-interacting protein kinases.[10,12,13] Though crystallographic data for LRRK2 kinase domain have yet to be published, the crystal structure of the slime mold *Dictyostelium discoideum* Roco4 kinase domain may provide insight.[14] With a kinase domain that is ~30% identical and ~50% similar to that of LRRK2, as well as a similar domain structures, the *D. discoideum* Roco4 kinase domain may be a suitable crystallography surrogate to help build an understanding of LRRK2 kinase structure and mechanism.

Primary sequence alignment, homology modeling, and surrogate X-ray crystal structures agree that G2019S is the Gly in the conserved DFG motif (DYG in LRRK2) at the beginning of the activation loop in the LRRK2

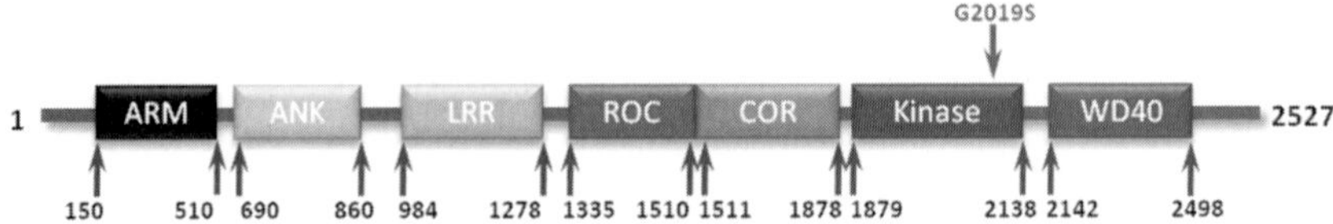

Figure 4.1 LRRK2 domain structure and location of the most common pathogenic mutation.

kinase domain (Fig. 4.2). Modeling of the G2019S mutation suggests that the change from a flexible Gly residue to a Ser causes the activation loop to be less flexible due to an increase in H-bonding interactions with D2017, E1920, or other nearby residues, increasing the population of conformations in the active state.[15–18] The X-ray crystal structure of Roco4 with the mutation that corresponds to LRRK2 G2019S (shown to have increased kinase activity compared to WT) was solved and revealed a H-bonding interaction which corresponds to G2019S and Q1919 in the αC-helix in LRRK2.[14] In addition, the double mutants in Roco4 corresponding to G2019S/Q1919A in LRRK2, in which the Ser is unable to make a H-bonding interaction to the αC-helix to stabilize the active conformation, have nearly wild-type activity. Metadynamics simulations combined with kinetics studies suggested that the energy barrier to achieve the inactive DYG-out conformation in LRRK2 is much higher for the G2019S mutant than for WT.[15] Together, this evidence suggests that an additional H-bond in the G2019S mutant stabilizes the active kinase conformation, causing an increase in kinase activity. This kinase gain of

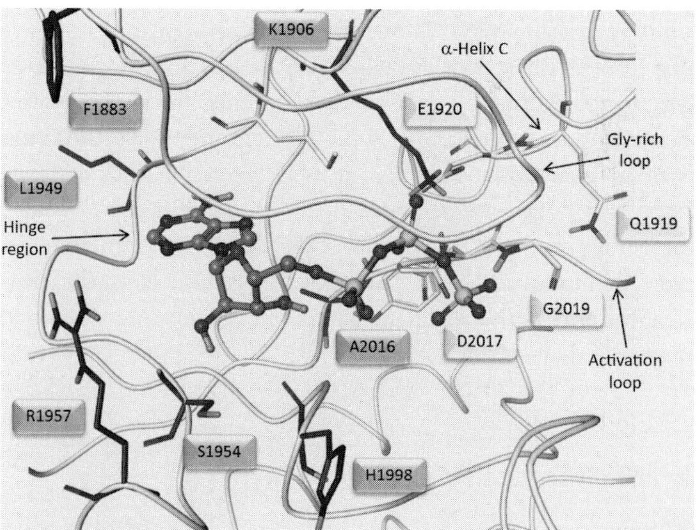

Figure 4.2 LRRK2 ATP-binding site. ATP (cyan (dark gray in the print version)) is shown docked into an LRRK2 homology model based on the X-ray crystal structure of JAK3 with CP-690550 (PDB ID: 3LXK). Location of the G2019S mutation and hypothesized interacting residues are highlighted in orange (light gray in the print version). Residues that have been targeted for LRRK2 specificity are highlighted in purple (dark gray in the print version).

function, determined using recombinant or isolated protein, has given rise to the working hypothesis that increased kinase function is somehow responsible for the PD phenotype.

Understanding LRRK2's role in normal and pathological biological pathways is still at a relatively early stage. For example, the endogenous substrate(s) for the kinase domain is (are) not known, but LRRK2 does autophosphorylate and recent data have demonstrated an increase in phosphorylation at S1292 by the pathological mutations.[6] There are additional phosphorylation sites[19] on LRRK2, including S910 and S935 and, while not autophosphorylation sites, they are clearly sensitive to LRRK2 kinase inhibitors and have been used extensively in cellular and *in vivo* studies (*vide infra*). Further, there has been significant recent progress identifying interaction partners[20] with a full understanding of LRRK2's function and the proteins that interact within different cell types remaining as an area of active research. The signaling pathways through which LRRK2 elicits its actions and others that it potentially modulates are emerging,[21] and its role in PD pathophysiology, while not definitive, is also developing.

Transgenic animals have also been studied in an attempt to gain greater insight into LRRK2 biology. In all cases, the animals are viable and exhibit normal life spans. While these animals appear to not have functional impairment, they do exhibit two distinct peripheral phenotypes: (1) the kidneys present with a dark color that show inclusion bodies upon microscopic examination and (2) type II pneumocytes of the lungs show changes in lamellar body morphology with altered surfactant secretion.[22] This latter phenotype has recently been observed with 7- and 29-day NHP toxicology studies.[23] It remains to be determined whether KO phenotypes are a consequence of a total lack of LRRK2 from the embryonic stage to adult and if the drug-mediated phenotypes are chemotype- or mechanism based.

3. MEDICINAL CHEMISTRY

LRRK2 kinase inhibitors that have been disclosed to date are ATP competitive. As one might expect, the LRRK2 ATP-binding site can accommodate known core scaffolds that incorporate 1-, 2-, and potentially 3-point hinge-binding motifs. Additionally, these compounds appear to prefer the DFG-in conformation, which may reflect the nature of the gain of function the pathophysiological mutations impart.

3.1. LRRK2 Patent Space Analysis

The first set of LRRK2 kinase inhibitors investigated were compounds repurposed from other kinase programs (*vide infra*). A handful of LRRK2-selective kinase inhibitors have been published in the primary literature; however, they are typically examples from a greater set of LRRK2 compounds claimed in the LRRK2 patent literature. The Markush structures for this set of kinase inhibitors are presented in Fig. 4.3. This represents a data set of over 4000 compounds from a variety of chemotypes displaying an equally diverse set of actual and calculated physicochemical properties.

Comparing this set of compounds, as a whole, to the data for CNS drugs and marketed kinases (Fig. 4.4) provides an overview of the LRRK2-patented chemical matter as it relates to physicochemical space. The box plots provide a summary of each property showing the data distribution with

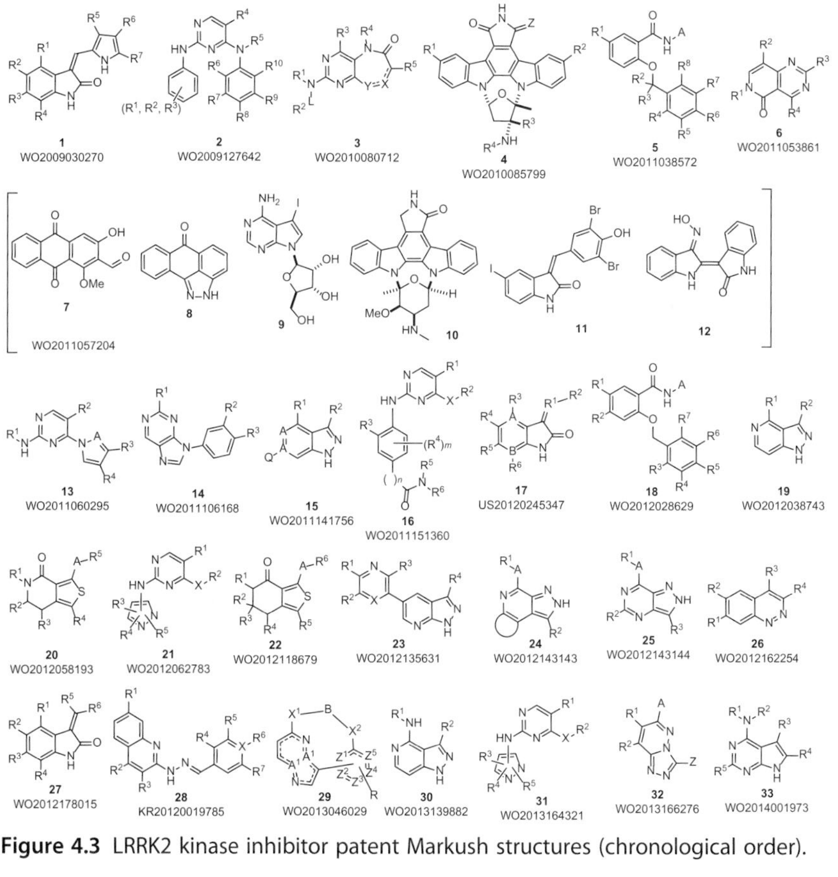

Figure 4.3 LRRK2 kinase inhibitor patent Markush structures (chronological order).

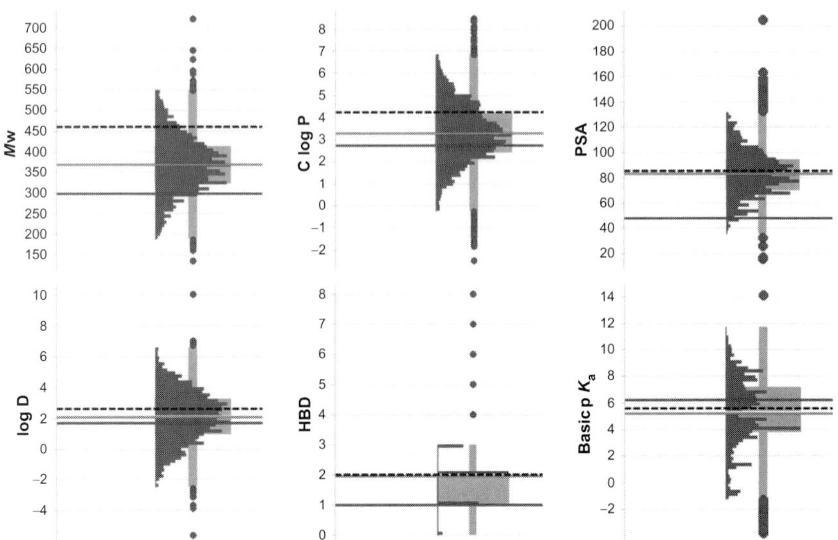

Figure 4.4 Physicochemical analysis of LRRK2 patent chemical matter (4389 compounds). Each box overlays the corresponding distribution for the descriptor on its box plots. Average: $M_W = 368.7$; $C\log P = 3.26$; PSA = 83.3 Å; $\log D = 2.09$; HBD = 1.9; basic $pK_a = 5.17$. The dots represent outliers and the lines denote: LRRK2 data set average (red (light gray in the print version)); CNS drug set average (green (dark gray in the print version)); marketed kinase drugs average (black dashed).

the average (red line; dark gray in the print version), the CNS drug set average (green line; dark gray in the print version), and the marketed kinase set (black dashed line). It becomes readily apparent that the current marketed kinase space is distinct and does not overlap very well with the CNS space defined by approved CNS drugs. While the marketed kinases are primarily for oncology indications, it does provide guidance for the medicinal chemist in how to adapt kinase design for a CNS target. Since the LRRK2 chemical matter defines a space that resides somewhere between these two areas, it remains to be determined how far the shift toward CNS space will need to be for an efficacious LRRK2 kinase inhibitor. The observation that very few of the compounds reported have any brain penetration may be providing an early indication of the direction for compound design.

3.2. Chemical Scaffolds

Twenty-eight patent applications of small molecules (Fig. 4.2), to date, represent roughly 10 unique chemical scaffolds. Through the use of structure–activity relationships, X-ray crystal structures, and homology models, specific residues in the LRRK2 ATP-binding pocket have been targeted for inhibitor interactions to improve potency and selectivity for LRRK2 kinase. Hydrophobic interactions with A2016 have been shown to be important for potency and selectivity of some LRRK2 kinase inhibitors.[24–27] Selectivity of these compounds is especially improved over kinases that have a more polar residue at that position—28% of kinases have a Ser or Thr at that position. Mutation of A2016 to Thr abolishes activity in some inhibitors,[24–26] but not others,[28–30] which is indicative of the inhibitor-binding modes.

The nonconserved residues L1949, S1954, R1957, and F1883 in the LRRK2 kinase domain (Fig. 4.2) were identified as potential selectivity handles due to their proximity to the ATP-binding pocket for a series of diaminopyrimidine compounds.[31–34] L1949 near the hinge region was targeted for selectivity since it is a smaller residue than Phe or Tyr, which are found in this position in approximately 60% of the kinome. Inhibitors with a nitrogen lone pair or substituents or that extend into this region can be accommodated with the Leu residue, but sterically clash with the larger Phe or Tyr. This strategy was used to obtain selectivity for LRRK2 over the JAK family of kinases and JAK2 was found to be a useful surrogate for predicting general kinome selectivity. Interactions with S1954 were targeted to achieve both general kinase selectivity and to overcome specific selectivity issues with the kinase TTK (MPS1). Approximately 55% of the kinome has a larger group at this position and approximately 50% has a negatively charged Asp or Glu residue in this position. Strategies to mitigate crossover to TTK, which has an Asp residue in this position, have included the introduction of small groups that would cause unfavorable steric and/or electrostatic interactions with the Asp side chain.[32,34] Selectivity of a series of 7-aryl-substituted quinoline derivatives may also be due to the placement of polar groups in the vicinity of the nonconserved S1954 and R1957 residues.[35] Interactions with the nonconserved R1957 and H1998 LRRK2 residues may also contribute to selectivity in a series of indolinone compounds.[27] Optimizing interactions with H1998 and the catalytic K1906 in this series of compounds was hypothesized to improve kinase selectivity, especially over RET kinase.

Alternative strategies used to obtain selectivity in kinases are through type II inhibitors, which stabilize an inactive DFG-out conformation, or type III inhibitors, which are not ATP competitive. Only a handful of LRRK2 inhibitors have been reported that fall into these categories.[15,36] These inhibitors were shown to be more potent against WT LRRK2 kinase compared to G2019S, supporting the hypothesis that the pathogenic

G2019S mutation stabilizes the active conformation of LRRK2.[15–18] This suggests a type II kinase inhibitor may not be the optimal approach for the treatment of PD due to the LRRK2 G2019S mutation. The X-ray crystal structure of Roco4 kinase domain cocrystallized with the inhibitor H1152, **35**, revealed two binding sites for this inhibitor: one in the expected ATP-binding site and the other close to the αC-helix.[14] The implications of this second binding site for LRRK2 are unclear, but suggest not completely ruling out the possibility of developing type III inhibitors for LRRK2 kinase.

3.2.1 Repurposed Kinase Inhibitors

Early work in the field of LRRK2 kinase inhibitor development sought to complement the range of biochemical tools that were becoming available[37] with small-molecule kinase inhibitors. By screening panels of commercial compounds, groups began to identify known kinase inhibitors that also displayed activity against LRRK2.[25,38,39] Initially identified compounds, using Lanthascreen detection and a surrogate peptide substrate (LRRKtide), included **7–9** and **11–12**. Staurosporine, **10**, was one of the most potent LRRK2 inhibitors reported (wt $IC_{50}=6$ nM); however, it is also highly promiscuous across the kinome.[40] Capitalizing on this observation, **4** illustrates how modifications to the bis-indole core were pursued to potentially address the selectivity issue. Sunitinib, **34**, is a potent LRRK2 inhibitor (wt $IC_{50}=37$ nM), but also displays similar promiscuity. Related indolidinones (**1**, **17**, and **27**) have been disclosed. Screening of a number of Rho-kinase inhibitors led to the identification of **35**.[25] Although it has a number of off-target effects, selectivity is much improved over **10** and **34**, although potency against LRRK2 was also reduced (wt $IC_{50}=244$ nM).

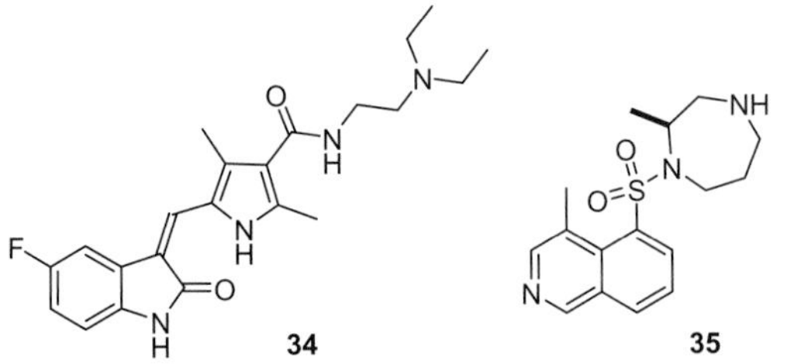

Although these early inhibitors are structurally different, they have in common potent activity across a range of kinases and thus it is difficult to separate effects at LRRK2 from off-target effects, especially when examined in isolation. These compounds proved useful for initial study of the function of LRRK2; however, it was clear that the field could benefit from more selective compounds.

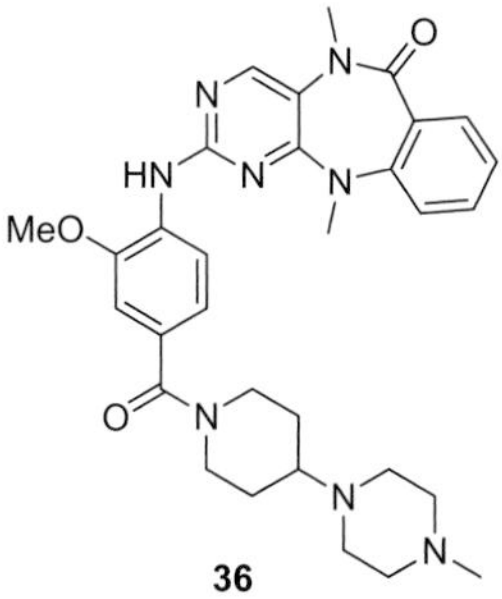

36 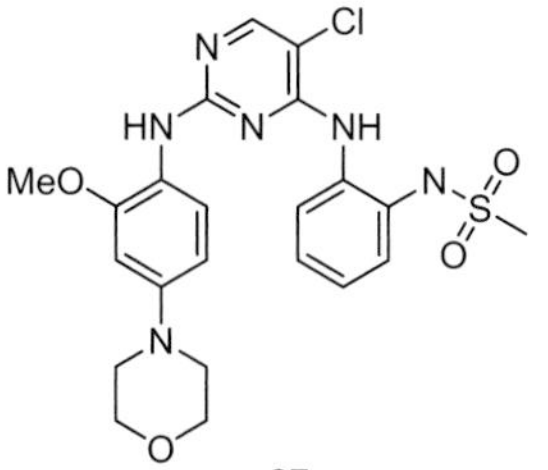 37 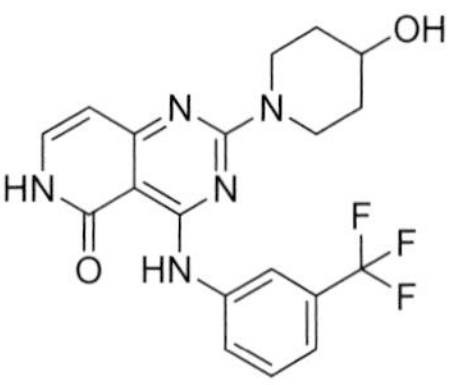 38

3.2.2 First-Generation LRRK2-Focused Kinase Inhibitors

In 2011, the first LRRK2-optimized kinase inhibitor LRRK2-IN-1, **36**, was reported (derived from **3**).[26] Due to its high potency against both wild-type and mutant LRRK2 (IC_{50} values of 13 and 6 nM, respectively, measured in cell-free enzymatic assays), and its improved selectivity across a range of kinases, it was clearly a vast improvement on previous compounds and was initially employed throughout the field as a selective LRRK2 inhibitor. While **36** (CNS MPO[41] = 4.52) is potent against both LRRK2 WT and G2019 mutant enzymes, it has been demonstrated to display moderate potency in a whole cell assay, 200–600 nM.[29,42] Further, high doses are required to observe any *in vivo* effects due to its low permeability and high levels of plasma protein binding. Additionally, no CNS effects were observed as **36** had low brain availability (unbound C_{brain}/unbound C_{plasma}). Although **36** demonstrated improved selectivity in comparison to previous LRRK2 inhibitors, it has some significant off-target potency, most notably displaying equipotent activity against ERK5 (BMK1, MAPK7), DCAMKL, PLK1, and PLK4. This off-target activity has been found to confound some of the biological assays in which the effects of **36** are attributed to LRRK2, in particular when assessing effects on neurite outgrowth and inflammation end points,[42] consequently, use of **36** has diminished.

3.2.3 Second-Generation LRRK2-Focused Kinase Inhibitors

As intensive research into the function and dysfunction of LRRK2 and its mutants continued, a number of other tool compounds and inhibitor series have been published from both academic and industrial laboratories. The most potent of these derived from **2**, in both enzyme (5–7 nM in WT and G2019S, respectively) and cell-based assays (attenuation of G2019S- and R1441C-induced neuronal injury and death in a concentration-dependent manner with an EC_{50} of <10 nM), was CZC-54252 (**37**) reported in 2011.[43] Unfortunately, **37** had disappointing selectivity across the kinome. Although CZC-54252 may prove a useful tool for cell-based assays, its poor CNS MPO[41] score (2.12) suggested it was outside of CNS space which will limit its application in *in vivo* models (confirmed with a reported brain availability of 0.05).

The diaminopyrimidine core has been extensively examined by several groups. In addition to compounds derived from **2** and **3** (*vide infra*), variations including analogs from **6**, **13**, **16**, **21**, and **31** have been reported. An early example is exemplified by **38** (CNS MPO[41] = 3.97), but more advanced analogs successfully balanced good physicochemical properties with high degrees of LRRK2 selectivity.[44,45] Focusing on diaminopyrimidine HTS hits, due to high ligand efficiency and reasonable PK properties, optimization of the hits using an LRRK2 homology model to identify both key interactions for LRRK2 potency and sites that may offer enhanced kinase selectivity yielded **39** (CNS MPO[41] = 5.37) with improved kinase selectivity over their initial hit, suitable *in vivo* exposures, and similar cellular activity (IC_{50} = 389 nM) to **36**. While disclosed in a patent the previous year, **39** was independently reported as HG-10-102-01 and reported to exhibit *in vivo* mouse brain activity.[31]

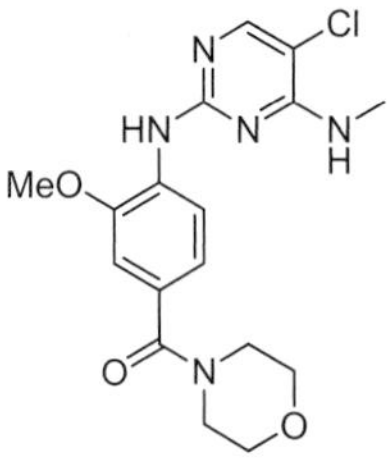

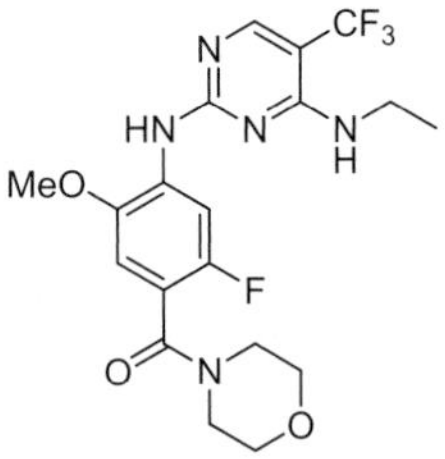

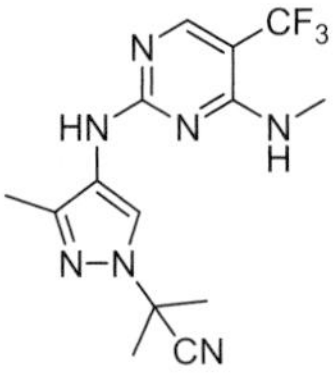

Having identified TTK (MPS1) as the major off-target interaction of this series, and significant safety concern, continued optimization using structure-based drug design initially arrived at **40** (CNS MPO[41] = 4.60), a compound having excellent selectivity for LRRK2.[28] The *in vivo* activity of this compound and related analogs was shown using dephosphorylation of S1292 as a measure of LRRK2 kinase activity. Although **40** is an excellent tool compound for probing the effect of LRRK2 inhibition *in vivo*, it has some major liabilities, thus further optimization was needed for progression toward a potential clinical candidate. As the diaminopyrimidine forms the key hinge interactions with LRRK2, optimization was focused around the side chain. The aniline motif, which could lead to idiosyncratic toxicity findings, was replaced with an aminopyrazole, and side chains to enhance aqueous solubility were incorporated.[32–34] This optimization leads to GNE-0877, **41** (CNS MPO[41] = 5.45), which has been progressed to preclinical safety studies.

3.2.4 Third-Generation LRRK2-Focused Kinase Inhibitors

Moving away from diaminopyrimidines, several other chemotypes have been reported. A series defined by **5** and **18**, while likely to be ATP competitive, do not contain a common hinge-binding motif.[34] This attribute may be responsible for the excellent kinase selectivity. GSK2578215A, **42**, is a representative analog from this series and showed a CNS MPO[41] = 3.82 (IC_{50} = 47 nM). It also suffered from poor rodent PK, with a relatively low oral bioavailability (%F = 12) and half-life of 1.1 h.[46] The lack of LRRK2 inhibition observed in brain, in comparison to kidney and spleen, suggested low levels of unbound compound in brain, potentially limiting the use of this series for *in vivo* models.

Quinoline derivatives, **26** and **28**, are known single-point hinge binders and have been found to be good LRRK2 kinase inhibitors.[30,47] Although the cinnoline variants[47] were potent LRRK2 inhibitors (wt IC_{50} = 7 nM), they were found to be promiscuous in a small kinase panel. By focusing the core to quinoline,[35] e.g., **43**, kinase selectivity was significantly improved and *in vivo* activity was demonstrated (CNS MPO[41] = 5.05).

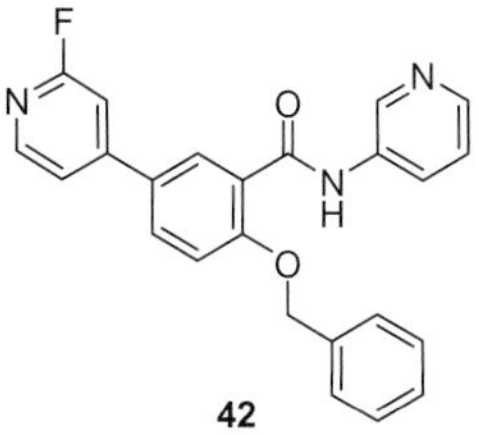

F N O N H O N NH CN N N N O N-N O N N N H

42 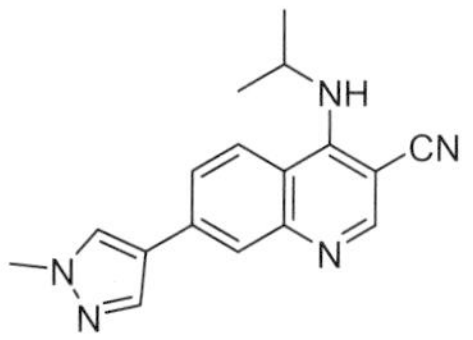 **43** 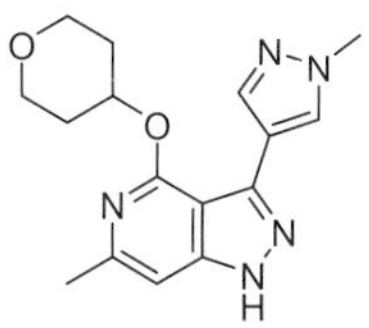 **44**

Pyrazolopyri(mi)dines, **15**, **19**, **23–25**, and **30**, have been well represented in the patent literature. For analogs, such as **44** (CNS MPO[41] = 5.83), potency is comparable to **36** with overall good ADME.[46] Alternate 6,5-fused ring systems, **14**, **20**, and **22**, have been reported; however, the pyrrolopyrimidine **33** appears to have the best alignment of properties. Specifically, **45** (CNS MPO[41] = 5.83) is one of the most active LRRK2 inhibitors *in vivo* (brain free drug wt $IC_{50}=15$ nM) along with high kinase selectivity.[48]

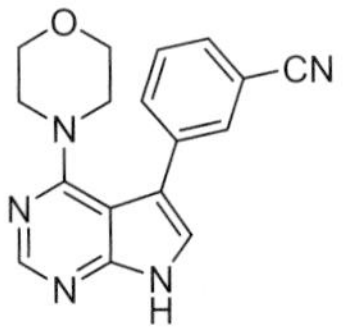

45 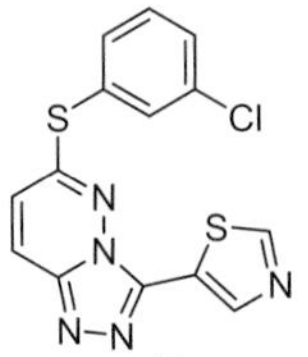 **46**

One final variation of the 6,5-fused ring system is illustrated by **32**. These compounds are also postulated to be single-point hinge binders, interacting through one of the triazole nitrogens. A representative compound from this series, **46** (CNS MPO[41] = 4.88), displayed excellent *in vitro* potency (wt $IC_{50}=31$ nM; G2019S $IC_{50}=8$ nM) against LRRK2 but this did not translate to cellular assays resulting in a 100-fold right shift.[49] As witnessed with other scaffolds (*vide supra*), compounds from this chemotype would make good tool compounds for probing LRRK2 *in vitro* biology.[50]

4. PRECLINICAL ANIMAL MODELS

Studies in *Drosophila melanogaster*[51] and *Caenorhabditis elegans*[52] have provided important insights into LRRK2 toxicity in these species. The LRRK2 transgenic mouse models created to date do not completely recapitulate the hallmarks of PD (i.e., dopaminergic neuronal loss, α-synuclein accumulation, the development of Lewy bodies, and behavioral phenotype),[53–55] and with the exception of pharmacodynamic end points, measuring reduction of LRRK2 phosphorylation at S935 and S1292 has not been used, to date, for any long-term LRRK2 inhibitor studies.[6] In contrast, viral overexpression of LRRK2 does result in dopaminergic neuronal loss,[39,56] but these models need to be fully validated with the more selective, brain-penetrant compounds described above.

Toxins such as 6-hydroxydopamine (6-OHDA) or 1-methyl-4-phenyl-1,2,3,6-tetra-hydropyridine (MPTP) have been effective in acutely inducing dopaminergic cell loss.[57] However, the LRRK2 knockout mouse displays the same sensitivity to MPTP as wild-type mice[58] suggesting that LRRK2 may not be involved in the events downstream of the mitochondrial toxicity in the mouse brain. There are no published reports on 6-OHDA *in vivo* models and LRRK2. The use of other toxins, such as lipopolysaccharide (LPS), may provide a more relevant model as LRRK2 kinase levels are increased by LPS *in vitro*[59,60] and *in vivo*.[60]

The lack of validated mammalian preclinical models of LRRK2 has been an impediment to the development of LRRK2 kinase inhibitors. At the same time, development of LRRK2 preclinical models has also been hampered by the lack of suitable tool compounds that possess the appropriate pharmacokinetic properties and safety profiles for long-term dosing studies that are likely to be required.

5. CONCLUSIONS

LRRK2 is a target for PD in which the biology is rapidly developing. Its complex protein domain structure adds to the challenge of understanding its cellular function and pathophysiological role. A thorough understanding of the protein domain organization and identification of interacting partners will be important in determining the underlying mechanism of LRRK2 pathophysiology. X-ray crystal structures of the LRRK2 enzymatic domains, including both WT and the pathogenic mutations, could contribute to an understanding of these questions and would be a key accomplishment to help drive inhibitor design.

The strength of research in the field of LRRK2, both academic and industrial, has generated a wealth of chemical tools for the study of LRRK2. Where the initially reported compounds had activity across multiple kinase targets, new inhibitors have a high degree of kinome selectivity along with improved physicochemical properties, providing opportunities to study inhibition of LRRK2 kinase activity from the isolated enzyme through to *in vivo* assays. Ultimately, this collection of inhibitors should allow researchers the flexibility to design experiments to probe LRRK2 function and develop LRRK2 efficacy models of PD.

REFERENCES

1. For a recent PD review Jellinger, K. A. *Expert. Rev. Neurother.* **2014**, *14*, 199.
2. Paisan-Ruiz, C.; Jain, S.; Evans, E. W.; Gilks William, P.; Simon, J.; van der Brug, M.; Lopez de Munain, A.; Aparicio, S.; Gil Angel, M.; Khan, N.; Johnson, J.; Martinez Javier, R.; Nicholl, D.; Carrera Itxaso, M.; Pena Amets, S.; de Silva, R.; Lees, A.; Marti-Masso Jose, F.; Perez-Tur, J.; Wood Nick, W.; Singleton Andrew, B. *Neuron* **2004**, *44*, 595.
3. Zimprich, A.; Biskup, S.; Leitner, P.; Lichtner, P.; Farrer, M.; Lincoln, S.; Kachergus, J.; Hulihan, M.; Uitti, R. J.; Calne, D. B.; Stoessel, A. J.; Pfeiffer, R. F.; Patenge, N.; Cabajal, I. C.; Vieregge, P.; Asmus, F.; Muller-Myhsok, B.; Dickson, D. W.; Meitinger, T.; Strom, T. M.; Wszolek, Z. K.; Gasser, T. *Neuron* **2004**, *44*, 601.

4. For a recent PD genetics review Labbe, C.; Ross, O. A. *Curr. Genomics* **2014**, *15*, 2.
5. Liao, J.; Wu, C.-X.; Burlak, C.; Zhang, S.; Sahm, H.; Wang, M.; Zhang, Z.-Y.; Vogel, K. W.; Federici, M.; Riddle, S. M.; Nichols, R. J.; Liu, D.; Cookson, M. R.; Stone, T. A.; Hoang, Q. Q. *Proc. Natl. Acad. Sci. U.S.A.* **2014**, *111*, 4055.
6. Sheng, Z.; Zhang, S.; Bustos, D.; Kleinheinz, T.; Le Pichon, C. E.; Dominguez, S. L.; Solanoy, H. O.; Drummond, J.; Zhang, X.; Ding, X.; Cau, F.; Song, Q.; Li, X.; Yue, Z.; van der Berg, M. P.; Burdick, D. J.; Gunzer-Toste, J.; Chen, H.; Liu, X.; Estrada, A. A.; Sweeney, Z. K.; Scearce-Levie, K.; Moffat, J. G.; Kirkpatrick, D. S.; Zhu, H. *Sci. Transl. Med.* **2012**, *4*, 164ra161.
7. www.uniprot.org/uniprot/Q5S007 (accessed 01.05.14).
8. Bosgraaf, L.; Van Haastert, P. J. M. *Biochim. Biophys. Acta* **2003**, *1643*, 5.
9. Marin, I.; van Egmond, W. N.; van Haastert, P. J. M. *FASEB J.* **2008**, *22*, 3103.
10. Marin, I. *Mol. Biol. Evol.* **2006**, *23*, 2423.
11. Mills, R. D.; Mulhern, T. D.; Cheng, H.-C.; Culvenor, J. G. *Biochem. Soc. Trans.* **2012**, *40*, 1086.
12. Manning, G.; Whyte, D. B.; Martinez, R.; Hunter, T.; Sudarsanam, S. *Science* **2002**, *298*, 1912.
13. Zhang, D.; Lin, J.; Han, J. *Cell. Mol. Immunol.* **2010**, 7, 243.
14. Gilsbach, B. K.; Ho, F. Y.; Vetter, I. R.; van Haastert, P. J. M.; Wittinghofer, A.; Kortholt, A. *Proc. Natl. Acad. Sci. U.S.A.* **2012**, *109*, 10322.
15. Liu, M.; Bender, S. A.; Cuny, G. D.; Sherman, W.; Glicksman, M.; Ray, S. S. *Biochemistry* **2013**, *52*, 1725.
16. Drolet, R. E.; Sanders, J. M.; Kern, J. T. *J. Neurogenet.* **2011**, *25*, 140.
17. Anand, V. S.; Braithwaite, S. P. *FEBS J.* **2009**, *276*, 6428.
18. Yun, H.; Heo, H. Y.; Kim, H. H.; DooKim, N.; Seol, W. *Bioorg. Med. Chem. Lett.* **2011**, *21*, 2953.
19. Dzamko, N.; Deak, M.; Hentati, F.; Reith, A. D.; Prescott, A. R.; Alessi, D. R.; Nichols, R. J. *Biochem. J.* **2010**, *430*, 405.
20. Beilina, A.; Rudenko, I. N.; Kaganovich, A.; Civiero, L.; Chau, H.; Kalia, S. K.; Kalia, L. V.; Lobbestrael, E.; Chia, R.; Ndukwe, K.; Ding, J.; Nalls, M. A.; International Parkinson's Disease Genomic Consortium; North American Brain Expression Consortium; Olszewski, M.; Hauser, D. N.; Kumaran, R.; Lozano, A. M.; Baikelandt, V.; Greene, L. E.; Taymans, J.-M.; Greggio, E.; Cookson, M. R. *Proc. Natl. Acad. Sci. U.S.A.* **2014**, *111*, 2626.
21. Berwick, D. C.; Harvey, K. *Trends Cell Biol.* **2011**, *21*, 257.
22. For a leading reference Miklavc, P.; Ehinger, K.; Thompson, K. E.; Hobi, N.; Shimshek, D. R.; Frick, M. *PLoS One* **2014**, *9*, e84926.
23. Fuji, R. Nonclinical Safety Studies of Selective LRRK2 Kinase Inhibitors. Presented at, In: *Seventh Annual Parkinson's Disease Therapeutics Conference, The New York Academy of Sciences, October 24*, 2013.
24. Deng, X.; Elkins Jonathan, M.; Zhang, J.; Yang, Q.; Erazo, T.; Gomez, N.; Choi Hwan, G.; Wang, J.; Dzamko, N.; Lee, J.-D.; Sim, T.; Kim, N.; Alessi Dario, R.; Lizcano Jose, M.; Knapp, S.; Gray Nathanael, S. *Eur. J. Med. Chem.* **2013**, *70C*, 758.
25. Nichols, R. J.; Dzamko, N.; Hutti, J. E.; Cantley, L. C.; Deak, M.; Moran, J.; Bamborough, P.; Reith, A. D.; Alessi, D. R. *Biochem. J.* **2009**, *424*, 47.
26. Deng, X.; Dzamko, N.; Prescott, A.; Davies, P.; Liu, Q.; Yang, Q.; Lee, J.-D.; Patricelli, M. P.; Nomanbhoy, T. K.; Alessi, D. R.; Gray, N. S. *Nat. Chem. Biol.* **2011**, 7, 203.
27. Troxler, T.; Greenidge, P.; Zimmermann, K.; Desrayaud, S.; Druckes, P.; Schweizer, T.; Stauffer, D.; Rovelli, G.; Shimshek, D. R. *Bioorg. Med. Chem. Lett.* **2013**, *23*, 4085.

28. Choi, H. G.; Zhang, J.; Deng, X.; Hatcher, J. M.; Patricelli, M. P.; Zhao, Z.; Alessi, D. R.; Gray, N. S. *ACS Med. Chem. Lett.* **2012**, *3*, 658.
29. Zhang, J.; Deng, X.; Choi, H. G.; Alessi, D. R.; Gray, N. S. *Bioorg. Med. Chem. Lett.* **2012**, *22*, 1864.
30. Reith, A. D.; Bamborough, P.; Jandu, K.; Andreotti, D.; Mensah, L.; Dossang, P.; Choi, H. G.; Deng, X.; Zhang, J.; Alessi, D. R.; Gray, N. S. *Bioorg. Med. Chem. Lett.* **2012**, *22*, 5625.
31. Chen, H.; Chan, B. K.; Drummond, J.; Estrada, A. A.; Gunzner-Toste, J.; Liu, X.; Liu, Y.; Moffat, J.; Shore, D.; Sweeney, Z. K.; Tran, T.; Wang, S.; Zhao, G.; Zhu, H.; Burdick, D. J. *J. Med. Chem.* **2012**, *55*, 5536.
32. Estrada, A. A.; Liu, X.; Baker-Glenn, C.; Beresford, A.; Burdick, D. J.; Chambers, M.; Chan, B. K.; Chen, H.; Ding, X.; DiPasquale, A. G.; Dominguez, S. L.; Dotson, J.; Drummond, J.; Flagella, M.; Flynn, S.; Fuji, R.; Gill, A.; Gunzner-Toste, J.; Harris, S. F.; Heffron, T. P.; Kleinheinz, T.; Lee, D. W.; Le Pichon, C. E.; Lyssikatos, J. P.; Medhurst, A. D.; Moffat, J. G.; Mukund, S.; Nash, K.; Scearce-Levie, K.; Sheng, Z.; Shore, D. G.; Tran, T.; Trivedi, N.; Wang, S.; Zhang, S.; Zhang, X.; Zhao, G.; Zhu, H.; Sweeney, Z. K. *J. Med. Chem.* **2012**, *55*, 9416.
33. Chan, B. K.; Estrada, A. A.; Chen, H.; Atherall, J.; Baker-Glenn, C.; Beresford, A.; Burdick, D. J.; Chambers, M.; Dominguez, S. L.; Drummond, J.; Gill, A.; Kleinheinz, T.; Le Pichon, C. E.; Medhurst, A. D.; Liu, X.; Moffat, J. G.; Nash, K.; Scearce-Levie, K.; Sheng, Z.; Shore, D. G.; Van de Poel, H.; Zhang, S.; Zhu, H.; Sweeney, Z. K. *ACS Med. Chem. Lett.* **2013**, *4*, 85.
34. Estrada, A. A.; Chan, B. K.; Baker-Glenn, C.; Beresford, A.; Burdick, D. J.; Chambers, M.; Chen, H.; Dominguez, S. L.; Dotson, J.; Drummond, J.; Flagella, M.; Fuji, R.; Gill, A.; Halladay, J.; Harris, S. F.; Heffron, T. P.; Kleinheinz, T.; Lee, D. W.; Pichon, C. E. L.; Liu, X.; Lyssikatos, J. P.; Medhurst, A. D.; Moffat, J. G.; Nash, K.; Scearce-Levie, K.; Sheng, Z.; Shore, D. G.; Wong, S.; Zhang, S.; Zhang, X.; Zhu, H.; Sweeney, Z. K. *J. Med. Chem.* **2014**, *57*, 921.
35. Garofalo, A. W.; Adler, M.; Aubele, D. L.; Brigham, E. F.; Chian, D.; Franzini, M.; Goldbach, E.; Kwong, G. T.; Motter, R.; Probst, G. D.; Quinn, K. P.; Ruslim, L.; Sham, H. L.; Tam, D.; Tanaka, P.; Truong, A. P.; Ye, X. M.; Ren, Z. *Bioorg. Med. Chem. Lett.* **2013**, *23*, 1974.
36. Liu, M.; Poulose, S.; Schuman, E.; Zaitsev, A. D.; Dobson, B.; Auerbach, K.; Seyb, K.; Cuny, G. D.; Glicksman, M. A.; Stein, R. L.; Yue, Z. *Anal. Biochem.* **2010**, *404*, 186.
37. Liou, G.-. Y.; Gallo, K. A. *Biochem. J.* **2009**, *424*, e1.
38. Covy, J. P.; Giasson, B. I. *Biochem. Biophys. Res. Commun.* **2009**, *378*, 473.
39. Lee, B. D.; Shin, J.-. H.; VanKampen, J.; Petrucelli, L.; West, A. B.; Ko, H. S.; Lee, Y.-. I.; Maguire-Zeiss, K. A.; Bowers, W. J.; Federoff, H. J.; Dawson, V. L.; Dawson, T. M. *Nat. Med.* **2010**, *16*, 998.
40. Davis, M. I.; Hunt, J. P.; Herrgard, S.; Ciceri, P.; Wodicka, L. M.; Pallares, G.; Hocker, M.; Treiber, D. K.; Zarrinkar, P. P. *Nat. Biotechnol.* **2011**, *29*, 1046.
41. Wager, T. T.; Hou, X.; Verhoest, P. R.; Villalobos, A. *ACS Chem. Neurosci.* **2010**, *1*, 435.
42. Luerman, G. C.; Nguyen, C.; Samaroo, H.; Loos, P.; Xi, H.; Hurtado-Lorenzo, A.; Needle, E.; Noell, G. S.; Galatsis, P.; Dunlop, J.; Geoghegan, K. F.; Hirst, W. D. *J. Neurochem.* **2014**, *128*, 561.
43. Ramsden, N.; Perrin, J.; Ren, Z.; Lee, B. D.; Zinn, N.; Dawson, V. L.; Tam, D.; Bova, M.; Lang, M.; Drewes, G.; Bantscheff, M.; Bard, F.; Dawson, T. M.; Hopf, C. *ACS Chem. Biol.* **2011**, *6*, 1021.
44. Lee, J.; Song, H.-J.; Koh, J. S.; Lee, H. K.; Kim, Y.; Chang, S.; Kim, H. W.; Lim, S.-H.; Choi, J.-S.; Lim, S.-H.; Kim, S.-W. Patent Application WO2011060295, 2011.

45. Baker-Glenn, C.; Burdick, D. J.; Chambers, M.; Chan, B. K.; Chen, H.; Estrada, A.; Guzner, J. L.; Shore, D.; Sweeney, Z. K.; Wang, S.; Zhao, G. Patent Application WO2011151360, 2011.
46. Nicols, P. L.; Eatherton, A. J.; Bamborough, P.; Jandu, K. J.; Phillips, O. J.; Andreotti, D. Patent Application WO2011038572, 2011.
47. Garofalo, A. W.; Adler, M.; Aubele, D. L.; Bowers, S.; Franzini, M.; Goldbach, E.; Lorentzen, C.; Neitz, R. J.; Probst, G. D.; Quinn, K. P.; Santiago, P.; Sham, H. L.; Tam, D.; Truong, A. P.; Ye, X. M.; Ren, Z. *Bioorg. Med. Chem. Lett.* **2013**, *23*, 71.
48. Galatsis, P.; Hayward, M. M.; Kormos, B. L.; Wager, T. T.; Zhang, L.; Stepan, A. F.; Henderson, J. L.; Kurumbail, R. G.; Verhoest, P. R. Patent Application WO2014001973, 2014.
49. Franzini, M.; Ye, X. M.; Adler, M.; Aubele, D. L.; Garofalo, A. W.; Gauby, S.; Goldbach, E.; Probst, G. D.; Quinn, K. P.; Santiago, P.; Sham, H. L.; Tam, D.; Truong, A. P.; Ren, Z. *Bioorg. Med. Chem. Lett.* **2013**, *23*, 1967.
50. Galatsis, P.; Henderson, J. L.; Kormos, B. L.; Han, S.; Kurumbail, R. G.; Wager, T. T.; Verhoest, P. R.; Noell, G. S.; Chen, Y.; Needle, E.; Berger, Z.; Steyn, S. J.; Houle, C.; Hirst, W. D. *Bioorg. Med. Chem. Lett.* **2014**, http://dx.doi.org/10.1016/j.bmcl.2014.07.052.
51. Hindle, S. J.; Elliott, C. J. *Autophagy* **2013**, *9*, 936.
52. Yao, C.; Johnson, W. M.; Gao, Y.; Wang, W.; Zhang, J.; Deak, M.; Alessi, D. R.; Zhu, X.; Mieyal, J. J.; Roder, H.; Wilson-Delfosse, A. L.; Chen, S. G. *Hum. Mol. Genet.* **2013**, *22*, 328.
53. Li, X.; Patel, J. C.; Wang, J.; Avshalumov, M. V.; Nicholson, C.; Buxbaum, J. D.; Elder, G. A.; Rice, M. E.; Yue, Z. *J. Neurosci.* **2010**, *30*, 1788.
54. Li, Y.; Liu, W.; Oo, T. F.; Wang, L.; Tang, Y.; Jackson-Lewis, V.; Zhou, C.; Geghman, K.; Bogdanov, M.; Przedborski, S.; Beal, M. F.; Burke, R. E.; Li, C. *Nat. Neurosci.* **2009**, *12*, 826.
55. Ramonet, D.; Daher, J. P.; Lin, B. M.; Stafa, K.; Kim, J.; Banerjee, R.; Westerlund, M.; Pletnikova, O.; Glauser, L.; Yang, L.; Liu, Y.; Swing, D. A.; Beal, M. F.; Troncoso, J. C.; McCaffery, J. M.; Jenkins, N. A.; Copeland, N. G.; Galter, D.; Thomas, B.; Lee, M. K.; Dawson, T. M.; Dawson, V. L.; Moore, D. J. *PLoS One* **2011**, *6*, e18568.
56. Dusonchet, J.; Kochubey, O.; Stafa, K.; Young, S. M., Jr.; Zufferey, R.; Moore, D. J.; Schneider, B. L.; Aebischer, P. *J. Neurosci.* **2011**, *31*, 907.
57. Tieu, K. *Cold Spring Harb. Perspect. Med.* **2011**, *1*, a009316.
58. Andres-Mateos, E.; Mejias, R.; Sasaki, M.; Li, X.; Lin, B. M.; Biskup, S.; Zhang, L.; Banerjee, R.; Thomas, B.; Yang, L.; Liu, G.; Beal, M. F.; Huso, D. L.; Dawson, T. M.; Dawson, V. L. *J. Neurosci.* **2009**, *29*, 15846.
59. Gardet, A.; Benita, Y.; Li, C.; Sands, B. E.; Ballester, I.; Stevens, C.; Korzenik, J. R.; Rioux, J. D.; Daly, M. J.; Xavier, R. J.; Podolsky, D. K. *J. Immunol.* **2010**, *185*, 5577.
60. Moehle, M. S.; Webber, P. J.; Tse, T.; Sukar, N.; Standaert, D. G.; DeSilva, T. M.; Cowell, R. M.; West, A. B. *J. Neurosci.* **2012**, *32*, 1602.

CHAPTER FIVE

Stimulating Neurotrophin Receptors in the Treatment of Neurodegenerative Disorders

Gunnar Nordvall, Pontus Forsell
AlzeCure Foundation, Karolinska Institutet Science Park, Huddinge, Sweden

Contents

1. INTRODUCTION

Neurons are the building blocks of both the central nervous system (CNS), including brain and spinal cord, and the peripheral nervous system, including the somatic, autonomic, and sensory nervous systems. There are several trophic factors that are necessary for the development and maintenance of neurons. Neurotrophins (NTs) are one class of trophic factors that promote the development, health, and survival of neurons, and thus they exert considerable control over the neuronal homeostasis.[1,2]

The lack of disease-modifying treatments for several neurodegenerative disorders, including Alzheimer's disease (AD), Parkinson's disease (PD), and Huntington's disease, has for several decades prompted the search for suitable ways to administer neuronal growth factors or for the identification of small molecules with neurotrophic support that will increase neuronal survival and function. The use of recombinant proteins such as nerve growth

Annual Reports in Medicinal Chemistry, Volume 49
ISSN 0065-7743
http://dx.doi.org/10.1016/B978-0-12-800167-7.00005-5

factor (NGF) or brain-derived growth factor (BDNF) is limited due to their poor pharmacokinetics and pleiotropic side effects.

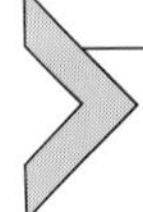

2. NTs AND NT RECEPTORS—STRUCTURE AND FUNCTION

The NTs, including NGF, BDNF, and NT-3–6, bind the tropomyosin-related kinase (Trk) family of receptor tyrosine kinases (TrkA, TrkB, and TrkC). In addition, each NT also binds to the common p75, a member of the tumor necrosis factor receptor superfamily. NGF binds with high affinity to TrkA, whereas BDNF and NT-4/5 bind to TrkB and NT-3 binds predominantly to TrkC, but it can also bind to TrkA and TrkB, albeit with less affinity. NTs are synthesized as precursor proteins, pro-NTs, which are cleaved to produce the mature form, a noncovalent-linked homodimer.

The functionality of NT receptors, Trks and p75, is quite complex.[3,4] Upon ligand binding, the Trks form homodimers leading to activation of the intracellular kinase domain and several autophosphorylation events of cytoplasmic tyrosine residues. In TrkA, this initiates a cascade of cell-signaling events, including phosphorylation of Y490, recruitment of SHC-1, concomitant activation of Ras-Raf-1 pathway, and phosphorylation of extracellular-regulated kinase (ERK)1/2. Phosphorylation of Y751 leads to activation of the phosphatidylinositol 3-kinase (PI3-kinase)/Akt pathways. Phosphorylation of Y785 on TrkA leads to the activation of phospholipase Cγ to generate diacylglycerol and inositol phosphates and the activation of protein kinase C pathway.[5] TrkB activation is mediated in much the same ways as that for TrkA, with main autophosphorylation taking place at Y516 and Y817 (human seq, corresponding to TrkA Y490 and 785).[6] The complexity of NT receptor signaling is likely to account for the diverse neuronal and non-neuronal actions that NTs exert.

The determination of the three-dimensional structure of NTs and their receptors has given further insight into the molecular interactions behind the activation events.[7–9] The NTs belong to the cystine knot superfamily, a large group of nonglobular proteins that consists of beta-sheets and that shares a conserved core of three intertwined disulfide bonds. The NTs exist as dimers in solution. The Trk receptors consist of two Ig-like domains, Ig-C1 (also known as D4) and Ig-C2 (D5), and a leucine-rich repeat domain that contains three leucine-rich repeats (LRRs) flanked by two cysteine-rich domains.

The structure of NGF in complex with the entire extracellular domain (ECD) of TrkA revealed that NGF only interacts with the Ig-C2 (D5) domain of TrkA.[10] The binding of NTs to their cognate receptor is mediated through two main regions, termed the conserved patch, which includes one of the beta-hairpin loops (L1) and residues from the four beta-strands that form the waist of NGF, which show high similarity between NTs and the specificity patch (residues 2–14 of NGF), which is an area with less sequence conservation that confers the selectivity exhibited by the neurothrophins. The structure of NGF bound to p75 has also been determined, showing that TrkA and p75 interaction with NGF is different, using nonoverlapping areas on NGF.[11]

3. ROLE OF NTs AND THEIR RECEPTORS IN NEURODEGENERATIVE DISORDERS

BDNF and NGF signaling is important to support neuronal survival, differentiation, neurogenesis, and synaptic plasticity. NGF and BDNF can have similar effects on neurons but at distinct sites due to differences in expression patterns of their receptors and/or accessory proteins. NGF has been demonstrated to promote the survival of basal forebrain cholinergic neurons (BFCNs), which are dysfunctional in the brain of patients with AD. These cholinergic neurons are highly dependent on NGF to maintain their biochemical and morphological phenotype. Further validation of the important role of NGF in preventing neurodegeneration comes from animal models studying the effect of chronic deprivation of NGF in AD11 transgenic mice, using expression of a recombinant neutralizing anti-NGF antibody postnatally.[12] Apart from inducing the expected cholinergic deficit in BFCNs, the mice also demonstrate a progressive neurodegeneration that recapitulates many aspects typical of AD neurodegeneration. BDNF has also been implicated in the pathophysiology of AD. The level of BDNF is reduced in the brain of patients with AD, suggesting that the lack of BDNF-mediated neurotrophic support may accelerate neurodegeneration.[13,14] Indeed, a growing body of experimental evidence suggests that increased BDNF signaling could potentially improve cognition in AD. The transplantation of stem cells into the brain of a triple-transgenic AD mouse model that expresses amyloid and tau pathology, i.e., the major neuropathological hallmarks of AD, resulted in improved cognition.[15] Depletion of BDNF blocked the beneficial effect of the neuronal transplantation, whereas ectopic expression of BDNF had a similar positive effect on cognition as the transplanted stem cells. These experiments strongly support enhanced BDNF signaling as a therapeutic mechanism in order to stimulate synaptic plasticity in AD.

Ectopic BDNF expression has also been shown to be beneficial during other neurodegenerative conditions. In animal models of PD, BDNF has been shown to promote the survival and/or prevent the degeneration of dopaminergic neurons in the substantia nigra.[16,17] Also, conditional ablation of BDNF in either the cortex or substantia nigra or ablation of TrkB in the striatum resulted in severely impaired striatal neuronal development.[18] Following their robust effects in animal models of neurodegeneration, NGF and BDNF have been evaluated for therapy for several human neurodegenerative diseases. However, mainly due to the short half-life in plasma of these proteins, the use of recombinant proteins has so far not proven very successful. One interesting exception is the local long-term delivery of NGF directly to the basal forebrain of patients with AD.[19] Although the clinical trial was conducted with only 10 patients and the primary objective of the trial was a safety evaluation, supportive biomarker data and cognitive tests suggested signs of improved neuronal function and cognition in response to the NGF treatment.

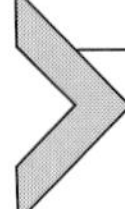

4. PHARMACOLOGICAL ACTIVATORS OF NT RECEPTORS

4.1. Peptidic Activators

The poor drug-like qualities of NGF and BDNF have spurred interest in making smaller peptides, which have been recently reviewed.[20] This has mainly been focused on making constrained ring closed peptides of one of the loop structures of the NTs. Constraining the conformation into a beta-turn appears essential to achieve activity, whereas more flexible peptides have been inactive.

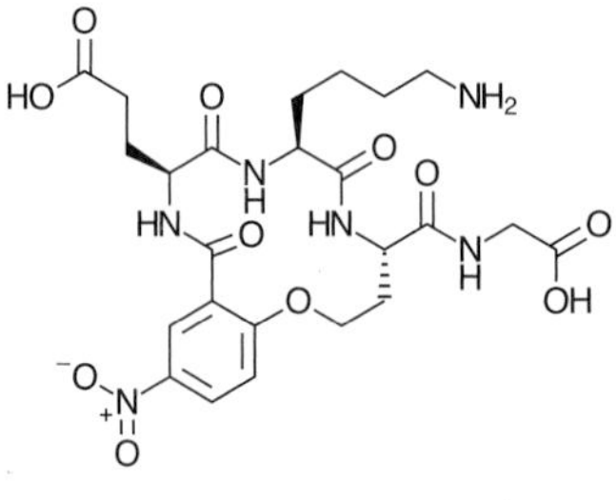

1 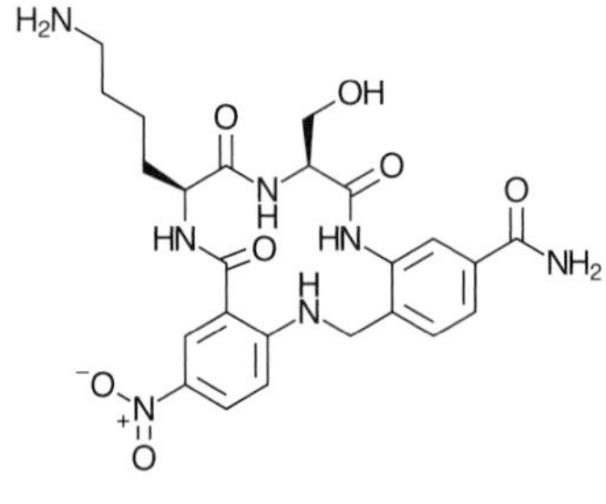 **2**

Based on pharmacophores derived from NGF peptide analogs and monoclonal antibodies (mAbs), a beta-turn library of about 60 members was synthesized.[21] From that effort, the cyclic peptide mimetic **1** (MIM-D3) was identified and demonstrated to be a proteolytically stable partial TrkA agonist. **1** was not able to block NGF binding to the ECD of TrkA; however, it appeared competitive with the TrkA-directed mAb 5C3, which has a binding epitope within the IgC2 domain of TrkA near the NGF-binding site. D3 was able to inhibit the interaction between the ECD of TrkA and 5C3 with an IC_{50} of 2 μM. In addition, the interaction appeared to be selective to TrkA, by not interacting with either $p75^{NTR}$ or TrkC. **1** alone was able to both increase the survival in primary neuronal cultures and induce differentiation. Interestingly, **1** was also able to potentiate the effects of NGF in these assays. **1** administered ICV via an osmotic minipump to cognitively impaired rats showed a long-lasting rescue of age-associated memory deficits.[22] The compound also increased the density of cortical cholinergic presynaptic boutons as well as the somal size of nucleus basalis neurons to the same extent as NGF. The poor blood–brain barrier (BBB) penetrance of **1** restricts its use in CNS disorders, and it is currently in phase 3 clinical trials for the treatment of dry eye syndrome. The related cyclic peptidomimic **2**[23] is reported to stimulate the TrkC receptor albeit with low potency, about 50 μM of compound **2** afforded Tyr phosphorylation to a similar extent as 100 pM NT-3.

Significant efforts have also been directed at producing agonistic mAbs acting at TrkA[24] and TrkB[25–28] receptors. Interestingly, some mAbs are displaceable with BDNF indicating a binding epitope partially overlapping with that of BDNF, whereas other mAbs are nondisplaceable by BDNF (e.g., 29D7). Thus, the TrkB receptor can be activated by ligands interacting with a site distinct from the BDNF-binding site.

4.2. Small-Molecule Activators

Even though a number of peptide derivatives and mAbs that mimic the effects of NTs have been identified, many of these are likely to suffer from many of the same drawbacks as the NTs themselves, such as poor stability in plasma and poor permeability especially into the CNS. A number of small-molecule activators of NTs have been identified using a combination of molecular modeling, virtual screens, and HTS assays. In addition, a number of natural products have been identified that are able to activate these receptors.

The quinoid structure **3** (L-783,281, demethylasterriquinone B-1 (DMAQB-1)), a fungal metabolite, was originally identified as an activator of the insulin receptor.[29] Further studies showed that it was also able to activate Trk receptors.[30] The compound acts as a partial agonist at both TrkA and TrkB receptors with a maximal stimulation of about 40% of that of NGF and BDNF, albeit at significantly lower potency (20 μM). In a Trk-transfected CHO cell lines, it was shown that **3** could stimulate both via the MAPK and PI-3-kinase pathways, to an extent similar to the cognate NTs. **3** (20 μM) was able to induce phosphorylation of Tyr 490. Using chimeric receptors, it was shown that **3** is likely to interact with the intracellular domain (ICD) of the TrkA receptor. It failed to demonstrate any neuronal survival or neurite outgrowth effects. The compound showed signs of cell toxicity that was probably compound related since a closely related non-active analog also showed toxicity. In order to identify nontoxic analogs, a library of ~300 asterriquinones and ~60 mono-indolylquinones was prepared.[31] Two compounds were identified with reduced toxicity and ability to activate TrkA receptors. Compound **4** had about 50% of the stimulatory effect of NGF (EC_{50} ~ 30 μM) inducing Y490, Erk, and Akt phosphorylation. Interestingly, **5** had 200% of the maximal NGF effect in CHO-TrkA cells at 30 μM and increased Y490 and Y674/675 phosphorylation. Compound **5** was unable to induce neuronal differentiation alone, but submaximal doses of **5** (1 μM) and NGF (10 ng/mL) increased the neuronal differentiation to the same level as 100 ng/mL NGF.

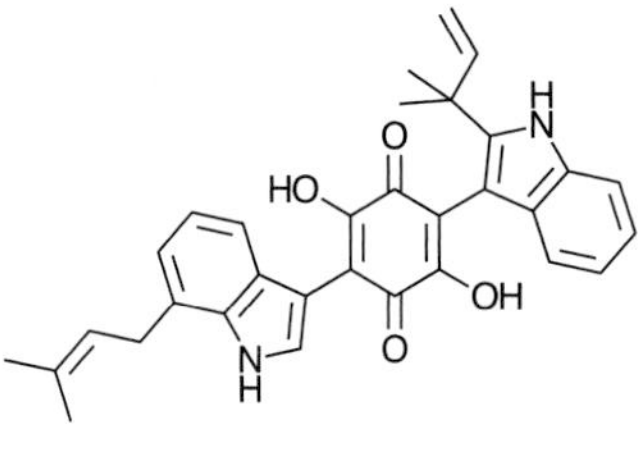

3 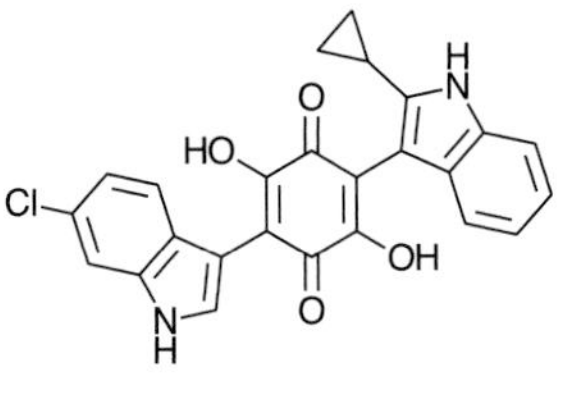 **4** 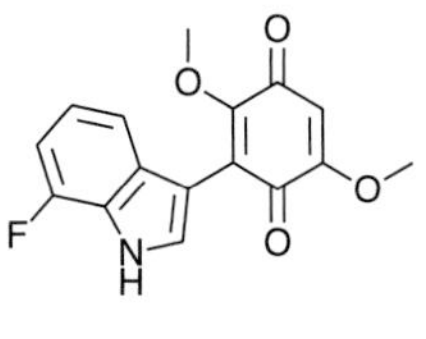 **5**

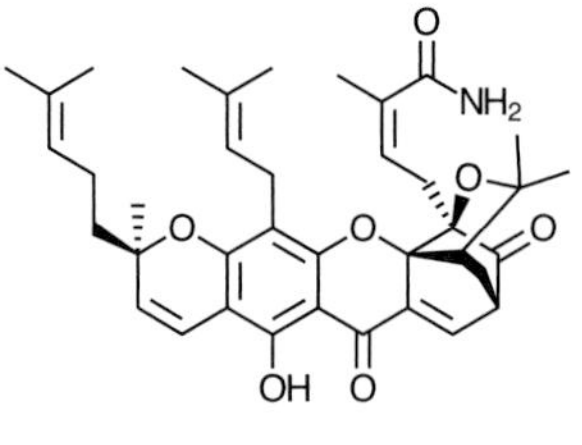

6 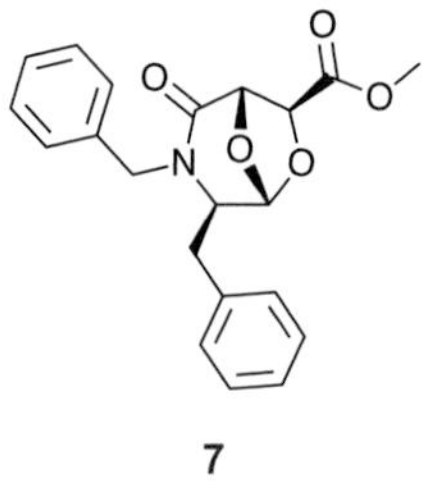 **7**

Gambogic amide (**6**), a derivative of Gambogic acid found in the resin exuded from the *Garcinia hanburryi* tree, was found to activate the TrkA receptor.[32] It selectively binds to the TrkA receptor, and not the TrkB or TrkC receptor, induces TrkA dimerization, and triggers phosphorylation of Tyr490 and Tyr751, thus eliciting MAPK and PI3-kinase/Akt activation. Using truncated receptor constructs, it was determined to interact with the cytoplasmic juxtamembrane domain of TrkA. The K_d for **6** to the TrkA receptor was determined to be 75 nM, whereas it had an EC_{50} of 5 nM in an assay for antiapoptotic activity in T17 cells treated with 1 μM staurosporine. It also showed greater protection than NGF against glutamate toxicity in hippocampal neurons. Finally, it was able to provoke neurite outgrowth in PC12 cells and attenuate kainate acid-induced neurotoxicity in C57BL/6 mice after s.c administration.

Based on suitable molecular shape and size for interaction with the TrkA-D5 domain, a small library of bicyclic compounds was selected for screening.[33] From this library, four compounds were identified that promoted cell proliferation in PC3 cells, among these MT-2 (**7**). Using tritiated [^{3}H]-MT-2, the binding was competitive with NGF and the K_d value was estimated to be 50–100 nM for the interaction with TrkA in NIH-3T3 cells. Docking studies and molecular modeling were used to identify a probable binding site for MT2 to the TrkA-D5 domain. The binding site was subsequently confirmed by mutagenesis of two key residues, T352 and F327, leading to almost complete loss of binding. MT2 added to PC12 cells caused phosphorylation of Tyr490, but minimal phosphorylation of Tyr674/675 and Tyr785. MT2 (10 μM) could, in an *in vitro* hippocampal model, protect neurons from Aβ-mediated death in NGF-deprived neurons, comparable to the effect attained by 2 nM NGF.

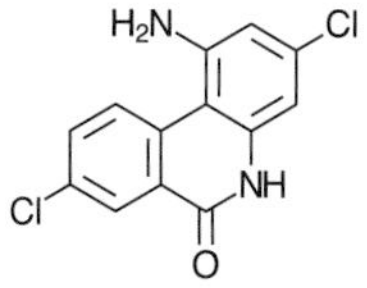

H_2N Cl NH Cl O HO H H O

8 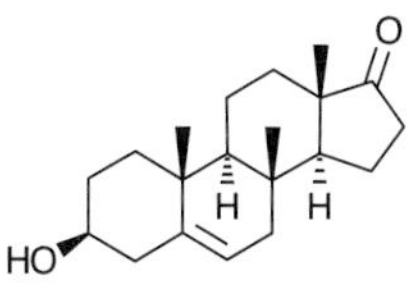 **9**

A number of phenanthridinone derivatives were claimed to potentiate the NGF-induced differentiation and neurite outgrowth in PC12 cells, exemplified with **8**.[34] In a cell survival assay, **8** potentiated the effect by NGF and NT-3 on the survival time of sensory neurons in the dorsal root ganglion of the rat spinal cord. The compound also potentiated the effect of NT-3 on Trk phosphorylation in rat hippocampal and striatal tissue slices.

Dehydroepiandrosterone (DHEA, **9**), a steroid produced in neurons and glia, has been shown to have neuroprotective effects. In HEK293 cells expressing TrkA or p75, [^{3}H]-DHEA displayed a binding affinity of 7 and 6 nM, respectively.[35] This binding was displaceable by treatment with NGF; however, **9** could not displace [^{125}I]-NGF. **9** induced phosphorylation of Tyr490 on TrkA and increased the phosphorylation of Akt and Erk in PC12 cells. In a NGF null mouse model, DHEA was able to reverse the loss of TrkA positive sensory neurons in the dorsal root ganglia.

A set of quaternary ammonium compounds exemplified by **10** were described as NGF agonists and were able to increase differentiation in PC12 cells and promote cell survival.[36] In a more recent patent application by the same group, a set of triglycine derivatives were described as NT agonists.[37] The compound, exemplified by **11** (G79, BN201), was reported to increase the differentiation in PC12 cells as well as improve cell survival.

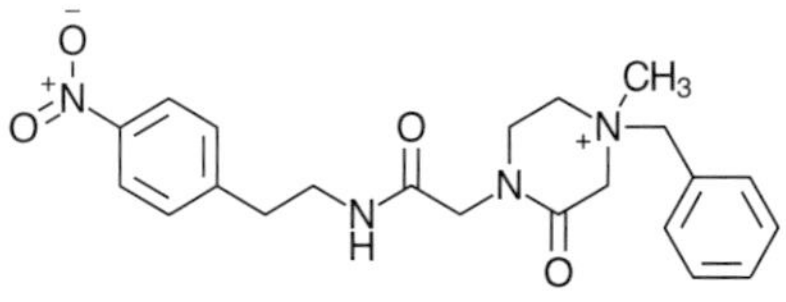

10 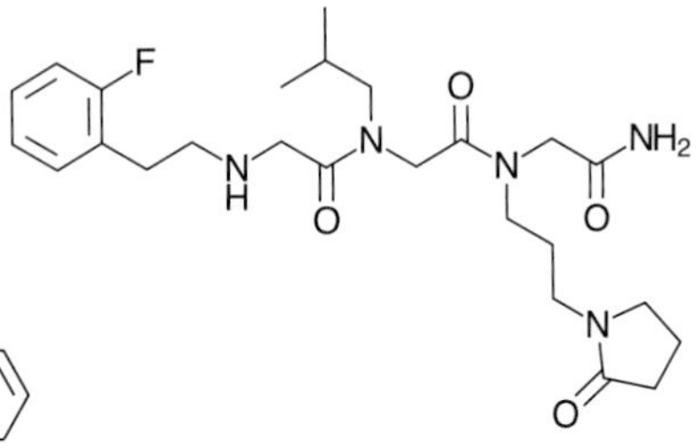 **11**

7,8-Dihydroxyflavone (**12**, 7,8-DHF) was identified after screening a diverse library of 2000 biologically active compounds using an apoptotic assay in a murine cell line.[38] The [^{3}H]7,8-DHF was shown to bind to the ECD of TrkB with a K_d of 320 nM, but not to TrkA. The binding appeared to be directed primarily to the cysteine cluster-2 domain. 7,8-DHF (5 mg/kg i.p.) induced TrkB phosphorylation in mouse brain,[38] primarily of Y816 (hY817) and Y706 (hY707).[39] Since 7,8-DHF was reported to enter the CNS, it has been extensively studied *in vivo*. It appears to exert neuroprotective roles in a variety of models of neurodegenerative disorders, notably 7,8-DHF has shown significant effects in several animal models of AD,[40–42] PD,[38] Huntington's disease,[43] and other models of learning and memory.[44]

Several close analogs of **12** have been identified, with improved potency, such as **13** that increased Akt about 40% more than 7,8-DHF at 10 nM in primary neurons, showed neuroprotective effects *in vitro* and *in vivo*, and had neurogenesis promoting effects in dentate gyrus after treating mice orally with 5 mg/kg once a day for 3 weeks.[45] Similarly, **14** improved neurogenesis in the hippocampus after 5 mg/kg once a day treatment for 3 weeks.[39]

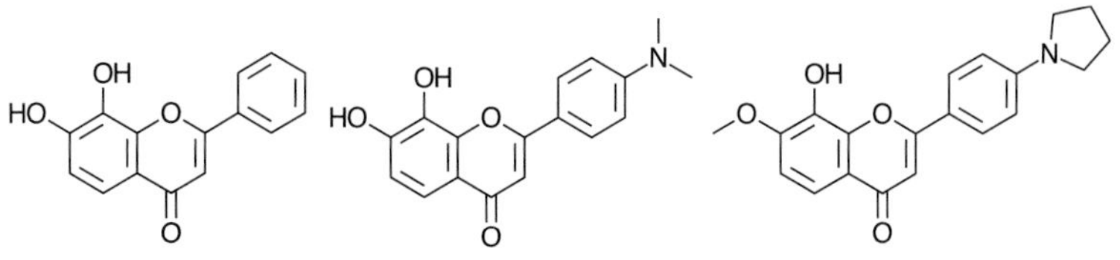

12 **13** **14**

12 has low *in vitro* Caco-2 permeability (Papp 1.5×10^{-6} cm/s) and also appears to have low oral bioavailability.[39] After treating mice with 50 mg/kg orally, the concentration of **12** in brain at 10 min was 50 ng/g of brain and 70 ng/mL in plasma.[39] A O-methylated metabolite of 7,8-DHF was also detected in brain and may contribute to TrkB receptor activation upon oral administration.[39]

12 was labeled to generate 4′-[^{18}F]-7,8-DHF (**15**) as a potential positron emission tomography (PET) ligand for the study of TrkB receptors *in vivo*.[46] Using *in vitro* autoradiography in rat brain sections, **15** showed high specific binding to TrkB receptors in cerebral cortex as well as hippocampus, caudate-putamen, and thalamus, which was displaceable by BDNF. However, a PET study in rats demonstrated a low BBB penetration of this PET tracer, thus making it unsuitable as a PET ligand.

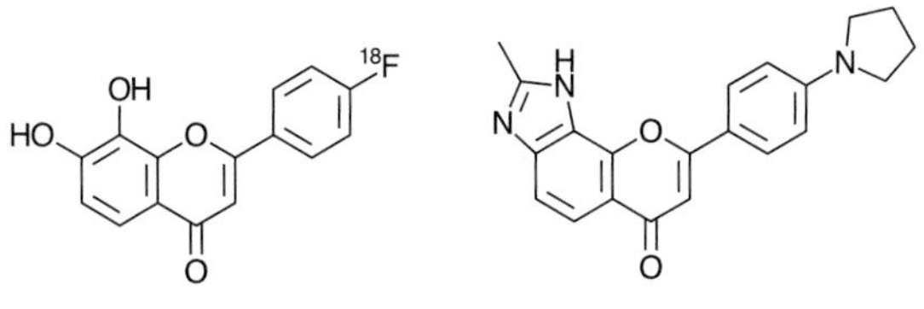

15 **16**

In an effort to improve the DMPK properties of **12**, the catechol motif was replaced with a benzimidazole (**16**).[47] This compound was able to increase TrkB Y816 phosphorylation and Akt activation, *in vitro* in primary neurons (at 500 nM) as well as *in vivo* in mouse brain after 2 h (1 mg/kg oral administration). The *in vivo* PK properties of the benzimidazole derivative **16** showed no improvement ($t_{1/2}$ 58 min, T_{max} 120 min, C_{max} 7 ng/mL, AUC_{inf} 1293 min ng/mL at 10 mg/kg p.o.)[47] compared to **12** ($t_{1/2}$ 134 min, T_{max} 10 min, C_{max} 70 ng/mL, AUC_{inf} 7515 min ng/mL at

50 mg/kg p.o.),[39] and the oral bioavailability of **16** was only 2% in mice.[47] *In vivo* effects in an animal model of depression of **16** were seen in after oral administration of 2.5 mg/kg to mice once a day for 3 weeks.

A pharmacophore model generated from loop 2 of BDNF was used to virtually screen a compound library of 1,000,000 compounds.[48] After computational filtering, seven compounds were evaluated for neurotrophic activity in a survival assay in E16 hippocampal neurons. Four compounds (**17–20**) were identified as partial agonists with EC_{50} values of 200–500 pM with a stimulatory effect of about 85% of that of BDNF. LM22A-4 (**20**) was shown to be selective versus TrkA and TrkC and against 57 other receptors in a panel screen. An increase in phosphorylation of Y515 (30% at 5 nM) and activation of Erk and Akt (50–70% at 0.1 nM), relative to unphosphorylated species, was seen after treatment of cultured E16 hippocampal neurons with **20**. The binding of **20** to TrkB was competitive with BDNF with an IC_{50} of approximately 47 nM in a fluorescence anisotropy assay. The compound showed poor BBB penetration and was therefore administered intranasally (0.22 mg/kg once daily) for 7 days to mice, which resulted in an increase in phosphorylation of Trk, Akt, and Erk in hippocampus and striatum. Intranasal administration of **20** (3.4 μg once daily) to rats for 2 weeks improved motor learning in a rotarod model after traumatic brain injury. A combination of intranasal and intraperitoneal administration of **20**, to increase brain concentrations, was shown to reduce motor impairment and neuropathology in mouse models of Huntington's disease.[49]

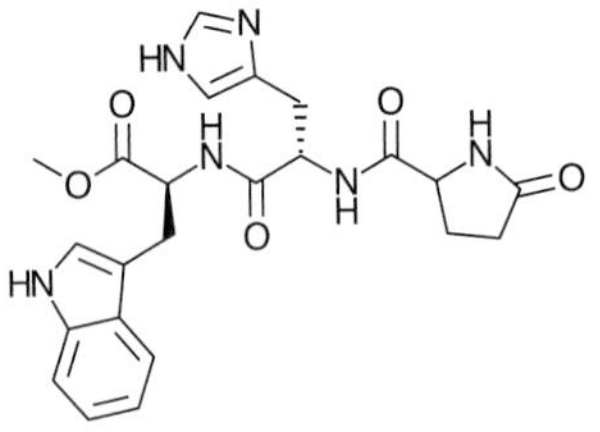

17

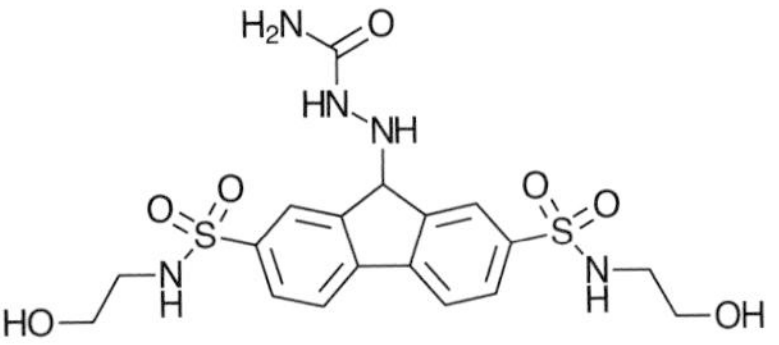

18

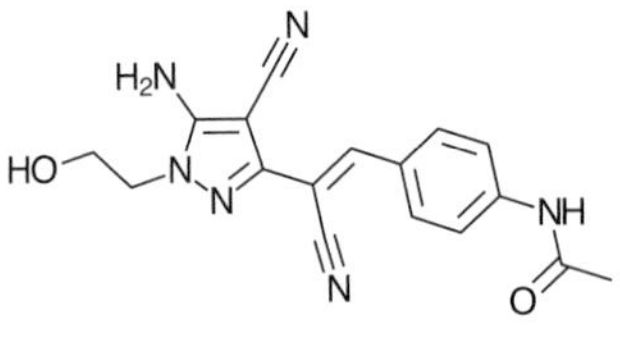

19

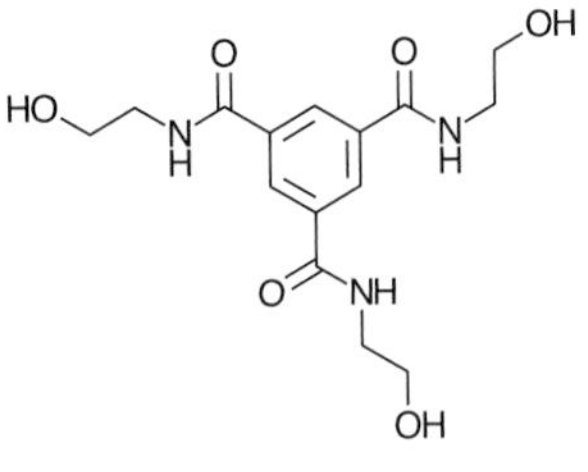

20

In a screening effort, the tetranortriterpenoid deoxygedunin (**21**) isolated from the Indian neem tree (*Azadirachta indica*) was identified as a TrkB agonist.[50] Using [^{3}H]deoxygedunin, it was shown that the compound bound to the TrkB ECD, but not to the ICD or to the TrkA, with a $K_d = 1.4$ μM. Deoxygedunin does not appear to be competitive with BDNF for binding to TrkB. Deoxygedunin (500 nM) activates TrkB Y817, Akt, and Erk as well as suppresses glutamate-provoked cell death in primary hippocampal neurons. It enters the CNS and had an antidepressant effect in the forced swim test (5 mg/kg for 5 days i.p.). Deoxygedunin has also been reported to promote axon regeneration in cut peripheral nerves.[51]

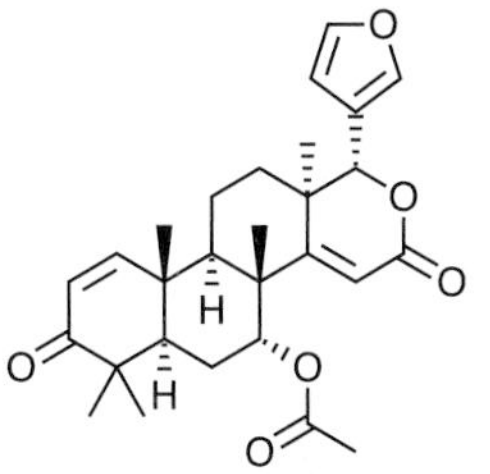

21

N-Acetylserotonin (**22**, NAS) and the more potent analog **23** (HIOC) selectively activate TrkB receptors.[52,53] HIOC increased TrkB phosphorylation in primary cortical neurons and increased Akt activation. At a concentration of 500 nM, **23** increased phosphorylation of TrkB to a level comparable to that produced by BDNF. It was also able to specifically activate TrkB receptors in mouse brain and protect neurons in a kainic acid excitotoxicity model.[53]

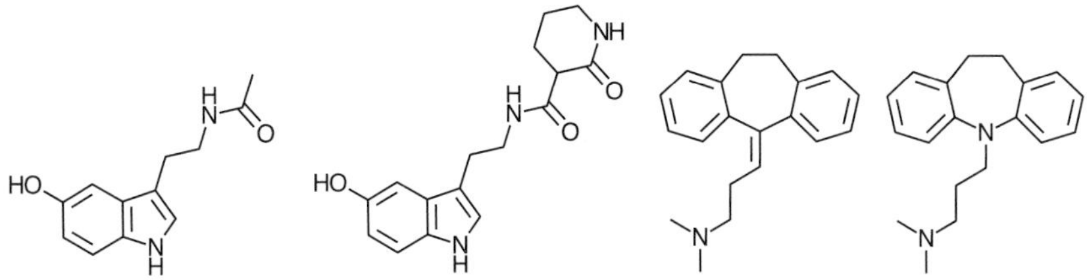

22 **23** **24** **25**

In a random screen effort, amitriptyline (**24**) was identified to protect both TrkA-expressing T17 cells, treated with staurosporine (EC_{50} 50 nM), and primary hippocampal neurons, treated with glutamate from apoptosis.[54] **24** activated phosphorylation of TrkA Y751 and Y785 but not of Y490. It also promotes Akt and Erk activation at a concentration of 250–500 nM. **24** (100 nM) induced neurite outgrowth, and 500 nM exhibited about 80% of the max stimulatory effect of NGF. In a binding assay, [^{3}H]-amitriptyline was determined to interact with the ECD of TrkA and TrkB, but not p75, with a K_d of 3 and 14 μM, respectively. By systematically deleting regions of the ECD, the binding site on TrkA was determined to be in the first leucine-rich region (LRR1) (aa 72–97). **24** was able to induce homodimerization of TrkA, similar to NGF, but interestingly also was able to induce heterodimerization between TrkA and TrkB. Amitriptyline increases TrkA expression and provokes TrkA and TrkB activation in hippocampus. Other antidepressants, like imipramine (**25**), have also been reported to active TrkB receptors.[55] **25** increased the phosphorylation of TrkB Y706/707 but not of Y515 in the prefrontal cortex of mice.

NT receptors have been shown to be transactivated by the adenosine A2 agonist CGS21680.[56] In addition, many small molecules have been reported to increase the level of BDNF expression or release. These observations make it important to fully define the specific mode of action of compounds stimulating TrkB receptors. In a recent publication, a number of small-molecule TrkB agonists were tested, such as NAS, amitriptyline, 7,8-DHF, and LM22A-4.[28] None of these compounds produced any activation of TrkB

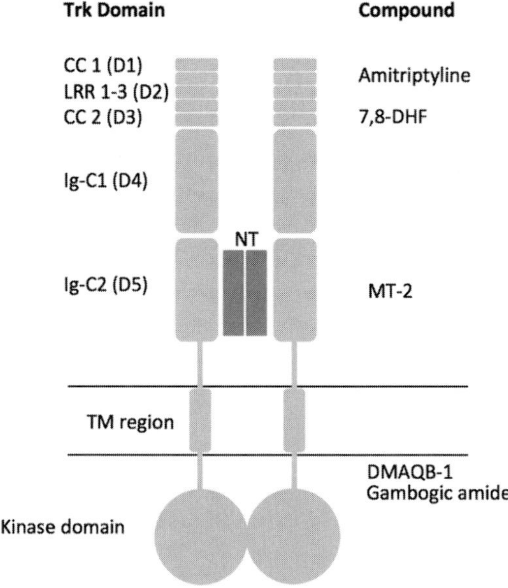

Figure 5.1 Schematic overview of regions of the Trk receptors where some of the small-molecule agonists are suggested to interact.

in a gene reporter assay or in an enzyme fragment complementation assay. In addition, there was no phosphorylation seen of TrkB, ERK, or AKT in CHO-K1 TrkB NFAT cell line or in primary corticostriatal cells. No increase in cell survival was seen in a primary neuronal system. This indicates that the activity may be assay and/or cell line dependent. Care must also be taken when evaluating compounds with multiple pharmacologies in complex systems or *in vivo* without a measure of drug exposure.

Notably, compounds identified to activate TrkA or TrkB receptor do not have a dimeric or symmetrical structure, indicating that they may activate the receptors differently than the dimeric NTs. Interestingly, activation of Trk receptors seems to be achievable by interacting with many different regions of the receptor, as indicated by the different binding sites identified for these ligands (see Fig. 5.1).

5. CONCLUSION

The activation of TrkA and TrkB seems to hold significant promise for the treatment of several neurodegenerative disorders such as AD, PD, and Huntington's disease. The NTs themselves are not suitable for peripheral administration, whereas local administration in the CNS shows promise. The identification of small-molecule stimulators of Trk receptors makes it possible not only to achieve high CNS exposure but also to achieve an increased selectivity than seen by the NTs.

The intracellular signaling pathways stimulated by activating NT receptors are complex. Interestingly, the pathways activated appear to be compound dependent. Thus, it appears possible to identify compounds that can specifically activate distinct neurotrophic signaling pathways, e.g., the neurotic pathway or the neurogenic pathway. In addition, it is possible to achieve selectivity for the TrkA or TrkB receptors against p75 and thereby reduce pleiotropic effects of NTs. This gives significant promise in the development of selective NT receptor agonists for the treatment of neurodegenerative disorders.

REFERENCES

1. Aloe, L.; Rocco, M. L.; Bianchi, P.; Manni, L. *J. Transl. Med.* **2012**, *10*, 239.
2. Lu, B.; Nagappan, G.; Guan, X.; Nathan, P. J.; Wren, P. *Nat. Rev. Neurosci.* **2013**, *14*, 401.
3. Longo, F. M.; Massa, S. M. *Nat. Rev. Drug Discov.* **2013**, *12*, 507.
4. Blum, R.; Konnerth, A. *Physiology* **2005**, *20*, 70.
5. Huang, E. J.; Reichardt, L. F. *Annu. Rev. Biochem.* **2003**, *72*, 609.
6. Minichiello, L. *Nat. Rev. Neurosci.* **2009**, *10*, 850.

7. Wiesmann, C.; Ultsch, M. H.; Bass, S. H.; de Vos, A. M. *Nature* **1999**, *401*, 184.
8. Ultsch, M. H.; Wiesmann, C.; Simmons, L. C.; Henrich, J.; Yang, M.; Reilly, D.; Bass, S. H.; de Vos, A. M. *J. Mol. Biol.* **1999**, *290*, 149.
9. Robinson, R. C.; Radziejewski, C.; Stuart, D. I.; Jones, E. Y. *Biochemistry* **1995**, *34*, 4139.
10. Wehrman, T.; He, X.; Raab, B.; Dukipatti, A.; Blau, H.; Garcia, K. C. *Neuron* **2007**, *53*, 25.
11. He, X.-L.; Garcia, K. C. *Science* **2004**, *304*, 870.
12. De Rosa, R.; Garcia, A. A.; Braschi, C.; Capsoni, S.; Maffei, L.; Berardi, N.; Cattaneo, A. *Proc. Natl. Acad. Sci. U.S.A.* **2005**, *102*, 3811.
13. Peng, S.; Wuu, J.; Mufson, E. J.; Fahnestock, M. *J. Neurochem.* **2005**, *93*, 1412.
14. Zhang, F.; Kang, Z.; Li, W.; Xiao, Z.; Zhou, X. *J. Clin. Neurosci.* **2012**, *19*, 946.
15. Blurton-Jones, M.; Kitazawa, M.; Martinez-Coria, H.; Castello, N. A.; Müller, F.-J.; Loring, J. F.; Yamasaki, T. R.; Poon, W. W.; Green, K. N.; LaFerla, F. M. *Proc. Natl. Acad. Sci. U.S.A.* **2009**, *106*, 13594.
16. Hyman, C.; Hofer, M.; Barde, Y. A.; Juhasz, M.; Yancopoulos, G. D.; Squinto, S. P.; Lindsay, R. M. *Nature* **1991**, *350*, 230.
17. Frim, D. M.; Uhler, T. A.; Galpern, W. R.; Beal, M. F.; Breakefield, X. O.; Isacson, O. *Proc. Natl. Acad. Sci. U.S.A.* **1994**, *91*, 5104.
18. Li, Y.; Yui, D.; Luikart, B. W.; McKay, R. M.; Li, Y.; Rubenstein, J. L.; Parada, L. F. *Proc. Natl. Acad. Sci. U.S.A.* **2012**, *109*, 15491.
19. Eriksdotter-Jönhagen, M.; Linderoth, B.; Lind, G.; Aladellie, L.; Almkvist, O.; Andreasen, N.; Blennow, K.; Bogdanovic, N.; Jelic, V.; Kadir, A.; Nordberg, A.; Sundström, E.; Wahlund, L.-O.; Wall, A.; Wiberg, M.; Winblad, B.; Seiger, A.; Almqvist, P.; Wahlberg, L. *Dement. Geriatr. Cogn. Disord.* **2012**, *33*, 18.
20. Skaper, S. D. *Curr. Pharm. Des.* **2011**, *17*, 2704.
21. Maliartchouk, S.; Feng, Y.; Ivanisevic, L.; Debeir, T.; Cuello, A. C.; Burgess, K.; Saragovi, H. U. *Mol. Pharmacol.* **2000**, *57*, 385.
22. Bruno, M. A.; Clarke, P.; Seltzer, A. *J. Neurosci.* **2004**, *24*, 8009.
23. Pattarawarapan, M.; Zaccaro, M. C.; Saragovi, U. H.; Burgess, K. *J. Med. Chem.* **2002**, *45*, 4387.
24. LeSauteur, L.; Maliartchouk, S.; Le Jeune, H.; Quirion, R.; Saragovi, H. U. *J. Neurosci.* **1996**, *16*, 1308.
25. Qian, M. D.; Zhang, J.; Tan, X.-Y.; Wood, A.; Gill, D.; Cho, S. *J. Neurosci.* **2006**, *26*, 9394.
26. Cazorla, M.; Arrang, J. M.; Prémont, J. *Br. J. Pharmacol.* **2011**, *162*, 947.
27. Bai, Y.; Xu, J.; Brahimi, F.; Zhuo, Y.; Sarunic, M. V.; Saragovi, H. U. *Invest. Ophthal. Vis. Sci.* **2010**, *51*, 4722.
28. Todd, D.; Gowers, I.; Dowler, S. J.; Wall, M. D.; McAllister, G.; Fischer, D. F.; Dijkstra, S.; Fratantoni, S. A.; van de Bospoort, R.; Veenman-Koepke, J.; Flynn, G.; Arjomand, J.; Dominguez, C.; Munoz-Sanjuan, I.; Wityak, J.; Bard, J. A. *PLoS One* **2014**, *9*, e87923.
29. Zhang, B.; Salituro, G.; Szalkowski, D.; Li, Z.; Zhang, Y.; Royo, I.; Vilella, D.; Díez, M. T.; Pelaez, F.; Ruby, C.; Kendall, R. L.; Mao, X.; Griffin, P.; Calaycay, J.; Zierath, J. R.; Heck, J. V.; Smith, R. G.; Moller, D. E. *Science* **1999**, *284*, 974.
30. Wilkie, N.; Wingrove, P. B.; Bilsland, J. G.; Young, L.; Harper, S. J.; Hefti, F.; Ellis, S.; Pollack, S. J. *J. Neurochem.* **2001**, *78*, 1135.
31. Lin, B.; Pirrung, M. C.; Deng, L.; Li, Z.; Liu, Y.; Webster, N. J. G. *J. Pharmacol. Exp. Ther.* **2007**, *322*, 59.
32. Jang, S.-W.; Okada, M.; Sayeed, I.; Xiao, G.; Stein, D.; Jin, P.; Ye, K. *Proc. Natl. Acad. Sci. U.S.A.* **2007**, *104*, 16329.
33. Scarpi, D.; Cirelli, D.; Matrone, C.; Castronovo, G.; Rosini, P.; Occhiato, E. G.; Romano, F.; Bartali, L.; Clemente, A. M.; Bottegoni, G.; Cavalli, A.; De Chiara, G.;

Bonini, P.; Calissano, P.; Palamara, A. T.; Garaci, E.; Torcia, M. G.; Guarna, A.; Cozzolino, F. *Cell Death Dis.* **2012**, *3*, e339.
34. Aoyagi, A.; Isono, F.; Fujii, M.; Ohono, A. WO00/63179, 2000.
35. Lazaridis, I.; Charalampopoulos, I.; Alexaki, V.-I.; Avlonitis, N.; Pediaditakis, I.; Efstathopoulos, P.; Calogeropoulou, T.; Castanas, E.; Gravanis, A. *PLoS Biol.* **2011**, *9*, e1001051.
36. Moreno, B.; Villoslada, P.; Messeguer, J.; Navarro, G.; Messeguer, A. WO2011024078, 2011.
37. Villoslada, P.; Messeguer, A. US2012/0052094, 2012.
38. Jang, S.-W.; Liu, X.; Yepes, M.; Shepherd, K. R.; Miller, G. W.; Liu, Y.; Wilson, W. D.; Xiao, G.; Blanchi, B.; Sun, Y. E.; Ye, K. *Proc. Natl. Acad. Sci. U.S.A.* **2010**, *107*, 2687.
39. Liu, X.; Qi, Q.; Xiao, G.; Li, J.; Luo, H. R.; Ye, K. *Pharmacology* **2013**, *91*, 185.
40. Devi, L.; Ohno, M. *Neuropsychopharmacology* **2011**, *37*, 434.
41. Zhang, Z.; Liu, X.; Schroeder, J. P.; Chan, C. B.; Song, M.; Yu, S. P.; Weinshenker, D.; Ye, K. *Neuropsychopharmacology* **2014**, *39*, 638.
42. Castello, N. A.; Nguyen, M. H.; Tran, J. D.; Cheng, D.; Green, K. N.; LaFerla, F. M. *PLoS One* **2014**, *9*, e91453.
43. Jiang, M.; Peng, Q.; Liu, X.; Jin, J.; Hou, Z.; Zhang, J.; Mori, S.; Ross, C. A.; Ye, K.; Duan, W. *Hum. Mol. Genet.* **2013**, *22*, 2462.
44. Zeng, Y.; Lv, F.; Li, L.; Yu, H.; Dong, M.; Fu, Q. *J. Neurochem.* **2012**, *122*, 800.
45. Liu, X.; Chan, C. B.; Jang, S.-W.; Pradoldej, S.; Huang, J.; He, K.; Phun, L. H.; France, S.; Xiao, G.; Jia, Y.; Luo, H. R.; Ye, K. *J. Med. Chem.* **2010**, *53*, 8274.
46. Bernard-Gauthier, V.; Boudjemeline, M.; Rosa-Neto, P.; Thiel, A.; Schirrmacher, R. *Bioorg. Med. Chem.* **2013**, *21*, 7816.
47. Liu, X.; Chan, C. B.; Qi, Q.; Xiao, G.; Luo, H. R.; He, X.; Ye, K. *J. Med. Chem.* **2012**, *55*, 8524.
48. Massa, S. M.; Yang, T.; Xie, Y.; Shi, J.; Bilgen, M.; Joyce, J. N.; Nehama, D.; Rajadas, J.; Longo, F. M. *J. Clin. Invest.* **2010**, *120*, 1774.
49. Simmons, D. A.; Belichenko, N. P.; Yang, T.; Condon, C.; Monbureau, M.; Shamloo, M.; Jing, D.; Massa, S. M.; Longo, F. M. *J. Neurosci.* **2013**, *33*, 18712.
50. Jang, S.-W.; Liu, X.; Chan, C. B.; France, S. A.; Sayeed, I.; Tang, W.; Lin, X.; Xiao, G.; Andero, R.; Chang, Q.; Ressler, K. J.; Ye, K. *PLoS One* **2010**, *5*, e11528.
51. English, A. W.; Liu, K.; Nicolini, J. M.; Mulligan, A. M.; Ye, K. *Proc. Natl. Acad. Sci. U.S.A.* **2013**, *110*, 16217.
52. Jang, S.-W.; Liu, X.; Pradoldej, S.; Tosini, G.; Chang, Q.; Iuvone, P. M.; Ye, K. *Proc. Natl. Acad. Sci. U.S.A.* **2010**, *107*, 3876.
53. Shen, J.; Ghai, K.; Sompol, P.; Liu, X.; Cao, X.; Iuvone, P. M.; Ye, K. *Proc. Natl. Acad. Sci. U.S.A.* **2012**, *109*, 3540.
54. Jang, S.-W.; Liu, X.; Chan, C. B.; Weinshenker, D.; Hall, R. A.; Xiao, G.; Ye, K. *Chem. Biol.* **2009**, *16*, 644.
55. Saarelainen, T.; Hendolin, P.; Lucas, G.; Koponen, E.; Sairanen, M.; MacDonald, E.; Agerman, K.; Haapasalo, A.; Nawa, H.; Aloyz, R.; Ernfors, P.; Castrén, E. *J. Neurosci.* **2003**, *23*, 349.
56. Lee, F. S.; Chao, M. V. *Proc. Natl. Acad. Sci. U.S.A.* **2001**, *98*, 3555.

SECTION 2

Cardiovascular and Metabolic Diseases

Section Editor: Robert L. Dow
Pfizer R&D, Cambridge, Massachusetts

CHAPTER SIX

Small-Molecule Modulators of GPR40 (FFA1)

Sean P. Brown, Joshua P. Taygerly
Amgen Inc., South San Francisco, California, USA

Contents

1. INTRODUCTION

Type II diabetics lose their ability to maintain glucose homeostasis due to eight metabolic defects including decreased insulin secretion, decreased incretin effects, increased lipolysis, increased glucose reabsorption, decreased glucose uptake, neurotransmitter dysfunction, increased hepatic glucose production, and increased glucagon secretion. Current drugs to treat type II diabetes improve glycemic control by restoring function in one or more of these metabolic defects. GPR40 is a G-protein-coupled receptor (GPCR), primarily expressed in pancreatic islets[1] and enteroendocrine cells.[2] Considerable interest has been focused on GPR40 as a novel therapeutic target for type II diabetes with the realization that activation of GPR40 leads to glucose-stimulated insulin secretion.[3] This alluring mechanism to treat type II diabetes presents the potential of little or no risk of hypoglycemia and the ability to affect multiple metabolic defects that contribute to the disease. This prospect has garnered a large amount interest from the pharmaceutical industry with six

Annual Reports in Medicinal Chemistry, Volume 49
ISSN 0065-7743
http://dx.doi.org/10.1016/B978-0-12-800167-7.00006-7

chemical entities entering clinical trials (TAK-875, AMG 837, LY2881835, JTT851, ASP5034, and P11187). In this chapter, we present the recent advances in small molecule agonists targeting GPR40.

2. RECENT DISCOVERIES IN GPR40 BIOLOGY

GPR40 was deorphanized in 2003 with the discovery that short and long-chain free fatty acids activate the cell surface seven-transmembrane domain receptor.[1] When activated in the β-cells of the pancreatic islets, GPR40 elicits increased insulin secretion only in the presence of elevated glucose levels (Fig. 6.1).[3] Upon activation of GPR40 in enteroendocrine L-cells, gut hormones glucagon-like peptide 1 (GLP-1) and glucose-dependent insulinotropic peptide (GIP) are released.[2] Both of these hormones are implicated in maintaining the incretin effect and GLP-1, in particular, has been identified as a drug target. Both GLP-1 stabilizers and GLP-1 degradation inhibitors (DPP-4 inhibitors) have been pursued by the pharmaceutical industry.[4] Two distinct classes of GPR40 agonists have been discovered, partial agonists and full agonists. Partial agonists have only been observed to activate GPR40 in pancreatic β-cells to secrete insulin, while full agonists have been shown to activate GPR40 in both β-cells and enteroendocrine L-cells to provide increased secretion of GIP and GLP-1 in addition to insulin.[5] Furthermore, it has been demonstrated that GPR40 possesses multiple allosteric binding sites that are distinct for partial and full agonists and that these sites display positive cooperativity upon ligand binding.[6]

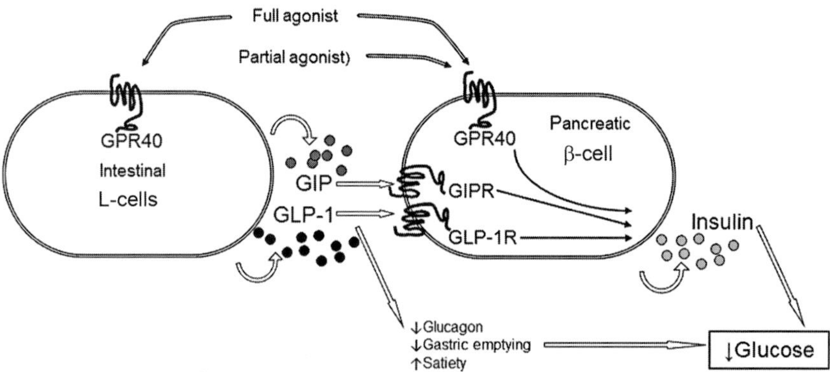

Figure 6.1 Scheme depicting the effects of partial and full GPR40 agonists on both intestinal L-cells and pancreatic β-cells. Both pathways lead to insulin release and glucose lowering.

3. GPR40 PARTIAL AGONISTS

3.1. TAK-875

To date, the small molecule that has progressed the farthest through human clinical trials is TAK-875 (fasiglifam, **3**). High-throughput screening of commercially available arylalkanoic acids revealed 3-phenylpropanoic acid to be a modest GPR40 agonist at 100 μM in a Ca^{2+} fluorescent imaging plate reader (FLIPR) activity in Chinese hamster ovary (CHO) cells transfected with human GPR40. A small synthetic effort identified benzyloxy-substituted analog **1** as a more potent agonist lead ($EC_{50}=510$ nM). Optimization led to potent agonist **2** ($EC_{50}=5.7$ nM), a ligand with moderate bioavailability ($F=21\%$) and high clearance (Cl = 900 mL/h/kg) in rats, resulting in low exposure ($AUC_{po}=0.25$ μg·h/mL).[7] Oxidation of the unsubstituted β-carbon of the phenylpropionic acid head group was implicated in the high clearance, and thus a series of benzofurans was designed to improve the pharmacokinetic (PK) parameters of this series by providing increased rigidity and increased steric hindrance around the metabolically labile β-carbon. Further optimization resulted in **3**, a GPR40 agonist with slightly diminished potency ($EC_{50}=14$ nM) *in vitro*, but with greatly improved PK parameters in rats ($F=76\%$; Cl = 34 mL/h/kg; $AUC_{po}=65$ μg·h/mL).[8] The sulfonylpropoxy functionality on **3** provides reduced lipophilicity and helps to impart high selectivity for GPR40 compared to other fatty acid receptors. Additional investigation revealed that **3** exerts partial agonist activity and can potentiate the agonist activity of endogenous GPR40 ligand γ-linolenic acid through allosteric binding.[9]

In preclinical pharmacological studies, **3** amplified glucose-stimulated insulin secretion (GSIS) in both rat insulinoma INS-1 833/15 β-cells and in primary rat islets while increasing intracellular inositol monophosphate and calcium concentrations.[10] Neither inhibition of GSIS nor apoptosis was observed in INS-1 β-cells incubated with **3** for extended time periods. This undesired lipotoxicity was observed previously in β-cells treated with endogenous fatty acids.[11] Also, **3** significantly increased plasma insulin levels and lowered plasma glucose levels in both the diabetic Zucker fatty rat model and the diabetic N-STZ-1.5 rat model at a 10 mg/kg dose level. Additionally, 6-week treatment of Zucker fatty rats with **3** at 10 mg/kg led to a significant decrease in glycosylated hemoglobin (HbA_{1c}) levels by −1.7%, and this effect was additively enhanced when codosing with metformin (50 mg/kg; HbA_{1c}: −2.4%).[12] Importantly, unlike other insulin secretagogues, **3** had no effect on insulin or plasma glucose levels in normal Sprague-Dawley rats even at a dose of 30 mg/kg, suggesting that **3** presents a low risk for hypoglycemia. Also, **3** was shown to have no effect on plasma glucose levels or insulin secretion in GPR40-knockout mice.[9]

The highly encouraging *in vitro* and *in vivo* profile of GPR40 activator **3** in preclinical models led to its advancement into human clinical trials. Compound **3** demonstrated a pharmacokinetic profile suitable for once a day dosing ($t_{½}$ = 28–36 h, with AUC and C_{max} linear up to a dose of 200 mg) in a single ascending dose study in healthy volunteers, and observed adverse events were mild. As expected, no glucose lowering was seen in this healthy cohort, as the mechanism of GPR40-dependent glucose lowering requires elevated glucose plasma levels.[13] In contrast, significant plasma glucose lowering was observed in a 2-week study in fasting diabetic patients (−60 mg/dL; −27% at day 13; 400 mg dose) and in an oral glucose tolerance test (OGTT) (−90 mg/dL, −40% at doses ≥100 mg). In a larger randomized double-blind trial of **3** versus placebo or glimepiride (1 mg), HbA_{1c} levels were robustly lowered (−1%) in diabetic patients treated with 100 mg of **3** once daily for 3 months. The observation of treatment-emergent hypoglycemic events in the **3** group (2%) was similar to placebo (3%) but significantly lower than those observed in the glimepiride group (19%).[14–16] On the basis of Phase II data, the pharmacodynamic relationship between HbA_{1c} levels and plasma **3** levels indicated an EC_{50} of 3.16 μg/mL.[17] In 2011, **3** entered Phase III trials, but these were discontinued in late 2013 because of clear signs of liver toxicity in patients. Whether the toxicity was specifically related to **3** or to the general mechanism was not disclosed.[18]

3.2. AMG 837

A high-throughput screen employing an aequorin-based calcium flux assay in transiently transfected CHO cells yielded phenylpropionic acid **4** (EC_{50} = 1.1 μM), a compound structurally related to phenylpropionic acid **1**, the lead structure from the TAK-875 program.[19] Optimization of substituents β to the carboxylic acid revealed that a (*S*)-propynyl moiety was preferred, leading to a significant improvement in potency

(**5**, $EC_{50}=0.06$ μM). The methoxyaryl region was further optimized to a *p*-CF_3-biphenyl, resulting in the discovery of clinical candidate AMG 837 (**6**) that displayed improved potency ($EC_{50}=0.03$ μM) and a good cross species pharmacokinetic profile (mouse, rat, dog, and cyno; Cl = 0.08–0.06 L/h/kg; $t_{1/2}=7$–28 h). Phenylpropionic acid 6 promoted insulin secretion from isolated mouse pancreatic islets only at glucose concentrations ≥8.3 mM ($EC_{50}=0.14$ μM), which implies a low risk of hypoglycemia.[20] After glucose challenge, **6** reduced glucose excursion in wild-type mice, but not in GPR40 knockout mice suggesting that **6** functions through activation of GPR40.[7] Efficacy following a glucose challenge was observed with a 0.3 mg/kg dose of **6** in both Sprague-Dawley rats with normal glycemic control and in Zucker fatty rats that display hyperglycemia (19% and 46% lower AUC, respectively).

GPR40 partial agonist **6** entered clinical trials in 2006, but development has been discontinued for undisclosed reasons. In an effort to provide a back-up molecule to AMG 837, efforts to increase the polar surface area of this class of agonist were undertaken.[21] This work culminated in the identification of AM-4668 (**7**), a GPR40 ligand that displays similar potency ($EC_{50}=0.036$ μM) but has significantly greater polar surface area (86 Å^2) compared to **6** (47 Å^2). Unlike **6**, compound **7** does not show appreciable affinity for rodent GPR40, so the glucose lowering potential of **6** was evaluated in knock-in mice in which the mouse GPR40 gene was replaced by the human gene. In this model, a 10 mg/kg dose provided 19% lowering of glucose AUC after OGTT.

4 EC_{50} = 1.1 μM → **5** EC_{50} = 0.06 μM → **6** EC_{50} = 0.03 μM → **7** EC_{50} = 0.04 μM

3.3. LY2881835

A structurally related compound, spirocycle LY2881835 (**8**), was identified from a medium-throughput screen followed by further structural modification. Phenylpropionic acid **8** induced GPR40 activation (92% of linoleic acid) with a potency of 233 nM in a calcium flux FLIPR assay in HEK293 cells transfected with human GPR40. Additionally, **8** stimulated GSIS in mouse Min6 cells and in mouse primary islets. *In vivo*, plasma glucose levels were lowered in a dose-dependent manner in both normal Sprague-Dawley mice and Zucker fatty rats treated with **8** in oral glucose tolerance tests.[22] Compound **8** progressed to Phase I clinical trials, but its development was discontinued for undisclosed reasons.[23]

8

4. GPR40 FULL AGONISTS

Reevaluation of the GPR40 agonists synthesized in the discovery of **6** in an assay with CHO cells transfected with GPR40 expression plasmid reduced from 5.0 to 0.05 μg (Fig. 6.2) exposed that **6** was in fact a partial agonist (EC_{50} = 0.06 μM, E_{max} = 20%) compared to DHA (**9**).[5] Interestingly, a structurally similar compound **10** (EC_{50} = 1.5 μM, E_{max} = 98%) displayed full agonist properties.[24] Separation of the enantiomers revealed that **11** retained full agonism (EC_{50} = 4.0 μM, E_{max} = 99%) while the corresponding enantiomer was a partial agonist (EC_{50} = 0.65 μM, E_{max} = 47%). Moving the ether linkage from the *para* to *meta* orientation

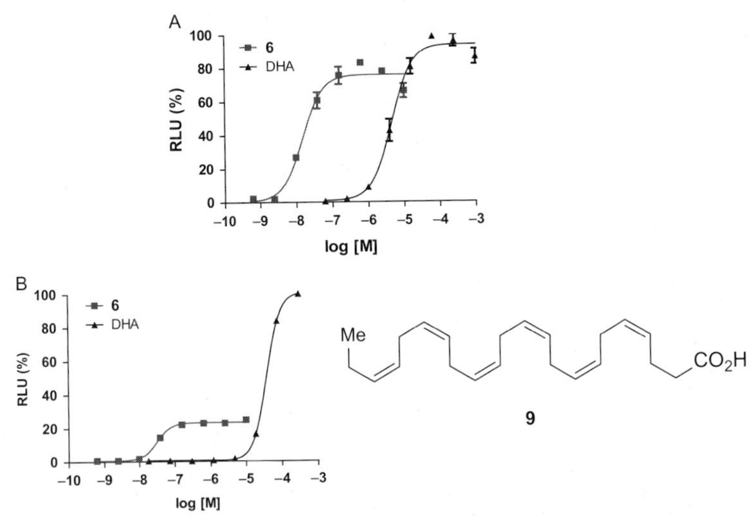

Figure 6.2 The effect of **6** and **9** in CHO cells transfected with (A) 5.0 μg and (B) 0.05 μg of GPR40 expression plasmid.

along with further optimization of the biaryl rings and the β-chiral substituent provided agonist **12** (EC_{50} = 0.16 μM, E_{max} = 100%). Competitive binding studies utilizing ^{3}H-**6** or ^{3}H-**12** showed that there are two distinct allosteric binding sites in addition to the orthosteric site.[6] One allosteric site binds chemotypes exemplified by **6** that give partial agonist activity, while the other binds chemotypes exemplified by **12** that give full agonist activity. Intriguingly, compounds **6** and **12** display positive cooperativity when coadministered with each other. Positive cooperativity is also observed when either **6** or **12** is coadministered with orthosteric binders α-linoleic acid or **9**. Additionally, full agonist **12** stimulates secretion of GLP-1 and GIP *in vitro* from isolated rat intestinal L-cells.

The pharmacokinetic properties of **12** were amenable to oral dosing allowing *in vivo* comparison to **6**. Compared to a maximum efficacious dose of **6** (60 mg/kg), **12** demonstrated improved glycemic control during OGTT in high-fat fed/STZ treated (HF/STZ) and BDF/diet-induced obesity (DIO) mouse models of type II diabetes. Significant increases in GLP-1 and GIP secretion were measured in HF/STZ mice treated with **12**, while partial agonist **6** did not show significant increases in GLP-1. This engagement of the incretin effect has not been demonstrated by other classes of GPR40 agonists, further illustrating the disparate pharmacology of GPR40 full agonists from partial agonists. The positive cooperativity of **6** and **12** that was shown *in vitro* was extended to *in vivo* studies where coadministration provided efficacy greater than either compound alone.[24]

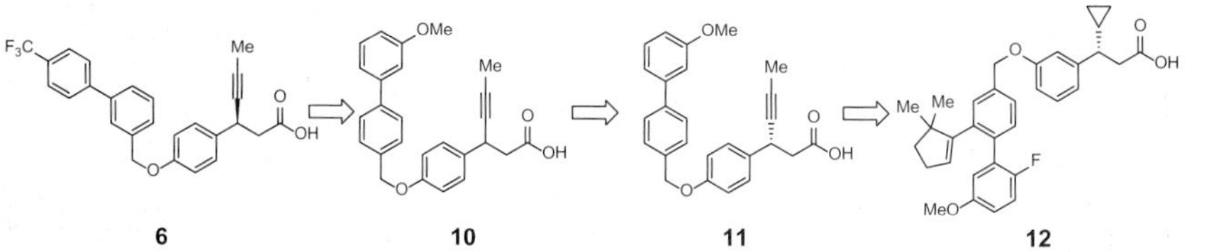

6 EC_{50} = 0.06 μM, E_{max} = 20%
10 EC_{50} = 1.5 μM, E_{max} = 98%
11 EC_{50} = 4.0 μM, E_{max} = 99%
12 EC_{50} = 0.16 μM, E_{max} = 100%

To improve the potency, specificity, and pharmacokinetic profile of **12**, spirocyclic constraints were introduced.[25] These modifications culminated in the discovery of spirocycle **13**, a ligand that demonstrated an improved selectivity profile on a panel of 101 GPCRs, ion channels, transporters, and enzymes at 10 μM. Compound 13 showed increased potency (EC_{50} = 0.081 μM, E_{max} = 101%) and half-life ($t_{1/2}$ = 4.2 h) in rats compared to **12**. In a HF/STZ mouse model of diabetes, **13** displayed greater reduction of glucose excursion after OGTT at 30 mg/kg than **12** at 60 mg/kg. Further efforts were focused on improving the drug-like properties and the pharmacokinetic profile of **12** by introducing polar functionalities.[26] A 3.3 unit reduction in $C\log P$ was achieved with compound **14** ($C\log P = 6.0$), while maintaining similar potency ($EC_{50} = 0.16$ μM, $E_{max} = 94\%$) to **12**. Both rat clearance and half-life were significantly improved in compound **14** (Cl = 0.06 L/h/kg; $t_{½} = 8.1$ h). Similar reductions in glucose excursion were observed with a 3 mg/kg dose of compound **14** as with 60 mg/kg of **12** after OGTT in the BDF/DIO mouse model of diabetes.

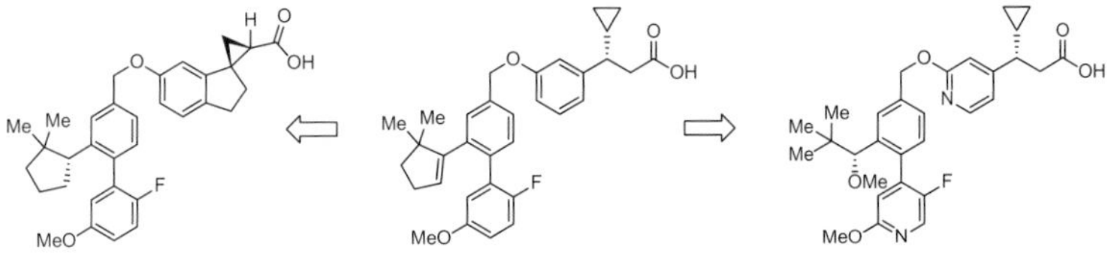

13
EC_{50} = 0.08 μM
E_{max} = 101%
Cl_{rat} = 0.25 L/h/kg
$t_{½}$ = 4.2 h

12
EC_{50} = 0.16 μM
E_{max} = 100%
Cl_{rat} = 0.91 L/h/kg
$t_{½}$ = 1.8 h

14
EC_{50} = 0.16 μM
E_{max} = 94%
Cl_{rat} = 0.06 L/h/kg
$t_{½}$ = 8.1 h

5. CONCLUSIONS

The clinically validated role of GPR40 in lowering HbA_{1c} by **3** has prompted excitement around the GPR40 partial agonist class of antidiabetics. However, given the role of GPR40 in sensing long-chain free fatty acids in the body, it is not surprising that many of the small-molecule partial agonists to date have possessed properties that fall outside of the range of typical drug-like molecules (such as $\log P < 5$).[27] Given that the Phase III clinical trial evaluating the safety and efficacy of **3** was halted due to hepatotoxicity and that three other clinical trials of GPR40 partial agonists were halted for undeclared reasons (**6**, **8**, and ASP5034), the next generation of GPR40 partial agonists will need to overcome the toxicity challenges that hindered the progress of the initial small molecules in this class. However, the robust clinical effects should ensure interest in this class going forward.

Additionally, the more recent class of GPR40 full agonists have the potential to provide even greater efficacy than GPR40 partial agonists by eliciting both insulin and incretin secretion. This class of GPR40 agonists has revealed that signaling through GPR40 is more complicated than originally believed and that our understanding of this therapeutically relevant receptor is incomplete. Further evaluation of the long-term safety and efficacy of these molecules will surely be of interest to the field.

REFERENCES

1. Itoh, Y.; Kawamata, Y.; Harada, M.; Kobayashi, M.; Fujii, R.; Fukusumi, S.; Ogi, K.; Hosoya, M.; Tanaka, Y.; Uejima, H.; Tanaka, H.; Maruyama, M.; Satoh, R.; Okubo, S.; Kizawa, H.; Komatsu, H.; Matsumura, F.; Noguchi, Y.; Shinohara, T.; Hinuma, S.; Fujisawa, Y.; Fujino, M. *Nature* **2003**, *422*(6928), 173.
2. Edfalk, S.; Steneberg, P.; Edlund, H. *Diabetes* **2008**, *57*(9), 2280.
3. Stoddart, L. A.; Smith, N. J.; Milligan, G. *Pharmacol. Rev.* **2008**, *60*(4), 405.
4. Drucker, D. J.; Nauck, M. A. *Lancet* **2006**, *368*, 1696.
5. Luo, J.; Swaminath, G.; Brown, S. P.; Zhang, J.; Guo, Q.; Chen, M.; Nguyen, K.; Tran, T.; Miao, L.; Dransfield, P. J.; Vimolratana, M.; Houze, J. B.; Wong, S.; Toteva, M.; Shan, B.; Li, F.; Zhuang, R.; Lin, D. C. H. *PLoS One* **2012**, 7(10), e46300.
6. Lin, D.; Guo, Q.; Luo, J.; Zhang, J.; Nguyen, K.; Chen, M.; Tran, T.; Dransfield, P. J.; Brown, S. P.; Houze, J.; Vimolratana, M.; Jiao, X. Y.; Wang, Y.; Birdsall, N. J. M.; Swaminath, G. *Mol. Pharmacol.* **2012**, *82*(5), 843.
7. Sasaki, S.; Kitamura, S.; Negoro, N.; Suzuki, M.; Tsujihata, Y.; Suzuki, N.; Santou, T.; Kanzaki, N.; Harada, M.; Tanaka, Y.; Kobayashi, M.; Tada, N.; Funami, M.; Tanaka, T.; Yamamoto, Y.; Fukatsu, K.; Yasuma, T.; Momose, Y. *J. Med. Chem.* **2011**, *54*(5), 1365.
8. Negoro, N.; Sasaki, S.; Mikami, S.; Ito, M.; Suzuki, M.; Tsujihata, Y.; Ito, R.; Harada, A.; Takeuchi, K.; Suzuki, M.; Miyazaki, J.; Santou, T.; Odani, T.; Kanzaki, N.; Funami, M.; Tanaka, T.; Kogame, A.; Matsunaga, S.; Yasuma, T.; Momose, Y. *ACS Med. Chem. Lett.* **2010**, *1*, 290.
9. Yabuki, C.; Komatsu, H.; Tsujihata, Y.; Maeda, R.; Ito, R.; Matsuda-Nagasumi, K.; Sakuma, K.; Miyawaki, K.; Kikuchi, N.; Takeuchi, K.; Habata, Y.; Mori, M. *PLoS One* **2013**, *8*(10), e76280.
10. Tsujihata, Y.; Ito, R.; Suzuki, M.; Harada, A.; Negoro, N.; Yasuma, T.; Momose, Y.; Takeuchi, K. *J. Pharmacol. Exp. Ther.* **2011**, *339*(1), 228.
11. Haber, E. P.; Ximenes, H. M.; Procopio, J.; Carvalho, C. R.; Curi, R.; Carpinelli, A. R. *J. Cell. Physiol.* **2003**, *194*(1), 1.
12. Ito, R.; Tsujihata, Y.; Matsuda-Nagasumi, K.; Mori, I.; Negoro, N.; Takeuchi, K. *Br. J. Pharmacol.* **2013**, *170*(3), 568.
13. Naik, H.; Vakilynejad, M.; Wu, J.; Viswanathan, P.; Dote, N.; Higuchi, T.; Leifke, E. *J. Clin. Pharmacol.* **2012**, *52*(7), 1007.
14. Burant, C. F.; Viswanathan, P.; Marcinak, J.; Cao, C.; Vakilynejad, M.; Xie, B.; Leifke, E. *Lancet* **2012**, *379*(9824), 1403.
15. Kaku, K.; Araki, T.; Yoshinaka, R. *Diabetes Care* **2013**, *36*(2), 245.
16. Leifke, E.; Naik, H.; Wu, J.; Viswanathan, P.; Demanno, D.; Kipnes, M.; Vakilynejad, M. *Clin. Pharmacol. Ther.* **2012**, *92*(1), 29.
17. Naik, H.; Lu, J.; Cao, C.; Pfister, M.; Vakilynejad, M.; Leifke, E. *CPT Pharmacometrics Sys. Pharmacol.* **2013**, *2*, e22.
18. *Nat. Rev. Drug Discov.* **2014**, *13*(2), 91.
19. Houze, J. B.; Zhu, L.; Sun, Y.; Akerman, M.; Qiu, W.; Zhang, A. J.; Sharma, R.; Schmitt, M.; Wang, Y.; Liu, J.; Liu, J.; Medina, J. C.; Reagan, J. D.; Luo, J.; Tonn, G.; Zhang, J.; Lu, J. Y.-L.; Chen, M.; Lopez, E.; Nguyen, K.; Yang, L.; Tang, L.; Tian, H.; Shuttleworth, S. J.; Lin, D. C. H. Bioorg. *Med. Chem. Lett.* **2012**, *22*(2), 1267.
20. Lin, D. C. H.; Zhang, J.; Zhuang, R.; Li, F.; Nguyen, K.; Chen, M.; Tran, T.; Lopez, E.; Lu, J. Y. L.; Li, X. N.; Tang, L.; Tonn, G. R.; Swaminath, G.; Reagan, J. D.; Chen, J.-L.; Tian, H.; Lin, Y.-J.; Houze, J. B.; Luo, J. *PLoS One* **2011**, *6*(11), e27270.
21. Liu, J.; Wang, Y.; Ma, Z.; Schmitt, M.; Zhu, L.; Brown, S. P.; Dransfield, P. J.; Sun, Y.; Sharma, R.; Guo, Q.; Zhuang, R.; Zhang, J.; Luo, J.; Tonn, G. R.; Wong, S.; Swaminath, G.; Medina, J. C.; Lin, D. C. H.; Houze, J. B. *ACS Med. Chem. Lett.* **2014**, http://dx.doi.org/10.1021/ml400501x.

22. Miller, A. R.; Briere, D.; Marcelo, M.; Rani, S.; Padmaja, R.; Wilbur, K.; Osborne, E.; Zink, R.; Jett, D.; Montrose-Rafizadeh, C.; Michail, M.; Bokvist, K.; Matti, P.; Hamdouch, C. In: *48th Annual EASD Meeting, Berlin* 2012.
23. http://clinicaltrials.gov/ct2/show/NCT01358981.
24. Brown, S. P.; Dransfield, P. J.; Vimolratana, M.; Jiao, X.; Zhu, L.; Pattaropong, V.; Sun, Y.; Liu, J.; Luo, J.; Zhang, J.; Wong, S.; Zhuang, R.; Guo, Q.; Li, F.; Medina, J. C.; Swaminath, G.; Lin, D. C. H.; Houze, J. B. *ACS Med. Chem. Lett.* **2012**, *3*, 726.
25. Wang, Y.; Liu, J.; Dransfield, P. J.; Zhu, L.; Wang, Z.; Du, X.; Jiao, X.; Su, Y.; Li, A.-r.; Brown, S. P.; Kasparian, A.; Vimolratana, M.; Yu, M.; Pattaropong, V.; Houze, J. B.; Swaminath, G.; Tran, T.; Nguyen, K.; Guo, Q.; Zhang, J.; Zhuang, R.; Li, F.; Miao, L.; Bartberger, M. D.; Correll, T. L.; Chow, D.; Wong, S.; Luo, J.; Lin, D. C. H.; Medina, J. C. *ACS Med. Chem. Lett.* **2013**, *4*(6), 551.
26. Du, X.; Dransfield, P. J.; Lin, D. C. H.; Wong, S.; Wang, Y.; Wang, Z.; Kohn, T.; Yu, M.; Brown, S. P.; Vimolratana, M.; Zhu, L.; Li, A.-R.; Su, Y.; Jiao, X.; Liu, J.; Swaminath, G.; Tran, T.; Luo, J.; Zhuang, R.; Zhang, J.; Guo, Q.; Li, F.; Connors, R.; Medina, J. C.; Houze, J. B. *ACS Med. Chem. Lett.* **2014**, http://dx.doi.org/10.1021/ml4005123.
27. Lipinski, C. A.; Lombardo, F.; Dominy, B. W.; Feeney, P. J. *Adv. Drug Deliv. Rev.* **2001**, *46*(1–3), 3.

CHAPTER SEVEN

Recent Advances in the Development of $P2Y_{12}$ Receptor Antagonists as Antiplatelet Agents

Allorie T. Caldwell, E. Blake Watkins
Department of Pharmaceutical Sciences, Union University School of Pharmacy, Jackson, Tennessee, USA

Contents

1. INTRODUCTION

$P2Y_{12}$ receptor antagonists have emerged as the current mode of treatment in conjunction with aspirin for acute coronary syndromes (ACS) and for patients undergoing percutaneous coronary intervention (PCI) with stenting through their ability to inhibit platelet aggregation.[1] The $P2Y_1$ and $P2Y_{12}$ receptors are members of the purinergic P2Y family of G-protein-coupled receptors and are located on the surface of platelets. Vascular damage initiates platelet activation[2] resulting in release of adenosine-5′-diphosphate (ADP, **1**) from dense granules. ADP binds to and activates the G_q-coupled purinergic $P2Y_1$ receptor resulting in increased intracellular calcium levels and initiation of rapidly reversible platelet aggregation. Successive binding of ADP to the $G\alpha_{i2}$-coupled $P2Y_{12}$ receptors inhibits adenylate cyclase resulting in a

Annual Reports in Medicinal Chemistry, Volume 49
ISSN 0065-7743
http://dx.doi.org/10.1016/B978-0-12-800167-7.00007-9

reduction of intracellular cyclic adenosine-3′,5′-monophosphate levels and subsequent amplification of stable platelet aggregation.[3–7] Successful platelet aggregation requires activation of both P2Y$_1$ and P2Y$_{12}$ receptors.[8] Antagonism of the P2Y$_{12}$ receptor results in inhibition of platelet aggregation and a reduction in occlusive thrombotic events.[9] A number of excellent reviews have recently appeared on the clinical use of P2Y$_{12}$ antagonists as antiplatelet agents.[10–13]

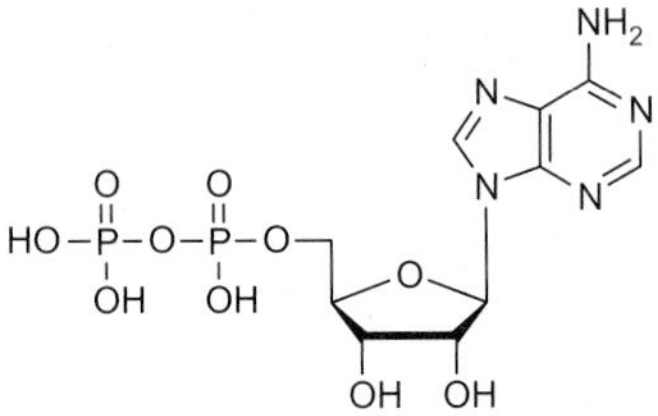

1

2. FDA-APPROVED P2Y$_{12}$ RECEPTOR ANTAGONISTS

Currently, two classes of P2Y$_{12}$ antagonists have been approved by the FDA for use in the US as antiplatelet agents. These consist of the irreversible thienopyridines and the more recently approved reversible ATP analogs. The thienopyridine class (Fig. 7.1) is composed of ticlopidine (**2**), clopidogrel (**3**), and prasugrel (**4**), while ticagrelor (**5**) is the only reversible ATP analog currently available for therapeutic use.[14]

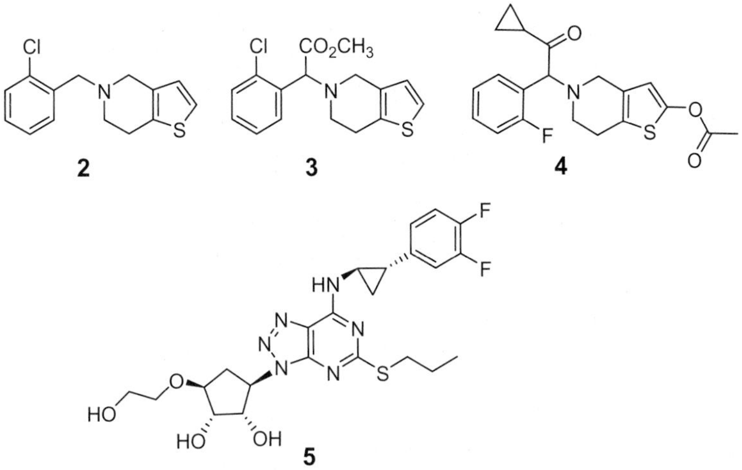

Figure 7.1 Currently FDA-approved P2Y$_{12}$ receptor antagonists.

2.1. Thienopyridines

The thienopyridines are prodrugs and, therefore, must undergo cytochrome P450 (CYP)-mediated conversion to the active metabolite prior to blocking the ADP-binding site of the P2Y$_{12}$ receptor. Hepatic activation is a two-step process involving multiple CYP isoforms (Fig. 7.2).[9] Ticlopidine and clopidogrel undergo an identical initial oxidative reaction leading to the formation of an α,β-unsaturated thiolactone (**6**). Prasugrel, on the other hand, due to the presence of the acetate moiety on the thiophene ring, is activated through esterase-mediated hydrolysis to a similar α,β-unsaturated thiolactone (**8**). The final step of metabolic activation is common to all thienopyridines and involves CYP-mediated, ring-opening hydrolysis of the thiolactone (**6** or **8**) to the active thiol (**7** or **9**). This active metabolite then forms an irreversible disulfide bond with a cysteine residue in the ADP-binding pocket.[4] The oligomeric receptor dissociates into dimers preventing ADP from binding to the receptor and preventing activation of the platelet in this manner.[15] The overall result is inhibition of platelet aggregation which prevents development of a thrombus at the site of vascular injury. Clopidogrel that is not rapidly converted to **7** is enzymatically hydrolyzed to the inactive carboxylic acid (~85% of a typical dose).[16]

The need for metabolic activation results in a delayed onset of action for this class of antiplatelet agents and increased inter-individual variability as a result of genetic polymorphism in the required CYP enzymes. Approximately 30% of Caucasians, 40% of African Americans, and 55% of East Asians carry a CYP2C19 reduced-function allele resulting in significantly lower plasma levels of the active metabolite. As a result, inhibition of platelets is reduced, posing an increased risk for myocardial infarction or stroke leading to death.[17] Irreversible antagonism of the P2Y$_{12}$ receptor also results

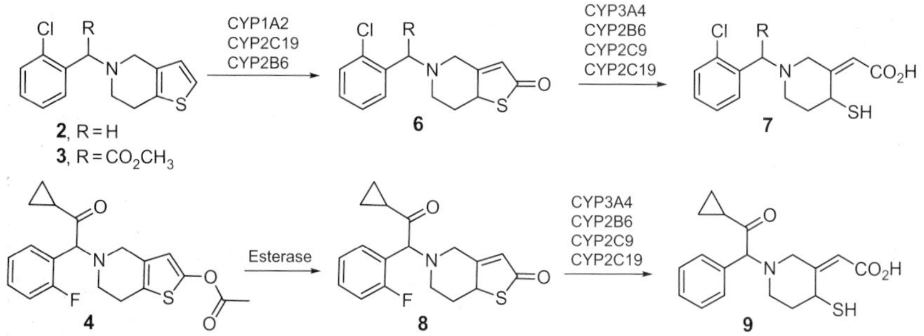

Figure 7.2 Metabolic activation of thienopyridines to the reactive thiols.[9]

in an increased risk of severe or even fatal bleeding events, since antagonism lasts throughout the life of the platelet. Of the thienopyridines, ticlopidine has largely been replaced by clopidogrel in treatment due to greater risk of bleeding, as well as a higher incidence of neutropenia and thrombocytopenia.[7,18,19]

Of the thienopyridines, **4** has a faster onset owing to the presence of a simple acetate ester which undergoes esterase-mediated hydrolysis to the intermediate thiolactone. Prasugrel was designed to forego the need for CYP-mediated oxidation of the thiophene ring to reduce inter-individual variability resulting in increased plasma concentrations of the active metabolite, more effective platelet inhibition and a reduction in the time to onset of action. Furthermore, prasugrel has a cyclopropyl ketone rather than an ester to provide metabolic stability leading to an increased concentration in the plasma.[20,21]

2.2. ATP Analogs

While ADP is the endogenous ligand for the $P2Y_{12}$ receptor, ATP acts as an antagonist. Ticagrelor belongs to the ATP analog class of reversible $P2Y_{12}$ antagonists and binds to a site distinct from the thienopyridines. Binding of ticagrelor causes a conformational change in the receptor thus prohibiting ADP binding and platelet activation.[22] Unlike the irreversible agents, the ATP analogs bind without the need for hepatic bioactivation, although the primary metabolite of ticagrelor is active with potency equal to the parent compound.[9] The overall result is a faster onset of action, less inter-individual variation in effectiveness, and greater control of platelet aggregation inhibition.[7,11,19]

3. NEW $P2Y_{12}$ RECEPTOR ANTAGONISTS

3.1. ATP Analogs

The search for additional reversible $P2Y_{12}$ antagonists with a reduced bleeding risk and greater potency continues. Several attempts have been made to improve upon existing FDA-approved agents. Cangrelor (**10**) is an intravenously administered, reversible ATP analog, which offers rapid onset of action plus rapid offset upon discontinuation of the drug. Cangrelor is rapidly metabolized to inactive products in plasma providing greater control over platelet aggregation.[12,14]

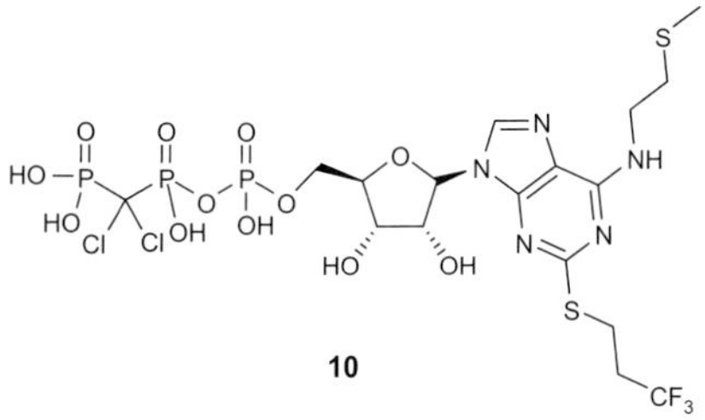

10

In an effort to improve upon the effectiveness and safety profile of **5**, a number of analogs with varying side chains were prepared based on the triazolopyrimidine scaffold.[23] Diversely substituted, hydroxy cyclopentyl side chains displayed a range of IC_{50} values of 0.22–16.3 μM in an ADP-induced human platelet-rich plasma (hPRP) assay (**5**, $IC_{50}=0.50$). Analog **11** emerged as the most likely candidate for further development having greater *in vitro* and *in vivo* antiplatelet activity. In rats dosed at 2.5 mg/kg, **11** showed 95% inhibition of platelet aggregation compared to 59% for **5**. Additionally, **11** displayed shorter bleeding time (9.76 min vs. >30 min at 2.5 mg/kg) in a rat tail bleeding test.[23]

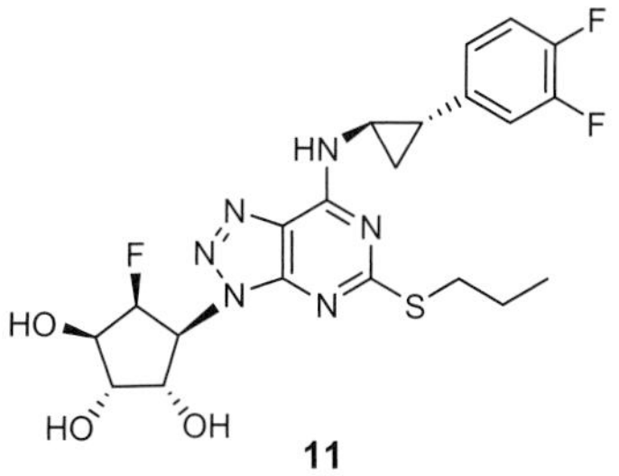

11

Additionally, a series of ticagrelor derivatives were synthesized making structural changes to the hydroxyethyl side chain as indicated below (**12**) to increase potency.[24] In general, potency in an ADP-induced platelet aggregation assay (PA) was lower for the ester/carbonate analogs (13.2–47.7%) compared to the parent compound (66.3%). Small alkyl or alkoxy substituents (R_1 or R_2) resulted in compounds with the highest activity (42.6–47.7%) indicating that hydrolysis to **5** was most likely required in order for inhibition of platelet aggregation to occur.[24]

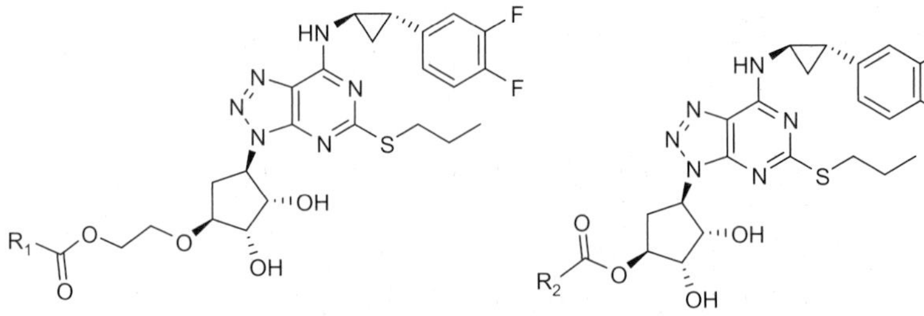

12

3.2. Thienopyridines

A number of additional P2Y_{12} antagonists have been synthesized as groups seek an agent capable of effectively inhibiting platelet aggregation void of the myriad of problems associated with current therapies. A series of single isomer (*S*)-clopidogrel and prasugrel hybrid molecules were developed in an effort to take advantage of the safety profile of **3**, as well as the added benefits of **4**, namely, the reduced dependence on CYP-mediated activation.[25] These agents belong to the thienopyridine class of antiplatelet agents and, therefore, must undergo activation to the active thiol in a fashion similar to **4**. Structural modifications (**13**) were made on the phenyl ring (R_1) and on the ester (R_2). SAR analysis, following a PA, demonstrated an inverse relationship between the size of R_2 and the antiplatelet activity. As the size of the ester increased, activity decreased seemingly in response to the reduction in the rate of hydrolysis by esterases to the intermediate thiolactone (R_1=Cl, R_2=Me, PA=34.6±13.5%; R_1=Cl, R_2=*tert*-amyl, PA=70.0±23.0%). Additional substituents at R_2 confirm this conclusion as the corresponding carbamates (i.e., R_1=Cl, R_2=NMe_2, PA=67.7±17.7%) and carbonates (i.e., R_1=Cl, R_2=Et, PA=53.0±11.6%) showed decreased effects on platelet aggregation.

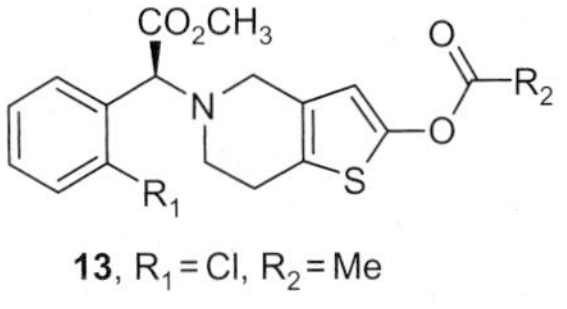

13, R_1=Cl, R_2=Me

To examine the stereochemical requirement for the P2Y_{12} receptor, researchers also examined the *R*-isomer (R_1=Cl, R_2=Me, PA=68.4±19.6%) and the racemic mixture (*R*/*S*) (R_1=Cl, R_2=Me,

PA = 63.7 ± 8.0%). Since activity resides in the *S*-isomer, the hybrid, vicagrel (**13**, R_1 = Cl, R_2 = Me) was developed as an enantiomerically pure agent. At 3 mg/kg in a rat model, **13** displayed potency (34.6 ± 13.5%) similar to **4** (32.1 ± 9.3%) but superior to **3** (73.7 ± 5.2%). In a pharmacokinetic rat model, **13** displayed a sixfold increase in bioavailability over **3**.[25]

3.3. Miscellaneous Scaffolds

3.3.1 Phenylpyrazole Glutamic Acid Piperazines

Utilizing a homology model of the $P2Y_{12}$ receptor, a series of compounds were designed to explore space requirements in order to maximize binding affinity [$K_i(P2Y_{12})$] and antiplatelet activity in a hPRP assay [IC_{50} (hPRP)].[26] The compounds were based on a phenylpyrazole glutamic acid piperazine backbone (**14**). Structural analysis of existing compounds using their homology model indicated a need for modification at the 5-position of the pyrazole core. As expected, substitutions at the 5-position displayed a significant impact on binding affinity and antiplatelet activity as functionality at this location, according to their model, filled a lipophilic region of the binding pocket. After exploring various groups, compound **14**, which contains a proline cyclobutyl amide at the C-5 position, displayed the highest binding affinity [$K_i(P2Y_{12})$ = 7.3 nM] and antiplatelet activity [IC_{50}(hPRP) = 9 nM], owing to its size, lipophilicity, and appropriate geometry. Additional modifications were made to the central amino acid portion and the piperazine region of the model compounds. While various polar substitutions were well tolerated in the replacement of glutamic acid, lipophilic side chains reduced binding affinity and lowered *in vitro* antiplatelet activity. Various substitutions for the piperazine fragment failed to yield a compound with potency greater than that of **14**.[26]

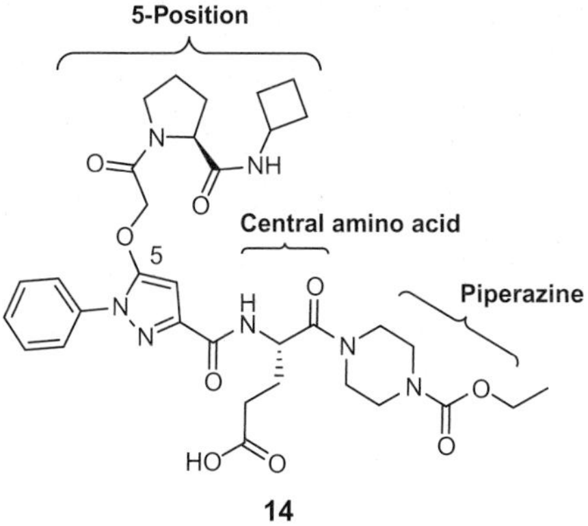

5-Position

Central amino acid

Piperazine

14

3.3.2 6-Aminonicotinates

A number of structurally similar compounds have been synthesized based on the lead compound (**15**) shown in Fig. 7.3 and tested for their *in vitro* binding affinities (IC_{50}) to the $P2Y_{12}$ receptor, as well as for their ability to inhibit fibrinogen-induced aggregation (IC_{50}) in a washed platelet assay (WPA).[27,28] The lead compound was discovered using high-throughput screening.[29] Evaluation revealed the need for a reduction in the lipophilic nature of the compounds to improve water solubility. Modification of the urea linkage to a sulfonylurea (**16**, $R_2 = Ph$) introduced improved water solubility owing to the lower pK_a value and increased ionization at pH 6.8. The sulfonylurea displayed a binding affinity of 0.18 μM and an IC_{50} of 1.1 μM in the WPA. Additional variations on rings A, B, and C were incorporated in order to better understand the SAR of this class of agents and to improve potency. Analogs containing a variety of C-ring substitutions were synthesized. Binding affinities ranged from 0.0063 to 0.090 μM, while WPA potency ranged from 0.0095 to 0.26 μM for the series containing a 5-chlorothienyl C-ring (**17**). A variety of substitutions of the B-ring showed the piperidine ring to be superior in terms of membrane permeability and potency. Modifications to the A-ring were limited to the ethyl ester since the primary metabolite of these compounds is the carboxylic acid, which lacks antiplatelet activity. Replacement of the ethyl ester with larger esters or ketones resulted in a significant decrease in binding affinity. A number of highly potent sulfonylureas were identified, but each of them failed to display the desired pharmacokinetic properties.[27]

In a continued effort to identify a $P2Y_{12}$ receptor antagonist possessing high potency and adequate physicochemical and pharmacokinetic properties, further investigations were undertaken. As a result, sulfonamide **18** emerged as a candidate drug for human clinical trials. Binding affinity was 11 nM while also displaying high water solubility (104 μM) and membrane permeability. The sulfonamide was a weak inhibitor of both CYP2C9 and CYP3A4, 3.3 and 20 μM, respectively. In a modified Folts Dog model, **18**

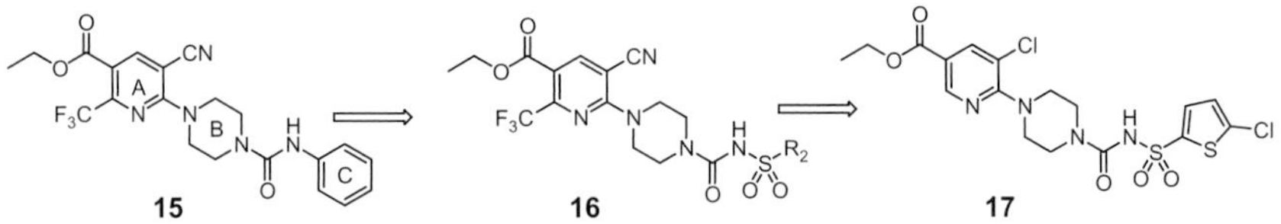

Figure 7.3 Evolution of lead compound to agents with improved pharmacokinetic properties.

was effective in inhibiting ADP-induced platelet aggregation [ED_{50} 3.0 μg/(kg × min)] and increasing blood flow in a dose-dependent manner. Additionally the *in vivo* therapeutic index (TI) was found to be ≥10, which compares favorably to the therapeutic indices for **5** (TI > 5.2) and **3** (TI = 2.3).[28]

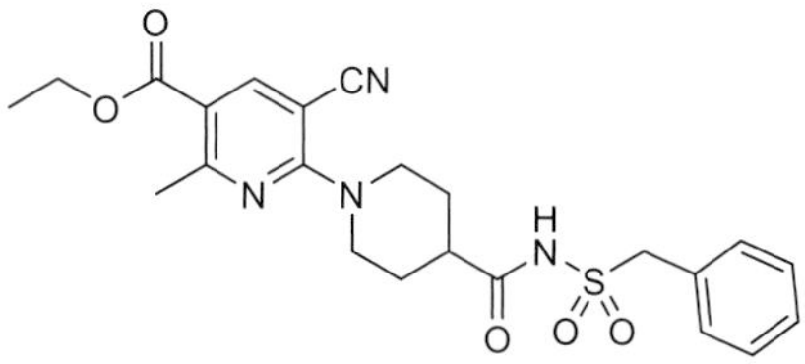

18 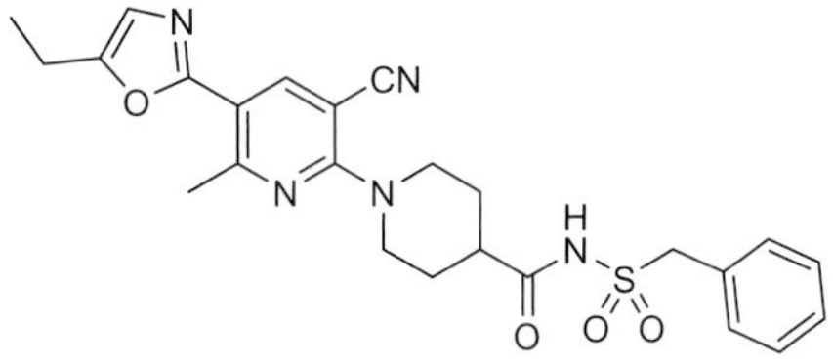 **19**

Esterase-mediated hydrolysis of **18** and related ethyl esters yields a carboxylic acid void of antiplatelet activity. In an effort to minimize metabolic susceptibility, a number of heterocyclic bioisosteres for the ethyl ester were examined. Most substitutions led to a decrease in potency, however, the 5-ethyloxazole (**19**) displayed binding affinity (IC_{50} = 0.18 μM) and WPA potency (0.2 μM) similar to **18** (binding affinity 0.011 μM) with a reduced rate of clearance. This general trend held for most 5-ethyloxazole analogs regardless of other structural modifications.[30]

4. CLINICAL APPLICATION

Four major trials have been published over the last 5 years regarding the advancements in antiplatelet use for patients with ACS. The first clinical trial of significance was the Trial to Assess Improvement in Therapeutic Outcomes by Optimizing Platelet Inhibition with Prasugrel–Thrombolysis in Myocardial Infarction (TRITON-TIMI) 38. This trial included over 13,000 patients with ACS who were scheduled for PCI. Data showed that prasugrel significantly reduced the rate of death when compared with clopidogrel. However, major bleeding was evaluated as a safety endpoint, and prasugrel demonstrated a higher risk of bleeding. After analysis of various subgroups, patients with a prior history of ischemic attacks, those over 65 years of age, and those with a body weight of less than 60 kg were found to be at highest risk for bleeding events. These findings lead to a contraindication in prior stroke patients and precautionary statements regarding the elderly and those with a low body weight. Of note, during subgroup analysis, diabetic patients, in particular, benefited more from prasugrel than clopidogrel. Diabetic patients in the prasugrel group had a greater relative risk reduction for the primary endpoint, as well as fewer myocardial infarctions during the follow-up period. Furthermore, no significant difference was found among this population with regards to bleeding risks.[31]

The Targeted Platelet Inhibition to Clarify the Optimal Strategy to Medically Manage Acute Coronary Syndromes study (TRILOGY-ACS) was conducted to evaluate the benefit of prasugrel versus clopidogrel in over 9000 patients who were not undergoing PCI and were managed medically. Also, the possibility of using a lower dose in elderly patients to curb bleeding risks was evaluated. The primary endpoint, composite death rates, was not significantly different between the two treatment groups after up to 30 months of follow-up. Interestingly, the bleeding risks between the two groups were not significantly different in this study group, in contrast to the results of the TRITON-TIMI 38 study. These results could be related to the dose adjustment for patients over 75 years of age and those under 60 kg of total body weight from 10 mg daily to 5 mg.[32]

In 2009, ticagrelor was approved by the FDA. The major phase III trial related to its approval was the study of Platelet Inhibition and Patient Outcomes (PLATO). Over 18,000 patients with ACS were randomized to receive either ticagrelor or clopidogrel. Unlike the patients in TRITON-TIMI 38, these patients were not required to be scheduled for a PCI. They could be managed medically like the patients in the TRILOGY-ACS trial. The primary end point of composite death due to vascular causes, myocardial infarction, or stroke was statistically lower in the ticagrelor group when compared with the clopidogrel group. As for safety endpoints, overall there was no difference in all major bleeding episodes between the two groups. However, ticagrelor had a higher incidence of noncoronary artery bypass graft-related major bleeding and a higher incidence of intracranial bleeding. As for other side effects, uric acid and serum creatinine levels increased slightly more with the use of ticagrelor versus clopidogrel, as did the incidence of dyspnea.[33]

Several papers have been published critiquing the methods and results of the PLATO trial. One issue raised was the fact that most of the benefit seen in the study corresponded to patient cases in Poland and Hungary. These areas were monitored by the study sponsors; whereas, regions such as the United States, Russia, and Georgia, which did not show the same benefit when evaluated independently, were monitored by a third party company. According to the authors of this critique, some of the endpoints were either not submitted for adjudication or were downgraded to a softer end point for ticagrelor but not for clopidogrel.[34] Another issue raised was related to discrepancies in the number of patients with an incomplete vital status follow-up, which could skew the study results.[35]

The most recent drug awaiting FDA approval is cangrelor, an intravenous and reversible $P2Y_{12}$ inhibitor. The Cangrelor versus Standard Therapy to Achieve Optimal Management of Platelet Inhibition (CHAMPION PHOENIX) trial compares the use of cangrelor prior to PCI to a loading dose of clopidogrel. Two prior CHAMPION studies, PCI and PLATFORM, were stopped prematurely due to lack of efficacy and an inability to meet power requirements. Also, some of the safety data showed a trend toward an increase in bleeding.[36,37] However, in the PHOENIX study, the primary endpoint was death from any cause, myocardial infarction, ischemia-driven revascularization or stent thrombosis in the 48 h after randomization. A statistically significant reduction in the endpoint was noted in the group using cangrelor versus clopidogrel. As for safety endpoints, bleeding occurred at a similar rate in both groups.[38]

5. CONCLUSIONS

Antagonism of the human $P2Y_{12}$ purinergic receptor has proven to be an effective means of managing patients with ACS and those undergoing PCI with stent placement through inhibition of platelet aggregation. Unfortunately, an effective $P2Y_{12}$ antagonist with a minimal bleeding risk remains elusive. As the landscape of the receptor becomes clear, more potent agents will be possible and additional work can be done to determine if separating bleeding characteristics from antiplatelet activity is feasible.

REFERENCES

1. Levine, G. N.; Bates, E. R.; Blankenship, J. C.; Bailey, S. R.; Bittl, J. A.; Cercek, B.; Chambers, C. E.; Ellis, S. G.; Guyton, R. A.; Hollenberg, S. M.; Khot, U. N.; Lange, R. A.; Mauri, L.; Mehran, R.; Moussa, I. D.; Mukherjee, D.; Nallamothu, B. K.; Ting, H. H. *Circulation* **2011**, *2011*(124), e574–e651.
2. Broos, K.; Feys, H. B.; De Meyer, S. F.; Vanhoorelbeke, K.; Deckmyn, H. *Blood Rev.* **2011**, *25*, 155–167.
3. Gachet, C. *Purinergic Signal* **2012**, *8*, 609–619.
4. Jacobson, K. A.; Deflorian, F.; Mishra, S.; Costanzi, S. *Purinergic Signal* **2011**, *7*, 305–324.
5. Storey, R. F. *Curr. Pharm. Des.* **2006**, *12*, 1255–1259.
6. Abbracchio, M. P.; Burnstock, G.; Boeynaems, J.-M.; Barnard, E. A.; Boyer, J. L.; Kennedy, C.; Knight, G. E.; Fumagalli, M.; Gachet, C.; Jacobson, K. A.; Weisman, G. A. *Pharmacol. Rev.* **2006**, *58*, 281–341.
7. Michelson, A. D. *Arterioscler. Thromb. Vasc. Biol.* **2008**, *28*, s33–s38.
8. Cattaneo, M. *Circulation* **2010**, *121*, 171–179.
9. Bhavaraju, K.; Mayanglambam, A.; Rao, K. A.; Kunapuli, S. P. *Curr. Opin. Drug Discov. Dev.* **2010**, *13*, 497–506.

10. Secco, G. G.; Parisi, R.; Mirabella, F.; Fattori, R.; Genoni, G.; Agostoni, P.; De Luca, G.; Marino, P. N.; Lupi, A.; Rognoni, A. *Cardiovasc. Hematol. Agents Med. Chem.* **2013**, *11*, 101–105.
11. Ferri, N.; Corsini, A.; Bellosta, S. *Drugs* **2013**, *73*, 1681–1709.
12. Barn, K.; Steinhubl, S. R. *Coron. Artery Dis.* **2012**, *23*, 368–374.
13. Liu, H.; Ge, H.; Peng, Y.; Xiao, P.; Xu, J. *Biophys. Chem.* **2011**, *155*, 74–81.
14. Kalra, K.; Franzese, C. J.; Gesheff, M. G.; Lev, E. I.; Pandya, S.; Bliden, K. P.; Tantry, U. S.; Gurbel, P. A. *Curr. Atheroscler. Rep.* **2013**, *15*, 1–13.
15. Savi, P.; Zachayus, J.-L.; Delesque-Touchard, N.; Labouret, C.; Herve, C.; Uzabigaga, M.-F.; Pereillo, J.-M.; Culouscou, J.-M.; Bono, F.; Ferrara, P.; Herbert, J.-M. *Proc. Natl. Acad. Sci. U.S.A* **2006**, *103*, 11069–11074.
16. Angiolillo, D. J.; Bhatt, D. L.; Gurbel, P. A.; Jennings, L. K. *Am. J. Cardiol.* **2009**, *103*, 40A–51A.
17. Mega, J. L.; Close, S. L.; Wiviott, S. D.; Shen, L.; Hockett, R. D.; Brandt, J. T.; Walker, J. R.; Antman, E. M.; Macias, W.; Braunwald, E.; Sabatine, M. S. *N. Engl. J. Med.* **2009**, *360*, 354–362.
18. Bertrand, M. E.; Rupprecht, H. J.; Urban, P.; Gershlick, A. H. *Circulation* **2000**, *102*, 624–629.
19. Porto, I.; Giubilato, S. *Expert Opin. Invest. Drugs* **2009**, *18*, 1317–1332.
20. Farid, N. A.; Smith, R. L.; Gillespie, T. A.; Rash, T. J.; Blair, P. E.; Kurihara, A.; Goldberg, M. J. *Drug Metab. Dispos.* **2007**, *35*, 1096–1104.
21. Brandt, J. T.; Payne, C. D.; Wiviott, S. D.; Weerakkody, G.; Farid, N. A.; Small, D. S.; Jakubowski, J. A.; Naganuma, H.; Winters, K. J. *Am. Heart J.* **2007**, *153*, 66.e9–66.e16.
22. Van Giezen, J. J. J.; Nilsson, L.; Berntsson, P.; Wissing, B. M.; Giordanetto, F.; Tomlinson, W.; Greasley, P. J. *J. Thromb. Haemost.* **2009**, 7, 1556–1565.
23. Tu, W.; Fan, J.; Zhang, H.; Xu, G.; Liu, Z.; Qu, J.; Yang, F.; Zhang, L.; Luan, T.; Yuan, J.; Gong, A.; Sun, P.; Dong, Q. *Bioorg. Med. Chem. Lett.* **2014**, *24*, 141–146.
24. Zhang, H.; Liu, J.; Zhang, L.; Kong, L.; Yao, H.; Sun, H. *Bioorg. Med. Chem. Lett.* **2012**, *22*, 3598–3602.
25. Shan, J.; Zhang, B.; Zhu, Y.; Jiao, B.; Zheng, W.; Qi, X.; Gong, Y.; Yuan, F.; Lv, F.; Sun, H. *J. Med. Chem.* **2012**, *55*, 3342–3352.
26. Zech, G.; Hessler, G.; Evers, A.; Weiss, T.; Florian, P.; Just, M.; Czech, J.; Czechtizky, W.; Goerlitzer, J.; Ruf, S.; Kohlmann, M.; Nazare, M. *J. Med. Chem.* **2012**, *55*, 8615–8629.
27. Bach, P.; Bostroem, J.; Brickmann, K.; van Giezen, J. J. J.; Groneberg, R. D.; Harvey, D. M.; O'Sullivan, M.; Zetterberg, F. *Eur. J. Med. Chem.* **2013**, *65*, 360–375.
28. Bach, P.; Antonsson, T.; Bylund, R.; Bjoerkman, J.-A.; Oesterlund, K.; Giordanetto, F.; van Giezen, J. J. J.; Andersen, S. M.; Zachrisson, H.; Zetterberg, F. *J. Med. Chem.* **2013**, *56*, 7015–7024.
29. Bach, P.; Bostroem, J.; Brickmann, K.; van Giezen, J. J. J.; Hovland, R.; Petersson, A. U.; Ray, A.; Zetterberg, F. *Bioorg. Med. Chem. Lett.* **2011**, *21*, 2877–2881.
30. Bach, P.; Bostroem, J.; Brickmann, K.; Burgess, L. E.; Clarke, D.; Groneberg, R. D.; Harvey, D. M.; Laird, E. R.; O'Sullivan, M.; Zetterberg, F. *Future Med. Chem.* **2013**, *5*, 2037–2056.
31. Wiviott, S. D.; Bruanwald, E.; McCabe, C. H.; Montalescot, G.; Ruzylio, W.; Gottlieb, S.; Neumann, F.-J.; Ardissino, D.; De Servi, S.; Murphy, S. A.; Riesmeyer, J.; Weerakkody, G.; Gibson, C. M.; Antmann, E. M. *N. Engl. J. Med.* **2007**, *357*, 2001–2015.
32. Roe, M. T.; Armstrong, P. W.; Fox, K. A. A.; White, H. D.; Prabhakaran, D.; Goodman, S. G.; Cornel, J. H.; Bhatt, D. L.; Clemmensen, P.; Martinez, F.; Ardissino, D.; Nicolau, J. C.; Boden, W. E.; Gurbel, P. A.; Ruzyllo, W.; Dalby, A. J.; McGuire, D. K.; Leiva-Pons, J. L.; Parkhomenko, A.; Gottlieb, S.;

Topacio, G. O.; Hamm, C.; Pavlides, G.; Goudev, A. R.; Oto, A.; Tseng, C.-D.; Merkely, B.; Gasparovic, V.; Corbalan, R.; Cinteza, M.; McLendon, R. C.; Winters, K. J.; Brown, E. B.; Lokhnygina, Y.; Aylward, P. E.; Huber, K.; Hochman, J. S.; Ohman, E. M. *N. Engl. J. Med.* **2012**, *367*, 1297–1309.

33. Wallentin, L.; Becker, R. C.; Budaj, A.; Cannon, C. P.; Emanuelsson, H.; Held, C.; Horrow, J.; Husted, S.; James, S.; Katus, H.; Mahaffey, K. W.; Scirica, B. M.; Skene, A.; Steg, P. G.; Storey, R. F.; Harrington, R. A. *N. Engl. J. Med.* **2009**, *361*, 1045–1057.
34. DiNicolantonio, J. J.; Tomek, A. *Int. J. Cardiol.* **2013**, *168*, 4076–4080.
35. DiNicolantonio, J. J.; Tomek, A. *Int. J. Cardiol.* **2013**, *169*, 145–146.
36. Harrington, R. A.; Stone, G. W.; McNulty, S.; White, H. D.; Lincoff, A. M.; Gibson, C. M.; Pollack, C. V., Jr.; Montalescot, G.; Mahaffey, K. W.; Kleiman, N. S.; Goodman, S. G.; Amine, M.; Angiolillo, D. J.; Becker, R. C.; Chew, D. P.; French, W. J.; Leisch, F.; Parikh, K. H.; Skerjanec, S.; Bhatt, D. L. *N. Engl. J. Med.* **2009**, *361*, 2318–2329.
37. Bhatt, D. L.; Lincoff, A. M.; Gibson, C. M.; Stone, G. W.; McNulty, S.; Montalescot, G.; Kleiman, N. S.; Goodman, S. G.; White, H. D.; Mahaffey, K. W.; Pollack, C. V., Jr.; Manoukian, S. V.; Widimsky, P.; Chew, D. P.; Cura, F.; Manukov, I.; Tousek, F.; Zubairjafar, M.; Arneja, J.; Skerjanec, S.; Harrington, R. A. *N. Engl. J. Med.* **2009**, *361*, 2330–2341.
38. Bhatt, D. L.; Stone, G. W.; Mahaffey, K. W.; Gibson, M.; Steg, P. G.; Hamm, C. W.; Price, M. J.; Leonardi, S.; Gallup, D.; Bramucci, E.; Radke, P. W.; Widimsky, P.; Tousek, F.; Tauth, J.; Spriggs, D.; McLaurin, B. T.; Angiolillo, D. J.; Genereux, P.; Liu, T.; Prats, J.; Todd, M.; Skerjanec, S.; White, H. D.; Harrington, R. A. *N. Engl. J. Med.* **2013**, *368*, 1303–1313.

CHAPTER EIGHT

Current Approaches to the Treatment of Atrial Fibrillation

Heather Finlay
Research and Development, Bristol-Myers Squibb, Princeton, New Jersey, USA

Contents

1. INTRODUCTION

Atrial fibrillation (AF) is the most common cardiac arrhythmia affecting an estimated population of 33 million worldwide with an associated prevalence of approximately 1%.[1] AF is frequently asymptomatic; however, AF is the strongest associated factor for increased risk of stroke (5-fold) and has an associated 2-fold increase in mortality.[2] The prevalence of AF increases significantly with age and concomitant to an aging population, the estimated global market will increase from 6 billion USD in 2012 to 15 billion USD in 2019.[3]

Clinical use of currently available anticoagulants is determined by assessing the level of stroke risk according to the clinical prediction rules $CHADS_2$ or CHA_2DS_2 score (C, congestive heart failure; H, hypertension; A, age greater than 75; D, diabetes; S, prior stroke or TIA or thromboembolism; one point

Annual Reports in Medicinal Chemistry, Volume 49
ISSN 0065-7743
http://dx.doi.org/10.1016/B978-0-12-800167-7.00008-0

each unless indicated) and available anti-arrhythmic drugs by evaluating known toxicity followed by efficacy.[4] Recent advances in the treatment of AF include the approval of a series of novel anticoagulants which inhibit thrombin or Factor Xa.[5] The option to more effectively anticoagulate reduces the risk of stroke significantly; however, AF patients remain in abnormal sinus rhythm. There is demonstrated benefit to restoring and maintaining normal sinus rhythm in AF patients in addition to anticoagulation therapy and this is the goal of concurrent anti-arrhythmic therapy.[6] Anti-arrhythmic agents with improved safety and efficacy profiles have been long-standing targets in the pharmaceutical industry and recent advances toward these goals are summarized in this review which covers compounds and series disclosed from 2010 to 2014.

2. ATRIAL-SELECTIVE AGENTS VERSUS NON-SELECTIVE AGENTS

2.1. Current Standard of Care

In addition to conversion to normal sinus rhythm by surgical ablation, anti-arrhythmic drugs are prescribed for the termination of AF, as agents for the maintenance of normal sinus rhythm and as rate control agents. Rate control agents, for example, β-blockers or calcium channel inhibitors, lower the heart rate (HR) by effecting re-entry currents by increasing the degree of block at the atrioventricular node leaving the patient in irregular sinus rhythm. Rhythm control agents, for example, potassium channel inhibitors, lead to cardioversion and restoration of normal sinus rhythm. Dronedarone was approved for AF in 2009 and is currently the most widely prescribed drug accounting for 24% of all global prescriptions in 2012 and 2013. Although not approved for AF, the first generation drug in this class, amiodarone, is currently prescribed for 21% of all global AF prescriptions.[7] Amiodarone and dronedarone inhibit ion channels I_{Na}, I_{CaL}, I_{Kr}, I_{Kur}, I_{to}, I_{KAch}, and I_f in addition to antagonism of both α and β receptors.[8] Drondedarone has a significantly reduced half-life, does not require an in-patient loading dose and has an improved cardiovascular side effect profile relative to amiodarone. It is the only anti-arrhythmic drug to demonstrate a stroke risk reduction in AF patients; however, due to the broad ion channel activity, dronedarone is contraindicated in patients with unstable or severe heart failure and the recent incidence of hepatotoxicity has limited more extensive use in AF patients.[9] Sotalol was first approved in 1992 for ventricular arrhythmia, but more recently approved for the treatment of AF in 2000. Sotalol is an I_{Kr} and β-blocker and is generally well tolerated as a first line therapy for patients with coronary artery disease.[10] Drugs which inhibit I_{Kr} are initially administered in a hospital setting due to mechanism based prolongation of QTc and the associated risk of the ventricular arrhythmia, Torsades de pointes. Additional drug therapies for maintenance of normal sinus rhythm include propafenone (I_{Na} and β-blocker, 14% of global prescriptions) and flecainide (I_{Na} inhibitor, 14% of global prescriptions), the I_{Kr} inhibitor dofetilide (US only) and vernakalant (I_{Kur} and I_{Na}, intravenous, Europe only).[11–14] Propafenone and flecainide are contraindicated in patients with coronary heart disease. Vernakalant is only approved for the termination of AF given the side effects of hypotension in addition to QTc prolongation.

2.2. Mechanism and Atrial-Specific Ion Channels

Non-selective agents have the associated risks of off-target toxicities in addition to potential cardiovascular side effects. Anti-arrhythmic agents which inhibit ion channels expressed in both the ventricle and atrium such as I_{Na}, I_{Kr}, and I_{to} can effect ventricular conduction and/or repolarization leading to QRS widening, PR increase and/or QT prolongation. As illustrated with most of the current standard of care agents, drugs that effect electrical activity in the ventricle are pro-arrhythmic and cause the life-threatening arrhythmia Torsades de pointes.[15] In the last decade, significant efforts have focused on selective inhibition of ion channels that are functionally expressed in the atrium and not the ventricle, since these mechanisms would be expected to have an improved safety profile. The repolarization voltage gated I_{Kur} channel is encoded by the human gene $K_V1.5$ and the depolarization ligand gated I_{KAch} channel is encoded by the human gene $K_V3.1$ and $K_V3.4$.[16,17] Both I_{Kur} and I_{KAch} are atrial-specific ion channels. Inhibition of I_{Kur} or I_{KAch} leads to an increase in atrial effective refractory period (AERP) in preclinical models and should result in maintenance of normal sinus rhythm and a reduced AF burden without being pro-arrhythmic in the ventricle. The clinical and preclinical updates summarized below will focus on I_{Kur} and I_{KAch} as the principal biological targets and efforts to identify candidates with the appropriate selectivity, efficacy, and safety to advance to clinical proof-of-concept.

3. CLINICAL UPDATES

3.1. Non-selective Agents

The most recently approved anti-arrhythmic agent for AF is vernakalant (Brinavess™, Kynapid™, **1**; Fig. 8.1), which was approved for acute cardioversion of AF in the EU in 2010 as an intravenous agent.[18] The US

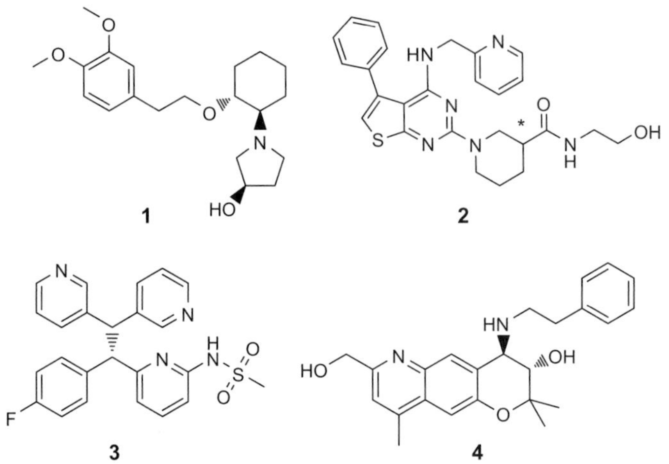

Figure 8.1 Recent clinical compounds.

development of the intravenous form was suspended in 2012, as was the oral form, although several commercial agreements were announced in 2013 with multiple partners.[19]

3.2. Atrial-Selective Agents

3.2.1 I_{Kur} Inhibitors

Efforts are on going to obtain clinical proof-of-concept that inhibition of I_{Kur} would have anti-arrhythmic efficacy in humans, however, to date this has not yet been achieved. The Phase I safety and exposure data was recently disclosed for the selective I_{Kur} inhibitor XEN-D0101.[20] In normal healthy volunteers monitored by continuous electrocardiography (ECG) and external Holter monitor, (dosed up to 300 mg, plasma concentration 3000 ng/mL total drug level) XEN-D0101 did not increase QT interval, induce AF or ventricular arrhythmia. Based on a recent patent disclosure, the structure of XEN-D0101 is likely the *R* or *S* antipode of **2**.[21] The *in vitro* potencies for the *R* and *S* antipodes of **2**, Fig. 8.1, are 5 and 27 nM respectively, with the lead compound showing selectivity for I_{Kur} over *h*ERG (275-fold) and an IC_{50} of 11 nM for inhibition of I_{Kur} current in dissociated human atrial myocytes *ex vivo*. Phase I data for a second-generation compound, XEN-D0103 (structure not disclosed) was subsequently detailed and Phase II studies in paroxysmal AF are planned.[22] XEN-D0103 was well tolerated in the single and multiple ascending dose studies and ECG monitoring demonstrated no effect on QTc prolongation.[23] Another I_{Kur} agent in the early development phase is BMS-919373 (structure not disclosed), which has entered Phase I single ascending and multiple ascending dose studies.[24]

The first reported clinical data for the selective agent MK-0448 (**3**; Fig. 8.1) in normal healthy volunteers with implanted quadripolar catheters did not demonstrate an increase in AERP when dosed *i.v.* despite robust effects in preclinical vagal nerve stimulated canine models.[25] It is not clear whether the degree of cardiac remodeling which occurs in the patients with AF is relevant to the potential efficacy of I_{Kur} inhibitors as measured by changes in AERP. In addition, the efficacy of I_{Kur} inhibitors to terminate AF, or the potential benefit in terms of maintenance of normal sinus rhythm or reduction in AF burden, remains to be determined as compounds advance to Phase II studies.

3.2.2 I_{KAch} Inhibitors

The potent and selective I_{KAch} inhibitor NTC-801(**4**; Fig. 8.1) demonstrated significant increases in AERP and AF termination efficacy in the preclinical vagal nerve stimulated canine model in addition to reduced AF inducibility in the dual paced canine model.[26] NTC-801 advanced to a Phase II study in the EU in 2011 to assess safety, tolerability, and the effects on AF burden in patients with paroxysmal AF and permanent pacemaker.[27] The I_{KAch} inhibitor AZD-2927 also advanced to Phase II clinical trials in 2011 as an *i.v.* administration to assess atrial and ventricular refractoriness in patients undergoing an invasive electrophysiological procedure.[28] There is currently no additional preclinical or clinical data published for AZD-2927. It has also been disclosed that Phase I studies for the selective agent XEN-R0702 (structure not disclosed) are expected to initiate in the near future.[29]

4. PRECLINCAL ADVANCES

4.1. I_{Kur} Inhibitors

4.1.1 Thienopyrimidines and Thienopyrazoles

Multiple closely related series have been disclosed as I_{Kur} inhibitors over the last decade and a comprehensive 2009 review included examples and series summaries to that date.[30] Subsequent to this review, a lead thienopyrimidine (**2**; Fig. 8.1) and a series of C4 analogs in the thienopyrimidine series (**5**; Fig. 8.2), in addition to a closely related thienopyrazole series (**6**; Fig. 8.2) were disclosed.[21,31,32] Examples **5** and **6** are representative of these series, with both possessing $K_V1.5$ $IC_{50} < 500$ nM.

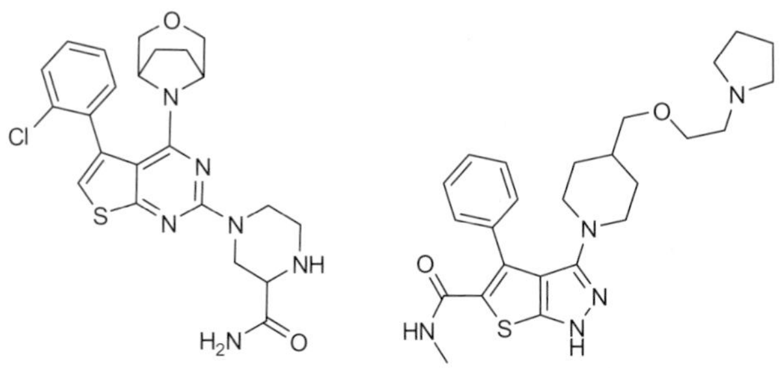

Figure 8.2 Examples of thienopyrimidine and thienoindazoles.

4.1.2 Imidazolidinones

The design and synthesis of imidazolidinone-based I_{Kur} blockers is an active area of disclosure with multiple series published in 2009.[33,34] Compound **7**, Fig. 8.3, and closely related compound **8**, Fig. 8.3, had the highest percent inhibition of $K_v1.5$ at 93% and 97%, respectively, at 1 μM; *in vivo* data was not included. KVI-020 (**9**; Fig. 8.3) was disclosed as the clinical lead for this series.[35] KVI-020 has a $K_V1.5$ IC_{50} of 480 nM with excellent selectivity versus *h*ERG, Na1.5, $Ca_V1.3$, $K_V1.1$, and $K_V4.3$. In the beagle AERP model, KVI-020 demonstrated a dose-dependant increase in AERP of 12% ($C_{total\ plasma}$ 4.6 μM) and 17% ($C_{total\ plasma}$ 23 μM) without effects on ventricular effective refractory period (VERP). Additional data in the canine AF model showed decreased success rate for AF induction, decreased AF average duration and a 52% reduction in total AF burden as measured by total time in AF at a dose of 3 mg/kg.

4.1.3 Indazole and Pyrrolopyrimidines

Multiple patent applications covering closely related indazole series published from 2010 to 2013. Extensive potency optimization and SAR for achieving selectivity over *h*ERG has been disclosed including the example compounds, **10**, Fig. 8.4, and **11**, Fig. 8.4, with efficacy in the canine AERP model of 22% increase in right AERPs for compound **10** and 20% increase for compound **11** when dosed *i.v.* at 1 mg/kg.[36,37] The recent form patent application in this series suggests that compound **12**, Fig. 8.4, is potentially the clinical lead in this series.[38] Additional lead compounds in the pyrrolopyrimidine series have been published including compound **13**, Fig. 8.4, which demonstrated 20% AERP increase in the rabbit pharmacodynamic model ($C_{total\ plasma}$ 0.7 μM).[39] The C6 triazole compound **14**, Fig. 8.4, demonstrated 25% decrease in VERP in the rat pharmacodynamic

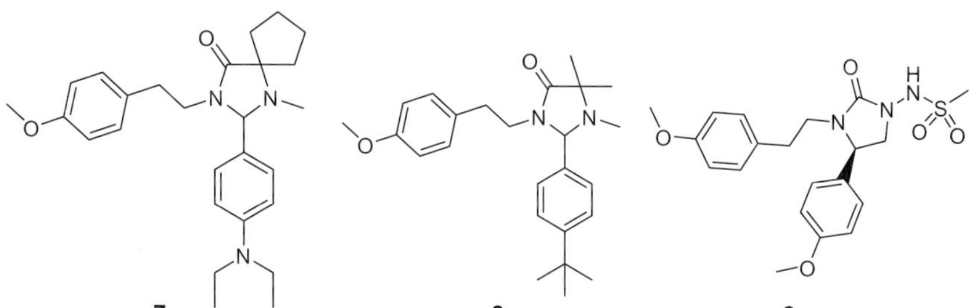

Figure 8.3 Examples of imidazolones.

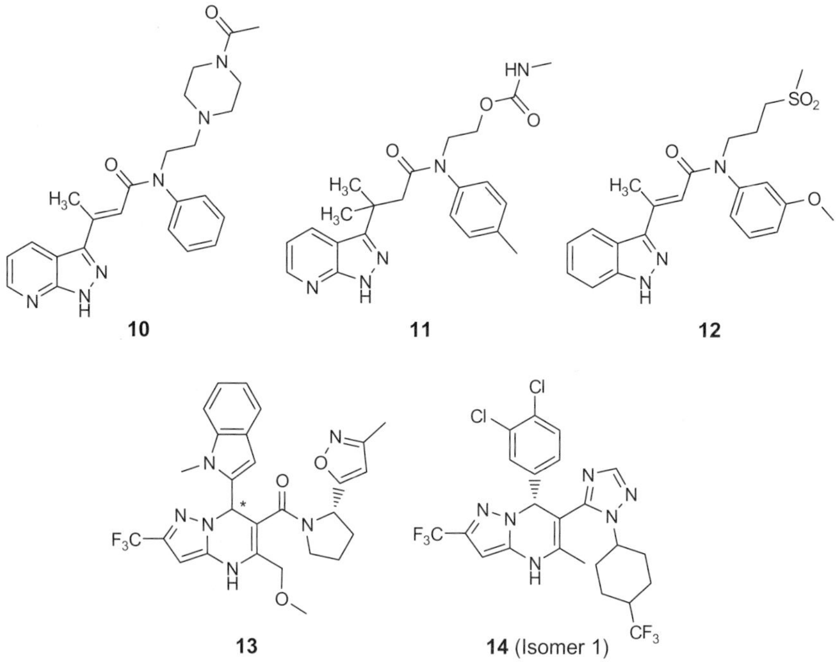

Figure 8.4 Examples of indazole and pyrrolopyrimidines.

model ($C_{\text{total plasma}}$ 1.6 μM) concomitant with a 25% decrease in blood pressure (BP), a 6% increase in HR, and 8% decrease in QT and was not advanced further.[40]

4.1.4 Phenylsulfonamides

The most widely disclosed series from 2009 to 2013 has been the phenylsulfonamides. Example compounds **15**, **16**, **17**, **18**, and **19** (Fig. 8.5)

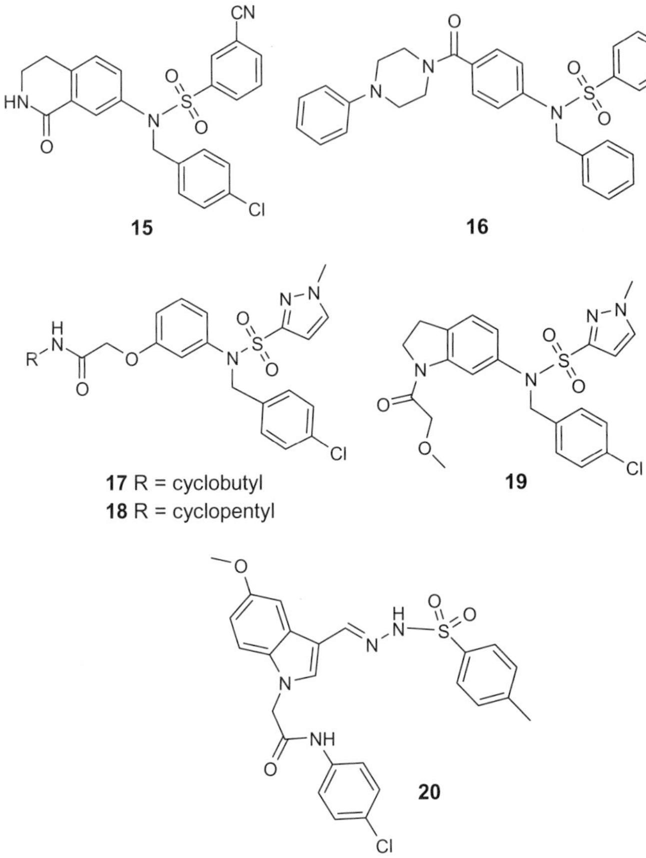

Figure 8.5 Examples of phenylsulfonamides.

are closely related analogs with potent *in vitro* $K_v 1.5$ inhibition from 95% to 100% at 1 μM.[41–45] Examples in this series were equipotent *in vitro* for $K_V1.3$, but no additional selectivity or *in vivo* data was included. A recent phenylsulfonamide literature disclosure covers the optimization of a lead compound RH01617 by pharmacophore mapping to identify compound **20**, Fig. 8.5, with $K_V 1.5$ IC_{50} potency of 0.54 μM.[46] Limited SAR was disclosed on improving selectivity versus *h*ERG. Isolated rat heart data was included demonstrating robust increase on AERP and no significant effect on VERP. Since $K_v 1.5$ is functionally expressed in both the atrium and ventricle in rats, an increase in VERP would be expected in this model despite selectivity versus *h*ERG.

4.1.5 Phenylcyclohexanes and *gem*-Dimethyl Isoindolinone

There have been multiple recent presentations and literature disclosures covering the phenylcyclohexanes which were first disclosed in 2002 as $K_v1.3$ inhibitors.[47] Recently, the *in vitro* profile and efficacy for **21**, Fig. 8.6, was presented.[48] Compound **21** showed 38% increase in VERP in the rat PD model at 3 mg/kg *i.v.* ($C_{max\ total\ plasma}$ 8.3 μM) without significant effects on BP or HR. As the series evolved to increase solubility, the anisole amide was replaced by heterocycles and the *gem*-dimethyl analog, **22**, Fig. 8.6, was disclosed, including AERP data in a rabbit PD model. Following administration of **22**, AERP increased 12% at 1 mg/kg *i.v.* ($C_{max\ total\ plasma}$ 1.2 μM) without significant effects on BP, HR, or VERP, however, at 3 mg/kg ($C_{max\ total\ plasma}$ 24 μM and C_{brain} 3.4 μM), seizures were observed.[49] The further optimized analog **23**, Fig. 8.6, has recently been disclosed to possess a robust 22% AERP increase in the rabbit model at 3 mg/kg *i.v.* ($C_{max\ total\ plasma}$ 2.4 μM) without significant effects on BP, HR, or VERP.[50]

4.1.6 Benzodiazepines

There were two related patent application disclosures in a novel benzodiazepine series in 2011–2012. The first patent application covered over 1000 examples and included *in vitro* $K_V1.5$ data for 30 examples, the most potent being **24**, Fig. 8.7, with a $K_V1.5$ IC_{50} 0.12 μM.[51] The second application covered compounds with $K_V1.5$ and $K_V3.1/K_V3.4$ *in vitro* data for 30 examples. The most potent compound disclosed was **25**, Fig. 8.7, which was equipotent for $K_V1.5$ and $K_V3.1/K_V3.4$ with IC_{50} values of 0.15 μM.[52] *In vivo* data was not included for examples in the series, however, compounds that inhibit both $K_V1.5$ and $K_V3.1/K_V3.4$ could offer a differentiated profile.

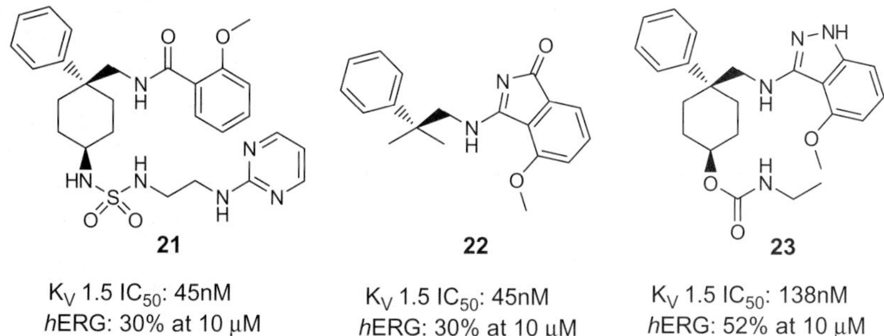

Figure 8.6 Examples of phenylcyclohexanes and gem-dimethyl isoindolinone.

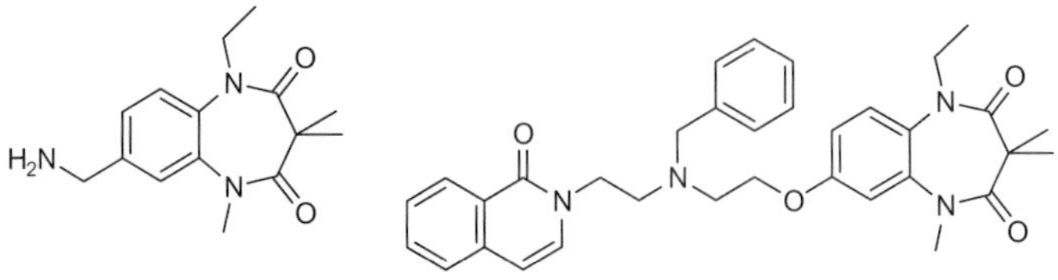

Figure 8.7 Examples of benzodiazepines.

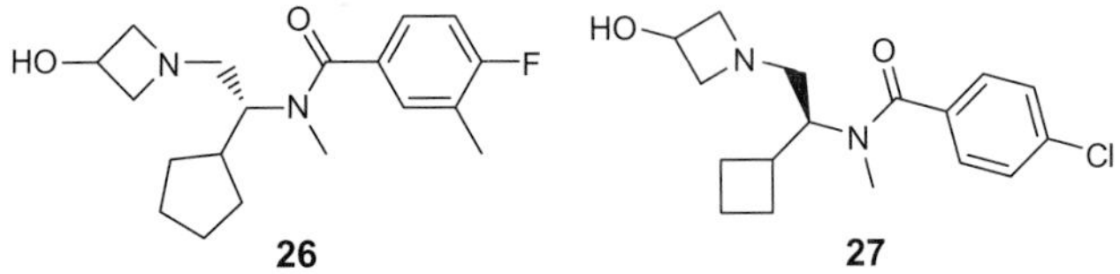

Figure 8.8 Benzamides.

4.2. I_{KAch} Inhibitors

4.2.1 Benzamides

Recently, a patent application published disclosing a novel benzamide series as I_{KAch} inhibitors. Compound **26**, Fig. 8.8, is the most potent example from an extensive series with I_{KAch} IC_{50} of 34 nM.[53] Closely related analog, **27**, Fig. 8.8, was also exemplified with I_{KAch} potency of IC_{50} of 72 nM and with efficacy in the rabbit model to affect an increase in right AERP of 20 ms at $C_{plasma\ free}$ 0.2 μM.

5. CONCLUSION

Despite recent significant advances in the series of novel anticoagulants with an approved indication for AF, there continues to be an unmet need for rhythm control agents with an improved safety and efficacy profile. Targeting selective inhibition of atrial-specific ion channel targets such as I_{Kur} and I_{KAch} continues to be an active area of research as indicated by the multiple disclosures covering diverse chemical series in both the clinical and preclinical arena. It is anticipated that multiple companies will advance compounds into proof-of-concept studies to determine whether these atrial-selective mechanisms have the potential to be useful therapies and generate clinical proof-of-concept for this approach. If successful, these atrial-selective agents offer a potentially safer alternative to the currently used non-selective agents and an opportunity to combine them with novel anticoagulant therapies to treat AF patients more safely and effectively.

REFERENCES

1. Chugh, S. S.; Havmoeller, R.; Narayanan, K.; Singh, D.; Rienstra, M.; Benjamin, E. J.; Gillum, R. F.; Kim, Y.; McAnulty, J. H.; Zheng, Z.; Forouzanfar, M. H.; Naghavi, M.; Mensah, G. A.; Ezzati, M.; Murray, C. J. L. *Circulation* **2013**, *129*, 837.
2. Zimetbaum, P. *Circulation* **2012**, *125*, 381.
3. Transparency Market Research. *Atrial Fibrillation Market (Pharmacological and Non-Pharmacological Cardioversion, Radiofrequency, Cryo, Laser and Microwave Catheter Surgery, Anti-Coagulant and Anti-Arrhythmic Drugs) Global Industry Share, Growth, Trends and Forecast 2013-2019.* www.transparencymarketresearch.com/atrial-fibrillation-market.html.
4. Camm, A. J.; Kirchhof, P.; Lip, G. Y. H.; Schotten, U.; Savelieva, I.; Ernst, S.; Van Gelder, I. C.; Al-Attar, N.; Hindricks, G.; Prendergast, B.; Heidbuchel, H.; Alfieri, O.; Angelini, A.; Atar, D.; Colonna, P.; De Caterina, R.; De Sutter, J.; Goette, A.; Gorenek, B.; Heldal, M.; Hohloser, S. H.; Kolh, P.; Le Heuzey, J.; Ponikowski, P.; Rutten, F. H. *Eur. Heart J.* **2010**, *31*, 2369.
5. Bassand, J. *Europace* **2012**, *14*, 312.
6. Hohnloser, S. H.; Crijns, H. J. G. M.; van Eickels, M.; Gaudin, C.; Page, R. L.; Torp-Pedersen, C.; Connolly, S. J. *N. Engl. J. Med.* **2009**, *360*, 668.
7. Worldwide Sales 2012 ($1.5 billion) and Q3 YTD 2013. *Cegedim GERS data for France and IMS MIDAS data for rest of the world.*
8. Sun, W.; Sarma, J. S.; Singh, B. N. *Circulation* **2009**, *100*(22), 2276.
9. U.S. FDA. *FDA drug safety communication: Severe liver injury associated with the use of dronedarone (marketed as Multaq)*; Safety Announcement; 14 Jan 2011.
10. Waldo, A.; Camm, A.; deRuyter, H.; Friedman, P.; MacNeil, D.; Pauls, J.; Pitt, B.; Pratt, C.; Schwartz, P.; Veltri, E. *Lancet* **1996**, *348*(9019), 7.
11. Burashnikov, A.; Belardinelli, L.; Antzelevitch, C. *J. Pharmacol. Exp. Ther.* **2012**, *340*(1), 161.
12. Echt, D.; Liebson, P.; Mitchell, L.; Peters, R.; Obias-Manno, D.; Barker, A.; Arensberg, D.; Baker, A.; Friedman, L.; Greene, H. *N. Engl. J. Med.* **1991**, *324*(12), 781.
13. Banchs, J. E.; Wolbrette, D. L.; Samii, S. M.; Penny-Peterson, E. D.; Patel, P. P.; Young, S. K.; Gonzalez, M. D.; Naccarelli, G. V. *J. Interv. Card. Electrophysiol.* **2008**, *23*(2), 111.
14. Naccarelli, G. V.; Wolbrette, D. L.; Samii, S.; Banchs, J. E.; Penny-Peterson, E.; Stevenson, R.; Gonzalez, M. D. *Drugs Today* **2008**, *44*(5), 325.
15. Elming, H.; Brendorp, B.; Pehrson, S.; Pedersen, O. D.; Kober, L.; Torp-Petersen, C. *Expert Opin. Drug Saf.* **2004**, *3*(6), 559.
16. Tamargo, J.; Caballero, R.; Gomez, R.; Delpon, E. *Expert Opin. Investig. Drugs* **2009**, *18*(4), 399.
17. Ehrlich, J. R.; Biliczki, P.; Hohnloser, S. H.; Nattel, S. *J. Am. Coll. Cardiol.* **2008**, *51*(8), 787.
18. Buccelletti, F.; Iacomini, P.; Botta, G.; Marsiliani, D.; Carroccia, A.; Silveri, N. G.; Franceschi, F. *J. Clin. Pharmacol.* **2012**, *52*(12), 1872.
19. Cardiome Press Release. http://www.cardiome.com; 19 Feb 2014.
20. Ford, J.; Milne, J.; Wettwer, E.; Christ, T.; Rogers, M.; Sutton, K.; Madge, D.; Virag, L.; Jost, N.; Horvath, Z.; Matschke, K.; Varro, A.; Ravens, U. *J. Cardiovasc. Pharmacol.* **2013**, *61*(5), 408.
21. John, D. E.; Ford, J. Patent Application WO 2012/131379, 2012.
22. Xention press release. http://www.xention.com; 10 Feb 2013.
23. Milne, J. T.; Madge, D. J.; Ford, J. W. *Drug Discov. Today* **2012**, *17*, 654.
24. World Health Organization. *International Clinical Trials Registry Platform*; Trial Reference ACTRN12611000373976.

25. Pavri, B. B.; Greenberg, H. E.; Kraft, W. K.; Lazarus, N.; Lynch, J. J.; Salata, J. J.; Bilodeau, M. T.; Regan, C. P.; Stump, G.; Fan, L.; Mehta, A.; Wagner, J. A.; Gutstein, D. E.; Bloomfield, D. *Circ. Arrhythm. Electrophysiol.* **2012**, *5*, 1193.
26. Machida, T.; Hashimoto, N.; Kuwahara, I.; Ogino, Y.; Matsuura, J.; Yamamoto, W.; Itano, Y.; Zamma, A.; Matsumoto, R.; Kamon, J.; Kobayashi, T.; Ishiwata, N.; Yamashita, T.; Ogura, T.; Nakaya, H. *Circ. Arrhythm. Electophysiol.* **2011**, *4*, 94.
27. www.Clinical-Trials.gov; Identifier NCT01356914.
28. www.Clinical-Trials.gov; Identifier NCT01396226.
29. Madge, D. In: *22nd International Symposium on Medicinal Chemistry, Berlin, Germany*, 2012.
30. Bilodeau, M.; Trotter, W. B. *Curr. Top. Med. Chem.* **2009**, *9*(5), 436.
31. Madge, D.; Chan, F.; John, D. E.; Edwards, S. D.; Blunt, R.; Hartzoulakis, B.; Brown, L. Patent Application WO 13072694, 2013.
32. John, D. E.; Edwards, S. D.; Hartzoulakis, B. Patent Application WO 13072693, 2013.
33. Blass, B. E.; Janusz, J. M.; Ridgeway, J. M.; Wu, S. Patent Application WO 09079630, 2009.
34. Blass, B. E.; Janusz, J. M.; Wu, S.; Ridgeway, J. II; Coburn, K.; Lee, W.; Fluxe, A. J.; White, R. E.; Jackson, C. M.; Fairweather, N. Patent Application WO 09079624, 2009.
35. Blass, B.; Fensome, A.; Trybulski, E.; Magolda, R.; Gardell, S. J.; Liu, K.; Samuel, M.; Feingold, I.; Huselton, C.; Jackson, C. M.; Djandjighian, L.; Ho, D.; Hennan, J.; Janusz, J. M. *J. Med. Chem.* **2009**, *52*, 6531.
36. Yamaguchi, T.; Kawanishi, H.; Ushirogochi, H.; Takahashi, T.; Takebe, T. Patent Application WO 09041559, 2009.
37. Yamaguchi, T.; Ushirogochi, H.; Hirai, M.; Imanishi, Y. Patent Application WO 10110414, 2010.
38. Yamaguchi, T.; Ishige, T. Patent Application WO 13103004, 2013.
39. Finlay, H. J.; Lloyd, J.; Vaccaro, W.; Kover, A.; Yan, L.; Bhave, G.; Prol, J.; Huynh, T.; Bhandaru, R.; Caringal, Y.; DiMarco, J.; Gan, J.; Harper, T.; Huang, C.; Conder, M.; Sun, H.; Levesque, P.; Blanar, M.; Atwal, K.; Wexler, R. *J. Med. Chem.* **2012**, *55*(7), 3036.
40. Finlay, H. J.; Jiang, J.; Caringal, Y.; Kover, A.; Conder, M.; Xing, D.; Levesque, P.; Harper, T.; Hsueh, M.; Atwal, K.; Blanar, M.; Wexler, R.; Lloyd, J. *Bioorg. Med. Chem. Lett.* **2013**, *23*(6), 1743.
41. Hamlyn, R. J.; Madge, D.; Mulla, M. Patent Application WO 10139953, 2010.
42. John, D. E.; Mushtaq, M.; Hamlyn, R. J.; Jones, S. M.; Pollard, D. L.; Hartzoulakis, B.; Payne, H.; Madge, D.; Ford, J. Patent Application WO 2010023445, 2010.
43. John, D. E.; Mushtaq, M.; Hamlyn, R. J.; Jones, S. M.; Hartzoulakis, B.; Madge, D.; Ford, J. Patent Application WO 2010023446, 2010; Patent Application WO 010023448, 2010.
44. John, D. E.; Mushtaq, M.; Hamlyn, R. J.; Garrett, S. L.; Hartzoulakis, B.; Madge, D.; Ford, J. Patent Application WO 2010023448, 2010.
45. Madge, D.; Mulla, M. Patent Application WO 10139967, 2010.
46. Guo, X.; Yang, Q.; Xu, J.; Chu, H.; Yu, P.; Zhu, J.; Wei, J.; Chen, W.; Zhang, Y.; Zhang, X.; Sun, H.; Tang, Y.; You, Q. *Bioorg. Med. Chem. Lett.* **2013**, *21*, 6466.
47. Miao, S.; Bo, J.; Garcia, M. L.; Goulet, J. L.; Hong, X. J.; Kaczorowski, G. J.; Kayser, F.; Koo, G. C.; Kotliar, A.; Schmalhofer, W. A.; Shah, K.; Sinclair, P. J.; Slaughter, R. S.; Springer, M. S.; Staruch, M. J.; Tsou, N. N.; Wong, F.; Parsons, W. H.; Rupprecht, K. M. *Bioorg. Med. Chem. Lett.* **2003**, *13*, 1161.
48. Li, L.; Jeon, Y. T.; Johnson, J. A.; Conder, M.; Xing, D.; Sun, H.; Li, D.; Levesque, P.; Harper, T. W.; Lloyd, J. In: *244th ACS National Meeting & Exposition, Philadelphia, PA, USA*, 2011, Abstract MEDI-384.

49. Johnson, J. A.; Xu, N.; Jeon, Y.; Finlay, H. J.; Kover, A.; Li, L.; Sun, H.; Li, D.; Levesque, P.; Conder, M.; Hsueh, M.; Harper, T. W.; Wexler, R.; Lloyd, J. In: *242nd ACS National Meeting & Exposition, Denver, CO, USA*, 2011, Abstract MEDI-148.
50. Johnson, J. A.; Xu, N.; Jeon, Y.; Finlay, H. J.; Kover, A.; Conder, M.; Sun, H.; Li, D.; Levesque, P.; Hsueh, M.; Harper, T.; Wexler, R. R.; Lloyd, J. *Bioorg. Med. Chem. Lett.* **2014**, *24*(14), 3108.
51. Oshima, K.; Matsumura, S.; Yamabe, H.; Isono, N.; Takemura, N.; Taira, S.; Oshiyama, T.; Menjo, Y.; Nagase, T.; Ueda, M. Patent Application JP 2012184225, 2012.
52. Oshima, K.; Oshiyama, T.; Taira, S.; Menjo, Y.; Yamabe, H.; Matsumura, S.; Ueda, M.; Koga, Y.; Tai, K.; Nakayama, T. Patent Application JP 2011063589, 2011.
53. Bostroem, J.; Granberg, K.; Emtenaes, H.; Mogemark, M.; Llinas, A. Patent Application WO2012074469, 2012.

SECTION 3

Inflammation/ Pulmonary/ GI Diseases

Section Editor: David S. Weinstein
Bristol-Myers Squibb R&D, Princeton, New Jersey

CHAPTER NINE

Advances in the Discovery of Small-Molecule IRAK4 Inhibitors

John Hynes Jr.*, Satheesh K. Nair†
*Bristol-Myers Squibb, Princeton, New Jersey, USA
†BMS Biocon Research Center, Bangalore, India

Contents

1. INTRODUCTION

Interleukin receptor-associated kinase 4 (IRAK4) is a member of the IRAK family of intracellular serine–threonine kinases consisting of IRAK1, IRAK2, IRAKM or 3, and IRAK4.[1] IRAK4 signals downstream of the pathogen sensing toll-like receptors (TLRs) and innate/adaptive immune signaling IL-1Rs. The proximal location of IRAK4 to these key immune signaling receptors has generated significant interest in therapeutic targeting of IRAK4 for autoimmune and inflammatory diseases.

Annual Reports in Medicinal Chemistry, Volume 49
ISSN 0065-7743
http://dx.doi.org/10.1016/B978-0-12-800167-7.00009-2

2. RATIONALE FOR TARGETING IRAK4 IN INFLAMMATORY DISEASES

2.1. TLR and IL-1R Signaling

A combination of innate and adaptive immune responses provides the initial and sustained defense against bacterial and viral infections. Central to the innate response are pattern recognition receptors (PRRs) which detect pathogen-associated molecular patterns (PAMPs) with foreign microbes. PRRs are important in chronic inflammatory conditions such as rheumatoid arthritis and lupus by recognizing the necrotic cell by-products generated during the course of these diseases. These damage-associated molecular pathogens (DAMPs) activate PRRs to propagate the proinflammatory condition and disease progression. The toll-like receptors (TLR1–10) are a family of PRRs that recognize a variety of bacterial and viral PAMPs and endogenous DAMPs. They signal through a complex cascade of adaptor proteins and kinases leading to NF-κB activation and cytokine expression. One of the predominant adaptor proteins required for human TLR signaling is myeloid differentiation primary-response gene 88 (MyD88). All human TLRs except for TLR3 signal through this protein (TLR4 has both MyD88-dependent and -independent signaling pathways). Following TLR activation and MyD88 recruitment to the toll-IL-1R domain, IRAK4 is recruited to the signaling complex and interacts with MyD88 via the kinase death domain (DD; Fig. 9.1).[2] IRAK4 activation leads to the recruitment and phosphorylation of IRAK1 or IRAK2 (not shown) which in turn leads to MAP kinase/IKK activation and proinflammatory cytokine production. In addition to being downstream of TLR signaling, the MyD88–IRAK4-dependent pathway is also required for signaling by the family of IL-1 receptors after their activation by proinflammatory cytokines IL-1, IL-8, and IL-33. The central role of IRAK4 in TLR/IL-1R signaling makes this an attractive target for blocking these proinflammatory-mediating receptors.

2.2. Human IRAK4-Deficient Patients

Clinically, IRAK4 deficiency is characterized by an increased susceptibility to infection (*Streptococcus pneumoniae* and *Staphylococcus aureus*) at a young age. Prophylactic treatment with antibiotics is recommended and infection rates decrease with age, resolving around adolescence.[3] Similar clinical observations were noted in MyD88-deficient patients.[4] The resolution of infection

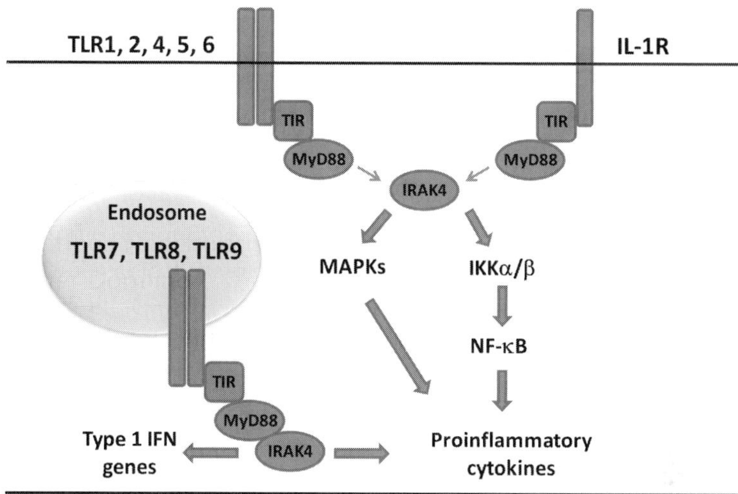

Figure 9.1 Schematic of IRAK4 signaling. (See the color plate.)

risk with age provides reason to believe that therapeutic IRAK4 inhibition will not be associated with unacceptable infection risk. Consistent with the central role, IRAK4 plays in TLR signaling, peripheral blood monocytes (PBMCs) from IRAK4-deficient patients were not responsive to TLR agonists (except TLR3). Additionally, lymphocyte subsets (B, T, and NK cells) from IRAK4-deficient patients were not responsive to TLR activation. Finally, fibroblasts from these patients did not respond to IL-1β stimulation.

2.3. Genetic Validation

IRAK4-deficient or kinase dead knockin (KDKI) mice demonstrate significantly reduced levels of cytokine production when stimulated with TLR2/4 and 9 ligands *in vivo*.[5] In KDKI mice, protection against disease progression in models of atherosclerosis,[6] multiple sclerosis,[7] and rheumatoid arthritis have been described.[8] A host of diseases—atherosclerosis, rheumatoid arthritis,[9] psoriasis,[10] systemic lupus erythematosis,[11] and inflammatory bowel disease[12]—have been linked to the MyD88-dependent TLRs. Therefore, it is reasonable to believe that IRAK4 inhibition would have a therapeutic benefit in these diseases.

2.4. TLR and IL-1-Targeted Therapies

As of this writing, no compounds directly targeting IRAK4 in human disease have been reported. However, blockade of TLR and IL-1R signaling upstream of IRAK4 has been evaluated clinically. In psoriatic patients, TLR7/9 blockade employing the oligonucleotide antagonist IMO-3100 provided a clinical benefit after 4 weeks.[13] A subsequent molecule (IMO-8400) targeting TLR7/8/9[14] has recently completed a 12-week psoriasis trial, meeting its primary objective demonstrating safety and tolerability in addition to the secondary objective of efficacy.[15] Biologics targeting IL-1 including anakinra (IL-1R antagonist), rilonacept (high-affinity fusion protein), and canakinumab (anti-IL-1β antibody) have been successfully developed for the treatment of diseases linked to inflammasome activation or autoinflammatory disorders.[16] Overall, these combined results implicate a critical role of mammalian TLR/IL-1R signaling in autoimmune diseases which would benefit from additional therapeutic options.

2.5. IRAK4 and Cancer

Activating pro-oncogenic MyD88 mutations (L265P) have been implicated in a variety of leukemias including activated B-cell-like diffuse large B-cell lymphoma (ABC-DLBCL),[17] chronic lymphocytic leukemia,[18] and Waldenström's macroglobulinemia.[19] Recently, synergistic effects of IRAK4 inhibition and B-cell signaling blockade have been shown to reduce the viability of cells derived from ABC-DLBCL patients[20] and inhibit tumor growth in murine xenograft models.[21] Melanoma cell lines also express high levels of IRAK4, and a combination of vinblastine and an IRAK4 inhibitor (dosed intratumorally) inhibited tumor growth relative to vinblastine or IRAK4 intervention alone in mice. Median survival was improved (38 days) over vinblastine alone (22 days).[22] These reports suggest an emerging role of aberrant immune system signaling in tumor progression and the potential for IRAK4 inhibition in cancer.

3. IRAK4 STRUCTURE

The IRAKs have a kinase domain (KD), a proline–serine–threonine-rich domain (ProST), and an N-terminal DD, the latter of which is critical for cytosolic adaptor protein interactions. In addition, IRAK1, 2, and M have a C-terminal domain which facilitates downstream interactions with TNF receptor-associated factor 6.[23] Structurally, the IRAK family is unique due to a tyrosine gatekeeper residue (Tyr262 for IRAK4) in the inhibitor-binding pocket.[24] Within the KD, IRAK4 contains an N-terminal extension (AAs 184–189, PISVGG) that forms a solvent exposed loop in close proximity to the αDE loop.[25] This extension, commonly referred to as the Schellman loop, provides a narrow cleft for inhibitor binding which is not present in other kinases.[26] These two unique features within the ATP-binding region of the kinase may provide engaging prospects for selective inhibitor design.

4. RECENT MEDICINAL CHEMISTRY EFFORTS

4.1. Benzimidazoles

The first reports of small-molecule IRAK4 inhibitors focused on medicinal chemistry efforts optimizing early screen leads derived from substituted benzimidazoles such as **1** (IRAK4 IC_{50} 4 μM).[27] Efforts to identify key IRAK4 interacting pharmacophores in this series revealed the C2 *N*-aroyl linker, preferably 3′-nitrobenzoyl, as optimal. Sulfonamide, urea, and methylene replacements for the amide lead were not potent implicating the amide carbonyl as the hinge binder. Hydroxypropyl (**2**, IC_{50} 0.15 μM) or morpholinoethyl (**3**, IC_{50} 0.2 μM) substitution at N1 provided key enzyme activity advances. Substitution of the nitrobenzamide did not improve IRAK4 inhibition. Selectivity over IRAK1 was not observed in this series (e.g., **3**, IRAK1 IC_{50} 0.3 μM).

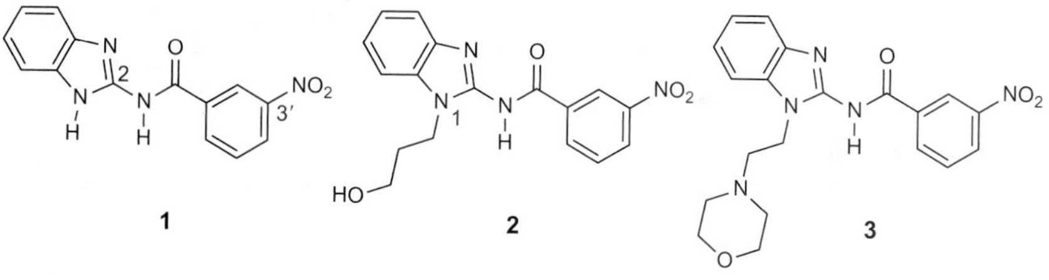

Subsequent efforts in this series provided the first inhibitor-IRAK4 protein crystal structure.[24] Inhibitor **4**[28] (Fig. 9.2) forms a hydrogen bond with the hinge backbone NH of Met265 via the benzamide carbonyl. The structure reveals a potential π-stacking interaction with the nitrophenyl group and the IRAK family-specific gatekeeper residue Tyr262. The benzamide nitro group forms a weak hydrogen bond with Lys213 in the back pocket of the kinase. The benzimidazole C5 substituent is directed through the narrow cleft provided by the Schellman loop and αDE loop to engage Arg273 via the carbonyl. Finally, the N1 hydroxypropyl engages Asp272.

Isosteric replacement of the amide with an indazole led to a series of compounds with IC_{50}s between 0.01 and 10 μM.[29] A significant subset of the compounds contain a 4-methoxypyridyl (**5**) or 2-methoxyphenyl (**6**) group at the 5′ position of the indazole. Benzimidazole N1 alkylation

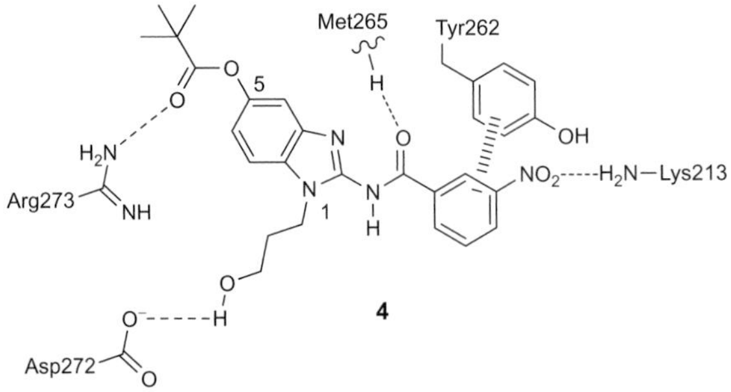

Figure 9.2 Binding interactions of IRAK4 with inhibitor 4.

was limited to cyclohexyl or 4-hydroxycyclohexyl groups. Broader substitutions are reported at the benzimidazole 5 position and polar groups such as hydroxymethyl or small alkyl amides (**7**) are tolerated. Selectivity over other kinases was not reported.

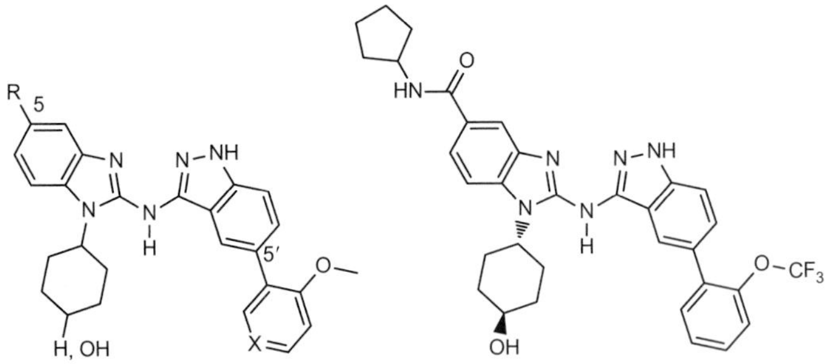

5, X = N, 9 examples
6, X = CH, 3 examples

7

Benzimidazole-substituted aminopyridines have also been disclosed as IRAK4 inhibitors.[30] Within this series, appreciable phosphoinositide-dependent kinase 1 (PDK1) and IRAK1 activity is highlighted. According to the inventors, combined PDK1 and IRAK inhibition may result in synergistic inhibition against tumors and autoimmune diseases. Halogen and alkoxy group substitution on the benzimidazole ring appear to be preferred in the presence of the pyrazolo piperidine (**8–12**, IRAK4 IC_{50} 0.001–0.1 μM). IRAK4 cell activity (assay conditions not reported) for three compounds indicate IC_{50}s between 0.1 and 10 μM. Thiophene amides in place of the C5 pyrazole led to pan PDK1/IRAK1/IRAK4 inhibitors such as **13** ($IC_{50} \leq 0.1$ μM against all three kinases).

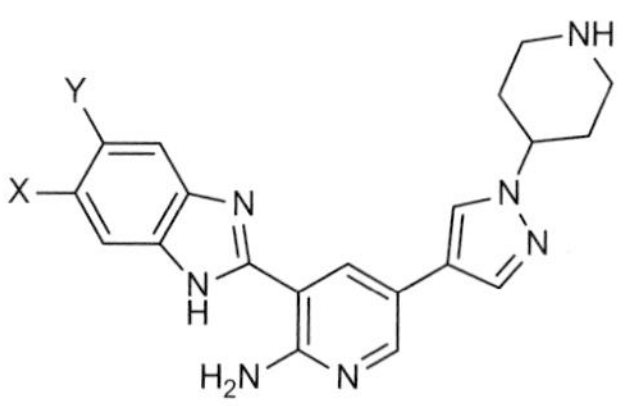

8, X = H, Y = H
9, X = H, Y = Cl
10, X = H, Y = F
11, X = H, Y = OMe
12, X = F, Y = OMe

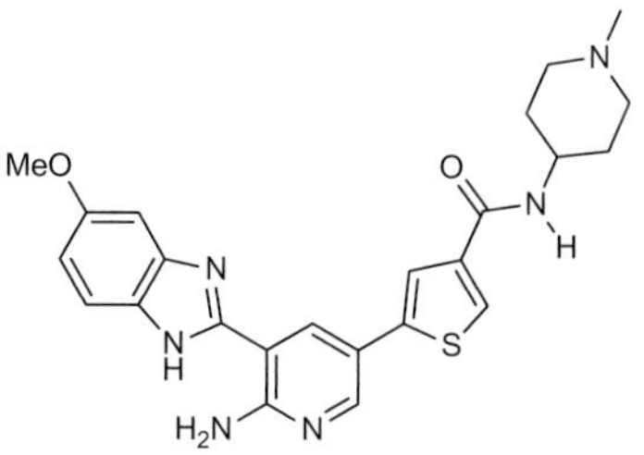

13

4.2. Thiazole, Pyridyl, and Oxazole Amides

Thiazole amide **14** (IC_{50} 2.8 μM) was identified during a screen for small-molecule inhibitors of IRAK4.[31] Modification of the aniline portion revealed 2-methoxy-4-morpholino aniline as a suitable replacement (**15**, IC_{50} 0.16 μM). Subsequently, the thiazole core was replaced with a pyridine that led to the identification of the pyrazole-substituted pyridine **16** (IC_{50} 0.02 μM).

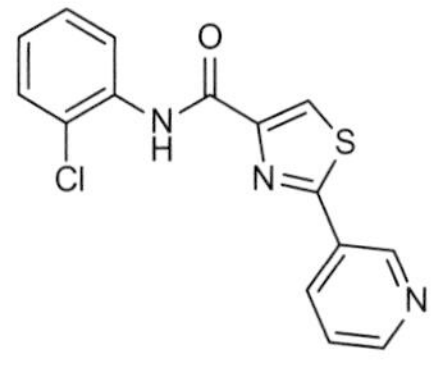

14 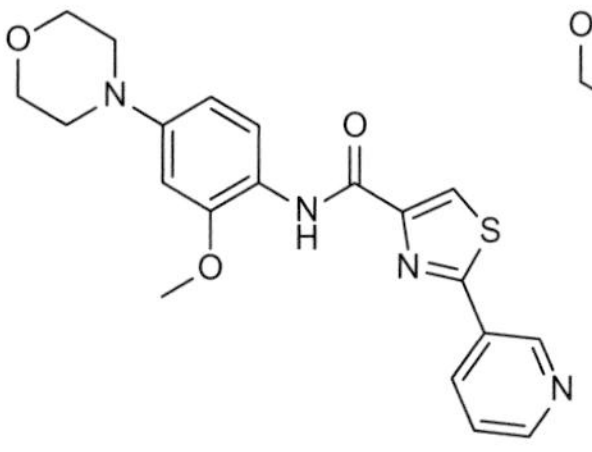 **15** 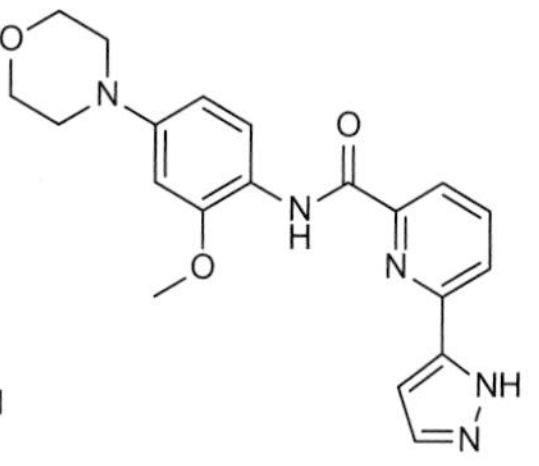 **16**

An additional thiazole amide **17** (IRAK4 43% inhibion at 1 μM) was modified by replacement of the aniline amide with a substituted pyrazole.[32] Pyrazole N-alkylation and conversion of the thiazole C2 phenyl to a 4-pyridyl afforded the potent inhibitor **18** (IC_{50} 11 nM). Time-dependent CYP3A4 inhibition was prevalent in the thiazole analogs. Compound **18** had appreciable CYP1A2, 2C9, and 2C19 inhibitory activity (IC_{50}s 0.4, 2.8, and 9.4 μM, respectively) which was modulated favorably at the pyridine via incorporation of a 3-methyl group (CYP1A2, 2C9, and 2C19 IC_{50}s >10 μM). Replacement of the thiazole with an oxazole eliminated CYP3A4 inhibition. Finally, replacement of the *N*-ethyl substituent on the pyrazole with 4-tetrahydropyran attenuated CYP1A induction to provide the lead compound AS2444697 **19** (IRAK4 IC_{50} 21 nM).

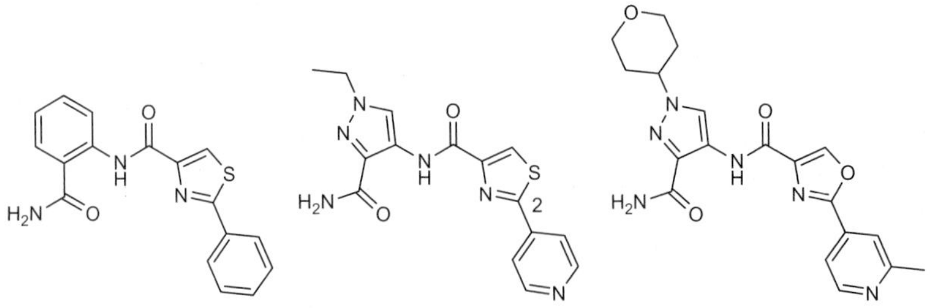

17 18 19

19 was selective against 146 kinases with exceptions being platelet-derived growth factor receptor (PDGFR)α, β, TrkA, TrkC, CLK1, ITK, and FLT3 (<10-fold selectivity against all seven). Cellular IC_{50}s measuring IRAK4 engagement downstream of the IL-1 receptor (IL-1β-stimulated IL-6) and TLR4 (lipopolysaccharide-stimulated TNFα) were 250 and 47 nM, respectively.[33] *In vivo*, AS2444697 was efficacious in the rat adjuvant-induced arthritis (ED_{50} 2.7 mg/kg, BID, PO) and the rat collagen-induced arthritis (ED_{50} 1.6 mg/kg, BID, PO) disease models. Good bioavailability was seen in rat (*F*% 50) and dog (*F*% 78) pharmacokinetic studies.

4.3. Pyrazolo and Thiophene Fused Pyrimidine Amides

Pyrazolo[1,5-*a*]pyrimidine-3-carboxamides have been disclosed as potent IRAK4 inhibitors.[34] Amides were predominately substituted 3-chlorophenyl groups (94 examples) such as **20** (IC_{50} 31 nM). Alternatively, quinoline-6-yl amides (41 examples) could be used which improved IRAK4 inhibition in some cases (**21** IC_{50} 2 nM). The pyrazolo[1,5-*a*] pyrimidine-3-carboxamide could be replaced with a thieno[3,2-*d*]pyrimidine-7-carboxamide with a moderate loss in biochemical activity at IRAK4 (**22** IC_{50} 12 nM). At least 10-fold selectivity over IRAK-1 was reported for select compounds.

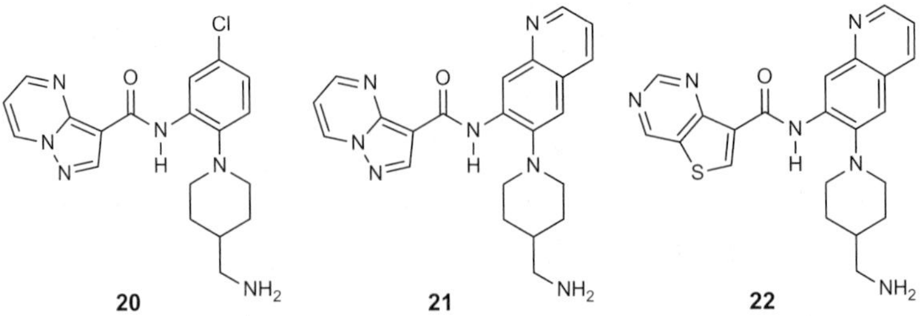

20 21 22

A related pyrazolo[1,5-*a*]pyrimidine-3-carboxamide series which replaced the aniline with a disubstituted pyrazole was disclosed.[35] A variety of pyrazole *N*-substitutions were explored with a preference for the *para*-tolyl group. Pyrazole 3-substitution varies significantly including saturated heterocycles (**23**) and phenyl (**24**) groups. Activity against IRAK4 (IC_{50} < 100 nM) is highlighted for key compounds in the application.

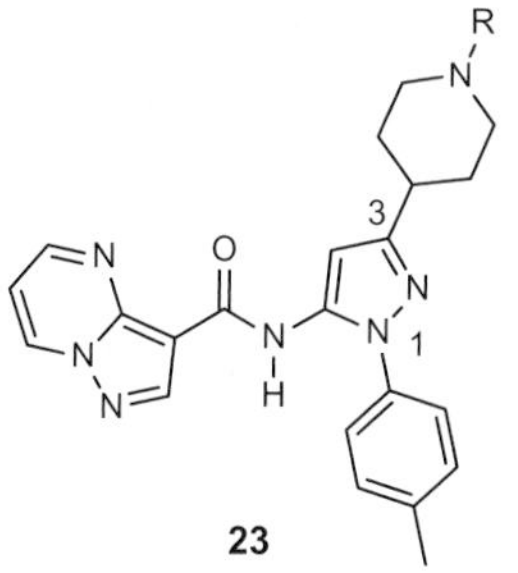

23 **24**

R = H, alkyl, cycloalkyl, CF_3CH_2, SO_2Me

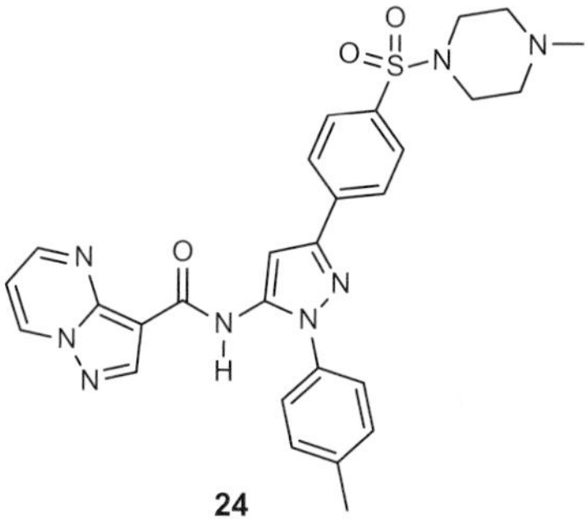

4.4. Pyridyl Amines

A counter screen of JNK1 inhibitors identified 2-aminopyrimidine-4-imidazo[1,2-*a*]pyridyl amide **25** as a moderately potent (IC_{50} 216 nM) IRAK4 inhibitor.[36] A sixfold increase in IRAK4 biochemical potency was realized upon conversion to the pyridine **26** (IC_{50} 35 nM). Benzimidazole analog **27** (IC_{50} 70 nM) was subjected to more extensive structure activity relationship (SAR) investigation around the benzimidazole ring leading to **28** (IC_{50} 1 nM).[37] Incorporating the chloro substitution into the original imidazo[1,2-*a*]pyridine series provided **29** (IRAK4 IC_{50} of 1 nM).

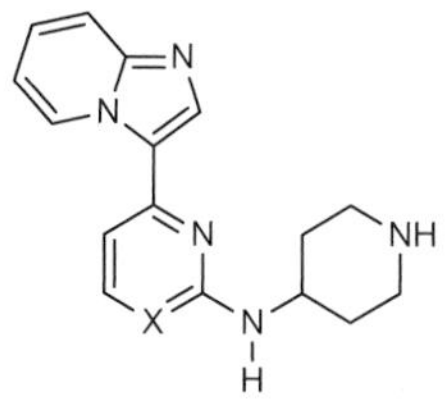

25, X = N
26, X = CH

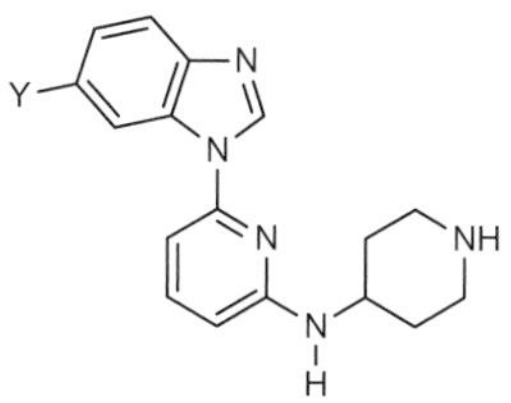

27, Y = H
28, Y = Cl

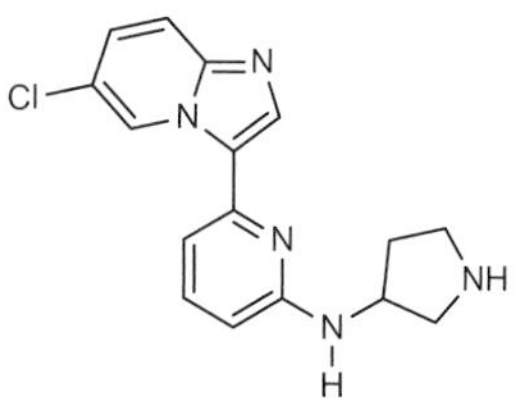

(±)-**29**

A series 2,4-diaminopyridines containing a C5 heterocycle have also been disclosed as potent IRAK4 inhibitors. 2-Aminobenzothiazole pyridine inhibitors such as triazole **30** were potent IRAK4 inhibitors.[38] Extension to thiazole[39] and oxazole[40]-based C5 heterocycles provided additional IRAK4 inhibitors.

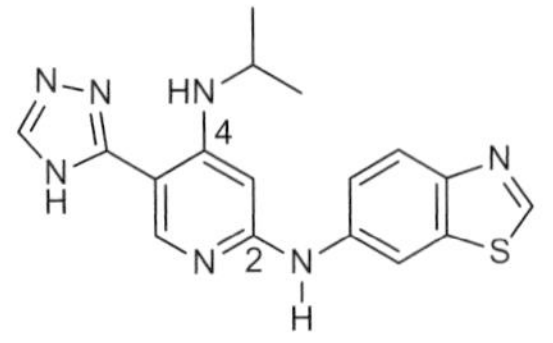

30

Additional 2-heteroaromatic amino pyridines are described in a recent patent application.[41] The predominant 2-substitution included benzothiazole (**31**), 7-aza-benzothiazole (**32**), and imidazo[1,2-*a*]pyridine (**33**). Pyridine C4 substitution and pendant C2 heterocycle substitution were both diverse. LG0224912 (IRAK4 IC_{50} 0.7 nM, structure not disclosed) may be from this chemical series.[42] *In vitro*, LG0224912 inhibits IL-1β-induced IRAK1 phosphorylation in a concentration-dependent manner. IL-1β- and LPS-induced IL-6 secretion in human PBMCs was potently inhibited with IC_{50}s of 84 and 194 nM, respectively. Translation to *in vivo* pharmacodynamic activity resulted in an oral ID_{50} of 25 mg/kg (IL-1β-induced IL-6). In the mouse collagen-induced arthritic disease (CIA) model, LG0224912 showed robust inhibition of clinical disease progression with once daily oral doses (30 mg/kg). A 10-fold lower dose was ineffective in this model.

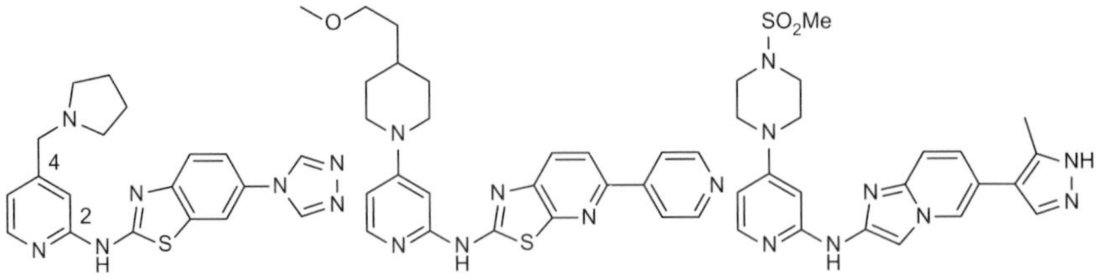

31 **32** **33**

4.5. 6,5-Fused Tricyclic Thienopyrimidines and Related Heterocycles

A recent patent application[43] and conference disclosure[44] highlighted fused thienopyrmidines ND-2110, **34** (IRAK4 K_i 7 nM) and ND-2158, **35**

(IRAK4 K_i 1 nM) as potent and efficacious IRAK4 inhibitors in animal models of inflammation. In a variety of IRAK4-dependent cellular assays, **35** inhibited TLR4-, TLR7/8-, and TLR9-mediated TNFα production with IC_{50}s between 90 and 160 nM. Cellular potency for **34** was reduced relative to **35**. In human whole blood, however, comparable potency in an R848 (TLR7/8-dependent)-induced TNFα production assay (IC_{50} 540–770 nM) was obtained for both compounds. In a rat model of acute cytokine production, **34** inhibited TLR4-dependent TNFα release with an oral ED_{50} of 7.5 mg/kg. Additionally, robust and dose-dependent efficacy in a murine model of human gout (monosodium urate-induced cellular infiltration) was observed with **34** (oral $ED_{50} \sim 30$ mg/kg). Both **34** and **35** were studied in an imiquimod-induced psoriasis model with twice daily intraperitoneal (IP) dosing. Clinical scores were comparable to the steroid control (dexamethasone) after 10 days. Finally, **34** blocked disease progression in the mouse CIA model (37% inhibition at 30 mg/kg, PO, BID). Inhibitor **35** was also evaluated in this model with IP dosing at 30 mg/kg to provide comparable results.

Inhibitor **35** has been studied against hematological tumors containing activating mutations of MyD88. Synergistic blockade *in vivo* with the marketed irreversible Bruton's tyrosine kinase inhibitor, ibrutinib, has recently been reported.[21]

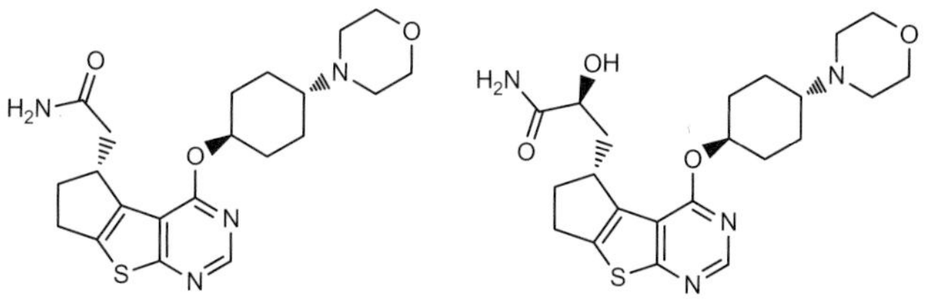

34 **35**

Three additional patent applications highlight the continued interest in this series. The fused cyclopentyl ring can be replaced with aromatic rings to provide benzo[4,5]thieno[2,3-*d*]pyrimidine analogs such as **36** ($IC_{50} \leq 0.1$ μM).[45] Second, the pyrimidine ring N3 was removed to provide 6,7-dihydro-5H-cyclopenta[4,5]thieno[2,3-*b*]pyridine-derived IRAK4 inhibitors such as **37** ($IC_{50} \leq 5$ μM).[46] Finally, the doubly annulated thiophene was replaced with a fluoropyrrole as in **38** ($IC_{50} \leq 0.1$ μM).[47] Additional compounds replacing the thiophene sulfur with an oxygen are also claimed.

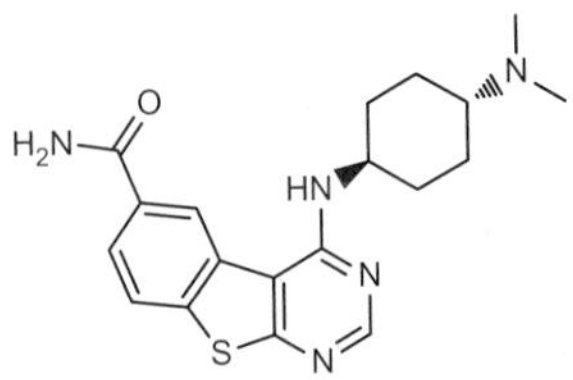

36 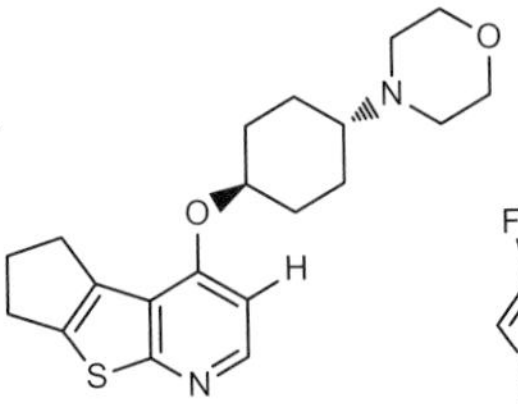 37 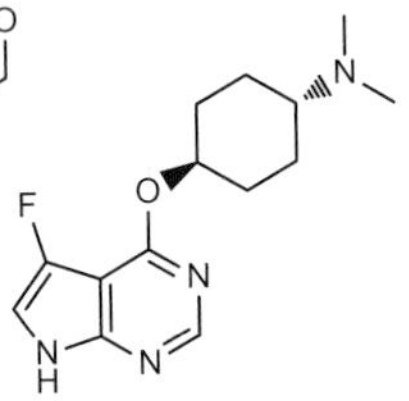 38

4.6. Other Heterocyclic Cores

Substituted aminopyrimidine-4-ones such as **39** (IRAK4 IC_{50} 11 nM) have been reported.[48] The C6 amino cyclopentane triol was extensively used in combination with the benzothiazole group at position 5 of the pyrimidinone. The triol could be replaced with *R*-3-amino piperidine (not shown) in many cases. An additional disclosure revealed the related pyrimidine series (**40** IC_{50} 0.2 nM).[49]

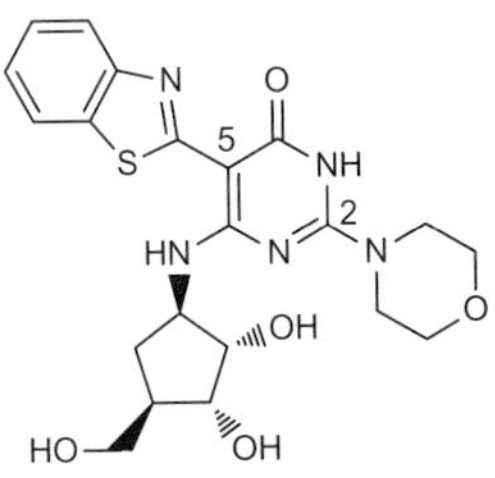

39 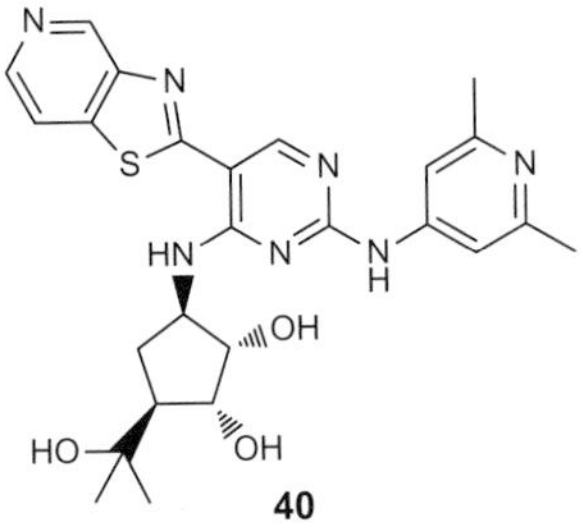 40

N-2-Morpholino benzothiazole (**41**) or benzothiazole (**42**) IRAK4 inhibitors with *in vivo* activity have been claimed.[50] Substitution at C6 is predominately a 2,6-disubstituted acyl pyridine as in **43** (IRAK4 $IC_{50} < 50$ nM) and C5 substitution varies from small alkyl amines to hydrogen, pyrrolidines, piperidines, and phenyl. *In vivo*, **43** inhibited LPS-induced serum TNFα in rats (53% inhibition at 30 mg/kg, PO). Doses of 3 and 10 mg/kg were ineffective. Recently reported X-ray cocrystal structures indicate that this series binds to the hinge region via the central amide carbonyl.[51]

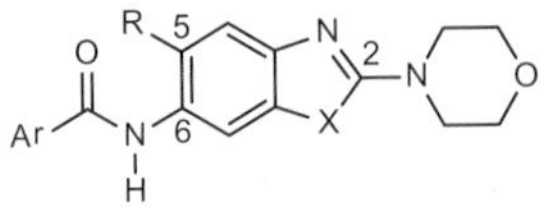

41, X = O
42, X = S

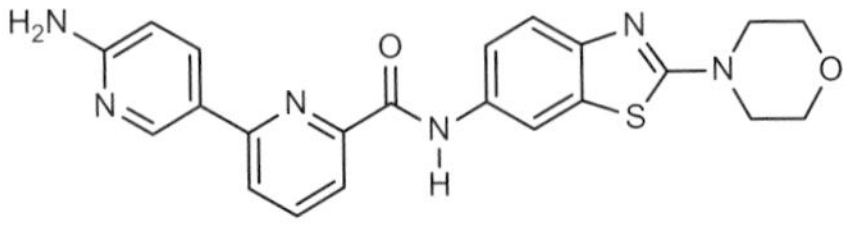

43

Indazolyl triazole **44** and related derivatives have been disclosed as IRAK4 and IRAK1 inhibitors.[52] Reported biochemical potencies for IRAK4 and IRAK1 for **44** are between 0.1 and 1 μM. Comparable potency is observed in THP-1 cells measuring IL-1β-induced IL-8 production. In human PBMCs, IL-1β-induced IL-6 production is inhibited in the IC_{50} range of 0.2–2 μM. In mice, **44** inhibited LPS-induced production of TNFα and IL-6 in a dose-dependent manner. Maximal inhibition of both cytokines was reportedly achieved at oral doses of 60 mg/kg.

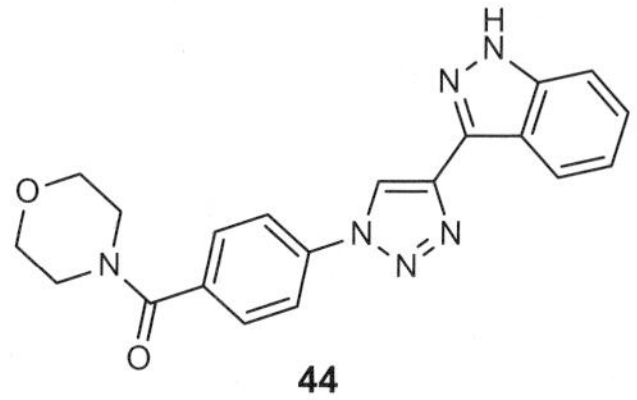

44

Compound **45** (IRAK4 $IC_{50} < 100$ nM) is representative of a series of pyrazolophenyl-substituted pyrimidines recently disclosed.[53] Most examples varied in the substituent of the nitrogen of the pyrazole which is directly attached to the pyrimidine ring, with polar groups such as pendant amides and saturated oxygen- and nitrogen-containing heterocycles explored.

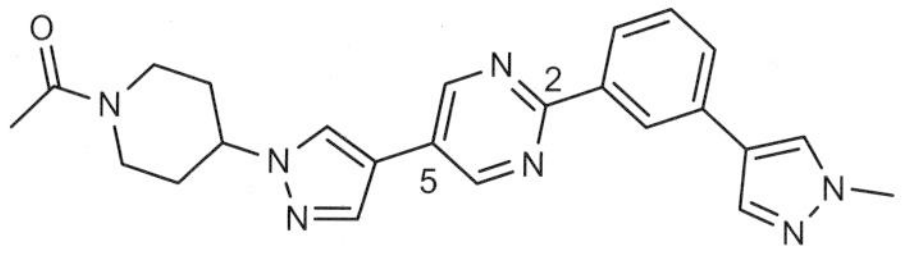

45

A structurally unique series of IRAK4 inhibitors were realized after isolation of an active impurity in a high-throughput screen.[54] The indolo[2,3-*c*] quinoline, PF-05387252 **46**, was potent against IRAK4 (IC_{50} 0.22 μM) with moderate whole blood activity (IC_{50} 4.5 μM). Efficacy in a murine LPS-induced TNFα model was obtained at 100 mg/kg PO. Concerns were noted due to vascular endothelial growth factor and PDGFR activity, and it was also highlighted that the compounds were DNA intercalators.

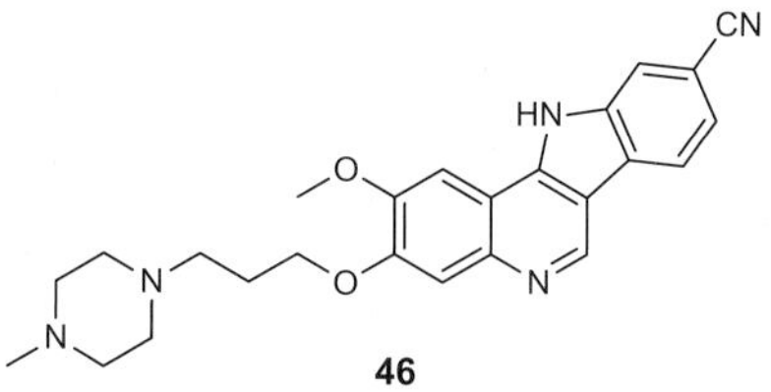

46

5. CONCLUSIONS

The potential for downstream inhibition of innate immune system processes by IRAK4 inhibition has generated significant interest in this target for the treatment of autoimmune disorders. Blockade of IL-1 receptor signaling, clinically validated with biologic therapies at the receptor and ligand level, has demonstrated promise for inflammatory diseases. It is anticipated that IRAK4 inhibition will complement those therapies. Additionally, inhibiting IRAK4 downstream of oncogenically active MyD88 mutations may provide for an additional mechanism to target these aggressive diseases. As evidenced by the rise in number of published patent applications claiming IRAK4 inhibitors over the 2008–2014 time period, the potential benefit these molecules may bring to the patient is being increasingly recognized. Although key concerns regarding innate immune system modulation remain, the development of a safe and efficacious IRAK4 inhibitor for the treatment of TLR/IL-1R-mediated diseases holds promise.

REFERENCES

1. Li, S.; Strelow, A.; Fontana, E. J.; Wesche, H. *PNAS* **2002**, *99*, 5567.
2. Lin, S.; Lo, Y.; Wu, H. *Nature* **2010**, *465*, 885.
3. Ku, C.-L.; Bernuth, H.; Picard, C.; Zhang, S.-Y.; Chang, H.-H.; Yang, K.; Chrabieh, M.; Issekutz, A. C.; Cunningham, C. K.; Gallin, J.; Holland, S. M.; Roifamn, C.; Ehl, S.; Smart, J.; Tang, M.; Barrat, F. J.; Levy, O.; McDonald, D.; Day-Good, N. K.; Miller, R.; Takada, H.; Hara, T.; Al-Hajjar, S.; Al-Ghonaium, A.; Speert, D.; Sanlaville, D.; Li, X.; Geissman, F.; Vivier, E.; Marodi, L.; Garty, B.-Z.; Chapel, H.; Rodriguez-Gallego, C.; Bossuyt, X.; Abel, L.; Puel, A.; Casanova, J.-L. *J. Exp. Med.* **2007**, *204*, 2407.
4. Picard, C.; Casanova, J.-L.; Puel, A. *Clin. Microbiol. Rev.* **2011**, *24*, 490.
5. Lye, E.; Dhanji, S.; Calzascia, T.; Elford, A. R.; Ohashi, P. S. *Eur. J. Immunol.* **2008**, *38*, 870.
6. Kim, T. W.; Febbraio, M.; Robinet, P.; DuGar, B.; Greene, D.; Cerny, A.; Latz, E.; Gilmour, R.; Staschke, K.; Chisolm, G.; Fox, P. L.; DiCorleto, P. E.; Smith, J. D.; Li, X. *J. Immunol.* **2011**, *186*, 2871.
7. Staschke, K. A.; Dong, S.; Saha, J.; Zhao, J.; Brooks, N. A.; Hepburn, D. L.; Xia, J.; Gulen, M. F.; Kang, Z.; Altuntas, C. Z.; Tuohy, V. K.; Gilmour, R.; Li, X.; Na, S. *J. Immunol.* **2009**, *183*, 568.
8. Koziczak-Holbro, M.; Littlewood-Evans, A.; Pöllinger, B.; Kovarik, J.; Dawson, J.; Zenke, G.; Burkhart, C.; Müller, M.; Gram, H. *Arthritis Rheum.* **2009**, *60*, 1661.
9. Goh, F. G.; Midwood, K. S. *Rheumatology* **2012**, *51*, 7.
10. Miller, L. S. *Adv. Dermatol.* **2008**, *24*, 71.
11. Kim, W.-U.; Sreih, A.; Bucula, R. *Autoimmun. Rev.* **2009**, *8*, 204.
12. Cario, E. *Inflamm. Bowel Dis.* **2010**, *16*, 1583.
13. Kimball, A. B.; Krueger, J.; Sullivan, T.; Arbeit, R. D. In: *International Investigative Dermatology Edinburgh, Edinburgh, Scotland, May 8–11*, 2013.

14. Bhagat, L.; Jiang, W.; Yu, D.; Arbeit, R. D.; Sullivan, T. *FOCIS* 2013, *2013, Boston, MA. Presentation Number S.85.*
15. Press Release. *Idera Pharmaceuticals.* 2014March 28www.iderapharm.com.
16. Dinarello, C. A.; Simon, A.; van der Meer, J. W. M. *Nat. Rev. Drug Discov.* **2012**, *11*, 633.
17. Ngo, V. N.; Young, R. M.; Schmitz, R.; Jhavar, S.; Xiao, W.; Lim, K. H.; Kohlhammer, H.; Xu, W.; Yang, Y.; Zhao, H.; Shaffer, A. L.; Romesser, P.; Wright, G.; Powell, J.; Rosenwald, A.; Muller-Hermelink, H. K.; Ott, G.; Gascoyne, R. D.; Connors, J. M.; Rimsza, L. M.; Campo, E.; Jaffe, E. S.; Delabie, J.; Smeland, E. B.; Fisher, R. I.; Braziel, R. M.; Tubbs, R. R.; Cook, J. R.; Weisenburger, D. D.; Chan, W. C.; Staudt, L. M. *Nature* **2011**, *470*, 115.
18. Puente, X. S.; Pinyol, M.; Quesada, V.; Conde, L.; Ordóñez, G. R.; Villamor, N.; Escaramis, G.; Jares, P.; Beà, S.; González-Díaz, M.; Bassaganyas, L.; Baumann, T.; Juan, M.; López-Guerra, M.; Colomer, D.; Tubío, J. M.; López, C.; Navarro, A.; Tornador, C.; Aymerich, M.; Rozman, M.; Hernández, J. M.; Puente, D. A.; Freije, J. M.; Velasco, G.; Gutiérrez-Fernández, A.; Costa, D.; Carrió, A.; Guijarro, S.; Enjuanes, A.; Hernández, L.; Yagüe, J.; Nicolás, P.; Romeo-Casabona, C. M.; Himmelbauer, H.; Castillo, E.; Dohm, J. C.; de Sanjosé, S.; Piris, M. A.; de Alava, E.; San Miguel, J.; Royo, R.; Gelpí, J. L.; Torrents, D.; Orozco, M.; Pisano, D. G.; Valencia, A.; Guigó, R.; Bayés, M.; Heath, S.; Gut, M.; Klatt, P.; Marshall, J.; Raine, K.; Stebbings, L. A.; Futreal, P. A.; Stratton, M. R.; Campbell, P. J.; Gut, I.; López-Guillermo, A.; Estivill, X.; Montserrat, E.; López-Otín, C.; Campo, E. *Nature* **2011**, *475*, 101.
19. Treon, S. P.; Xu, L.; Yang, G.; Zhou, Y.; Liu, Z.; Cao, Y.; Sheehy, P.; Manning, R. J.; Patterson, C. J.; Tripsas, C.; Arcaini, L.; Pinkus, G.; Rodig, S. J.; Sohani, A. R.; Harris, N. L.; Laramie, J. M.; Skifter, D. A.; Lincoln, S. E.; Hunter, Z. R. *NEJM* **2012**, *367*, 826.
20. Chaudhary, D.; Wood, N.; Romero, D. L.; Robinson, S. D.; Greenwood, J. R.; Shelley, M.; Morin, M.; Kapeller, R.; Westlin, W. F. In: *55th Annual Meeting of the American Society of Hematology, New Orleans, LA; Abstract 3833*, 2013.
21. Kelly, P. N.; Chaudhary, D.; Young, R. M.; Shaffer, A.; Robinson, S.; Romero, D. L.; Kapeller, R.; Staudt, L. M. In: *American Association for Cancer Research, San Diego, CA; Abstract LB-112* 2014.
22. Srivastava, R.; Geng, D.; Liu, Y.; Zheng, L.; Li, Z.; Joseph, M. A.; McKenna, C.; Bansal, N.; Ochoa, A.; Davila, E. *Cancer Res.* **2012**, *72*, 6209.
23. Flannery, S.; Bowie, A. G. *Biochem. Pharmacol.* **2010**, *80*, 1981.
24. Wang, Z.; Liu, J.; Sudom, A.; Ayres, M.; Li, S.; Wesche, H.; Powers, J. P.; Walker, N. P. C. *Structure* **2006**, *14*, 1835.
25. Kuglstatter, A.; Villsenor, A. G.; Shaw, D.; Lee, S. W.; Tsing, S.; Niu, L.; Song, K. W.; Barnett, J. W.; Browner, M. F. *J. Immunol.* **2007**, *178*, 2641.
26. Wang, Z.; Wesche, H.; Stevens, T.; Walker, N.; Yeh, W.-C. *Curr. Top. Med. Chem.* **2009**, *9*, 724.
27. Powers, J. P.; Li, S.; Jaen, J. C.; Liu, J.; Walker, N. P. C.; Wang, Z.; Wesche, H. *Bioorg. Med. Chem. Lett.* **2006**, *16*, 2842.
28. Frenkel, A. D.; Lively, S. E.; Powers, J. P.; Smith, A.; Sun, D.; Tomooka, C.; Wang, Z. Patent Application US 2007/0037803, 2007.
29. Guckian, K.; Jewell, C.; Conlon, P.; Lin, E. Y. S.; Chan, T. US Patent 8,293,923, 2012.
30. Calderini, M.; Wucherer-Plietker, M.; Graedler, U.; Esdar. C. Patent Application US 2012/0115861, 2012.
31. Buckley, G. M.; Gowers, L.; Higueruelo, A. P.; Jenkins, K.; Mack, S. R.; Morgan, T.; Parry, D. M.; Pitt, W. R.; Rausch, O.; Richard, M. D.; Sabin, V.; Fraser, J. L. *Bioorg. Med. Chem. Lett.* **2008**, *18*, 3211.

32. Inami, H.; Mizutani, T.; Watanabe, J.; Usuda, H.; Nagashima, S.; Ito, T.; Aoyama, N.; Kontani, T.; Hayashida, H.; Terasawa, T.; Moritomo, A.; Ishikawa, T.; Hayashi, K.; Takeuchi, M.; Ohta, M. In: *242nd ACS National Meeting, Denver, CO; Abstract MEDI-329* 2011.
33. Ishikawa, T.; Imamura, E.; Mizuhara, H.; Iwaoka, H.; Evelyn, E.; Inami, H.; Mizutani, T.; Watanabe, J.; Usuda, H.; Nagashima, S.; Ito, T.; Kontani, T.; Shimizu, Y.; Mutoh, S. In: *ACR/ARHP Meeting, Chicago, IL; Abstract 1000* 2011.
34. Arora, N.; Chen, S.; Hermann, J.; Kuglstatter, A.; Labadie, S. S.; Lin, C.; Lucas, M. C.; Moore, A. G.; Papp, E.; Talamas, F. X.; Wanner, J.; Zhai, Y. Patent Application WO 2012/007375, 2012.
35. McElroy, W. T.; Li, G.; Ho, G. D.; Tan, Z.; Paliwal, S.; Seganish, W. M.; Tulshian, D.; Lampe, J.; Methot, J. L.; Zhou, H.; Altman, M. D.; Zhu, L. Patent Application WO 2012/129258, 2012.
36. Buckley, G. M.; Ceska, T. A.; Fraser, J. L.; Gowers, L.; Groom, C. R.; Higueruelo, A. P.; Jenkins, K.; Mack, S. R.; Morgan, T.; Parry, D. M.; Pitt, W. R.; Rausch, O.; Richard, M. D.; Sabin, V. *Bioorg. Med. Chem. Lett.* **2008**, *18*, 3291.
37. Buckley, G. M.; Fosbeary, R.; Fraser, J. L.; Gowers, L.; Higueruelo, A. P.; James, L. A.; Jenkins, K.; Mack, S. R.; Morgan, T.; Parry, D. M.; Pitt, W. R.; Rausch, O.; Richard, M. D.; Sabin, V. *Bioorg. Med. Chem. Lett.* **2008**, *18*, 3656.
38. Dodd, D.; Mussari, C. P.; Bhide, R. S.; Nair, S. K.; Paidi, V. R.; Kumar, S. R.; Banerjee, A.; Sistla, R.; Pitts, W. J.; Hynes, J. Patent Application WO 2013/106614, 2013.
39. Paidi, V. R.; Kumar, S. R.; Nair, S. K.; Banerjee, A.; Sistla, R.; Pitts, W. J.; Hynes, J. Patent Application WO 2013/106641, 2013.
40. Paidi, V. R.; Kumar, S. R.; Banerjee, A.; Nair, S. K.; Sistla, R.; Pitts, W. J.; Hynes, J. Patent Application WO 2013/106612, 2013.
41. Ho, K.-K.; Diller, D.; Letourneau, J. J.; McGuinness, B. F.; Cole, A. G.; Rosen, D.; van Oeveren, C. A.; Pickens, J. C.; Zhi, L.; Shen, Y.; Pedram, B. Patent Application WO 2012/068546, 2012.
42. Vajda, E. J.; Lin, T. H.; Wang, B.; Ho, K.-K.; van Oeveren, A.; McGuinness, B.; Letourneau, J.; Lee, Y.-H.; Rungta, D.; Zhi, L.; Marschke, K. B. In: ACR/ARHP Meeting, Chicago, IL; Abstract 836, 2011.
43. Harriman, G. C.; Wester, R. C.; Romero, D. L.; Robinson, S.; Shelley, M.; Wessel, M. D.; Greenwood, J. R.; Masse, C. E. Patent Application WO 2013/106535, 2013.
44. Divya Chaudhary, D.; Robinson, D.; Masse, C. E.; Wessel, M. D.; Watts, S.; Greenwood, J.; Shelley, M.; Brewer, M.; Harriman, G.; Frye, L. L.; Wester, R. T.; Kapeller, R.; Romero, D. In: ACR/ARHP Meeting, Washington, DC; Abstract 1062, 2012.
45. Romero, D. L.; Robinson, S.; Wessel, M. D.; Greenwood, J. R. Patent Application WO 2014/011902, 2014.
46. Harriman, G. C.; Wester, R. C.; Romero, D. L.; Masse, C. E.; Robinson, S.; Greenwood, J. R. Patent Application WO 2014/011906, 2014.
47. Harriman, G. C.; Romero, D. L.; Masse, C. E.; Robinson, S.; Wessel, M. D.; Greenwood, J. R. Patent Application WO 2014/011911, 2014.
48. Seganish, W. M.; Brumfield, S. N.; Lim, J.; Matasi, J. J.; McElroy, W. T.; Tulshian, D. B.; Lavey, B. J.; Altman, M. D.; Gibeau, C. R.; Lampe, J. W.; Methot, J.; Zhu, L.; Patent Application WO 2013/066729, 2013.
49. Seganish, W. M.; McElroy, W. T.; Brumfield, S.; Herr, J. R.; Yet, L.; Yang, J.; Harding, J. P.; Ho, G. D.; Tulshian, D. B.; Yu, W.; Wong, M. K. C.; Lavey, B. J.; Kozlowski, J. A. Patent Application WO 2014/058685, 2014.
50. Anima, B.; Hosohalli, S.; Subhendu, M. Patent Application WO 2013/042137, 2013.
51. Samajdar, S. In: *CHI's 9th Annual Drug Discovery Chemistry, San Diego, CA* 2014.

52. Jorand-Lebrun, C.; Crosignani, S.; Dorbais, J.; Grippie-Vallotton, T.; Pretre, A. Patent Application WO 2012/084704, 2012.
53. Crosignani, S.; Jorand-Lebrun, C.; Gerber, P.; Muzerelle, M. Patent Application WO 2014/008992, 2014.
54. Tumey, L. N.; Rao, V.; Bhagirath, N.; Subrath, J.; Boschelli, D. H.; Bennett, E.; Shim, J.; Goodwin, D.; Lin, L.-L.; Telliez, J.-B.; Murphy, E.; Shen, M. In: *241st ACS National Meeting, Anaheim, CA; Abstract MEDI-3* 2011.

CHAPTER TEN

H_4 Receptor Antagonists and Their Potential Therapeutic Applications

David Burns, Niu Shin, Ravi Jalluri, Wenqing Yao, Peggy Scherle
Incyte Corporation, Wilmington, Delaware, USA

Contents

1. INTRODUCTION

1.1. Histamine Receptor Family

Histamine (2-(imidazole-4-yl) ethylamine) is a biogenic amine that is produced by decarboxylation of L-histidine. Mast cell, basophils, and neurons in the stomach and brain are among the cell types where histamine is synthesized and stored. Histamine exerts its biological functions by signaling through four different G-protein-coupled receptors, the histamine H_1, H_2, H_3, and H_4 receptors. These receptors share relatively low amino acid

Annual Reports in Medicinal Chemistry, Volume 49
ISSN 0065-7743
http://dx.doi.org/10.1016/B978-0-12-800167-7.00010-9

sequence homology with each other, with the H_4R showing 29%, 31%, and 43% identity with the H_1R, H_2R, and H_3R, respectively, suggesting that they may have evolved from different ancestral genes.[1,2] Also, the affinity with which histamine binds to these four different receptors varies significantly, with K_i values ranging from 5 to 10 nM for the H_3R and H_4R to 2–10 μM for the H_1R and H_2R.[3] The H_1R and H_2R have proven to be excellent antihistamine targets for relieving allergic responses, and gastric acid reflux, respectively. The H_3R antagonists are currently in clinical trials for different neurological disorders including narcolepsy, hypersomnia, and schizophrenia. The H_4R inhibitors are also in early stages of clinical trials for allergic and autoimmune inflammatory diseases and will be the focus of this review.

1.2. Expression and Function of the Histamine H_4 Receptor

Several lines of evidence suggest that the H_4 receptor plays a role in immunomodulation. In addition to its restricted expression in various hematopoietic cells[1,4,5], the H_4R mediates histamine-induced chemotaxis of eosinophils,[6,7] mast cells,[8] dendritic cells,[9–11] and possibly NK cells and monocytes.[10] It cooperates with the H_2R in controlling IL-16 release from $CD8^+$ T cells.[12] Activation of the H_4R was found to suppress CCL2 and IL-12 production from monocytes and dendritic cells, repectively,[9,13] promote LTB_4 release from mast cells,[14] and increase IL-17 production from memory Th17 cells.[15] In a dendritic and T cell coculture system, blockade of the H_4R on dendritic cells led to decreases in cytokine and chemokine production and limited their ability to induce a Th2 response in T cells.[16] Detection of the H_4R in different locations of the central and peripheral nervous system suggest that the receptor may serve functions beyond immunomodulation.

2. ANTAGONISTS OF THE H_4 RECEPTOR

2.1. Indole and Benzimidazole Amide Ligands

Due to the high degree of homology between H_4R and H_3R, the initial ligands that were used to study H_4R perturbation were known H_3R ligands that were based on the native histamine ligand and in most cases these ligands behaved as agonists.[17] **1** (JNJ7777120) was the first selective non-imidazole H_4R antagonist with >1000-fold selectivity over the other histamine subtypes and has become one of the most utilized tool compounds for studying H_4R anatagonism.[18] Poor metabolic stability of **1** due to rapid *N*-demethylation of the piperazine has precluded its development. Efforts to obtain compounds with improved properties by scaffold hopping led to benzimidazole **2** (JNJ10191584) which had improved solubility and liver microsome stability; although the *in vivo* half-life in rodents was still limited ($t_{1/2}=1.0$ h).[19] Bioisosteric replacement of the metabolically labile *N*-methyl piperazine of **1** with a bicyclic octahydropyrrolo[3,4-*c*]pyrrole moiety (**3**) resulted in increased metabolic stability in both rat and human *in vitro*.[20] To further increase the metabolic stability, the indole was replaced with the more polar benzimidazole resulting in decreased potency even after extensive SAR studies of varying substitution on the benzimidazole (**4**). A serendipitous discovery led to the identification of the amidine **5** (PF-2988403) which restored potency while retaining metabolic stability.

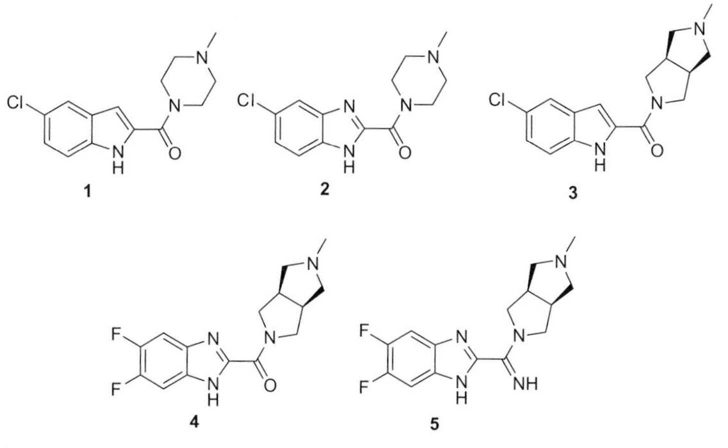

2.2. Dibenzodiazepine, Quinoxalinone, and Quinazoline Ligands

The clozapine analog **6** (VUF6884) and **1** were used in an *in silico* flexible alignment model to generate a series of small heterocyclic fragments that were elaborated upon and optimized to afford **7** (VUF10214) and **8** (VUF10148).[21] Extrapolation of these two lead compounds utilizing a scaffold hopping approach led to the discovery of **9** (VUF10499 and VUF10497) which were potent human H_4R inverse agonists.[22]

These compounds also possess considerable affinity for the human histamine H_1R and therefore represent a novel class of dual action H_1R/H_4R ligands.

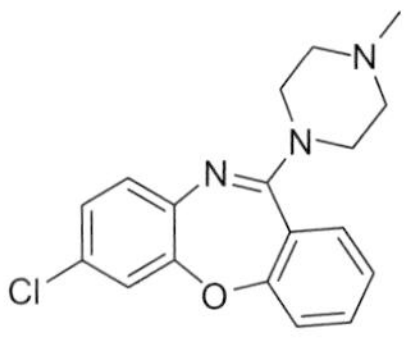

6

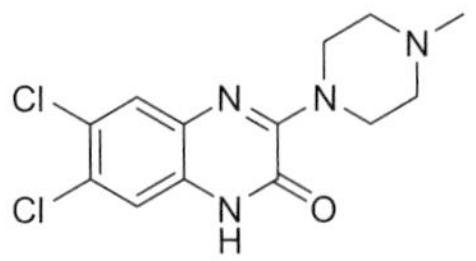

7

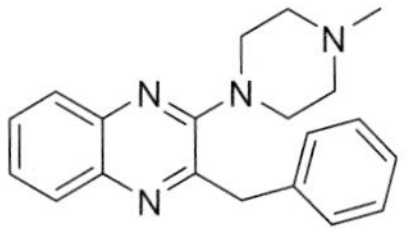

8

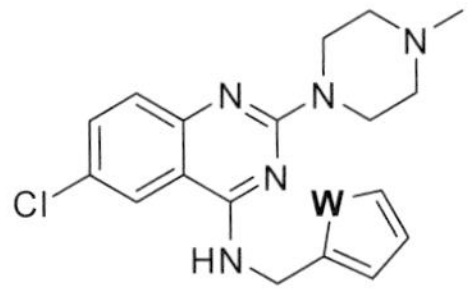

9

VUF10499 (W: O) VUF10497 (W: S)

2.3. Pyrimidine-Based Ligands

Pyrimidine-based ligands have been one of the most extensively mined scaffolds for H_4R affinity.[23–25] The following discussion will highlight historically important compounds, although the main emphasis will be on recently disclosed compounds. In 2005, 2-aminopyrimidines were first disclosed as H_4 receptor antagonists, which led to the expedient expansion of this class of ligands by multiple groups almost simultaneously.[26,27] Following extensive SAR studies based on a lead compound from a high-throughput screen (HTS), it was found that the 6-position was most amenable to substitution with a variety of nonpolar substituents such as aryl and alkyl groups, which led to the discovery of **10**.[28] A second group found that the 6-position could be substituted with fused heterobicycles, such as isoquinolines as exemplified by **11** (INCB38579).[29] **11** was shown to act as an antagonist with single digit nanomolar potency and at least 80-fold selectivity over the other human histamine H_1, H_2, and H_3 receptors.[30] **11** had good pharmacokinetic (PK) properties in both mice and rats and was shown to be efficacious in a histamine-induced itch model in mice and carrageenan-induced acute inflammatory pain model in rats as described below. In addition, **11** demonstrated a reduction in inflammatory pain experienced by both rats and mice in formalin-induced pain models. A third group, discovered 6-alkyl- and 6-cycloalkyl-2,4-diaminopyrimidine compounds using **1** as a lead and incorporating SAR findings from a structurally distinct class of tricyclic pyrimidines.[31] Following extensive SAR and *in vivo* studies **12** (JNJ39758979) was selected as a clinical candidate. **12** was withdrawn during phase II trials due to two patients contracting drug-induced agranulocytosis.[31]

In 2007, triamino pyrimidines were disclosed as modulators of the histamine H_4 receptor activity, which was quickly followed by several groups publishing on this class of compounds.[23–25,32] Following extensive SAR studies and optimization **13** (PF-3893787) emerged as a clinical lead and was evaluated for effects on lung function in asthmatic subjects who were presented with an allergen challenge. The results of this clinical proof of concept study are pending.[33,34] The SAR of three triaminoheteroaryl scaffolds were explored and it was found that the general hierarchy in potency was triamino pyridines > 1,3-pyrimidines > 1,5-pyrimidines with **14** having subnanomolar potencty.[35] Interestingly, the functional activity against the H_4R was dependent upon both the nature of the heteroaryl core and the diamino group with the 3-methylamino pyrrolidine acting as an antagonist across all three heteroaryl cores and the methyl piperazine only acting as a partial antagonist and agonist when attached to the 1,3-pyrimidine core and as a partial agonist for the other two heteroaryl cores.

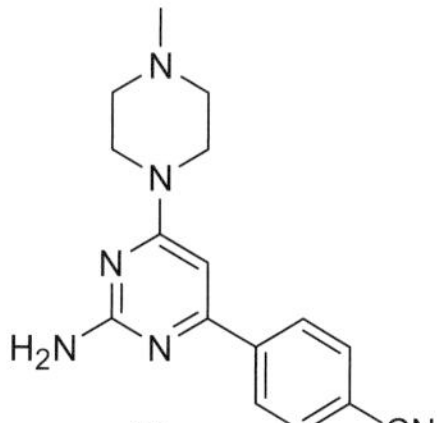

N N N N H_2N CN **10**

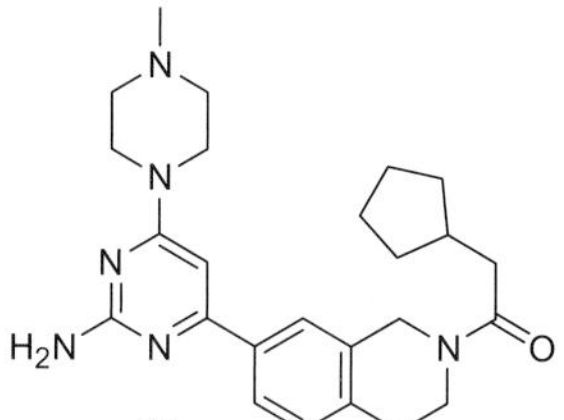

N N N N H_2N N O **11**

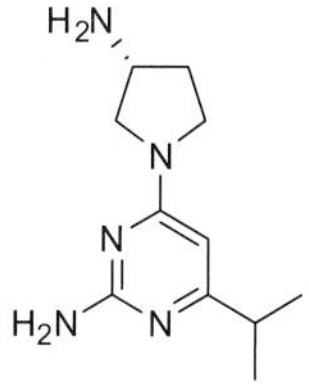

H_2N N N N H_2N **12**

HN N N N H_2N N H **13**

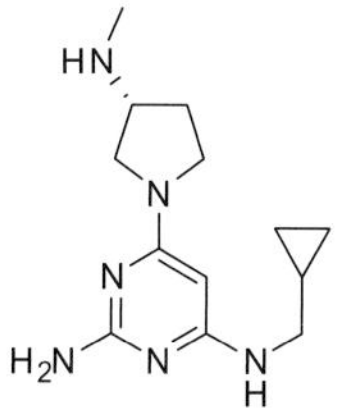

HN N N H_2N N H **14**

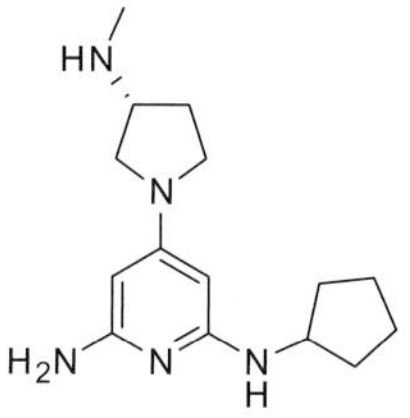

A ligand-based virtual screening was performed using **1** as a reference ligand providing two hits that were found to have low micromolar H_4R binding affinity, namely a 3*H*-[1,2,3]triazolo[4,5-*d*]pyrimidine (**15**) and a related thieno[2,3-d]pyrimidine.[36] In an effort to decipher the pharmacophore, the fused bicycle of **15** was dissected into two hetero-monocyclic components **16** and **17**. It was found that the pyrimidine compound **17** was 84-fold more potent than **16**, which was attributed to it having both a hydrogen bond donor and acceptor group analogous to the amide indole portion of **1**. Insertion of the 2-amino group on the pyrimidine ring resulted in a fourfold increase in potency and further optimization of the substitution on the aromatic ring led to compounds such as **18**. Interestingly, it was found that the substitution pattern on the phenyl ring dictated how the compound behaved in a functional binding assay. Generally speaking, substitution at the ortho and para positions of the phenyl ring resulted in partial agonists and substitution at the meta position resulted in an inverse agonist. Compound **18** exhibited improved reduction in constitutive activity in a GTPγS efficacy model in comparison to **1**, despite having a modest H_4R binding affinity (282 nM).

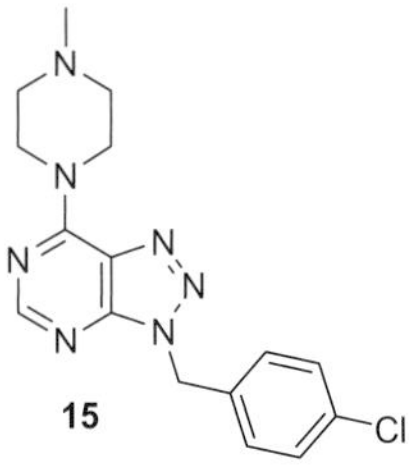

15 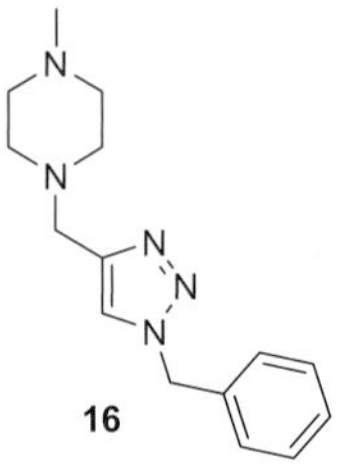 **16** 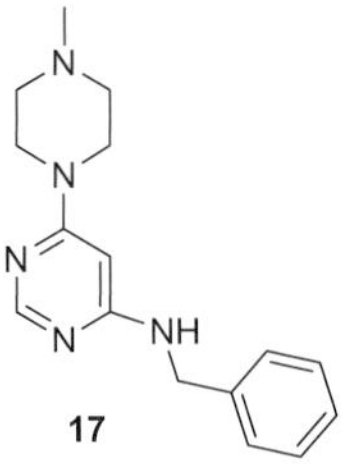 **17**

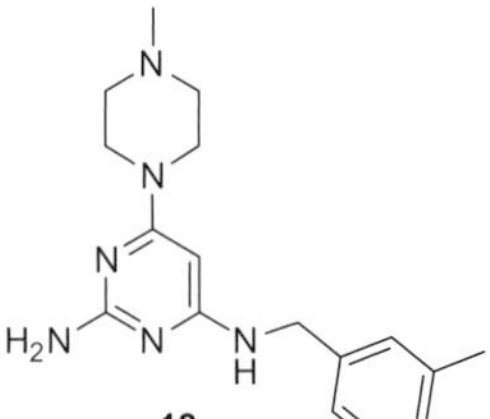

18

In 2010, a new class of triamino pyrimidines H_4R antagonists were disclosed that contained a heterocycle in the 4-position, most notably a 1,2,3,4-tetrahydroisoquinoline (THIQ) with substitution at the C7 position.[37] The THIQ segment was designed based on the hybridization of previously disclosed compounds **19** and **20**.[32,38] **21** (INCB37690) was identified as an early lead from this series. Following a 5 mg/kg oral dosing in rats, **21** was determined to have a half-life of 8.6 h, a high volume of distribution (16.4 L/kg), relatively low clearance (~33% hepatic blood flow), and was completely bioavailable. The major liability for this compound was high hERG inhibition and a low H_4R/H_3R selectivity. Both of these issues could be addressed by modification of the THIQ ring substituents, such as installation of a hydroxyl group on the 3-methyl substituent or the replacement of the 7-pyridyl ring with a saturated heterocycle, as in **22**.

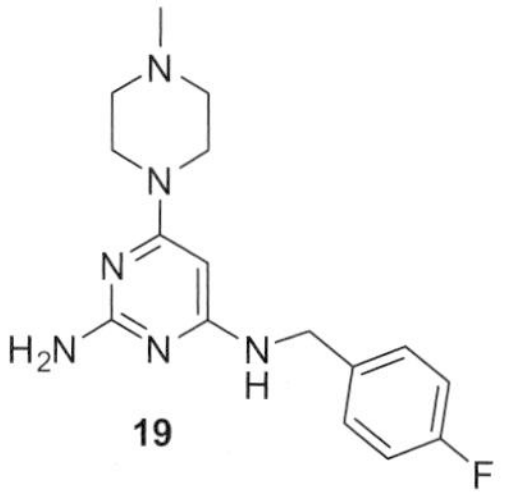

19 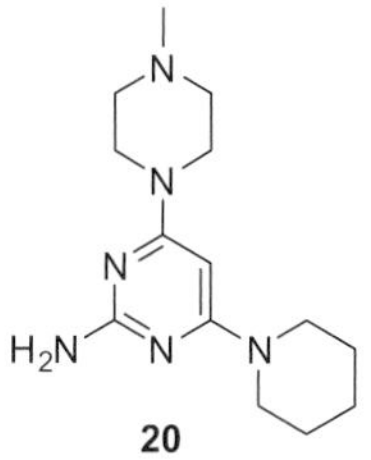 **20** 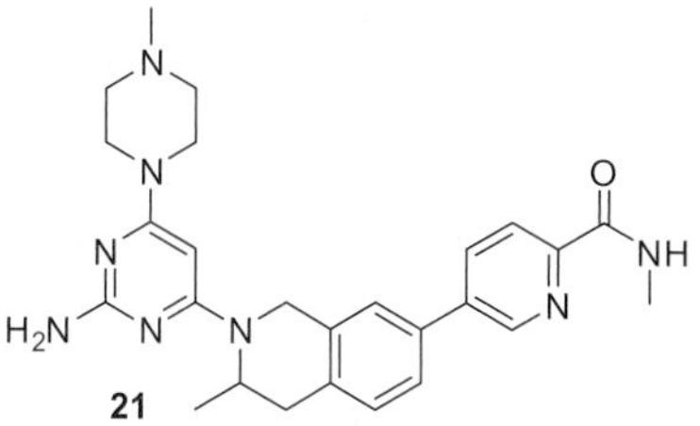 **21** 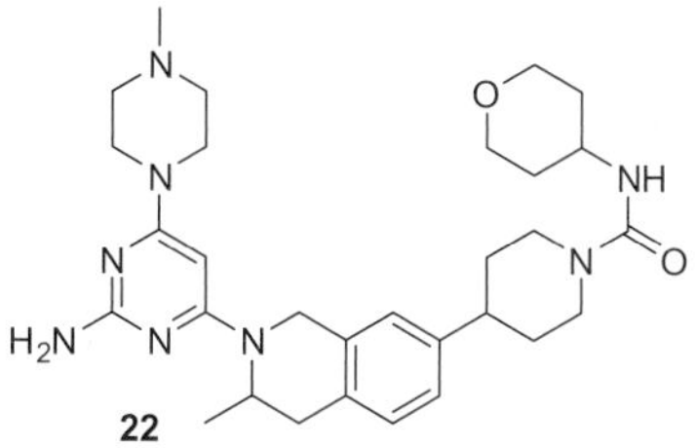 **22**

Patent applications have published from two groups that claim 2-aminoquinazolines as potent hH_4R antagonists.[29,39] Both groups displayed potent examples with substitution at the 7-position of the quinazoline ring, one having heteroaromatic groups (e.g., **23**) and the other having secondary amides (e.g., **24**). A systematic evaluation of the importance of the three nitrogens of the 2-amino quinazoline ring system was carried out, indicating that a 640-fold increase in binding affinity was observed in the evolution of 1-(4-methylpiperazin-1-yl)isoquinoline to 4-(4-methylpiperazin-1-yl)quinazolin-2-amine.[40] SAR studies also revealed the preference for substitution at the 7-position of the quinazoline ring versus the 8-position to fill a hydrophobic pocket and resulted in the discovery of VUF11489, **25**, which exhibited a 255-fold selectivity over hH_3R and had good oral bioavailability (47%) in the mouse. Despite this favorable profile, **25** had three major issues: (1) high hERG binding, (2) metabolically labile *N*-methyl piperazine, (3) the furan group was associated with undesired metabolic activation and covalent binding to proteins. These issues were resolved by the introduction of an additional nitrogen to the quinazoline to form a pyrido[3,2-*d*]pyrimidine, replacement of the *N*-methylpiperazine with 3-methylamino azetidine, and replacement of the tetrahydrofuran group with an alkyl or cycloalkyl group to afford compounds such as **26**.[41] It was found that the hERG binding was significantly reduced while maintaining good affinities to both human and rodent histamine receptors, with a fourfold higher residence time in the hH_4 receptor in comparison to **25**.

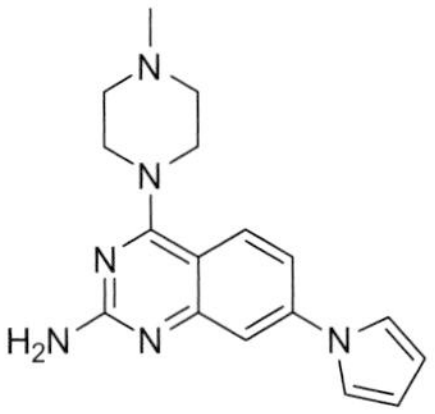

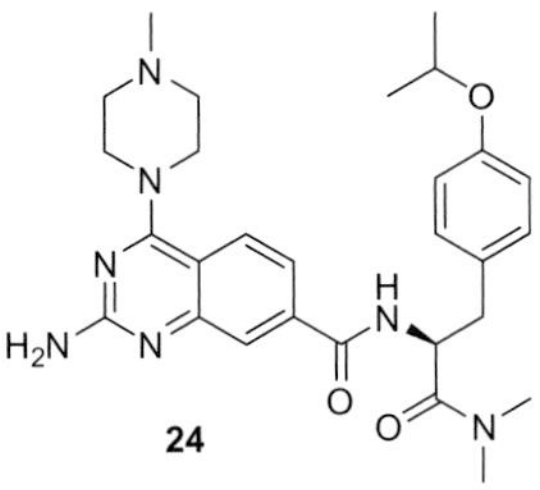

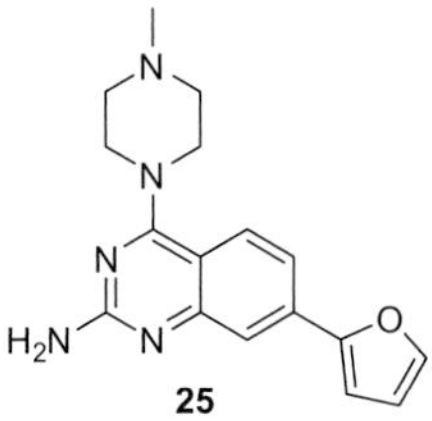

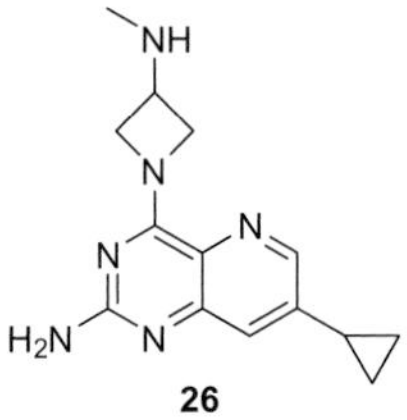

Despite the potent H_4R antagonist activity *in vitro* and *in vivo*, **10** suffered from rapid metabolism via demethylation and had significant off-target affinity for $5\text{-}HT_{1a}$ and $5\text{-}HT_{1d}$ receptors.[42] It was postulated that both the overall PK and selectivity profiles could be improved upon by restricting the rotation around the two aromatic rings by forming a ring junction between the carbon at C5 of the pyrimidine ring and the ortho position of the phenyl ring. A seven-membered annulation was found to form optimal ring size with regards to acting as a potent antagonist although there was no detectable H_4R functional activation across multiple species. Replacement of the *N*-methyl piperazine with 3-amino-pyrolidine afforded **27** (A943931) which had reduced *in vitro* metabolism, improved PK properties, higher H_4 selectivity, and *in vivo* antagonism. The same group expanded upon the success of the rotationally constrained 2,4-diamino-5,6-disubstituted pyrimidines by preparing tetracyclic fused systems with a high level of saturation as exemplified by **28** (A987306).[43] **28** displayed a similar *in vitro* and *in vivo* profile to **27**, with a notably high potency in a pain assay in rats, blocking carrageenan-induced thermal hyperalgesia with an ED_{50} of 42 μmol/kg.

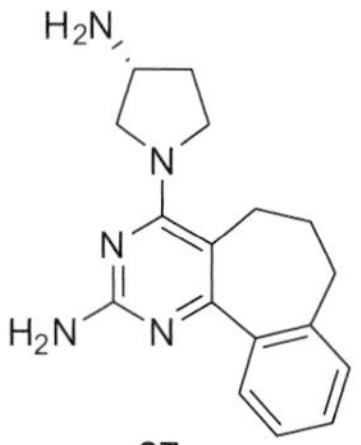

In 2009, furo[3,2-*d*]pyrimidine derivatives were disclosed and from this patent UR-63325 emerged as a clinical candidate and was the first in man H_4R antagonist that was studied for the treatment of asthma and allergic rhinitis.[44] 29 (UR-63325) was safe and well tolerated in phase I studies. A proof of concept phase IIa trial was completed in 2011 in which symptomatic relief in otherwise healthy allergic rhinitis patients was measured in response to **29** and nasal allergen challenge (results pending).[45]

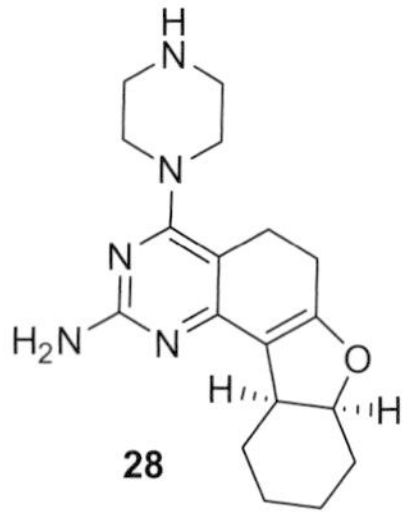

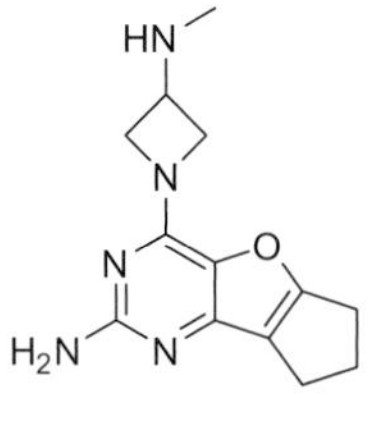

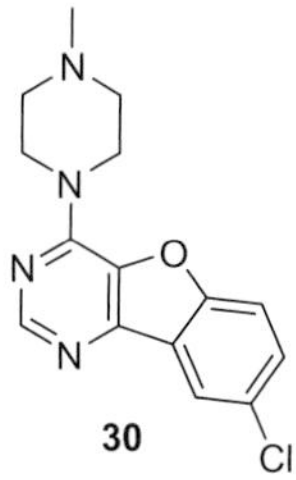

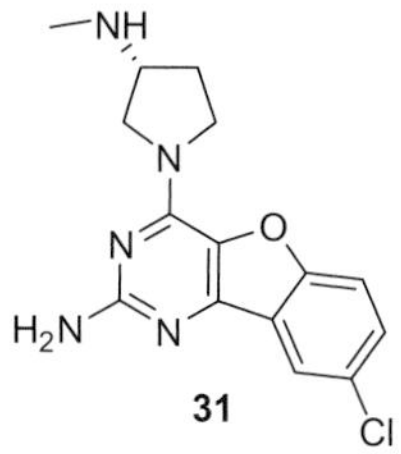

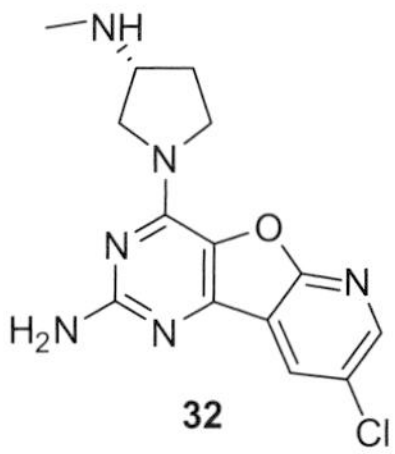

4-Amino-benzofuro[3,2-*d*]pyrimidines, such as **30**, emerged from HTS as a novel scaffold for optimization.[46] It was revealed that microsomal stability could be improved upon by replacement of the piperazine group with *N*-methylaminopyrrolidine and the binding affinity was increased by the introduction of the amino group at the 2-position of the pyrimidine ring leading to the discovery of **31**. Compound **31** is reported to be a potent inverse agonist with excellent selectivity against other histamine subtypes and a reasonable PK profile in dog (36% bioavailability) and cyno (99% bioavailability). Another group subsequently reported the same class of compounds from HTS and noted that the tricyclic lead **30** is a constrained version of their previously discovered **1** and **2**. **31** was also reported in the context of this work.[47] Despite high potency and improved microsomal stability, compounds related to **31** suffered from low aqueous solubility and an unacceptable level of hERG inhibition in patch clamp assays. In an effort to address these issues, the aryl ring was replaced with a pyridyl ring, leading to the discovery of JNJ40279486, **32**, which had improved solubility and was devoid of hERG inhibition. In addition, **32** had good metabolic stability and bioavailability in rat and dog (91% for both).

3. ROLE OF THE HISTAMINE H_4 RECEPTOR IN DISEASE MODELS

3.1. Acute Inflammation and Inflammatory Pain

The effect of H_4R inhibition has been studied in preclinical models of acute inflammation. In the zymosan-induced pleurisy and peritonitis models in mice, pretreatment of the animals with H_4R antagonist **1** reduced neutrophil influx, leukotriene B4, and prostaglandin D2 levels in pleural fluid and peritoneal lavage.[8,14,48] In a carrageenan-induced pleurisy model in rats, **1** also markedly reduced recruitment of leukocytes and levels of proinflammatory prostaglandins (PGE2 and 6-keto PGF1a) and cytokines (IL-1β and TNF-α) in the pleural exudates.[49] **1** and VUF6002 significantly reduced paw edema induced by carrageenan injection in rats.[50] In addition, two different H_4R antagonists, **1** and JNJ28307474, reduced LPS-induced TNF level in plasma that was produced in liver resident macrophages in mice and this production was also reduced in H_4R-deficient mice.[51]

Expression of the histamine H_4R within the central and peripheral nervous systems is consistent with its potential role in the excitation of certain neurons. Recently, **1**, A-943931, **11**, and VUF6002 were all shown to be efficacious in reducing acute inflammatory pain induced by carrageenan in thermal and mechanical hyperalgesia models and by formalin in the flinch model in rats.[30,42,50] **1** also exhibited robust antinociceptive activity in a complete Freund's adjuvant-induced persistent inflammatory pain model and effectively reversed monoiodoacetate-induced osteoarthritic joint pain. Interestingly, it also reduced neuropathic pain in the spinal nerve ligation and sciatic nerve constriction injury models.[52] It is not known at this time whether the site of action is peripheral or central.

3.2. Rheumatoid Arthritis

The H_4R had been reported to be expressed in synovial cells of rheumatoid arthritis patients. It was thus hypothesized that histamine contributes to the occurrence of rheumatoid arthritis through activation of the H_4R. This has been borne out by a recent study in which both H_4R-deficient mice and mice treated with JNJ28307474 exhibited reduced arthritis disease severity in both collagen antibody-induced arthritis and collagen-induced arthritis models.[53]

3.3. Asthma

As is the case for rheumatoid arthritis, histamine has long been associated with asthma; patients with asthma have higher levels of histamine associated with increased airway inflammation and obstruction.[54] But the traditional antihistamines are not effective in the treatment of asthma,[55] which, together with functional expression of the H_4R in immune cells and the high affinity of histamine for H_4R relative to H_1R and H_2R (*vide supra*), makes the histamine H_4R the next logical target to investigate for this disease. The first study on the role of the H_4R in an asthma model was done using the ovalbumin (OVA) lung inflammation model in mice several years ago.[16] In this study, mice with H_4R deficiency or those treated with **1** during either challenge or sensitization phase exhibited decreased lung inflammation, with fewer infiltrating lung eosinophils and lymphocytes and weaker Th2 responses. Further, blockade of the H_4R on dendritic cells led to a decrease in cytokine and chemokine production and an inability to induce Th2 responses. Subsequently, the same research group demonstrated that in a subchronic airway inflammation model, treatment of the mice with **1** not only led to reduced airway inflammation but also airway remodeling and dysfunction.[56] **1** and UR-60427 have also been shown to be efficacious in OVA-induced asthma models in rats and guinea pigs.[57,58]

3.4. Pruritis

Histamine has also long been recognized as a mediator of pruritis in humans. In mice, the pruritic response induced by histamine or histamine H_4R agonists can be reduced by H_4R deficiency or H_4R antagonists like **1**, and almost completely eliminated by combined inhibition of the H_1R and H_4R.[59,60] **1** can also significantly reduce an allergic itch response induced by different haptens in mice.[61,62] Interestingly, despite the fact that mast cells are the major source of histamine and express the H_4R, the histamine H_4R-mediated pruritus was shown to be independent of mast cells or other hematopoietic cells and was hypothesized to result from its action on peripheral neurons in the skin.[62]

4. CLINICAL DEVELOPMENT OF H_4 RECEPTOR ANTAGONISTS

Three histamine H_4R inhibitors have entered clinical trials. JNJ38518168 (structure undisclosed) is being tested for safety, efficacy, PK, and pharmacodynamics in randomized, double-blind, placebo-controlled multicenter trials in symptomatic adult subjects with uncontrolled, persistent asthma, or active rheumatoid arthritis despite concomitant methotrexate therapy. Trials have been completed to evaluate safety, pharmacodynamics and efficacy of **13** on allergen-induced airway response in mild asthmatic subjects. The third compound, **25**, offers good safety, PK, and pharmacodynamics. The proof of concept study for this compound in allergic patients with allergic rhinitis induced by nasal challenge has also been completed. Efficacy data for these three compounds are not available at this time.

5. CONCLUSIONS

The generally restricted expression of the histamine H_4R in immune cells together with data from numerous preclinical models have supported the critical role of the H_4R in inflammatory diseases including rheumatoid arthritis, asthma, pruritis, and inflammatory pain. H_4R antagonists from multiple chemical scaffolds have been identified with improved PK and pharmacodynamics properties. The first H_4R antagonists have entered into clinical trials and data are expected soon to further establish the importance of targeting the H_4R as a novel therapeutic in inflammation.

REFERENCES

1. Morse, K. L.; Behan, J.; Laz, T. M.; West, R. E., Jr.; Greenfeder, S. A.; Anthes, J. C.; Umland, S.; Wang, Y.; Hipkin, R. W.; Gonsiorek, W.; Shin, N.; Gustafson, E. L.; Qiao, X.; Wang, S.; Hedrick, J. A.; Greene, J.; Bayne, M.; Monsma, F. J., Jr. *J. Pharmacol. Exp. Ther.* **2001**, *296*, 1058.
2. Nguyen, T.; Shapiro, D. A.; George, S. R.; Setola, V.; Lee, D. K.; Cheng, R.; Rauser, L.; Lee, S. P.; Lynch, K. R.; Roth, B. L.; O'Dowd, B. F. *Mol. Pharmacol.* **2001**, *59*, 427.
3. Fung-Leung, W.-P.; Thurmond, R. L.; Ling, P.; Karlsson, L. *Curr. Opin. Investig. Drugs* **2004**, *5*, 1174.
4. Zhu, Y.; Michalovich, D.; Wu, H.-L.; Tan, K. B.; Dytko, G. M.; Mannan, I. J.; Boyce, R.; Alston, J.; Tierney, L. A.; Li, X.; Herrity, N. C.; Vawter, L.; Sarau, H. M.; Ames, R. S.; Davenport, C. M.; Hieble, J. P.; Wilson, S.; Bergsma, D. J.; Fitzgerald, L. R. *Mol. Pharmacol.* **2001**, *59*, 434.

5. Lippert, U.; Artuc, M.; Grutzkau, A.; Babina, M.; Guh, S.; Haase, I.; Blaschke, V.; Zachmann, K.; Knosalla, M.; Middel, P.; Kruger-Krasagakis, S.; Henz, B. M. *J. Invest. Dermatol.* **2004**, *123*, 116.
6. O'Reilly, M.; Alpert, R.; Jenkinson, S.; Gladue, R. P.; Foo, S.; Trim, S.; Peter, B.; Trevethick, M.; Fidock, M. *J. Recept. Signal Transduction* **2002**, *22*, 431.
7. Ling, P.; Ngo, K.; Nguyen, S.; Thurmond, R. L.; Edwards, J. P.; Karlsson, L.; Fung-Leung, W.-P. *Br. J. Pharmacol.* **2004**, *142*, 161.
8. Thurmond, R. L.; Desai, P. J.; Dunfourd, P. J.; Fung-Leung, W.-P.; Hofstra, C. L.; Jiang, W.; Nguyen, S.; Riley, J. P.; Sun, S.; Williams, K. N.; Edwards, J. P.; Karlsson, L. *J. Pharmacol. Exp. Ther.* **2004**, *309*, 404.
9. Gutzmer, R.; Diestel, C.; Mommert, S.; Kother, B.; Stark, H.; Wittmann, M.; Werfel, T. *J. Immunol.* **2005**, *174*, 5224.
10. Damaj, B. B.; Becerra, C. B.; Esber, H. J.; Wen, Y.; Maghazachi, A. A. *J. Immunol.* **2007**, *179*, 7907.
11. Baumer, S.; Wendorff, S.; Gutzmer, R.; Werfel, T.; Dijkstra, D.; Chazot, P.; Stark, H.; Kietzmann, M. *Allergy* **2008**, *63*, 1387.
12. Gantner, F.; Sakai, K.; Tusche, M. W.; Cruikshank, W. W.; Center, D. M.; Bacon, K. B. *J. Pharmacol. Exp. Ther.* **2002**, *303*, 300.
13. Dijkstra, D.; Leurs, R.; Chazot, P.; Shenton, F. C.; Stark, H.; Werfel, T.; Gutzman, R. *J. Allergy Clin. Immunol.* **2007**, *120*, 300.
14. Takeshita, K.; Sakai, K.; Bacon, K. B.; Gantner, F. *J. Pharmacol. Exp. Ther.* **2003**, *307*, 1072.
15. Mommert, S.; Gschwandtner, M.; Koether, B.; Gutzmer, R.; Werfel, T. *Am. J. Pathol.* **2012**, *180*, 177.
16. Dunford, P. J.; O'Donnell, N.; Riley, J. P.; Williams, K. N.; Karlsson, L.; Thurmond, R. L. *J. Immunol.* **2006**, *176*, 7062.
17. Leurs, R.; Chazot, P. L.; Shenton, F. C.; Lim, H. D.; de Esch, I. J. *Br. J. Pharmacol.* **2009**, *1*, 14.
18. Jablonowski, J. A.; Grice, C. A.; Chai, W.; Dvorak, C. A.; Venable, J. D.; Kwok, A. K.; Ly, K. S.; Wei, J.; Baker, S. M.; Desai, P. J.; Jiang, W.; Wilson, S. J.; Thurmond, R. L.; Karlsson, L.; Edwards, J. P.; Lovenberg, T. W.; Carruthers, N. I. *J. Med. Chem.* **2003**, *46*, 3957.
19. Venable, J. D.; Cai, H.; Chai, W.; Dvorak, C. A.; Grice, C. A.; Jablonowski, J. A.; Shah, C. R.; Kwok, A. K.; Ly, K. S.; Pio, B.; Wei, J.; Desai, P. J.; Jiang, W.; Nguyen, S.; Ling, P.; Wilson, S. J.; Dunford, P. J.; Thurmond, R. L.; Lovenberg, T. W.; Karlsson, L.; Carruthers, N. I.; Edwards, J. P. *J. Med. Chem.* **2005**, *48*, 8289.
20. Lane, C. A.; Hay, D.; Mowbray, C. E.; Paradowski, M.; Selby, M. D.; Swain, N. A.; Williams, D. H. *Bioorg. Med. Chem. Lett.* **2012**, *22*, 1156.
21. Smits, R. A.; Lim, H. D.; Hanzer, A.; Zuiderveld, O. P.; Guaita, E.; Adami, M.; Coruzzi, G.; Leurs, R.; de Esch, I. J. P. *J. Med. Chem.* **2008**, *51*, 2457.
22. Smits, R. A.; de Esch, I. J. P.; Zuiderveld, O. P.; Broeker, J.; Sansuk, K.; Guaita, E.; Coruzzi, G.; Adami, M.; Haaksma, E.; Leurs, R. *J. Med. Chem.* **2008**, *51*, 7855.
23. Engelhardt, H.; Smits, R. A.; Leurs, R.; Haaksma, E.; de Esch, I. J. P. *Curr. Opin. Drug Discovery Dev.* **2009**, *12*, 628.
24. Lazewska, D.; Kiec-Kononowicz, K. *Front. Biosci. Schol.* **2012**, *4*, 967.
25. Kiss, R.; Keserű, G. M. *Expert Opin. Ther. Pat.* **2012**, *22*, 205.
26. Sato, H.; Fukushima, K.; Shimazaki, M.; Urbahns, K.; Sakai, K.; Gantner, F.; Bacon, K. WO2005/014556, 2005.
27. Sato, H.; Tanaka, K.; Shimazaki, M.; Urbahns, K.; Sakai, K.; Gantner, F.; Bacon, K. WO2005/054239, 2005.
28. Altenbach, R. J.; Adair, R. M.; Bettencourt, B. M.; Black, L. A.; Fix-Stenzel, S. R.; Gopalakrishnan, S. M.; Hsieh, G. C.; Liu, H.; Marsh, K. C.; McPherson, M. J.; Milicic, I.; Miller, T. R.; Vortherms, T. A.; Warrior, U.; Wetter, J. M.; Wishart, N.;

Witte, D. G.; Honore, P.; Esbenshade, T. A.; Hancock, A. A.; Brioni, J. D.; Cowart, M. D. *J. Med. Chem.* **2008**, *51*, 6571.
29. Zhou, J.; Maduskuie, T. P.; Qian, D.; Yao, W. WO2010/108059, 2010.
30. Shin, N.; Covington, M.; Bian, D.; Zhuo, J.; Bowman, K.; Li, Y.; Soloviev, M.; Qian, D.-Q.; Feldman, P.; Leffet, L.; He, X.; Wang, K. H.; Krug, K.; Bell, D.; Czerniak, P.; Hu, Z.; Zhao, H.; Zhang, J.; Yeleswaram, S.; Yao, W.; Newton, R.; Scherle, P. *Eur. J. Pharmacol.* **2012**, *657*, 47.
31. Savall, B. M.; Chavez, F.; Tays, K.; Dunford, P. J.; Cowden, J. M.; Hack, M. D.; Wolin, R. L.; Thurmond, R. L.; Edwards, J. P. *J. Med. Chem.* **2014**, *57*, 2429–2439.
32. Carceller Gonzalez, E.; Salas Solana, J.; Soliva Soliva, R.; Medina, F.; Eva. M.; Marti Via, J. WO2007/031529, 2007.
33. Bell, S. A.; Lane, C. A. L.; Mowbray, C.; Selby, M. D.; Swain, N. A.; Williams, D. H. WO2007/072163, 2007.
34. Mowbray, C. E.; Bell, A. S.; Clarke, N. P.; Collins, M.; Jones, R. M.; Lane, C. A. L.; Liu, W. L.; Newman, S. D.; Paradowski, M.; Schenck, E. J.; Selby, M. D.; Swain, N. A.; Williams, D. H. *Bioorg. Med. Chem. Lett.* **2011**, *21*, 6596.
35. Meduna, S. P.; Savall, B. M.; Cai, H.; Edwards, J. P.; Thurmond, R. L.; McGovern, P. M. *Bioorg. Med. Chem. Lett.* **2011**, *21*, 3113.
36. Sander, K.; Kottke, T.; Tanrikulu, Y.; Proschak, E.; Weizel, L.; Schneider, E. H.; Seifert, R.; Schneider, G.; Stark, H. *Bioorg. Med. Chem.* **2009**, *17*, 7186.
37. Zhang, C.; Qian, D.; Zhuo, J.; Yao, W. WO2010/075270, 2010.
38. Raphy, G.; Watson, R. J.; Hannah, D.; Pegurier, C.; Ortmans, I.; Lock, C. J.; Knight, R. L.; Owen, D. A. WO2008/031556, 2008.
39. Smits, R. A.; De Esch, I. J. P.; Leurs, R. WO2010/146173, 2010.
40. Smits, R. A.; Lim, H. D.; van der Meer, T.; Kuhne, S.; Bessembinder, K.; Zuiderveld, O. P.; Wijtmans, M.; de Esch, I. J.; Leurs, R. *Bioorg. Med. Chem. Lett.* **2012**, *22*, 461.
41. Andaloussi, M.; Lim, H. D.; van der Meer, T.; Sijm, M.; Poulie, C. B.; de Esch, I. J.; Leurs, R.; Smits, R. A. *Bioorg. Med. Chem. Lett.* **2013**, *23*, 2663.
42. Cowart, M. D.; Altenback, R. J.; Liu, H.; Hsieh, G. C.; Drizin, I.; Milicic, I.; Miller, T. R.; Witte, D. G.; Wishart, N.; Fix-Stenzel, S. R.; McPherson, M. J.; Adair, R. M.; Wetter, J. M.; Bettencourt, B. M.; Marsh, K. C.; Sullivan, J. P.; Sullivan, J. P.; Honore, P.; Esbenshade, T. A.; Brinoi, J. D. *J. Med. Chem.* **2008**, *51*, 6547.
43. Liu, H.; Altenbach, R. J.; Carr, T. L.; Chandran, P.; Hsieh, G. C.; Lewis, G. R.; Manelli, A. M.; Milicic, I.; Marsh, K. C.; Miller, T. R.; Strakhova, M. I.; Vortherms, T. A.; Wakefield, B. D.; Wetter, J. M.; Witte, D. G.; Honore, P.; Esbenshade, T. A.; Brioni, J. D.; Cowart, M. D. *J. Med. Chem.* **2008**, *51*, 7094.
44. Carceller Gonzalez, E.; Medina Fuentes, E. M.; Marti Via, J.; Virgili Bernado, M. WO2009/056551, **2009**.
45. Salcedo, C.; Pontes, C.; Merlos, M. *Front. Biosci. Elite* **2013**, *5*, 178.
46. Cramp, S.; Dyke, H. J.; Higgs, C.; Clark, D. E.; Gill, M.; Savy, P.; Jennings, N.; Price, S.; Lockey, P. M.; Norman, D.; Porres, S.; Wilson, F.; Jones, A.; Ramsden, N.; Mangano, R.; Leggate, D.; Andersson, M.; Hale, R. *Bioorg. Med. Chem. Lett.* **2010**, *20*, 2516.
47. Savall, B. M.; Gomez, L.; Chavez, F.; Curtis, M.; Meduna, S. P.; Kearney, A.; Dunford, P.; Cowden, J.; Thurmond, R. L.; Grice, C.; Edwards, J. P. *Bioorg. Med. Chem. Lett.* **2011**, *21*, 6577.
48. Strakhova, M. I.; Cuff, C. A.; Manelli, A. M.; Carr, T. L.; Witte, D. G.; Baranowski, J. L.; Vortherms, T. A.; Miller, T. R.; Rundell, L.; McPherson, M. J.; Adair, R. M.; Brito, A. A.; Bettencourt, B. M.; Yao, B. B.; Wetter, J. M.;

Marsh, K. C.; Liu, H.; Cowart, M. D.; Brioni, J. D.; Esbenshade, T. A. *Br. J. Pharmacol.* **2009**, *157*, 44.
49. Pini, A.; Somma, T.; Formicola, G.; Lucarini, L.; Bani, D.; Thurmond, R.; Masnini, E. *Curr. Pharm. Des.* **2014**, *20*, 1338.
50. Coruzzi, G.; Adami, M.; Guaita, E.; de Esch, I. J. P.; Leurs, R. *Eur. J. Pharmacol.* **2007**, *563*, 240.
51. Cowden, J. M.; Yu, F.; Challapalli, M.; Huang, J.-F.; Kim, S.; Fung-Leung, W.-P.; Ma, J. Y.; Riley, J. P.; Zhang, M.; Dunford, P. J.; Thurmond, R. L. *Inflammation Res.* **2013**, *62*, 599.
52. Hsieh, G. C.; Chandran, P.; Salyers, A. K.; Pai, M.; Zhu, C. Z.; Wensink, E. J.; Witte, D. G.; Miller, T. R.; Mikusa, J. P.; Baker, S. J.; Wetter, J. M.; Marsh, K. C.; Hancock, A. A.; Cowart, M. D.; Esbenshade, T. A.; Brion, J. D.; Honore, P. *Pharmacol. Biochem. Behav.* **2010**, *95*, 41.
53. Cowden, J. M.; Yu, F.; Banie, H.; Farahani, M.; Ling, P.; Nguyen, S.; Riley, J. P.; Zhang, M.; Zhu, J.; Dunford, P. J.; Thurmond, R. L. *Ann. Rheum. Dis.* **2013**, *73*, 600.
54. Jarjour, N.; Calhoun, W.; Schwartz, L.; Busse, W. *Am. Rev. Respir. Dis.* **1991**, *144*, 83.
55. Van Ganse, E.; Kaufman, L.; Derde, M. P.; Yernault, J. C.; Delaunois, L.; Vincken, W. *Eur. Respir. J.* **1997**, *10*, 2216.
56. Cowden, J. M.; Riley, J. P.; Ma, J. Y.; Thurmond, R. L.; Dunford, P. J. *Respiratory Res.* **2010**, *11*, 86.
57. Alfon, J.; Ardanaz, N.; Gil-Torregrosa, B.; Fernandez, A.; Balsa, D.; Carceller, E.; Gomez, L.; Merlos, M.; Cortijo, J.; Morceillo, E.; Bartroli, X. *Inflammation Res.* **2010**, *59*(Suppl. 2), S199.
58. Somma, T.; Cinci, L.; Formicola, G.; Pini, A.; Thurmond, R.; Ennis, M.; Bani, D.; Masini, E. *Br. J. Pharmacol.* **2013**, *170*, 200.
59. Bell, J. K.; McQueen, D. S.; Rees, J. L. *Br. J. Pharmacol.* **2004**, *142*, 374–380.
60. Dunford, P. J.; Williams, K. N.; Desai, P. J.; Larlsson, L.; McQueen, D.; Thurmond, R. L. *J. Allergy Clin. Immunol.* **2007**, *119*, 176.
61. Robach, K.; Wendorf, S.; Sander, K.; Stark, H.; Gutzmer, R.; Werfel, T.; Kietzmann, M.; Baumer, W. *Exp. Dermatol.* **2008**, *18*, 57.
62. Cowden, J. M.; Zhang, M.; Dunford, P. J.; Thurmond, R. L. *J. Invest. Dermatol.* **2010**, *130*, 1023.

CHAPTER ELEVEN

Urate Crystal Deposition Disease and Gout—New Therapies for an Old Problem

Jean-Luc Girardet, Jeffrey N. Miner
Ardea Biosciences, San Diego, California, USA

Contents

1. INTRODUCTION

Gout is an inflammatory arthritic condition resulting from monosodium urate crystal deposition (UCD) in joints and tissues, which develop because of high levels of serum uric acid. Patients suffering from gout often have a genetic defect in the renal transport of uric acid, resulting in high uric acid levels, termed hyperuricemia. Diets rich in purines, which are quickly and entirely converted to uric acid, and drugs that impair excretion of uric acid (e.g., thiazides) contribute to hyperuricemia. In some individuals, uric acid reaches levels above 6.8 mg/dL, leading to precipitation in tissues and organs, forming urate crystal aggregates called tophi. These monosodium urate (MSU) tophi are highly inflammatory and can lead to the occurrence of a gout flare. These flares are extremely painful, and it has been reported by patients that even air blowing over the joint can cause excruciating pain. The presence of deposits of uric acid on joints, cartilage, tendons, muscles, and the inflammatory milieu surrounding the deposit causes significant damage, bony erosions, and other sequelae.[1]

Annual Reports in Medicinal Chemistry, Volume 49
ISSN 0065-7743
http://dx.doi.org/10.1016/B978-0-12-800167-7.00011-0

Gout is highly prevalent, occurring in almost 6% of men and 2% of women in the United States, making it the most common form of inflammatory arthritis.[2] The gender difference is likely due to the uricosuric effects of estrogen, since gout prevalence increases in women after menopause.[2] The incidence of hyperuricemia (serum uric acid level of >7.0 mg/dL for men and >5.7 mg/dL for women) is much higher than that of gout, ranging between 20% and 25% of the population worldwide.

UCD disease is a chronic condition that may result in gout flare, but in the majority of patients remains occult for many years.[3] While acute gout flare is the most conspicuous consequence of high uric acid levels, there may be other more insidious processes at work under these conditions. Uric acid levels have been shown in many different studies to be independently and positively associated with the risk of a number of additional illnesses. These range from stroke and myocardial infarction to hypertension, obesity, and diabetes.[4] In these cases, it is unclear as to whether high uric acid is causative or simply a tightly associated marker of the disease process. Possible mechanisms include direct effects of uric acid on the vasculature and activation of inflammatory danger signal pathways.[5] Long-term interventional studies using effective uric acid lowering therapy will likely be required to fully understand the role of uric acid in these other diseases.

Hyperuricemia can be exacerbated by a high dietary purine intake, overproduction, and intestinal underexcretion; however, the primary cause is inefficient renal uric acid excretion.[6] In healthy individuals, uric acid in the kidney is filtered through the glomerulus and then reabsorbed in the proximal tubule. In persons with hyperuricemia or gout, uric acid transporter numbers and/or activity may be increased, resulting in inappropriate greater reabsorption of uric acid. This mechanism prevents efficient excretion of uric acid, raising serum levels.[7] Drugs used to treat hyperuricemia primarily include inhibitors of xanthine oxidase (XO), which block the production of uric acid. In addition, agents that enhance what could be described as uricuresis, a therapeutic increase in uric acid excretion from the kidney, are also utilized. These uricuretic agents produce analogous effects to those of diuretics (drugs that inhibit water reabsorption in the kidney) and glucuretics (drugs that inhibit glucose reabsorption in the kidney) in hypertensive and diabetic patients, respectively. These approaches can be used individually as monotherapy and in combination, blocking both production and reabsorption of uric acid to lower serum uric acid very efficiently.

2. THERAPEUTICS FOR GOUT BY CLINICAL MANIFESTATION

Gout treatment includes short-term approaches for acute gout attacks and long-term approaches for treating hyperuricemia. There are also treatment regimens to block the occurrence of flare during the initiation of uric acid lowering therapy. These prophylactic treatments are often the same as those for acute gout flare, although often at lower dosages and for longer duration (6 months is recommended).[8]

2.1. Gout Flares

Acute gout therapy focuses on the rapid inhibition of the pain and inflammation resulting from the inflammatory response to monosodium UCD. Most commonly prescribed in the United States are nonsteroidal anti-inflammatory drugs (NSAIDs), colchicine, and corticosteroids. In addition, canakinumab, an anti-IL-1β antibody, is approved in Europe for the treatment of acute gout flare.

2.1.1 Nonsteroidal Anti-Inflammatory Drugs

NSAIDs have been shown to be effective in the treatment of acute gout flare, while not treating the underlying crystal burden.[9] These compounds act through inhibition of cyclooxygenase-1 and -2 enzymes, resulting in the decreased production of a variety of inflammatory mediators. They are used both prophylactically and in acute therapy during flare. These compounds exhibit significant gastrointestinal side effects when dosed long term in some patients, which can limit their use in the prophylaxis of flares. The NSAIDs most commonly utilized include indomethacin (**1**), naproxen (**2**), and occasionally ibuprofen (**3**).

2.1.2 Colchicine

Colchicine (**4**) is a long-standing medication, first identified in the leaves of the autumn crocus (*Colchicum autumnale*), which were used medicinally in ancient Egypt for its laxative properties around 1500 BC.[10] Eventually recognized for its antiarthritic properties,[10] colchicine is widely prescribed for gout and likely works by binding tubulin in target lymphocytes (neutrophils). This prevents cellular migration and causes a failure to respond normally to typical stimuli. It is effective in both preventing acute gout flare during prophylaxis and treating symptoms after they occur.[11]

2.1.3 Glucocorticoids

Glucocorticoids are particularly effective anti-inflammatory agents with a long history of use in gout and other inflammatory diseases. They are efficacious during gout flares as oral agents and as intra-articular, injectable preparations. Glucocorticoids are not as commonly used as colchicine and NSAIDs in uncomplicated gout, likely because of a concern about the relative safety of glucocorticoids. However, it has been suggested that gout patients may safely tolerate low-dose steroids.[12]

Commonly used corticosteroids for gout include prednisone (**5**), more commonly in Europe, prednisolone (**6**), and triamcinolone (**7**). They inhibit a wide range of inflammatory gene expression through their agonist activity on the glucocorticoid receptor (GR). Upon binding glucocorticoids, GR translocates into the nucleus of target cells affecting the expression of various genes. Key inflammatory genes that are inhibited by the protein–ligand complex include those for proinflammatory cytokines (e.g., interleukins [ILs] and tumor necrosis factor), selectins, collagenases, and matrix metalloproteinases. The ubiquitous expression of the GR combined with the broad volume of distribution for steroidal drugs ensures activity throughout the body, a critical property because gout flares can occur in many different joints.[13]

2.1.4 IL-1 Blockade

IL-1β is responsible for the inflammatory response in gouty arthritis. Newer agents include a series of IL-1 inhibitors that have shown efficacy in acute gout flare and in prophylaxis. These injectable agents provide powerful inhibitory activity, but do show evidence for injection site reactions and administration is not straightforward. The table below outlines the data and usage for three protein-based therapeutics that block IL-1β activity. Canakinumab is the only agent that has obtained regulatory approval, and this approval was limited to the European Union (EU).[14]

Drug (type)	Gout prophylaxis Hyperuricemia	Acute flare treatment Gouty arthritis
Anakinra (IL-1R antagonist)	No planned or ongoing trails	Human Pilot Study Completed (2007) (reg# ISRCTN10862635)
Canakinumab (IL-1β mab)	Ph2 study complete (reg# NCT00819585)	EU Approval US unapproved (reg# NCT01080131)
Rilonacept (IL-1R decoy)	Ph3 Not approved (reg# NCT01459796)	Marginal efficacy (reg# NCT00855920)

An oral agent (AC-201) that reduces IL-1β expression through the inhibition of inducible nitric oxide synthase has been developed. AC-201 (structure not disclosed) is in Phase 2 trials for gout prophylaxis.[15] Additionally, AC-201 has also been demonstrated to have uric acid-lowering effects in clinical trials. This interesting addition provides both anti-inflammatory activity and beneficial impact on the underlying problem with gout for prophylaxis. Increased infections and reduced tumor surveillance are among the major concerns.

2.1.5 Phosphodiesterase-4

Apremilast (**8**) is a phosphodiesterase-4 inhibitor that is claimed to have anti-inflammatory activity and was tested in a trial for acute gout flare. This approach may be useful in patients who are intolerant to colchicine or NSAIDs.[16] The primary endpoints for the planned trial included subject-reported number of swollen, tender, and warm joints, as well as self-reported pain, global assessment, and functional status. This study was withdrawn prior to initiation, and it is unclear if the sponsors are continuing with this program. Apremilast has recently received approval for the treatment of psoriatic arthritis.[16]

2.1.6 *Anti-C5a Antibody*

MSU crystals activate complement as part of the inflammatory cascade during a gout flare. C5a, a highly potent inflammatory complement peptide, is the key player in MSU crystal-induced IL-1β secretion through its interaction with C5aR. C5a dose-dependently induces pro-IL-1β expression in human primary monocytes. It has been proposed that antibodies blocking C5a function could provide a new therapeutic strategy for gout because of the strong participation of the inflammasome in the mechanism.[17]

2.1.7 *CXCR2*

Neutrophils are the single most prevalent cell type involved in the inflammatory response to MSU crystals. The influx of these critical cells into the synovium is almost entirely dependent on the function of the CXCR2 (IL-8) receptor and is considered critical in the pathogenesis of gout. This receptor mediates the activity of the neutrophil-related chemokines, CXCL1 and CXCL2. In an air-pouch model, these chemokines and their interactions with the receptor were found to be essential for the inflammatory response to urate crystals.[18] Furthermore, in a mouse model of intra-articular injection of MSU crystals, treatment with ladarixin (DF-2162, **9**), an allosteric inhibitor of CXCR2, significantly decreased the neutrophil influx into the joint.[19]

2.2. Hyperuricemia

The treatment of gout must also include approaches to address chronically high levels of uric acid. Lowering uric acid levels below 6 mg/dL long term is known to prevent future gout flare and reduce the size of and possibly eliminate tophi.[20] If treatment is aggressive enough, and uric acid levels remain below 6 mg/dL over the long term, it is possible to completely control the disease. Gout patients typically suffer from a failure to excrete sufficient quantities of uric acid, and uricuretic drugs have been developed with the aim to effectively improve excretion rates. These drugs include both older uricuretic agents and newer selective uric acid reabsorption inhibitors (SURIs) in development. In addition, it is also possible to reduce serum uric acid levels by inhibiting the production of uric acid using XO inhibitors. These drugs are effective at lowering uric acid if dosed at high enough levels to obtain normal serum uric acid in patients. There has been reluctance to dose escalate allopurinol because of concerns about tolerability and potential side effects. Drugs targeting the two mechanisms (production and excretion) can be used together, and combination therapy inhibiting production as well as increasing excretion can be very effective at lowering uric acid and reducing tophi size.[21]

Patients with severely debilitating and disfiguring tophaceous gout often do not achieve target serum uric acid levels with XO inhibitors and may continue to experience frequent gout flares, so they are placed on injectable uricase therapy. Uricases are enzymes that are capable of degrading uric acid into much more soluble allantoin.[22] However, these agents generate significant neutralizing antibody responses and occasional severe reactions, limiting their use to only a subset of the most severely affected patients.

2.2.1 Drugs Blocking Uric Acid Production

XO inhibitors block the production of uric acid by blocking the enzyme responsible for the oxidation of hypoxanthine and xanthine (Fig. 11.1).

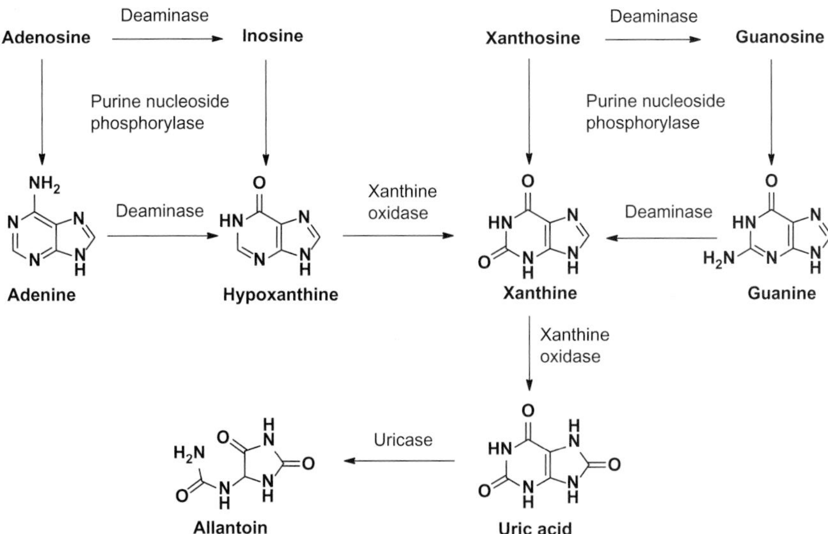

Figure 11.1 Purine catabolism.

2.2.1.1 XO Inhibition

Allopurinol (Zyloprim, generic, **10**) was the first marketed XO inhibitor. It was originally designed as an antitumor agent,[23] but was never approved for this use. Instead, it found an application for the treatment of gout and hyperuricemia and was approved by the U.S. FDA in 1966. Dosage forms available are 100 and 300 mg tablets, and the approval allows dosing up to 900 mg per day in the most extreme cases, although this is seldom done, and most physicians will prescribe a 300-mg daily dose.[24] Allopurinol is a purine analog with moderate activity against human XO. Its major metabolite is oxypurinol, which is also an XO inhibitor. Likely because of the purine-like structure of the parent molecule and its metabolites, these compounds inhibit other enzymes involved in purine and pyrimidine metabolism in addition to XO. This raises the specter of off-target effects. The most serious side effect for allopurinol is the occurrence of Stevens Johnson's syndrome, a rare but life-threatening immune reaction to the drug. A more common and less serious rash may also lead to discontinuation of the use of allopurinol in some patients. Comparative, cohort, and observational studies have shown that only about one-half of patients treated with allopurinol had their serum uric acid reach levels below the target of 6 mg/dL.[25]

Despite this, it took more than 30 years for a second XO inhibitor to come to market, with the approval of febuxostat (Uloric, Adenuric, **11**) in 2009. Febuxostat is presented as a second-generation XO inhibitor that has a better tolerability profile and is more efficacious than allopurinol. Its structure is not related to any natural purine base, and as such, it is more selective than allopurinol. Two doses of febuxostat have been approved in the United States (40 and 80 mg, both administered QD), while a higher 120 mg daily dose has been approved in Europe. Febuxostat can be given to patients who are renally impaired or have allopurinol hypersensitivity syndrome.[26] There have been rare reports of hypersensitivity reactions,[27] including Stevens Johnson's syndrome with febuxostat as well as cases of hepatic injury or failure, for liver injury; it is advised that febuxostat treatment should be stopped if no alternative etiology can be found.

Topiroxostat (FYX-051, Topiloric, Uriadec, **12**) is the third approved XO inhibitor, but, to date, this approval was only received in Japan in 2013. It is structurally more related to febuxostat than to allopurinol. It is dosed twice a day, and the maximum daily dose is 160 mg. Another structurally related inhibitor is niraxostat (Y-700, **13**). A Phase 2 trial was conducted, but the development of this compound appears to have been discontinued since 2009. LC-350189 is another XO listed as being in Phase

1 clinical development in South Korea,[28] but few details about this drug have emerged.

Additional XO inhibitors have been described in patent applications,[29] but to date, this research has not generated additional clinical development compounds.

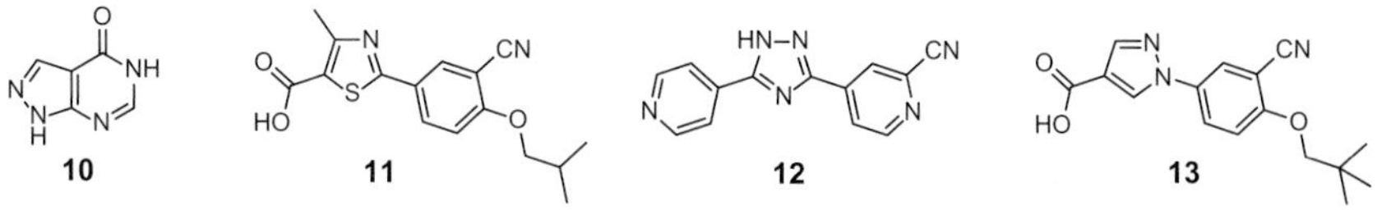

2.2.1.2 Purine Nucleoside Phosphorylase Inhibition

Other published approaches to lowering production of uric acid include the inhibition of purine nucleoside phosphorylase (PNP), just upstream from XO (Fig. 11.1). Blockade of this enzyme reduces the substrates available to XO to produce uric acid. PNP inhibitor ulodesine (BCX4208, **14**) is currently in development for chronic gout.[30] This drug has been associated with dose-dependent reduction in plasma xanthine and hypoxanthine levels, confirming its mode of action. It was studied in a Phase 2b trial in combination with allopurinol, and after 52 weeks of treatment with doses of 5, 10, and 20 mg daily, the response rates were 45%, 47%, and 64%, respectively, compared with 19% for placebo.

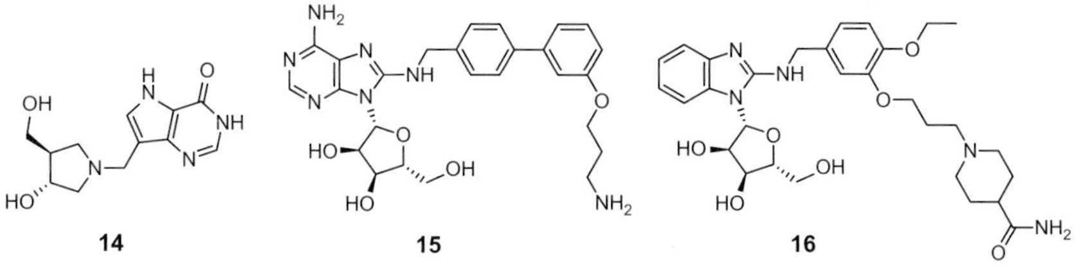

2.2.1.3 Concentrative Nucleoside Transporter Type 2

In addition, blockade of purine absorption has been proposed as a pathway to reduce serum uric acid levels. Concentrative Nucleoside Transporter Type 2 (CNT2) is responsible for absorption of purines from the intestine, and inhibitors are being developed to block the action of this intestinal transporter. Since dietary purines are entirely converted into uric acid in the intestine,[31] blockade of this source of uric acid could lower serum uric acid significantly. CNT2 inhibitors, KGO-2142 (**15**) and KGO-2173 (**16**), are being investigated for the potential treatment of gout and hyperuricemia. These compounds have been reported to selectively inhibit CNT2 without affecting CNT1 and CNT3 and to inhibit dietary RNA-induced hyperuricemia in Cebus monkeys.[32]

2.2.2 Drugs Increasing Uric Acid Excretion

A large number of drugs are known to reduce serum uric acid levels by increasing the excretion of uric acid in the urine.[33] Interestingly, most of these agents do not have an indication for the treatment of gout. These drugs include losartan (angiotensin II receptor inhibitor—hypertension, **17**), amlodipine (calcium channel blocker—hypertension, **18**), atorvastatin (HMG-CoA reductase inhibitor—high cholesterol, **19**), and guaifenesin (expectorant, **20**). A smaller number are indicated for gout, but each has a profile that limits clinical utility. These compounds include probenecid[34] (**21**), sulfinpyrazone (**22**), and benzbromarone[35] (**23**) (all organic ion transporter inhibitors) and are indicated for the treatment of hyperuricemia or, in the case of sulfinpyrazone, gout, and hyperuricemia. Probenecid and sulfinpyrazone are not selective for uric acid transporters in the kidney, and they tend to be associated with either significant drug–drug interactions or other adverse events associated with their lack of selectivity, which significantly limits their use overall.[36] Benzbromarone exhibits increased selectivity for uric acid transporters compared with the previous agents. However, benzbromarone was withdrawn from most markets after reports of serious hepatotoxicity which was unrelated to its mechanism of action.[37] It is still used, however, in some markets such as Japan and in parts of Europe. These drugs were discovered more than 50 years ago, and for many years, uric acid transport inhibitors were not actively targeted for drug discovery.

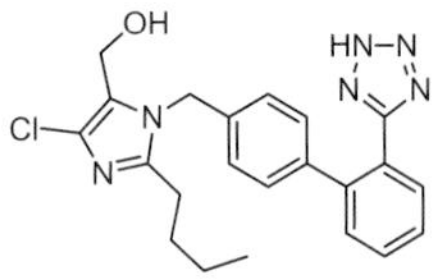

17 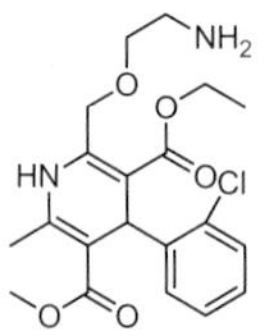 **18** 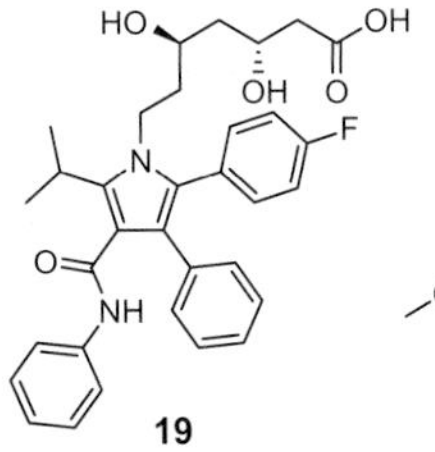 **19** 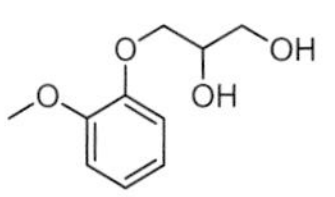 **20**

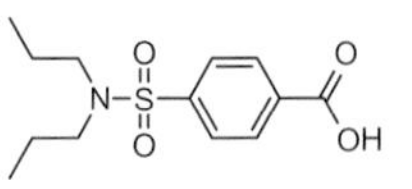

21 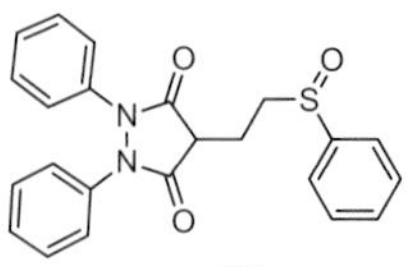 **22** 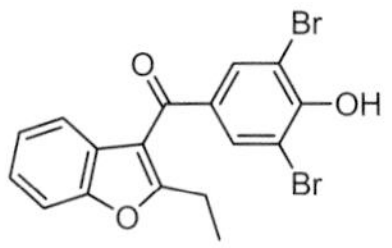 **23**

In 2008, the serum uric acid reduction effect of RDEA594 (lesinurad, **24**), a newly discovered SURI, was reported.[38] Lesinurad was shown to selectively inhibit transporters of uric acid without affecting other generic ion transporters like OAT1 and OAT3. Lesinurad was later determined to have a primary mode of action involving the inhibition of URAT1 in the kidney. Lesinurad is currently undergoing Phase 3 clinical trials for the chronic treatment of gout, in combination with XO inhibitor (either allopurinol or febuxostat). The lesinurad doses studied are 200 and 400 mg once daily. Results of these trials are expected in 2014.

Work has also been reported on the development of RDEA3170 (undisclosed structure), a second-generation SURI that is currently undergoing Phase 2 trials at doses ranging from 5 to 12.5 mg once daily.[39]

Levotofisopam (**25**) is another drug in development that increases the excretion of uric acid. Its precise mechanism of action has not been elucidated, but this drug increases uric acid excretion at the doses studied (50 mg, three times daily) in Phase 1/2. Other drugs, such as URC-102[40] and KUX-1151, are undergoing early Phase 1 clinical development. Interestingly, KUX-1151 is described as a dual inhibitor, acting on XO to reduce production of uric acid and acting as an uricuretic drug.[32,41]

Arhalofenate (**26**), another dual mechanism-of-action drug in development,[42] is reported as having an anti-inflammatory effect in addition to its uricuretic effect. This drug is a PPAR-gamma partial agonist that was first developed for type 2 diabetes, but upon observing a strong uricuretic effect, the drug was investigated for gout, and as of 2013, it was undergoing Phase 2 studies in this therapeutic area.

Another clinical study result showing reduced serum uric acid and urinary uric acid levels was recently reported for a sodium channel inhibitor (**27**), which was found to have micromolar inhibition of URAT1.

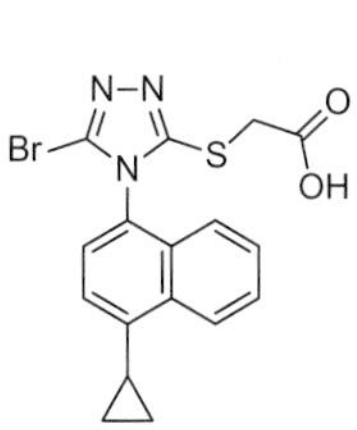

24 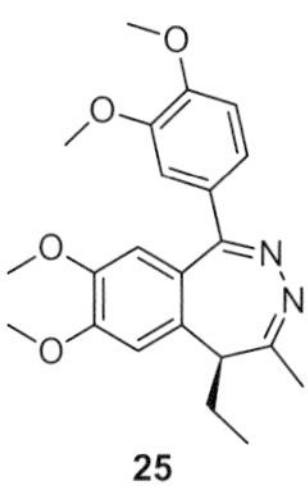 **25** 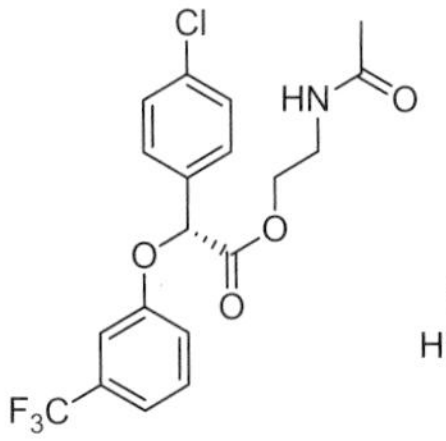 **26** 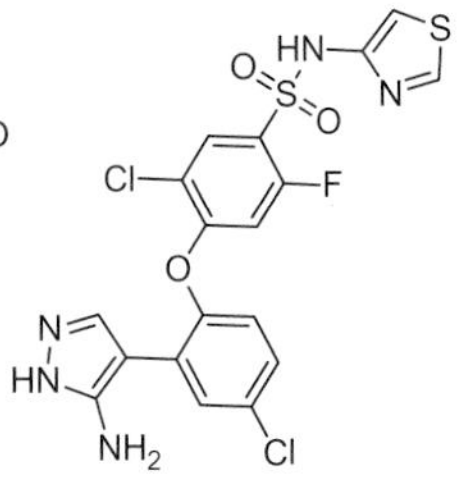 **27**

2.2.3 Drugs Catalyzing Uric Acid Metabolism

As presented in Section 1, humans are the only mammals entirely lacking uricase, an enzyme capable of metabolizing uric acid into the much more soluble allantoin. Uricases are extremely effective at reducing uric acid and can depress levels down to below 1 mg/dL in treated patients. This results in rapid dissolution of tophi and significant improvement in function after long-term dosing. There is a large, poorly understood increase in the rate of flare with rapid lowering of serum uric acid; however, after chronic dosing, the flare rate decreases.[43] Uricases are also used for tumor lysis syndrome, also known as chemotherapy-induced secondary hyperuricemia. Rapid degradation of liquid tumors occasionally results in an extremely high uric acid level with subsequent renal damage. Several companies investigated the use of recombinant porcine-like uricases to catalyze the oxidation of uric acid to 5-hydroxyisourate in patients. The first of these uricases approved for the reduction of serum uric acid was rasburicase, which was launched in Europe in 2001 and in the United States in 2002. It is an injectable given intravenously, and its indication is currently limited to cancer patients expected to suffer from tumor lysis and subsequent serum uric acid elevation. Rasburicase can rapidly return uric acid levels to a normal range in these patients, preventing renal damage. It is given as a daily infusion for up to 5 days.

A second recombinant uricase named pegloticase was approved in 2010 in the United States and 2013 in Europe. This drug is also administered intravenously, but because it is pegylated, it can be administered every two weeks. It is approved for use in chronic gout patients refractory to conventional therapy. Infusion-site reactions are the most common adverse events reported for this drug, and unfortunately, over 40% of patients develop neutralizing antibodies against the foreign enzyme.[44]

3. CONCLUSIONS

The long hiatus in gout therapy drug development is coming to a close. Increased understanding of the burden of gout is driving the need for new therapies for this disease. In addition, the incidence of gout has increased dramatically over recent years as a result of an aging population and increases in gout risk factors such as hypertension and obesity.[45]

As a result, within the last 10 years, there is clearly increased research activity in hyperuricemia and gout. The publication rate for articles describing work in this area has risen extremely quickly over the last decade. The activity noted in the research literature may suggest that additional therapeutic modalities are going to be in development in the future. The results of the ongoing clinical research into the genetics of gout may provide additional targets for both the treatment of chronic hyperuricemia and the treatment of acute gout flare. There is still a significant unmet medical need for new agents not only to be capable of safely reducing uric acid levels in gout patients, but, more importantly, to demonstrate actual clinical benefit by reducing tophi formation and the incidence of flare.

Furthermore, the importance of uric acid in gout, as well as its potential in directly contributing to the development of a variety of cardiovascular diseases, is increasingly being recognized. This is an area of ongoing work, which will determine if uric acid is a causal agent of cardiovascular disease or merely a marker.

REFERENCES

1. McQueen, F. M.; Doyle, A.; Reeves, Q.; Gao, A.; Tsai, A.; Gamble, G. D.; Curteis, B.; Williams, M.; Dalbeth, N. *Rheumatology (Oxford)* **2014**, *53*(1), 95.
2. Zhu, Y.; Pandya, B. J.; Choi, H. K. *Arthritis Rheum.* **2011**, *63*(10), 3136.
3. Roddy, E.; Mallen, C. D.; Doherty, M. *BMJ* **2013**, *347*, f5648.
4. Grassi, D.; Desideri, G.; Di Giacomantonio, A. V.; Di Giosia, P.; Ferri, C. *High Blood Press. Cardiovasc. Prev.* **2014** Feb 20, http://dx.doi.org/10.1007/s40292-014-0046-3. [Epub ahead of print].
5. Rock, K. L.; Kataoka, H.; Lai, J.-J. *Nat. Rev. Rheumatol.* **2012**, 1.
6. Boss, G. R.; Seegmiller, J. E. N. *Eng. J. Med.* **1979**, *300*(26), 1459.
7. Bobulescu, I. A.; Moe, O. W. *Adv. Chronic Kidney Dis.* **2012**, *19*(6), 358.
8. Crittenden, D. B.; Pillinger, M. H. *Bull. Hosp. Jt. Dis.* **2013**, *71*(3), 189.
9. Shrestha, M.; Morgan, D. L.; Moreden, J. M.; Singh, R.; Nelson, M.; Hayes, J. *Ann. Emerg. Med.* **1995**, *26*(6), 682.
10. Graham, W.; Roberts, J. B. *Ann. Rheum. Dis.* **1953**, *12*(1), 16.
11. Terkeltaub, R. A.; Furst, D. E.; Bennett, K.; Kook, K. A.; Crockett, R. S.; Davis, M. W. *Arthritis Rheum.* **2010**, *62*(4), 1060.
12. Cronstein, B. N.; Sunkureddi, P. J. *Clin. Rheumatol.* **2013**, *19*(1), 19.
13. Ramamoorthy, S.; Cidlowski, J. A. *Endocr. Dev.* **2013**, *24*, 41.
14. (a) Schlesinger, N.; De Meulemeester, M.; Pikhlak, A.; Yucel, A. E.; Richard, D.; Murphy, V.; Arulmani, U.; Sallstig, P.; So, A. *Arthritis Res. Ther.* **2011**, *13*(2), R53; (b) Schlesinger, N.; Mysler, E.; Lin, H. Y.; De Meulemeester, M.; Rovensky, J.; Arulmani, U.; Balfour, A.; Krammer, G.; Sallstig, P.; So, A. *Ann. Rheum. Dis.* **2011**. *70*(7), 1264; (c) Terkeltaub, R. A.; Schumacher, H. R.; Carter, J. D.; Baraf, H. S.; Evans, R. R.; Wang, J.; King-Davis, S.; Weinstein, S. P. *Arthritis Res. Ther.* **2013**, *15*(1), R25; (d) So, A.; De Smedt, T.; Revaz, S.; Tschopp, J. Arthritis Res. Ther. 2007, *9*(2), R28.
15. http://www.clinicaltrials.gov/ct2/show/NCT01712204.
16. Kavanaugh, A.; Mease, P. J.; Gomez-Reino, J. J.; Adebajo, A. O.; Wollenhaupt, J.; Gladman, D. D.; Lespessailles, E.; Hall, S.; Hochfeld, M.; Hu, C.; Hough, D.; Stevens, R. M.; Schett, G. *Ann. Rheum. Dis.* **2014**, *73*(6), 1020–1026.
17. Samstad, E. O.; Niyonzima, N.; Nymo, S.; Aune, M. H.; Ryan, L.; Bakke, S. S.; Lappegard, K. T.; Brekke, O. L.; Lambris, J. D.; Damas, J. K.; Latz, E.; Mollnes, T. E.; Espevik, T. *J. Immunol.* **2014**, *192*(6), 2837.

18. Terkeltaub, R.; Baird, S.; Sears, P.; Santiago, R.; Boisvert, W. *Arthritis Rheum.* **1998**, *41*(5), 900.
19. Amaral, F. A.; Costa, V. V.; Tavares, L. D.; Sachs, D.; Coelho, F. M.; Fagundes, C. T.; Soriani, F. M.; Silveira, T. N.; Cunha, L. D.; Zamboni, D. S.; Quesniaux, V.; Peres, R. S.; Cunha, T. M.; Cunha, F. Q.; Ryffel, B.; Souza, D. G.; Teixeira, M. M. *Arthritis Rheum.* **2012**, *64*(2), 474.
20. Khanna, D.; Fitzgerald, J. D.; Khanna, P. P.; Bae, S.; Singh, M. K.; Neogi, T.; Pillinger, M. H.; Merill, J.; Lee, S. *Arthritis Care Res.* **2012**, *64*(10), 1431.
21. Perez-Ruiz, F.; Calabozo, M.; Pijoan, J. I.; Herrero-Beites, A. M.; Ruibal, A. *Arthritis Rheum.* **2002**, *47*(4), 356.
22. George, R. L., Jr.; Sundy, J. S. *Drugs Today (Barc.)* **2012**, *48*(7), 441.
23. Robins, R. K. *J. Am. Chem. Soc.* **1956**, *78*, 784.
24. Chung, Y.; Lu, C. Y.; Graham, G. G.; Mant, A.; Day, R. O. Intern. Med. J. *38*(6), 388.
25. (a) Becker, M. A.; Schumacher, H. R.; MacDonald, P. A.; Lloyd, E.; Lademacher, C. *J. Rheumatol.* **2009**, *36*(6), 1273; (b) Becker, M. A.; Schumacher, H. R., Jr.; Wortmann, R. L.; MacDonald, P. A.; Eustace, D.; Palo, W. A.; Streit, J.; Joseph-Ridge, N. N. *Engl. J. Med.* **2005**, *353*(23), 2450; (c) Wei, L.; Mackenzie, I. S.; Chen, Y.; Struthers, A. D.; MacDonald, T. M. *Br. J. Clin. Pharmacol.* **2011**, *71*(4), 600; (d) Becker, M.; Fitz-Patrick, D.; Storgard, C.; Cravets, M.; Baumgartner, S. *ACR/ARHP abstract 1187*. 2013.
26. Chohan, S. *J. Rheumatol.* **2011**, *38*(9), 1957.
27. Abeles, A. M. *J. Rheumatol.* **2012**, *39*(3), 659.
28. http://www.clinicaltrials.gov/ct2/show/NCT01361646.
29. Kumar, R.; Darpan; Sharma, S.; Singh, R. *Expert Opin. Ther. Pat.* **2011**, *21*(7), 1071.
30. Gras, J. *Drugs Fut.* **2014**, *39*(2), 123.
31. (a) Zöllner, N. *Proc. Nutr. Soc.* **1982**, *41*(3), 329; (b) Ho, C. Y.; Miller, K. V.; Savaiano, D. A.; Crane, R. T.; Ericson, K. A.; Clifford, A. J. *J. Nutr.* **1979**, *109*(8), 1377.
32. Hiratochi, M.; Tatani, K.; Shimizu, K.; Kuramochi, Y.; Kikuchi, N.; Kamada, N.; Itoh, F.; Isaji, M. *Eur. J. Pharmacol.* **2012**, *690*, 183.
33. Bach, M. H.; Simkin, P. A. *Curr. Opin. Rheumatol.* **2014**, *26*(2), 169.
34. Mason, R. M. *Ann. Rheum. Dis.* **1954**, *13*(2), 120.
35. Sinclair, D. S.; Fox, I. H. *J. Rheumatol.* **1975**, *2*(4), 437.
36. Pittman, J. R.; Bross, M. H. *Am. Fam. Physician* **1999**, *59*(7), 1799.
37. Kaufmann, P.; Torok, M.; Hanni, A.; Roberts, P.; Gasser, R.; Krahenbuhl, S. *Hepatology* **2005**, *41*(4), 925.
38. Fleischmann, R.; Kerr, B.; Yeh, L. T.; Suster, M.; Shen, Z.; Polvent, E.; Hingorani, V.; Quart, B.; Manhard, K.; Miner, J. N.; Baumgartner, S. *Rheumatology (Oxford)* **2014**, http://dx.doi.org/10.1093/rheumatology/ket487.
39. http://www.clinicaltrials.gov/ct2/show/NCT02078219.
40. http://www.clinicaltrials.gov/ct2/show/NCT01953497.
41. http://www.kissei.co.jp/e_contents/rd/pipeline/index.html.
42. Edwards, N. L.; So, A. *Rheum. Dis. Clin. North Am.* **2014**, *40*(2), 375.
43. Perez-Ruiz, F. *Rheumatology (Oxford)* **2009**, *48*(Suppl. 2), ii9–ii14.
44. Lipsky, P. E.; Calabrese, L. H.; Kavanaugh, A.; Sundy, J. S.; Wright, D.; Wolfson, M.; Becker, M. A. *Arthritis Res. Ther.* **2014**, *16*(2), R60.
45. Saag, K. G.; Choi, H. *Arthritis Res. Ther.* **2006**, *8*(Suppl. 1), S2.

SECTION 4

Oncology

Section Editor: Shelli R. McAlpine
School of Chemistry, University of New South Wales,
Sydney, Australia

CHAPTER TWELVE

p53–MDM2 and MDMX Antagonists

Constantinos Neochoritis*, Natalia Estrada-Ortiz*, Kareem Khoury†, Alexander Dömling*,†

*Department for Drug Design, University of Groningen, Groningen, The Netherlands
†Carmolex BV, Groningen, The Netherlands

Contents

1. INTRODUCTION

p53, described for the first time in 1979, is considered to be the cellular gatekeeper or guardian for cell division and growth[1,2] and was the first

Annual Reports in Medicinal Chemistry, Volume 49
ISSN 0065-7743
http://dx.doi.org/10.1016/B978-0-12-800167-7.00012-2

tumor-suppressor gene to be identified. p53 is mutated in over 50% of human cancers. These mutations are related to the DNA-binding domain and preclude p53 from acting as a transcription factor.[3] In other cases, the activity of p53 is inhibited either by binding to viral proteins or by alterations in other genes that code for proteins interacting with or linked to the function of p53 such as mouse double minute MDM 2 and MDMX. The level of p53 in cells is subject to tight control, and its main nonredundant regulators are MDM2 and MDMX (MDM2 and MDM4, HDM2, and HDMX in humans), which cooperate with each other to regulate p53 in different flanks.[4,5] Therefore, blocking the interaction between wild-type p53 and its negative regulators MDM2 and MDMX has become an important approach in oncology for restoring p53s antitumor activity. Interestingly, based on the multiple disclosed compound classes and cocrystal structures, p53–MDM2 is perhaps the most studied and top targeted protein–protein interaction.[6,7] Furthermore, several compounds have proceeded into phase I clinical trials (*vide infra*).

1.1. Importance of p53/MDM2/MDMX in Tumor Suppression (p53 Pathway Regulation)

The *MDM2* gene was found to be upregulated in tumors by gene amplification, increased transcription levels, and enhanced translation by approximately 7%, with the highest frequency observed in soft tissue sarcomas (20–30%), osteosarcomas (16%), and oesophageal carcinomas (13%).[8] Simultaneous mutation of p53 and amplification of *MDM2* do not generally occur within the same tumor, suggesting that *MDM2* amplification is an effective means for the inactivation of p53-promoted tumorigenesis.[8] MDMX is as critical as MDM2 in repressing p53s function, as it can both downregulate p53 and act as an upregulator for MDM2.[9] MDMX does not have E3 ligase activity, but its binding with MDM2 increases the rates of ubiquitination made by MDM2; this binding is crucial for the suppression of p53 activity during embryonic development.[10,11] Recent reports highlight the differential activities of MDM2 and MDMX,[12] sparking interest in selective MDMX inhibitors or dual action MDM2/MDMX inhibitors.[13,14]

1.2. The p53/MDM2/MDMX Interaction (Crystal Structures)

MDM2 has a deep hydrophobic pocket in which the p53 protein binds as an alpha helix. This deep MDM2 cleft holds p53's "hotspot triad" made up of p53's Trp23, Leu26, and Phe19 (Fig. 12.15A). In fact, most of the binding energy of the p53/MDM2 interaction resides in these three amino acids (PDB: 1YCR).[15,16] The cocrystal structures of MDMX-p53 show that the binding site is very similar to that of the MDM2–p53 binding site; however, the Met53 and Tyr99 residues of MDMX bulge into the hydrophobic pocket making it smaller and slightly different in shape (PDB: 3DAB and 3DAC).[15]All well-characterized antagonists compete with p53 in the MDM2 or MDMX cleft.

2. MDM2 ANTAGONISTS

2.1. Nutlin-Type Compounds

In 2004, the first class of potent and specific MDM2 small-molecule inhibitors was published.[17] These inhibitors, which were the first described small-molecule inhibitors of the p53–MDM2 interaction constituting a real breakthrough, were based on 1,2,4,5-tetrasubstituted 4,5-*cis*-imidazolines, also known as Nutlins (Fig. 12.1). Extensive chemical modifications of the initial lead compound yielded Nutlin-3 (**2**), which is an active enantiomer. Nutlin-3 blocks the MDM2–p53 protein–protein interaction with an IC_{50} value of 90 nM. A cocrystal structure of Nutlin-2 (**1**) complexed with MDM2 (PDB: 1RV1; Fig. 12.1C) shows that the two bromo-substituted phenyl rings of Nutlin-2 occupy the Trp23 and Leu26 pockets, while the ethyl substituent on the third phenyl group of Nutlin-2 is lodged into the Phe19 pocket. Overlay of Nutlin-2 on the p53 peptide bound to MDM2 shows the two phenyl rings, and the ethoxyl group of Nutlin-2 nicely superimpose on the three key hydrophobic binding residues of the p53 peptide. Nutlin-3 was active in several cell lines and showed antitumor activity in a mouse xenograft model.[17]

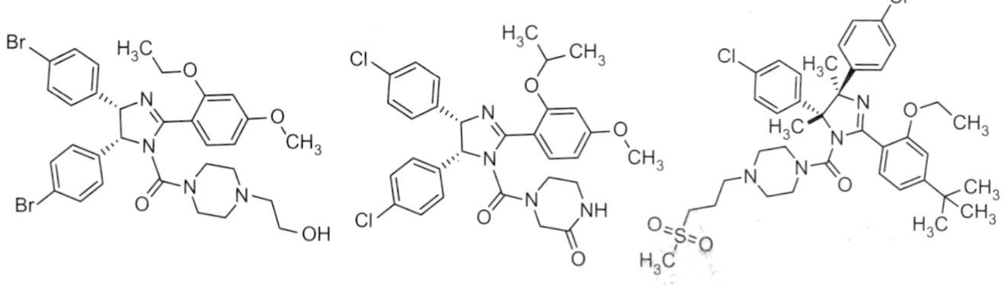

1 (Nutlin-2) **2 (Nutlin-3a)** **3 (RG7112, RO5045337)**

Figure 12.1 The potent Nutlin compounds entered to clinical trials.

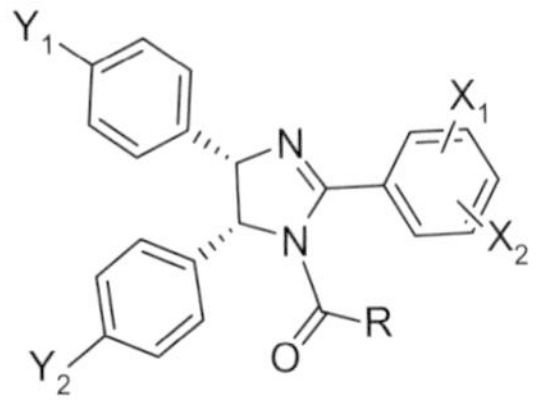

In 2008, advanced Nutlin compound **3** (RG7112, RO5045337, PDB: 4IPF)[18] (Fig. 12.1) entered clinical trials for different types of cancers. In a liposarcoma clinical study, only modest clinical benefit was seen and adverse reactions were greater than expected from preclinical data.[19]

There are four key features of the modification of Nutlin-3a:

- Dimethyl groups were introduced into the imidazole ring of the Nutlins to prevent oxidation.
- Isopropyl ether was replaced with ethyl ether to reduce molecular weight while retaining the same efficiency for hydrophobic interaction.
- Methylphenyl ether, which was found to be the metabolic soft spot, was replaced with a *tert*-butyl group to decrease the metabolic liability.
- Methyl sulfonyl, as a polar group, was added to the urea part to improve MDM2 binding as well as pharmacokinetic (PK) properties.

With these modifications, RG7112 showed enhanced binding affinity to MDM2 with $K_d = 10.7$ nM, three times more potent than Nutlin-3a in inhibition of cell growth. Another MDM2 inhibitor, RO5503781 (RG7388, compound 31), has recently completed phase I clinical trials for patients with advanced malignancies, except leukemia and is currently undergoing phase I clinical trials in patients with acute myelogenous leukemia (ClinicalTrials.gov Identifier: NCT01773408 and NCT01462175).[5,20,21]

2.2. Imidazoles

The crystallization of imidazole compound **4** (WK23, PDB: 3LBK; Fig. 12.15D) with MDM2 and its analogue, **5** (WK298, PDB: 3LBJ[5]) with MDMX, allowed studying of the differences in the binding site of MDM2 and MDMX. Although these compounds have some detectable affinity for MDMX, their affinity is greater for MDM2 due to the surrounding amino acids at the Leu26 pocket making the MDMX binding region more open to solvent and thereby decreasing the affinity of the imidazoles.[5,22] Compound **5** shows a K_i value of 119 nM with MDM2 and 11 μM with MDMX, respectively. A different class of compounds was synthesized placing a planar aromatic ring in a van der Waals contact with the side chain of Val93, as represented by compound **6** (PDB: 4DIJ) with $IC_{50} = 3.8$ μM (Fig. 12.2).

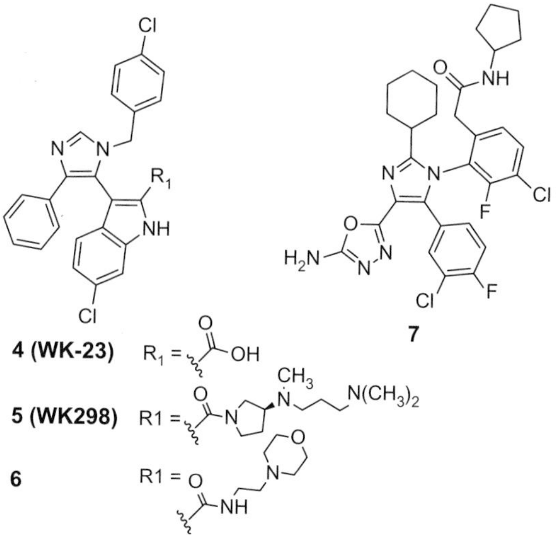

Figure 12.2 Imidazole scaffolds cocrystallized in the binding site of MDM2.

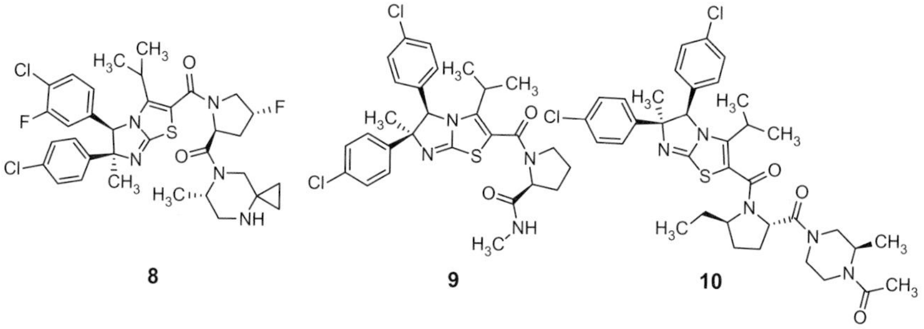

Figure 12.3 Imidazothiazole derivatives inhibiting the p53–MDM2 interaction.

Moreover, compound **7** has an IC_{50} value of 2 nM to MDM2 and also binds to MDMX with low micromolar affinity.[18] Further optimization gave compounds with IC_{50} values between 0.9 and for MDM2.[23]

2.3. Imidazothiazoles

MDM2 inhibitors, starting from the core structure of the Nutlins, have also been designed. Compound **8** with the structural core of a bicyclic dihydroimidazothiazole (Fig. 12.3) has an IC_{50} of 1.2 nM.[24] According to a recent review,[25] the latter patent is believed to contain Daiichi's phase I candidate DS-3032b. Potent p53–MDM2 interaction inhibitors possessing the dihydroimidazothiazole scaffold, aiming to mimic the mode of interaction between MDM2 and the Nutlins such as compounds **9** (PDB: 3VZV; Fig. 12.15E)[26] and **10** (PDB: 3W69)[27] with an IC_{50} of 1.1 and 0.026 μM, respectively, were also reported. Although in general not more potent than the Nutlins they add substantially more molecular weight due to the condensed heavy atom heteroring and additional substituents.

2.4. Benzodiazepines

In 2005, a class of benzodiazepine (BZD) compounds were reported as MDM2 inhibitors.[28,29] Optimization of the substituents on the three phenyl rings resulted in compound **11** with $IC_{50}=0.22$ μM and $K_d=80$ nM. The crystal structure of the complex formed by MDM2 and BZD **11** (*S*,*S*-isomer) was obtained (PDB: 1T4E; Fig. 12.15F) at a resolution of 2.7 Å.[30] Interactions between BZD **11** and MDM2 closely mimic the three key p53-binding residues that interact with MDM2. The iodobenzene ring occupies the Phe19 pocket, the *p*-chlorophenyl occupies the Trp23 pocket, and the other *p*-chlorophenyl, at the α-position relative to the carboxylic acid group, occupies the Leu26 pocket. Further optimization resulted in BZD **12** (reducing epimerization issues compared with the stereogenic center at

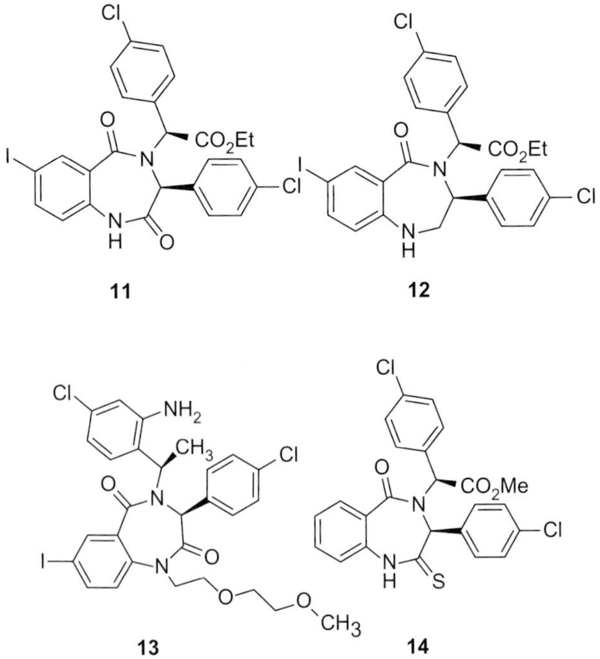

Figure 12.4 1,4-Benzodiazepines as inhibitors.

the α-carbonyl position of **11**) and **13** (introducing a hydrogen bond to Val93) with IC_{50} values of 0.49 μM and 394 nM, respectively (Fig. 12.4). Despite significant chemistry efforts, MDM2 inhibitory activity could not be sufficiently improved and antitumor activity in animal models of human cancer was unsatisfactory, precluding advancement into clinical development.[5] Compound **14**, a 2-thiobenzodiazepine derivative, was related as inhibitor by a recent patent with an IC_{50} of 3.18 μg/mL against human osteosarcoma U-2 OS cell line (wild-type p53).[31]

2.5. Spirooxindoles

Based on the insight that an oxindole group can mimic the Trp23 moiety, spirooxindole-containing natural products were identified and docked into the MDM2 pocket. Computational docking using natural products that contained the spirooxindole moiety led to modified spirooxindole small molecules using an additional phenyl moiety to fit into the Phe19 pocket and an isopropyl group to mimic Leu26.[16,32] Compounds **15** (RO-2468), **16** (RO-5353, PDB: 4LWV), and **17** (RO-8994) were the most potent compounds divulged in patents, with IC_{50} values in the low nanomolar range. Optimization of the primary scaffold produced compounds with K_i values in the low μM and nM range (**18–20**), giving promising results in *in vitro* and *in vivo* assays with compounds **18** and **19** showing K_i values of 8.5 μM and 5 nM, respectively.[16,32–34] A more advanced compound (**20**, $K_i = 0.44$ nM) demonstrated improved oral bioavailability in rats and succeeded in two xenograft models imparting complete tumor regression.[35] Its analogue MI-773 (SAR405838) proceeded into phase I clinical development (Fig. 12.5).[5]

A series of compounds (e.g., **21**), discovered via high-throughput screening assays, based on Nutlins and other spirooxindoles were developed with favorable results in homogeneous time-resolved fluorescence assays showing IC_{50} values in the nanomolar range (PDB: 4JVR; Fig. 12.15G).[36] Recently, a spirooxindole–pyrroloimidazole derivative **22** has been described showing an IC_{50} of 6.8 μM (Fig. 12.5).[37]

2.6. Isoindolones

In 2005, the design of a series of MDM2 inhibitors containing an isoindolinone scaffold guided by computational docking studies was reported.[38] Extensive modifications of the isoindolinones led to compound **23**, the most potent compound in this set, with an IC_{50} value of 0.17 μM in an enzyme-linked immunosorbent assay-binding assay. Computational

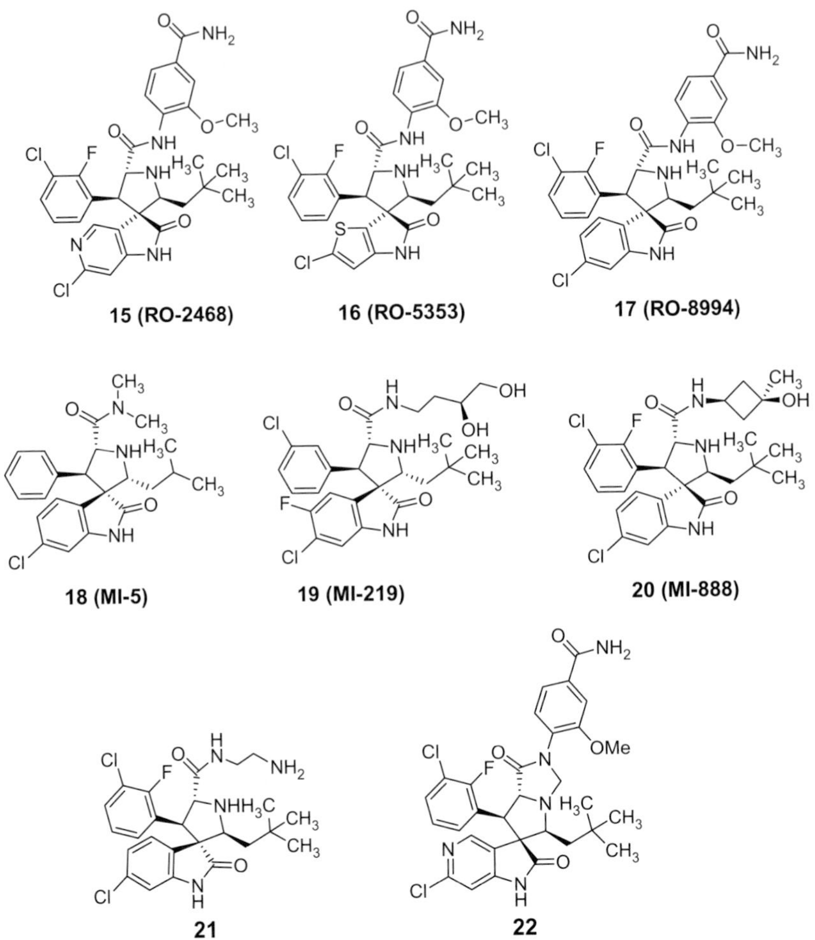

Figure 12.5 Spirooxindole derivatives as inhibitors for the p53–MDM2.

docking suggested that the 4-chlorophenyl group occupies the Leu26 pocket and the isoindolinone scaffold occupies the Trp23 pocket. The other two functional groups of compound **23** are thought to undergo polar interactions with MDM2. In US patent 8058269,[39] an isoindolone anticancer agent **24** is mentioned which shows an IC_{50} of 1.2 nM binding to MDM2 (Fig. 12.6).

2.7. Indole-2-Carboxylic Acid Derivatives

Active compounds were reported based on anchoring of a 6-chloro-indole moiety designed through special computational software[40] and synthesized

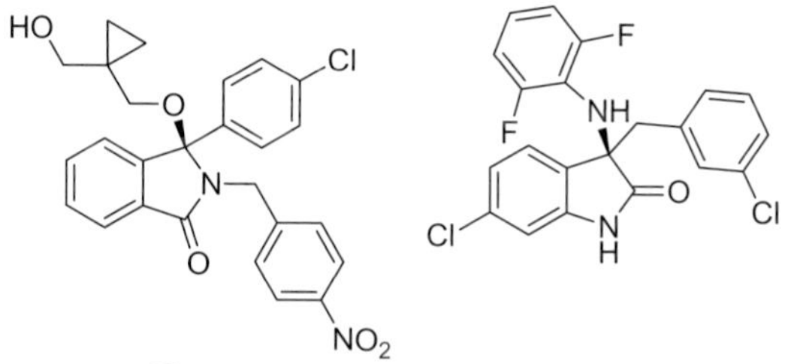

23 24

Figure 12.6 Isoindolones as potent inhibitors.

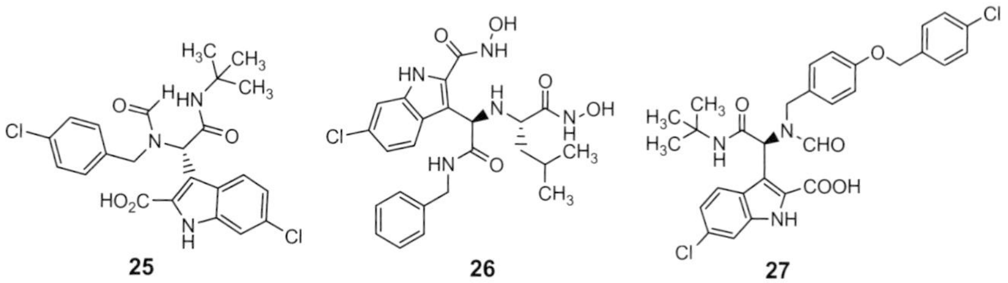

25 26 27

Figure 12.7 Indole-2-carboxylic acid derivatives showing potency against p53–MDM2/X interaction.

through Ugi multicomponent chemistry.[41] The most potent compounds are **25** (PDB: 3TJ2), **26** (PDB: 4MDQ), and **27** (PDB: 4MDN; Fig. 12.15J)[42] with IC_{50} values of 400 nM, 1.2 μM, and 600 nM, respectively. Compounds **25** and **26** mimic three distinct amino acids of p53 (Phe19, Trp23, and Leu26), but compound **27** induced an additional hydrophobic pocket on the MDM2 surface and unveiled for the first time a four-point binding mode (Fig. 12.7).[42–44]

2.8. Pyrrolidinones

Novel derivatives of pyrrolidinones and pyrrolidines were recently described. The pyrrolidinone ring is fused with other heterocyclic systems like pyrroles, pyrazoles, and imidazoles. Compound **28** exhibited IC_{50} values of 9.5 nM and 27.8 μM for MDM2 and MDMX, respectively.[45] The imidazopyrrolidinone derivative **29** has shown inhibition with an IC_{50} of 0.17 nM (MDM2) and 0.468 μM (MDMX),[46] and a pyrazolopyrrolidinone derivative **30** was active with IC_{50} values of 7.8 nM and 4.16 μM for MDM2 and MDMX, respectively (Fig. 12.8).[47]

2.9. Pyrrolidines

Following the identification of RG7112, researchers reported the discovery and characterization of a second-generation clinical MDM2 inhibitor, compound **31** (RG7388, RO5503781PDB: 4JSC; Fig. 12.15H), with superior potency and selectivity. This pyrrolidine derivative effectively activates the p53 pathway, leading to cell cycle arrest and/or apoptosis in cell lines expressing wild-type p53 and tumor growth inhibition or regression of osteosarcoma xenografts in nude mice. RG7388 is undergoing clinical investigation in solid and hematological tumors.[48] Other pyrrolidine derivatives with activity were recently described, substituted pyrrolidine-2-carboxamide

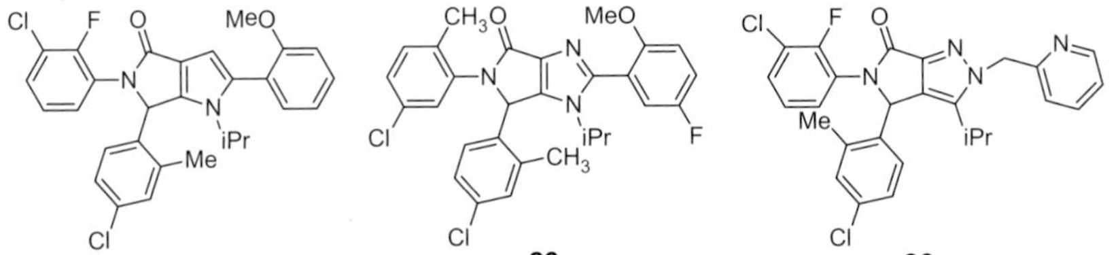

28 **29** **30**

Figure 12.8 Pyrrolidinones as novel scaffolds for p53–MDM2 inhibition.

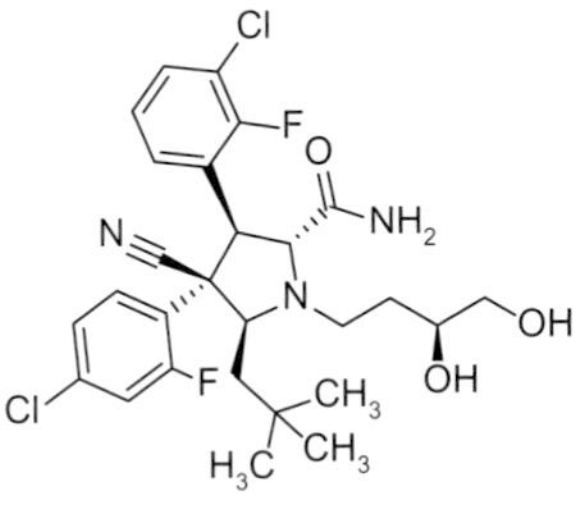

31(RG7388)

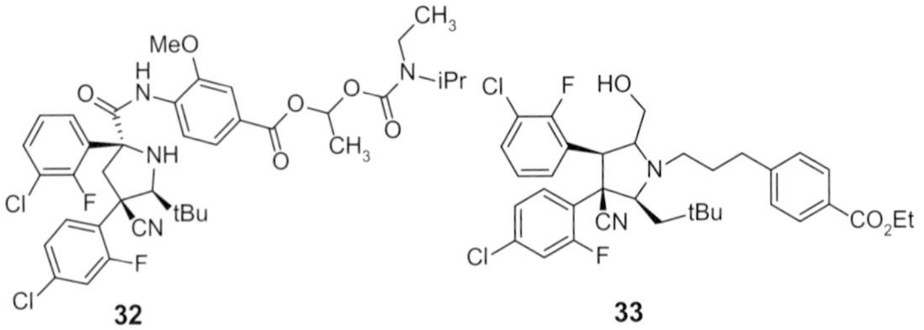

32 **33**

Figure 12.9 Pyrrolidine derivatives screened for the suppression of p53–MDMX binding.

32[49] with an IC_{50} value of 0.0296 μM in MDM2 and cyanohydroxymethylphenyl pyrrolidine **33** (Fig. 12.9).[50]

2.10. Isoquinolines and Piperidinones

Substituted isoquinolinones and piperidinones were reported as inhibitors of MDM2 and MDMX.[51] The most potent compound **34** (Fig. 12.10) has IC_{50} values of 0.8 nM and 2.1 μM for MDM2 and MDMX, respectively.

Recently, scientists have reported the structure-based design of a new class of potent MDM2 inhibitors.[52] Compound **35** (PDB: 2LZG; Fig. 12.15I; IC_{50}=34 nM) has a *p*-chloro-substituted phenyl ring that occupies the Trp23 pocket, an *m*-chloro substituted phenyl ring which sits in the Leu26 pocket, and a cyclopropyl group that occupies the Phe19 pocket. Replacement of the cyclopropyl group in compound **35** by a chiral *tert*-butyl 2-butanoate yielded piperidinone **36**, which is eight times more

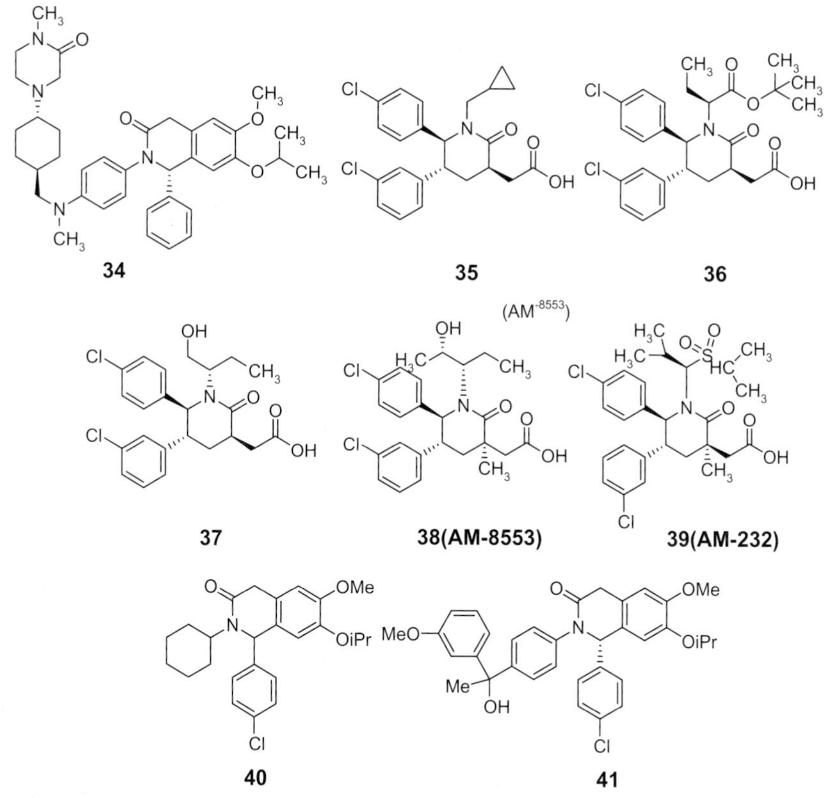

Figure 12.10 Various isoquinolines and piperidinones as inhibitors.

potent than **35** to MDM2 (PDB: 4ERE). Further modification of the *tert*-butyl ester of **36** gave compound **37** (PDB: 4HBM), which after optimization led to compound **38** (AM-8553, PDB: 4ERF), the most potent compound in this set with an IC_{50} of 1.1 nM to MDM2 (Fig. 12.10). In a mouse SJSA-1 tumor xenograft model, oral administration of AM-8553 at 200 mg/kg once daily imparted partial tumor regression. Further development of AM-8553 led to the discovery of compound **39** (AM-232) with $K_i = 0.045$ nM and $IC_{50} = 9.1$ nM in SJSA-1 cell line, and *in vivo* antitumor activity in the SJSA-1 osteosarcoma xenograft model ($ED_{50} = 9.1$ mg/kg). Main difference to AM-8553 is the sulfonamide moiety, which interacts with glycine in a shallow, underutilized cleft of the MDM2 surface. AM-232 is currently being evaluated in human clinical trials.[53] Recent patents related to cyclohexyl isoquinoline compound **40** and hydroxyl isoquinoline **41** are also capable of inhibiting the interaction between p53–MDM2/X with IC_{50} of 0.894 μM (MDM2), 44.25 μM (MDMX), and 0.432 μM (MDM2), respectively.[54,55]

2.11. Peptides

Small synthesized peptides have shown to be about 100-fold more active toward MDM2 than native p53,[56] however have suffered from low cell permeability. When permeabilities were greatly improved, activities were not high enough to justify further investigation (**42** peptide SAH-p53-86, PDB: 3V3B; Fig. 12.15B).[20,57]

Recently, a new stapled peptide ATSP-7041 was synthesized (Fig. 12.11), which activates the p53 pathway in tumors *in vitro* and *in vivo*, with $K_i = 0.9$ and 6.8 nM for MDM2 and MDMX, respectively. These stapled peptides have shown improved bioavailability compared with previously described peptides and induce p53-dependent apoptosis and inhibit cell proliferation in multiple MDM2- and MDMX-overexpressing tumors in cell-based models (PDB: 4N5T cocrystallized with MDMX).[58]

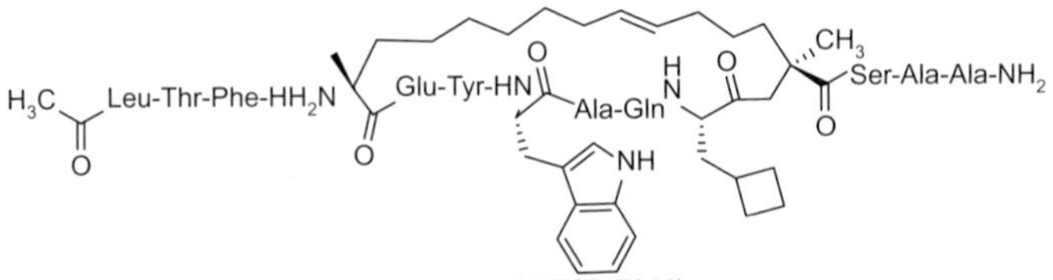

Figure 12.11 The stapled peptide ATSP-7041.

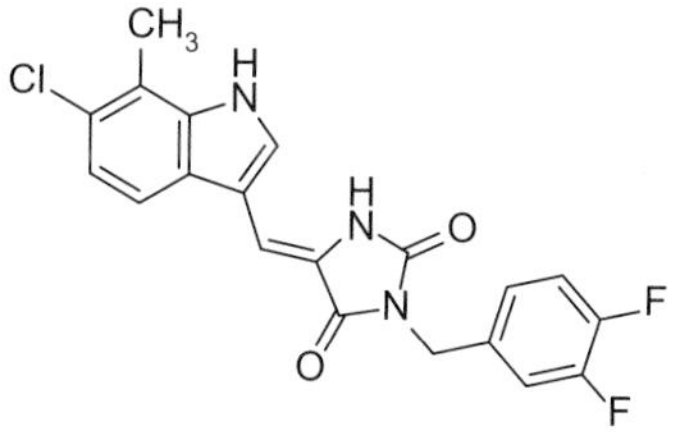

43(RO-2443)

Figure 12.12 A novel indolyl hydantoin scaffold as a potent inhibitor.

2.12. Miscellaneous Compounds

A series of indolyl hydantoin compounds **43** (RO-2443; Fig. 12.12) emerged as potent, dual MDM2/MDMX inhibitors with MDM2 and MDMX IC_{50} values of 33 nM and 41 nM, respectively.[59] RO-2443 showed high potency (ligand efficiency = 0.36) and is thought to bind to at least two of the three subpockets in the MDM2 and MDMX binding sites. Crystallographic and biochemical investigations suggested inhibition of homodimerization and heterodimerization as a mode of action (PDB: 3VBG).

Few novel tetrasubstituted morpholinone derivatives were described. The most potent morpholinone analogue **44** (PDB: 4JV7) was cocrystallized with MDM2 (IC_{50} = 1 μM). The observed binding mode prompted a bidirectional optimization approach: The first strategy relied on the optimization of the contacts of the substituents of the morpholinone with MDM2 while maintaining the observed crystallographic binding mode. As a result, the potent MDM2 inhibitor **45** was identified. The second strategy involved induction of an alternative, more efficient binding mode by systematically modifying the substituents of the morpholinone ring; thus, compound **46** was discovered. Cocrystallization of compound **47** (PDB: 4JWR) provided further evidence for the binding mode of this scaffold (Fig. 12.13).[36] In 2014, a novel MDM2 inhibitor, compound **48** (AM-8735), was described based on a morpholinone core, with IC_{50} = 0.4 nM, cellular potency (SJSA-1 EdU IC_{50} = 25 nM), along with desirable PK properties. The compound also shows antitumor activity in the SJSA-1 osteosarcoma xenograft model with an ED_{50} of 41 mg/kg.[60] Furthermore, novel thiophene derivative **49** was described as an antitumor agent, inhibiting the p53–MDM2 interaction.[61,62]

A series of substituted piperidines were synthesized with the two strongest binding compounds **50** and **51** with IC_{50} values against MDM2 of 0.02 and 0.271 μM, respectively, as determined by the fluorescence polarization (FP) assay.[63]

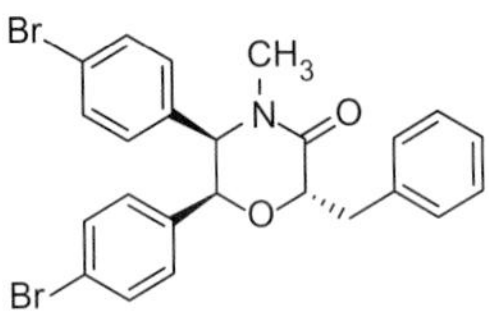

44 (4JV7)

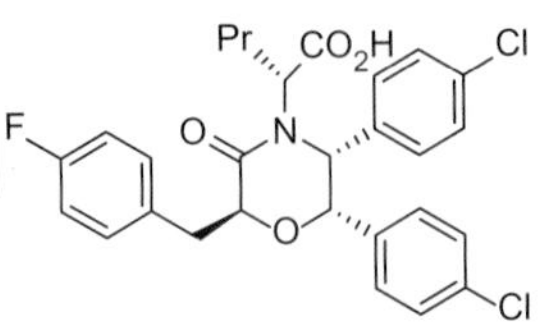

45

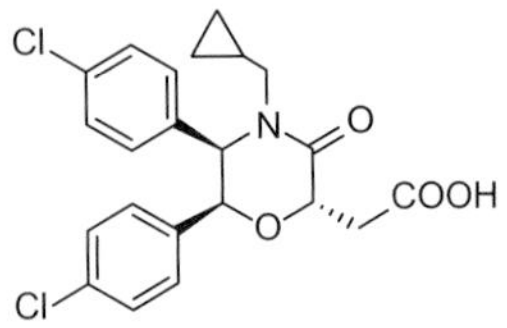

46

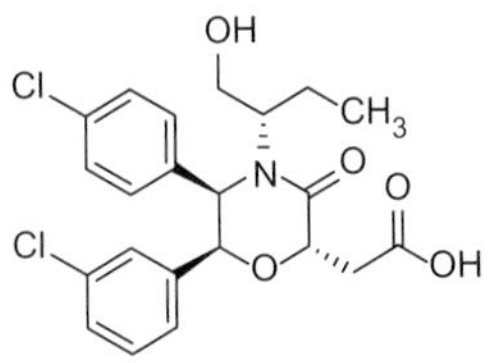

47(4JWR)

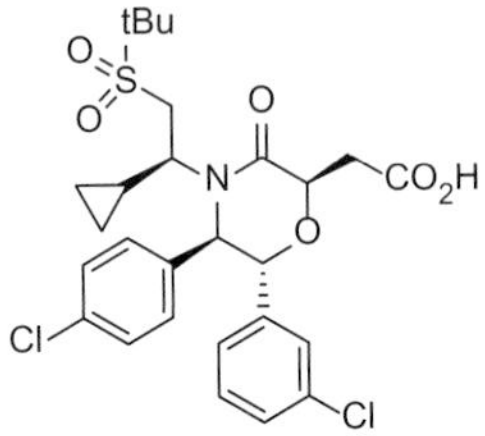

48 (AM-8735)

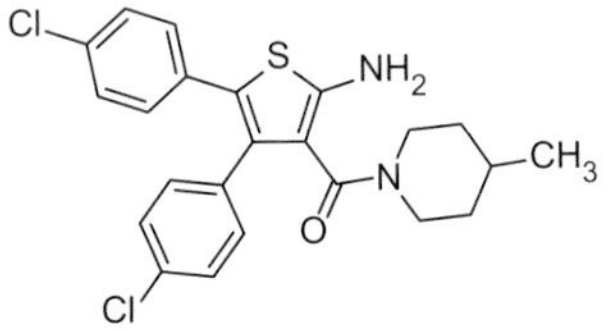

49

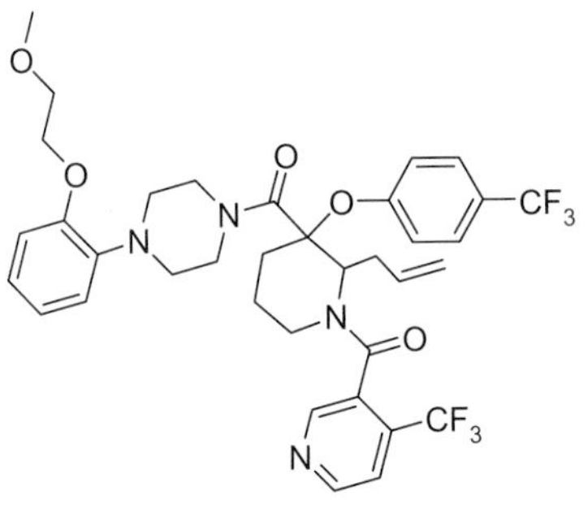

50

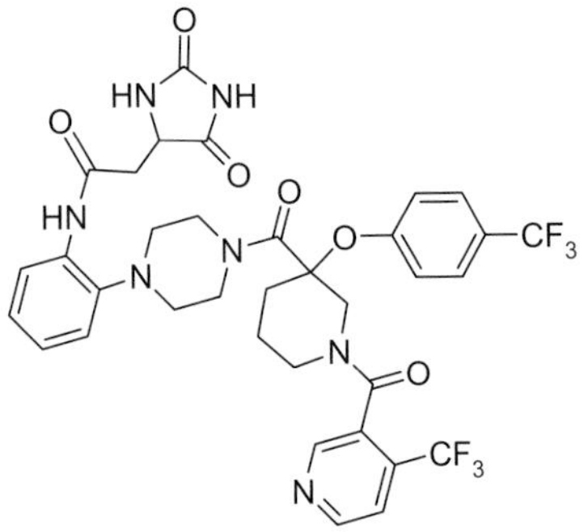

51

Figure 12.13 Various scaffolds including morpholinones and piperidines showing inhibition.

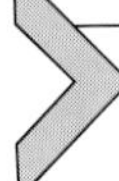

3. MDMX ANTAGONISTS

3.1. Imidazoles

Imidazole compounds **4** and **5** showing dual affinity for both MDM2 and MDMX were developed. Cocrystal analysis of WK298 bound to MDMX showed a similar binding mode to that of p53/MDMX. WK298 fills the three key sub-binding pockets corresponding to p53's Trp23, Phe19, and Leu26. Additionally, the oxygen of the 2-carboxamide forms a hydrogen bond to MDMX's His54 (PDB: 3LBJ). The amino acid differences between MDM2 and MDMX (L45M and H95P) form a distinct sub-binding pocket which can explain the difference in the bindings of small molecules to the two proteins.[5,22]

3.2. Miscellaneous Compounds

Indolyl hydantoin compound **43** showed inhibitory activity against MDM2 and MDMX with IC_{50} of 33 and 41 nM, respectively; however, its poor water solubility precluded testing in cells (PDB: 3U15). Chemical optimization led to compound **52** (RO-5963) showing IC_{50} values of 17 and 24 nM for MDM2 and MDMX, respectively, with improved solubility.[59] A library of 40,000 compounds was screened to identify MDMX inhibitors leading to compound **53** which was found to have an MDMX binding affinity IC_{50} of 0.5 μM in the surface plasmon resonance assay and 1.2 μM in the fluorescence correlation spectroscopy assay. Additionally, it had a cell growth inhibitory GI_{50} of 1.6 μM in the MV4;11 leukemia cell line, 5.5 μM in the p53-deficient HI299 cell line, and 9.8 μM in WI-38 cultured normal human cells (Fig. 12.14).[64]

A peptide *pDI* was identified using phage display that selects for sequences with maximal binding to MDM2 and MDMX.[65] pDI is 300-fold more potent than p53 peptide in binding to MDM2 or MDMX (PDB: 3JZO). Additionally, a single and a quadruple mutant peptide was developed with improved affinities for MDM2 and MDMX (PDB: 3JZP and 3JZQ).[66] The PMI peptide (TSFAEYWNLLSP), a duodecimal peptide inhibitor, has shown to compete with p53 for MDM2 and MDMX binding at an affinity of roughly two orders of magnitude higher then native p53 and roughly fivefold higher then *pDI*[67] (PDB: 3EQY).

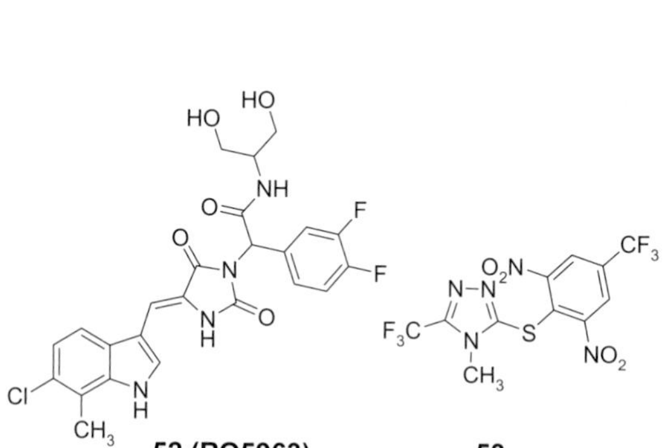

Figure 12.14 Hydantoin and triazole derivatives as potent inhibitors of p53–MDMX.

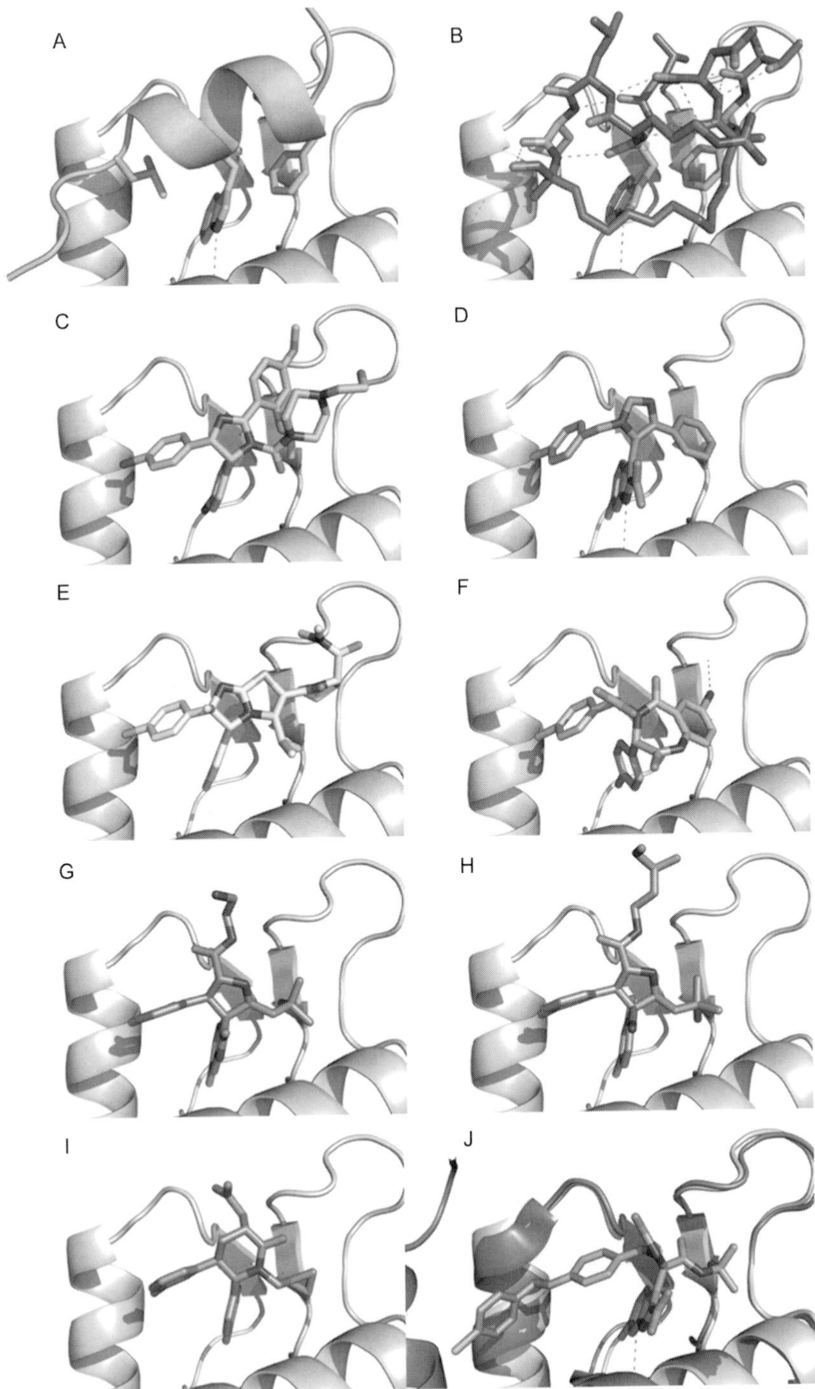

Figure 12.15 See legend on the opposite page.

4. CONCLUSION

More than 50% of human cancer presents mutation in p53 gene. These mutations are associated with the DNA binding domain and make p53 unable to act as a transcription factor. In other cases, the lack of activity of p53 is either due to its binding to viral proteins or due to alterations in other genes that code for proteins interacting with or linked to the function of p53 such as MDM2 and MDMX. For this reason, the p53 pathway is currently a major target in oncology. While there are several p53/MDM2 antagonists in early clinical trials, it was recently recognized that both the related proteins MDM2 and MDM4 must be efficiently inhibited to induce apoptosis in a multitude of cancers. In this chapter, an extensive overview of the novel scaffolds, inhibiting the interaction of p53–MDM2/X, was described. The compounds were classified in order to facilitate the comparison between the various structures.

Figure 12.15 Alignment of different chemotypes cocrystallized with MDM2 superimposed onto PDB: 1YCR. The majority of X-ray structures show only side-chain rotations and small movements of the receptor α-helices. (A) p53 peptide (cyan α-helix with anchoring amino acids $F^{19}W^{23}L^{26}$ rendered as sticks) in the MDM2 binding pocket (gray secondary structure, PDB: 1YCR). W^{23} of p53 forms a hydrogen bond to MDM2's L^{54}. (B) Stapled peptide **42** closely mimics the p53 peptide's anchoring amino acids (pink sticks) and backbone and has an additional hydrocarbon staple (blue stick, PDB: 3V3B) which exhibits multiple contacts to the MDM2 receptor. (C) Nutlin derivative **1** (green sticks) with two hydrophobic *p*-halophenyl substituents occupying the W^{23} and L^{26} pockets and an *o*-ethoxy phenyl moiety targeting the F^{19} pocket (PDB: 1RV1). (D) Tri-substituted indoloimidazole **4** exhibiting an indole substituent closely mimicking p53's W^{23} and two phenyl groups occupying the L^{26} and F^{19} pockets. (PDB: 3LBK). (E) Imidazothiazole **9** exhibiting two hydrophobic *p*-halophenyl substituents occupying the W^{23} and L^{26} pockets and an *iso*-butyric side chain targeting the F^{196} pocket (PDB: 3VZV). (F) Benzodiazepindione **11** with two hydrophobic *p*-halophenyl substituents occupying the W^{23} and L^{26} pockets and a 7-I benzene forming a halogen bond to the backbone carbonyl of MDM2's Q^{72} (PDB: 1T4E). (G) Spirocyclic oxindole **21** featuring an indole mimicking p53's W^{23}, a hydrophobic *p*-halobenzene substituent sitting in the L^{26} pocket, and a bulky alkyl chain introduced F^{19} pocket (PDB: 4JVR). (H) Cyanopyrrolidine **31** featuring two hydrophobic phenyl groups in the W^{23} and L^{26} pockets, a bulky alkyl chain in the F19 pocket, and a solubilizing diol-side chain (PDB: 4JSC). (I) Piperidinone **35** featuring two hydrophobic phenyl groups and a cyclopropyl methyl alkyl chain targeting the $L^{26}W^{23}F^{19}$ pockets, respectively (PDB: 1RV1). (J) Indole-2-carboxylic acid **27** showing an unusual binding mode inducing a never before seen deep binding pocket by ordering the otherwise disordered N-terminus of MDM2 (blue α-helix at the left side). The phenyl group of the extended benzylphenol can be regarded as a fourth pharmacophore element. The indole moiety mimics the W^{23} and the *tert*-butyl group occupies the F^{19} pocket (PDB: 4MDN). (See the color plate.)

REFERENCES

1. Lane, D. P. *Nature* **1992**, *358*, 15.
2. Levine, A. J. *Cell* **1997**, *88*, 323.
3. Hainaut, P.; Hollstein, M. In: *p53 and Human Cancer: The First Ten Thousand Mutations*; Vande Woude, George F.; Klein, George Eds.; Advances in Cancer Research, Vol. 77; Academic Press, 1999; pp 81–137.
4. Vogelstein, B.; Lane, D.; Levine, A. J. *Nature* **2000**, *408*, 307.
5. Zhao, Y.; Bernard, D.; Wang, S. *BioDiscovery* **2013**; *8*, 1.
6. Popowicz, G. M.; Dömling, A.; Holak, T. A. *Angew. Chem. Int. Ed. Engl.* **2011**, *50*, 2680.
7. Khoury, K.; Holak, T. A.; Dömling, A. p53/MDM2 Antagonists: Towards Non-genenotoxic Anticancer Treatments. In: *Protein-Protein Interactions in Drug Discovery*; Methods and Principles in Medicinal Chemistry, Wiley-VCH, Weinheim, Germany; Dömling, A. Ed.; Wiley-VCH Verlag GmbH & Co. KGaA, 2013; pp 129–164.
8. Momand, J.; Zambetti, G. P. *J. Cell. Biochem.* **1997**, *64*, 343.
9. Levine, A. J.; Hu, W.; Feng, Z. *Cell Death Differ.* **2006**, *13*, 1027.
10. Huang, L.; Yan, Z.; Liao, X.; Li, Y.; Yang, J.; Wang, Z.-G.; Zuo, Y.; Kawai, H.; Shadfan, M.; Ganapathy, S.; Yuan, Z.-M. *Proc. Natl. Acad. Sci. U.S.A.* **2011**, 12001–12006.
11. Linares, L. K.; Hengstermann, A.; Ciechanover, A.; Müller, S.; Scheffner, M. *Proc. Natl. Acad. Sci. U.S.A.* **2003**, *100*, 12009.
12. Marine, J.-C. W.; Dyer, M. A.; Jochemsen, A. G. *J. Cell Sci.* **2007**, *120*, 371.
13. Khoury, K.; Popowicz, G. M.; Holak, T. A.; Dömling, A. *MedChemComm* **2011**, *2*, 246.
14. Gembarska, A.; Luciani, F.; Fedele, C.; Russell, E. A.; Dewaele, M.; Villar, S.; Zwolinska, A.; Haupt, S.; de Lange, J.; Yip, D.; Goydos, J.; Haigh, J. J.; Haupt, Y.; Larue, L.; Jochemsen, A.; Shi, H.; Moriceau, G.; Lo, R. S.; Ghanem, G.; Shackleton, M.; Bernal, F.; Marine, J.-C. *Nat. Med.* **2012**, *18*, 1239.
15. Popowicz, G. M.; Czarna, A.; Holak, T. A. *Cell Cycle* **2008**, *7*, 2441.
16. Ding, K.; Lu, Y.; Nikolovska-Coleska, Z.; Qiu, S.; Ding, Y.; Gao, W.; Stuckey, J.; Krajewski, K.; Roller, P. P.; Tomita, Y.; Parrish, D. A.; Deschamps, J. R.; Wang, S. *J. Am. Chem. Soc.* **2005**, *127*, 10130.
17. Vassilev, L. T.; Vu, B. T.; Graves, B.; Carvajal, D.; Podlaski, F.; Filipovic, Z.; Kong, N.; Kammlott, U.; Lukacs, C.; Klein, C.; Fotouhi, N.; Liu, E. A. *Science* **2004**, *303*, 844.
18. Vu, B.; Wovkulich, P.; Pizzolato, G.; Lovey, A.; Ding, Q.; Jiang, N.; Liu, J.-J.; Zhao, C.; Glenn, K.; Wen, Y.; Tovar, C.; Packman, K.; Vassilev, L.; Graves, B. *ACS Med. Chem. Lett.* **2013**, *4*, 466.
19. Ray-Coquard, I.; Blay, J.-Y.; Italiano, A.; Le Cesne, A.; Penel, N.; Zhi, J.; Heil, F.; Rueger, R.; Graves, B.; Ding, M.; Geho, D.; Middleton, S. A.; Vassilev, L. T.; Nichols, G. L.; Bui, B. N. *Lancet Oncol.* **2012**, *13*, 1133.
20. Wang, S.; Zhao, Y.; Bernard, D.; Aguilar, A.; Kumar, S. In: *Protein-Protein Interactions*; Wendt, M. D. Ed.; Topics in Medicinal Chemistry; Springer: Berlin Heidelberg, 2012; pp 57–79.
21. Zak, K.; Pecak, A.; Rys, B.; Wladyka, B.; Dömling, A.; Weber, L.; Holak, T. A.; Dubin, G. *Expert Opin. Ther. Pat.* **2013**, *23*, 425.
22. Popowicz, G. M.; Czarna, A.; Wolf, S.; Wang, K.; Wang, W.; Dömling, A.; Holak, T. A. *Cell Cycle* **2010**, *9*, 1104.
23. Furet, P.; Chène, P.; De Pover, A.; Valat, T. S.; Lisztwan, J. H.; Kallen, J.; Masuya, K. *Bioorg. Med. Chem. Lett.* **2012**, *22*, 3498.
24. Uoto, K.; Kawato, H.; Sugimoto, Y.; Naito, H.; Miyazaki, M.; Taniguchi, T.; Aonuma, M. Patent Application WO 2009/151069 A1, 2009.
25. Khoo, K. H.; Verma, C. S.; Lane, D. P. *Nat. Rev. Drug Discov.* **2014**, *13*, 217.

26. Miyazaki, M.; Naito, H.; Sugimoto, Y.; Kawato, H.; Okayama, T.; Shimizu, H.; Miyazaki, M.; Kitagawa, M.; Seki, T.; Fukutake, S.; Aonuma, M.; Soga, T. *Bioorg. Med. Chem. Lett.* **2013**, *23*, 728.
27. Miyazaki, M.; Naito, H.; Sugimoto, Y.; Yoshida, K.; Kawato, H.; Okayama, T.; Shimizu, H.; Miyazaki, M.; Kitagawa, M.; Seki, T.; Fukutake, S.; Shiose, Y.; Aonuma, M.; Soga, T. *Bioorg. Med. Chem.* **2013**, *21*, 4319.
28. Parks, D. J.; Lafrance, L. V.; Calvo, R. R.; Milkiewicz, K. L.; Gupta, V.; Lattanze, J.; Ramachandren, K.; Carver, T. E.; Petrella, E. C.; Cummings, M. D.; Maguire, D.; Grasberger, B. L.; Lu, T. *Bioorg. Med. Chem. Lett.* **2005**, *15*, 765.
29. Raboisson, P.; Marugán, J. J.; Schubert, C.; Koblish, H. K.; Lu, T.; Zhao, S.; Player, M. R.; Maroney, A. C.; Reed, R. L.; Huebert, N. D.; Lattanze, J.; Parks, D. J.; Cummings, M. D. *Bioorg. Med. Chem. Lett.* **2005**, *15*, 1857.
30. Grasberger, B. L.; Lu, T.; Schubert, C.; Parks, D. J.; Carver, T. E.; Koblish, H. K.; Cummings, M. D.; LaFrance, L. V.; Milkiewicz, K. L.; Calvo, R. R.; Maguire, D.; Lattanze, J.; Franks, C. F.; Zhao, S.; Ramachandren, K.; Bylebyl, G. R.; Zhang, M.; Manthey, C. L.; Petrella, E. C.; Pantoliano, M. W.; Deckman, I. C.; Spurlino, J. C.; Maroney, A. C.; Tomczuk, B. E.; Molloy, C. J.; Bone, R. F. *J. Med. Chem.* **2005**, *48*, 909.
31. Zhang, W.; Miao, Z.; Zhuang, C.; Sheng, C.; Zhu, L.; Zhang, Y.; Yao, J.; Guo, Z. Patent Application CN 102321034 A, 2012.
32. Ding, K.; Lu, Y.; Nikolovska-Coleska, Z.; Wang, G.; Qiu, S.; Shangary, S.; Gao, W.; Qin, D.; Stuckey, J.; Krajewski, K.; Roller, P. P.; Wang, S. *J. Med. Chem.* **2006**, *49*, 3432.
33. Liu, J.-J.; Zhang, J.; Zhang, Z. WO/2011/101297, 2011.
34. Bartkovitz, D. J.; Chu, X.-J.; Ding, Q.; Graves, B. J.; Jiang, N.; Zhang, J.; Zhang, Z. WO/2011/067185, 2011.
35. Zou, P.; Zheng, N.; Yu, Y.; Yu, S.; Sun, W.; McEachem, D.; Yang, Y.; Yu, L. X.; Wang, S.; Sun, D. *J. Pharm. Pharm. Sci.* **2012**, *15*, 265.
36. Gonzalez-Lopez de Turiso, F.; Sun, D.; Rew, Y.; Bartberger, M. D.; Beck, H. P.; Canon, J.; Chen, A.; Chow, D.; Correll, T. L.; Huang, X.; Julian, L. D.; Kayser, F.; Lo, M.-C.; Long, A. M.; McMinn, D.; Oliner, J. D.; Osgood, T.; Powers, J. P.; Saiki, A. Y.; Schneider, S.; Shaffer, P.; Xiao, S.-H.; Yakowec, P.; Yan, X.; Ye, Q.; Yu, D.; Zhao, X.; Zhou, J.; Medina, J. C.; Olson, S. H. *J. Med. Chem.* **2013**, *56*, 4053.
37. Bartkovitz, D. J.; Chu, X. J.; Ding, Q.; Jiang, N.; Liu, J. J.; Zhang, Z. Patent Application US 2013/053410 A1, 2013.
38. Hardcastle, I. R.; Ahmed, S. U.; Atkins, H.; Calvert, A. H.; Curtin, N. J.; Farnie, G.; Golding, B. T.; Griffin, R. J.; Guyenne, S.; Hutton, C.; Källblad, P.; Kemp, S. J.; Kitching, M. S.; Newell, D. R.; Norbedo, S.; Northen, J. S.; Reid, R. J.; Saravanan, K.; Willems, H. M. G.; Lunec, J. *Bioorg. Med. Chem. Lett.* **2005**, *15*, 1515.
39. Chen, L.; Yang, S.; Zhang, J.; Zhang, Z. Oxindole derivatives. US 8058269 B2, 2011.
40. Koes, D.; Khoury, K.; Huang, Y.; Wang, W.; Bista, M.; Popowicz, G. M.; Wolf, S.; Holak, T. A.; Dömling, A.; Camacho, C. J. *PLoS One* **2012**, 7, e32839.
41. Huang, Y.; Wolf, S.; Koes, D.; Popowicz, G. M.; Camacho, C. J.; Holak, T. A.; Dömling, A. *ChemMedChem* **2012**, 7, 49.
42. Bista, M.; Wolf, S.; Khoury, K.; Kowalska, K.; Huang, Y.; Wrona, E.; Arciniega, M.; Popowicz, G. M.; Holak, T. A.; Dömling, A. *Structure* **2013**, *21*, 2143.
43. Dömling, A. Patent Application WO 2012/033525 A3, 2012.
44. Huang, Y.; Wolf, S.; Beck, B.; Köhler, L.-M.; Khoury, K.; Popowicz, G. M.; Goda, S. K.; Subklewe, M.; Twarda, A.; Holak, T. A.; Dömling, A. *ACS Chem. Biol.* **2014**, *9*, 802–8011.
45. Cotesta, S.; Furet, P.; Guagnano, V.; Holzer, P.; Kallen, J.; Mah, R.; Masuya, K.; Schlapbach, A.; Stutz, S.; Vaupel, A. Patent Application US 2013/317024 A1, 2013.

46. Furet, P.; Guagnano, V.; Holzer, P.; Kallen, J.; Liao, L.; Mah, R.; Mao, L.; Masuya, K.; Schlapbach, A.; Stutz, S. Patent Application US 2014/0011798 A1, 2014.
47. Furet, P.; Guagnano, V.; Holzer, P.; Mah, R.; Masuya, K.; Schlapbach, A.; Stutz, S.; Vaupel, A. Patent Application WO 2013/080141 A1, 2013.
48. Ding, Q.; Zhang, Z.; Liu, J.-J.; Jiang, N.; Zhang, J.; Ross, T. M.; Chu, X.-J.; Bartkovitz, D.; Podlaski, F.; Janson, C.; Tovar, C.; Filipovic, Z. M.; Higgins, B.; Glenn, K.; Packman, K.; Vassilev, L. T.; Graves, B. *J. Med. Chem.* **2013**, *56*, 5979.
49. Bartkovitz, D. J.; Chu, X. J.; Vu, B. T.; Zhao, C.; Fishlock, D. Patent Application US 2013/0244958 A1, 2013.
50. Liu, J. J.; Ross, T. M. Patent Application US 2012/0149660 A1, 2012.
51. Berghausen, J.; Buschmann, N.; Furet, P.; Gessier, F.; Hergovich Lisztwan, J.; Holzer, P.; Jacoby, E.; Kallen, J.; Masuya, K.; Pissot Soldermann, C.; Ren, H.; Stutz, S. Patent Application WO 2011/076786 A1, 2011.
52. Rew, Y.; Sun, D.; Gonzalez-Lopez De Turiso, F.; Bartberger, M. D.; Beck, H. P.; Canon, J.; Chen, A.; Chow, D.; Deignan, J.; Fox, B. M.; Gustin, D.; Huang, X.; Jiang, M.; Jiao, X.; Jin, L.; Kayser, F.; Kopecky, D. J.; Li, Y.; Lo, M.-C.; Long, A. M.; Michelsen, K.; Oliner, J. D.; Osgood, T.; Ragains, M.; Saiki, A. Y.; Schneider, S.; Toteva, M.; Yakowec, P.; Yan, X.; Ye, Q.; Yu, D.; Zhao, X.; Zhou, J.; Medina, J. C.; Olson, S. H. *J. Med. Chem.* **2012**, *55*, 4936.
53. Sun, D.; Li, Z.; Rew, Y.; Gribble, M.; Bartberger, M. D.; Beck, H. P.; Canon, J.; Chen, A.; Chen, X.; Chow, D.; Deignan, J.; Duquette, J.; Eksterowicz, J.; Fisher, B.; Fox, B. M.; Fu, J.; Gonzalez, A. Z.; Gonzalez-Lopez De Turiso, F.; Houze, J. B.; Huang, X.; Jiang, M.; Jin, L.; Kayser, F.; Liu, J. J.; Lo, M.-C.; Long, A. M.; Lucas, B.; McGee, L. R.; McIntosh, J.; Mihalic, J.; Oliner, J. D.; Osgood, T.; Peterson, M. L.; Roveto, P.; Saiki, A. Y.; Shaffer, P.; Toteva, M.; Wang, Y.; Wang, Y. C.; Wortman, S.; Yakowec, P.; Yan, X.; Ye, Q.; Yu, D.; Yu, M.; Zhao, X.; Zhou, J.; Zhu, J.; Olson, S. H.; Medina, J. C. *J. Med. Chem.* **2014**, *57*, 1454.
54. Holzer, P.; Masuya, K.; Furet, P.; Kallen, J.; Stutz, S.; Buschmann, N. Patent Application WO 2012/175520 A1, 2012.
55. Holzer, P.; Masuya, K.; Guagnano, V.; Furet, P.; Kallen, J.; Stutz, S. Patent Application WO 2012175487 A1, 2012.
56. García-Echeverría, C.; Chène, P.; Blommers, M. J. J.; Furet, P. *J. Med. Chem.* **2000**, *43*, 3205.
57. Baek, S.; Kutchukian, P. S.; Verdine, G. L.; Huber, R.; Holak, T. A.; Lee, K. W.; Popowicz, G. M. *J. Am. Chem. Soc.* **2012**, *134*, 103.
58. Chang, Y. S.; Graves, B.; Guerlavais, V.; Tovar, C.; Packman, K.; To, K.-H.; Olson, K. A.; Kesavan, K.; Gangurde, P.; Mukherjee, A.; Baker, T.; Darlak, K.; Elkin, C.; Filipovic, Z.; Qureshi, F. Z.; Cai, H.; Berry, P.; Feyfant, E.; Shi, X. E.; Horstick, J.; Annis, D. A.; Manning, A. M.; Fotouhi, N.; Nash, H.; Vassilev, L. T.; Sawyer, T. K. *Proc. Natl. Acad. Sci. U.S.A.* **2013**, 201303002.
59. Graves, B.; Thompson, T.; Xia, M.; Janson, C.; Lukacs, C.; Deo, D.; Lello, P. D.; Fry, D.; Garvie, C.; Huang, K.-S.; Gao, L.; Tovar, C.; Lovey, A.; Wanner, J.; Vassilev, L. T. *Proc. Natl. Acad. Sci. U.S.A.* **2012**, *109*, 11788.
60. Gonzalez, A. Z.; Eksterowicz, J.; Bartberger, M. D.; Beck, H. P.; Canon, J.; Chen, A.; Chow, D.; Duquette, J.; Fox, B. M.; Fu, J.; Huang, X.; Houze, J. B.; Jin, L.; Li, Y.; Li, Z.; Ling, Y.; Lo, M.-C.; Long, A. M.; McGee, L. R.; McIntosh, J.; McMinn, D. L.; Oliner, J. D.; Osgood, T.; Rew, Y.; Saiki, A. Y.; Shaffer, P.; Wortman, S.; Yakowec, P.; Yan, X.; Ye, Q.; Yu, D.; Zhao, X.; Zhou, J.; Olson, S. H.; Medina, J. C.; Sun, D. *J. Med. Chem.* **2014**, *57*, 2963.
61. Czarna, A.; Beck, B.; Srivastava, S.; Popowicz, G. M.; Wolf, S.; Huang, Y.; Bista, M.; Holak, T. A.; Dömling, A. *Angew. Chem. Int. Ed. Engl.* **2010**, *49*, 5352.

62. Wang, W.; Hu, Y.; Hu, C.; Sheng, R.; Liu, T.; He, Q.; Vojtesek, B. Patent Application CN 102503930 A, 2012.
63. Bogen, S. L.; Ma, Y.; Wang, Y.; Lahue, B. R.; Nair, L. G.; Shizuka, M.; Voss, M. E.; Kirova-Snover, M.; Pan, W.; Tian, Y.; Kulkarni, B. A.; Gibeau, C. R.; Liu, Y.; Scapin, G.; Rindgen, D.; Doll, R. J.; Guzi, T. J.; Hicklin, D. J.; Nomeir, A.; Seidel-Dugan, C., Jr.; G. W. S.; Maccoss, M. Patent Application WO 2011/046771 A1, 2011.
64. Tsuganezawa, K.; Nakagawa, Y.; Kato, M.; Taruya, S.; Takahashi, F.; Endoh, M.; Utata, R.; Mori, M.; Ogawa, N.; Honma, T.; Yokoyama, S.; Hashizume, Y.; Aoki, M.; Kasai, T.; Kigawa, T.; Kojima, H.; Okabe, T.; Nagano, T.; Tanaka, A. *J. Biomol. Screen.* **2013**, *18*, 191.
65. Hu, B.; Gilkes, D. M.; Chen, J. *Cancer Res.* **2007**, *67*, 8810.
66. Phan, J.; Li, Z.; Kasprzak, A.; Li, B.; Sebti, S.; Guida, W.; Schönbrunn, E.; Chen, J. *J. Biol. Chem.* **2010**, *285*, 2174.
67. Li, C.; Pazgier, M.; Li, C.; Yuan, W.; Liu, M.; Wei, G.; Lu, W.-Y.; Lu, W. *J. Mol. Biol.* **2010**, *398*, 200.

CHAPTER THIRTEEN

Modulators of Atypical Protein Kinase C as Anticancer Agents

Jonathan R.A. Roffey*, **Gregory R. Ott**†
*Cancer Research Technology, Discovery Laboratories, Wolfson Institute for Biomedical Research, University College London, London, United Kingdom
†Discovery & Product Development, Teva Global R&D, Teva Pharmaceuticals, West Chester, Pennsylvania, USA

Contents

1. INTRODUCTION

1.1. Overview of Protein Kinase C Isoforms

Since the discovery that protein kinase C (PKC) acted as the major signal-transducer of the tumor-promoting phorbol esters, the PKC multigene

Annual Reports in Medicinal Chemistry, Volume 49
ISSN 0065-7743
http://dx.doi.org/10.1016/B978-0-12-800167-7.00013-4

family of serine/threonine kinases have been defined as key regulators in a multitude of signal transduction pathways that impinge on diverse cellular processes.[1,2] The PKC family comprises of 12 distinct gene products sharing a similar C-terminal serine/threonine kinase domain (AGC class) that are divided into four distinct subfamilies related to the same single-yeast gene product.[3] This subdivision is based on sequence homology, function, and cofactors required for catalytic regulation. Thus, the classical (PKCα, PKCβ, PKCγ),[4] the novel (PKCδ, PKCε, PKCη, and PKCθ),[5] the atypical (PKCι, PKCζ),[6,7] and the PKN isoforms (PKN1, PKN2, PKN3)[8,9] make up the PKC family.

2. ATYPICAL PROTEIN KINASE C ISOFORMS

In mammals, there are two highly homologous aPKC isoforms, PKCι (PKCλ in murine)[10] and PKCζ, whereas lower organisms such as *Caenorhabditis elegans* and *Drosophila melanogaster* have only a single aPKC isoform. The mammalian aPKC isoforms exhibit 73% sequence homology at the amino acid level, and in the kinase domain, this increases to 88% sequence homology and 94% sequence similarity; the nucleotide-binding cavity differs in only a single amino acid between these isoforms.

2.1. aPKC Activation Mechanisms

The atypical PKCs are structurally and functionally distinct from the other PKC isoforms in that they are unresponsive to phorbol esters and their catalytic activity is not dependent upon diacylglycerol, Ca^{2+} ions, or phosphatidylserine.[11] A sequence of allosteric inputs serves to process the aPKCs from an inhibited basal state to a fully activated conformation.[2] The first step is driven through what is termed priming phosphorylation on the C-terminal kinase domain by upstream master regulator kinases.[12,13] The aPKC kinase domain is stabilized and fully activated by this priming phosphorylation on the activation loop (T403/410) and C-terminal turn motif (T555/556) in PKCι and PKCζ, respectively. Catalytic activity is maintained in a latent catalytically competent closed conformer through intramolecular binding of a regulatory pseudosubstrate sequence positioned in the N-terminal regulatory domain. This sequence resembles a true substrate but lacks a phosphorylatable residue, serving to maintain the kinase in an autoinhibited state until release. The kinase localization, function, and activity are regulated through a Phox-Bem (PB1) domain in the aPKC N-terminal regulatory domain that mediates protein–protein interactions with other PB1 domain-containing proteins triggering allosteric activation to ensure that substrate phosphorylation occurs at the correct place and time.[14] A well-characterized PB1 domain-containing interaction partner is the adaptor protein partitioning-defective protein 6 (Par-6).[15] Depletion of Par-6 leads to aPKC loss of function, and conversely, overexpression of Par-6 results in a gain of aPKC function.[16,17] Par-6 has been demonstrated to activate aPKC through a steric release mechanism in which PB1 domain heterodimerization leads to release of aPKC autoinhibition due to the close proximity of the aPKC PB1 domain to its pseudosubstrate sequence.[18]

2.2. aPKC Structure

No full-length high-resolution structures of an aPKC isoform have been determined; however, a number of groups have published the structures of PKCι kinase domain in Apo, ATP-bound, and inhibitor-bound structures.[19–21] To date, no structure of PKCζ kinase domain has been reported. The PKCι kinase domain tertiary structure is similar to other AGC kinases with the catalytic nucleotide-binding site situated between a small N-terminal lobe consisting of β-sheets and a larger predominantly α-helical C-terminal lobe (Fig. 13.1A).[22] A notable feature in these structures is the glycine-rich loop (GXGXXA) that is peculiar in that it has alanine at the last position instead of glycine as in the other PKC family members. Threonine 403 in the activation loop is phosphorylated, and the phosphate group makes contact to conserved residues, bringing the loop into a fixed conformation necessary for catalytic activity. The threonine residue at position 555 in the C-terminal turn motif is also phosphorylated and serves to stabilize the kinase domain by fixing the C-terminus at the top of the upper lobe to the kinase core. In the ATP-bound structure, the phenyl side chain of the Phe543 residue protrudes into the ATP-binding pocket to make van der Waals interactions with the adenine moiety of the bound nucleotide (Fig. 13.1B). This residue forms part of the Asn-Phe-Asp or NFD motif that is conserved in the AGC kinase superfamily.[23] In a partially activated full-length PKCβ structure, an equivalent NFD motif Phe residue has been implicated in allosteric activation in which the phenyl side chain is held out of the ATP-binding site by the C1 domain preventing full catalytic activation until diacylglycerol activation.[23] In both the Apo- and ATP-bound PKCι structures, the NFD motif forms an ordered α-helix. Interestingly, in the CRT0066854 liganded structure, the side-chain phenyl ring in the PKCι NFD motif is displaced from the nucleotide-binding pocket by an

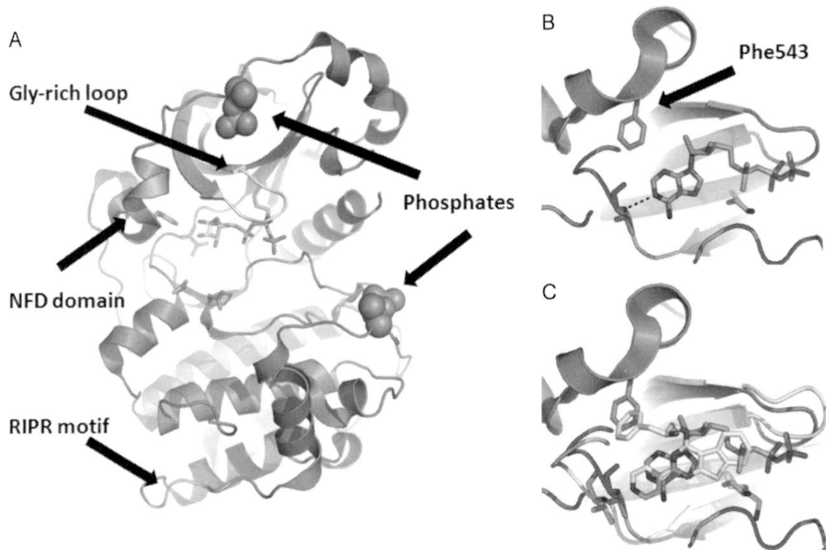

Figure 13.1 Structures of kinase domain (KD)-PKCι. (A) Structure of ATP bound to kinase domain PKCι, highlighting the gly-rich loop (yellow), NFD domain (green), RIPR motif (cyan) and the priming phosphorylation sites (space fill). (B) Structure of ATP bound to the nucleotide-binding cleft, key hydrogen bonding interactions represented by dashed lines along with Phe543 in the NFD motif. (C) Superposition of CRT0066854 (yellow) on the ATP-bound enzyme structure (magenta). Upon binding of CRT0066854 to the nucleotide-binding cleft, F543 is ejected from the active site caused by a steric clash with the ligand phenyl ring moiety (yellow). *Figures adapted from research originally published in* The Biochemical Journal, *see ref. 21. © The Biochemical Society.* (See the color plate.)

equivalent phenyl group in the inhibitor, leading to a Phe543-NFD out conformation and a disordering of NFD motif (Fig. 13.1C).[21]

2.3. aPKC Function

The catalytic activity of both the aPKC isoforms has emerged as central players in the evolutionarily conserved mechanisms that govern the establishment and maintenance of cellular polarity.[24] Cell polarity proteins regulate a diverse array of biological process such as asymmetric cell division, proliferation, migration, and morphogenesis, the biological process that causes an organism to develop its shape.[25] Apical–basal cell polarity is the asymmetric distribution of cellular components between the outward facing apical and the basolateral membrane surfaces. The apical membrane is involved in absorption and excretion, whereas the basolateral membrane attaches the epithelial tissue to the connective tissue. Normal epithelial cell polarity is established and maintained by a dynamic interplay between three critical protein complexes that engage in mutually competitive interactions to sort proteins between the apical and basolateral domains. Thus, the apically located Par and crumbs complexes and the basolateral domain scribble complex are conserved from yeast to mammals. The Par complex consists of aPKC, the multidomain scaffolding protein Par-3 and the adaptor Par-6, which promotes the establishment of the apical–basal membrane. The crumbs complex specifies the apical membrane and comprises the transmembrane protein Crumbs, Pals1, and PATJ. The basolateral domain is established by the scribble complex comprising of three tumor suppressor proteins: lethal giant larvae (lgl), Discs Large, and scribble. Polarity complex interplay through mutual antagonism serves to regulate their cellular location. For example, aPKC can phosphorylate the scribble complex component lgl, leading to its inactivation and translocation from the basolateral domain to the cytoplasm. Conversely, lgl can disrupt the Par complex by competing with Par-3 for aPKC–Par-6 binding, sequestering the Par complex from the apical membrane.[26,27]

3. DISEASE LINKAGE OF ATYPICAL PKCs

3.1. Oncology

Both aPKC isoforms have been implicated in the initiation and development of many cancers making a strong case for the inhibition of the aPKC kinases as a novel cancer therapy.[28] Loss of PKCι catalytic activity inhibits carcinogenesis in different mouse models,[29–33] whereas overexpression of PKCι is observed in wide range of human malignancies.[34–38] Moreover, overexpression of PKCι is strongly correlated to tumor specific-gene copy number amplification events and is positively associated with poor prognosis and aggressive disease.[30,39–41] An emerging body of evidence implicates the dysregulation of the conserved polarity complexes in cancer, and loss of apical–basal polarity is a hallmark of aggressive cancers.[42,43] A frequent mechanism by which human tumors lose polarity is through loss of Par-3, and as a consequence, PKCι is frequently mislocalized within the cytoplasm and nucleus of transformed cells and human tumors.[44] Furthermore, in ovarian and breast cancers, transformed growth is correlated with mislocalized PKCι and deregulated epithelial cell polarity.[45,46] Moreover, PKCι has been demonstrated to cooperate with oncogenic Ras and Her2 to promote loss of polarized epithelial morphogenesis (dysplasia).[47] Further evidence supporting the importance of polarity loss in tumorigenesis has come from a recent finding identifying a cancer associated mutation in a substrate-specific recruitment motif of PKCι, leading to a disruption in the polarizing activity of PKCι.[48] Accumulating evidence supports the existence of cancer stem cells in driving tumor initiation, progression, and metastasis.[49] The genes that control epithelial cell polarity also regulate the symmetry of cell divisions in stem cells.[50] Loss of asymmetric cell divisions in stem or progenitor cells can lead to increased proliferation and expansion through symmetric cell divisions. PKCι has been shown to control epidermal homeostasis and stem cell fate through regulation of cell division orientation and through the promotion of symmetric divisions, and as such, PKCι would be expected to drive expansion of a more undifferentiated stem cell-like population. In line with these findings, PKCι activity has been implicated as a central player in lung cancer stem cell expansion downstream of mutant Ras.[51] In addition, PKCι is required for Ras-mediated colon carcinogenesis.[52] Furthermore, PKCι maintains a tumor-initiating cell phenotype that is required for ovarian tumorigenesis[53] and maintains cancer stem cell-like properties of lung squamous cell carcinoma.[54] Finally, evidence is beginning to emerge of a distinct role for PKCι as an oncogenic kinase in Hedgehog (HH) signaling. In this context, PKCι has been implicated in tumors with cell-intrinsic HH signaling such as basal cell carcinoma[55] as well as tumors with autocrine/paracrine HH signaling such as NSCLC.[54]

Despite the high degree of similarity between aPKC isoforms, the role of PKCζ in cancer is distinct from that of PKCι and a significant body of evidence exists to implicate PKCζ in cancer disease pathophysiology. PKCζ expression is highly upregulated in prostate carcinoma progression from benign hyperplasia to metastatic disease.[56,57,41] Furthermore, enhanced PKCζ activity and spontaneous development of tumors in the prostate and endometrium were observed in genetically engineered mouse model containing a homozygous deletion of Par4, a negative regulator of PKCζ.[58,59] PKCζ activity has been linked to resistance against a wide range of cytotoxic and genotoxic agents,[60–62] and PKCζ activity has been linked to chemoresistance in tamoxifen-resistant estrogen receptor-positive breast cancers,[63,64] suggesting that inhibition of PKCζ activity may have beneficial therapeutic effects by acting as a chemosensitizer to a wide array of commonly used chemotoxic agents.

3.2. Metabolic Diseases

Differing roles for aPKCs in metabolic signaling pathways have been reported within discrete tissues such as muscle and liver,[65] though reversal of abnormalities in lipid and carbohydrate signaling have been reported with inhibitors of PKCι in diabetic hepatocytes[66] as well as in rodent models.[67] Inhibition of PKCζ has been shown to restore breakdown of outer blood–retinal barriers in diabetic retinopathy.[68] Furthermore, aPKCs have been found to be key regulatory molecules in the VEGF, TNF, CCL2, and thrombin-induced dysfunction in the permeability of the blood–retinal barrier and thus modulation of aPKCs may be effective at controlling retinal edema and inflammation.[69]

3.3. Other Indications

Outside the oncology/metabolic disease focus, aPKC function and/or dysregulation has been implicated in pathologies from learning and memory to infectious diseases and immunology/inflammation. aPKCs have been shown to affect polarity in neurons and neuronal progenitor cells.[70,71] Recently, PKCι/λ and PKCζ/PKMζ were shown to have potentially mutually antagonistic regulatory functions in the modulation of neuronal polarity.[72] Furthermore, conflicting reports concerning the role of aPKCs on learning and memory have appeared.[73–75] Altered Wnt/aPKC and Wnt/Ryk signaling have been found in SOD1(G93A) mice, an important animal model for the pathogenesis of ALS.[76] More recently, aPKCs have been shown to phosphorylate HIV-1 Gag and have been implicated as regulatory kinases for incorporation of Vpr into HIV-1 virions; importantly, inhibition of aPKC suppressed infectivity of HIV-1 in macrophages.[77] In addition, aPKCs have been implicated in the regulation of NF-κB signaling, thus potentially acting as modulators of immune response and inflammation.[78] Furthermore, PKCι regulates the activation of key transcription factors—NF-κB, NFAT, GAT2, and STAT6—necessary for optimal T-helper cell (Th2) proliferation and cytokine production, and is implicated in allergic airway inflammation *in vivo*.[79]

In the context of this summary of aPKC activation, there are multiple points to consider for therapeutic intervention. These include ATP and protein substrate-binding competitive inhibitors as well as PB1 domain inhibitors and modulators of aPKC membrane association. The underlying evidence supporting modulation of aPKCs as effective regulators of disease state is quite compelling, and thus, small-molecule inhibitors have increased in prevalence recently. The following small-molecule aPKC inhibitors acting in a variety of modalities as noted above have been reported in the literature.

4. NON-ATP-BINDING SITE INHIBITORS

4.1. PB1 Domain of aPKC (Gold Complexes)

The gold-containing compounds sodium aurothiomalate (ATM, **1**) and aurothioglucose (ATG, **2**) were identified from a roughly 1000-compound library of known drugs using a FRET assay that looked for interaction between PB1 domains of PKCι and the complexation partner PAR6.[80] Since only atypical isoforms of the PKC family contain the PB1 domain, agents that disrupt signaling through this mechanism should be specific for atypical versus classical and novel PKCs. Furthermore, by avoiding interaction with the ATP-binding site, selectivity against the roughly 500 known kinases should also be achieved. Binding of **1** and **2** was in the low micromolar range to PKCι and both displayed activity in a cellular assay measuring PKCι-dependent Rac1 activation. Additionally, activity in the disruption of tumor cell growth in soft agar and *in vivo* tumor growth inhibition was found. Since these are marketed drugs for rheumatoid arthritis (RA), ATM quickly advanced to a Phase 1 trial in NSCLC.[81] Recently, auranofin (**3**), another known gold-containing drug for RA, similar in structure to ATG, demonstrated comparable *in vitro/in vivo* activity as **1** and **2** and was advanced to the clinic in NSCLC.[53]

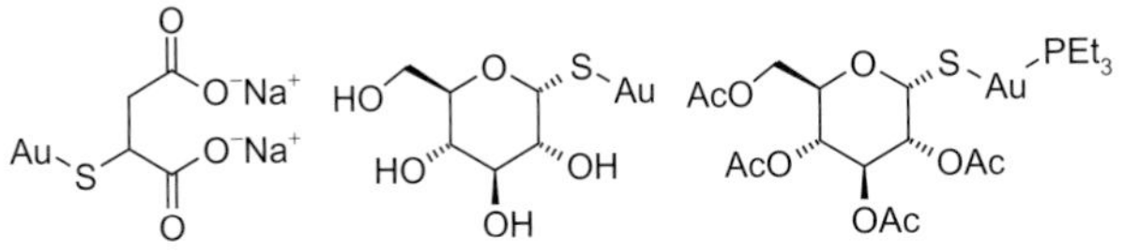

4.2. C-terminal Lobe of the Catalytic Domain of aPKC

ICA-1 (**4**) was first synthesized as a carbocyclic derivative of the imidazole nucleotide analogue AICAR (**5**).[82] These carbon variants were envisioned to act as inhibitors and/or substrates of enzymes in the *de novo* nucleotide biosynthetic pathway and were of interest as antiproliferative agents. It was not until much later that ICA-1 was found to be a selective inhibitor of PKCι.[83] It purportedly binds to residues 469–475 in the C-terminal lobe as suggested by molecular modeling. It was found to be a competitive inhibitor of PKCι, inhibiting the phosphorylation of myelin basic protein by recombinant PKCι by 45% at 5 μM. It showed no inhibitory effect on PKCζ in the same assay. Using a γ-32P-ATP PIP3-induced phosphotransferase assay, the IC_{50} of ICA-1 was found to be 10 nM using recombinant PKCι. Furthermore, ICA-1 at 0.1 μM inhibited the *in vitro* proliferation of

BE(2)-C neuroblastoma cells by 58% and induced apoptosis but had no effect on normal neuronal cells.[83]

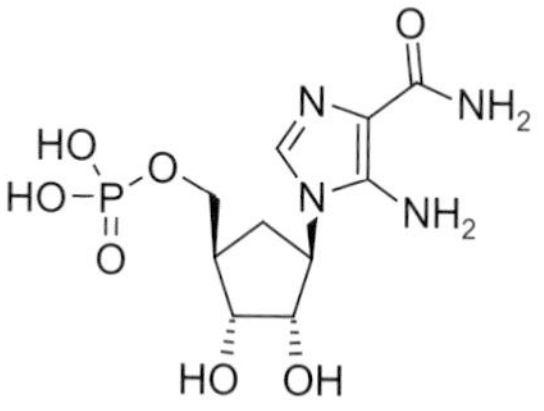

4

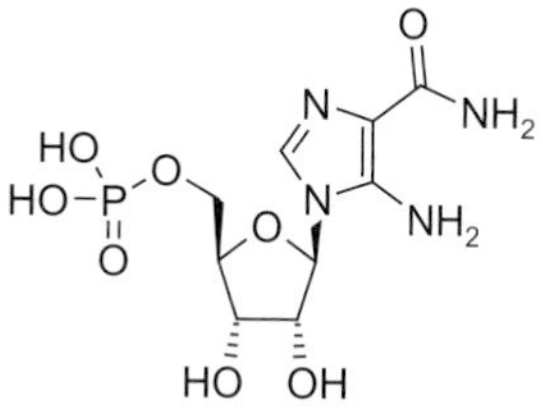

5

The natural product pachastrissamine (Jaspine B, **6**), isolated from the Okinawan sponge *Pachastrissa* sp., demonstrated cytotoxic effects against several cancer cell lines with potencies in the submicromolar range.[84] **6** and the other possible stereoisomers had inhibitory activity against the sphingosine kinases 1 and 2 (SphK1 and SphK2) with **7** and **8** being the most potent of the congeners. Since *N,N*-dimethylsphingosine (**9**), was shown to be an inhibitor of both PKC and sphingosine kinases, **6–8** were screened against novel, classical, and atypical PKC isoforms. Complete inhibition of PKCζ/ι at 10 μM was observed; only modest effects were noted for the novel and classical isoforms.[85] Screening at lower concentrations revealed about 50% inhibition at 3 μM and no inhibition at 1 μM. It was conjectured that since ceramides (**10**) bind to the aPKC isoforms at the atypical C1 domain, the binding of **6–8** would occur in this region as well.

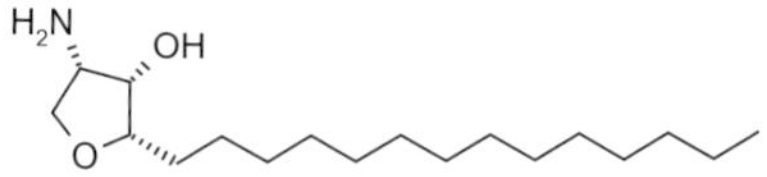

6

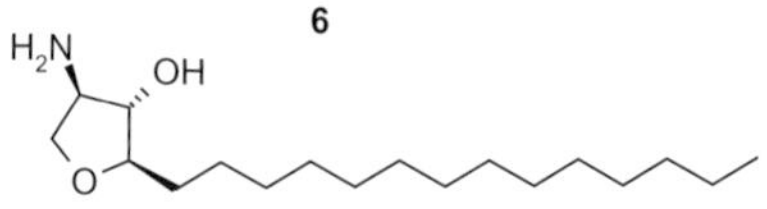

7

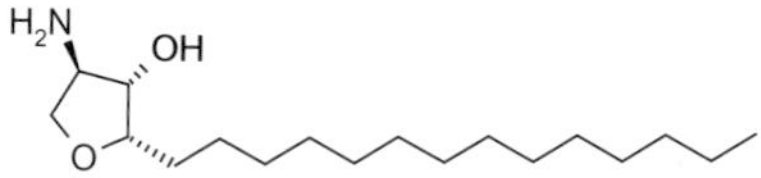

8

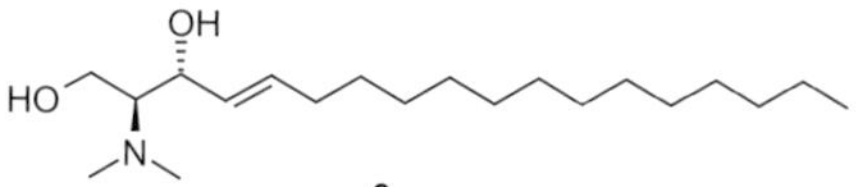

9

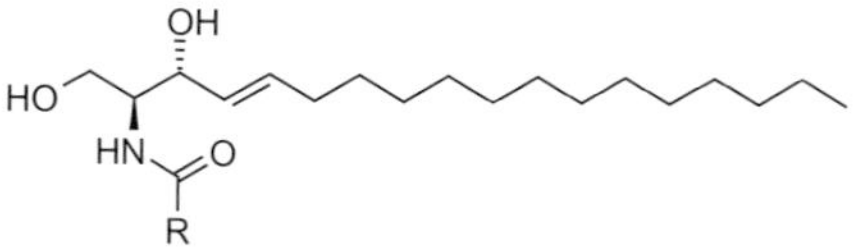

10, R = Fatty acid

4.3. aPKC Pseudosubstrate Binding Site

The PKCζ inhibitory peptide (**11**, ZIP) has been a widely used probe to study the atypical PKCζ (and PKMζ) signaling pathways. Encompassing the pseudosubstrate sequence, the 13-mer ZIP peptide (Myr-Ser-Ile-Tyr-Arg-Arg-Gly-Ala-Arg-Arg-Trp-Arg-Lys-Leu) has been myristoylated at the N-terminus to improve cell penetration.[86]

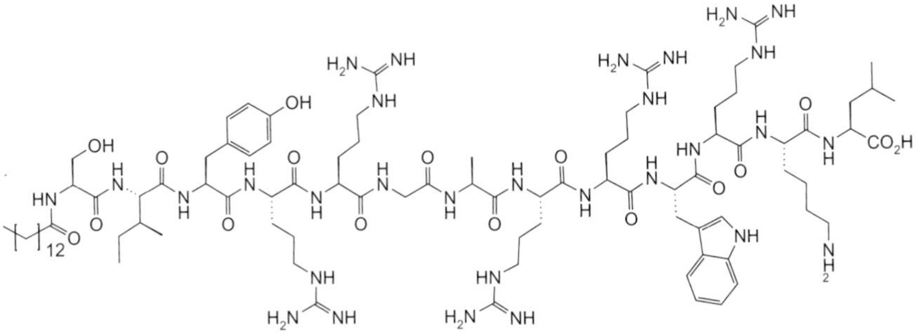

11

4.4. Allosteric PIF-1 Domain Binding

A series of allosteric inhibitors of PKCζ were identified whose mechanism of action involves PIF-1 site binding and regulation of activity through the C1 domain.[87] More detailed SAR for this work revealed isosteric replacement of the acetylated anilino moiety of **12** with a benzimidazole provided modestly active PKCζ inhibitors.[88] The more potent analogues from this series (**13**, **14**) displayed IC_{50}s of 18 and 33 μM against recombinant PKCζ. These analogues also inhibited NF-κB activation in U937 cells at 12 and 14 μM, respectively. Despite the close structural relationship, **14** was more selective against a panel of AGC kinases. **14** was also selective over novel and classical PKCs as well as PKCι.

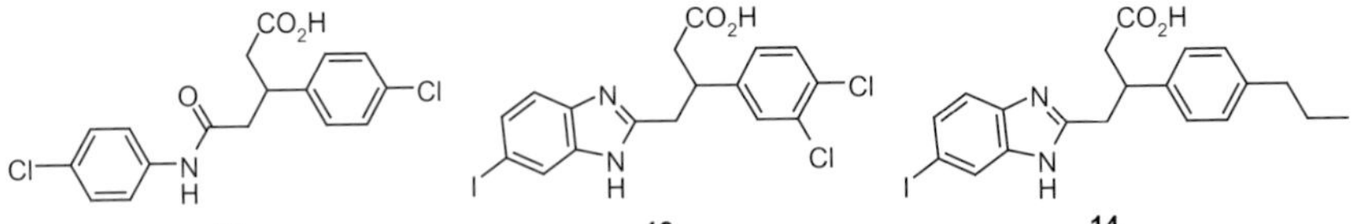

12 **13** **14**

4.5. Undefined Binding Modes

A series of phenylthiophenes were identified from a library screen for their ability to inhibit atypical PKCs. In addition, they prevented both VEGF- and TNF-induced retinal permeability in the rodent retina without inducing apoptosis or adverse retinal pathology.[68] A more detailed SAR account of this work demonstrated that inhibitors **15–17** displayed IC_{50} values of 1–6 μM against PKCζ and PKCι, and were *noncompetitive* with respect to ATP.[89]

The cellular EC_{50}s for TNF-induced NF-κB activation were 1–3 nM and the EC_{50}s for TNF/VEGF-induced endothelial permeability were between 1 and 280 nM. No rationale was given for this order of magnitude difference in potencies though it was stated that these inhibitors did not appear to have off-target activity affecting the canonical VEGF signaling pathways.

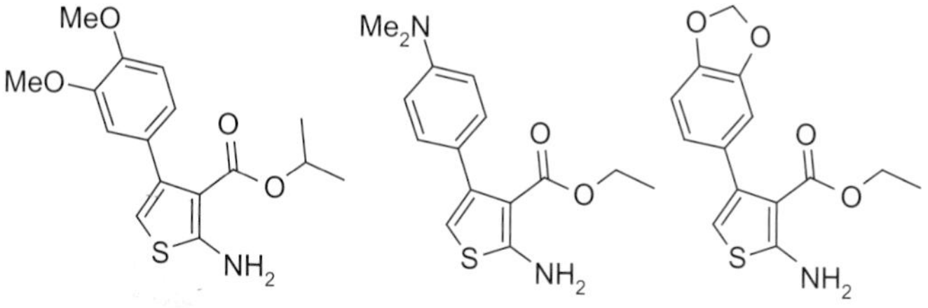

MeO MeO O O S NH2 **15**

Me2N O O S NH2 **16**

O O O O S NH2 **17**

5. ATP-BINDING SITE INHIBITORS

The maleimide substructure is a well-known pan-PKC inhibitor motif, though it is more notable for novel and classical PKC inhibition than atypical PKC inhibition.[90,91] Despite the marginal potency of **18** (PKCζ $IC_{50}=5.8$ μM),[92] a co-crystal structure using the kinase domain of PKCι has been reported.[19] The closely related analogue **19**, also a pan-PKC inhibitor, displayed more robust activity against aPKCs with a PKCζ $IC_{50}=60$ nM.[93]

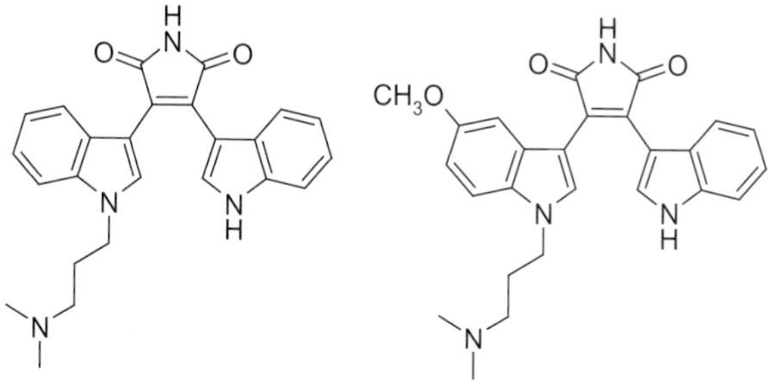

H N O O N N H N **18**

H N O O CH3O N N H N **19**

The speciosterosulfates A–C (**20–22**) were isolated from the marine sponge *Spheciospongia* sp., harvested near the Phillipines.[94] These new sterolsulfates, isolated along with the known topsentiasterol sulfate **23**, were found to inhibit PKCζ. In an ATP-competitive enzyme assay, **20–23** were shown to inhibit PKCζ with IC_{50}s of 1.59, 0.53, 0.11, and 1.21 μM, respectively. Further investigation of these compounds in a cell-based screen measuring inhibition of NF-κB activation, gave EC_{50}s between 12 and 64 μM.

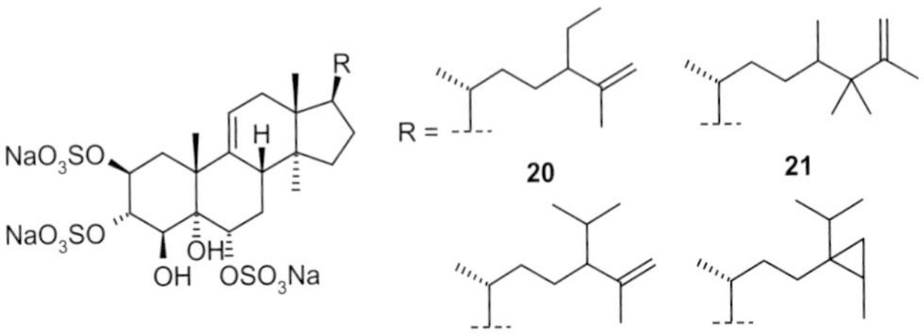

A series of compounds containing an indazole–benzimidazole motif were found to be PKCζ inhibitors.[95] Representative inhibitors **24** and **25** displayed IC_{50}s of 22 and 5 nM against PKCζ, though **24** was not PKC-isoform selective. **25** was modestly selective (10-fold) for the other atypical isoform PKCι, and greater than 700-fold for novel and classical isoforms. However, **25** displayed poor kinome selectivity when screened at 10 μM against a 40-kinase panel.

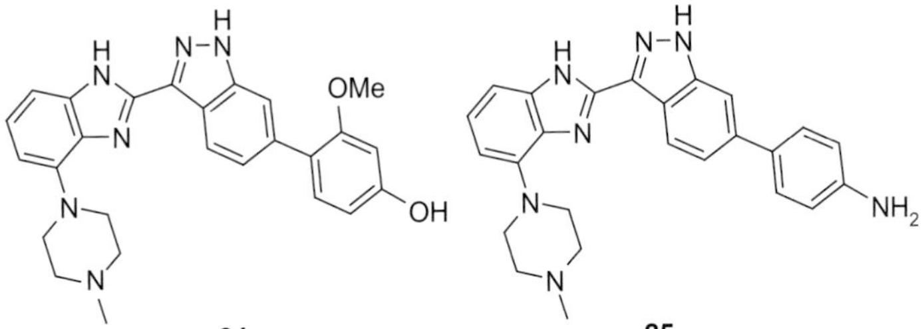

A high-throughput screen of PKCι kinase domain activity identified a series of ATP- competitive thieno[2,3-*d*]pyrimidine analogues based upon screening hit **26**.[21] An optimized compound, **27**, did not exhibit selectivity for PKCζ, consistent with the high level of conservation within the atypical PKC kinase domain. There was a marked stereochemical preference for PKCι potency with an approximately 20-fold eudysmic ratio for the more active *S*-enantiomer. Compound **27** reduced the phosphorylation of the direct PKCι substrate, LLGL, in a cellular assay. In phenotypic assays, **27** decreased non-adherent cell growth, inhibited polarized epithelial morphogenesis, and impeded directed cell migration. Furthermore, in a model of deregulated apicobasolateral polarity, **27** reversed a mutant Ras transformed amorphous growth phenotype back to a regular spherical structure resembling the nontransformed parental line, consistent with inhibition of atypical PKCs.[48]

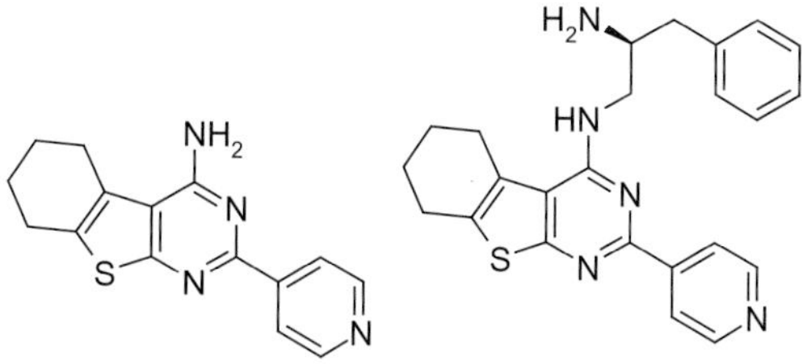

26 **27**

Pyrrole amide PKCzI257.3 (**28**, PKCζ $IC_{50}=28\ \mu M$) was identified by screening a several hundred compound library.[96] Counterscreening against PKCα demonstrated approximately sevenfold selectivity. No direct cytotoxicity was observed at 60 μM in an MTT assay. Assessing functional activity at roughly the PKCζ IC_{50} in MDA-MB-231 breast carcinoma cell line showed that chemotaxis was inhibited as well as cell migration in a scratch assay.

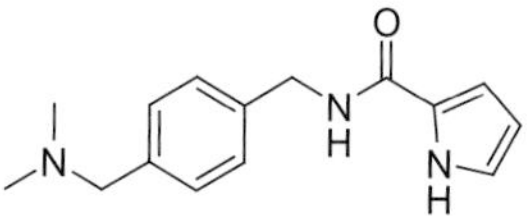

28

Another screening campaign led to a series of 3-hydroxy-2-(3-hydroxyphenyl)-4H-1-benzopyran-4-ones active against PKCζ.[97] Compounds **30–33** displayed IC_{50}s between 2 and 10 μM. When screened at 10 μM against PKA, PKBα, PKCβ2, and PKCε, some selectivity was observed. Molecular modeling suggested that these inhibitors bind to the ATP-cleft of the kinase through a keto-hydroxyl motif in multiple conformations.

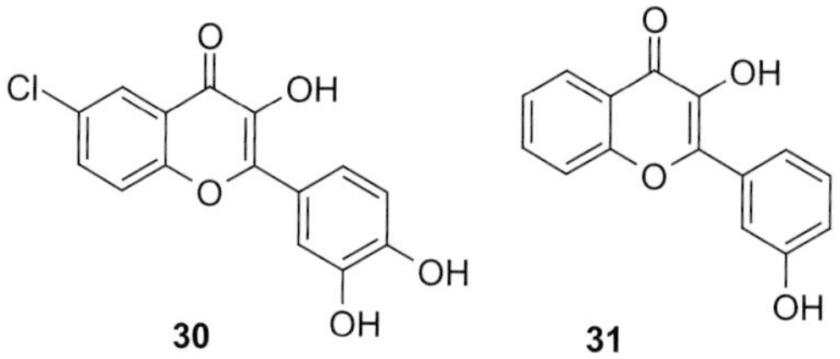

30 **31**

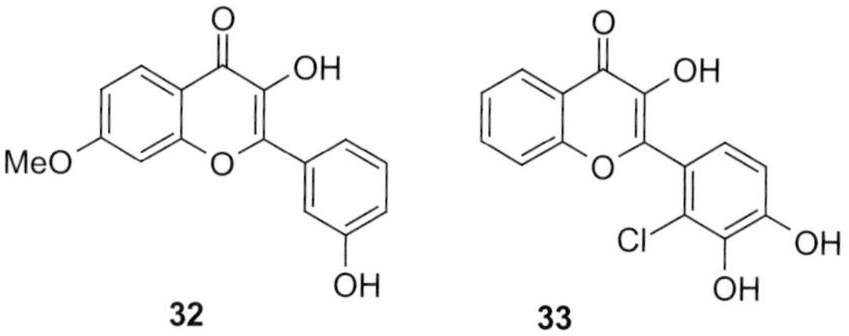

32 **33**

6. CONCLUSIONS

Atypical PKC isoforms are critical regulators of a variety of cellular processes including polarity, migration, proliferation, and survival. Both isoforms have been implicated in the initiation and development of many cancers providing significant support toward the development of a small-molecule inhibitor of the aPKC kinases for therapeutic application in oncology. Through detailed understanding of aPKC regulatory and activation mechanisms, multiple strategies for modulators of aPKC activity are being progressed.

REFERENCES

1. Castagna, M.; Takai, Y.; Kaibuchi, K.; Sano, K.; Kikkawa, U.; Nishizuka, Y. *J. Biol. Chem.* **1982**, *257*, 7847–7851.
2. Rosse, C.; Linch, M.; Kermorgant, S.; Cameron, A. J. M.; Boeckeler, K.; Parker, P. J. *Nat. Rev. Mol. Cell Biol.* **2010**, *11*, 103–112.
3. Mellor, H.; Parker, P. J. *Biochem. J.* **1998**, *332*, 281–292.
4. Takai, Y.; Kishimoto, A.; Iwasa, Y.; Kawahara, Y.; Mori, T.; Nishizuka, Y. *J. Biol. Chem.* **1979**, *254*, 3692–3695.
5. Ono, Y.; Fujii, T.; Oigita, K.; Kikkawa, U.; Igarashi, K.; Nishizuka, Y. *J. Biol. Chem.* **1988**, *263*, 6927–6932.
6. Selbie, L. A.; Schmitz-Peiffer, C.; Sheng, Y.; Biden, T. J. *J. Biol. Chem.* **1993**, *268*, 24296–24302.
7. Ono, Y.; Fujii, T.; Oigita, K.; Kikkawa, U.; Igarashi, K.; Nishizuka, Y. *Proc. Natl. Acad. Sci. U.S.A.* **1989**, *86*, 3099–3103.
8. Mukai, H.; Ono, Y. *Biochem. Biophys. Res. Commun.* **1994**, *199*, 897–904.
9. Palmer, R. H.; Ridden, J.; Parker, P. J. *Eur. J. Biochem.* **1995**, *227*, 344–351.
10. Akimoto, K.; Mizuno, K.; Osadal, S.; Hirail, S.; Tanuma, S.; Suzuki, K.; Ohno, S. *J. Biol. Chem.* **1994**, *269*, 12677–12683.
11. Pu, Y.; Peach, M. L.; Garfield, S. H.; Wincovitch, S.; Marquez, V. E.; Blumberg, P. M. *J. Biol. Chem.* **2006**, *281*, 33773–33788.
12. Le Good, J. A.; Ziegler, W. H.; Parekh, D. B.; Alessi, D. R.; Cohen, P.; Parker, P. J. *Science* **1998**, *281*, 2042–2045.
13. Facchinetti, V.; Ouyang, W.; Wei, H.; Soto, N.; Lazorchak, A.; Gould, C.; Lowry, C.; Newton, A. C.; Mao, Y.; Miao, R. Q.; Sessa, W. C.; Qin, J.; Zhang, P.; Su, B.; Jacinto, E. *EMBO J.* **2008**, *27*, 1932–1943.
14. Moscat, J.; Diaz-Meco, M. T.; Wooten, M. W. *Cell Death Differ.* **2009**, *16*, 1426–1437.
15. Atwood, S. X.; Chabu, C.; Penkert, R. R.; Doe, C. Q.; Prehoda, K. E. *J. Cell Sci.* **2007**, *120*, 3200–3206.
16. Petronczki, M.; Knoblich, J. A. *Nat. Cell Biol.* **2001**, *3*, 43–49.
17. David, D. J. V.; Tishkina, A.; Harris, T. J. C. *Development* **2010**, *137*, 1645–1655.
18. Graybill, C.; Wee, S.; Atwood, S. X.; Prehoda, K. E. *J. Biol. Chem.* **2012**, *287*, 21003–21011.
19. Messerschmidt, A.; Macieira, S.; Velarde, M.; Bädeker, M.; Benda, C.; Jestel, A.; Brandstetter, H.; Neuefeind, T.; Blaesse, M. *J. Mol. Biol.* **2005**, *352*, 918–931.
20. Takinura, T.; Kamata, K.; Fukasawa, K.; Ohsawawa, H.; Komatani, H.; Yoshizumi, T.; Kotani, H.; Iwasawa, Y. *Acta. Cryst.* **2010**, *D66*, 577–583.

21. Kjaer, S.; Linch, M.; Purkiss, A.; Kostelecy, B.; Knowles, P. P.; Rosse, C.; Riou, P.; Soudy, C.; Kaye, S.; Patel, B.; Soriano, E.; Murry-Rust, J.; Barton, C.; Dillon, C.; Roffey, J.; Parker, P. J.; McDonald, N. Q. *Biochem. J.* **2013**, *451*, 329–342.
22. Pearce, L. R.; Komander, D. K.; Alessi, D. R. *Nat. Rev. Mol. Cell Biol.* **2010**, *11*, 9–22.
23. Leonard, T. A.; Rozycki, B.; Saidi, L. F.; Hummer, G.; Hurley, J. H. *Cell* **2011**, *144*, 55–66.
24. Suzuki, A.; Ohno, S. *J. Cell Sci.* **2006**, *119*, 979–987.
25. Muthuswamy, S. K.; Xue, B. *Annu. Rev. Cell Dev. Biol.* **2012**, *28*, 599–625.
26. Betschinger, J.; Mechtler, K.; Knoblich, J. A. *Nature* **2003**, *422*, 326–330.
27. Plant, P. J.; Fawcett, J. P.; Lin, D. C. C.; Holdorf, A. D.; Binns, K.; Kulkarin, S.; Pawson, T. *Nat. Cell Biol.* **2003**, *5*, 301–308.
28. Parker, P. J.; Justilien, V.; Riou, P.; Linch, M.; Fields, A. P. *Biochem. Pharmacol.* **2014**, *88*, 1–11.
29. Zhang, Z.; Anastasiadis, P. Z.; Liu, Y.; Thompson, A. E.; Fields, A. P. *J. Biol. Chem.* **2004**, *279*, 22118–22123.
30. Regala, R. P.; Weems, C.; Jamieson, L.; Khoor, A.; Edell, E. S.; Lohse, C. M.; Fields, A. P. *Cancer Res.* **2005**, *65*, 8905–8911.
31. Frederick, L. A.; Matthews, J. A.; Jamieson, L.; Justilien, V.; Thompson, E. A.; Radisky, D. C.; Fields, A. P. *Oncogene* **2008**, *27*, 4841–4853.
32. Murray, N. R.; Weems, J.; Braun, U.; Leitges, M.; Fields, A. P. *Cancer Res.* **2009**, *69*, 656–662.
33. Scotti, M. L.; Bamlet, W. R.; Smyrk, T. C.; Fields, A. P.; Murray, N. R. *Cancer Res.* **2010**, *70*, 2064–2074.
34. Zhang, L.; Huang, J.; Yang, N.; Liang, S.; Barchetti, A.; Giannakakis, A.; Cadungog, M. G.; O'Brien-Jenkins, A.; Massobrio, M.; Roby, K. F.; Katsaros, D.; Gimotty, P.; Butzow, R.; Weber, B. L.; Coukos, G. *Cancer Res.* **2006**, *66*, 4627–4635.
35. Erdogan, E.; Klee, E. W.; Thompson, A. E.; Fields, A. P. *Clin. Cancer Res.* **2009**, *15*, 1527–1533.
36. Takagawa, R.; Akimoto, K.; Ichikawa, Y.; Akiyama, H.; Kojima, Y.; Ishiguro, H.; Inayama, Y.; Aoki, I.; Kunisaki, C.; Endo, I.; Nagashima, Y.; Ohno, S. *Ann. Surg. Oncol.* **2010**, *17*, 81–88.
37. Kikuchi, K.; Soundararajan, A.; Zarzabal, L. A.; Weems, C. R.; Nelon, L. D.; Hampton, S. T.; Michalek, J. E.; Rubin, B. P.; Fields, A. P.; Keller, C. *Oncogene* **2013**, *32*, 286–295.
38. Fields, A. P.; Regala, R. P. *Pharmacol. Res.* **2007**, *55*, 487–497.
39. Yang, Y.; Chu, J.; Luo, M.; Wu, Y.; Zhang, Y.; Feng, Y.; Shi, Z.; Xu, X.; Han, Y.; Cai, Y.; Dong, J.; Zhan, Q.; Wu, M.; Wang, M. *Genes Chromosomes Cancer* **2008**, *47*, 127–136.
40. Beroukhim, R.; Mermel, C. H.; Porter, D.; Wei, G.; Raychaudhuri, S.; Donovan, J.; Barretina, J.; Boehm, J. S.; Dobson, J.; Urashima, M.; McHenry, K. T.; Pinchback, R. M.; Ligon, A. H.; Cho, Y.-J.; Haery, L.; Greulich, H.; Reich, M.; Winckler, W.; Lawrence, M. S.; Weir, B. A.; Tanaka, K. E.; Chiang, D. Y.; Bass, A. J.; Loo, A.; Hoffman, C.; Prensner, J.; Liefeld, T.; Gao, Q.; Yecies, D.; Signoretti, S.; Maher, E.; Kaye, R. J.; Sasaki, H.; Tepper, J. E.; Fletcher, J. A.; Tabernero, J.; Baselga, J.; Tsao, M.-S.; Demichelis, F.; Rubin, M. A.; Janne, P. A.; Daly, M. J.; Nucera, C.; Levine, R. L.; Ebert, B. L.; Gabriel, S.; Rustgi, A. K.; Antonescu, C. R.; Ladanyi, M.; Letai, A.; Garraway, L. A.; Loda, M.; Beer, D. G.; True, L. D.; Okamoto, A.; Pomeroy, S. L.; Singer, S.; Golub, T. R.; Lander, E. S.; Getz, G.; Sellers, W. R.; Meyerson, M. *Nature* **2010**, *463*, 899–905.
41. Yao, S.; Bee, A.; Brewer, D.; Dodson, A.; Beesley, C.; Ke, Y.; Ambroisine, A.; Fisher, F.; Møller, H.; Dickinson, Y.; Gerard, P.; Lian, L. Y.; Risk, J.; Lane, B.; Smith, P.; Reuter, V.; Berney, D.; Gosden, C.; Scardino, P.; Cuzick, J.; Djamgoz, M. B. A.; Cooper, C.; Foster, C. S. *Genes Cancer* **2010**, *1*, 444–464.

42. Dow, L. E.; Humbert, P. O. *Int. Rev. Cyt.* **2007**, *262*, 253–302.
43. Martin-Belmonte, F.; Perez-Moreno, M. *Nat. Rev. Cancer* **2012**, *12*, 23–38.
44. Du, G.; Wang, J.; Lu, J.; Li, Q.; Ma, C.; Du, J.; Zou, S. *Ann. Surg. Oncol.* **2009**, *16*, 1578–1586.
45. Eder, A. M.; Sui, X.; Rosen, D. G.; Nolden, L. K.; Cheng, K. W.; Lahad, J. P.; Kango-Singh, M.; Lu, K. H.; Warneke, C. L.; Atkinson, E. N.; Bedrosian, I.; Keyomarsi, K.; Kuo, W.; Gray, J. W.; Yin, J. C. P.; Liu, J.; Halder, G.; Mills, G. B. *Proc. Natl. Acad. Sci. U.S.A* **2005**, *102*, 12519–12524.
46. Kojima, Y.; Akimoto, K.; Nagashima, Y.; Ishiguro, H.; Shirai, S.; Chishima, T.; Ichikawa, Y.; Ishikawa, Y.; Sasaki, T.; Kubota, Y.; Inayama, Y.; Aoki, I.; Ohno, S.; Shimada, H. *Human Path.* **2008**, *39*, 824–831.
47. Linch, M.; Sanz-Garcia, M.; Rosse, C.; Riou, P.; Peel, N.; Madsen, C. D.; Sahai, E.; Downward, J.; Khwaja, A.; Dillon, C.; Roffey, J.; Cameron, A. J. M.; Parker, P. J. *Carcinogenesis* **2014**, *35*, 396–406.
48. Linch, M.; Sanz-Garcia, M.; Soriano, E.; Zhang, Y.; Riou, P.; Rosse, C.; Cameron, A.; Knowles, P.; Purkiss, A.; Kjaer, S.; McDonald, N. Q.; Parker, P. J. *Sci. Signalling* **2013**, *6*, 1–11.
49. O'Connor, M. L.; Xiang, D.; Shigdar, S.; Macdonald, J.; Li, Y.; Wang, T.; Pu, C.; Wang, Z.; Qiao, L.; Duan, W. *Cancer Lett.* **2014**, *344*, 180–187.
50. Januschke, J.; Go, C. *Oncogene* **2008**, *27*, 6994–7002.
51. Regala, R. P.; Davis, R. K.; Kunz, A.; Khoor, A.; Leitges, M.; Fields, A. P. *Cancer Res.* **2009**, *69*, 7603–7611.
52. Murray, N. R.; Jamieson, L.; Yu, W.; Zhang, J.; Gökmen-Polar, Y.; Sier, D.; Anastasiadis, P.; Gatalica, Z.; Thompson, A. E.; Fields, A. P. *J. Cell Biol.* **2004**, *164*, 797–802.
53. Wang, Y.; Hill, K. S.; Fields, A. P. *Mol. Cancer Res.* **2013**, *11*, 1624–1635.
54. Justilien, V.; Walsh, M. P.; Ali, S. A.; Thompson, E. A.; Murray, N. R.; Fields, A. P. *Cancer Cell* **2014**, *25*, 139–151.
55. Atwood, S. X.; Li, M.; Lee, A.; Tang, J. Y.; Oro, A. E. *Nature* **2013**, *494*, 484–488.
56. Rhodes, D. R.; Kalyana-Sundaram, S.; Mahavisno, V.; Varambally, R.; Yu, J.; Briggs, B. B.; Barrette, T. R.; Anstet, M. J.; Kincead-Beal, C.; Kulkarni, P.; Varambally, S.; Ghoshy, D.; Chinnaiyan, A. M. *Neoplasia* **2007**, *9*, 166–180.
57. Yao, S.; Ireland, J. S.; Bee, A.; Beesley, C.; Forootan, S. S.; Dodson, A.; Dickinson, T.; Gerard, P.; Lian, L. Y.; Risk, J. M.; Smith, P.; Malki, M. I.; Ke, Y.; Cooper, C. S.; Gosden, C.; Foster, C. S. *J. Cancer Educ.* **2012**, *107*, 388–399.
58. Garcia-Cao, I.; Duran, A.; Collado, M.; Carrascosa, M. J.; Martin-Caballero, J.; Flores, J. M.; Diaz-Meco, M. T.; Moscat, J.; Serrano, M. *EMBO J.* **2005**, *6*, 577–583.
59. Joshi, J.; Fernandez-Marcos, P. J.; Galvez, A.; Amanchy, R.; Linares, J. F.; Duran, A.; Pathrose, P.; Leitges, M.; Canamero, M.; Collado, M.; Salas, C.; Serrano, M.; Moscat, J.; Diaz-Meco, M. T. *EMBO J.* **2008**, *27*, 2181–2193.
60. Filomenko, R.; Poirson-Bichat, F.; Billerey, C.; Belon, J.-P.; Garrido, C.; Solary, E.; Bettaieb, A. *Cancer Res.* **2002**, *62*, 1815–1821.
61. Plo, I.; Hernandez, H.; Kohlhagen, G.; Lautier, D.; Pommier, Y.; Laurent, G. *J. Biol. Chem.* **2002**, *277*, 31407–31415.
62. Bezombes, C.; De Thonel, A.; Apostolou, A.; Louat, T.; Jaffrezou, J. P.; Laurent, G.; Quillet-Mary, A. *Mol. Pharmacol.* **2002**, *62*, 1446–1455.
63. Yi, P.; Feng, Q.; Amazit, L.; Lonard, D. M.; Tsai, S. Y.; Tsai, M. J.; O'Malley, B. W. *Mol. Cell* **2008**, *29*, 465–476.
64. Iorns, E.; Lord, C. J.; Ashworth, A. *Biochem. J.* **2009**, *417*, 361–370.
65. Farese, R. V.; Sajan, M. P. *Curr. Opin. Lipidol.* **2012**, *23*, 175–181.
66. Sajan, M. P.; Farese, R. V. *Diabetologia* **2012**, *55*, 1446–1457.
67. Sajan, M. P.; Nimal, S.; Mastorides, S.; Acevedo-Duncan, M.; Kahn, C. R.; Fields, A. P.; Braun, U.; Leitges, M.; Farese, R. V. *Metab. Clin. Exp.* **2012**, *61*(4), 459–469.

68. Omri, S.; Behar-Cohen, F.; Rothschild, P.-R.; Gélizé, E.; Jonet, L.; Jeanny, J.-C.; Omri, B.; Crisanti, P. *PLoS One* **2013**, *8*(11), e81600.
69. Titchenell, P. M.; Lin, C. M.; Keil, J. M.; Sundstrom, J. M.; Smith, C. D.; Antonetti, D. A. *Biochem. J.* **2012**, *446*, 455.
70. Henrique, D.; Schweisguth, F. *Curr. Opin. Genet. Dev.* **2003**, *13*, 341–350.
71. Ghosh, S.; Marquardt, T.; Thaler, J. P.; Carter, N.; Andrews, S. E.; Pfaff, S. L.; Hunter, T. *Proc. Natl. Acad. Sci. U.S.A* **2008**, *105*, 335–340.
72. Parker, S. S.; Mandell, E. K.; Hapak, S. M.; Maskaykina, I. Y.; Kusne, Y.; Kim, J.-Y.; Moy, J. K.; St. John, P. A.; Wilson, J. M.; Gothard, K. M.; Price, T. J.; Ghosh, S. *Proc. Natl. Acad. Sci. U.S.A* **2013**, *110*, 14450–14455.
73. Wu-Zhang, A. X.; Schramm, C. L.; Nabavi, S.; Malinow, R.; Newton, A. C. *J. Biol. Chem.* **2012**, *287*, 12879–12885.
74. Volk, L. J.; Bachman, J. L.; Johnson, R.; Yu, Y.; Huganir, R. L. *Nature* **2013**, *493*, 420–423.
75. Lee, A. M.; Kanter, B. R.; Wang, D.; Lim, J. P.; Zou, M. E.; Qiu, C.; McMahon, T.; Dadgar, J.; Fischbach-Weiss, S. C.; Messing, R. O. *Nature* **2013**, *493*, 416–419.
76. Tury, A.; Tolentino, K.; Zhou, Y. *Dev. Neurobiol.* **2014**, *74*, 839–850.
77. Kudoh, A.; Takahama, S.; Ode, H.; Yokoyama, M.; Okayama, A.; Ishikawa, A.; Miyakawa, K.; Matsunaga, S.; Kimura, H.; Sugiura, W.; Sato, H.; Hirano, H.; Ohno, S.; Yamamoto, N.; Ryo, A. *Retrovirology* **2014**, *11*, 1–16.
78. Diaz-Meco, M. T.; Moscat, J. *Immunol. Rev.* **2012**, *246*, 154–167.
79. Yang, J. Q.; Leitges, M.; Duran, A.; Diaz-Mecom, M. T.; Moscat, J. *Proc. Natl. Acad. Sci. U.S.A* **2009**, *106*, 1099–1104.
80. Stallings-Mann, M.; Jamieson, L.; Regala, R. P.; Weems, C.; Murray, N. R.; Fields, A. P. *Cancer Res.* **2006**, *66*, 1767–1774.
81. Mansfield, A. S.; Fields, A. P.; Jatoi, A.; Qi, Y.; Adjei, A. A.; Erlichman, C.; Molina, J. R. *Anticancer Drugs* **2013**, *24*, 1079–1083.
82. Schmitt, L.; Caparelli, C. A. *Nucleosides Nucleotides* **1995**, *14*, 1929–1945.
83. Pillai, P.; Desai, S.; Patel, R.; Sajan, M.; Farese, R.; Ostrov, D.; Duncan, M.-A. *Int. J. Biochem. Cell Biol.* **2011**, *43*, 784–794.
84. Kuroda, I.; Musman, M.; Ohtani, I.; Ichiba, T.; Tanaka, J.; Garcia-Gravalos, D.; Higa, T. *J. Nat. Prod.* **2002**, *65*, 1505.
85. Yoshimitsu, Y.; Oishi, S.; Miyagaki, J.; Inuki, S.; Ohno, H.; Fujii, N. *Bioorg. Med. Chem.* **2011**, *19*, 5402–5408.
86. Laudanna, C.; Mochly-Rosen, D.; Liron, T.; Constantin, G.; Butcher, E. C. *J. Biol. Chem.* **1998**, *273*, 30306–30315.
87. Lopez-Garcia, L. A.; Schulze, J. O.; Fröhner, W.; Zhang, H.; Weber, N.; Navratil, J.; Amon, S.; Hindie, V.; Zeuzem, S.; Jørgensen, T. J. D.; Alzari, P. M.; Neimanis, S.; Engel, M.; Biondi, R. M. *Chem. Biol.* **2011**, *18*, 1463–1473.
88. Fröhner, W.; Lopez-Garcia, L. A.; Neimanis, S.; Weber, N.; Navratil, J.; Maurer, F.; Stroba, A.; Zhang, H.; Biondi, R. M.; Engel, M. *J. Med. Chem.* **2011**, *54*, 6714–6723.
89. Titchenell, P. M.; Hollis Showalter, H. D.; Pons, J.-F.; Barber, A. J.; Jin, Y.; Antonetti, D. A. *Bioorg. Med. Chem. Lett.* **2013**, *23*, 3034–3038.
90. Davis, P. D.; Hill, C. H.; Keech, E.; Lawton, G.; Nixon, J. S.; Sedgwick, A. D.; Wadsworth, J.; Westmacott, D.; Wilkinson, S. E. *FEBS Lett.* **1989**, *259*, 61–63.
91. Toullec, D.; Pianetti, P.; Coste, H.; Bellevergue, P.; Grand-Perret, T.; Ajakane, M.; Baudet, V.; Boissin, P.; Boursier, E.; Loriolle, F.; Duhamel, L.; Charon, D.; Kirilovsky, J. *J. Biol. Chem.* **1991**, *266*, 15771–15781.
92. Martiny-Baron, G.; Kazanietz, M. G.; Mischak, H.; Blumberg, P. M.; Kochs, G.; Hug, H.; Marme, D.; Schachtel, C. *J. Biol. Chem.* **1993**, *268*, 9194–9197.
93. Gschwendt, M.; Dieterich, S.; Rennecke, J.; Kittstein, W.; Mueller, H. J.; Johannes, F. J. *FEBS Lett.* **1996**, *392*, 77–80.

94. Whitson, E. L.; Bugni, T. S.; Chockalingam, P. S.; Concepcion, G. P.; Harper, M. K.; He, M.; Hooper, J. N. A.; Mangalindan, G. C.; Ritacco, F.; Ireland, C. M. *J. Nat. Prod.* **2008**, *71*, 1213–1217.
95. Trujillo, J. I.; Kiefer, J. R.; Huang, W.; Thorarensen, A.; Xing, L.; Caspers, N. L.; Day, J. E.; Mathis, K. J.; Kretzmer, K. K.; Reitz, B. A.; Weinberg, R. A.; Stegeman, R. A.; Wrightstone, A.; Christine, L.; Compton, R.; Li, X. *Bioorg. Med. Chem. Lett.* **2009**, *19*, 908–911.
96. Wu, J.; Zhang, B.; Wu, M.; Li, H.; Niu, R.; Ying, G.; Zhang, N. *Invest. New Drugs* **2010**, *28*, 268–275.
97. Yuan, L.; Seo, J. S.; Kang, N. S.; Keinan, S.; Steele, S. E.; Michelotti, G. A.; Wetsel, W. C.; Beratan, D. N.; Gong, Y. D.; Lee, T. H.; Hong, J. *Mol. BioSyst.* **2009**, *5*, 927–930.

SECTION 5

Infectious Diseases

Section Editor: William J. Watkins
Gilead Sciences, Inc., Foster City, California

CHAPTER FOURTEEN

Advancement of Cell Wall Inhibitors in *Mycobacterium tuberculosis*

Pravin S. Shirude*, Monalisa Chatterji, Shridhar Narayanan, Pravin S. Iyer
Department of Medicinal Chemistry and Biosciences, AstraZeneca India Pvt. Ltd., Bangalore, India

Contents

1. INTRODUCTION

Despite decades of scientific progress, tuberculosis (TB) still remains a major global health problem. In 2012, an estimated 8.6 million new cases of active TB were reported and 1.3 million people died from the disease.[1] The past two decades have seen the emergence of multidrug-resistant TB (MDR-TB) and, more recently, extensively drug-resistant TB (XDR-TB).[2] There is an urgent need for drugs that can shorten the duration of TB chemotherapy and are active against drug-resistant strains of *Mycobacterium tuberculosis* (Mtb).[3] It is imperative to discover and develop potent small-molecule inhibitors of Mtb that have a novel mode of action and are active against both drug-sensitive and drug-resistant TB.

*Present address: Sai Life Sciences Ltd, Pune, India.

Annual Reports in Medicinal Chemistry, Volume 49
ISSN 0065-7743
http://dx.doi.org/10.1016/B978-0-12-800167-7.00014-6

The unique and complex cell wall of Mtb is associated with its pathogenicity.[4] First-line and second-line TB drugs include compounds that inhibit some part of Mtb cell wall synthesis or metabolism. The cell wall of Mtb consists of a mycolic acid layer connecting to a peptidoglycan (PG) layer through an arabinogalactan polysaccharide (Fig. 14.1). The mycolate–arabinogalactan complex is covalently linked to the PG layer and is essential for cell viability.[5–7] Although the cell wall of Mtb remains an active area of interest for novel TB drug discovery, it is not a favored target for the identification of drugs that acts against drug-resistant TB. This perception may have arisen from the belief that Mtb exists in two distinct populations, replicating and nonreplicating, with reduced cell wall activity in dormant stages.[9] Isoniazid (INH) is a well-known inhibitor of InhA, a key bacterial enzyme in the FAS-II (fatty acid biosynthesis) system involved in cell wall synthesis,[10] and is highly bactericidal against replicating bacteria, with a minimum inhibitory concentration of 0.05 μg/ml and a high therapeutic margin. However, it has much reduced activity against nonreplicating Mtb (>10 μg/ml).[11] The emerging literature evidence suggests that the basal "cell wall core" activity is likely to be maintained throughout the different stages of infection of Mtb *in vivo*[12,13] and hence points toward the potential of cell wall inhibitors targeting both replicating and nonreplicating Mtb. The demonstration of efficacy against nonreplicating bacteria by penicillins and clavulanate combinations further illustrates the importance of PG, an integral part of the cell wall.[14] Moreover, the synergistic effects of the benzothiazinone BTZ043 with TMC207[15] and SQ109 with TMC207, INH, and rifampin[16,17] suggest that cell wall inhibitors contribute toward

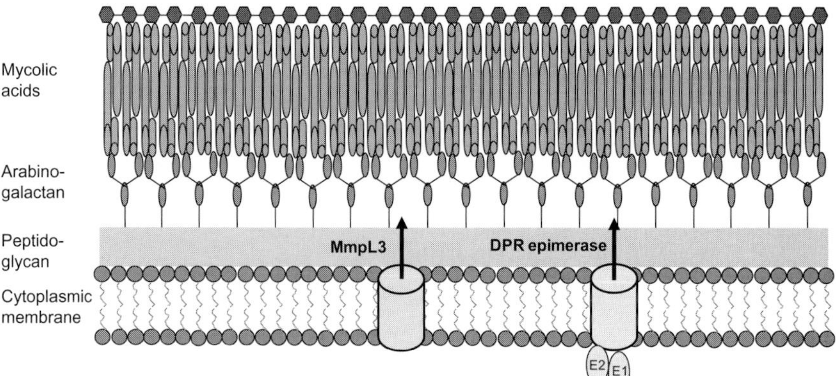

Figure 14.1 Schematic showing the cell wall of *Mycobacterium tuberculosis*.[8](See the color plate.)

improved efficacy in combination. These recent developments have spurred interest in the cell wall and its components for novel TB drug discovery. This review broadly describes recent advances in targeting the cell wall of Mtb, with specific focus on InhA, decaprenylphosphoryl-β-D-ribose-2′-epimerase (DprE1), MmpL3, and PG biosynthesis.

2. CELL WALL INHIBITORS

2.1. InhA

Fatty acids cannot be scavenged from the host and must be synthesized *de novo*, and thus are essential for bacterial growth.[18,19] The mycobacterial fatty acid biosynthesis pathway (FAS-II) is a clinically validated target for TB, but remains relatively unexploited.[20] The FAS-II pathway elongates fatty acids provided by the FAS-I pathway[18,19] and is highly conserved across many pathogens. FAS-II is a disaggregated system in which each reaction, performed by individual enzymes, is fundamentally distinct from the multienzyme FAS-I complex found in mammals. Mycolic acids are crucial components of the cell wall and serve as the first protective barrier for the organism. Pathways for their synthesis are therefore both essential in Mtb and sufficiently dissimilar from human systems to be attractive targets for treatment of TB.[20] The last reaction in each round of elongation involves the reduction of an enoyl-acyl carrier protein (enoyl-ACP) by the enoyl-ACP reductase. In Mtb, this protein is InhA, encoded by *inhA*.[19,21]

The clinical validation of InhA is based on the success of INH in treating TB patients.[21,22,11] INH is a prodrug that is activated by the catalase-peroxidase KatG to form an isonicotinoyl radical. This reacts with nicotinamide adenine dinucleotide (NAD) to form a covalent adduct **1** (INH-NAD, Fig. 14.2),[22] which inhibits InhA. It is known that drugs such as ethionamide and propionamide also target InhA via adduct formation with NAD.[10] Similar to INH, diazaborines have also been reported to form a boronate-NAD adduct that competes kinetically with the cofactor NADH.[23] A large majority of INH-resistant Mtb clinical isolates have mutations in KatG.[22] Additionally, mutations in the *inhA* coding region or *inhA* promoter region that confer resistance to INH have been reported.[22] These observations have led many TB drug discovery groups to identify direct inhibitors of InhA that do not require activation by KatG, and thus to identify novel compounds that are active against INH-resistant TB.[11] Interestingly, the natural product pyridomycin (**2**) also targets InhA via direct competitive inhibition of NADH binding to InhA.[24] Pyridomycin shows

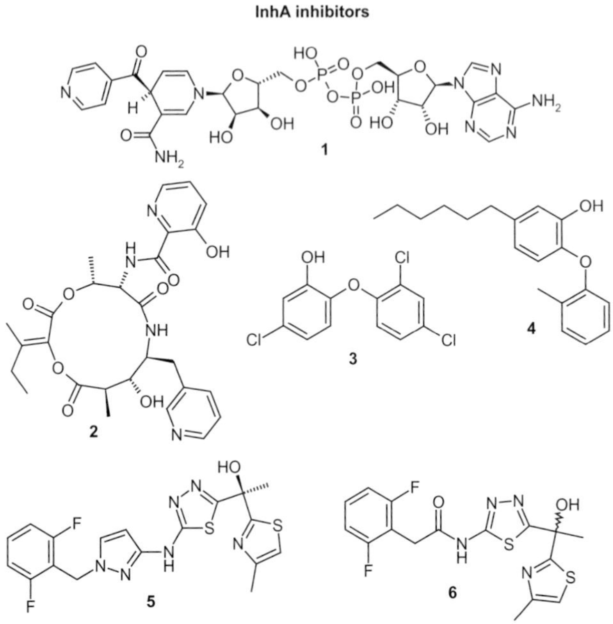

InhA inhibitors

Figure 14.2 Chemical structures of key InhA inhibitors. **1**, Isoniazid-NAD adduct (INH-NAD); **2**, pyridomycin; **3**, triclosan; **4**, PT70; **5**, **6**, methyl thiazole derivatives.

activity against an INH-resistant Mtb strain in which the resistance results from a mutation in KatG. This suggests that pyridomycin may occupy the entire site that binds to the cofactor and INH-NAD adduct.[24] In principle, direct inhibition of InhA might also be achieved through binding allosterically or within the substrate binding site. However, no allosteric inhibitors of InhA have been reported and only triclosan (**3**) (a highly effective broad-spectrum antibacterial agent that targets FabI, the non-mycobacterial equivalent of InhA[25]) is known to compete with the substrate of InhA. The weak inhibition of **3** results from direct binding to the NAD^+ binary complex of the enzyme.[25,26] The discovery of PT70[27] (**4**), a highly ligand efficient analog of triclosan, reinforces the validity of this approach. However, direct inhibition of InhA has been a considerable challenge due to the poor translation of enzyme inhibition to cellular potency. A number of different scaffolds targeting InhA have been identified, but most lack cellular activity despite potent enzyme inhibition.[28]

The recent disclosure of a series of methyl thiazoles (e.g., **5**) suggests that cellular potency against Mtb can be achieved with potent enzyme inhibition, while retaining desirable physicochemical properties.[28] More recent efforts addressing the mode of action of this series helped to understand the connection between enzyme inhibition and cellular potency and led to analog (**6**) with reduced cytochrome P450 (CYP) inhibition and an improved safety profile.[28] The scaffold exhibits a unique mechanism of InhA inhibition, involving a "Y158-out" conformation of the protein that accommodates an unionized inhibitor. The demonstration of a new hydrophilic interaction with protein residue M98 permits incorporation of motifs that modify physicochemical properties and improve cellular activity. Interestingly, the methyl thiazole prefers the NADH-bound (as against the NAD^+-bound) form of the enzyme.[28]

2.2. DprE1

DprE1 is involved in the biosynthesis of decaprenylphosphoryl-β-D-arabinofuranose (DPA), a precursor of mycobacterial cell wall arabinan by using decaprenylphosphoryl-β-D-ribose as a substrate.[29] This reaction is catalyzed by the heteromeric enzyme decaprenylphosphoribose 2′-epimerase (DprE), via a sequential oxidation–reduction mechanism involving decaprenylphosphoryl-2-keto-β-D-erythro-pentofuranose as an intermediate. DprE1 is composed of two proteins, encoded by the *dprE1* and *dprE2* genes. DprE1 is an FAD-containing oxidoreductase, while DprE2 is an NADH-dependent reductase.[30,31] The recent descriptions of the DprE1 crystal structures from *Mycobacterium smegmatis* (Msm) and Mtb[32–34] have contributed to a better understanding of this mechanism of inhibition. The structures of Msm and Mtb DprE1 enzymes are very similar, with 83% sequence identity, and the active site is fully conserved. The analyses of ligand-bound structures of DprE1 from both organisms revealed no major conformational change in the active site upon binding of different chemical classes, including both covalent and noncovalent inhibitors.

The identification of the nitrobenzothiazinones (BTZ043, **7** and PBTZ169, **8**) (Fig. 14.3) as covalent inhibitors of DprE1 that kill Mtb *in vitro*, *ex vivo*, and in mouse models of TB confirms the validity of this target for discovery of novel TB drugs.[31,32] Dinitrobenzenes (DNBs) (**9**) are also known to inhibit DPA synthesis.[35] Strains with a mutation in *dprE1* (BTZ043 resistant mutants) were resistant to DNBs, supporting the hypothesis that DprE1 could be the target for this class.[35] A recent screen of the NIAID library against Mtb led to the identification of a triazole derivative (377790, **10**).[36] Triazoles are thought to inhibit DprE1 by the formation of a specific covalent bond with Cys387, similar to BTZ043, following

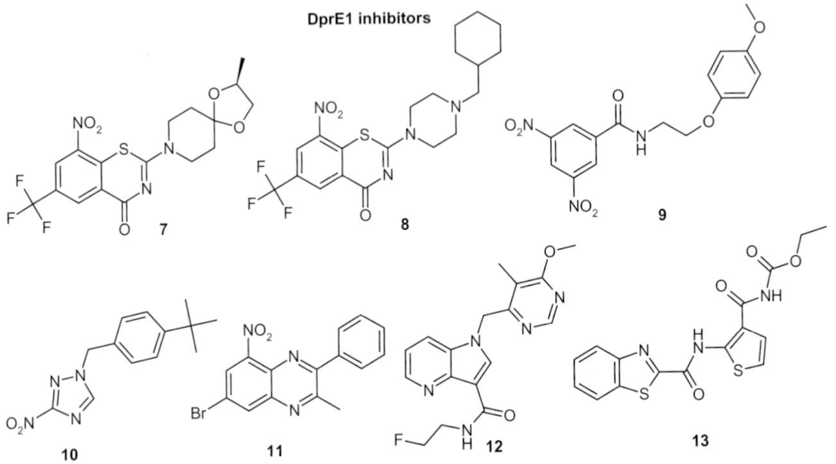

DprE1 inhibitors

Figure 14.3 Chemical structures of key DprE1 inhibitors. **7**, BTZ-043; **8**, PBTZ169; **9**, DNB; **10**, 377790; **11**, VI9376; **12**, azaindole; **13**, TCA-01.

reduction of the nitro group to a nitroso-derivative. The quinoxaline VI-9376 (**11**) is structurally related to BTZ compounds and the nitro group at the 5-position of the quinoxaline scaffold is essential for activity.[37] Similar to DNBs, **11** was found to inhibit DprE1 in the same manner as BTZ043. Recent efforts have led to the identification of the new promising molecules **12** and **13** (Fig. 14.3), which have demonstrated efficacy *in vivo* through noncovalent inhibition of DprE1.[38,39] Overall, data suggests that *in vivo* efficacy can be achieved by both covalent and noncovalent inhibition of DprE1.

2.3. MmpL3

As described above, the biosynthesis of mycolic acids is complex and involves the FAS-I and FAS-II pathways. Following biosynthesis, the mycolic acids are transported to their site on the cell wall where they are attached to the peptidoglycan–arabinogalactan complex.[40,6,41] Although the biosynthesis of mycolic acids is well characterized, the details of their intracellular transport are not well established.[6] Bacterial proteins that are involved in resistance, nodulation, and cell division are called resistance-nodulation-cell division proteins; 13 of these are found in Mtb, where they are referred to as mycobacterial membrane protein large (MmpL) proteins and are thought to encode lipid transporters in Mtb. It is believed that MmpL3 is a transporter of mycolic acids.[40] MmpL3 has also been implicated in heme transport in Mtb,[42] and has been shown to be essential for growth. Small-molecule Mtb inhibitors that target MmpL3 have been identified through cell-based screens (Fig. 14.4).

One of the first inhibitors identified in a cell-based screen is the adamantyl-urea AU1235 (**14**), which targets MmpL3 as confirmed by mutant generation coupled with whole-genome sequencing.[43] A recent report on the optimization of this series demonstrated improvement of solubility, microsomal stability & permeability, increased selectivity over human soluble epoxide hydrolase enzymes and improved anti-TB potencies.[44] The pyrrolo-methyl-piperazine derivative BM212 (**16**) was reported to inhibit both drug-sensitive and drug-resistant strains of Mtb, including XDR.[45] A close analog of BM212 with improved drug-like properties exhibited potent antitubercular activity when tested in an acute murine infection model.[46] MmpL3 was identified as the cellular target for BM212 via mutant generation followed by whole-genome sequencing.[46] SQ109[47] (**15**), an analog of ethambutol, has completed a Phase IIa trial

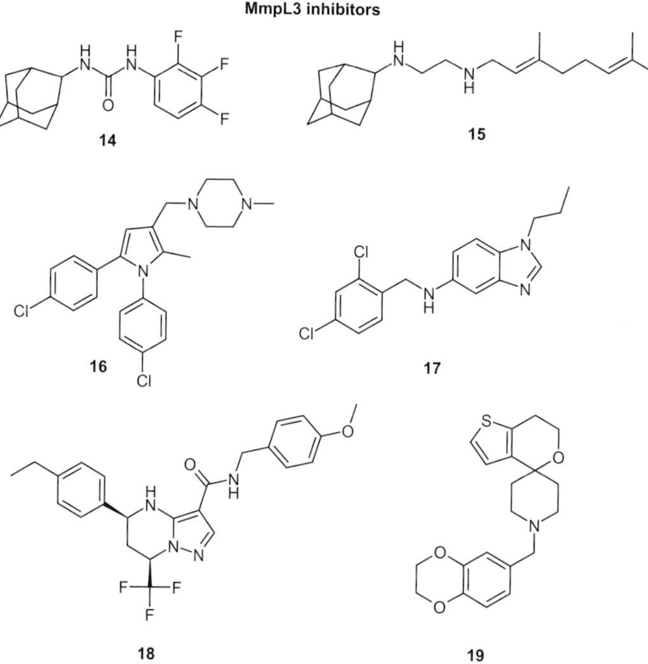

Figure 14.4 Chemical structures of key MmpL3 inhibitors. **14**, AU1235; **15**, SQ-109; **16**, BM212; **17**, C215; **18**, THPP; **19**, spiros.

for the treatment of TB.[48] Interestingly, it is now known that it has different mode of action than ethambutol. Although attempts to generate spontaneous mutants of SQ109 were unsuccessful, mutants generated using SQ109 analogs had single-nucleotide polymorphisms in MmpL3, suggesting it to be the cellular target of SQ109.[49] Other recent anti-TB compounds obtained from whole-cell screens include imidazoles[35] (C215, **17**), tetrahydropyrazolo [1,5-1a] pyrimidine-3-carboxamides (**18**) and *N*-benzyl-6′-7′-dihydrospiro [piperidine-4,4′-thienol [3,2-c] pyrans] (Spiros, **19**).[50] All these compounds have been shown to target MmpL3 using similar methods to those described above.[35,50] Thus, MmpL3 offers new avenues to target Mtb and these inhibitors may form the basis of new therapies to address MDR and XDR-TB.

2.4. Peptidoglycan Synthesis

The well-known class of β-lactam antibiotics act by inhibiting the final transpeptidation step in the synthesis of the PG catalyzed by L,D-transpeptidases. Despite their broad-spectrum success, β-lactams have not yet found use as TB drugs. β-Lactams as a class include penicillin derivatives (penams), cephalosporins (cephems), monobactams, and carbapenems.[51] Unfortunately, resistance to all is widespread, mediated primarily through expression of a collection of enzymes known as β-lactamases that efficiently hydrolyse the β-lactam to an inactive metabolite.[52] One successful strategy to circumvent this resistance combines the β-lactam antibiotic with an agent capable of inhibiting β-lactamases. Mtb also possesses a strong, constitutive β-lactamase activity that renders it resistant to β-lactams.[53,54] There is a report of off-label use of the β-lactam and β-lactamase inhibitor (BL-BLI) combination of amoxicillin (**20**) and clavulanic acid (**23**) (Fig. 14.5) in the successful treatment of two patients with MDR-TB.[55] However, amoxicillin-clavulanic acid saw limited success in several other clinical trials.[56,14] Further, a recent study demonstrated that meropenem (**21**)/clavulanate (**23**) had potent activity against drug-resistant TB and anaerobic models of persistent TB infection.[57] Moreover, multiple *in vitro* studies have reported enhanced potency against Mtb when treated with a combination of BL-BLI rather than β-lactams alone.[58,59] In an animal model of TB infection, meropenem and imipenem (**22**) showed moderate bactericidality in combination with clavulanate.[60–62] These findings have led to clinical trials for meropenem/clavulanate against XDR-TB, with encouraging results.[63–65]

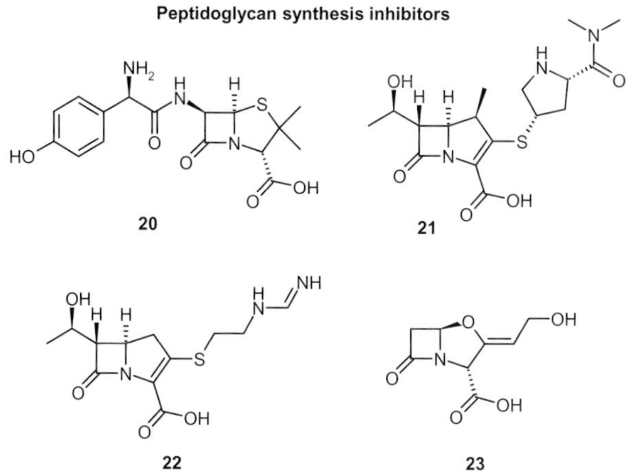

Peptidoglycan synthesis inhibitors

20 21 22 23

Figure 14.5 Chemical structures of key peptidoglycan synthesis inhibitors. **20**, Amoxicillin; **21**, meropenem; **22**, imipenem; **23**, clavulanic acid.

2.5. Emerging Targets

In the recent past, new targets involved in cell wall function or synthesis have been closely investigated. A few of the interesting targets that are required for the biosynthesis or function of mycolic acids are β-ketoacyl-acyl carrier protein synthase (KasAB and FabH), acyl-AMP ligase (Fad32), PimA, polyketide synthase (Psk13) and methyl transferase (PcaA). Small-molecule inhibitors of FabH[66] and KasAB[67] have been identified previously. Recently, inhibitors for FadD32 with anti-TB activity have also been identified.[68]

3. CONCLUSIONS

Even though many of the current first- and second-line anti-TB drugs target cell wall biosynthesis, most of the newly discovered compounds continue to emphasize the cell wall as the target of choice for Mtb. The recent understanding of "cell wall core" activity in the different stages of infection *in vivo* has highlighted the importance of cell wall inhibitors for targeting both replicating and nonreplicating Mtb. Importantly, drug-to-target screening approaches have identified novel targets impacting major cell wall components (e.g., DprE1, MmpL3, and others) in addition to better known targets (InhA, PG synthesis). Novel cell wall inhibitors provide a potential new therapy to address MDR and XDR-TB in patients who are resistant to existing first- and second-line drugs.

REFERENCES

1. World Health Organization. *Global Tuberculosis Report*, 2013.
2. Zumla, A.; Raviglione, M.; Hafner, R.; von Reyn, C. F. *New Engl. J. Med.* **2013**, *368*, 745.
3. Gandhi, N. R.; Shah, N. S.; Andrews, J. R.; Vella, V.; Moll, A. P.; Scott, M.; Weissman, D.; Marra, C.; Lalloo, U. G.; Friedland, G. H. *Am. J. Respir. Crit. Care Med.* **2010**, *181*, 80.
4. Hett, E. C.; Rubin, E. J. *Microbiol. Mol. Biol. Rev.* **2008**, *72*, 126.
5. Brennan, P. J. *Tuberculosis* **2003**, *83*, 91–97.
6. Takayama, K.; Wang, C.; Besra, G. S. *Clin. Microbiol. Rev.* **2005**, *18*, 81.
7. Kaur, D.; Guerin, M. E.; Skovierova, H. *Adv. Appl. Microbiol.* **2009**, *69*, 23.
8. Jackson, M.; McNeil, M. R.; Brennan, P. J. *Future Microbiol.* **2013**, *8*, 855.
9. Mitchison, D.; Davies, G. *Int. J. Tuberc. Lung Dis.* **2012**, *16*, 724.
10. Banerjee, A.; Dubnau, E.; Quemard, A.; Balasubramanian, V.; Um, K. S.; Wilson, T.; Collins, D.; de Lisle, G.; Jacobs, W. R., Jr. *Science* **1994**, *263*, 227.
11. Vilcheze, C.; Jacobs, W. R., Jr. *Annu. Rev. Microbiol.* **2007**, *61*, 35.
12. Glickman, M. S.; Cox, J. S.; Jacobs, W. R., Jr. *Mol. Cell* **2000**, *5*, 717.
13. Gupta, R.; Lavollay, M.; Mainardi, J. L.; Arthur, M.; Bishai, W. R.; Lamichhane, G. *Nat. Med.* **2010**, *16*, 466.
14. Hugonnet, J. E.; Tremblay, L. W.; Boshoff, H. I.; Barry, C. E., 3rd.; Blanchard, J. S. *Science* **2009**, *323*, 1215.
15. Lechartier, B.; Hartkoorn, R. C.; Cole, S. T. *Antimicrob. Agents Chemother.* **2012**, *56*, 5790.
16. Reddy, V. M.; Einck, L.; Andries, K.; Nacy, C. A. *Antimicrob. Agents Chemother.* **2010**, *54*, 2840.
17. Chen, P.; Gearhart, J.; Protopopova, M.; Einck, L.; Nacy, C. A. *J. Antimicrob. Chemother.* **2006**, *58*, 332.
18. McDevitt, D.; Rosenberg, M. *Trends Microbiol.* **2001**, *9*, 611.
19. Payne, D. J.; Warren, P. V.; Holmes, D. J.; Ji, Y.; Lonsdale, J. T. *Drug Discov. Today* **2001**, *6*, 537.
20. Lu, H.; Tonge, P. J. *Acc. Chem. Res.* **2008**, *41*, 11.
21. Lamichhane, G. *Trends Mol. Med.* **2011**, *17*, 25.
22. Rawat, R.; Whitty, A.; Tonge, P. J. *Proc. Natl. Acad. Sci. U.S.A* **2003**, *100*, 13881.
23. Davis, M. C.; Franzblau, S. G.; Martin, A. R. *Bioorg. Med. Chem. Lett.* **1998**, *8*, 843.
24. Hartkoorn, R. C.; Sala, C.; Neres, J.; Pojer, F.; Magnet, S.; Mukherjee, R.; Uplekar, S.; Boy-Röttger, S.; Altmann, K. H.; Cole, S. T. *EMBO Mol. Med.* **2012**, *4*, 1032.
25. Rafferty, J. B.; Simon, J. W.; Baldock, C.; Artymiuk, P. J.; Baker, P. J.; Stuitje, A. R.; Slabas, A. R.; Rice, D. W. *Structure* **1995**, *3*, 927.
26. Kuo, M. R.; Morbidoni, H. R.; Alland, D.; Sneddon, S. F.; Gourlie, B. B.; Staveski, M. M.; Leonard, M.; Gregory, J. S.; Janjigian, A. D.; Yee, C.; Musser, J. M.; Kreiswirth, B.; Iwamoto, H.; Perozzo, R.; Jacobs, W. R.; Sacchettini, J. C.; Fidock, D. A. *J. Biol. Chem.* **2003**, *278*, 20851.
27. Pan, P.; Tonge, P. J. *Curr. Top. Med. Chem.* **2012**, *12*, 672.
28. Shirude, P. S.; Madhavapeddi, P.; Naik, M.; Murugan, K.; Shinde, V.; Radha, N.; Bhat, J.; Kumar, A.; Hameed, S.; Holdgate, G.; Davies, G.; McMiken, H.; Hegde, N.; Ambady, A.; Venkatraman, J.; Panda, M.; Bandodkar, B.; Sambandamurthy, V. K.; Read, J. A. *J. Med. Chem.* **2013**, *56*, 8533.
29. Trefzer, C.; Rengifo-Gonzalez, M.; Hinner, M. J.; Schneider, P.; Makarov, V.; Cole, S. T.; Johnsson, K. *J. Am. Chem. Soc.* **2010**, *132*, 13663.
30. Mikušová, K.; Huang, H.; Yagi, T.; Holsters, M.; Vereecke, D.; D'Haeze, W.; Scherman, M. S.; Brennan, P. J.; McNeil, M. R.; Crick, D. C. *J. Bacteriol.* **2005**, *187*, 8020.

31. Makarov, V.; Manina, G.; Mikušová, K.; Möllmann, U.; Ryabova, O.; Saint-Joanis, B.; Dhar, N.; Pasca, M. R.; Buroni, S.; Lucarelli, A. P.; Milano, A.; De Rossi, E.; Belanova, M.; Bobovska, A.; Dianiskova, P.; Kordulakova, J.; Sala, C.; Fullam, E.; Schneider, P.; McKinney, J. D.; Brodin, P.; Christophe, T.; Waddell, S.; Butcher, P.; Albrethsen, J.; Rosenkrands, I.; Brosch, R.; Nandi, V.; Bharath, S.; Gaonkar, S.; Shandil, R. K.; Balasubramanian, V.; Balganesh, T.; Tyagi, S.; Grosset, J.; Riccardi, G.; Cole, S. T. *Science* **2009**, *324*, 801.
32. Makarov, V.; Lechartier, B.; Zhang, M.; Neres, J.; Van der Sar, A. M.; Raadsen, S. A.; Hartkoorn, R. C.; Ryabova, O. B.; Vocat, A.; Decosterd, L. A.; Widmer, N.; Buclin, T.; Bitter, W.; Andries, K.; Pojer, F.; Dyson, P. J.; Cole, S. T. *EMBO Mol. Med.* **2014**, published online.
33. Neres, J.; Pojer, F.; Molteni, E.; Chiarelli, L. R.; Dhar, N.; Boy-Röttger, S.; Buroni, S.; Fullam, E.; Degiacomi, G.; Lucarelli, A. P.; Read, R. J.; Zanoni, G.; Edmondson, D. E.; De Rossi, E.; Pasca, M. R.; McKinney, J. D.; Dyson, P. J.; Riccardi, G.; Mattevi, A.; Cole, S. T.; Binda, C. *Sci. Transl. Med.* **2012**, *4*, 150ra121.
34. Batt, S. M.; Jabeen, T.; Bhowruth, V.; Quill, L.; Lund, P. A.; Eggeling, L.; Alderwick, L. J.; Futterer, K.; Besra, G. S. *Proc. Natl. Acad. Sci. U.S.A* **2012**, *109*, 11354.
35. Christophe, T.; Jackson, M.; Jeon, H. K.; Fenistein, D.; Contreras-Dominguez, M.; Kim, J.; Genovesio, A.; Carralot, J. P.; Ewann, F.; Kim, E. H.; Lee, S. Y.; Kang, S.; Seo, M. J.; Park, E. J.; Skovierová, H.; Pham, H.; Riccardi, G.; Nam, J. Y.; Marsollier, L.; Kempf, M.; Joly-Guillou, M. L.; Oh, T.; Shin, W. K.; No, Z.; Nehrbass, U.; Brosch, R.; Cole, S. T.; Brodin, P. *PLoS Pathog.* **2009**, *5*, e1000645.
36. Stanley, S. A.; Grant, S. S.; Kawate, T.; Iwase, N.; Shimizu, M.; Wivagg, C.; Silvis, M.; Kazyanskaya, E.; Aquadro, J.; Golas, A.; Fitzgerald, M.; Dai, H.; Zhang, L.; Hung, D. T. *ACS Chem. Biol.* **2012**, 7, 1377137.
37. Magnet, S.; Hartkoorn, R. C.; Székely, R.; Pató, J.; Triccas, J. A.; Schneider, P.; Szántai-Kis, C.; Orfi, L.; Chambon, M.; Banfi, D.; Bueno, M.; Turcatti, G.; Kéri, G.; Cole, S. T. *Tuberculosis* **2010**, *90*, 354.
38. Wang, F.; Sambandan, D.; Halder, R.; Wang, J.; Batt, S. M.; Weinrick, B.; Ahmad, I.; Yang, P.; Zhang, Y.; Kim, J.; Hassani, M.; Huszar, S.; Trefzer, C.; Ma, Z.; Kaneko, T.; Mdluli, K. E.; Franzblau, S.; Chatterjee, A. K.; Johnson, K.; Mikusova, K.; Besra, G. S.; Fütterer, K.; Jacobs, W. R., Jr.; Schultz, P. G. *Proc. Natl. Acad. Sci. U.S.A.* **2013**, *110*, E2510.
39. Shirude, P. S.; Shandil, R.; Sadler, C.; Naik, M.; Hosagrahara, V.; Hameed, S.; Shinde, V.; Bathula, C.; Humnabadkar, V.; Kumar, N.; Reddy, J.; Panduga, V.; Sharma, S.; Ambady, A.; Hegde, N.; Whiteaker, J.; McLaughlin, B.; Gardner, H.; Madhavapeddi, P.; Ramachandran, V.; Kaur, P.; Narayan, A.; Guptha, S.; Awasthy, D.; Narayan, C.; Mahadevaswamy, J.; KG, V.; Ahuja, V.; Srivastava, A.; KR, P.; Bharath, S.; Kale, R.; Ramaiah, M.; Roy Choudhury, N.; Sambandamurthy, V. K.; Solapure, S.; Iyer, P. S.; Narayanan, S.; Chatterji, M. J. *Med. Chem.* **2013**, *56*, 9701.
40. Varela, C.; Rittmann, D.; Singh, A.; Krumbach, K.; Bhatt, K.; Eggeling, L.; Besra, G. S.; Bhatt, A. *Chem. Biol.* **2012**, *19*, 498.
41. Rayasam, G. V. *Expert Opin. Ther. Targets* **2014**, *18*, 247.
42. Tullius, M. V.; Harmston, C. A.; Owens, C. P.; Chim, N.; Morse, R. P.; McMath, L. M.; Iniguez, A.; Kimmey, J. M.; Sawaya, M. R.; Whitelegge, J. P.; Horwitz, M. A.; Goulding, C. W. *Proc. Natl. Acad. Sci. U.S.A* **2011**, *108*, 5051.
43. Grzegorzewicz, A. E.; Pham, H.; Gundi, V. A.; Scherman, M. S.; North, E. J.; Hess, T.; Jones, V.; Gruppo, V.; Born, S. E.; Korduláková, J.; Chavadi, S. S.; Morisseau, C.; Lenaerts, A. J.; Lee, R. E.; McNeil, M. R.; Jackson, M. *Nat. Chem. Biol.* **2012**, *8*, 334.
44. North, E. J.; Scherman, M. S.; Bruhn, D. F.; Scarborough, J. S.; Maddox, M. M.; Jones, V.; Grzegorzewicz, A.; Yang, L.; Hess, T.; Morisseau, C.; Jackson, M.; McNeil, M. R.; Lee, R. E. *Bioorg. Med. Chem.* **2013**, *21*, 2587.

45. La Rosa, V.; Poce, G.; Canseco, J. O.; Buroni, S.; Pasca, M. R.; Biava, M.; Raju, R. M.; Porretta, G. C.; Alfonso, S.; Battilocchio, C.; Javid, B.; Sorrentino, F.; Ioerger, T. R.; Sacchettini, J. C.; Manetti, F.; Botta, M.; De Logu, A.; Rubin, E. J.; De Rossi, E. *Antimicrob. Agents Chemother.* **2012**, *56*, 324.
46. Poce, G.; Bates, R. H.; Alfonso, S.; Cocozza, M.; Porretta, G. C.; Ballell, L.; Rullas, J.; Ortega, F.; De Logu, A.; Agus, E.; La Rosa, V.; Pasca, M. R.; De Rossi, E.; Wae, B.; Franzblau, S. G.; Manetti, F.; Botta, M.; Biava, M. *PLoS One* **2013**, *8*, e56980.
47. Jia, L.; Tomaszewski, J. E.; Hanrahan, C.; Coward, L.; Noker, P.; Gorman, G.; Nikonenko, B.; Protopopova, M. *Br. J. Pharmacol.* **2005**, *144*, 80.
48. http://clinicaltrials.gov/ct2/show/NCT01218217?term=&rank=1.
49. Tahlan, K.; Wilson, R.; Kastrinsky, D. B.; Arora, K.; Nair, V.; Fischer, E.; Barnes, S. W.; Walker, J. R.; Alland, D.; Barry, C. E., 3rd.; Boshoff, H. I. *Antimicrob. Agents Chemother.* **2012**, *56*, 1797.
50. Remuiñán, M. J.; Pérez-Herrán, E.; Rullás, J.; Alemparte, C.; Martínez-Hoyos, M.; Dow, D. J.; Afari, J.; Alemparte, C.; Martínez-Hoyos, M.; Dow, D. J.; Afari, J.; Mehta, N.; Esquivias, J.; Jiménez, E.; Ortega-Muro, F.; Fraile-Gabaldón, M. T.; Spivey, V. L.; Loman, N. J.; Pallen, M. J.; Constantinidou, C.; Minick, D. J.; Cacho, M.; Rebollo-López, M. J.; González, C.; Sousa, V.; Angulo-Barturen, I.; Mendoza-Losana, A.; Barros, D.; Besra, G. S.; Ballell, L.; Cammack, N. *PLoS One* **2013**, *8*, e60933.
51. Bush, K.; Macielag, M. J. *Expert Opin. Ther. Pat.* **2010**, *20*, 1277.
52. Kong, K. F.; Schneper, L.; Mathee, K. *APMIS* **2010**, *118*, 1.
53. Kasik, J. E. *The Am. Rev. Respir. Dis.* **2005**, *91*, 117.
54. Flores, A. R.; Parsons, L. M.; Pavelka, M. S. *Microbiology* **2005**, *151*, 521.
55. Nadler, J. P.; Berget, J.; Nord, J. A.; Cofiky, R.; Saxena, M. *Chest* **1991**, *99*, 1025.
56. Chambers, H. F.; Kogacöz, T.; Sipit, T.; Turner, J.; Hopewell, P. C. *Clin. Infect. Dis.* **1998**, *26*, 874.
57. Donald, P. R.; Sirgel, F. A.; Venter, A.; Parkin, D. P.; Van de Wal, B. W.; Barendse, A.; Smit, E.; Carman, D.; Talent, J.; Maritz, J. *Scand. J. Infect. Dis.* **2001**, *33*, 466.
58. Prabhakaran, K.; Harris, E. B.; Randhawa, B.; Adams, L. B.; Williams, D. L.; Hastings, R. C. *Microbios* **1993**, *76*, 251.
59. Gonzalo, X.; Drobniewsk, F. *J. Antimicrob. Chemother.* **2013**, *68*, 366.
60. Veziris, N.; Truffot, C.; Mainardi, J.; Jarlier, V. *Antimicrob. Agents Chemother.* **2011**, *55*, 2597.
61. England, K.; Boshoff, H. I. M.; Arora, K.; Weiner, D.; Dayao, E.; Schimel, D.; Laura, E.; Barry, C. E., 3rd. *Antimicrob. Agents Chemother.* **2012**, *56*, 3384.
62. Solapure, S.; Dinesh, N.; Shandil, R.; Ramachandran, V.; Sharma, S.; Bhattacharjee, D.; Ganguly, S.; Reddy, J.; Ahuja, V.; Panduga, V.; Parab, M.; Vishwas, K. G.; Kumar, N.; Balganesh, M.; Balasubramanian, V. *Antimicrob. Agents Chemother.* **2013**, *57*, 2506.
63. Payen, M.; De Wit, S.; Martin, C.; Sergysels, R.; Muylle, I.; Van Laethem, Y.; Clumeck, N. *Int. J. Tuberc. Lung Dis.* **2012**, *16*, 558.
64. Lorenzo, S. D.; Alffenaar, J. W.; Sotgiu, G.; Centis, R.; D'Ambrosio, L.; Tiberi, S.; Bolhuis, M. S.; van Altena, R.; Viggiani, P.; Piana, A.; Spanevello, A.; Migliori, G. B. *Eur. Respir. J.* **2013**, *41*, 1386.
65. Dauby, N.; Muylle, I.; Mouchet, F.; Sergysels, R.; Payen, M. C. *Pediatr. Infect. Dis. J.* **2011**, *30*, 812.
66. Zhang, X. H.; Yu, H.; Zhong, W.; Wang, L. L.; Song, L. *Chin. J. Chem.* **2009**, *20*, 1019.
67. Kremer, L.; Douglas, J. D.; Baulard, A. R.; Morehouse, C.; Guy, M. R.; Alland, D.; Lynn, G.; Dover, L. G.; Lakey, J. H.; Jacobs, W. R.; Brennan, P. J.; Minnikin, D. E.; Besra, G. S. *J. Biol. Chem.* **2000**, *275*, 16857.
68. Stanley, S. A.; Kawate, T.; Iwase, N.; Shimizu, M.; Clatworthy, A. E.; Kazyanskaya, E.; James, C.; Sacchettini, J. C.; Ioergere, T. R.; Siddiqif, N. A.; Minamif, S.; Aquadroa, J. A.; Granta, S. S.; Rubin, E. J.; Hung, D. T. *Proc. Natl. Acad. Sci. U.S.A* **2013**, *110*, 11565.

CHAPTER FIFTEEN

Nucleosides and Nucleotides for the Treatment of Viral Diseases

Michael J. Sofia
OnCore Biopharma, Inc. and The Blumberg Institute, Doylestown, Pennsylvania, USA

Contents

1. INTRODUCTION

Nucleosides and nucleotides have played an integral role in the treatment of viral diseases. For patients with human immunodeficiency virus (HIV), they have proven to be the backbone in number of combination regimens. Currently, nucleos(t)ides are the preferred option and standard of care for treating patients infected with hepatitis B virus (HBV) and they are emerging as a key component in therapies to treat hepatitis C virus (HCV) infection. They also play a central role in the management of other viral infections such as those caused by herpes viruses (HSV-1 and HSV-2), varicella zoster virus, Epstein–Barr virus, and cytomegalovirus.[1,2] The attractiveness of a nucleos(t)ide strategy in the development of therapeutics for viral diseases stems from the fact that all viruses require a polymerase for either DNA or RNA replication. Although viral polymerases are frequently differentiated from host polymerases, the development of a nucleos(t)ide as a treatment for a viral disease requires careful attention to toxicity associated with host replication processes. Particular attention must be given to

Annual Reports in Medicinal Chemistry, Volume 49
ISSN 0065-7743
http://dx.doi.org/10.1016/B978-0-12-800167-7.00015-8

assessment of mitochondrial toxicity, bone marrow cellular toxicity, and the inhibition of host DNA polymerases.

Another factor that must be considered when developing a nucleos(t)ide inhibitor pertains to nucleos(t)ide metabolic activation. It is the nucleotide triphosphate analog, as the functional substrate for the viral polymerase, that becomes incorporated into the growing RNA or DNA chain, typically leading to a chain termination event and ultimately an end to viral replication. Consequently, the efficiency by which a nucleos(t)ide gets converted to the active triphosphate and the concentration and half-life of the triphosphate within the cell are important factors in how effective the nucleos(t)ide is as an inhibitor of viral replication. In general, the first phosphorylation step is the most discriminating among the three needed to generate the active triphosphate. In cases where the nucleoside itself is not a good substrate for the kinase involved in the initial phosphorylation step, delivery of the monophosphate is desired, but this typically requires the use of prodrug technology to mask the unfavorable characteristics of the phosphate group and facilitate permeability.[3] Consequently, nucleotide prodrug strategies have seen much use in the development of nucleotides to treat viral diseases. This review will focus on recent developments in the search for nucleoside and nucleotide therapies for the treatment of HIV, HBV, HCV, and Dengue virus.

2. HUMAN IMMUNODEFICIENCY VIRUS

HIV is a retrovirus that infects approximately 35 million individuals worldwide. HIV requires a RNA-dependent DNA polymerase or reverse transcriptase (RT) for replication of the viral genome. A number of nucleoside and nucleotide RT inhibitors (**1**–**8**) have been approved for the treatment of HIV infection.[4,5] Several have been coformulated into fixed-dose combinations and in some cases coformulated with other HIV replication inhibitors affording convenient therapeutic regimens that have become the standard in highly active antiretroviral therapy (HAART). These fixed-dose combinations include Combivir®, Trizivir®, Epzicom®, Truvada®, Atripla®, Stribild®, and Complera®. Truvada®, Atripla®, Complera®, and Stribild® include the nucleoside emtricitabine (FTC, **4**) and the acyclic nucleotide tenofovir diisoproxil fumarate (TDF, **2**) while Combivir™, Trizivir®, and Epzicom® comprise a two- or three-drug combination comprising the nucleosides zidovudine (AZT, **3**), lamivudine (3TC, **5**), and/or abacavir (ABC, **8**).[4] The success of HAART has made HIV a manageable disease and has dramatically increased the life expectancy of those infected with this once terminal illness. However, even with this success the search continues for new agents that address the needs of a chronically infected population where resistance, side effects due to long-term use, and drug–drug interactions are increasingly prevalent especially within an aging HIV population.

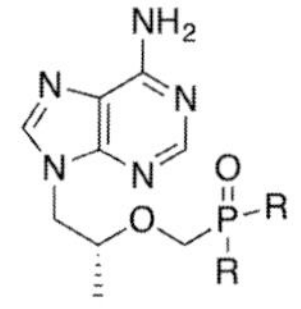

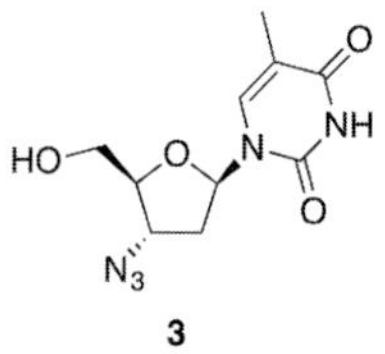

In an effort to increase targeted exposure of tenofovir (TFV) in peripheral blood mononuclear cells (PBMCs) and consequently reduce the renal toxicity associated with TDF, a phosphoramidate prodrug, tenofovir alafenamide (TAF) (GS-7340, **9**), was developed.[6] TAF was determined to be 400-fold more potent than TFV (**1**) in PBMCs and cleavage to TFV was mediated by lysosomal cathepsin A, which is highly expressed in these cells. The use of this prodrug approach resulted in an enhanced exposure ratio of the parent nucleoside TFV in PBMCs relative to plasma and led to higher efficacy. Viral load declines of −1.57 and −1.71 $\log_{10}$ IU/mL at doses of 50 and 100 mg, respectively, were observed in a previously untreated HIV-infected patient population treated for 14 days compared to a viral load decline of −0.94 $\log_{10}$ IU/mL for those receiving TDF (300 mg).[7] Subsequent monotherapy clinical studies have confirmed the superiority of TAF over TDF and demonstrated improvement in both efficacy and safety parameters at a substantially lower dose. TAF is currently in phase II/III clinical development as part of a fixed-dose combination protocol.

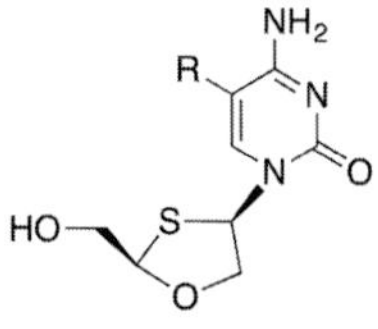

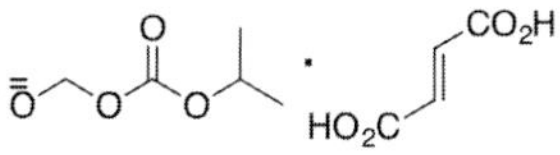

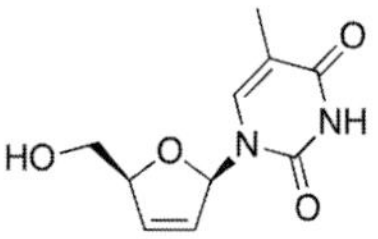

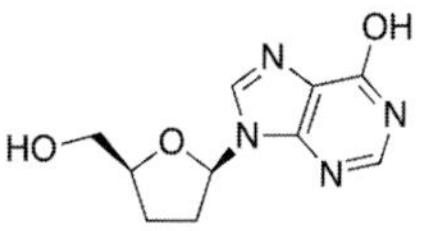

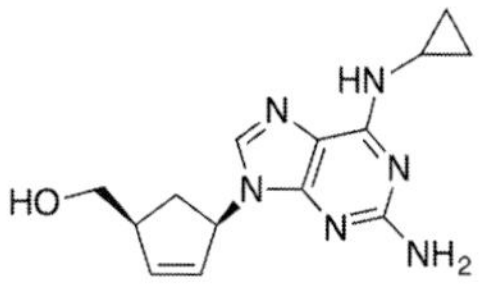

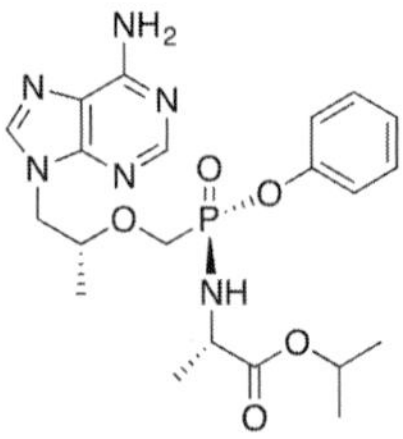

9

A search for a nucleotide phosphonate that would provide an improved resistance profile over existing nucleos(t)ides and that would exhibit a better safety profile relative to host DNA polymerases led to the 2′-F-2′,3′-dideoxydidehydro-adenosine derivative **10**.[6] This phosphonate nucleoside showed an improved resistance profile across a wide range of resistance mutations relative to all nucleos(t)ides in clinical use. In order to improve cell permeability properties and uptake into lymphoid cells, phosphoramidate prodrugs were evaluated as was previously demonstrated in the case of **9**. Ultimately, the *in vivo* profile and PBMC loading characteristics led to the identification of GS-9131 (**11**) as a lead development candidate.[6] *In vitro* and *in vivo* studies showed that **11** had a reduced potential for renal accumulation relative to TDF and no significant renal findings were observed in 28-day toxicity studies in multiple species. GS-9131 entered clinical development but there have been no recent reports on its progress.

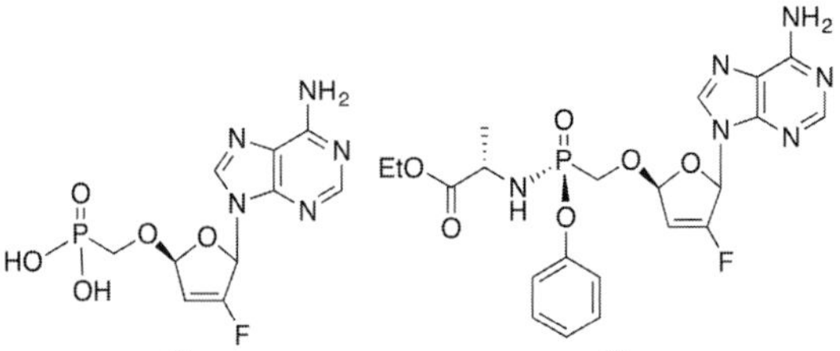

10 **11**

Another attempt to improve on the profile of TFV led to the hexadecyloxypropyl prodrug CMX157 (**12**).[8,9] This TFV lipid conjugate was designed to take advantage of generalized lipid uptake pathways and bypass the organic ion transporter mechanism that contributes to renal toxicity problems seen with TDF. CMX157 was active against many HIV-1 subtypes with EC_{50}s ranging from 0.20 to 7.2 nM, but the activity was affected by the presence of human serum. Significant activity was retained against an HIV-1 panel representing a number of nucleoside RT-resistant mutations. Low cytotoxicity and comparable bone marrow toxicity were seen with CMX157 relative to TFV. In a 7-day toxicity study in rats, CMX157 given orally presented no observed toxicity with a NOAEL of >100 mg/kg/day. Clinical progress has not been reported.

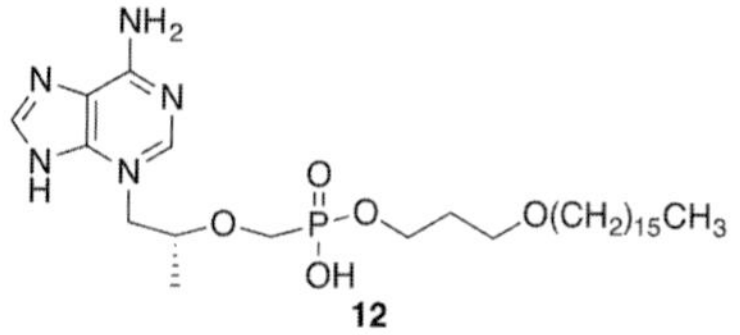

NH_2 N N N N H O P O OH O $O(CH_2)_{15}CH_3$

12

A variety of fatty acid lipid–nucleoside conjugates have been developed to both enhance membrane transport and introduce functionality that provides a second anti-HIV mechanism of action.[10–12] The second mechanism of action would result from the fatty acid inhibition of *n*-myristoyl transferase, which is involved in the myristoylation of several proteins in the HIV life cycle. Various fatty acid conjugates of AZT (**13D**), d4T (**14A–D**), and 3TC (**15B–D** and **16B** and **C**) were prepared.[10–12] In each case, the lipid conjugates were shown to be more potent *in vitro* and lacked cytotoxicity. Another approach investigated 1,3-diacylglycerol conjugates of AZT 5′-monophosphate in which the conjugate **17** demonstrated anti-HIV activity similar to the parent nucleoside AZT but suffered from chemical and enzymatic instability.[13]

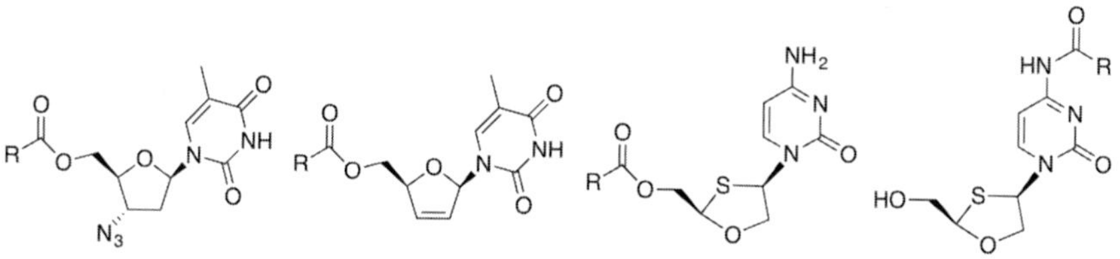

13D
(EC_{50} = 0.00031 μM)

14A,B,C,D
(EC_{50} = 0.002–14 μM)

15B,C,D
(EC_{50} = 0.5–2.3 μM)

16B,C
(EC_{50} = 5.3–10.9 μM)

A R = $CH_3(CH_2)_{12}$-
B R = $N_3(CH_2)_{11}$-
C R = $CH_3CH_2S(CH_2)_{11}$-
D R = $CO_2H(CH_2)_8$-

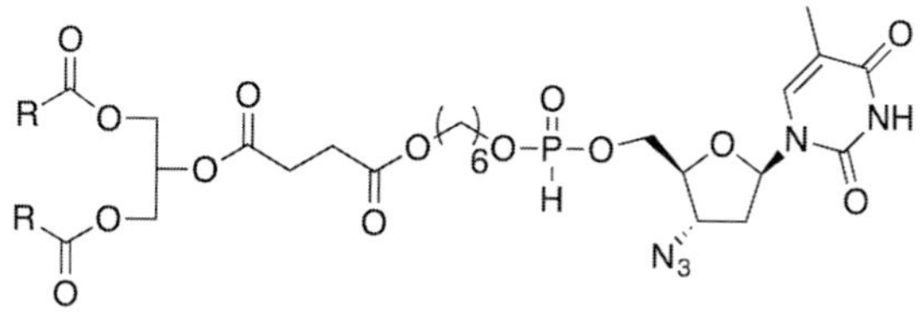

R = C_5H_{11}

17

An extensive SAR study of 4′-substituted nucleosides as inhibitors of HIV-1 has occurred over the years with the observation that 4′-ethynyl substitution provides the best potency versus cytotoxicity ratio.[14,15] Several such derivatives have progressed, including the thymidine derivative BMS-986001 (festinavir, **18**) and 2′-deoxy-4′-*C*-ethynyl-2-fluoroadenosine (EFdA, **19**).[16–18] BMS-986001 was reported to have an $EC_{50}=0.25$ μM, and with no cytotoxicity in CEM cells and no significant decrease in mitochondrial DNA or cell protein up to 200 μM appears less toxic than d4T (**6**). It showed equivalent activity against wild-type HIV-1 and HIV-1 with K65R and Q151M RT mutations, but reduced potency against multidrug-resistant isolates.[16] In a phase I clinical trial, **18** was well tolerated at doses up to 900 mg qd and subsequently showed efficacy (max −1.36 log IU/mL viral load) when dosed over 10 days at doses up to 600 mg qd.[7]

The observations that 4′-ethynyl purine derivatives were more potent as inhibitors of HIV-1 than their corresponding pyrimidine analogs, that guanosine and inosine bases led to cytotoxic nucleoside derivatives, and that substitution at the C-2 position of the adenosine base prevented deamination by adenosine deaminase ultimately led to the identification of **19** as a potent ($EC_{50}=0.068$ nM) noncytotoxic inhibitor of HIV-1.[14] **19** also showed activity against M184V and multidrug-resistant HIV-1 strains and had minimal effect on human mitochondrial DNA polymerase γ, with an $EC_{50}=10$ μM, or human DNA polymerases α and β. An *in vivo* study in two SIV-infected macaques with advanced AIDS showed that **19** was highly effective, demonstrating 3–4 log decreases in levels of virus within 1 week of treatment.[18] No signs of drug toxicity were observed over 6 months of continuous therapy with concomitant sustained viral suppression. Clinical data have yet to be reported.

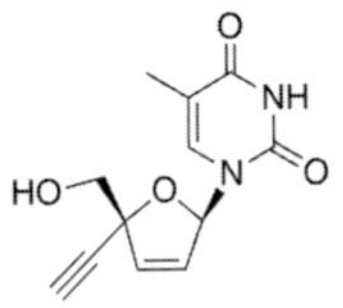

18 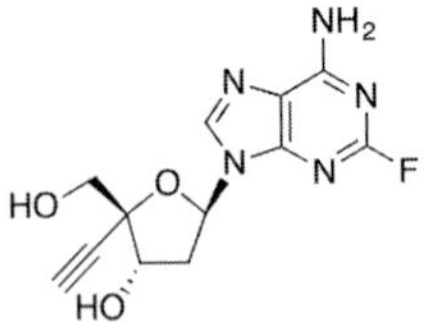 **19**

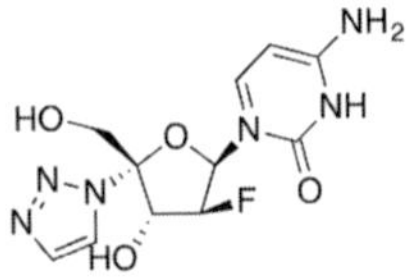

20

Another series of 4′-substituted nucleosides is represented by the 4′-triazoles. Introduction of a 4′-triazole moiety onto a

2′-deoxy-2′-fluoro-β-D-arabinofuranosylcytidine nucleus produced several potent inhibitors of HIV-1.[19] The most potent of these was the unsubstituted 4′-triazole derivative **20** ($EC_{50}=0.09$ μM).

Several dioxolane nucleoside mimetics have been shown to be potent inhibitors of HIV-1. The most advanced of these is (−)-β-D-2,6-diaminopurine dioxolane (amdoxovir, **21**), which is a prodrug of 9-(β-D-1,3-dioxolan-4-yl)-guanine (DXG, **22**).[7] Amdoxovir is in phase ll clinical development for the treatment of HIV-1 infection. Adenosine deaminase converts amdoxovir, via hydrolysis of the C-6 amino group, to DXG, which is then converted to the triphosphate active metabolite. DXG is active *in vitro* against M184V/I and thymidine analog HIV-1-resistant mutations but is less potent against K65R and L74V mutations. Amdoxovir administered to HIV-1-infected patients at 500 mg bid for 10 days resulted in a −1.00 $\log_{10}$ reduction in viral load, and a −2.00 $\log_{10}$ viral load reduction was observed in combination with zidovudine (200 mg bid). All adverse events were reported to be mild or moderate.[7]

(−)-β-D-(2*R*,4*R*)-Dioxolane-thymine (DOT, **23**) is a dioxolane nucleoside thymidine mimetic that was shown *in vitro* to be active against both wild-type ($EC_{50}=6.5$ μM) and clinically significant nucleoside-resistant HIV-1 subtypes.[20] However, DOT is a poor substrate in the first step of the phosphorylation cascade to the active triphosphate. An extensive SAR study was undertaken to assess phosphoramidate prodrugs identifying several such as **24** ($EC_{50}=0.23$ μM) having submicromolar activity.[20] Similarly, phosphoramidate prodrugs of 6-substituted-2-H-purine dioxolanes were investigated as double prodrugs that would afford dioxolane-A monophosphate **25**.[21] The most potent and least cytotoxic analog was **26** ($EC_{50}=0.086$ μM).

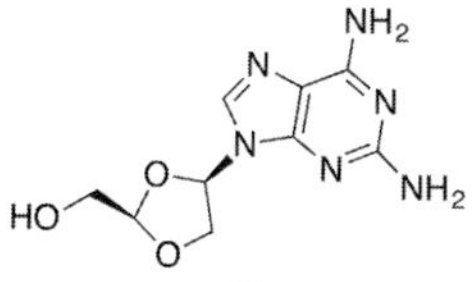

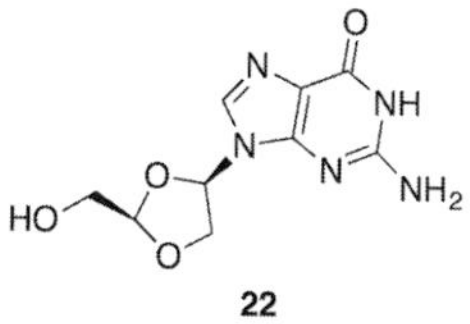

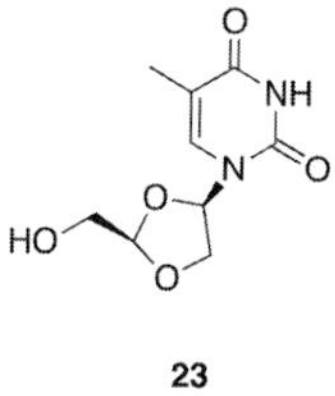

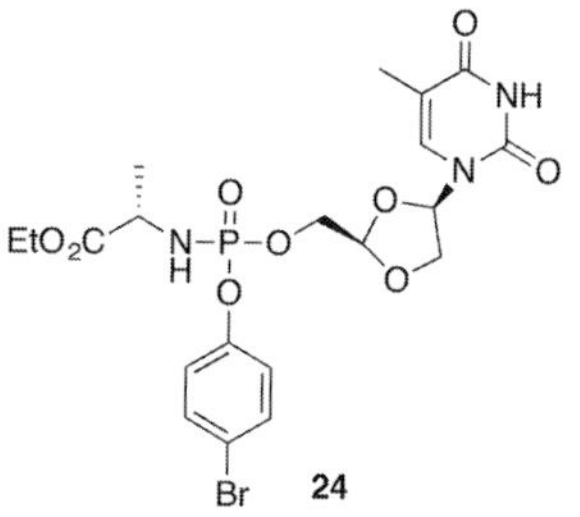

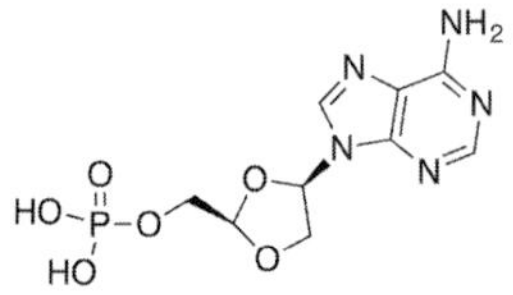

25

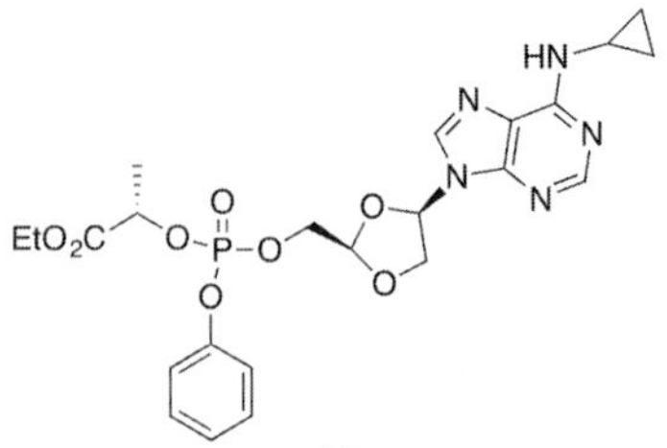

26

Thiophosphonate derivatives **27** and **28** of the acyclic nucleosides adefovir (**29**) and TFV were prepared and shown to inhibit HIV-1 replication with potency similar to adefovir and TFV.[22] They were also equally effective against a broad variety of clinical isolates *in vitro*.

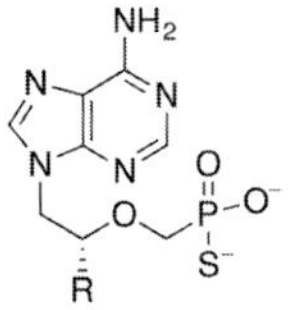

27 R = H
28 R = CH_3

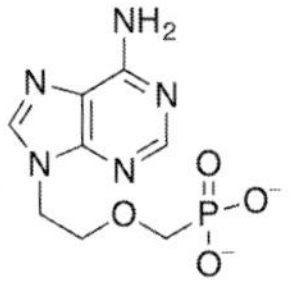

29

3. HEPATITIS B VIRUS

HBV is a DNA virus in the *Hepadnaviridae* family. It is estimated that 400 million individuals are infected with HBV worldwide. The current standard of care for treatment of HBV is long-term nucleos(t)ide therapy. The nucleos(t)ides approved for treating HBV infection include lamivudine, adefovir dipivoxil (**30**), entecavir (**31**), telbivudine (**32**), and TDF. Entecavir and TDF are the most widely prescribed of these agents. Long-term use of entecavir leads to resistance in a significant patient population and TDF is associated with nephrotoxicity and bone loss.[23,24] However, continued use of nucleos(t)ide therapy has been associated with reduction in liver fibrosis demonstrating that suppression of viral replication has positive long-term value.[25,26]

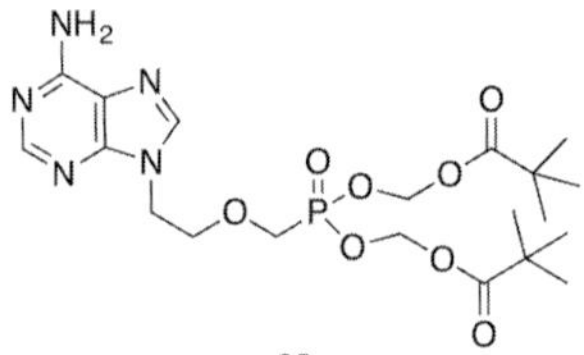

30 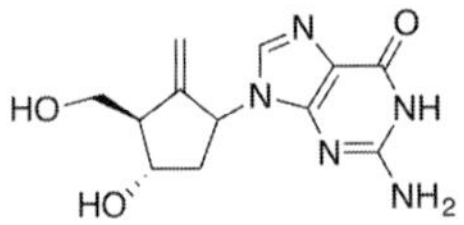 **31** 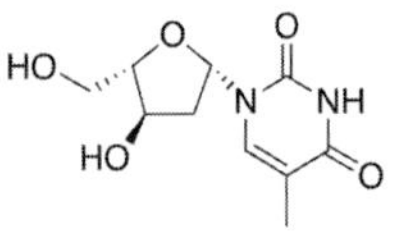 **32**

Even with the success of existing nucleos(t)ide HBV therapy, work has continued in an effort to identify novel inhibitors that may provide additional benefit relative to the existing agents, and several of the anti-HIV agents mentioned above have also been assessed for us in treating HBV infection.[27] Relative to TDF, TAF delivers high levels of TFV to hepatocytes as well as PBMCs and is under development for the treatment of HBV as well as HIV infection.[28] In a 28-day phase I study in HBV-infected patients at doses of 8–120 mg, TAF produced approximately a 2.5 log IU/mL decline in viral load. This effect is more pronounced than that seen with at 300 mg dose of TDF.[28] In addition, TAF doses of 25 mg or lower reduced plasma levels of TFV to less than 8% of those seen with a 300 mg dose of TDF and generated smaller decreases in creatinine clearance.

Recently, preparation of the 2′-fluoro-6′-methylene-carbocyclic adenosine (FMCA, **33**) ($EC_{50}=0.55$ μM), which borrowed the 6′-methylene-carbocyclic nucleus of entecavir, led to a potent inhibitor of HBV replication that was also active against the lamivudine–entecavir-resistant clone (L180M + M204V + S202G).[29] Furthermore, preparation of the corresponding 5′-phosphoramidate prodrug **34** resulted in a compound that was 10-fold more potent than **33** against both the wild-type ($EC_{50}=0.62$ μM) and resistant mutant ($EC_{50}=0.054$ μM).[29]

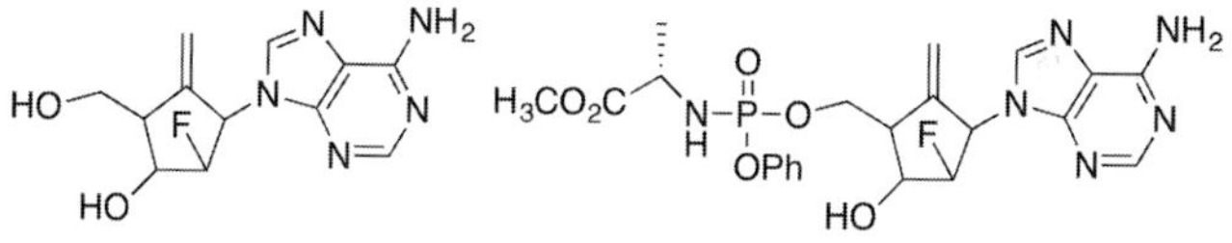

33 **34**

A series of 5′-carboxylic acid nucleosides was investigated for their anti-HBV activity with the rationale that the 5′-carboxyl group could behave as an isosteric replacement for the 5′-hydroxyl group of the ribose nucleus. Several modestly active inhibitors (**35**–**37**) were identified, with EC_{50}s ranging from 5 to 10 μg/mL.[30]

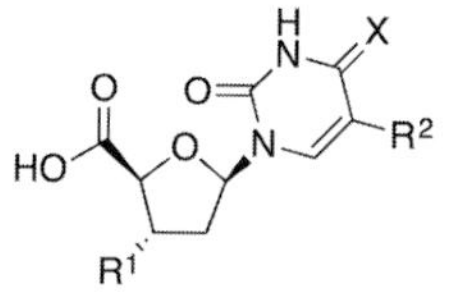

35 R^1 = H, R^2 = H, X = NH
36 R^1 = N_3, R^2 = H, X = O
37 R^1 = N_3, R^2 = CH_3, X = O

2′-β-F-4′-Substituted cytidine derivatives (e.g., **38**) in which the 4′-substituent was a 4-substituted-1,2,3-triazole moiety were shown to exhibit potent activity in HepG2.2.15 cells.[19] These nucleosides demonstrated the ability to reduce viral HBsAg and HBeAg production without accompanying cytotoxicity.

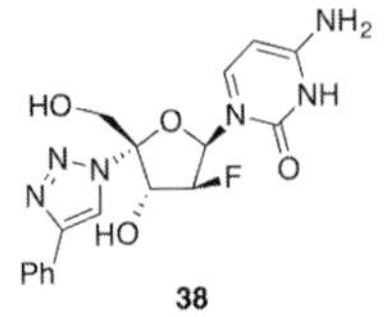

38

(EC_{50} = 0.1μM, HBsAg; EC_{50} = 0.25μM, HBeAg)

The combination of acyclovir **39** and silatrane led to conjugate **40** with the rationale that silatrane could promote activation and proliferation of T lymphocytes and enhance the weak activity of the nucleoside.[31] The conjugate **40** had positive effects on the secretion of HBsAg and HBeAg in both cell culture and in a HBV transgenic mouse model. In addition, this conjugate produced increases in both mRNA levels and serum levels of both IFN-γ and IL-2 *in vivo*. Increases in T lymphocyte subgroups were also observed in transgenic mice treated with **40**.

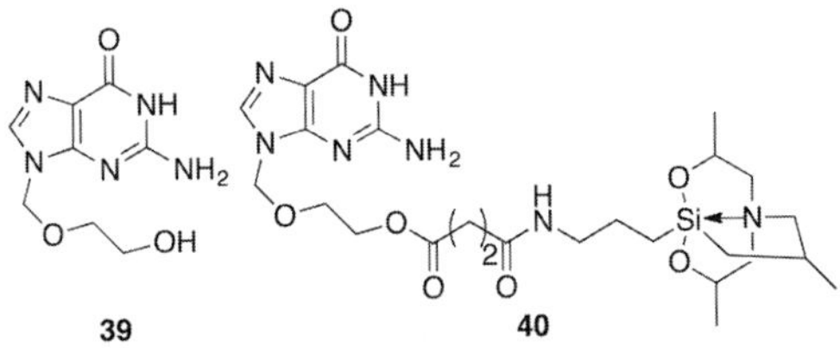

39 40

Clevudine (**41**) is a nucleoside approved in South Korea for treatment of HBV-infected patients.[32] However, clinical studies in the United States were terminated because of drug-related myopathy.[32] Clevudine has the unusual ability to maintain viral suppression for an extended period of time. Unlike other nucleoside HBV replication inhibitors, clevudine is not considered to be an obligate chain terminator because it maintains a 3′-OH group. Recent studies have shown that clevudine triphosphate is a non-competitive inhibitor of HBV DNA elongation, binding to the HBV polymerase active site so as to induce distortions that prevent its incorporation into the growing DNA strand.[33,34] Further studies showed that clevudine is unique in demonstrating synergistic effects in combination with any of the other HBV polymerase inhibitors currently approved for therapy. These results illustrate the potential clinical value of an analog that lacks the toxicity associated with clevudine itself.

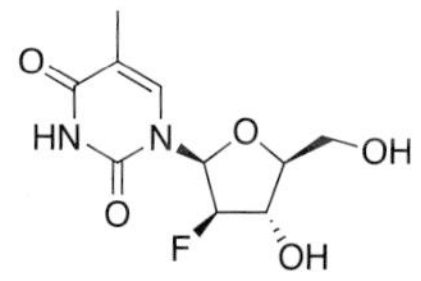

41

In addition to their reported anti-HIV activity, compounds **12** (EC_{50} = 1.6 μM),[8] **26** (EC_{50} = 0.8 μM),[21] **27** (EC_{50} = 3.0 μM),[22] and **28** (EC_{50} = 2.8 μM)[22] were also reported to demonstrate inhibition of HBV replication.

4. HEPATITIS C VIRUS

The development of nucleoside and nucleotide inhibitors of the HCV RNA-dependent RNA polymerase (RdRp) has continued since 2004 with no less than 16 agents entering clinical development (Table 15.1).[35,36] In the last several years, a number of extensive reviews covering HCV polymerase inhibitors have appeared.[35,36] Recently, the uridine nucleotide prodrug sofosbuvir (SOF, **42**)[37] became the first nucleos(t)ide approved by both the FDA and EU regulatory authorities for the treatment of HCV patients infected with genotype (GT) 1, 2, 3, and 4 virus, and in clinical trials it also showed efficacy against all relevant HCV GTs (1–6).[38,39] Its approval marked the first introduction of an all-oral interferon (IFN)-free regimen to treat patients suffering from HCV infection. SOF + ribavirin (RBV, **43**) administered orally to HCV patients infected with GT-2 or -3 virus for 12 or 24 weeks demonstrated high cure rates (sustained virological response 12 weeks after discontinuation of therapy, SVR12).[40] In the GT-2 patient population, SOF + RBV was shown to eradicate the virus (SVR12) in 93% of treatment naïve and IFN-intolerant, ineligible or unwilling patients. The GT-3 patient population required a longer duration of therapy (24 weeks) to achieve high cure rates (SVR12 = 84%). For GT-1 and -4 patients, the combination of SOF + IFN/RBV was required to achieve high SVR12 rates (GT-1 = 89% and GT-4 = 96%).[39] In a small cohort of GT-5 and -6 patients, the cure rates were reported to be 100%. Even in difficult-to-treat patient populations such as African Americans with varying stages of liver fibrosis and generally unfavorable IL28B allele frequency, HIV–HCV-coinfected patients, and liver transplant patients with recurrent HCV, high cure rates were achieved.[40–43] IFN-free combinations of SOF, either with the NS5A inhibitors daclatasvir or ledipasvir or with the protease inhibitor

Table 15.1 Nucleoside and nucleotide HCV polymerase inhibitors that have entered clinical development

Compound	Class	Nucleobase	Phase of development	Development status	Issues
NM283	2′-Methyl	Cytosine	II	Discontinued	GI toxicity
R1626	4′-Azido	Cytosine	II	Discontinued	Hematological toxicity
PSI-6130	2′-Methyl-2′-F	Cytosine	I	Discontinued	PK
Mericitabine (RG7128)	2′-Methyl-2′-F	Cytosine	III	Active	–
Sofosbuvir (GS-7977 and PSI-7977)	2′-Methyl-2′-F	Uracil	Marketed	FDA and EU approved	–
GS-938/PSI-352938	2′-Methyl-2′-F	Guanine	II	Discontinued	Liver enzyme elevation
IDX-184	2′-Methyl	Guanine	II	Discontinued	Potential cardio toxicity
INX-189	2′-Methyl	Guanine	II	Discontinued	Cardio toxicity
RG7348	4′-Azido	Uracil	I	Discontinued	Efficacy
ALS-2158	2′-Methyl-?	Adenine	I	Discontinued	Efficacy
VX-135/ALS-220	2′-Methyl-?	Uracil	II	Active	–
ACH-3422	2′-Methyl-?	Uracil	II	Active	–
GS-6620	2′-Methyl	Adenine	I	Discontinued	PK
IDX-20963	2′-Methyl-?	Uracil	IND	Hold	Preclinical toxicity
IDX-21437	?	Uracil	I/II	Active	–
IDX-21459	?	Uracil	I	Active	–

simeprevir (with or without RBV), demonstrated that simple two-drug combinations of direct acting antivirals were very effective at producing high cure rates (SVR12 95–100%).[44–46] The NDA for the two-drug fixed-dose combination of SOF and ledipasvir was recently submitted for regulatory approval. Resistance to SOF has not emerged as a clinical concern, primarily because the NS5B S282T mutant that is less susceptible *in vitro* is not found in pretreated patients and is also quite unfit.[47,48]

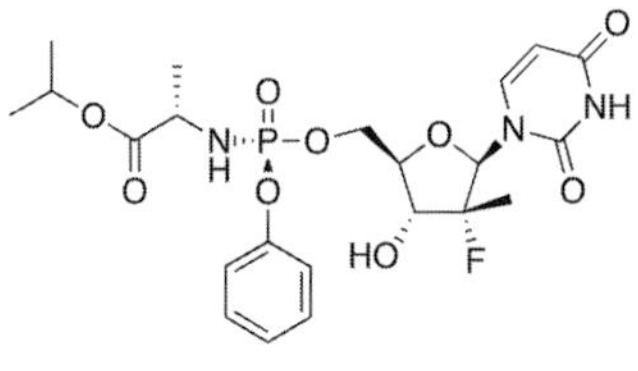

42
(EC_{90} = 0.42 μM)

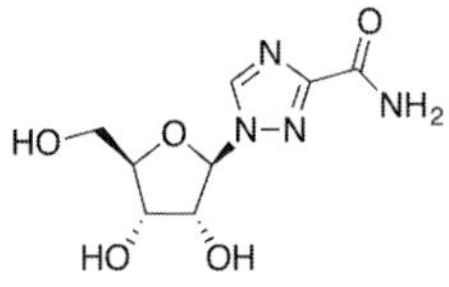

43

The success of SOF has prompted a flurry of activity around 2′-methyluridine nucleotide prodrugs (Table 15.1).[35,36] Although the detailed structures of these agents have yet to be disclosed, the 2′-methyl characteristic can be inferred from their reported resistance profile. A number are in active clinical development, but IDX20963 (Table 15.1) is currently on clinical hold because of a reported preclinical toxicity signal.

Attempts to identify new nucleos(t)ide agents to treat HCV have persisted in an effort to capitalize on the pan-genotypic character and high barrier to resistance exhibited by the class. A significant effort was focused on purine nucleotides because of early work that showed them to be exceptionally potent in the whole cell replicon assay. This led to the investigation of a number of guanosine nucleotide analogs. PSI-352938 (**44**) was the first in a class of cyclic phosphate prodrugs and also employed a double prodrug strategy.[49,50] The cyclic phosphate prodrug release mechanism was shown to be mediated by CYP3A4.[51] PSI-352938 demonstrated modest replicon activity (EC_{90} = 1.37 μM) but was shown to be active against the NS5B S282T mutation known to confer resistance to 2′-F-2′-*C*-methyl pyrimidine nucleosides; three other mutations (S15G, C223H, and V321I) were required to confer a high level of resistance.[50] In a phase I clinical study, PSI-352938 achieved exceptional efficacy (−3.94 $\log_{10}$ IU/mL viral load decline) when administered to HCV GT-1 patients.[52] In a proof of concept combination study, administration with SOF over 14 days led to a −5.5 $\log_{10}$ reduction in viral load, with 94% of patients being below the limits of detection.[53] Unfortunately, liver enzyme elevations in patients receiving

PSI-352938 led to discontinuation of a subsequent study. Additional SAR work showed that improved replicon potency could be achieved through optimization of cyclic phosphate prodrug substituents, as seen for compound **45**.[54] However, further development has not been reported.

Other guanosine nucleotide prodrugs have been investigated, also employing the phosphoramidate prodrug moiety in an attempt to leverage liver targeting. Thus, PSI-353661 (**46**) demonstrated potent inhibition in the replicon assay ($EC_{90}=0.008$ μM) and a novel resistance profile similar to PSI-352938, but was never progressed into clinical development.[55,50] The structurally related 2′-*C*-methylguanosine 5′-phosphoramidate prodrugs IDX-184 (**47**)[56,57] and INX-08189 (BMS-986094, **48**),[58–60] each producing an identical triphosphate, were progressed into the clinic, but severe cardiovascular toxicity associated with INX-08189 resulted in discontinuation of development for both compounds.[61] The severe nature of the cardiovascular toxicity seen with INX-08189 seems to have curtailed the interest in developing a guanosine nucleoside for treating HCV patients.

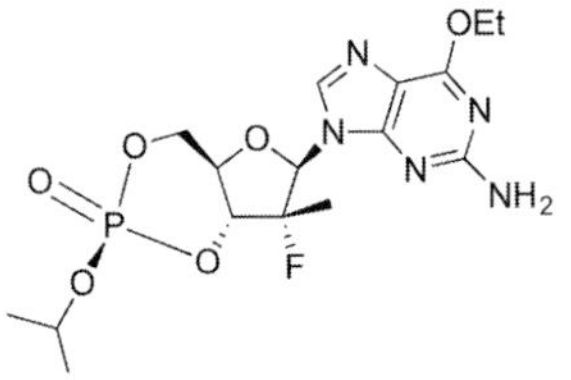

44
($EC_{90}=1.37$ μM)

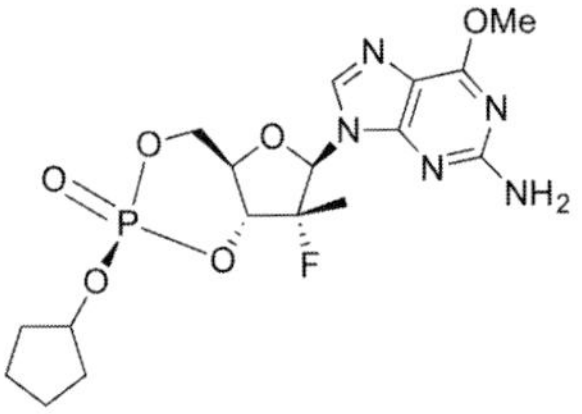

45
($EC_{50}=0.10$ μM)

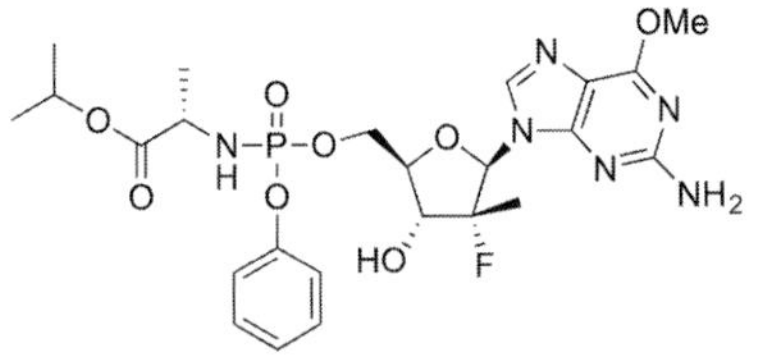

46
($EC_{90}=0.008$ μM)

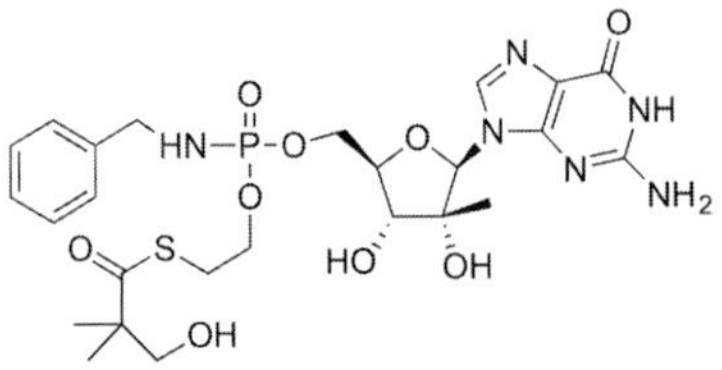

47
($EC_{50}=0.4$ μM)

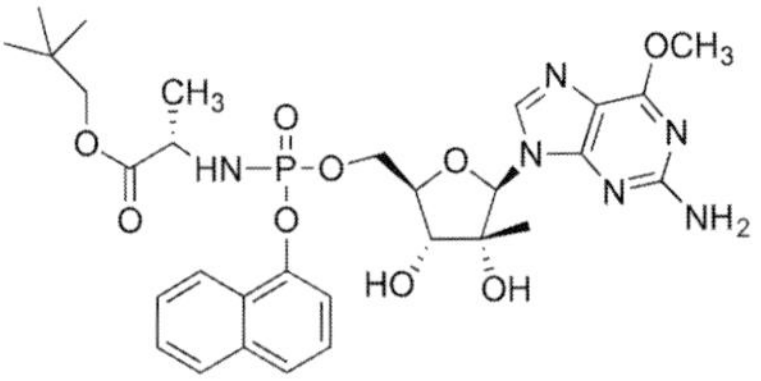

48
($EC_{50}=0.010$ μM)

C-Nucleosides, which comprise a furanose sugar moiety connected to a non-natural heterocyclic base via a carbon–carbon bond, are known to be more resistant to enzymatic and hydrolytic cleavage than their *N*-nucleoside counterparts. To investigate the potential for *C*-nucleosides as inhibitors of HCV replication, several reports have investigated both purine and pyrimidine *C*-nucleoside derivatives (**49–55**).[62,63] In general, these derivatives were less active as inhibitors of HCV replication in the whole cell replicon assay than their *N*-nucleoside derivatives. Only the adenosine analogs **49** ($EC_{50}=1.28$ μM) and **50** ($EC_{50}=1.98$ μM) demonstrated inhibition of HCV replication with their corresponding triphosphates showing inhibition of the viral polymerase. Compound **50** was efficiently phosphorylated in primary human hepatocytes, had high oral bioavailability in dogs (70%) and rats (50%), and showed superior liver triphosphate levels (two- to five-fold) in hamsters, relative to its *N*-nucleoside counterpart, when dosed orally.[62] Despite a favorable *in vitro* toxicity profile, adverse events, including death, were observed when **50** was dosed orally in rats.[63]

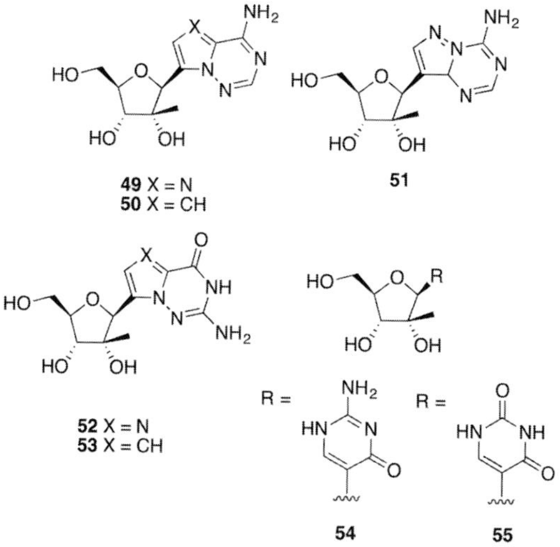

49 X = N
50 X = CH

51

52 X = N
53 X = CH

54

55

A unique series of 1′-substituted *C*-nucleosides possessing a 4-aza-7,9-dideaza-adenosine base **56** were shown to have modest activity against HCV with improved potency upon preparation of their *S*-acyl-2-thioethyl prodrugs **57**.[64] However, they also demonstrated substantial cytotoxicity. Preparation of the 2′-*C*-methyl-1′-cyano derivatives eliminated the cytotoxicity and subsequent development of the 5′-phosphoramidate prodrug of the 1′-cyano analog resulted in the clinical candidate GS-6620 **58**.[65] Like many other HCV nucleos(t)ide inhibitors, GS-6620 contained a

2′-*C*-methylribosyl core and maintained a pan-genotypic profile. In a phase I human clinical study, GS-6620 demonstrated efficacy in HCV-infected patients, but a variable intra- and interpatient PK and PD profile ultimately led to termination of its development.[66]

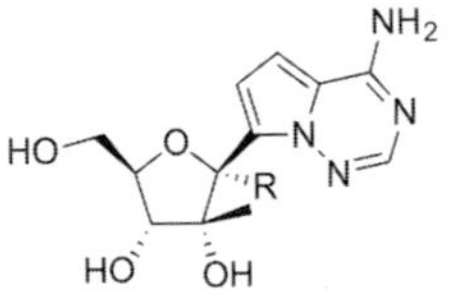

R = CN (EC_{50} = 4.1 μM), CH_3, vinyl, ethynyl

56

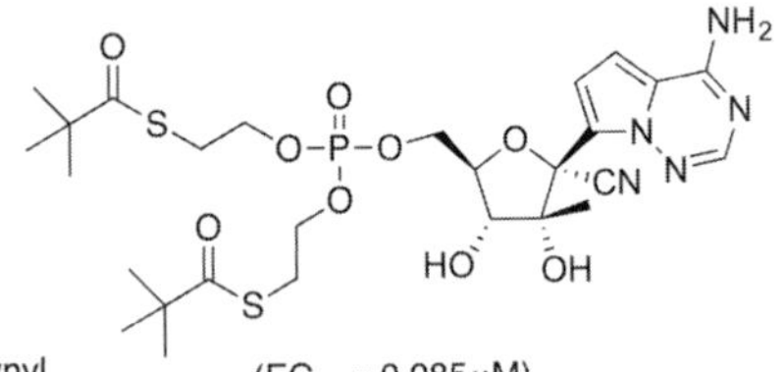

(EC_{50} = 0.085 μM)

57

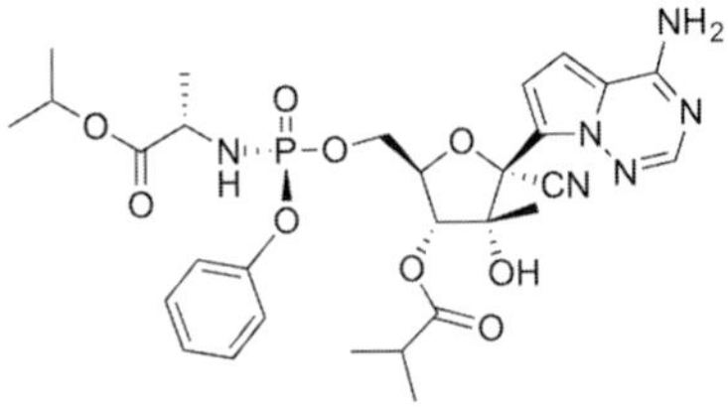

(EC_{50} = 0.36 μM)

58

In attempts to identify a novel class of nucleosides that inhibit HCV, several groups searched for 2′-*C*-methyl motif replacements. This search led to the identification of the 2′-cyclopropyl and 2′-spiro oxetane nucleosides and their corresponding 5′-phosphate prodrugs.[67–70] In the 2′-cyclopropyl series, TMC647078 (**59**) was active in the replicon assay (EC_{50} = 7.3 μM) across all GTs, but exhibited reduced activity against NS5B S282T.[67,68] Low plasma exposure in rats led to the preparation of the ester prodrugs **60** and **61**, which resulted in reduced replicon activity (EC_{50} = 42.8 and 20.3 μM, respectively) but improved *in vivo* exposure relative to the parent nucleoside. No further development of this series has been reported.

Investigation of a series of 2′-spironucleosides bearing either a 2′-oxetane or 2′-tetrahydrofuran moiety (**62**) identified several oxetanes as inhibitors of the HCV polymerase but only weakly active as inhibitors of HCV replication in the replicon assay.[69,70] Subsequent preparation of phosphoramidate nucleotide prodrug variants (**63**) resulted in increased whole cell potency. Several compounds were shown to generate substantial levels of the active triphosphate when incubated with primary human hepatocytes or *in vivo* in rat liver. The guanosine derivative **64** also was shown to be equipotent in replicons having either the wild-type or S282T mutant polymerase.

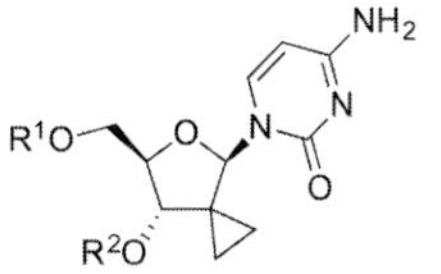

59 R^1, R^2 = H, (EC_{50} = 7.3μM)
60 R^1 = H, R^2 = $C(O)CH(CH_3)_2$ (EC_{50} = 42.8μM)
61 R^1, R^2 = $C(O)CH(CH_3)_2$ (EC_{50} = 20.3μM)

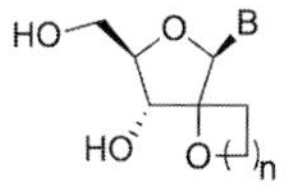

n = 1, 2
B = Uracil, cytosine, guanine
(EC_{50} = 14.9μM - >100μM)
62

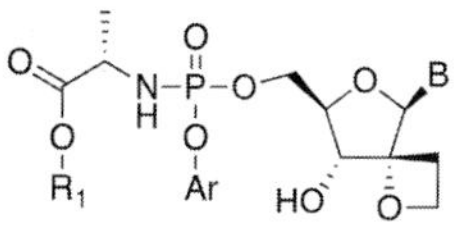

B = Cytosine, uracil
(EC_{50} = 0.74 - >96μM)
63

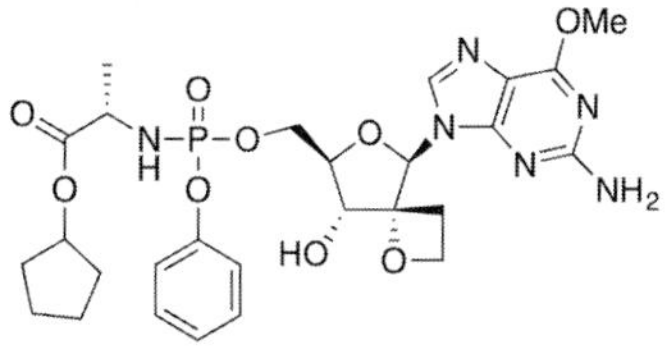

(EC_{50} = 0.45μM)
64

The pursuit of nucleosides containing novel bases has seen limited success. 7-Heterocyclic substituted 7-deaza-adenine derivatives containing either the 2′-F-2′-*C*-methyl or 2′-OH-2′-*C*-methyl substitution (**65**, **66**) demonstrated modest HCV replicon potency and modest liver triphosphate levels when administered *in vivo* to rats.[71] Preparation of phosphate prodrugs (**67**) proved beneficial only for the derivative with the 2′-OH-2′-*C*-methyl substitution.[71] In a similar fashion, an extensive series of 6-substituted-7-heteroaryl-7-deazapurine ribonucleosides (**68–70**) were studied, but replicon activity was generally correlated with cytotoxicity.[72] However, in a few examples, low micromolar EC_{50}s were observed with minimal cytotoxicity.

Another attempt to identify novel base containing HCV nucleos(t)ide inhibitors was inspired by Janus linear tricyclic nucleosides (**71–73**).[73] Unfortunately, neither of these nucleosides nor a representative phosphoramidate prodrug demonstrated anti-HCV activity. Most of the activity for these compounds was ascribed to cytotoxicity.

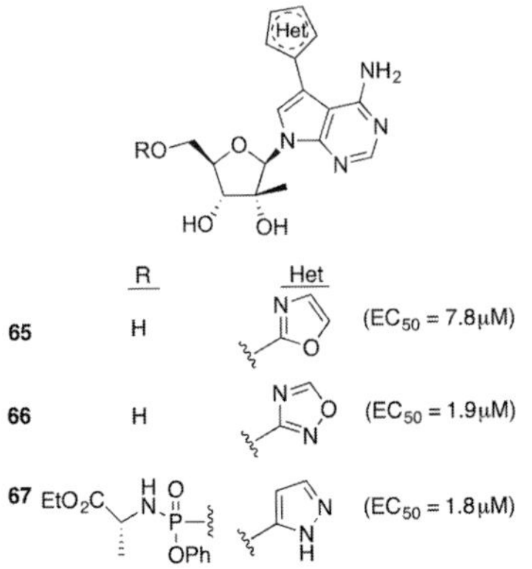

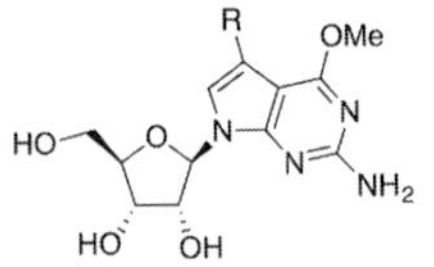

	R	Het	
65	H	(oxazolyl)	(EC_{50} = 7.8μM)
66	H	(oxadiazolyl)	(EC_{50} = 1.9μM)
67	EtO_2C…N(H)–P(O)(OPh)–	(pyrazolyl)	(EC_{50} = 1.8μM)

68 R = furan-2-yl (EC_{50} = 2.44μM)
69 R = thiophen-2-yl (EC_{50} = 6.07μM)
70 R = benzofuran-2-yl (EC_{50} = 8.98μM)

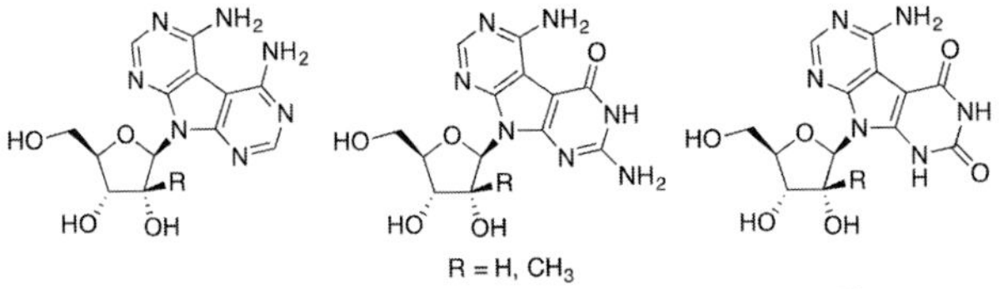

71 **72** (R = H, CH_3) **73**

Several carbocyclic nucleos(t)ides were prepared in an attempt to identify analogs that are more metabolically stable than ribose derivatives. The direct carbocyclic analog of SOF, **74**,[74] produced an inactive compound and several 7-substituted-7-deaza carbocyclic nucleosides **75**[75] were shown to be only weekly active, thus questioning the viability of this strategy.

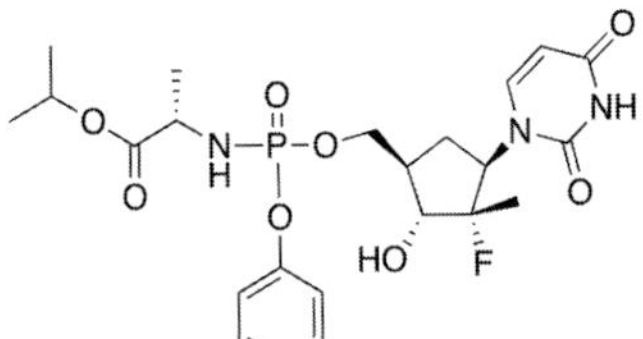

74

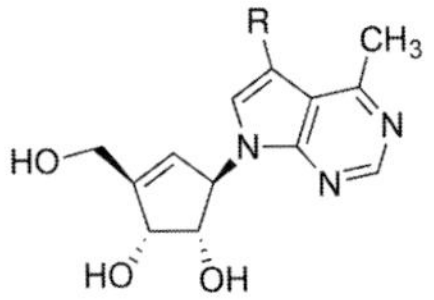

R = H, F, Cl, Br, I, vinyl, ethynyl
(EC_{50} = 28 - >100μM)
75

A variation of the proven phosphoramidate nucleotide prodrug strategy, phosphorodiamidates, was applied to 2′-methylguanosine derivatives.[58] The replacement of the aryloxy phosphate substituent with a second amino acid was intended to eliminate both the toxicological concern of releasing a phenol on promoiety decomposition and the chirality at phosphorus. This strategy produced a number of potent inhibitors of HCV replication (e.g., **76**) and demonstrated that significant liver levels of the corresponding nucleoside triphosphate could be achieved *in vivo* in rats.

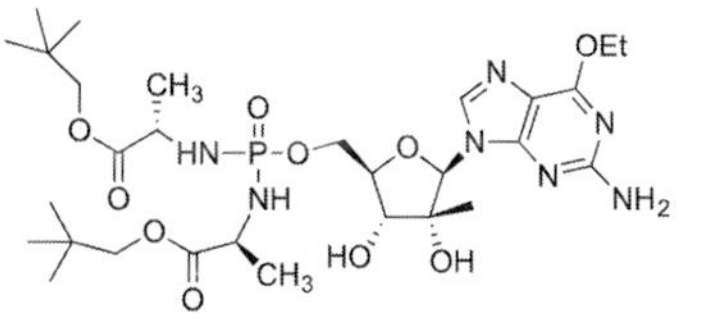

($EC_{50} = 0.04\mu M$)
76

Further investigation of the previously reported 2′-substituted 4′-azidocytidine series resulted in identification of the 4′-azido-2′-deoxy-2′-*C*-methylcytidine derivative **77** and its prodrugs as inhibitors of HCV replication.[76] The 3′,5′-diisobutryate ester prodrug TMC649128 (**78**) was subsequently identified as a clinical development candidate with 65% oral bioavailability when dosed in rats.

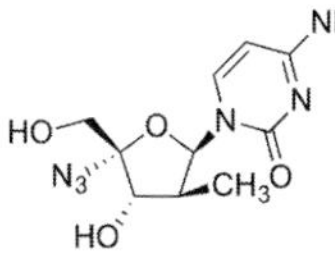

($EC_{50} = 1.2\mu M$)
77

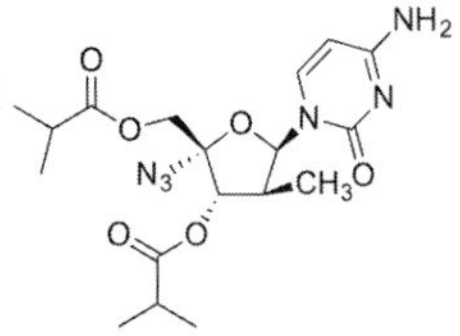

($EC_{50} = 6.6\mu M$)
78

A nucleoside phosphonate approach, in which the phosphonate moiety represents a bioisostere of the monophosphate, investigated cytidine, uridine, and adenosine analogs **79** and **80**.[77] No replicon activity was observed, despite good NS5B enzyme inhibitory activity for the diphosphate derivatives. Preparation of the diamidate prodrugs (e.g., **81**) did lead to whole cell activity but these prodrugs also displayed increased cytotoxicity.[77]

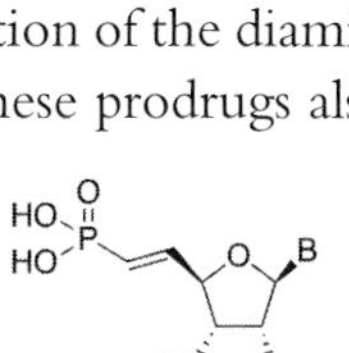

B = Cytosine (C) ($IC_{50} = 1.9\mu M$, $EC_{50} = 76\mu M$)
Uracil (U) ($IC_{50} = 31\mu M$, $EC_{50} = >250\mu M$)
Adenine (A) ($IC_{50} = 2.1\mu M$, $EC_{50} = 132\mu M$)
79

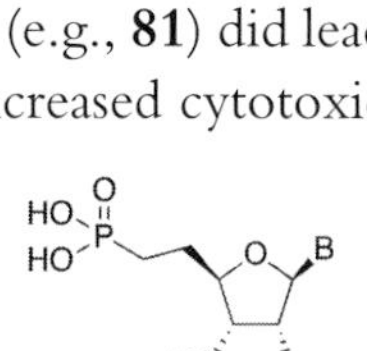

B = Adenine (A) ($IC_{50} = 4.6\mu M$, $EC_{50} = >500\mu M$)
80

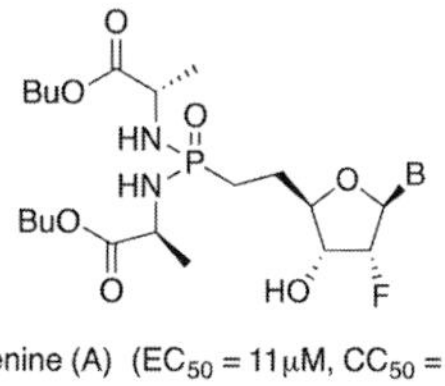

B = Adenine (A) ($EC_{50} = 11\mu M$, $CC_{50} = 24–36\mu M$)
81

5. DENGUE VIRUS

Like HCV, Dengue virus (DENV) is a member of the *Flaviviridae* family of positive RNA viruses. Within the genus are other viruses that have a significant impact on global human health that include West Nile virus (WNV), yellow fever virus (YFV), Japanese encephalitis virus (JEV), and tick-borne encephalitis virus (TBEV). There are vaccines for YFV, JEV, and TBEV but none exist for DENV or WNV. In addition, no direct acting antiviral treatments exist for DENV or WNV.

Although the structures of the RdRps in the *Flaviviridae* family have a low sequence homology, they all share the right-handed polymerase architecture featuring fingers, palm, and thumb subdomains.[78] In addition, these viral RNA polymerases also possess a relatively conserved palm domain that contains the active site motif GDD. One difference between the HCV and DENV and WNV RdRps is that HCV has a monofunctional RdRp, whereas DENV and WNV have both RdRp and methyltransferase activities. Because of the similarities, much of the work done in the HCV arena has been parlayed into the search for nucleos(t)ide inhibitors of DENV and WNV.[79,80]

Several examples of 2′-substituted nucleosides have been identified as inhibitors of DENV-2.[81,82] The preferred 2′-substituents, when combined with an adenine base, appear to be methyl (**82**, $EC_{50}=1.12$ μM) and ethynyl (**83**, $EC_{50}=1.41$ μM). Replacement of the natural adenine base with 7-deaza-adenosine in conjunction with the 2′-*C*-ethynyl substitution gave NITD008 (**84**), which showed only a modest increase in potency ($EC_{50}=0.7$ μM) but with reduced cytotoxicity.[83] NITD008 was also shown to be active against all DENV serotypes as well as WNV and HCV. Resistant virus did not readily emerge in cell culture. *In vivo* studies in a Dengue lethal mouse model demonstrated that using either single or multiple doses, **84** suppressed peak viremia, decreased cytokine elevation, and prevented death even when given after establishment of viral infection. Preparation of the diisobutyrate ester prodrug (NITD203, **86**) of the 7-acetamide-7-deaza-adenosine derivative (NITD449, **85**) led to a compound that was equipotent across DENV serotypes ($EC_{50}=0.54$–0.71 μM) with improved bioavailability.[84] *In vivo*, **86** was shown to be efficacious in a mouse viremia model with an NOAEL of 10 mg/kg/day in rats.

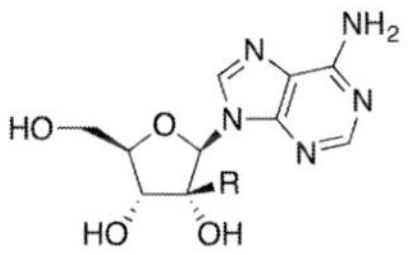

82 R = CH_3 (EC_{50} = 1.12 μM)
83 R = C≡CH (EC_{50} = 1.14 μM)

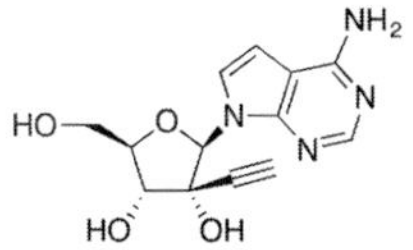

(EC_{50} = 0.7 μM)
84

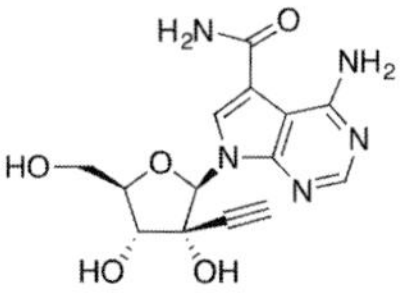

(EC_{50} = 2.61 μM DENV-2 NCG)
85

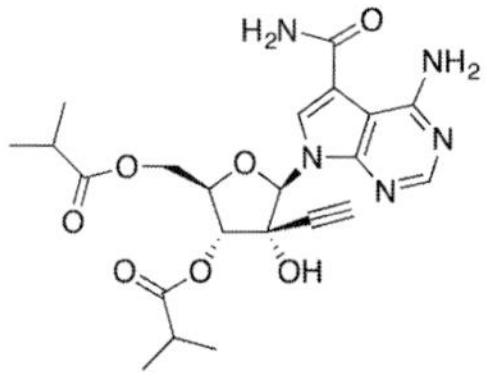

(EC_{50} = 0.69 μM DENV-2 NCG)
86

Another series of deaza-adenosine analogs were evaluated as inhibitors of DENV-2.[85] These compounds were designed based on the known nucleoside analogs tubercidin (**87**) and 6-methyl-9-β-D-ribofuranosylpurine (**88**), whose utility is limited by their cytotoxicity profile. Of the series (**89–92**), only the 6-methyl-7-deaza-adenosine derivative **91** (IC_{50} = 0.88 μM) showed activity in a DENV-2 replicon assay without cytotoxicity. Further evaluation demonstrated that **91** was a highly potent DENV-2 replication inhibitor in both a plaque reduction assay (IC_{50} = 0.062 μM) and by real-time PCR (IC_{50} = 0.039 μM). However, mechanistic studies indicate that **91** does not function as a inhibitor of the RdRp or as a chain terminator. Rather, its activity may be due to some effect on RNA–RNA interactions, RNA–protein interactions, or action on a host cell target.

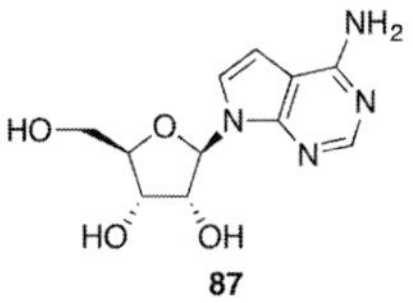

87

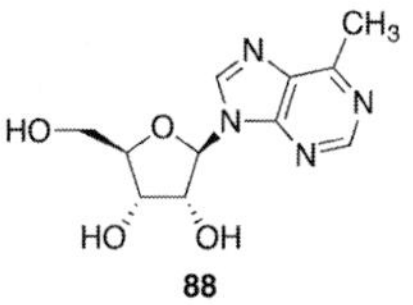

88

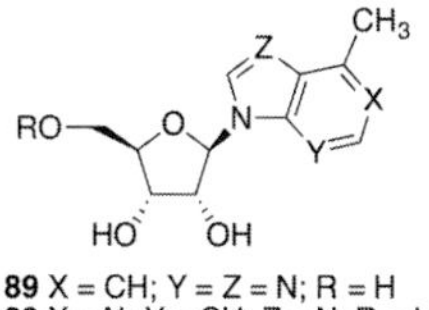

89 X = CH; Y = Z = N; R = H
90 X = N; Y = CH; Z = N; R = H
91 X = Y = N; Z = CH; R = H
92 X = Y = N; Z = CH; R = Me

A screen of lipophilic uridine analogs identified a series of triphenylmethyl containing derivatives **93** and **94** with antiviral activity.[86] Several demonstrated activity against DENV, but most of the compounds in the series were plagued by significant cytotoxicity.

93 R_1 = Tr, R_2 = Tr, R_3 = H (EC_{50} = 1.75 μM)
94 R_1 = Tr, R_2 = H, R_3 = Tr (EC_{50} = 30 μM)

6. CONCLUSION

Historically, nucleosides and nucleotides have played a central role in the treatment of viral diseases, and their significance remains undiminished in contemporary clinical applications. The search for new nucleoside and nucleotide therapies has encompassed both novel ribose and nucleobase moieties. These discovery efforts have also utilized prodrug strategies to a larger extent than in any other field of drug discovery. Even with the vast exploration of nucleoside and nucleotide chemical space in the past, efforts continue to investigate new chemical matter because of the broad utility and success of these molecules in the treatment of viral diseases.

REFERENCES

1. De Clercq, E., Ed., Vol. 67; *Antiviral Agents*; Academic Press: New York, 2013.
2. Jordheim, L. P.; Durantel, D.; Zoulim, F.; Dumontet, C. *Nat. Rev.* **2013**, *12*, 447.
3. Hecker, S. J.; Erion, M. D. *J. Med. Chem.* **2008**, *51*, 2328.
4. Schinazi, R. F.; Hernandez-Santiago, B. I.; Hurwitz, S. J. *Antiviral Res.* **2006**, *71*, 322.
5. De Clercq, E. *Adv. Pharmacol.* **2013**, *67*, 317.
6. Mackman, R. L. In: Barrish, J. C.; Carter, P. H.; Cheng, P. T. W.; Zahler, R. Eds.; Accounts in Drug Discovery: Case Studies in Medicinal Chemistry; Royal Society of Chemistry: Cambridge, UK, 2011; p 215.
7. Hurwitz, S. J.; Schinazi, R. F. *Drug Discov. Today Technol.* **2012**, *9*, e183.
8. Painter, G. R.; Almond, M. R.; Trost, L. C.; Lampert, B. M.; Neyts, J.; De Clercq, E.; Korba, B. E.; Aldern, K. A.; Beadle, J. R.; Hostetler, K. Y. *Antimicrob. Agents Chemother.* **2007**, *51*, 3505.
9. Lanier, E. R.; Ptak, R. G.; Lampert, B. M.; Keilholz, L.; Hartman, T.; Buckheit, R. W., Jr.; Mankowski, M. K.; Osterling, M. C.; Almond, M. R.; Painter, G. R. *Antimicrob. Agents Chemother.* **2010**, *54*, 2901.
10. Agarwal, H. K.; Chhikara, B. S.; Hanley, M. J.; Ye, G.; Doncel, G. F.; Parang, K. *J. Med. Chem.* **2012**, *55*, 4861.
11. Pemmaraju, B.; Agarwal, H. K.; Oh, D.; Buckheit, K. W.; Buckheit, R. W., Jr.; Tiwari, R.; Parang, K. *Tetrahedron Lett.* **2014**, *55*, 1983.

12. Agarwal, H. K.; Loethan, K.; Mandal, D.; Doncel, G. F.; Parang, K. *Bioorg. Med. Chem. Lett.* **2011**, *21*, 1917.
13. Shastina, N. S.; Maltseva, T. Y.; D'yakova, L. N.; Lobach, O. A.; Chataeva, M. S.; Nosik, D. N.; Shvetz, V. I.; Russian, J. *Bioorg. Chem.* **2013**, *39*, 161.
14. Ohrui, H. *Proc. Jpn. Acad. Ser. B Phys. Biol. Sci.* **2011**, *87*, 53.
15. Hayakawa, H.; Kohgo, S.; Kitano, K.; Ashida, N.; Kodama, E.; Mitsuya, H.; Ohrui, H. *Antivir. Chem. Chemother.* **2004**, *15*, 169.
16. Wang, F.; Flint, O. P. *Antimicrob. Agents Chemother.* **2013**, *57*, 6205.
17. Dutschman, G. E.; Grill, S. P.; Gullen, E. A.; Haraguchi, K.; Takeda, S.; Tanaka, H.; Baba, M.; Cheng, Y. C. *Antimicrob. Agents Chemother.* **2004**, *48*, 1640.
18. Murphey-Corb, M.; Rajakumar, P.; Michael, H.; Nyaundi, J.; Didier, P. J.; Reeve, A. B.; Mitsuya, H.; Sarafianos, S. G.; Parniak, M. A. *Antimicrob. Agents Chemother.* **2012**, *56*, 4707.
19. Wu, J.; Yu, W.; Fu, L.; He, W.; Wang, Y.; Chai, B.; Song, C.; Chang, J. *Eur. J. Med. Chem.* **2013**, *63*, 739.
20. Wang, P.; Rachakonda, S.; Zennou, V.; Keilman, M.; Niu, C.; Bao, D.; Ross, B. S.; Furman, P. A.; Otto, M. J.; Sofia, M. J. *Antivir. Chem. Chemother.* **2012**, *22*, 217.
21. Bondada, L.; Detorio, M.; Bassit, L.; Tao, S.; Montero, C. M.; Singletary, T. M.; Zhang, H.; Zhou, L.; Cho, J. H.; Coats, S. J.; Schinazi, R. F. *ACS Med. Chem. Lett.* **2013**, *4*, 747.
22. Barral, K.; Weck, C.; Payrot, N.; Roux, L.; Durafour, C.; Zoulim, F.; Neyts, J.; Balzarini, J.; Canard, B.; Priet, S.; Alvarez, K. *Eur. J. Med. Chem.* **2011**, *46*, 4281.
23. Grimm, D.; Thimme, R.; Blum, H. E. *Hepatol. Int.* **2011**, *5*, 644.
24. Borgia, G.; Gentile, I. *Curr. Med. Chem.* **2006**, *13*, 2839.
25. Chang, T. T.; Liaw, Y. F.; Wu, S. S.; Schiff, E.; Han, K. H.; Lai, C. L.; Safadi, R.; Lee, S. S.; Halota, W.; Goodman, Z.; Chi, Y. C.; Zhang, H.; Hindes, R.; Iloeje, U.; Beebe, S.; Kreter, B. *Hepatology* **2010**, *52*, 886.
26. Marcellin, P.; Gane, E.; Buti, M.; Afdhal, N.; Sievert, W.; Jacobson, I. M.; Washington, M. K.; Germanidis, G.; Flaherty, J. F.; Schall, R. A.; Bornstein, J. D.; Kitrinos, K. M.; Subramanian, G. M.; McHutchison, J. G.; Heathcote, E. J. *Lancet* **2013**, *381*, 468.
27. Geng, C. A.; Wang, L. J.; Guo, R. H.; Chen, J. J. *Mini Rev. Med. Chem.* **2013**, *13*, 749.
28. Gane, E. J.; Agarwal, K.; Fung, S. K.; Nguyen, T. T.; Cheng, W.; Sicard, E.; Ryder, S. D.; Flaherty, J. F.; Lawson, E.; Zhao, S.; Subramanian, M.; McHutchinson, J. G.; Foster, G. R. In: *24th Conference of the Asian Pacific Association for the Study of the Liver, Brisbane, Australia*, 2014.
29. Rawal, R. K.; Singh, U. S.; Chavre, S. N.; Wang, J.; Sugiyama, M.; Hung, W.; Govindarajan, R.; Korba, B.; Tanaka, Y.; Chu, C. K. *Bioorg. Med. Chem. Lett.* **2013**, *23*, 503.
30. Shakya, N.; Vedi, S.; Liang, C.; Agrawal, B.; Tyrrell, D. L.; Kumar, R. *Bioorg. Med. Chem. Lett.* **2012**, *22*, 6475.
31. Han, A.; Li, L.; Qing, K.; Qi, X.; Hou, L.; Luo, X.; Shi, S.; Ye, F. *Bioorg. Med. Chem. Lett.* **2013**, *23*, 1310.
32. Jang, J. H.; Kim, J. W.; Jeong, S. H.; Myung, H. J.; Kim, H. S.; Park, Y. S.; Lee, S. H.; Hwang, J. H.; Kim, N.; Lee, D. H. *J. Viral Hepat.* **2011**, *18*, 84.
33. Jones, S. A.; Murakami, E.; Delaney, W.; Furman, P.; Hu, J. *Antimicrob. Agents Chemother.* **2013**, *57*, 4181.
34. Chong, Y.; Chu, C. K. *Bioorg. Med. Chem. Lett.* **2002**, *12*, 3459.
35. Sofia, M. J.; Chang, W.; Furman, P. A.; Mosley, R. T.; Ross, B. S. *J. Med. Chem.* **2012**, *55*, 2481.
36. Schinazi, R.; Halfon, P.; Marcellin, P.; Asselah, T. *Liver Int.* **2014**, *34*(Suppl. 1), 69.

37. Sofia, M. J.; Bao, D.; Chang, W.; Du, J.; Nagarathnam, D.; Rachakonda, S.; Reddy, P. G.; Ross, B. S.; Wang, P.; Zhang, H. R.; Bansal, S.; Espiritu, C.; Keilman, M.; Lam, A. M.; Steuer, H. M.; Niu, C.; Otto, M. J.; Furman, P. A. *J. Med. Chem.* **2010**, *53*, 7202.
38. Jacobson, I. M.; Gordon, S. C.; Kowdley, K. V.; Yoshida, E. M.; Rodriguez-Torres, M.; Sulkowski, M. S.; Shiffman, M. L.; Lawitz, E.; Everson, G.; Bennett, M.; Schiff, E.; Al-Assi, M. T.; Subramanian, G. M.; An, D.; Lin, M.; McNally, J.; Brainard, D.; Symonds, W. T.; McHutchison, J. G.; Patel, K.; Feld, J.; Pianko, S.; Nelson, D. R. *N. Engl. J. Med.* **2013**, *368*, 1867.
39. Lawitz, E.; Mangia, A.; Wyles, D.; Rodriguez-Torres, M.; Hassanein, T.; Gordon, S. C.; Schultz, M.; Davis, M. N.; Kayali, Z.; Reddy, K. R.; Jacobson, I. M.; Kowdley, K. V.; Nyberg, L.; Subramanian, G. M.; Hyland, R. H.; Arterburn, S.; Jiang, D.; McNally, J.; Brainard, D.; Symonds, W. T.; McHutchison, J. G.; Sheikh, A. M.; Younossi, Z.; Gane, E. J. *N. Engl. J. Med.* **2013**, *368*, 1878.
40. Koff, R. S. *Aliment. Pharmacol. Ther.* **2014**, *39*, 478.
41. Osinusi, A.; Meissner, E. G.; Lee, Y. J.; Bon, D.; Heytens, L.; Nelson, A.; Sneller, M.; Kohli, A.; Barrett, L.; Proschan, M.; Herrmann, E.; Shivakumar, B.; Gu, W.; Kwan, R.; Teferi, G.; Talwani, R.; Silk, R.; Kotb, C.; Wroblewski, S.; Fishbein, D.; Dewar, R.; Highbarger, H.; Zhang, X.; Kleiner, D.; Wood, B. J.; Chavez, J.; Symonds, W. T.; Subramanian, M.; McHutchison, J.; Polis, M. A.; Fauci, A. S.; Masur, H.; Kottilil, S. *JAMA* **2013**, *310*, 804.
42. Charlton, M.; Gane, E.; Manns, M. P.; Brown, R. S.; Curry, M. P.; Kwo, P.; Fontana, R. J.; Gilroy, R.; Teperman, L.; Muir, A.; McHutchinson, J. G.; Symonds, W.; Denning, J.; McNair, L.; Arterburn, S.; Terrault, N.; Samuel, D.; Forns, X. In: *64th Annual Meeting of the American Association for the Study of Liver Diseases, Washington, DC* 2013, Abstract LB-2.
43. Curry, M. P.; Forns, X.; Chung, R. T.; Terrault, N.; Brown, R. S.; Fenkel, J. M.; Gordon, F.; O'Leary, J.; Kuo, A.; Schiano, T.; Everson, G.; Schiff, E.; Befeler, A.; McHutchinson, J. G.; Symonds, W.; Denning, J.; McNair, L.; Arterburn, S.; Moonka, D.; Gane, E.; Afdhal, N. In: *64th Annual Meeting of the American Association for the Study of Liver Diseases, Washington, DC*, 2013, Abstract 213.
44. Fontana, R. J.; Hughes, E. A.; Bifano, M.; Appelman, H.; Dimitrova, D.; Hindes, R.; Symonds, W. T. *Am. J. Transplant.* **2013**, *13*, 1601.
45. Gane, E. J.; Stedman, C. A.; Hyland, R. H.; Ding, X.; Svarovskaia, E.; Subramanian, G. M.; Symonds, W. T.; McHutchison, J. G.; Pang, P. S. *Gastroenterology* **2014**, *146*, 736.
46. Sulkowski, M. S.; Gardiner, D. F.; Rodriguez-Torres, M.; Reddy, K. R.; Hassanein, T.; Jacobson, I.; Lawitz, E.; Lok, A. S.; Hinestrosa, F.; Thuluvath, P. J.; Schwartz, H.; Nelson, D. R.; Everson, G. T.; Eley, T.; Wind-Rotolo, M.; Huang, S. P.; Gao, M.; Hernandez, D.; McPhee, F.; Sherman, D.; Hindes, R.; Symonds, W.; Pasquinelli, C.; Grasela, D. M. *N. Engl. J. Med.* **2014**, *370*, 211.
47. Lam, A. M.; Espiritu, C.; Bansal, S.; Micolochick Steuer, H. M.; Niu, C.; Zennou, V.; Keilman, M.; Zhu, Y.; Lan, S.; Otto, M. J.; Furman, P. A. *Antimicrob. Agents Chemother.* **2012**, *56*, 3359.
48. Tong, X.; Le Pogam, S.; Li, L.; Haines, K.; Piso, K.; Baronas, V.; Yan, J. M.; So, S. S.; Klumpp, K.; Najera, I. *J. Infect. Dis.* **2014**, *209*, 668.
49. Reddy, P. G.; Bao, D.; Chang, W.; Chun, B. K.; Du, J.; Nagarathnam, D.; Rachakonda, S.; Ross, B. S.; Zhang, H. R.; Bansal, S.; Espiritu, C. L.; Keilman, M.; Lam, A. M.; Niu, C.; Steuer, H. M.; Furman, P. A.; Otto, M. J.; Sofia, M. J. *Bioorg. Med. Chem. Lett.* **2010**, *20*, 7376.
50. Lam, A. M.; Espiritu, C.; Bansal, S.; Micolochick Steuer, H.; Zennou, V.; Otto, M. J.; Furman, P. A. *J. Virol.* **2011**, *85*, 12334.

51. Niu, C.; Tolstykh, T.; Bao, H.; Park, Y.; Babusis, D.; Lam, A. M.; Bansal, S.; Du, J.; Chang, W.; Reddy, P. G.; Zhang, H. R.; Woolley, J.; Wang, L. Q.; Chao, P. B.; Ray, A. S.; Otto, M. J.; Sofia, M. J.; Furman, P. A.; Murakami, E. *Antimicrob. Agents Chemother.* **2012**, *56*, 3767.
52. Rodriguez-Torres, M.; Lawitz, E.; Denning, J. M.; Cornpropst, M. T.; Albanis, E.; Symonds, W. T.; Berrey, M. M. In: *46th Annual Meeting of the European Association for the Study of the Liver, Berlin, Germany* 2011, Abstract 1235.
53. Lawitz, E.; Rodriguez-Torres, M.; Denning, J. M.; Cornpropst, M. T.; Clemons, D.; McNair, L.; Berrey, M. M.; Symonds, W. T. In: *46th Annual Meeting of the European Association for the Study of the Liver, Berlin, Germany* 2011, Abstract 1370.
54. Du, J.; Bao, D.; Chun, B. K.; Jiang, Y.; Reddy, P. G.; Zhang, H. R.; Ross, B. S.; Bansal, S.; Bao, H.; Espiritu, C.; Lam, A. M.; Murakami, E.; Niu, C.; Micolochick Steuer, H. M.; Furman, P. A.; Otto, M. J.; Sofia, M. J. *Bioorg. Med. Chem. Lett.* **2012**, *22*, 5924.
55. Chang, W.; Bao, D.; Chun, B.-K.; Naduthambi, D.; Nagarathnam, D.; Rachakonda, S.; Reddy, P. G.; Ross, B. S.; Zhang, H.-R.; Bansal, S.; Espiritu, C. L.; Keilman, M.; Lam, A. M.; Niu, C.; Steuer, H. M.; Furman, P. A.; Otto, M. J.; Sofia, M. J. *ACS Med. Chem. Lett.* **2011**, *2*, 130.
56. Zhou, X. J.; Pietropaolo, K.; Chen, J.; Khan, S.; Sullivan-Bolyai, J.; Mayers, D. *Antimicrob. Agents Chemother.* **2011**, *55*, 76.
57. Lalezari, J.; Asmuth, D.; Casiro, A.; Vargas, H.; Lawrence, S.; Dubuc-Patrick, G.; Chen, J.; McCarville, J.; Pietropaolo, K.; Zhou, X. J.; Sullivan-Bolyai, J.; Mayers, D. *Antimicrob. Agents Chemother.* **2012**, *56*, 6372.
58. McGuigan, C.; Madela, K.; Aljarah, M.; Bourdin, C.; Arrica, M.; Barrett, E.; Jones, S.; Kolykhalov, A.; Beilman, B.; Bryant, K. D.; Ganguly, B.; Gorovits, E.; Henson, G.; Hunley, D.; Hutchins, J.; Muhammad, J.; Obikhod, A.; Patti, J. M.; Walters, C. R.; Wang, J.; Vernachio, J. H.; Ramamurty, C. V. S.; Battina, S. K.; Chamberlain, S. *J. Med. Chem.* **2011**, *54*, 8632.
59. Vernachio, J. H.; Bleiman, B.; Bryant, K. D.; Chamberlain, S.; Hunley, D.; Hutchins, J.; Ames, B.; Gorovits, E.; Ganguly, B.; Hall, A.; Kolykhalov, A.; Liu, Y.; Muhammad, J.; Raja, N.; Walters, C. R.; Wang, J.; Williams, K.; Patti, J. M.; Henson, G.; Madela, K.; Aljarah, M.; Gilles, A.; McGuigan, C. *Antimicrob. Agents Chemother.* **2011**, *55*, 1843.
60. Patti, J. M.; Matson, M.; Goehlecke, B.; Barry, A.; Wensel, E.; Pentikis, H.; Alam, J.; Henson, G. In: *46th Annual Meeting of the European Association for the Study of the Liver, Berlin, Germany* 2011, Abstract 460.
61. Arnold, J. J.; Sharma, S. D.; Feng, J. Y.; Ray, A. S.; Smidansky, E. D.; Kireeva, M. L.; Cho, A.; Perry, J.; Vela, J. E.; Park, Y.; Xu, Y.; Tian, Y.; Babusis, D.; Barauskus, O.; Peterson, B. R.; Gnatt, A.; Kashlev, M.; Zhong, W.; Cameron, C. E. *PLoS Pathog.* **2012**, *8*, e1003030.
62. Cho, A.; Zhang, L.; Xu, J.; Babusis, D.; Butler, T.; Lee, R.; Saunders, O. L.; Wang, T.; Parrish, J.; Perry, J.; Feng, J. Y.; Ray, A. S.; Kim, C. U. *Bioorg. Med. Chem. Lett.* **2012**, *22*, 4127.
63. Draffan, A. G.; Frey, B.; Pool, B.; Gannon, C.; Tyndall, E. M.; Lilly, M.; Francom, P.; Hufton, R.; Halim, R.; Jahangiri, S.; Bond, S.; Nguyen, V. T. T.; Jeynes, T. P.; Wirth, V.; Luttick, A.; Tilmanis, D.; Thomas, J. D.; Pryor, M.; Porter, K.; Morton, C. J.; Lin, B.; Duan, J.; Kukolj, G.; Simoneau, B.; McKercher, G.; Lagace, L.; Amad, M. A.; Bethell, R. C.; Tucker, S. P. *ACS Med. Chem. Lett.* **2014**, *5*(6), 679–684. http://dx.doi.org/10.1021/ml500077j (online early access).
64. Cho, A.; Saunders, O. L.; Butler, T.; Zhang, L.; Xu, J.; Vela, J. E.; Feng, J. Y.; Ray, A. S.; Kim, C. U. *Bioorg. Med. Chem. Lett.* **2012**, *22*, 2705.
65. Cho, A.; Zhang, L.; Xu, J.; Lee, R.; Butler, T.; Metobo, S.; Aktoudianakis, V.; Lew, W.; Ye, H.; Clarke, M.; Doerffler, E.; Byun, D.; Wang, T.; Babusis, D.; Carey, A. C.;

German, P.; Sauer, D.; Zhong, W.; Rossi, S.; Fenaux, M.; McHutchison, J. G.; Perry, J.; Feng, J.; Ray, A. S.; Kim, C. U. *J. Med. Chem.* **2014**, *57*, 1812.
66. Lawitz, E.; Hill, J.; Marbury, T.; Hazan, L.; Gruener, D.; Webster, L.; Majauskas, R.; Morrison, R.; DeMicco, M.; German, P.; Stefanidis, D.; Svarovskaia, E.; Arterburn, S.; Ray, A.; Rossi, S.; McHutchinson, J. G.; Rodriguez-Torres, M. In: *47th Annual Meeting of the European Association for the Study of the Liver, Barcelona, Spain*, 2012.
67. Jonckers, T. H. M.; Lin, T.-I.; Buyck, C.; Lachau-Durand, S.; Vandyck, K.; Van Hoff, S.; Vandekerckhove, L. A. M.; Hu, L.; Berke, J. M.; Vijgen, L.; Dillen, L. L. A.; Cummings, M. D.; de Kock, H.; Nilsson, M.; Sund, C.; Rydegard, C.; Samuelsson, B.; Rosenquist, A.; Fanning, G.; Van Emelen, K.; Simmen, K.; Raboisson, P. *J. Med. Chem.* **2010**, *53*, 8150.
68. Berke, J. M.; Vijgen, L.; Lachau-Durand, S.; Powdrill, M. H.; Rawe, S.; Sjuvarsson, E.; Eriksson, S.; Gotte, M.; Fransen, E.; Dehertogh, P.; Van den Eynde, C.; Leclercq, L.; Jonckers, T. H.; Raboisson, P.; Nilsson, M.; Samuelsson, B.; Rosenquist, A.; Fanning, G. C.; Lin, T. I. *Antimicrob. Agents Chemother.* **2011**, *55*, 3812.
69. Du, J.; Chun, B. K.; Mosley, R. T.; Bansal, S.; Bao, H.; Espiritu, C.; Lam, A. M.; Murakami, E.; Niu, C.; Micolochick Steuer, H. M.; Furman, P. A.; Sofia, M. J. *J. Med. Chem.* **2014**, *57*, 1826.
70. Jonckers, T. H.; Vandyck, K.; Vandekerckhove, L.; Hu, L.; Tahri, A.; Van Hoof, S.; Lin, T. I.; Vijgen, L.; Berke, J. M.; Lachau-Durand, S.; Stoops, B.; Leclercq, L.; Fanning, G.; Samuelsson, B.; Nilsson, M.; Rosenquist, A.; Simmen, K.; Raboisson, P. *J. Med. Chem.* **2014**, *57*, 1836.
71. Di Francesco, M. E.; Avolio, S.; Pompei, M.; Pesci, S.; Monteagudo, E.; Pucci, V.; Giuliano, C.; Fiore, F.; Rowley, M.; Summa, V. *Bioorg. Med. Chem.* **2012**, *20*, 4801.
72. Naus, P.; Caletkova, O.; Konecny, P.; Dzubak, P.; Bogdanova, K.; Kolar, M.; Vrbkova, J.; Slavetinska, L.; Tloust'ova, E.; Perlikova, P.; Hajduch, M.; Hocek, M. *J. Med. Chem.* **2014**, *57*, 1097.
73. Zhou, L.; Amblard, F.; Zhang, H.; McBrayer, T. R.; Detorio, M. A.; Whitaker, T.; Coats, S. J.; Schinazi, R. F. *Bioorg. Med. Chem. Lett.* **2013**, *23*, 3385.
74. Liu, J.; Du, J.; Wang, P.; Nagarathnam, D.; Espiritu, C.; Bao, H.; Murakami, E.; Furman, P. A.; Sofia, M. J. *Nucleosides Nucleotides Nucleic Acids* **2012**, *31*, 277.
75. Thiyagarajan, A.; Salim, M. T.; Balaraju, T.; Bal, C.; Baba, M.; Sharon, A. *Bioorg. Med. Chem. Lett.* **2012**, *22*, 7742.
76. Nilsson, M.; Kalayanov, G.; Winqvist, A.; Pinho, P.; Sund, C.; Zhou, X. X.; Wahling, H.; Belfrage, A. K.; Pelcman, M.; Agback, T.; Benckestock, K.; Wikstrom, K.; Boothee, M.; Lindqvist, A.; Rydegard, C.; Jonckers, T. H.; Vandyck, K.; Raboisson, P.; Lin, T. I.; Lachau-Durand, S.; de Kock, H.; Smith, D. B.; Martin, J. A.; Klumpp, K.; Simmen, K.; Vrang, L.; Terelius, Y.; Samuelsson, B.; Rosenquist, S.; Johansson, N. G. *Bioorg. Med. Chem. Lett.* **2012**, *22*, 3265.
77. Parrish, J. P.; Lee, S. K.; Boojamra, C. G.; Hui, H.; Babusis, D.; Brown, B.; Shih, I. H.; Feng, J. Y.; Ray, A. S.; Mackman, R. L. *Bioorg. Med. Chem. Lett.* **2013**, *23*, 3354.
78. Noble, C. G.; Shi, P. Y. *Antiviral Res.* **2012**, *96*, 115.
79. Parkinson, T.; Pryde, D. C. *Future Med. Chem.* **2010**, *2*, 1181.
80. Green, J.; Bandarage, U.; Luisi, K.; Rijnbrand, R. In: *Annual Reports in Medicinal Chemistry*; Desai, M. C. Ed.; Vol. 48; Academic Press: New York, 2012; p 297.
81. Chen, Y. L.; Yin, Z.; Duraiswamy, J.; Schul, W.; Lim, C. C.; Liu, B.; Xu, H. Y.; Qing, M.; Yip, A.; Wang, G.; Chan, W. L.; Tan, H. P.; Lo, M.; Liung, S.; Kondreddi, R. R.; Rao, R.; Gu, H.; He, H.; Keller, T. H.; Shi, P. Y. *Antimicrob. Agents Chemother.* **2010**, *54*, 2932.

82. Latour, D. R.; Jekle, A.; Javanbakht, H.; Henningsen, R.; Gee, P.; Lee, I.; Tran, P.; Ren, S.; Kutach, A. K.; Harris, S. F.; Wang, S. M.; Lok, S. J.; Shaw, D.; Li, J.; Heilek, G.; Klumpp, K.; Swinney, D. C.; Deval, J. *Antiviral Res.* **2010**, *87*, 213.
83. Yin, Z.; Chen, Y. L.; Schul, W.; Wang, Q. Y.; Gu, F.; Duraiswamy, J.; Kondreddi, R. R.; Niyomrattanakit, P.; Lakshminarayana, S. B.; Goh, A.; Xu, H. Y.; Liu, W.; Liu, B.; Lim, J. Y.; Ng, C. Y.; Qing, M.; Lim, C. C.; Yip, A.; Wang, G.; Chan, W. L.; Tan, H. P.; Lin, K.; Zhang, B.; Zou, G.; Bernard, K. A.; Garrett, C.; Beltz, K.; Dong, M.; Weaver, M.; He, H.; Pichota, A.; Dartois, V.; Keller, T. H.; Shi, P. Y. *Proc. Natl. Acad. Sci. U.S.A.* **2009**, *106*, 20435.
84. Chen, Y. L.; Yin, Z.; Lakshminarayana, S. B.; Qing, M.; Schul, W.; Duraiswamy, J.; Kondreddi, R. R.; Goh, A.; Xu, H. Y.; Yip, A.; Liu, B.; Weaver, M.; Dartois, V.; Keller, T. H.; Shi, P. Y. *Antimicrob. Agents Chemother.* **2010**, *54*, 3255.
85. Wu, R.; Smidansky, E. D.; Oh, H. S.; Takhampunya, R.; Padmanabhan, R.; Cameron, C. E.; Peterson, B. R. *J. Med. Chem.* **2010**, *53*, 7958.
86. Chatelain, G.; Debing, Y.; De Burghgraeve, T.; Zmurko, J.; Saudi, M.; Rozenski, J.; Neyts, J.; Van Aerschot, A. *Eur. J. Med. Chem.* **2013**, *65*, 249.

CHAPTER SIXTEEN

Advances in Inhibitors of Penicillin-Binding Proteins and β-Lactamases as Antibacterial Agents

Yong He, Jonathan Lawrence, Christopher Liu, Ning Yin
Discovery Chemistry, Cubist Pharmaceuticals, Lexington, Massachusetts, USA

Contents

1. INTRODUCTION

The continued rise of multidrug resistant bacterial infections poses a significant threat to public health.[1] A 2012 study by the World Health Organization (WHO) found that drug-resistant pathogens cause approximately 400,000 infections in Europe each year leading to 25,000 deaths.[2] Likewise, the Centers for Disease Control and Prevention (CDC) has recently pronounced that carbapenem-resistant *Enterobacteriaceae* (CRE) is an "urgent hazard level" threat (i.e., a highest level threat to human health) and multidrug-resistant (MDR) *Pseudomonas aeruginosa* and extended-spectrum-β-lactamase (ESBL)-producing *Enterobacteriaceae* "serious hazard level" threats (i.e., the second highest level).[3] These and other bacterial threats have prompted the pharmaceutical industry to reinvest in the

Annual Reports in Medicinal Chemistry, Volume 49
ISSN 0065-7743
http://dx.doi.org/10.1016/B978-0-12-800167-7.00016-X

discovery of new antibiotics with increased potency and expanded spectrum against these lethal pathogens. Of the many validated antibacterial targets, the penicillin binding proteins (PBPs) remain the most clinically successful and best understood protein target area.[4–6] PBPs catalyze the cross-linking of lipopolysaccharide monomers, giving strength to the bacterial cell wall. Disruption of this cross-linking weakens the wall and leads to cell lysis; with no human homologues, this makes them very attractive targets for drug discovery and development.[7–10] A variety of PBP isoforms exist across bacterial strains and are most commonly classified by their molecular weight (i.e., PBP 1a having the highest weight followed by 1b, 2, 3, etc).[11] Although no single PBP is required for bacterial viability, PBPs 1, 2, and 3 are often referred to as "essential" because inhibition of *at least* one of these isoforms is required for antibiotic activity. Likewise, it is not necessary to inhibit all PBPs to have a clinically meaningful antibiotic effect.

A second key aspect driving the development of new PBP inhibitors is the rise of β-lactamase resistance mechanisms, especially in Gram-negative organisms.[12,13] β-Lactam antibiotics were among the first antibiotics discovered and are the most commonly used drug class in history. In response, evolutionary pressures have caused bacteria to express special enzymes called β-lactamases which hydrolyze the reactive β-lactam ring and render the antibiotic inert. Researchers have responded in turn by developing β-lactamase inhibitors (BLIs) which, when co-dosed with β-lactam antibiotics, protect them from hydrolysis.[14] The constant détente between antibiotics and resistance mechanisms puts relentless pressure on researchers to deliver subsequently more powerful antibiotics and BLIs. Although novel therapeutics targeting Gram-negative bacteria have recently been treated by this publication,[15] this review focuses specifically on the most recent advances in the development of β-lactam and non-β-lactam PBP inhibitors as well as development of inhibitors targeting the serine and metallo-β-lactamase enzymes.

2. PBP INHIBITORS

2.1. β-Lactam-Based Inhibitors

Although it has been more than 70 years since the discovery of penicillin, β-lactams remain the only class of PBP inhibitors on the market and the most widely used antibiotics throughout the world. β-Lactams are typically classified by their core ring structure as a cephem, **1**; a penem, **2**; or a monobactam, **3**. Recent β-lactam discovery research has focused on improving the spectrum, potency against resistant pathogens, and pharmacokinetic properties of all three classes.

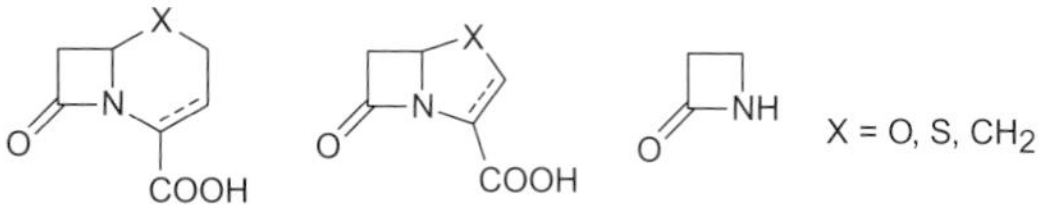

X = O, S, CH_2

1 **2** **3**

2.1.1 Cephems: Cephalosporins

The third- and fourth-generation cephalosporins (e.g., ceftazidime and cefepime, respectively) demonstrate potent activity against Gram-negative bacteria.[16–18] However, increasing resistance to cephalosporins in *P. aeruginosa* has become a serious problem worldwide.[19] The most advanced cephalosporin currently in clinical development is ceftolozane **4** (previously CXA-101 or FR264205), a novel extended-spectrum cephalosporin with excellent *in vitro* and *in vivo* activities against *P. aeruginosa*, including multidrug-resistant clinical isolates (i.e., isolated from infected patients).[20] The spectrum coverage of ceftolozane against Gram-positive and Gram-negative bacteria is generally comparable to that of third- and fourth-generation cephalosporins, but with improved anti-*Pseudomonal* activity. Combining ceftolozane (**4**) with the BLI tazobactam (**5**) further improves its spectrum to include most ESBL-producing *Enterobacteriaceae*.[21] Currently, a ceftolozane/tazobactam coformulation is under development for the treatment of complicated urinary tract infections, complicated intra-abdominal infections, and ventilator-associated pneumonia.

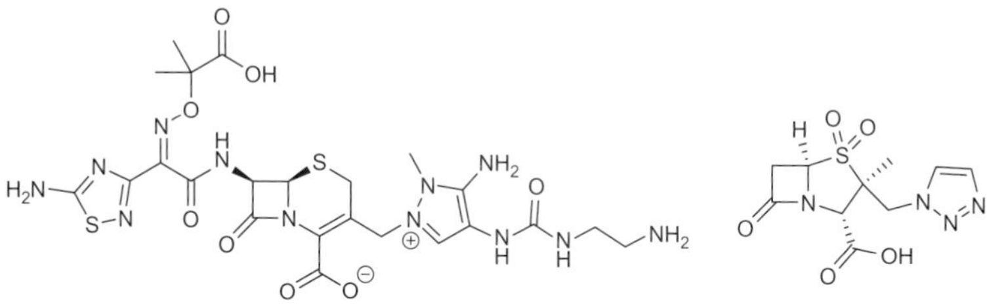

4 **5**

The cephalosporins ceftaroline, **6**, and ceftobiprole, **7**, are unusual members of the family in that they are active against methicillin-resistant *Staphylococcus aureus*.[22,23] Unfortunately, their *Pseudomonas* coverage is poor, especially against MDR strains. In 2013, a series of cephalosporin derivatives bearing imidazolium groups at the C-3′ position were disclosed.[24] An exemplified compound **8** displayed excellent broad spectrum activity encompassing Gram-positive bacteria such as methicillin-susceptible *S. aureus* (minimum inhibitory concentration $MIC_{90}=1$ μg/mL) and methicillin-resistant *S. aureus* (MRSA, $MIC_{90}=8$ μg/mL), as well as Gram-negative bacteria such as ESBL-producing *Klebsiella pneumoniae* ($MIC_{90}=0.25$ μg/mL) and *P. aeruginosa* ($MIC_{90}=4$ μg/mL).

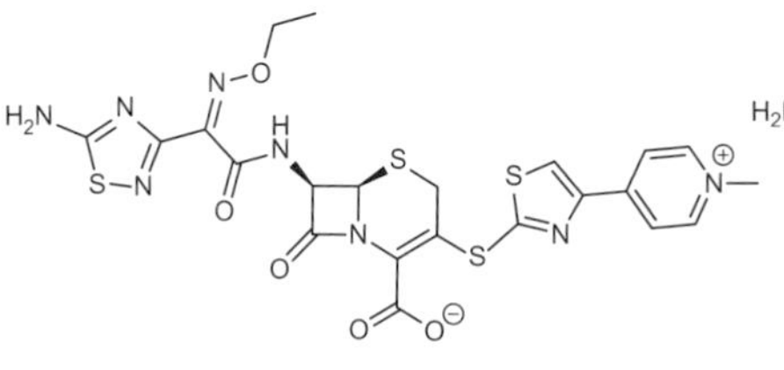

6 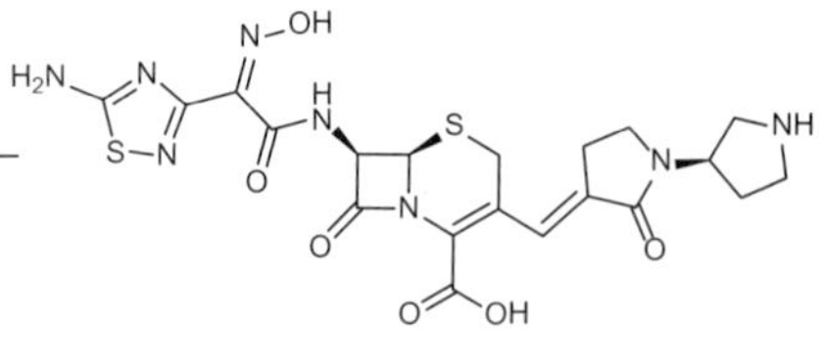 **7**

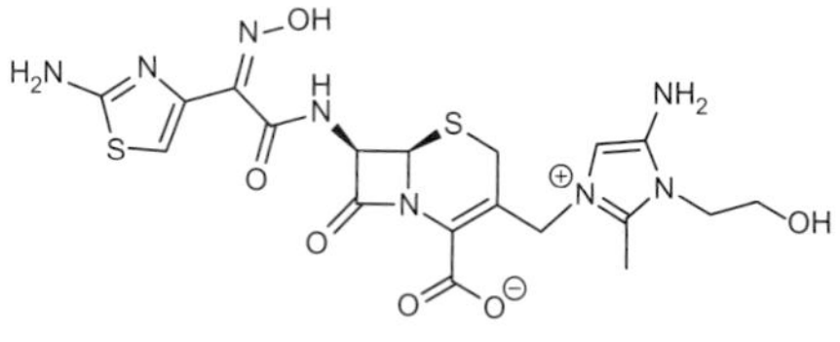

8

Compound **9** is representative of a series of novel cephalosporins with siderophore functionality aimed at improving Gram-negative coverage.[25] Siderophores are iron-binding moieties added to antibiotics designed to exploit iron-scavenging bacterial pathways to enhance cell penetration via a transporter mechanism.[26] Compound **9** showed improved antibacterial activity against a panel of bacterial isolates such as *P. aeruginosa*, *K. pneumoniae*, *Acinetobacter baumannii*, and *Escherichia coli* as compared to ceftazidime. In a model of systemic lethal infection by ceftazidime-resistant *P. aeruginosa*, compound **9** demonstrated excellent *in vivo* efficacy with a survival ED_{50} of 10 mg/kg (SC, BID) in a whole-body mouse infection model.

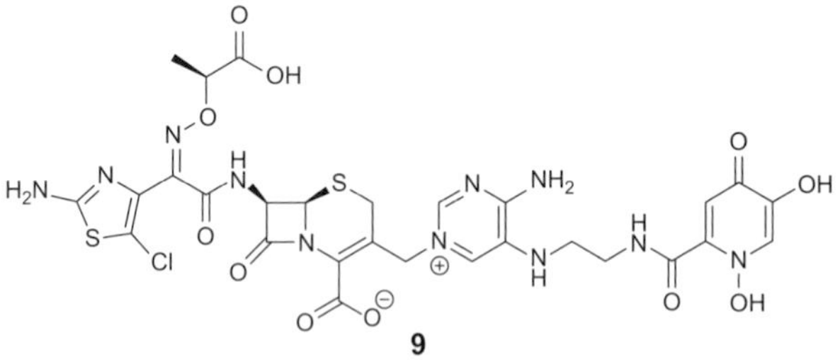

9

2.1.2 Penems: Carbapenems

The emergence of MDR Gram-negative bacteria such as carbapenem-resistant *A. baumannii* and carbapenem-resistant *K. pneumoniae* has become a considerable public health concern. FSI-1671, **10**, has been recently disclosed as a new carbapenem antibiotic possessing excellent potency against MDR Gram-negative bacteria. It displayed MIC_{90} values of 4 μg/mL against multidrug-resistant *A. baumannii*, which was further reduced to 1 μg/mL when combined with sulbactam (6 μg/mL). FSI-1671 also showed excellent potency against β-lactamase-producing organisms.[27]

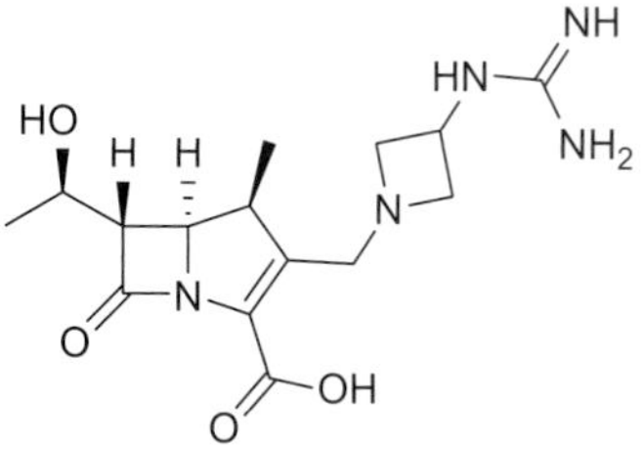

10

Aiming to improve the pharmacokinetic properties of carbapenems, a new series of derivatives containing an aryl sulfonamide moiety was reported. Representative penem **11** showed antibacterial activity against *S. aureus*, *S. pneumoniae*, and *K. pneumoniae* comparable to ertapenem and a longer half-life in rats (i.v. $t_{1/2}=1.26$ h vs. $t_{1/2}=0.36$ h).[28]

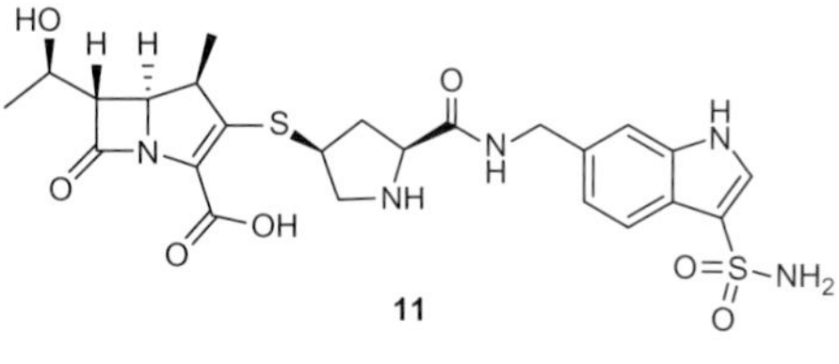

11

All four carbapenems approved for clinical use in the United States are administered via intravenous infusion due to poor oral absorption.[29] A series of ertapenem prodrugs were disclosed in a recent publication of which only bis-(5-methyl-2-oxo-1,3-dioxol-4-yl)methyl (medoxomil) ester prodrug **12** was shown to be rapidly hydrolyzed in rat, human, dog, and monkey plasma. Ertapenem diethyl ester prodrug **13** was slowly hydrolyzed in dog plasma, but showed the best *in vivo* dog pharmacokinetic profile with >31% oral bioavailability by intraduodenal dosing.[30]

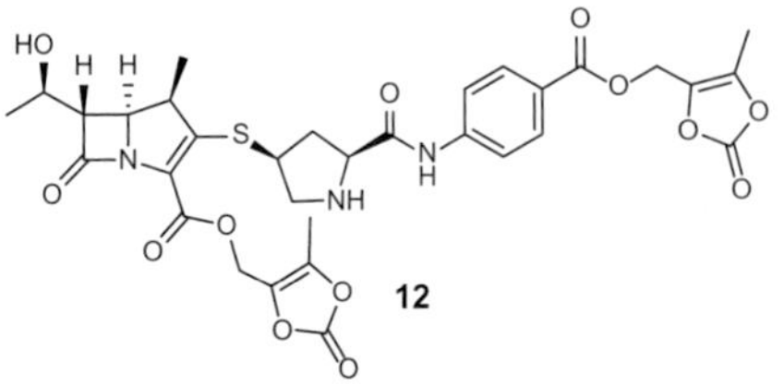

12 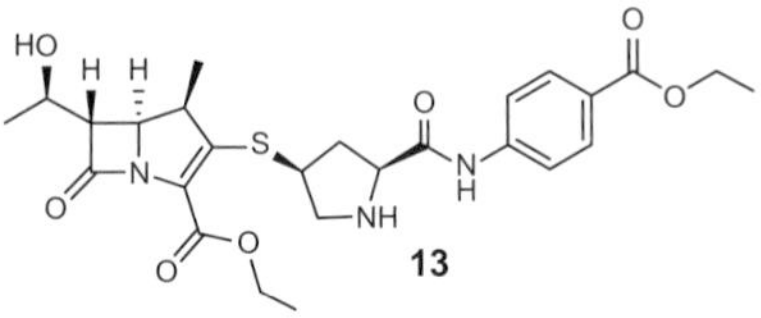 **13**

2.1.3 Monobactams

The strategy to develop novel monobactams with a siderophore approach has also received significant attention.[26] The monobactam–siderophore conjugates have promising *in vitro* profiles, especially against MDR Gram-negative strains. MB-1, **14**, a pyridone-conjugated monobactam, exhibited excellent *in vitro* activity against a broad panel of MDR Gram-negatives including *P. aeruginosa* when tested using standard assay conditions. However, further research demonstrated that the *in vivo* efficacy of MB-1 did not correlate with the *in vitro* MIC: MB-1 was not efficacious against several clinical strains despite having low MICs. The authors proposed an adaptation-type resistance mechanism to explain these observations.[31,32]

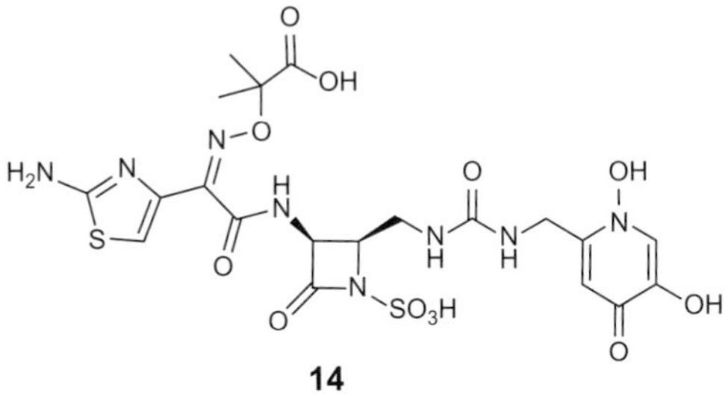

14

A recent patent disclosed a series of amidine-substituted monobactam derivatives. Representative compounds such as **15** displayed potent antibacterial activity against *E. coli*, *K. pneumoniae*, *P. aeruginosa*, and *A. baumannii*.[33]

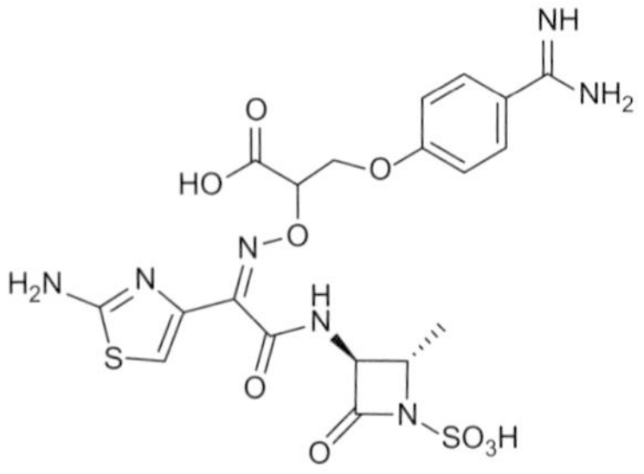

15

2.2. Non-β-Lactam PBP Inhibitors

Some resistance mechanisms have been addressed through the development of more advanced β-lactam antibiotics, and others through coadministration with a BLI. Unfortunately, resistance is still on the rise and the expression and lateral transfer of newer resistance-conferring genes threatens to render β-lactam antibiotics useless. The discovery of non-β-lactam PBP inhibitors has become a high priority, as these molecules could retain the clinical utility of this mode of action but circumvent the established resistance mechanisms deriving from hydrolysis of the β-lactam ring.

2.2.1 Covalent Inhibitors

The diazabicyclooctane (DBO) scaffold arose through rational design of a β-lactam mimic.[34] While formative analogs had no inherent antibacterial activity, they did act as potent inhibitors of class A, class C, and some class D β-lactamases (*vida infra*).

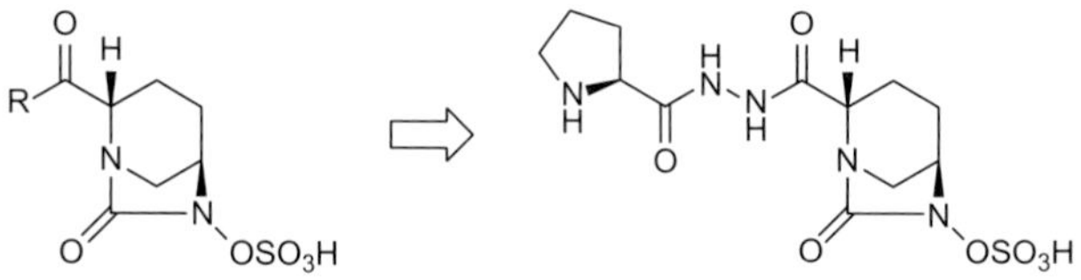

16 **17**

In 2013, a patent disclosed a series of DBO analogs that possess inherent antibacterial activity.[35] Dihydrazide DBOs such as **17** achieve killing against a variety of bacteria, including species that express class A, class B, and extended-spectrum β-lactamases (*vida infra*) such as *E. coli*, *P. aeruginosa*, and *K. pneumoniae*. The authors do not disclose the molecular target of these analogs; the similarity of the scaffold to known BLIs and the broad spectrum of activity suggest that this class of DBOs covalently inhibit PBPs as well as act as BLIs.

There are many examples of boronic acid derivatives acting as transition state mimics for serine hydrolases.[36–38] This functional group has also found application as inhibitors of β-lactamases and PBPs. As with the DBO scaffold, successes in creating covalent reversible inhibitors of β-lactamases informed subsequent extension of the paradigm to PBPs. Glycine-based boronic acids such as **18** have recently been designed using structural biology information in multiple PBP isoforms and species (PBP1b and PBP2xR6 from *S. pneumoniae*, and R39, a class C low-molecular mass PBP from *Actinomadura*).[39] The authors show that both the boronic acid and the amino-methyl spacer to aromatic substituents are required for inhibition. However, none of the reported compounds had antibacterial activity as measured by MIC assays.

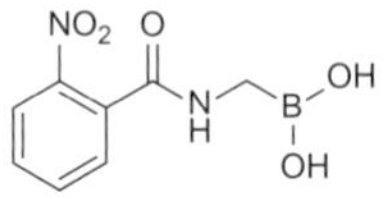

18

2.2.2 Noncovalent Inhibitors

Efforts have also been made to identify noncovalent inhibitors of PBPs. One such class of molecules, the 4-quinolines, arose from a computational, fragment-based screening approach undertaken in *E. coli* PBP5 (a low-molecular weight PBP nonessential for antibacterial activity).[40] Computational docking of the initial lead **19** suggested additional analogs and the researchers found that these new analogs—exemplified by **20**—bind *E. coli* and *Bacillus subtilis* PBPs 1a/1b, 2, 3, and 4. Notably, the parent 4-quinoline **21** did not show any measurable binding. These analogs showed no antibacterial activity against either the Gram-positive or Gram-negative pathogens tested; the authors concluded that the observed affinity for essential PBPs was not sufficiently strong to elicit a measurable MIC.

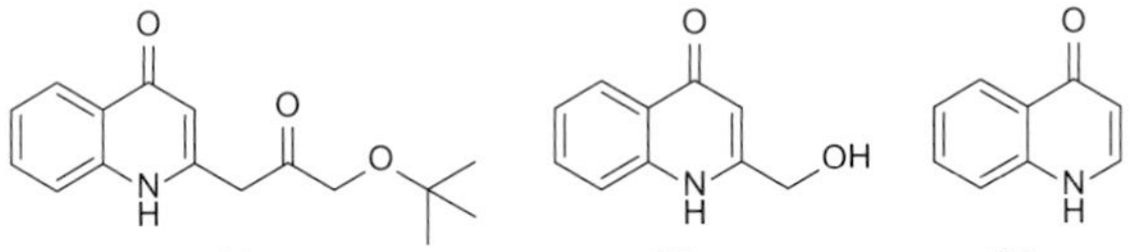

19 **20** **21**

Other efforts utilized mutated PBP isoforms from resistant bacteria as the basis for discovering new noncovalent compounds; the focus on mutant PBPs aids in the identification of compounds that display activity against previously resistant bacteria. An SAR study of anthranilic acid and naphthalene-sulfonamide derivatives revealed weak inhibitors of PBP2a from a MRSA strain, PBP1b from *S. pneumoniae*, and PBP5 from *Enterococcus faecium*.[4] The most potent examples from each class, **22** and **23**, showed little to no antibacterial activity against most Gram-positive or Gram-negative organisms. Although **22** was weakly active (MIC = 2–4 μg/mL) against MRSA, it displayed similar activity in strains both expressing and lacking PBP2a. The authors suggest this activity may thus result not from PBP inhibition, but a different bacterial target.

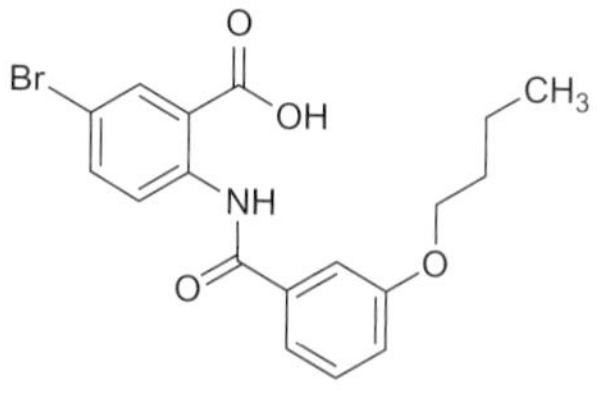

22 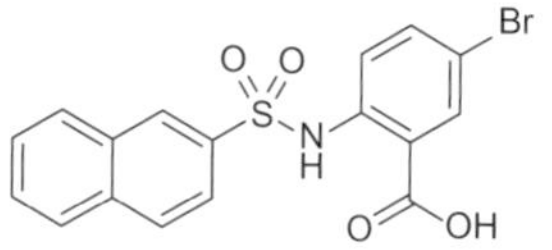 **23**

A high-throughput screen of 50,080 compounds against PBP2 of penicillin-susceptible *Neisseria gonorrhoeae* uncovered seven compounds that showed target binding as well as antibacterial activity against susceptible and penicillin or cephalosporin-resistant strains.[41] These compounds spanned a variety of chemical classes and had MIC values ranging from 2 to 32 μg/mL. The authors observed little correlation between PBP2 IC_{50} and MIC, and given the moderate levels of PBP inhibition, lack of knowledge about potency in other essential PBPs, and the resistance mechanisms in *N. gonorrhoeae* (upregulation of efflux pumps, altered porin expression), alternate mechanisms of action could not be ruled out as a reason for the observed MICs.

3. β-LACTAMASE INHIBITORS

The most prevalent mechanism of bacterial resistance to β-lactam antibiotics is the production of β-lactamase enzymes.[42] β-Lactamases are classified according to their hydrolytic mechanism: those that require an active site serine for acylation/deacylation (class A, C, and D serine-dependent β-lactamases)[43] and those that require one or more zinc ions (class B metallo-β-lactamase). Although some progress has been made on developing inhibitors of either serine-β-lactamases or metallo-β-lactamases, no inhibitors that cover all classes have been reported to date.

3.1. Serine β-Lactamase Inhibitors

3.1.1 DBO-Based BLIs

Avibactam, **24**, is the first non-β-lactam BLI advanced to the clinic in the last three decades.[44] It exhibits an excellent spectrum, covering essentially all class A and class C β-lactamases, including ESBLs and *Klebsiella*-produced carbapenemases (KPCs), which—despite the name—can now be found in many other Gram-negative strains. Avibactam is currently in clinical development in combination with ceftazidime and ceftaroline for the treatment of various serious bacterial infections.[45] It represents a marked improvement over current marketed BLIs (i.e., tazobactam, sulbactam, and clavulanic acid), which only show activity against class A enzymes, excluding carbapenemases such as KPCs. Most interestingly, a recent disclosure shows that avibactam is a reversible covalent inhibitor and thus has a unique mechanism of inhibition among all BLIs. As shown in Scheme 16.1, avibactam forms a noncovalent complex with the enzyme first, then acylates the enzyme to give a covalent complex. Subsequent deacylation of the covalent complex proceeds through reversible cyclization and regeneration of the active inhibitor rather than undergoing hydrolysis or rearrangement like many other BLIs.[46,47]

Since the discovery of avibactam, the DBO scaffold[34] has functioned as a starting point for the development of next generation serine BLIs. MK-7655, **25**, a piperidine amide derivative of avibactam, is under clinical development paired with imipenem.[48] Several new series of DBO-based BLIs have been disclosed in 2013. For example, modification of the amide side chain yielded several alternatives[49–52] including hydroxymates, **26**; hydrazides, **27**; and heteroaryl derivatives, **28**. Core-modified DBOs have also been disclosed. The olefin-containing inhibitors exemplified by **29** have been reported to show good enzymatic activity against class A, class C, and to some extent class D enzymes, and showed a synergistic effect when paired with selected β-lactam antibiotics.[53]

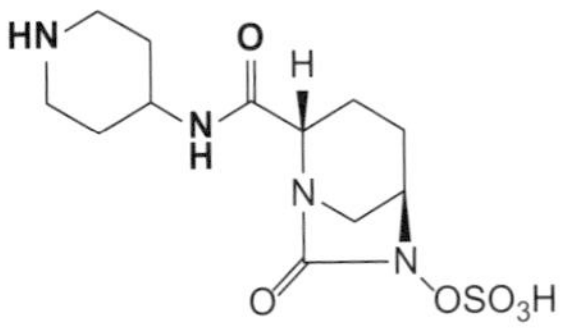

25 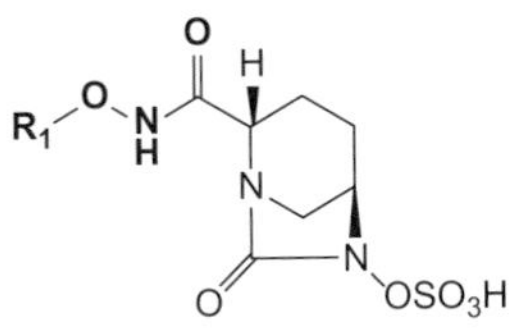 **26** 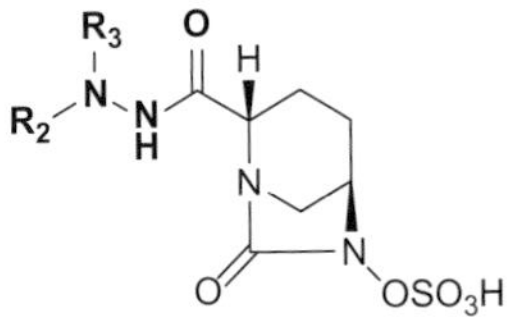 **27**

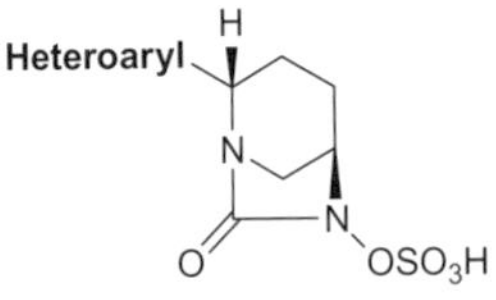

28 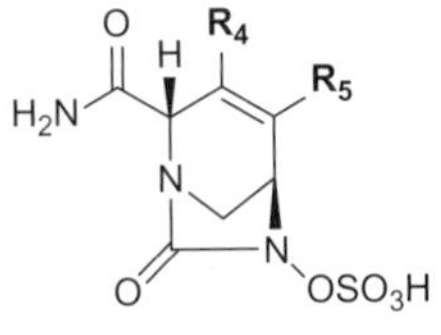 **29**

It is worth noting that these inhibitors do not have activity against class B metallo-β-lactamases (*vida infra*) and have no meaningful inherent antibacterial activity, although some have idiosyncratic activity against selected strains.

Scheme 16.1 Covalent inhibition of beta-lactamase enzymes by avibactam.

3.1.2 *Boronic Acid-Based BLIs*

Boronic acid derivatives have previously been identified as class A and C BLIs. The trivalent boron acts as an electrophile and forms a tetrahedral adduct with the catalytic serine, closely resembling the transition state found in the hydrolytic mechanism. RPX-7009, **30**, is the first boron-based BLI that has advanced to clinical testing.[54] It is currently under phase I clinical development in combination with biapenem. It displays excellent activity against class A and C enzymes including KPCs, but no meaningful activity against class B and D enzymes.

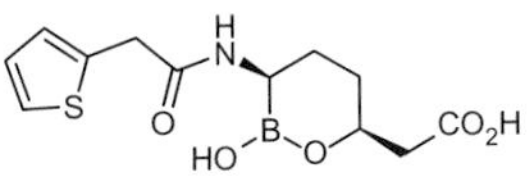

30 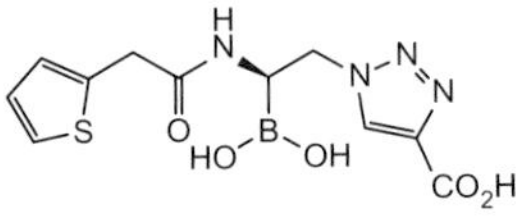 **31** 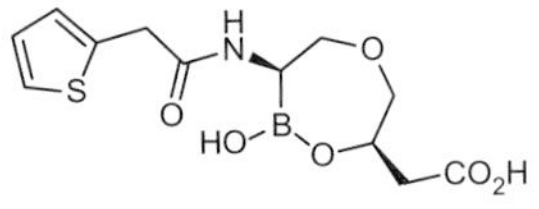 **32**

Several acyclic and cyclic boronic acid/ester-based BLIs were disclosed in 2013. For example, the 1,2,3-traizolecarboxylic acid substituted amido boronic acid derivatives represented by **31** have been reported to inhibit class A (CTX-M-15, SHV-12, KPC-2), class C, and some class D (Oxa-10) enzymes and potentiate monobactam antibiotics aztreonam and tigemonam against β-lactamase-producing strains.[55] Similar triazole boronic acid/ester analogs have similar β-lactamase activity and potentiate ampicillin.[56]

Heterocyclic boronic acid derivatives exemplified by **32** have been reported to potentiate tigemonam against strains expressing class A (ESBLs and KPCs), class C, and some class D enzymes.[57]

3.2. Metallo-β-Lactamase Inhibitors

The class B, or metallo-β-lactamases (MBLs), represent a unique subfamily of resistance enzymes. The MBLs rely on zinc ions trapped in the catalytic site to form an active hydroxide ion to serve as the key nucleophile in β-lactam hydrolysis.[58] MBLs are classified into three families based on DNA sequence: B1, B2, and B3. MBLs in the B1 and B3 family have a single catalytic zinc and are broadly active β-lactamases while MBLs in the B2 family have two catalytic zinc atoms in the active site, but are solely active against carbapenems.[59]

MBLs are particularly potent resistance enzymes capable of hydrolyzing all known classes of β-lactam antibiotics, with the exception of monobactams. In addition, the genes that encode for MBLs (e.g., IMP, VIM, and NDM) are often colocated with genes that encode for resistance to aztreonam, the only clinically used monobactam, rendering some MBL-expressing bacteria effectively untreatable. Unlike other β-lactamases, there are no clinically approved inhibitors of MBLs. Because of the unique promiscuity[60] and potency of MBLs, the development of effective metallo-β-lactamase inhibitors (MBLIs) is a high priority research topic. Recent research has focused on identifying novel chemical matter as a starting point for SAR studies. Although a number of publications on this subject exist, many researchers have struggled to demonstrate synergistic behavior at concentrations consistent with on-target activity.

3.2.1 Dicarboxylate Inhibitors

The most in-depth studies published in the last 2 years have focused on dicarboxylate-containing MBLIs. ME1071, **33**, showed mixed results when a constant concentration of ME1071 (32 μg/mL) was combined with different β-lactam antibiotics and tested against IMP- and VIM-expressing *P. aeruginosa* strains. When tested against 166 IMP-1-expressing *Pseudomonas* strains, a combination of imipenem and ME1071 reduced the number of resistant strains by 50% where resistance was defined by the CLSI 2010 breakpoint of 8 μg/mL. Against eight VIM-2-expressing strains, the percentage of resistant strains dropped by less than 25%, but the combination did cause a fourfold improvement in the MIC_{50} value.[61] Other combinations led the authors to conclude that ME1071 was most effective when combined with carbapenem antibiotics. Later testing with biapenem, doripenem, imipenem, and meropenem showed a dose-dependent improvement in MIC values across 20 *Enterobacteriaceae* expressing NDM, IMP, or VIM, but no improvement against 5 NDM-*Acinetobacter* strains irrespective of the carbapenem.[62] *In vivo* pneumonia protection studies in mice demonstrated that 100 mg/kg of ME1071 combined with 100 mg/kg biapenem was markedly more effective than biapenem alone (90% survival vs. 40%) against a BLA-2 *P. aeruginosa* strain.[63] Further advanced analogs were patented later in 2013.[64]

Another potent dicarboxylate recently disclosed is the 3-amino-phthalic acid derivative **34**. Representative of a larger group of compounds, structure **34** demonstrated 2.7 μM IC_{50} for IMP-1 and was also able to restore the activity of carbapenems against IMP-1 *P. aeruginosa* strains. When combined with biapenem, this inhibitor reduced the MIC against *Pseudomonas* from 64–128 to 1 μg/mL.[65] The *N*-heterocyclic dicarboxylic acids typified by **35** also showed low micromolar to high nanomolar K_i

values against the CcrA (B1), ImiS (B2), and L1 (B3) MBLs. *In vitro* assays against *E. coli* strains expressing those MBLs showed slight improvements in MIC (one- to eightfold) when combined with either cefazolin or imipenem.[66]

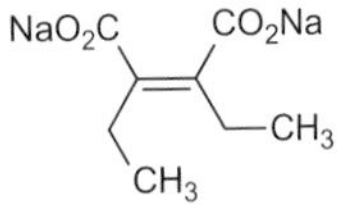

33 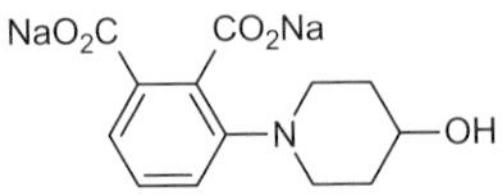 **34** 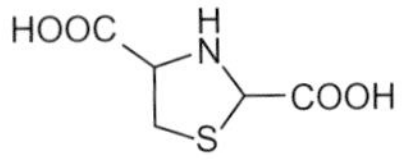 **35**

3.2.2 Thiolate-Based Inhibitors

In the last 2 years, a number of thiolate-based MBLIs have been disclosed. Sulfur is a common feature of many published MBLIs; indeed, one recent publication provided the X-ray structure of known competitive inhibitor mercaptoacetate **36** cocrystallized with SMB-1, a B3 MBL.[67] Other thiolates published[66,68–70] includes 5-(4-pyridyl)-2-mercapto-1,3,4-thiadiazoles, **37**; 3-mercapto-1,2,4-triazoles, **38**, along with their immediate synthetic precursors the acylsemicarbazides, **39**; thiopyrimidine, **40**; and the 4,5-dihydrothiazole carboxylic acid, **41**. All of the structures listed have been characterized as binders of MBLs at μM concentrations; some have shown weak inhibitory activity in enzyme-based assays, but none have been tested in whole-cell antibacterial synergy assays.

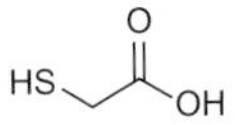

36 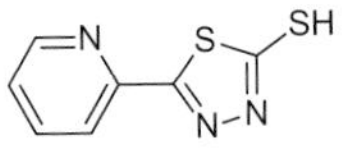 **37** 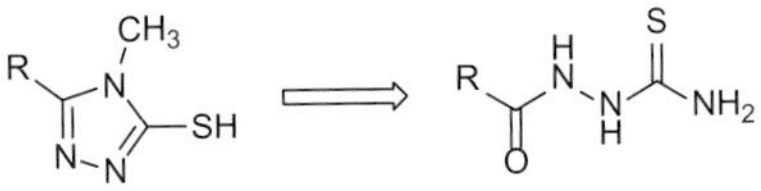 **38** **39**

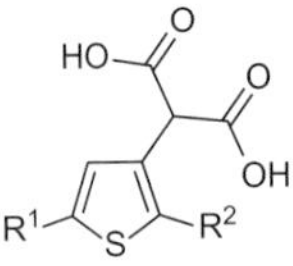

40 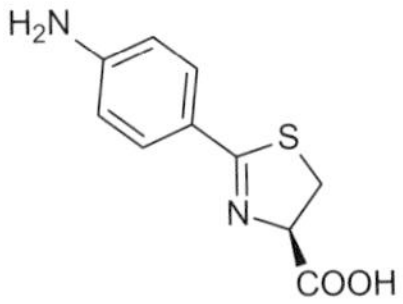 **41**

3.2.3 Other Small Molecule Inhibitors

Publications on recent efforts to identify small molecule inhibitors outside of the structural classes mentioned have been limited. One recent example disclosed a substrate-mimicking strategy to identify β-phospholactam **42** (stereochemistry unspecified) as a potential inhibitor of a variety of MBLs across all three structural classes. After a 30-min incubation of the MBLs with **42** at 100 μM concentration, inhibition levels ranging from 53% to 94% were reported against B1 and B3 MBLs.[71] Aminoimidazole **43** improved the MIC of imipenem against an NDM-1-expressing *K. pneumoniae* from 64 to 4 μg/mL when co-dosed at 30 μM. Subsequent bacterial membrane analysis showed significant permeabilization at that concentration although less potent compounds demonstrated similar disruptions. The authors suggest that **43** may act through multiple mechanisms.[72]

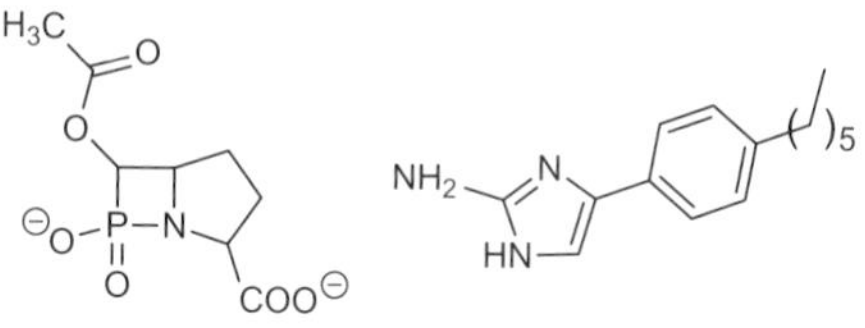

42 **43**

4. CONCLUSIONS AND OUTLOOK

Given the successful track record of β-lactam antibiotics, PBPs and β-lactamases continue to be critical targets for the development of new antibacterial agents to combat evolving drug-resistant bacterial infections. Despite the great clinical need, developments in these and other antibiotic classes have been slow. A survey of the literature dated to 2013 demonstrates that advancements are being made—both clinically and preclinically—toward novel PBP and BLI drugs. The discovery of non-β-lactam PBP inhibitors and BLIs such as the DBOs and boronic acids represent significant breakthroughs; data suggest these have the ability to overcome some existing resistance mechanisms. Similarly, a new generation of cephalosporin and carbapenem β-lactam antibiotics expands the spectrum of coverage in order to combat drug-resistant bacterial infections.

REFERENCES

1. Arias, C. A.; Murray, B. E. *N. Engl. J. Med.* **2009**, *360*, 439.
2. Grayson, M. L.; Heymann, D.; Pittet, D. *The Evolving Threat of Antimicrobial Resistant: Options for Action*; World Health Organization: Geneva, Switzerland, 2012; p1.
3. Centers for Disease Control. *Antibiotic Resistance Threats in the United States, 2013*, Centers for Disease Control: Atlanta, GA, 2013; p 112.
4. Leonard, D. A.; Bonomo, R. A.; Powers, R. A. *Acc. Chem. Res.* **2013**, *46*, 2407.
5. Elander, R. P. *Appl. Microbiol. Biotechnol.* **2003**, *61*, 385.
6. Anderson, R. J.; Groundwater, P. W.; Todd, A.; Worsley, A. J. *Antibacterial Agents*. John Wiley & Sons, Ltd.: Chichester, UK, 2012; p 261.

7. Bugg, T. D. H.; Braddick, D.; Dowson, C. G.; Roper, D. I. *Trends Biotechnol.* **2011**, *29*, 167.
8. Matteï, P.-J.; Neves, D.; Dessen, A. *Curr. Opin. Struct. Biol.* **2010**, *20*, 749.
9. Sauvage, E.; Kerff, F.; Terrak, M.; Ayala, J. A.; Charlier, P. *FEMS Microbiol. Rev.* **2008**, *32*, 234.
10. Drawz, S. M.; Bonomo, R. A. *Clin. Microbiol. Rev.* **2010**, *23*, 160.
11. Curtis, N. A.; Orr, D.; Ross, G. W.; Boulton, M. G. *Antimicrob. Agents Chemother.* **1979**, *16*, 325.
12. Fisher, J. F.; Meroueh, S. O.; Mobashery, S. *Chem. Rev.* **2005**, *105*, 395.
13. Nicolas-Chanoine, M. H. *Int. J. Antimicrob. Agents* **1996**, 7(Suppl. 1), S21.
14. Bebrone, C.; Lassaux, P.; Vercheval, L.; Sohier, J.-S.; Jehaes, A.; Sauvage, E.; Galleni, M. *Drugs* **2010**, *70*, 651.
15. Obrecht, D.; Bernardini, F.; Dale, G.; Dembowsky, K. In: *Annual Reports in Medicinal Chemistry*; Macor, J. E., Ed.; 46; AcademicPress : Burlington, 2011; p245.
16. Roberts, J. A.; Webb, S. A. R.; Lipman, J. *Int. J. Antimicrob. Agents* **2007**, *29*, 117.
17. Bingen, E.; Bidet, P.; D'humières, C.; Sobral, E.; Mariani-Kurkdjian, P.; Cohen, R. *Antimicrob. Agents Chemother.* **2013**, *57*, 2437.
18. Wüst, J.; Frei, R. *Clin. Microbiol. Infect. Off. Publ. Eur. Soc. Clin. Microbiol. Infect. Dis.* **1999**, *5*, 262.
19. Strateva, T.; Yordanov, D. *J. Med. Microbiol.* **2009**, *58*, 1133.
20. Takeda, S.; Nakai, T.; Wakai, Y.; Ikeda, F.; Hatano, K. *Antimicrob. Agents Chemother.* **2007**, *51*, 826.
21. Livermore, D. M.; Mushtaq, S.; Ge, Y. *J. Antimicrob. Chemother.* **2010**, *65*, 1972.
22. Saravolatz, L. D.; Stein, G. E.; Johnson, L. B. *Clin. Infect. Dis. Off. Publ. Infect. Dis. Soc. Am.* **2011**, *52*, 1156.
23. Dauner, D. G.; Nelson, R. E.; Taketa, D. C. *Am. J. Health Syst. Pharm.* **2010**, *67*, 983.
24. Zhang, M.; Xie, X. Patent application CN 102070656 A, 2011.
25. Cho, Y. L.; Yun, J. Y.; Chae, S. E.; Park, C..; Lee, H. S.; Oh, K. M.; Kang, H. J.; Kang, D. H.; Yang, Y. J.; Kwonj, H. J.; Park, T. K.; Woo, S. H.; Kim, Y. Z. US 8329684, 2012.
26. Braun, V.; Braun, M. *Curr. Opin. Microbiol.* **2002**, *5*, 194.
27. Joo, H. Y.; Kim, D. I.; Kowalik, E.; Li, Y.; Mao, S.; Hager, H. W.; Choi, W. B. FSI-1671, A Novel Anti-Acinetobacter carbapenem. In: *In Vitro Activities of FSI-1671 and FSI-1671/Sulbactam against MDR-A. baumannii, 53rd ICAAC, September 10–13*, 2013.
28. Sun, L. CN 102584827, 2012.
29. Zhanel, G. G.; Wiebe, R.; Dilay, L.; Thomson, K.; Rubinstein, E.; Hoban, D. J.; Noreddin, A. M.; Karlowsky, J. A. *Drugs* **2007**, *67*, 1027.
30. Singh, S. B.; Rindgen, D.; Bradley, P.; Cama, L.; Sun, W.; Hafey, M. J.; Suzuki, T.; Wang, N.; Wu, H.; Zhang, B.; Wang, L.; Ji, C.; Yu, H.; Soll, R.; Olsen, D. B.; Meinke, P. T.; Nicoll-Griffith, D. A. *ACS Med. Chem. Lett.* **2013**, *4*, 715.
31. Brown, M. F.; Mitton-Fry, M. J.; Arcari, J. T.; Barham, R.; Casavant, J.; Gerstenberger, B. S.; Han, S.; Hardink, J. R.; Harris, T. M.; Hoang, T.; Huband, M. D.; Lall, M. S.; Lemmon, M. M.; Li, C.; Lin, J.; McCurdy, S. P.; McElroy, E.; McPherson, C.; Marr, E. S.; Mueller, J. P.; Mullins, L.; Nikitenko, A. A.; Noe, M. C.; Penzien, J.; Plummer, M. S.; Schuff, B. P.; Shanmugasundaram, V.; Starr, J. T.; Sun, J.; Tomaras, A.; Young, J. A.; Zaniewski, R. P. *J. Med. Chem.* **2013**, *56*, 5541.
32. McPherson, C. J.; Aschenbrenner, L. M.; Lacey, B. M.; Fahnoe, K. C.; Lemmon, M. M.; Finegan, S. M.; Tadakamalla, B.; O'Donnell, J. P.; Mueller, J. P.; Tomaras, A. P. *Antimicrob. Agents Chemother.* **2012**, *56*, 6334.
33. Klenke, B.; Wiegand, I.; Schiffer, G.; Broetz-Oesterhelt, H.; Maiti, S. N.; Khan, J.; Reddy, A.; Yang, Z.; Hena, M.; Jia, G.; Ligong, O.; Liang, H.; Yip, J.; Gao, C.; Tajammul, S.; Mohammad, R.; Biswajeet, G. WO 2013110643, 2013.

34. Coleman, K. *Curr. Opin. Microbiol.* **2011**, *14*, 550–555.
35. Patel, M. V.; Deshpande, P. K.; Bhawasar, S.; Bhagwat, S.; Jafri, M. A.; Mishra, A.; Pavase, L.; Gupta, S.; Kale, R.; Joshi, S. WO 2013030733, 2013.
36. Smoum, R.; Rubinstein, A.; Dembitsky, V. M.; Srebnik, M. *Chem. Rev.* **2012**, *112*, 4156.
37. Bachovchin, D. A.; Cravatt, B. F. *Nat. Rev. Drug Discov.* **2012**, *11*, 52.
38. Baker, S. J.; Tomsho, J. W.; Benkovic, S. J. *Chem. Soc. Rev.* **2011**, *40*, 4279.
39. Zervosen, A.; Bouillez, A.; Herman, A.; Amoroso, A.; Joris, B.; Sauvage, E.; Charlier, P.; Luxen, A. *Bioorg. Med. Chem.* **2012**, *20*, 3915.
40. Shilabin, A. G.; Dzhekieva, L.; Misra, P.; Jayaram, B.; Pratt, R. F. *ACS Med. Chem. Lett.* **2012**, *3*, 592.
41. Fedarovich, A.; Djordjevic, K. A.; Swanson, S. M.; Peterson, Y. K.; Nicholas, R. A.; Davies, C. *PLoS One* **2012**, 7, e44918.
42. Fisher, J. F.; Meroueh, S. O.; Mobashery, S. *Chem. Rev.* **2005**, *105*, 395.
43. Bush, K.; Jacoby, G. A.; Medeiros, A. A. *Antimicrob. Agents Chemother.* **1995**, *39*, 1211.
44. Stachyra, T.; Levasseur, P.; Péchereau, M.-C.; Girard, A.-M.; Claudon, M.; Miossec, C.; Black, M. T. *J. Antimicrob. Chemother.* **2009**, *64*, 326.
45. Pucci, M. J.; Bush, K. *Clin. Microbiol. Rev.* **2013**, *26*, 792.
46. Ehmann, D. E.; Jahić, H.; Ross, P. L.; Gu, R.-F.; Hu, J.; Kern, G.; Walkup, G. K.; Fisher, S. L. *Proc. Natl. Acad. Sci.* **2012**, *109*, 11663.
47. Lahiri, S. D.; Mangani, S.; Durand-Reville, T.; Benvenuti, M.; De Luca, F.; Sanyal, G.; Docquier, J.-D. *Antimicrob. Agents Chemother.* **2013**, *57*, 2496.
48. Blizzard, T. A.; Chen, H.; Kim, S.; Wu, J.; Bodner, R.; Gude, C.; Imbriglio, J.; Young, K.; Park, Y.-W.; Ogawa, A.; Raghoobar, S.; Hairston, N.; Painter, R. E.; Wisniewski, D.; Scapin, G.; Fitzgerald, P.; Sharma, N.; Lu, J.; Ha, S.; Hermes, J.; Hammond, M. L. *Bioorg. Med. Chem. Lett.* **2014**, *24*, 780.
49. Maiti, S. N.; Nguyen, D.; Khan, J.; Ling, R. US 2013225554, 2013.
50. Abe, T.; Furuuchi, T.; Sakamaki, Y.; Inamura, S.; Morinaka, A. WO 2013180197, 2013.
51. Bhagwat, S.; Deshpande, P. K.; Bhawasar, S.; Patil, V. J.; Tadiparthi, R.; Pawar, S. S.; Jadhav, S. B.; Dabhade, S. K.; Deshmukh, V. V.; Dhond, B.; Birajdar, S.; Shaikh, M. U.; Dekhane, D.; Patel, P. A. WO 2013030735, 2013.
52. Gu, Y. G.; He, Y.; Yin, N.; Alexander, D. C.; Cross, J. B.; Busch, R.; Dolle, R. E.; Metcalf, C. A., III US 2013296555, 2013.
53. Mcguire, H.; Bist, S.; Bifulco, N.; Zhao, L.; Wu, Y.; Huynh, H.; Xiong, H.; Comita-Prevoir, J.; Dussault, D.; Geng, B.; Chen, B.; Durand-Reville, T.; Guler, S. WO 2013150296, 2013.
54. Livermore, D. M.; Mushtaq, S. *J. Antimicrob. Chemother.* **2013**, *68*, 1825.
55. Reddy, R.; Glinka, T.; Totrov, M.; Hecker, S. US 2013316978, 2013.
56. Prati, F.; Caselli, E. WO 2013053372, 2013.
57. Reddy, R.; Boyer, S.; Totrov, M.; Hecker, S. WO 2013033461, 2013.
58. Palzkill, T. *Ann. N.Y. Acad. Sci.* **2013**, *1277*, 91.
59. King, D. T.; Strynadka, N. C. *Future Med. Chem.* **2013**, *5*, 1243.
60. Saini, A. *Adv. Biol. Chem.* **2012**, *02*, 323.
61. Ishii, Y.; Eto, M.; Mano, Y.; Tateda, K.; Yamaguchi, K. *Antimicrob. Agents Chemother.* **2010**, *54*, 3625.
62. Livermore, D. M.; Mushtaq, S.; Morinaka, A.; Ida, T.; Maebashi, K.; Hope, R. *J. Antimicrob. Chemother.* **2013**, *68*, 153.
63. Yamada, K.; Yanagihara, K.; Kaku, N.; Harada, Y.; Migiyama, Y.; Nagaoka, K.; Morinaga, Y.; Nakamura, S.; Imamura, Y.; Miyazaki, T.; Izumikawa, K.; Kakeya, H.; Hasegawa, H.; Yasuoka, A.; Kohno, S. *Int. J. Antimicrob. Agents* **2013**, *42*, 238.

64. Akihiro, M.; Kazunori, M.; Takashi, I. D. A.; Muneo, H.; Mototsugu, Y.; Takao, A. B. E. WO 2013015388, 2013.
65. Hiraiwa, Y.; Morinaka, A.; Fukushima, T.; Kudo, T. *Bioorg. Med. Chem.* **2013**, *21*, 5841.
66. Feng, L.; Yang, K.-W.; Zhou, L.-S.; Xiao, J.-M.; Yang, X.; Zhai, L.; Zhang, Y.-L.; Crowder, M. W. *Bioorg. Med. Chem. Lett.* **2012**, *22*, 5185.
67. Wachino, J.; Yamaguchi, Y.; Mori, S.; Kurosaki, H.; Arakawa, Y.; Shibayama, K. *Antimicrob. Agents Chemother.* **2013**, *57*, 101.
68. Faridoon; Hussein, W. M.; Vella, P.; Islam, N. U.; Ollis, D. L.; Schenk, G.; McGeary, R. P. *Bioorg. Med. Chem. Lett.* **2012**, *22*, 380.
69. Hussein, W. M.; Fatahala, S. S.; Mohamed, Z. M.; McGeary, R. P.; Schenk, G.; Ollis, D. L.; Mohamed, M. S. *Chem. Biol. Drug Des.* **2012**, *80*, 500.
70. Chen, P.; Horton, L. B.; Mikulski, R. L.; Deng, L.; Sundriyal, S.; Palzkill, T.; Song, Y. *Bioorg. Med. Chem. Lett.* **2012**, *22*, 6229.
71. Yang, K.-W.; Feng, L.; Yang, S.-K.; Aitha, M.; LaCuran, A. E.; Oelschlaeger, P.; Crowder, M. W. *Bioorg. Med. Chem. Lett.* **2013**, *23*, 5855.
72. Worthington, R. J.; Bunders, C. A.; Reed, C. S.; Melander, C. *ACS Med. Chem. Lett.* **2012**, *3*, 357.

SECTION 6

Topics in Biology

Section Editor: John Lowe
JL3Pharma LLC, Stonington, Connecticut

CHAPTER SEVENTEEN

Tumor Microenvironment as Target in Cancer Therapy

Reuven Reich*, Claudiu T. Supuran†, Eli Breuer*

*Institute for Drug Research, School of Pharmacy, The Hebrew University of Jerusalem, P.O. Box 12065, Jerusalem, Israel

†Laboratorio di Chimica Bioinorganica, Polo Scientifico, Universita degli Studi di Firenze, Sesto Fiorentino, Florence, Italy

Contents

ABBREVIATIONS

ATX autotaxin
CAI carbonic anhydrase inhibitor
CAIX carbonic anhydrase type IX
CAXII carbonic anhydrase type XII
CPO carbamoyl phosphonates
MMP matrix metalloproteinase
PEDF pigment epithelium-derived factor
VEGF vascular endothelial growth factor

1. INTRODUCTION

The role of the microenvironment during the initiation and progression of carcinogenesis is of critical importance, both for understanding

Annual Reports in Medicinal Chemistry, Volume 49
ISSN 0065-7743
http://dx.doi.org/10.1016/B978-0-12-800167-7.00017-1

cancer biology, as well as exploiting this new knowledge for improved diagnostics and therapeutics at the cellular and molecular level. The microenvironment provides essential cues for the maintenance of cancer stem cells/ cancer-initiating cells and provides the "soil" for the seeding of cancer cells at metastatic sites. Numerous studies demonstrate a positive correlation between tumor-promoting enzymes, angiogenesis, carcinoma-associated fibroblasts, and inflammatory-infiltrating cells and poor disease outcome, emphasizing the clinical relevance of the tumor microenvironment to tumor behavior. Thus, the dynamic and reciprocal interactions between tumor cells and cells of the tumor microenvironment orchestrate events critical to tumor progression toward metastasis, and many cellular and molecular elements of the microenvironment are attractive targets for novel therapeutic approaches.

Several molecules including growth factors (e.g., vascular endothelial growth factor (VEGF), connective tissue growth factor (CTGF)), growth factor receptors (endoglin (CD105), VEGFRs), adhesion molecules, and enzymes (ectonucleotidases, ATX, CAIX, CAXII, fibroblast activation protein a (FAPa), matrix metalloproteinases (MMPs), prostate specific membrane antigen (PSMA), urokinase-type plasminogen activator (uPA)) are induced or upregulated in the tumor microenvironment, but are otherwise under tight expression regulation in normal tissues. Consequently, these molecules can be targeted pharmacologically by certain inhibitors as well as by active and passive immunotherapy. Several parameters of the tumor microenvironment, such as hypoxia, inflammation, and angiogenesis, play a critical role in tumor aggressiveness and treatment response. The present summary deals with one of the most dynamic aspects of the microenvironment, namely tumor-promoting enzymes.

2. CANCER-PROMOTING ENZYMES AND INHIBITORS

2.1. MMPs and Their Inhibitors

The MMP family is composed of about 26 zinc endopeptidases, which collectively have the capacity to degrade all the major components of the extracellular matrix. The invasive nature of malignant tumors has long been associated with the ability to degrade collagens, just as the resistance of cartilage to tumor invasion is associated with the presence of collagenase inhibitors. However, it was not until Liotta recognized the importance of basement membrane (type IV collagen) degradation in delineating the invasive and metastatic potential of carcinoma that the enzymatic activities associated with cancer cells became better defined.[1] These investigators identified and purified the type IV collagenase that became the second member of the MMP family (MMP-2). Over the years, the MMP family has expanded to include 26 zinc-dependent endopeptidases (MMPs), many of which were first identified by their overexpression in tumor cells.[2] Although MMPs have roles in many pathologies, this chapter is mainly concerned with cancer[3] and tumor cell dissemination[4] leading to the formation of metastases.[5,6] Despite more than two decades of work in designing MMP inhibitors, none has proven to be clinically useful. One limitation of this work was that the design of the inhibitor candidates was based on X-ray crystallographic structures obtained by the interactions between the water-soluble enzymes and insoluble inhibitor candidates such as hydroxamic acids. Despite *in vitro* subnanomolar IC_{50} values, these inhibitors invariably failed *in vivo*, either because of painful side effects or due to lack of *in vivo* activity.[7] Papers describing such inhibitors are still being published. Some of these hydroxamate MMP inhibitor clinical candidates are described below.[8,9] In addition, some of the specific MMP inhibitors described below are oriented toward specific pathological conditions such as COPD and RA. These targets are also associated with the tumor microenvironment, and hence these new compounds might also be used for treating cancer.

2.1.1 Matrix Metalloproteinase-2 and Its Inhibitors

Matrix metalloproteinase-2 (MMP-2) formally known as gelatinase A or 72 kDa type IV collagenase, is secreted as an inactive proenzyme and requires activation by limited proteolysis. The enzyme is involved in numerous physiological and pathological conditions such as endometrial menstrual breakdown, regulation of vascularization, and the inflammatory response. The main interest in this enzyme came about by the discovery of its role in tumor progression about three decades ago. The understanding of its role in severe pathological conditions was followed by an intensive search for efficient inhibitors however, although no clinical candidates have resulted from this work, as discussed below.

Naturally occurring endogenous protein and a nonfunctional serine protease inhibitor (serpin) pigment epithelium-derived factor (PEDF), has been proposed for cancer therapy partly due to its ability to regulate specific MMPs central to cancer progression. PEDF has been found to specifically downregulate membrane-type I matrix metalloproteinase (MT1-MMP) and furthermore, potentially MMP-2, two of the most commonly implicated MMPs in neoplasia.[10]

Doxycycline, a tetracycline antibiotic, the only kind of MMP inhibitor licensed for clinical use,[11] was recently used to treat rabbit eyes after exposure to sulfur mustard vapor. Tear fluid and cornea samples have showed elevated MMP-2 and MMP-9 (a closely related family member previously called gelatinase B) activities in all corneas during the acute injury. Treating the sulfur mustard-induced ocular injury by doxycycline as a preventive therapy in different phases ameliorated the delayed pathology.[12]

The latest contribution in a series of reports on thiirane based MMP inhibitors describes prodrugs intended for targeting the brain. Two derivatives, a carbamate and a urea proved to be selective and potent inhibitors of MMP-2, and showed potential for intervention of MMP-2-dependent diseases such as brain metastasis.[13]

Bisphosphonate anticancer agents acting as MMP inhibitors have been described earlier and some are used in cancer treatment.[14] A recent report describes novel "bone seeking" bisphosphonates as MMP inhibitors (BP-MMPIs), capable of being targeted and overcoming undesired side effects of broad spectrum MMPIs.[15]

2.1.2 Matrix Metalloproteinase-12 and Its Inhibitors

Matrix metalloproteinase-12 or macrophage metalloelastase (EC 3.4.24.65) was cloned in 1993. As with the other members of the MMP family, this enzyme is secreted as a proenzyme that has to be proteolytically activated by certain serine proteases such as plasmin and thrombin. MMP-12 has a broad substrate range. In addition to its activity against elastin, MMP-12 has been shown to cleave human CXC-chemokines such as CXCL1, CXCL2, CXCL3, CXCL5, and CXCL8. It also degrades collagen type IV, fibronectin, laminin, vitronectin, and certain proteoglycans. Although it seems to be mainly involved in COPD, MMP-12 has also been reported to be involved in malignant progression in colorectal,[16,17] nonsmall cell lung cancer,[18] prostate cancer,[19] hepatocarcinoma,[20] and head and neck squamous cell carcinomas.[21]

Astra Zeneca with Danish and Dutch laboratories jointly developed an oral MMP-9/12 inhibitor, AZD1236, **1**, which is a potent, orally bioavailable MMP inhibitor that has demonstrated selectivity for human MMP-9 and MMP-12 *in vitro* compared with reference compounds. **1** dosed orally at 75 mg twice daily was generally well tolerated over 6 weeks in patients with moderate-to-severe COPD. No clinical efficacy of AZD1236 was demonstrated in this short-term signal-searching study, although possible evidence of an impact on desmosine may suggest the potential value of selective inhibitors of MMPs in the treatment of COPD in longer term trials.[22]

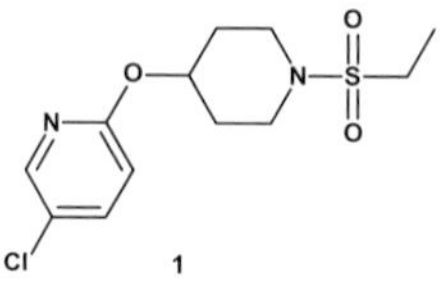

1 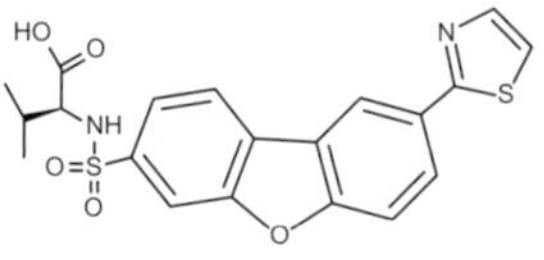 2

Another inhibitor developed jointly by Pfizer and NiKem Research, **2**, exhibits excellent oral efficacy in MMP-12 induced ear-swelling inflammation and lung inflammation mouse models, and it has been successfully advanced into development track status.[23]

A well-known French-Italian group has been occupied by the elusive problem of designing phosphinic peptide inhibitors toward matrix metalloprotease-12 (MMP-12) for about a decade. The present chapter is focused on crystallographic and thermodynamic studies.[24] Comparing phosphinic versus hydroxamate inhibitors reveals that the chelation of the zinc ion is slightly different, leading the inhibitor backbone to adopt a position in which the hydrogen bonding with the MMP-12 active site is less favorable in phosphinic inhibitor **3** while maintaining high affinity.

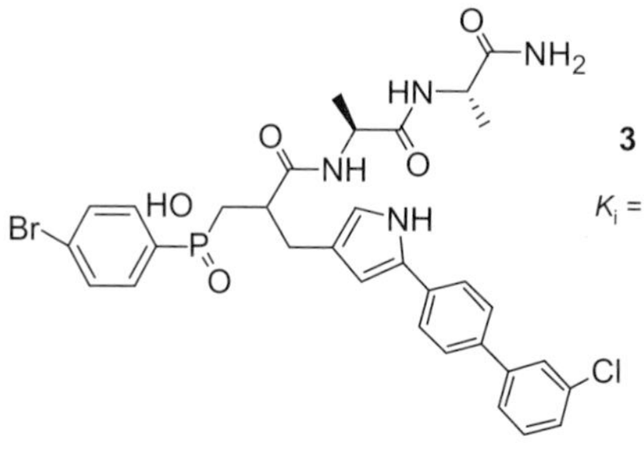

3 K_i = 0.28 nM 4

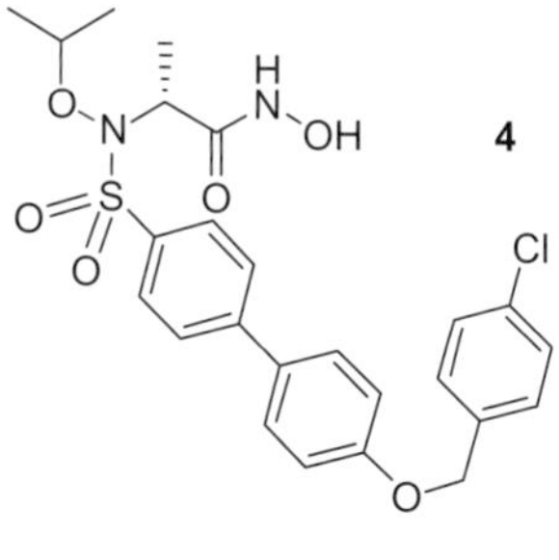

2.1.3 Matrix Metalloproteinase-13 and Its Inhibitors

Matrix metalloproteinase-13, previously known as collagenase-3, was cloned in 1994 and is secreted as a proenzyme that has to be proteolytically activated. MMP-13 is tightly bound to tissues and utilizes heparan sulfate proteoglycans as extracellular docking molecules. MMP-13 also has wide substrate specificity, as it degrades collagen types I–IV but preferentially hydrolyzes soluble collagen type II.[25] The enzyme plays a significant role in cartilage collagen degradation during osteoarthritis and COPD[26]; however, it is also associated with certain tumors and its presence correlates with tumor aggressiveness and the invasive capacity of tumors such as squamous cell carcinoma,[27] gastric cancer,[28] colorectal cancer,[19] papillary thyroid carcinomas,[29] and breast cancer.[30]

From a series of MMP-12/MMP-13 dual-target inhibitors based on sulfonamidohydroxamates, **4** was found to possess the most favorable characteristics which made it a valid dual-target inhibitor for the treatment of RA and a promising candidate for future *in vivo* testing in mice models.[31]

In the design of the MMP-13 inhibitor AZD6605, **5**, a reverse hydroxamate was replaced by a hydantoin zinc-binding group followed by optimization of MMP-13 inhibitory potency, and solubility properties, while maintaining good selectivity over MMP-14. **5** was advanced to clinical trials.[32]

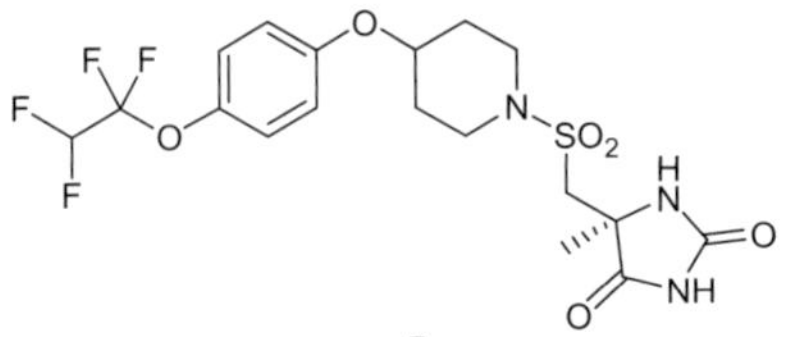

5 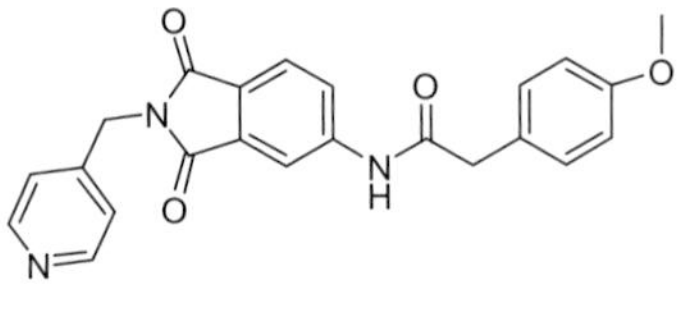 **6**

Authors from Zurich University approached the problem of designing a new inhibitor by analyzing co-crystal structures of the target molecule in order to define the pharmacophore, which led to a potent inhibitor, **6**, which is selective versus MMP-2, MMP-12, and MMP-14.[33]

2.1.4 Tumor Necrosis Factor-A-Converting Enzyme and Its Inhibitors

Disintegrin and metalloproteinases (ADAMs) are a gene family of proteins that share the metalloproteinase domain with MMPs. They are structurally classified into two groups: the membrane-anchored ADAM and ADAM with thrombospondin motifs (ADAMTS).

ADAM metallopeptidase domain 17 (ADAM17), also known as TACE (*tumor necrosis factor-α-converting enzyme*), is involved in the processing of tumor necrosis factor alpha (TNF-α) at the surface of the various cells by cleaving the Ala76–Val77 amide bond of the membrane-bound form of the cytokine. ADAM17 was cloned in 1997 and found to be expressed in virtually all cells, while the active ADAM17 on the cell surface is mainly found during inflammation and cancer.[34] TNF-α itself is involved in numerous cellular functions including inflammatory conditions and tumor progression. One of the strategies to prevent its undesirable functions was to prevent its shedding from cellular membranes by inhibiting its activator, ADAM17. Over the years, a significant number of TACE inhibitors have been reported in the literature which have yet to be approved for clinical application.[35]

2.2. Regulation of Microenvironment pH

Almost a century ago, Warburg discovered that tumor cells prefer glycolytic metabolism even in the presence of sufficient oxygen, thereby creating acidic and hypoxic conditions.[36] Further, this condition creates an advantage for tumor cells to proliferate and disseminate and even survive chemotherapy. To avoid intracellular and extracellular damage due to acidification, a number of mechanisms are activated within the tumor cells. Among the mechanisms detected are the upregulation of proton transporters such as the V-ATPase, the Na^+/H^+ exchanger and the various carbonic anhydrases.[37] Increased activity of these transporters restores the intracellular pH and regulates the extracellular pH, to give a local advantage to tumor cells. The relatively acidic microenvironment promotes tumor cell invasiveness and enhanced secretion of exosomes.[38] Further, the acidic microenvironment reduces normal cell viability and prevents a normal T cell immune response.[39] Therefore, interference with these mechanisms may present a novel approach for exerting a holistic suppression of tumor progression.

2.2.1 H⁺ Pumps and Transporters

The presence of enhanced expression of V-ATPase in tumor cell membranes has been demonstrated recently and bafilomycin, a V-ATPase inhibitor, reduced tumor growth in an orthotopic murine model.[40] Since V-ATPase and $(H^+ + K^+)$ ATPase in the stomach share significant similarities, proton-pump inhibitors have been examined in recent clinical trials as potential modulators of the tumor microenvironment.[41,42] Such inhibitors might act as pH modulators and both affect tumor cells and restore chemotherapy sensitivity and the restoration of the immune response of T cells.[43]

2.2.2 Tumor-Associated Carbonic Anhydrases

α-Carbonic anhydrases (CAs, EC 4.2.1.1) are widespread metalloenzymes in vertebrates, and 16 isozymes have been characterized to date. CAs are very efficient catalysts for the reversible hydration of carbon dioxide to bicarbonate and protons ($CO_2 + H_2O \leftrightarrow HCO_3^- + H^+$).[44–46]

Many CA isoforms are involved in critical physiologic processes such as respiration and acid–base regulation, electrolyte secretion, bone resorption, calcification and biosynthetic reactions which require bicarbonate as a substrate (lipogenesis, gluconeogenesis, and ureagenesis).[44–46] Only two CA isozymes (CAIX and CAXII) are predominantly associated with, and overexpressed in, many tumors and have been shown to be involved in critical processes connected with cancer progression and response to therapy.[45,47–49]

CAIX expression is confined to a few normal tissues (stomach and body cavity lining), but it is ectopically induced and highly overexpressed in many solid tumor types through strong transcriptional activation by hypoxia mediated by the hypoxia inducible factor 1 transcription factor.[45,47,48] CAIX and CAXII are transmembrane isoforms with an extracellular active site, but the catalytic activity of CAXII is lower compared to CAIX.[45] Since CO_2 is the main byproduct of all oxidative processes, the CAs play a fundamental role in acid–base equilibria in all systems, including tumors.[45] While there are many known CA inhibitors (CAIs), only those specific for CAIX (or/and XII) have shown a significant antitumor effect.

Primary sulfonamides, sulfamates, and sulfamides act as CAIs.[44,45] Recently, a large number of such compounds were specifically designed for targeting the tumor-associated CA isoforms.[50–52]

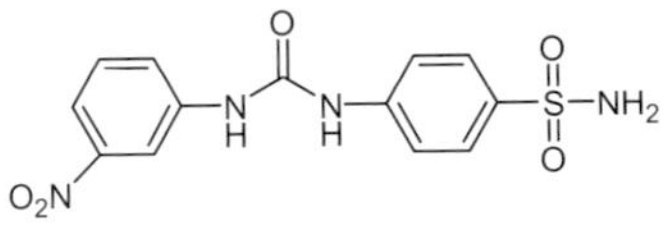

7 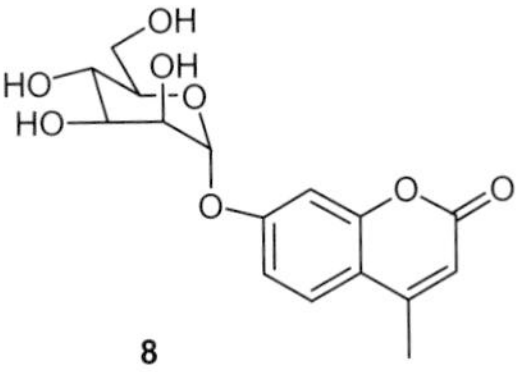 **8**

4T1 mouse metastatic breast cancer cells were shown to be capable of inducing CAIX overexpression in hypoxia, resulting in spontaneous lung metastasis formation.[53,54] The ureido-sulfonamide inhibitor **7** (a low nanomolar CAIX inhibitor) at doses of 30–45 mg/kg strongly inhibited both the growth of primary tumors and metastases in this animal model.[54] Similar results have been observed with the glycosyl coumarin inhibitor **8**.[54] Thus, both sulfonamide- and coumarin-based CAIs targeting CAIX were effective in inhibiting the growth of both primary tumors and metastases in an animal model of disease.[54] This work suggests that CAIX and XII are validated drug targets for developing novel antitumor drugs, which interfere with the pH regulation in cancer cells.

Recently, a structure–activity relationship (3D-QSAR) study for a series of CAIX inhibitors using comparative molecular field analysis and comparative molecular similarity indices analysis techniques with the help of SYBYL 7.1 software for the development of improved hCAIX inhibitors was developed.[55]

2.3. Ectonucleotidases

Ectonucleotidases are families of nucleotide-metabolizing enzymes that are expressed on the plasma membrane and have externally oriented active sites. These ectoenzymes operate in concert or consecutively and metabolize nucleotides to the respective nucleoside analogs. There is growing evidence that adenosine receptors could be promising therapeutic targets in wide range of conditions, including cerebral and cardiac ischemic diseases, sleep disorders, immune and inflammatory disorders, and cancer. After more than three decades of medicinal chemistry research, a considerable number of selective agonists and antagonists of adenosine receptors have been discovered, and some have been clinically evaluated, although none has yet received regulatory approval.[56,57]

There are four major families of ectonucleotidases described so far.[58] There are many ectoenzymes present on the cell surface which are involved in extracellular hydrolysis pathways leading to nucleotides and to other products. These include the ENTPDase, ectonucleoside triphosphate diphosphohydrolase family/CD39; the ENPP, ectonucleotide pyrophosphatase/phosphodiesterase family, ecto-5′-nucleotidase/CD73, and alkaline phosphatases. In addition, ectonucleoside diphosphokinase can interconvert extracellular nucleotides, and ATP can serve as a cosubstrate of ectoprotein kinase in the phosphorylation of surface-located proteins.

2.3.1 Autotaxin

Autotaxin (ATX) is an extracellular hydrolase, the activity of which originates from a threonine residue and from two zinc ions in the active site.[59,60] It deserves special attention because of its potency and activity due to being ENPP2, which was discovered as an autocrine motility factor that has been implicated in cell growth, motility, blood vessel formation and cancer progression.[61] As stated earlier, ATX belongs to the ENPP family of enzymes and is the only family member that displays lysophospholipase D (lysoPLD) activity, which hydrolyzes lysophosphatidyl choline to the bioactive lysophosphatidic acid.[62] The latter is involved in diverse malignancies and processes such as migration, proliferation, and differentiation.[63] There is a report of potent *in vivo* ATX inhibitors based on long chains linked to aromatic phosphonic acids.[64] Compounds **9** and **10** are the most potent of these inhibitors.

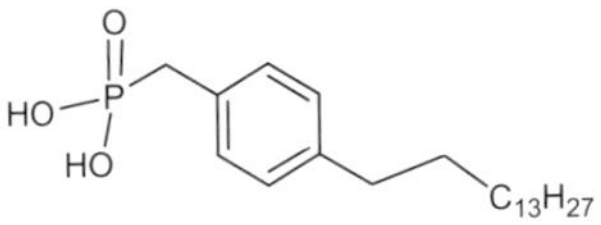

HO—P(=O)(HO)—CH₂—C₆H₄—CH₂CH₂—$C_{13}H_{27}$

9

HO—P(=O)(HO)—CH₂—naphthyl—CH=CH—$C_{11}H_{23}$

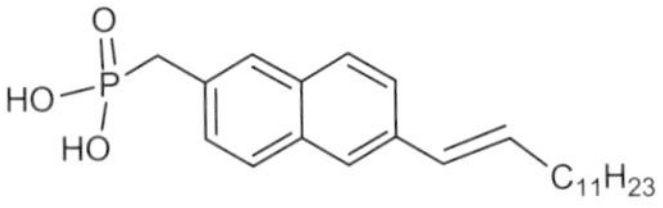

10

2.3.2 CD73

CD73, an ecto-5′-nucleotidase/ecto-5′-NT, is overexpressed in many types of human and mouse tumor types and is implicated in the control of tumor progression. It has been shown that stromal CD73 promotes tumor growth in a T cell-dependent manner and that the optimal antitumor effect of CD73 blockade requires inhibiting both tumor and stromal CD73.[65] These findings suggest that both tumor and stromal CD73 cooperatively protect tumors from incoming antitumor T cells and show the potential of targeting CD73 as a cancer immunotherapy strategy. CD73 is involved in the survival, proliferation, and invasion of gliomas through its enzymatic action. APCP (α,β-methylene ADP), a CD73 inhibitor, causes significant reduction in glioma cell proliferation.[66] The same inhibitor or anti-CD73 monoclonal antibodies have reported to significantly reduce tumor growth and metastasis, which was promoted by CD73.[67]

CD73 also plays a key role in the generation of the extracellular immune-suppressor adenosine. While its protumorigenic effects have thus far been attributed to adenosine-mediated immunosuppression, it has been demonstrated that CD73 also promotes tumor angiogenesis via enzymatic and non-enzymatic functions. Tumor levels of VEGF were reduced and tumor angiogenesis suppressed in a breast cancer mouse model following treatment with a monoclonal antibody targeted against CD73. These findings highlight a previously unknown mechanism of action for anti-CD73 therapy and further support its clinical development for cancer treatment.[68]

2.3.3 CD39

A number of recent papers indicate synergism of CD39, a nucleotidase triphosphate dephosphorylase (ENTPD), and CD73, which may lead to treatment of malignant glioma and other immunosuppressive diseases.[69] Another report indicates the possibility of improved targeted therapy of ovarian cancer by blocking adenosine-dependent immune evasion with anti-CD39 and anti-CD73 antibodies A1 and 7G2.[70]

2.3.4 *Miscellaneous*

A number of papers report on small molecule ectonucleotidase inhibitors. One group identified naphthalenesulfonic acids as efficient inhibitors of ecto-5′-nucleotidases. The most potent compound had an IC_{50} in the low micromolar range.[71] A very recent article reports a sulfoanthraquinone inhibitor complexed with NTPDase2, the crystal structure of which has been determined.[72] Another aspect worthy of mention is a recent article reporting a new sensitive ecto-5′-nucleotidase assay for compound screening, as shown below.[73]

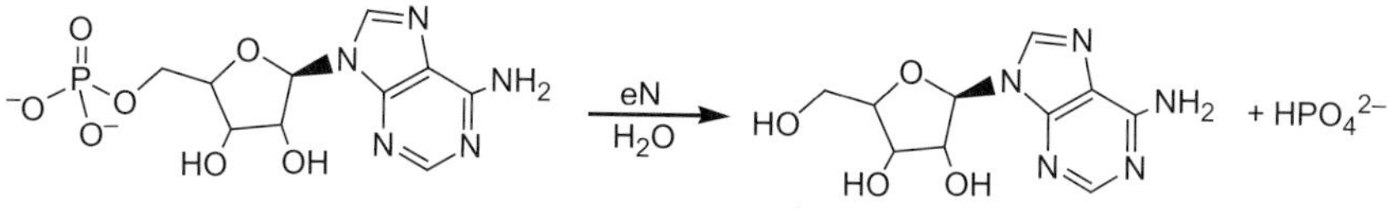

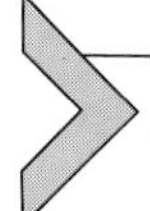

3. CARBAMOYLPHOSPHONATES: INHIBITORS OF EXTRACELLULAR ZINC-ENZYMES

Medicinal chemists had failed worldwide for over two decades in the design of clinically useful MMP inhibitors. Dozens of compounds have been synthesized, many of them very potent (mostly water insoluble and/or neutral) molecules, with high affinity to the enzymes yielding many very informative X-ray crystallographic structures which provided a mechanistic understanding of inhibition but no new drugs. Another cancer supporting hydrolase enzyme in the tumor microenvironment is autotaxin (ATX) that has also attracted the synthesis of inhibitor candidates, e.g. boronic acids, phosphates and phosphonates, but yielded no *in vivo* active compounds.[74]

A new approach using ionizable acids soluble in aqueous solutions at physiological pH is based on the assumption that if both the inhibitor and the target enzyme are dissolved in the same solution, the result will be enzymatic inhibition. Previous work indicated that water solubility for an inhibitor is not enough, because neutral molecules can partition between two phases resulting in inefficient inhibition. This approach led to the design and synthesis of dozens of carbamoyl phosphonates (CPOs) of varying structures ranging from aliphatic,[75] alicyclic,[76] and aromatic structures.[77] The *in vitro* test model for these compounds was tumor cell invasion through a reconstituted basement membrane (Matrigel), which depends on the ability of the cells to locally degrade collagen IV present in the Matrigel. This degradation is dependent on MMP-2. If a compound showed significant potency in preventing tumor cell invasion in a dose-dependent manner, then it was tested for antimetastatic activity in a murine melanoma model. In this model, mice were injected i.v. with B16F10 cells and treated daily for 3 weeks with the various compounds; values reflect percent decrease in lung metastases relative to control. The activities of these new CPOs against MMP-2, CAIX, and ATX are shown in Table 17.1.[78]

As can be seen in Table 17.1, compound **11** is too short to inhibit ATX, but **12** (JS-403) is a potent inhibitor of ATX. The three longest CPOs were found to be active ATX inhibitors, the two longest ones **13** (TCH-18) and **14** (TCH-23) are inactive toward MMP-2.

Table 17.2 shows *in vivo* results for two compounds, **12** and **13**, in the previously mentioned murine melanoma model. Compound **12** has been examined both when administered i.v. and PO, while **13** has been administered only orally. The results show a decrease of the metastases to a very high extent relative to control, both for **12**, which inhibits three enzymes (MMP-2–CA–ATX) and for **13**, which inhibits only ATX and CA.

Table 17.1 Test results of previously synthesized CPO inhibitors[77] examined as potential ATX inhibitors.

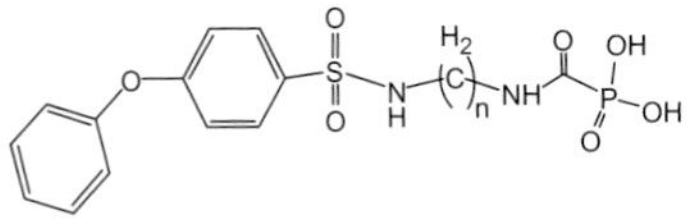

No.	Symbols	CH_2 chain length, *n*	ATX IC_{50} μM	MMP-2 IC_{50} μM	hCAIX IC_{50} μM
11	JS-389	*5*	>100	4	6.3
12	**JS-403**	***6***	**1.5**	**4**	**7.0**
13	**TCH-18**	7	**1.5**	**>100**	**0.85**
14	**TCH-23**	*8*	**1.5**	**>100**	**0.68**

Compounds **12, 13** and **14** in this table have been identified as ATX inhibitors.

Table 17.2 *In vivo* effect of selected CPO inhibitors on metastasis formation in mice[a]

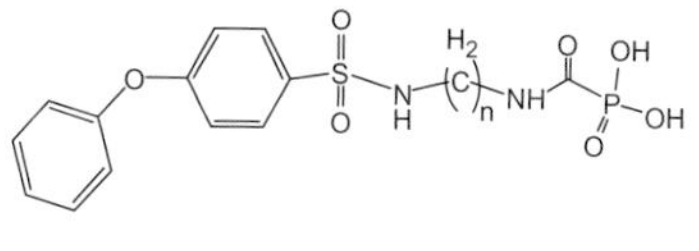

Compounds	Mode of introduction	CH_2 chain length, *n*	Dose 12.5 mg/kg	25 mg/kg
			Values reflect percent decrease in lung metastases relative to control	
12	IP	6	82	82
12	PO	6	61	71
13	PO	7	67	90

[a]Mice were injected i.v. with B16F10 cells and treated daily for 3 weeks by the drugs; values reflect percent decrease in lung metastases relative to control.

4. CONCLUSIONS AND OUTLOOK

A previous chapter suggested that water-soluble metalloenzymes, including MMPs, CAIX, and ATX, could be inhibited by cell-impermeable, water-soluble, charged, ionic molecules.[79] The present review demonstrates the feasibility of this prediction.

We expect that the coming years will see further discoveries of cancer related enzymes and their inhibitors. One kind might be newly discovered extracellular enzymes that support cancer, while the second will probably be more ectonucleotidases, and small molecule inhibitors of the enzymes mentioned. We hope that the outcome of these developments will lead to novel side effect-free cancer therapy.

ACKNOWLEDGMENTS

This work was supported in part by the Grass Center for Drug Design and Synthesis of Novel Therapeutics to E. B. and R. R. R. R. and E. B. are affiliated with the David R. Bloom Center of Pharmacy and the Brettler Center for Pharmacology, in the School of Pharmacy, Faculty of Medicine, The Hebrew University of Jerusalem.

REFERENCES

1. Liotta, L. A. *Am. J. Pathol.* **1984**, *117*, 339.
2. Shuman Moss, L. A.; Jensen-Taubman, S.; Stetler-Stevenson, W. G. *Am. J. Pathol.* **2012**, *181*, 1895.
3. Stellas, D.; Patsavoudi, E. *Anticancer Agents Med. Chem.* **2012**, *12*, 707.
4. Li, N. G.; Shi, Z.-H.; Tang, Y. P.; Duan, J. A. *Curr. Med. Chem.* **2009**, *16*, 3805.
5. Kessenbrock, K.; Plaks, V.; Werb, Z. *Cell* **2010**, *141*, 52.
6. Morrison, C. J.; Butler, G. S.; Rodriguez, D.; Overall, C. M. *Curr. Opin. Cell Biol.* **2009**, *21*, 645.
7. Renkiewicz, R.; Qiu, L.; Lesch, C.; Sun, X.; Devalaraja, R.; Cody, T.; Kaldjian, E.; Welgus, H.; Baragi, V. *Arthritis Rheum.* **2003**, *48*, 1742.
8. Romanchikova, N.; Trapencieris, P.; Zemıtis, J.; Turks, M. *J. Enzyme Inhib. Med. Chem.* **2013**, http://dx.doi.org/10.3109/14756366.2013.855207.
9. Fabre, B.; Filipiak, K.; Diaz, N.; Zapico, J. M.; Surez, D.; Ramos, A.; de Pascual-Teresa, B. *ChemBioChem* **2014**, *15*(3), 399–412. http://dx.doi.org/10.1002/cbic.201300698, and an earlier paper by this group.
10. Alcantara, M. B.; Dass, C. R. *J. Pharm. Pharmacol.* **2014**, *66*(7), 895–902.
11. Schwartz, J.; Holmuhamedov, E.; Zhang, X.; Lovelace, G. L.; Smith, C. D.; Lemasters, J. J. *Toxicol. Appl. Pharmacol.* **2013**, *273*, 172.
12. Horwitz, V.; Dachir, S.; Cohen, M.; Gutman, H.; Cohen, L.; Fishbine, E.; Brandeis, R.; Turetz, J.; Amir, A.; Gore, A.; Kadar, T. *Curr. Eye Res.* **2014**, *39*, 803.
13. Gooyit, M.; Song, W.; Mahasenan, K. V.; Lichtenwalter, K.; Suckow, M. A.; Schroeder, V. A.; Wolter, W. R.; Mobashery, S.; Chang, M. *J. Med. Chem.* **2013**, *56*, 8139.
14. Veerendhar, A.; Reich, R.; Breuer, E. C. R. *C.R. Chim.* **2010**, *13*, 1191.
15. Tauro, M.; Laghezza, A.; Loiodice, F.; Agamennone, M.; Campestre, C.; Tortorella, P. *Bioorg. Med. Chem.* **2013**, *21*, 6456.
16. Said, A. H.; Raufman, J. P.; Xie, G. *Cancers (Basel)* **2014**, *6*, 366.
17. van Nguyen, S.; Skarstedt, M.; Löfgren, S.; Zar, N.; Andersson, R. E.; Lindh, M.; Matussek, A.; Dimberg, J. *Anticancer Res.* **2013**, *33*, 3247.
18. Wen, Y.; Cai, L. *Zhongguo Fei Ai Za Zhi* **2014**, *17*, 30.
19. Larson, S. R.; Zhang, X.; Dumpit, R.; Coleman, I.; Lakely, B.; Roudier, M.; Higano, C. S.; True, L. D.; Lange, P. H.; Montgomery, B.; Corey, E.; Nelson, P. S.; Vessella, R. L.; Morrissey, C. *Prostate* **2013**, *73*, 932.
20. Ng, K. T.; Qi, X.; Kong, K. L.; Cheung, B. Y.; Lo, C. M.; Poon, R. T.; Fan, S. T.; Man, K. *Eur. J. Cancer* **2011**, *47*, 2299.
21. Kim, J. M.; Kim, H. J.; Koo, B. S.; Rha, K. S.; Yoon, Y. H. *Eur. Arch. Otorhinolaryngol.* **2013**, *270*, 1137.
22. Dahl, R.; Titlestad, I.; Lindqvist, A.; Wielders, P.; Wray, H.; Wang, M.; Samuelsson, V.; Mo, J.; Holt, A. *Pulm. Pharmacol. Ther.* **2012**, *25*, 169.
23. Wu, Y.; Li, J.; Wu, J.; Morgan, P.; Xu, X.; Rancati, F.; Vallese, S.; Raveglia, L.; Hotchandani, R.; Fuller, N.; Bard, J.; Cunningham, K.; Fish, S.; Krykbaev, R.; Tama, S.; Goldman, S. J.; Williams, C.; Mansour, T. S.; Saiah, E.; Sypek, J.; Li, W. *Bioorg. Med. Chem. Lett.* **2012**, *22*, 138.
24. Czarny, B.; Stura, E. A.; Devel, L.; Vera, L.; Cassar-Lajeunesse, E.; Beau, F.; Calderone, V.; Fragai, M.; Luchinat, C.; Dive, V. *J. Med. Chem.* **2013**, *56*, 1149.
25. Knäuper, V.; López-Otin, C.; Smith, B.; Knight, G.; Murphy, G. *J. Biol. Chem.* **1996**, *271*, 1544.
26. (a) Pelletier, J. P.; Boileau, C.; Boily, M.; Brunet, J.; Mineau, F.; Geng, C.; Reboul, P.; Laufer, S.; Lajeunesse, D.; Martel-Pelletier, J. *Arthritis Res. Ther.* **2005**, *7*, R1091; (b) Joronen, K.; Ala-aho, R.; Majuri, M. L.; Alenius, H.; Kähäri, V. M.; Vuorio, E. *Ann. Rheum. Dis.* **2004**, *63*, 656–664.

27. Meides, A.; Gutschalk, C. M.; Devel, L.; Beau, F.; Czarny, B.; Hensler, S.; Neugebauer, J.; Dive, V.; Angel, P.; Mueller, M. M. *Int. J. Cancer* **2014**, http://dx.doi.org/10.1002/ijc.28866 (Epub ahead of print).
28. Yang, G.; Ma, F.; Zhong, M.; Fang, L.; Peng, Y.; Xin, X.; Zhong, J.; Zhu, W.; Zhang, Y. *Mol. Med. Rep.* **2014**, *9*, 1371.
29. Wang, J. R.; Li, X. H.; Gao, X. J.; An, S. C.; Liu, H.; Liang, J.; Zhang, K.; Liu, Z.; Wang, J.; Chen, Z.; Sun, W. *Eur. Rev. Med. Pharmacol. Sci.* **2013**, *17*, 427.
30. Wang, L.; Wang, X.; Liang, Y.; Diao, X.; Chen, Q. *Acta Biochim. Pol.* **2012**, *59*, 593.
31. Santamaria, S.; Nuti, E.; Cercignani, G.; Marinelli, L.; La Pietram, V.; Novellino, E.; Rossello, A. *Biochem. Pharmacol.* **2012**, *84*, 813.
32. De Savi, C.; Waterson, D.; Pape, A.; Lamon, S.; Hadley, E.; Mills, M.; Page, K. M.; Bowyer, J.; Maciewicz, R. A. *Bioorg. Med. Chem. Lett.* **2013**, *23*, 4705.
33. Fischer, T.; Riedl, R. *ChemMedChem* **2013**, *8*, 1457.
34. (a) Scheller, J.; Chalaris, A.; Garbers, C.; Rose-John, S. *Trends Immunol.* **2011**, *32*, 380; (b) Kenny, P. A. *Differentiation* **2007**, *75*, 800.
35. Murumkar, P. R.; Giridhar, R.; Yadav, M. R. *Expert Opin. Drug Discov.* **2013**, *8*, 157.
36. Hsu, P. P.; Sabatini, D. M. *Cell* **2008**, *134*, 703.
37. Parks, S. K.; Chiche, J.; Pouysségur, J. *Nat. Rev. Cancer* **2013**, *13*, 611.
38. Federici, C.; Petrucci, F.; Caimi, S.; Cesolini, A.; Logozzi, M.; Borghi, M.; D'Ilio, S.; Lugini, L.; Violante, N.; Azzarito, T.; Majorani, C.; Brambilla, D.; Fais, S. *PLoS One* **2014**, *9*, e88193. http://dx.doi.org/10.1371/journal.pone.0088193, eCollection 2014.
39. Calcinotto, A.; Filipazzi, P.; Grioni, M.; Iero, M.; De Milito, A.; Ricupito, A.; Cova, A.; Canese, R.; Jachetti, E.; Rossetti, M.; Huber, V.; Parmiani, G.; Generoso, L.; Santinami, M.; Borghi, M.; Fais, S.; Bellone, M.; Rivoltini, L. *Cancer Res.* **2012**, *72*, 2746.
40. Xu, J.; Xie, R.; Liu, X.; Wen, G.; Jin, H.; Yu, Z.; Jiang, Y.; Zhao, Z.; Yang, Y.; Ji, B.; Dong, H.; Tuo, B. *Carcinogenesis* **2012**, *33*, 2432.
41. Spugnini, E. P.; Baldi, A.; Buglioni, S.; Carocci, F.; de Bazzichini, G. M.; Betti, G.; Pantaleo, I.; Menicagli, F.; Citro, G.; Fais, S. *J. Transl. Med.* **2011**, *9*, 221. http://dx.doi.org/10.1186/1479-5876-9-221.
42. Michel, V.; Licon-Munoz, Y.; Trujillo, K.; Bisoffi, M.; Parra, K. J. *Int. J. Cancer* **2013**, *132*, E1–E10.
43. Bellone, M.; Calcinotto, A.; Filipazzi, P.; De Milito, A.; Fais, S.; Rivoltini, L. *Oncoimmunology* **2013**, *2*, e22058.
44. Supuran, C. T. *Nat. Rev. Drug Discov.* **2008**, 7, 168.
45. Neri, D.; Supuran, C. T. *Nat. Rev. Drug Discov.* **2011**, *10*, 767.
46. Touisni, N.; Maresca, A.; McDonald, P. C.; Lou, Y.; Scozzafava, A.; Dedhar, S.; Winum, J.-Y.; Supuran, C. T. *J. Med. Chem.* **2011**, *54*, 8271.
47. Supuran, C. T. *Front. Pharmacol.* **2011**, *2*, 34.
48. Alterio, V.; Di Fiore, A.; D'Ambrosio, K.; Supuran, C. T.; De Simone, G. *Chem. Rev.* **2012**, *112*, 4421.
49. McIntyre, A.; Patiar, S.; Wigfield, S.; Li, J. L.; Ledaki, I.; Turley, H.; Leek, R.; Snell, C.; Gatter, K.; Sly, W. S.; Vaughan-Jones, R. D.; Swietach, P.; Harris, A. L. *Clin. Cancer Res.* **2012**, *18*, 3100.
50. Supuran, C. T.; McKenna, R. In: *Carbonic Anhydrase: Mechanism, Regulation, Links to Disease, and Industrial Applications*; McKenna, R. , Frost, S. , Eds.; Springer Verlag: Heidelberg, 2014; pp 291–323, Subcell. Biochem. 2014, 75, 291.
51. Lopez, M.; Salmon, A. J.; Supuran, C. T.; Poulsen, S.-A. *Curr. Pharm. Des.* **2010**, *16*, 3277.
52. Morris, J. C.; Chiche, J.; Grellier, C.; Lopez, M.; Bornaghi, L. F.; Maresca, A.; Supuran, C. T.; Pouyssegur, J.; Poulsen, S.-A. *J. Med. Chem.* **2011**, *54*, 6905.
53. Maresca, A.; Temperini, C.; Vu, H.; Pham, N. B.; Poulsen, S.-A.; Scozzafava, A.; Quinn, R. J.; Supuran, C. T. *J. Am. Chem. Soc.* **2009**, *131*, 3057.

54. Lou, Y.; McDonald, P. C.; Oloumi, A.; Chia, S. K.; Ostlund, C.; Ahmadi, A.; Kyle, A.; Auf dem Keller, U.; Leung, S.; Huntsman, D. G.; Clarke, B.; Sutherland, B. W.; Waterhouse, D.; Bally, M. B.; Roskelley, C. D.; Overall, C. M.; Minchinton, A.; Pacchiano, F.; Carta, F.; Scozzafava, A.; Touisni, N.; Winum, J. Y.; Supuran, C. T.; Dedhar, S. *Cancer Res.* **2011**, *71*, 3364.
55. Sethi, K. K.; Verma, S. M. *J. Enzyme Inhib. Med. Chem.* **2013**, *29*, 571.
56. Jacobson, K. A.; Gao, Z. G. *Nat. Rev. Drug Discov.* **2006**, *5*, 247.
57. Hasko, G.; Linden, J.; Cronstein, B.; Pacher, P. *Nat. Rev. Drug Discov.* **2008**, *7*, 759.
58. Zimmermann, H. *Drug Dev. Res.* **2001**, *52*, 44.
59. Stracke, M. L.; Krutzsch, H. C.; Unsworth, E. J.; Arestad, A.; Cioce, V.; Schiffmann, E.; Liotta, L. A. *J. Biol. Chem.* **1992**, *267*, 2524.
60. Sakagami, H.; Aoki, J.; Natori, Y.; Nishikawa, K.; Kakehi, Y.; Natori, Y.; Arai, H. *J. Biol. Chem.* **2005**, *280*, 23084.
61. Koike, S.; Keino-Masu, K.; Ohto, T.; Masu, M. *Genes Cells* **2006**, *11*, 133.
62. Tokumura, A.; Majima, E.; Kariya, Y.; Tominaga, K.; Kogure, K.; Yasuda, K.; Fukuzawa, K. *J. Biol. Chem.* **2002**, *277*, 39436.
63. Noguchi, K.; Herr, D.; Mutoh, T.; Chun, J. *Curr. Opin. Pharmacol.* **2009**, *9*, 15.
64. Gupte, R.; Patil, R.; Liu, J.; Wang, Y.; Lee, S. C.; Fujiwara, Y.; Fells, J.; Bolen, A. L.; Emmons-Thompson, K.; Yates, C. R.; Siddam, A.; Panupinthu, N. T.; Pham, T.-C.; Baker, D. L.; Parrill, A. L.; Mills, G. B.; Tigyi, G.; Miller, D. D. *ChemMedChem* **2011**, *6*, 922.
65. Wang, L.; Fan, J.; Thompson, L. F.; Zhang, Y.; Shin, T.; Curiel, T. J.; Zhang, B. *J. Clin. Invest.* **2011**, *121*, 2371.
66. Bavaresco, L.; Bernardi, A.; Braganhol, E.; Cappellari, A. R.; Rockenbach, L.; Farias, P. F.; Wink, M. R.; Delgado-Canedo, A.; Battastini, A. M. O. *Mol. Cell. Biochem.* **2008**, *319*, 61.
67. Zhang, B. *Oncoimmunology* **2012**, *1*, 67.
68. Allard, B.; Turcotte, M.; Spring, K.; Pommey, S.; Royal, I.; Stagg, J. *Int. J. Cancer* **2014**, *134*, 1466.
69. Xu, S.; Shao, Q.-Q.; Sun, J.-T.; Yang, N.; Xie, Q.; Wang, D.-H.; Huang, Q.-B.; Huang, B.; Wang, X.-Y.; Li, X.-G.; Qu, X. *Neuro Oncol.* **2013**, *15*, 1160.
70. Häusler, S. F. M.; Montalbán del Barrio, I.; Diessner, J.; Stein, R. G.; Strohschein, J.; Hönig, A.; Dietl, J.; Wischhusen, J. *Am. J. Transl. Res.* **2014**, *6*, 129.
71. Iqbal, J.; Saeed, A.; Raza, R.; Matin, A.; Hameed, A.; Furtmann, N.; Lecka, J.; Sévigny, J.; Bajorath, J. *Eur. J. Med. Chem.* **2013**, *70*, 685.
72. Zebisch, M.; Baqi, Y.; Schaefer, P.; Müller, C. E.; Straeter, N. *J. Struct. Biol.* **2014**, *185*, 336–341.
73. Freundlieb, M.; Zimmermann, H.; Müller, C. E. *Anal. Biochem.* **2014**, *446*, 53.
74. Albers, H. M. H. G.; Ovaa, H. *Chem. Rev.* **2012**, *112*, 2593.
75. Breuer, E.; Salomon, C. J.; Katz, Y.; Chen, W.; Lu, S.; Röschenthaler, G.-V.; Hadar, R.; Reich, R. *J. Med. Chem.* **2004**, *47*, 2826.
76. Reich, R.; Katz, Y.; Hadar, R.; Breuer, E. *Clin. Cancer Res.* **2005**, *11*, 3925.
77. Frant, J.; Veerendhar, A.; Chernilovsky, T.; Nedvetzki, S.; Vaksman, O.; Hoffman, A.; Breuer, E.; Reich, R. *ChemMedChem* **2011**, *6*, 1471.
78. Reich, R.; Hoffman, A.; Veerendhar, A.; Maresca, A.; Innocenti, A.; Supuran, C. T.; Breuer, E. *J. Med. Chem.* **2012**, *55*, 7875.
79. Breuer, E.; Frant, J.; Reich, R. *Expert Opin. Ther. Pat.* **2005**, *15*, 253.

Novel Screening Paradigms for the Identification of Allosteric Modulators and/or Biased Ligands for Challenging G-Protein-Coupled Receptors

Stephan Schann*, Pascal Neuville*,†, Michel Bouvier‡

*Domain Therapeutics SA, Bioparc, Boulevard Sebastien Brant, Strasbourg-Illkirch, France
†Domain Therapeutics NA, Inc., Neomed Institute, Montreal, Quebec, Canada
‡Department of Biochemistry, Institute for Research in Immunology and Cancer, University of Montréal, Montreal, Quebec, Canada

Contents

ABBREVIATIONS

Ab antibody
AM allosteric modulators
BL biased ligands
cAMP cyclic adenosine monophosphate
ERK extracellular signal-regulated kinase
FRET fluorescence resonance energy transfer
GFP green fluorescence protein
GPCR G-protein-coupled receptor

Annual Reports in Medicinal Chemistry, Volume 49
ISSN 0065-7743
http://dx.doi.org/10.1016/B978-0-12-800167-7.00018-3

HTS high-throughput screen
mGluR metabotropic glutamate receptor
NAMs negative allosteric modulators
PAMs positive allosteric modulators
SAMs silent allosteric modulators
SAR structure–activity relationship

1. INTRODUCTION

G-protein-coupled receptors (GPCRs) constitute a superfamily of more than 800 membrane proteins that recognize a highly diverse range of extracellular ligands to transmit a signal inside the cell. They have been a very successful drug target family, as nearly 30% of marketed drugs act through this class of receptors. But all these drugs target only about 50 members representing the "low-hanging fruit" GPCRs.

Until recently, the method of choice to develop a GPCR drug was the development of a small molecule or peptide agonist/antagonist, mimicking or competing with the endogenous ligand binding site (orthosteric binding site). Such molecules were classically discovered using either a radioactive binding assay or functional assays using a second messenger as a readout. These methods are limited in their ability to address the more challenging "high-hanging fruit" members of the GPCR tree, and recently multiple novel strategies aiming to tackle these difficult members have emerged. This chapter discusses these novel strategies, including allosteric modulators (AMs), biased ligands (BLs), and antibody (Ab)-based modulators.

AMs are by definition molecules interacting with a binding site distinct from the orthosteric one. They can be used therapeutically to modulate GPCR activity: positive/negative allosteric modulators (PAMs/NAMs) will, respectively, enhance/decrease orthosteric ligand affinity and/or activity, but will be devoid of activity in the absence of an orthosteric ligand. Some AMs called allosteric agonists can also activate the receptor in the absence of the endogenous ligand, and some others called ago-allosteric modulators encompass properties of both PAMs and allosteric agonists. The main advantages of AMs for drug discovery and development include the fact that they give access to molecules with better subtype selectivity profiles, more druggable chemistries for peptide, lipid, or class C GPCRs, and better selectivity of action.[1,2]

BLs on the other hand are ligands showing distinct potency and/or efficacy toward different signaling pathways engaged by a given receptor.[3–5]

This concept, also known as "functional selectivity," presents an evident advantage for the development of therapeutics with better safety profiles. Indeed, targeting only the pathway(s) involved in disease modification without affecting other functions of the receptor enables the development of drug candidates with reduced risk of adverse effects.

Finally, recent years have also shown a growing interest in the use of Abs to target GPCRs.[6,7] This is mainly the result of two breakthroughs in the domain: (1) the development of new technologies (novel antigens for their identification and the appearance of a novel Ab family) and (2) the discovery that Abs can exert agonist, AM and/or BL properties. Abs were also instrumental in revealing structural information for many GPCRs. Information has hitherto been lacking and now opens the door for structural drug discovery strategies with this family of receptors.[7]

All these paradigms illustrate the fact that we are now entering a novel era of GPCR drug discovery with new challenges to consider but also new promises for better drugs.

2. ALLOSTERIC MODULATORS

2.1. AM Screening: Challenges

Although rarely employed, equilibrium radioactive binding assays can be used for AM identification or characterization. They yield information about the AM cooperativity factor-α, a parameter quantifying the influence of the AM on the binding of the orthosteric probe. PAMs enhancing the orthosteric ligand binding will result in positive α values, whereas NAMs will lead to negative α values. However, these radioactive binding assays should be used with caution as NAMs with high negative α values could be mistaken as competitive orthosteric ligands, and in many cases, AMs can influence signaling with no detectable effect on ligand binding. Also, because of the probe dependency the choice of the radioligand selected for study may greatly impact the outcome.[8]

The method of choice for the identification of GPCR AMs is the use of functional assays. With these tests, the influence of the AMs on the endogenous ligand potency/efficacy is directly measured and quantified with the modulation factor-β.[8] The typical high-throughput screening (HTS) procedure for AM identification consists in combining single concentrations of both the endogenous agonist and the tested molecule. Two successive additions of the test compound and the endogenous agonist are performed on the cell system expressing the receptor of interest. With the first addition, the potential agonist activity (from orthosteric agonist, allosteric agonist, or ago-allosteric modulator) of the test compound is assessed. With the second one, PAM or NAM activity of the test compound is assessed in the presence of an EC_{10-20} or EC_{80} concentration of a known (preferably endogenous) agonist, respectively.[9]

Identification of GPCR AMs can, however, be complex. Indeed, so-called "context-dependent pharmacology" makes the task highly sensitive to the assay conditions and subtle changes can result in conflicting results.[1] For instance, "probe dependency" can have dramatic consequences as the same modulator can be characterized as a PAM against a particular orthosteric agonist and a NAM against another one.[10] This feature is of particular importance for receptor systems having multiple endogenous ligands. Moreover, AMs can influence the endogenous ligand potency, its efficacy, or both, and they can differently affect multiple signaling pathways engaged by a single endogenous ligand on a given GPCR. A complete characterization should therefore involve assessments of the AM on affinity and potency/efficacy of one or multiple endogenous ligand(s) and on one or multiple functional pathway(s). This can result in a complex picture not always easy to handle for medicinal chemists in the course of optimization programs.

2.2. AM Screening: Illustrative Examples

An illustration of an AM-HTS was recently published in the field of μ-opioid receptors.[11] Scientists from Bristol-Myers Squibb used a functional assay based on enzyme complementation to follow β-arrestin recruitment in U2OS cells. The screen was performed in two modes, in absence and presence of an EC_{10} of the endogenous ligand endomorphin-I, to detect both agonists and PAMs. This campaign resulted in the identification of **1** and **2** (Fig. 18.1) that showed no activity when tested alone, but increased endomorphin-I-induced β-arrestin recruitment. Follow-up characterization revealed that these two PAMs have respective EC_{50} values of 1 and 3 μM and shift the endomorphin-I EC_{10} response to 76% and 83% of the endogenous ligand maximal effect. Counterscreening with the δ-opioid receptor in the same cellular background and with the same β-arrestin readout demonstrated that the two PAMs were specific to the μ-opioid subtype. These PAMs were also evaluated in alternative functional tests and showed a similar PAM profile in a $G_{i/o}$ cAMP inhibitory paradigm in CHO cells as well as in a GTPγS binding assay run on more

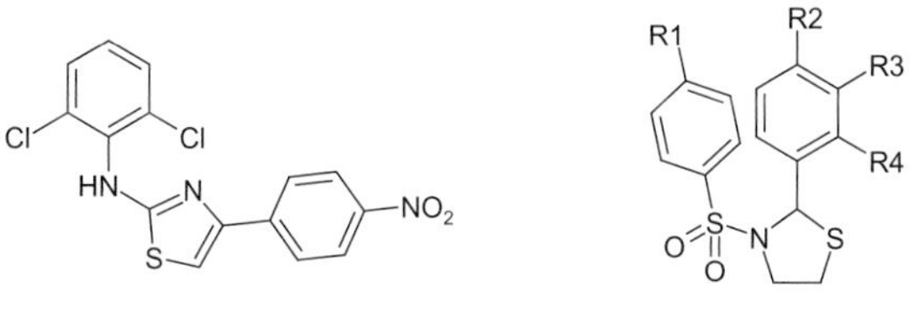

1

2 R1=Cl R2=OMe R3=Br R4=H
3 R1=Me R2=OMe R3=NO_2 R4=H
4 R1=Cl R2=Br R3=H R4=OMe

Figure 18.1 μ-Opioid receptor AMs described by BMS.

physiologically relevant mouse brain membranes. **2** was also tested in radioligand binding tests and showed different behaviors on DAMGO (a μ-opioid receptor selective agonist) and diprenorphine binding properties, increasing affinity for DAMGO but not for diprenorphine. This probe dependency, a typical hallmark of GPCR AMs, is an illustration of the limitation of binding studies for AM screening. Very interestingly, BMS scientists also reported in their publication that very close analogs of **2** such as **3** and **4** (Fig. 18.1) behaved as silent allosteric modulators (SAMs), inhibiting PAM activity of **2** without exerting any PAM or NAM activity. Again, this PAM to SAM switch within a same chemical family is also a hallmark of GPCR AMs.[12]

One of the standard advantages associated with GPCR AMs is their reduced propensity to induce desensitization or receptor-mediated side effects caused by long-term stimulation by orthosteric agonists and in the case of μ-opioid receptors the most common side effects including constipation or hyperalgesia. Therefore, **1** and **2** constitute prototypes for the development of novel categories of opioid ligands for pain management with an improved side-effect profile.[13]

A second illustrative example of AM-HTS was reported for the GLP-1R, a class B GPCR with screening and characterization studies toward the identification of small molecule modulators of the GLP-1R.[14] To that end, the authors chose a duplexed protocol for their campaign, where two cell lines containing the GLP-1R and the glucagon receptor were coplated. The primary HTS was based on a Ca^{2+}-based functional assay run in Chem-9 and Chem-1 cells expressing the GLP-1R and glucagon receptors, respectively. Tested molecules were evaluated in the absence or presence of an EC_{50} of the endogenous ligand to detect agonists, and PAMs or NAMs simultaneously. This protocol constitutes an innovative way to identify simultaneously both categories of GLP-1R AMs, with direct

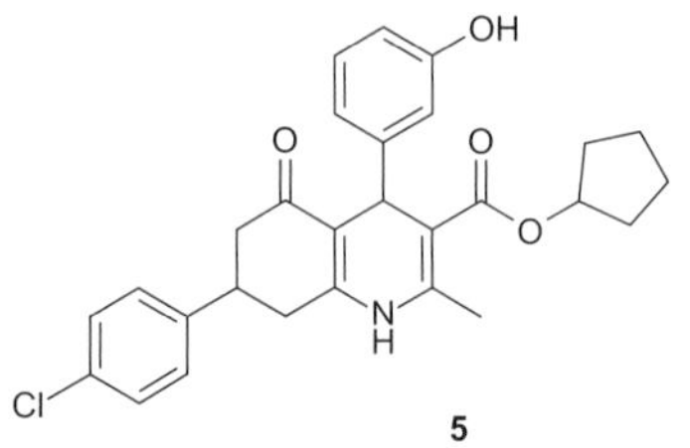

Figure 18.2 GLP-1R PAM described by Morris *et al.*

counter-screen information in a single screening effort. Follow-up characterization of identified hits involved a dose–response of AMs in the presence of a GLP-1 EC_{20} and EC_{80} for PAMs and NAMs, respectively, and a second functional test based on cyclic adenosine monophosphate (cAMP) accumulation in a more native cellular background, INS-1. Among the multiple classes of GLP-1R AMs reported, some showed a promising PAM profile with an absence of activity on the glucagon receptor and the melanocortin type 4 receptor (see compound **5** in Fig. 18.2). Similarly as with the μ-opioid receptor, a GLP-1R PAM constitutes a very promising approach to overcome limitations associated with the use of injectable orthosteric agonists, and numerous pharmaceutical companies are currently involved in small molecule drug discovery programs.[15] However, the complexity of the GLP-1R system with multiple endogenous ligands and multiple functional pathways stresses the need for a complete characterization of candidates.[16,17]

2.3. AM Screening: Novel Approaches

The inherent challenges of GPCR AMs such as context-dependent pharmacology have led to the development of new technological approaches for their identification and characterization.[1,18] Among them, fluorescence resonance energy transfer (FRET)-based binding assays have been proposed to circumvent system-dependent complexity.[19,20] One of these binding technologies named DTect-All™ has been proposed as a filter to specifically identify AMs.[1] DTect-All™ is based on a double labeling of the receptor with (1) a green fluorescence protein (GFP) link to the truncated N-terminus of the GPCR and (2) a fluorescent probe binding the receptor transmembrane part. When these two fluorophores are in close proximity, a FRET signal is detected and the screening consists of identifying molecules

6
mGluR2 binding: $K_i = 6.6\ \mu M$
mGluR2: no activity
mGluR3: no activity

7
mGluR2 binding: $K_i = 1\ \mu M$
mGluR2: NAM with $IC_{50} = 0.8\ \mu M$
mGluR3: PAM with $EC_{50} = 13.4\ \mu M$

8
mGluR2 binding: $K_i = 0.8\ \mu M$
mGluR2:NAM with $IC_{50} = 1\ \mu My$
mGluR3: PAM with $EC_{50} = 10.4\ \mu M$

Figure 18.3 Group II mGluR AMs described by Schann *et al.*

that will interfere with the FRET signal by changing the receptor conformation. An illustration of its use has been published for group II metabotropic glutamate receptor (mGluR2/3).[21] In this example, a binding assay for mGluR2 was engineered with a receptor having its N-terminal domain replaced by a GFP. The corresponding hybrid receptor no longer recognizes endogenous glutamate and, therefore, is well suited for the specific identification of AMs. Screening of a small compound collection led to the identification of the silent binder **6** (mGluR2 binder devoid of activity in a Ca^{2+}-based functional assay) that was changed with subtle chemical modifications into **7** or **8**, dual AMs showing NAM activity on mGluR2 and PAM activity on mGluR3 (Fig. 18.3).[21] Interestingly, it should be noted that the direct screen with the calcium assay would not have enabled the identification of the initial hit **6**.

3. BIASED LIGANDS

3.1. BL Identification: Characteristics

No specific technology has been developed so far for BL identification. In the few examples found in the literature, BLs were identified from an initial bias hypothesis that one pathway should be favored over a second one to improve the candidate behavior *in vivo* (see part 1 in Scheme 18.1). Under these circumstances, the HTS relies on the use of one or two functional assays, with the objective to identify ligands biased for a pathway A over a pathway B. Corresponding biased hits are then further optimized by medicinal chemists following potency/efficacy on both pathways in parallel (Scheme 18.1).

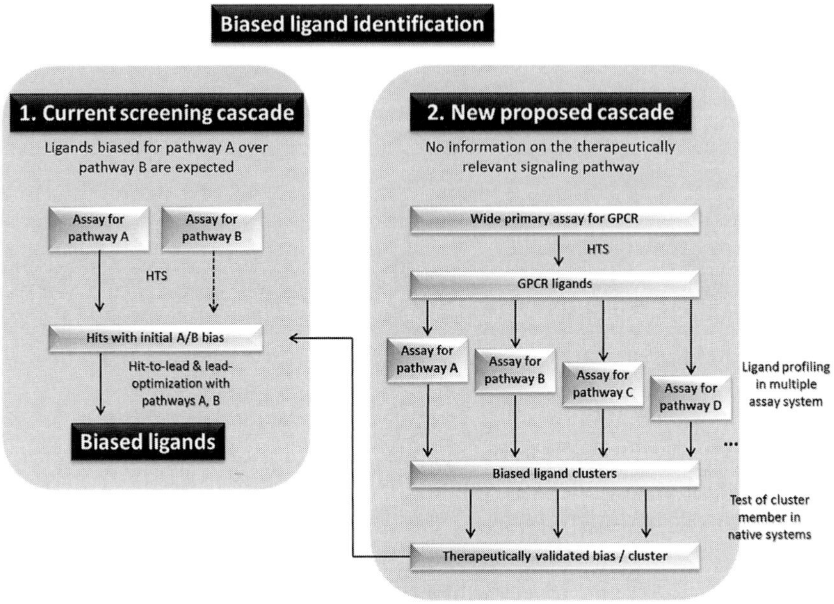

Scheme 18.1 Screening cascades for BL identification.

3.2. BL Identification: Illustrative Examples

A recent example of the identification of GPCR BLs relates to the μ-opioid receptor, which can engage both the G-protein Gi and β-arrestin-2 pathways.[22] It was proposed that the therapeutic analgesic response of morphine is mediated by the G-protein pathway, whereas tolerance, constipation, and respiratory depression, three well-known side effects of opioid-like drugs, are at least in part mediated by the β-arrestin-2 pathway.[23,24] Based on this information, scientists from Trevena, Inc. screened for μ-opioid agonists biased for the Gi pathway using morphine **9** as the reference "balanced compound." A HTS was conducted with the two corresponding assays and identified compound **10** (Fig. 18.4) as an initial hit. This compound exerted similar potency on the two pathways (pEC50 = 6.3 on Gi and pEC50 = 5.7 on β-arrestin-2) but is a partial agonist with stronger efficacy on the Gi pathway (74%) compared with the β-arrestin pathway (32%). Structure–activity relationship (SAR) optimization was conducted with parallel characterization on both pathways in the same cell system. Differential SARs were identified for each pathway, leading to the identification of biased candidates such as compound **11** (Fig. 18.3) that is currently being investigated in the clinic for postoperative

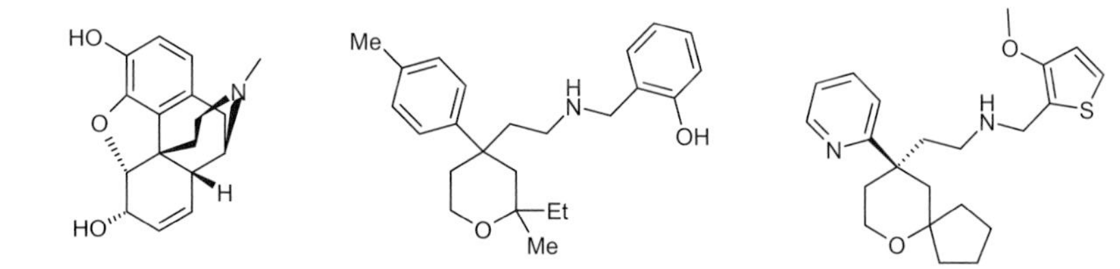

9	10	11
Morphine	Initial hit	TRV-130
Gi: pEC_{50}=7.4, efficacy=100%	Gi: pEC_{50}=6.3, efficacy=74%	Gi: pEC_{50}=8.1, efficacy=84%
b-arr: pEC_{50}=6.3, efficacy=100%	b-arr: pEC_{50}=5.7, efficacy=32%	b-arr: pEC_{50}=7.3, efficacy=15%

Figure 18.4 μ-Opioid receptor ligands.

pain.[25] The μ-opioid receptor is therefore at the forefront of innovative GPCR R&D approaches to identify both AM and BL for safer and more effective treatments for pain.[26]

The second example concerns the CXCR4 receptor, a therapeutic target for multiple indications including HIV and cancers. This receptor, like many GPCRs, can recruit several intracellular pathways upon activation, and scientists from Siena Biotech have conducted a screening program to identify new biased CXCR4 antagonists.[27] The strategy chosen was to conduct a primary HTS with a Gi-based assay in CHO cells, followed successively by a β-arrestin and an internalization test in C2C12 and U2OS cells, respectively. Here again, a native assay was incorporated in the screening cascade with a cAMP assay performed on the glioblastoma cell line GBMR16. The results of screening a 30,000 compound library reveal five distinct clusters of hits showing distinct signaling profiles with regard to the three studied pathways. Some hits, like compound **12** (Fig. 18.5), were antagonists of the Gi pathway only, showing no activity on the two other ones. These BLs can therefore be considered as NAMs for Gi and SAMs for β-arrestin and internalization as they bind the receptor without affecting these two specific pathways. Interestingly, other Gi antagonists illustrated by **13** were SAMs for β-arrestin and (allosteric) agonists for internalization and others, such as compound **14**, were NAMs for the three assays, showing activity similar to the reference antagonists AMD3100 or CTCE-9908. The next step is to determine the signaling profile that will offer the best equilibrium between therapeutic efficacy and the absence of side effects.

The third example is a report of BLs for the dopamine D_2 receptor.[28] The starting point for this program was an antipsychotic currently under clinical development, cariprazine **15** (Fig. 18.6). Again, two functional pathways were studied in this work, the cAMP one and the phosphorylation of

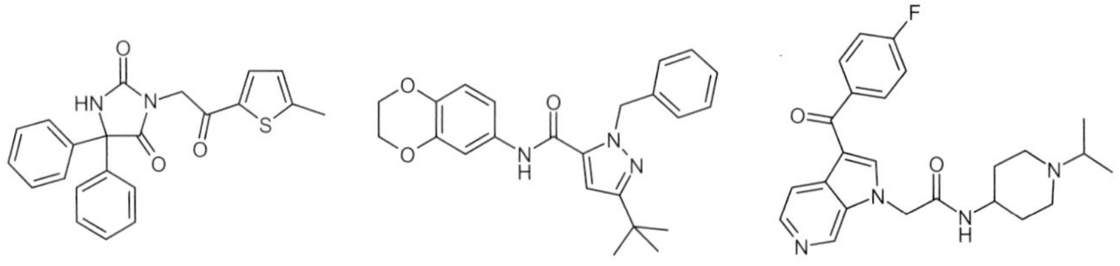

12
Gi assay: IC_{50}=2.76 μM
beta-arrestin: inactive
Internalization: inactive

13
Gi assay: IC_{50}=4 μM
beta-arrestin: inactive
Internalization: EC_{50}=16.7 μM

14
Gi assay: IC_{50}=2.53 μM
beta-arrestin: IC_{50}=3.2 μM
Internalization: IC_{50}=7.56 μM

Figure 18.5 Members of CXCR4-ligand clusters with different signaling profile.

extracellular signal-regulated kinase 1/2 (ERK1/2). The novel and interesting aspect of this report lies in the use of specific parameters allowing quantification of bias.[5,29] This is to our knowledge the first medicinal chemistry publication reporting an SAR effort guided by the transduction coefficient $\log(\tau/K_A)$. This parameter, derived from the operational model, constitutes a system-independent value to quantify agonist activity. τ refers to the agonist efficacy, whereas K_A represents the agonist equilibrium dissociation constant.[29] The authors also used $\Delta\Delta\log(\tau/K_A)$ to quantify bias between the two pathways for a given compound. This publication elegantly showed how subtle structural changes engender significant biases within the same chemical series. For instance, replacement of the urea moiety of cariprazine **15** with a NH-BOC group in **16** resulted in a meaningful reduction of cAMP activity, but almost no change in the pERK1/2 activation or in the binding affinity. This translated into a cAMP-pERK1/2 bias of 2.34 for **15** moving down to 0.71 for analog **16**. Similarly, changing one stereocenter on compound **17** to yield analog **18** resulted in an improved cAMP potency and in a complete loss of pERK1/2 activity. This simple modification, which does not affect affinity, resulted in a dramatic bias in the compound activity toward the cAMP pathway. Interestingly, a previous report described the identification of close analogs of the FDA-approved aripiprazole showing D_2-agonist activity with a bias for β-arrestin over cAMP inhibition.[30] Such BLs exerted antipsychotic activity in wild-type animals but not in β-arrestin knockout mice demonstrating that this pathway also plays a crucial role in D_2-mediated antipsychotic activity.

3.3. BL Identification: Novel Approaches

As of today, no functional assay fulfills the three criteria discussed by Kenakin in 2011 for an ideal primary BL HTS: (a) to be all encompassing,

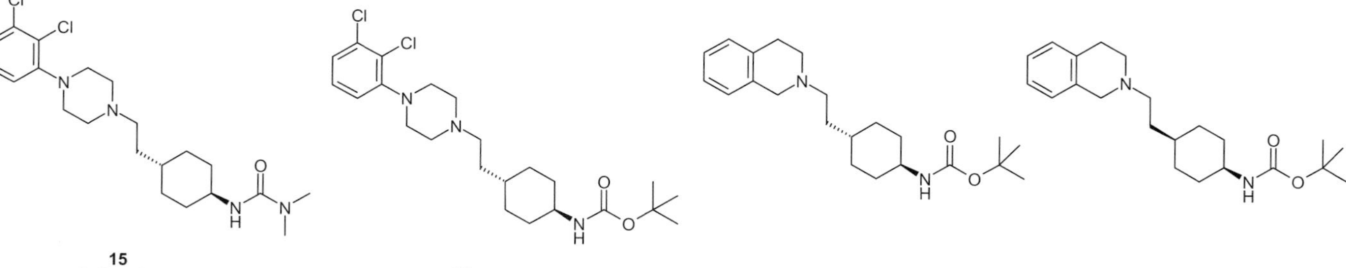

Figure 18.6 Dopamine D_2 receptor ligands.

(b) sensitive, and (c) simple to decode.[31] Moreover, the physiological relevance of assays selected for screening and optimization is sometimes questionable, and validation of signaling bias requires native system assays (ideally before initiating and) during medicinal chemistry efforts.

Consequently, a new screening cascade addressing the limitations of the current approach has been proposed (see right-hand side of Scheme 18.1).[31] A HTS should be conducted with a robust assay, as proximal to the receptor as possible, detecting the basic interaction of molecules with the GPCR to identify both biased and unbiased ligands. Then, hit profiling should be conducted in multiple assay systems covering all the complexity of receptor signaling to deliver clusters of ligands with different signaling profiles. The next step should consist of testing members of these clusters in integrated assays being as close as possible to the physiologically relevant outcome. This should deliver key validation information on both the chemical series and signaling pathways of importance to initiate and drive the optimization phases.

Such a strategy is used at Domain Therapeutics with the DTect-All™ technology as a proximal primary assay for ligand detection and BioSens-All™ technology for broad signaling characterization of the receptor binders.[1] BioSens-All™ consists of a combination of multiple BRET-based biosensor assays each following one specific signaling pathway.[32–36] They are performed in a homogenous high-throughput format in living cells of the same cellular background to avoid system-dependent bias observation.[1] They have been used recently to characterize the signaling properties of a new class of molecules, pepducins, targeting the chemokine CXCR4 receptor. This study revealed that a CXCR4-targeting pepducin acts as a biased agonist toward the Gi pathway without activating the G13 or β-arrestin pathways.[36] As a reflection of the clinical efficacy/side-effect profile, this information can be useful for the design of novel and safer agents.

4. Ab DISCOVERY AS NOVEL AMs/BLs

Abs have been used for many years to advance our understanding of biological targets and to propose novel therapeutic solutions. However, they have been rarely used with GPCRs, as these receptors present specific challenges for the identification of high affinity selective Abs: a limited surface area for Ab recognition, conformational dynamism of the receptor, and the challenge of producing and purifying sufficient quantities of the GPCR protein. In early 2014, only one GPCR Ab drug reached the market, mogamulizumab, targeting the CCR4 receptor, approved in Japan for the treatment of relapsed or refractory T cell leukemia and lymphoma.[37]

However, the above-mentioned challenge of GPCR conformational dynamism can also be seen as an advantage as one could consider developing Abs binding selectively to specific receptor states (agonist-bound, antagonist-bound, AM-bound, G-protein-coupled, β-arrestin-coupled, etc.). Therefore, Abs could also be considered as a novel strategy for the development of therapeutics that will elicit specific effects (agonist, PAM, NAM, biased agonist/antagonist) on specific signaling pathways, thus leading to therapeutics showing cell-specific actions, with all the safety advantages this might represent.

The most important technical aspect of Ab identification deals with the nature of the antigen. Indeed, the receptor antigen ideally needs to be in a pure, homogenous, and stable form, coherent with the native receptor structure. However, the use of a simple transfected cell line results in poor results as GPCR levels are rather low and the presence of multiple other membrane-bound proteins pollutes the Ab screening process. Among newly emerging technologies for Ab identification are the use of constrained GPCR peptide fragments,[38] peptide fragments conjugated to viral particles,[39] liposome-/exosome-/nanoparticles presenting a higher density of the pure receptor,[40,41] as well as access to purified GPCRs stabilized with specific mutations.[6] Moreover, the recent discovery of monomeric single chain Abs (nanobodies) has enabled the identification of therapeutic candidates of smaller size and enhanced stability.[42–44] Nanobodies that can bind distinct conformational states of a GPCR are also invaluable tools for the study of GPCR structures and functions.[45–48]

Thanks to these novel approaches, more and more programs involving the development of therapeutic GPCR-targeting Abs are currently reaching clinical development.[6] Interestingly, this renaissance allowed the discovery of novel categories of Abs showing AM or BL properties. For instance, scientists from Roche recently reported the discovery of the Ab MAB1/28 showing bias toward internalization versus G-protein activation of mGluR7.[49] The data showed that MAB1/28 behaves as an agonist with regard to internalization and mitogen-activated protein kinase/ERK pathway activation, and as an antagonist for L-AP4-induced cAMP inhibition. This clearly shows that this Ab is stabilizing a specific receptor conformation triggering a specific functional signature.[50] More recently, agonist monoclonal Abs of the β1-adrenergic receptor showing biased activity for the Gs pathway over β-arrestin have also been reported. These Abs were reported as acting only in dimeric form, suggesting they could work by promoting/stabilizing GPCR dimerization. One of these Abs was found to be active *in vivo* on heart rate, which is consistent with the expected action of a β1-adrenergic agonist.[51]

5. CONCLUSIONS

Recent novel paradigms are revitalizing the traditional field of GPCR drug discovery. They encompass new categories of ligands such as AMs, BLs, or Abs associated with advantages for the discovery of new therapeutics. They also represent new challenges, as their identification requires novel technological approaches that have not fully penetrated the pharmaceutical industry. Taken together, these new paradigms mark the beginning of a new GPCR drug discovery age.

REFERENCES

1. Schann, S.; Bouvier, M.; Neuville, P. *Drug Discov. Today Technol.* **2013**, *10*(2), e261.
2. Melancon, B. J.; Hopkins, C. R.; Wood, M. R.; Emmitte, K. A.; Niswender, C. M.; Christopoulos, A.; Conn, P. J.; Lindsley, C. W. *J. Med. Chem.* **2012**, *55*(4), 1445.
3. Rajagopal, S.; Rajagopal, K.; Lefkowitz, R. J. *Nat. Rev. Drug Discov.* **2010**, *9*(5), 373.
4. Galandrin, S.; Oligny-Longpré, G.; Bouvier, M. *Trends Pharmacol. Sci.* **2007**, *28*(8), 423.
5. Kenakin, T.; Christopoulos, A. *Nat. Rev. Drug Discov.* **2013**, *12*(3), 205.
6. Webb, D. R.; Handel, T. M.; Kretz-Rommel, A.; Stevens, R. C. *Biochem. Pharmacol.* **2013**, *85*(2), 147.
7. Congreve, M.; Dias, J. M.; Marshall, F. H. *Prog. Med. Chem.* **2014**, *53*, 1.
8. Wootten, D.; Christopoulos, A.; Sexton, P. M. *Nat. Rev. Drug Discov.* **2013**, *12*(8), 630.
9. Burford, N. T.; Watson, J.; Bertekap, R.; Alt, A. *Biochem. Pharmacol.* **2011**, *81*(6), 691.
10. Kenakin, T. *Mol. Interv.* **2004**, *4*(4), 222.
11. Burford, N. T.; Clark, M. J.; Wehrman, T. S.; Gerritz, S. W.; Banks, M.; O'Connell, J.; Traynor, J. R.; Alt, A. *Proc. Natl. Acad. Sci. U.S.A.* **2013**, *110*, 10830.
12. Wood, M. R.; Hopkins, C. R.; Brogan, J. T.; Conn, P. J.; Lindsley, C. W. *Biochemistry* **2011**, *50*(13), 2403.
13. Burford, N. T.; Traynor, J. R.; Alt, A. *Br. J. Pharmacol.* **2014**, http://dx.doi.org/10.1111/bph.12599.
14. Morris, L. C.; Days, E. L.; Turney, M.; Mi, D.; Lindsley, C. W.; Weaver, C. D.; Niswender, K. D. *J. Biomol. Screen* **2014**, *19*(6), 847–858. http://dx.doi.org/10.1177/1087057114520971.
15. Willard, F. S.; Bueno, A. B.; Sloop, K. W. *Exp. Diabetes Res.* **2012**, article ID 709893.
16. Wootten, D.; Savage, E. E.; Willard, F. S.; Bueno, A. B.; Sloop, K. W.; Christopoulos, A.; Sexton, P. M. *Mol. Pharmacol.* **2013**, *83*(4), 822.
17. Pabreja, K.; Mohd, M. A.; Koole, C.; Wootten, D.; Furness, S. G. *Br. J. Pharmacol.* **2014**, *171*(5), 1114.
18. Lütjens, R.; Perry, B.; Schelshorn, D.; Rocher, J. P. *Drug Discov. Today Technol.* **2013**, *10*(2), e253.
19. Emami-Nemini, A.; Roux, T.; Leblay, M.; Bourrier, E.; Lamarque, L.; Trinquet, E.; Lohse, M. J. *Nat. Protoc.* **2013**, *8*(7), 1307.

20. Daval, S. B.; Valant, C.; Bonnet, D.; Kellenberger, E.; Hibert, M.; Galzi, J.-L.; Ilien, B. *J. Med. Chem.* **2012**, *55*(5), 2125.
21. Schann, S.; Mayer, S.; Franchet, C.; Frauli, M.; Steinberg, E.; Thomas, M.; Baron, L.; Neuville, P. *J. Med. Chem.* **2010**, *53*(24), 8775.
22. Chen, X. T.; Pitis, P.; Liu, G.; Yuan, C.; Gotchev, D.; Cowan, C. L.; Rominger, D. H.; Koblish, M.; Dewire, S. M.; Crombie, A. L.; Violin, J. D.; Yamashita, D. S. *J. Med. Chem.* **2013**, *56*(20), 8019.
23. Bohn, L. M.; Leftkowitz, R. J.; Gainetdinov, R. R.; Peppel, K.; Caron, M. G.; Lin, F. T. *Science* **1999**, *286*(5449), 2495.
24. Raehal, K. M.; Walker, J. K.; Bohn, L. M. *J. Pharmacol. Exp. Ther.* **2005**, *314*(3), 1195.
25. Soergel, D. G.; Subach, R. A.; Sadler, B.; Connell, J.; Marion, A. S.; Cowan, C. L.; Violin, J. D.; Lark, M. W. *J. Clin. Pharmacol.* **2014**, *54*(3), 351.
26. Thompson, G.; Kelly, E.; Christopoulos, A.; Canals, M. *Br. J. Pharmacol.* **2014**, http://dx.doi.org/10.1111/bph.12600.
27. Castaldo, C.; Benicchi, T.; Otrocka, M.; Mori, E.; Pilli, E.; Ferruzzi, P.; Valensin, S.; Diamanti, D.; Fecke, W.; Varrone, M.; Porcari, V. *J. Biomol. Screen* **2014**, *19*(6), 859–869. http://dx.doi.org/10.1177/1087057114526283.
28. Shonberg, J.; Herenbrink, C. K.; López, L.; Christopoulos, A.; Scammells, P. J.; Capuano, B.; Lane, J. R. *J. Med. Chem.* **2013**, *56*(22), 9199.
29. Kenakin, T.; Watson, C.; Muniz-Medina, V.; Christopoulos, A.; Novick, S. *ACS Chem. Neurosci.* **2012**, *3*(3), 193.
30. Allen, J. A.; Yost, J. M.; Setola, V.; Chen, X.; Sassano, M. F.; Chen, M.; Peterson, S.; Yadav, P. N.; Huang, X. P.; Feng, B.; Jensen, N. H.; Che, X.; Bai, X.; Frye, S. V.; Wetsel, W. C.; Caron, M. G.; Javitch, J. A.; Roth, B. L.; Jin, J. *Proc. Natl. Acad. Sci. U.S.A.* **2011**, *108*(45), 18488.
31. Kenakin, T. J. *Pharmacol. Exp. Ther.* **2011**, *336*(2), 296.
32. Angers, S.; Salahpour, A.; Joly, E.; Hilairet, S.; Chelsky, D.; Dennis, M.; Bouvier, M. *Proc. Natl. Acad. Sci. U.S.A.* **2000**, *97*(7), 3684.
33. Galés, C.; Rebois, R. V.; Hogue, M.; Trieu, P.; Breit, A.; Hébert, T. E.; Bouvier, M. *Nat. Methods* **2005**, *2*(3), 177.
34. Galés, C.; Van Durm, J. J.; Schaak, S.; Pontier, S.; Percherancier, Y.; Audet, M.; Paris, H.; Bouvier, M. *Nat. Struct. Mol. Biol.* **2006**, *13*(9), 778.
35. Charest, P. G.; Terrillon, S.; Bouvier, M. *EMBO Rep.* **2004**, *6*(4), 334.
36. Quoyer, J.; Janz, J. M.; Luo, J.; Ren, Y.; Armando, S.; Lukashova, V.; Benovic, J. L.; Carlson, K. E.; Hunt, S. W., 3rd.; Bouvier, M. *Proc. Natl. Acad. Sci. U.S.A.* **2013**, *110*(52), 15088.
37. de Lartigue, J. *Drugs Today (Barc.)* **2012**, *48*, 655.
38. Misumi, S.; Nakayama, D.; Kusaba, M.; Iiboshi, T.; Mukai, R.; Tachibana, K.; Nakasone, T.; Umeda, M.; Shibata, H.; Endo, M.; Takamune, N.; Shoji, S. *J. Immunol.* **2006**, *176*(1), 463.
39. Sommerfelt, M. A. *Expert Opin. Ther. Pat.* **2009**, *19*(9), 1323.
40. Delcayre, A.; Shu, H.; Le Pecq, J. B. *Expert Rev. Anticancer Ther.* **2005**, *5*(3), 537.
41. Jones, J. W.; Greene, T. A.; Grygon, C. A.; Doranz, B. J.; Brown, M. P. *J. Biomol. Screen.* **2008**, *13*(5), 424.
42. Siontorou, C. G. *Int. J. Nanomedicine* **2013**, *8*, 4215.
43. Jähnichen, S.; Blanchetot, C.; Maussang, D.; Gonzalez-Pajuelo, M.; Chow, K. Y.; Bosch, L.; De Vrieze, S.; Serruys, B.; Ulrichts, H.; Vandevelde, W.; Saunders, M.; De Haard, H. J.; Schols, D.; Leurs, R.; Vanlandschoot, P.; Verrips, T.; Smit, M. J. *Proc. Natl. Acad. Sci. U.S.A.* **2010**, *107*(47), 20565.
44. Maussang, D.; Mujić-Delić, A.; Descamps, F. J.; Stortelers, C.; Vanlandschoot, P.; Stigter-van Walsum, M.; Vischer, H. F.; van Roy, M.; Vosjan, M.; Gonzalez-Pajuelo, M.; van Dongen, G. A.; Merchiers, P.; van Rompaey, P.; Smit, M. J. *J. Biol. Chem.* **2013**, *228*(41), 29562.

45. Rasmussen, S. G.; Choi, H. J.; Fung, J. J.; Pardon, E.; Casarosa, P.; Chae, P. S.; Devree, B. T.; Rosenbaum, D. M.; Thian, F. S.; Kobilka, T. S.; Schnapp, A.; Konetzki, I.; Sunahara, R. K.; Gellman, S. H.; Pautsch, A.; Steyaert, J.; Weis, W. I.; Kobilka, B. K. *Nature* **2011**, *469*(7329), 175.
46. Rasmussen, S. G.; DeVree, B. T.; Zou, Y.; Kruse, A. C.; Chung, K. Y.; Kobilka, T. S.; Thian, F. S.; Chae, P. S.; Pardon, E.; Calinski, D.; Mathiesen, J. M.; Shah, S. T.; Lyons, J. A.; Caffrey, M.; Gellman, S. H.; Steyaert, J.; Skiniotis, G.; Weis, W. I.; Sunahara, R. K.; Kobilka, B. K. *Nature* **2011**, *477*(7366), 549.
47. Kruse, A. C.; Ring, A. M.; Manglik, A.; Hu, J.; Hu, K.; Eitel, K.; Hübner, H.; Pardon, E.; Valant, C.; Sexton, P. M.; Christopoulos, A.; Felder, C. C.; Gmeiner, P.; Steyaert, J.; Weis, W. I.; Garcia, K. C.; Wess, J.; Kobilka, B. K. *Nature* **2013**, *504*(7478), 101.
48. Staus, D. P.; Wingler, L. M.; Strachan, R. T.; Rasmussen, S. G.; Pardon, E.; Ahn, S.; Steyaert, J.; Kobilka, B. K.; Lefkowitz, R. J. *Mol. Pharmacol.* **2014**, *85*(3), 472.
49. Ullmer, C.; Zoffmann, S.; Bohrmann, B.; Matile, H.; Lindemann, L.; Flor, P.; Malherbe, P. *Br. J. Pharmacol.* **2012**, *167*(7), 1448.
50. Audet, M.; Bouvier, M. *Cell* **2012**, *151*(1), 14.
51. Hutchings, C. J.; Cseke, G.; Osborne, G.; Woolard, J.; Zhukov, A.; Koglin, M.; Jazayeri, A.; Pandya-Pathak, J.; Langmead, C. J.; Hill, S.; Weir, M.; Marshall, F. H. *MAbs* **2013**, *6*(1), 246.

CHAPTER NINETEEN

Mer Receptor Tyrosine Kinase: Therapeutic Opportunities in Oncology, Virology, and Cardiovascular Indications

Xiaodong Wang*, Stephen Frye*,†

*Center for Integrative Chemical Biology and Drug Discovery, Division of Chemical Biology and Medicinal Chemistry, Eshelman School of Pharmacy, University of North Carolina at Chapel Hill, Chapel Hill, North Carolina, USA

†Department of Medicine, Lineberger Comprehensive Cancer Center, School of Medicine, University of North Carolina at Chapel Hill, Chapel Hill, North Carolina, USA

Contents

ABBREVIATIONS

ALL acute lymphoblastic leukemia
AML acute myeloid leukemia
DCs dendritic cells
Gas6 growth arrest-specific 6
GBM glioblastoma
IFNs interferons

Annual Reports in Medicinal Chemistry, Volume 49
ISSN 0065-7743
http://dx.doi.org/10.1016/B978-0-12-800167-7.00019-5

NK natural killer
NSCLC non-small cell lung cancer
PtdSer phosphatidylserine
RTK receptor tyrosine kinase
TKI tyrosine kinase inhibitor
TLRs toll-like receptors

1. INTRODUCTION

Mer receptor tyrosine kinase (RTK) is a member of the TAM (Tyro3, Axl, and Mer) family. TAM RTKs play no essential role in embryonic development; rather, they function as homeostatic regulators in adult tissues and organ systems that are subject to continuous challenge and renewal throughout life.[1] Deficiencies in TAM signaling contribute to chronic inflammatory and autoimmune disease model systems and potentially in humans, while aberrantly elevated TAM signaling in tumor cells is strongly associated with cancer progression, metastasis, and resistance to targeted therapies.[1]

Each member of the TAM family contains an extracellular domain, a transmembrane domain, and a conserved intracellular kinase domain.[2] The extracellular domain includes two N-terminal immunoglobulin (Ig)-like domains and two fibronectin type III (FNIII) domains. Growth arrest-specific 6 (Gas6) is the common biological ligand (using its carboxy termini to bind the Ig-like domain of TAM RTKs) shared by the TAM family. Other ligands, including Protein S,[3] Tubby,[4] TULP-1,[4] and Galectin-3[5] can also stimulate Mer. The intracellular kinase domain of the TAM family is quite dissimilar from other RTKs (the average sequence homology of the Mer kinase domain compared to other RTK families is 40%), making Mer an excellent candidate for selective targeting by small molecules.

Mer is normally expressed on platelets, macrophages, and a range of epithelial cells. Under normal physiologic conditions, Mer mediates the second, clot stabilization phase of platelet aggregation, macrophage, and epithelial cell clearance of apoptotic cells, modulation of macrophage cytokine synthesis, cell motility, and cell survival.[2a,6] Abnormal activation or overexpression of Mer RTK has been implicated in neoplastic progression of many human cancers and has been correlated with poorer prognosis.[2a,6] Stimulation of Mer can induce diverse cellular functions dependent on the ligand–receptor combination as well as the cell type and microenvironment. In this review, we will discuss four therapeutic opportunities for Mer tyrosine kinase inhibitors (TKIs).

2. MER BIOLOGICAL FUNCTION AND THERAPEUTIC OPPORTUNITIES

2.1. Mer's Role in Macrophages, Natural Killer, and Dendritic Cells

Apoptosis is an important process that multicellular organisms use to maintain healthy and normal cell populations. Clearance of apoptotic cells is critical in many biological processes, including wound healing, tissue development and homeostasis, lymphocyte maturation, and pathological responses such as inflammation.[7] Although a number of different types of phagocytes can ingest infectious microorganisms and particles, clearance of apoptotic cells is primarily mediated by macrophages and, to a lesser degree, dendritic cells (DCs). Although macrophages express all three TAM receptors, Mer is more central to this role in macrophages, whereas Axl is more prominent in DCs.[8]

While Mer can bind the phosphatidylserine ligand complex and link these to macrophages, it is not required for binding to apoptotic cells. However, it is essential for cell shape changes associated with rapid engulfment of the apoptotic material.[7,9] Gas6 bridges the externalized phosphatidylserine (PtdSer) on apoptotic cells and MerTK on the surface of macrophages; this complex ligand activates the intracellular TK activity. The tyrosine kinase domains of Mer activate downstream signaling events, including integrins such as $\alpha v\beta 5$, which lead to cytoskeletal changes necessary for phagocytic engulfment of apoptotic cells.[10] The Mer homozygous knockout produces macrophages deficient in the clearance of apoptotic thymocytes, which may ultimately induce autoimmunity.[7] In addition, abnormal regulation of cytokine release in the absence of Mer is responsible for development of autoimmunity. In DCs, Mer also provides negative feedback via binding of Protein S produced by activated T-cells and signaling through its tyrosine kinase domain to limit the magnitude of DC activation during inflammatory response.[11] In addition, all three TAM receptors are required for normal differentiation and functional maturation of natural killer (NK) cells.[12] They also dampen NK cell activity which is important in decreasing inflammatory reactions but is deleterious in cancer.[13] Furthermore, Mer is found to be critical for normal cytokine release from NK T-cells.[14] The normal function of TAMs is therefore to signal for clearance of apoptotic cells while actively signaling to prevent an inflammatory immune response.

The high levels of Mer expressed on tumor-infiltrating macrophages[15] may increase their efficiency at clearing apoptotic melanoma cells; thus, limiting the time during which antigens from those dying cells could trigger an immune response. This also decreases the innate immune response to tumor cells. This is consistent with the results from syngeneic mouse models of breast cancer, melanoma, and colon cancer; tumors grow slowly and are poorly metastatic in MerTK −/− mice.[16] Therefore, Mer inhibition could decrease macrophage-aided cell clearance, increase inflammatory cytokine release (IL-12), and stimulate innate and subsequent adaptive immune responses to the tumor in MerTK-expressing cancers, in addition to its direct effect on tumors which will be discussed in the next section.

2.2. Aberrant Expression of Mer in Hematological and Solid Tumors: A Dual Target for Anticancer Effects

Traditional cancer treatment involves highly cytotoxic chemotherapeutics, which target actively dividing cells nonspecifically and are thus associated with toxic side effects, resulting in long-term disabilities, especially for pediatric patients.[17] The selective targeting of oncogenic pathways in specific cancers has opened the possibility of increased efficacy and potentially milder side effects.[6a] However, with the exception of Gleevec, most targeted agents have resulted in measurable but modest improvement in patient outcomes,[18] suggesting that additional antitumor pathways such as immune activation need to be engaged[19]—this is the promise of a Mer TKI (Fig. 19.1).

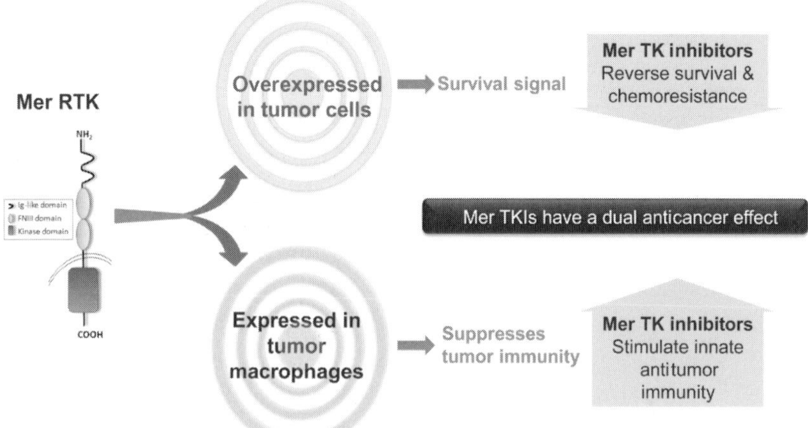

Figure 19.1 Mer RTK is a dual target in cancer. (See the color plate.)

Mer is overexpressed or ectopically expressed in various hematological and solid tumors. This results in the activation of several canonical oncogenic signaling pathways including the mitogen-activated protein kinase and phosphoinositide 3-kinase pathways, as well as regulation of signal transducer and activator of transcription family members, migration-associated proteins (the focal adhesion kinase and myosin light chain 2) and prosurvival proteins (survivin and Bcl-2).[20]

For example, Mer is ectopically expressed in both B- and T-cell acute lymphoblastic leukemia (ALL) but not in normal mouse- and human-circulating T- and B-lymphocytes.[21] Similarly, Mer is expressed in the majority of acute myeloid leukemia (AML) cell lines and patient samples but not in normal bone marrow myeloid precursor cells.[22] Similar findings have been reported in other solid tumors, including non-small cell lung carcinoma (NSCLC),[23] glioblastoma (GBM),[6b,24] and metastatic melanoma.[15a,25] This tumor-specific overexpression pattern may confer a large therapeutic window versus normal cells of the same lineage—at least in these tumor types.

Expression of a Mer transgene in hematopoietic cells leads to development of leukemia and/or lymphoma in mice.[26] When Mer is inhibited by sh-RNA knockdown in leukemia,[21c] NSCLC,[23] or GBM cells,[24a] they are more susceptible to apoptotic death and exhibit reduced colony formation in soft agar. In orthotropic ALL and AML xenograft mouse models, onset and progression of disease are delayed and survival is significantly increased in mice transplanted with Mer knockdown leukemia cell lines relative to mice transplanted with wild-type (WT) cell lines.[21c,22] Similarly, in a NSCLC, xenograft mouse model growth of tumors is markedly slowed by Mer knockdown. In addition, the combination of standard cytotoxic chemotherapies with sh-RNA-mediated Mer knockdown results in synergistic tumor cell killing. Taken together, these data demonstrate an important role for Mer in oncogenesis and chemoresistance, in multiple tumor types. Therefore, in addition to its role in suppressing antitumor innate immune responses, Mer provides a novel therapeutic target for the treatment of ALL, AML, NSCLC, and other Mer-expressing tumors.

2.3. The Role of TAM Family Kinases in Viral Immune Avoidance

After virus engagement, host pattern recognition receptors including TLRs, RIG-I-like receptors, and cytosolic DNA sensors[27] activate signal transduction pathways that induce type I interferons (IFNs), which stimulate the production of antiviral cellular restriction factors in order to control virus replication.[28] The TAM RTKs and their ligands Gas6 and Protein S negatively regulate the innate immune response to microbial infection through their signaling activity.[1] Many enveloped viruses, including West Nile (WNV), Dengue (DENV), HIV-1, Ebola, Marburg, Amapari, Tacaribe, Chikungunya, and Eastern Equine Encephalitis viruses, display PtdSer on the external leaflet of their membrane envelopes.[29] These viruses imitate apoptotic cells in a PtdSer-dependent process (apoptotic mimicry) to attach and gain access to cells.[29a,30] The TAM system facilitates this process through an interaction between these RTKs and virions that are opsonized with a TAM ligand. Because of the multimerization, concentration, and immobilization of Gas6 or Protein S on the virion surface, the ligand-coated virions also act as potent agonists to activate TAM receptors; and consequently, inhibit the type I IFN response in target cells.[31] On the other hand, TAM-deficient DCs are less susceptible to infection by retroviruses and flaviviruses due to elevated type I IFN production.[32] Taken together, these results demonstrate that TAM receptors are engaged by viruses in order to attenuate type I IFN signaling and avoid innate immunity. Therefore, TAMs represent potential antiviral targets.

2.4. Mer's Role in Coagulation: An Anticoagulation Target with Minimal Bleeding Liabilities

Antiplatelet compounds comprise an important family of drugs for cardiovascular diseases and for certain surgical procedures where a risk of stroke or thrombosis is prevalent.[33] Deficiency of any one of the TAM receptors protects mice against platelet-mediated life-threatening thrombosis in certain models due to the destabilization of platelet aggregates.[34] Mer knock-out mice have diminished platelet aggregation while maintaining normal bleeding times and coagulation parameters. Consequently, these mice are protected from thrombosis without increased spontaneous bleeding.[34,35] These observations indicate that small molecule Mer kinase inhibitors have potential as new antiplatelet drugs with decreased bleeding complications,[36] a profile that confers a major advantage over currently available antiplatelet therapies. As many cancer patients are prothrombotic, this activity is unlikely to be detrimental to oncology indications.

3. SMALL MOLECULE MER INHIBITORS

There are several reported small molecules having inhibitory activity against the TAM family. Some are current clinical agents with their primary targets outside of the TAM family; others are small molecules designed to target members of the TAM family. We will discuss these inhibitors in separate sections. Although the degree of intentionality in kinase targeting is not necessarily reflected in the kinome profiles of all agents, it does strongly bias the types of assays in which small molecules are profiled and therefore, the data available for evaluation.

3.1. Current Clinical Agents with TAM Family Activity

As shown in Fig. 19.2, BGB324 (R428, **1**) is the only intentionally TAM-targeted small molecule (selectively targeting Axl) currently in clinical development (completed Phase Ia clinical trial and planned Phase Ib clinical trial in AML and NSCLC).[37] The remaining compounds possess activity against the TAM family while their intended targets are other kinases (Fig. 19.2). For example, Crizotinib (PF-2341066, **2**) is an anticancer drug acting as an anaplastic lymphoma kinase and c-ros oncogene 1 (ROS1) inhibitor for NSCLC.[38] It is also very active against other kinases such as c-Met, Axl, TrkA, TrkB, Lck, and Mer with $IC_{50} < 10$ nM.[38] Similarly, Sunitinib (SU11248, **3**)[39] is a multitargeted RTK inhibitor for the treatment of renal cell carcinoma and imatinib-resistant gastrointestinal stromal tumor. Bosutinib (SKI-606, PF-5208763, **4**),[40] a dual Src/Abl inhibitor, is approved for the treatment of resistant Philadelphia chromosome-positive chronic myelogenous leukemia patients and Cabozantinib (XL184, **5**),[41] targeting c-Met and VEGFR2, has been used for treatment of medullary thyroid cancer. There are additional compounds currently at various stages of clinical trials (Fig. 19.2). Some are very potent against Mer such as Foretinib (XL880, **6**),[42] LY2801653 (**7**),[43] S49076 (**8**),[44] and AT9283 (**9**);[45] others only have moderate activity against Mer such as MK-2461 (**10**),[46] Lestaurtinib (**11**),[47] SU14813 (**12**),[48] R406 (**13**),[49] and BMS-777607 (**14**).[50] **14** has been used as a tool compound to demonstrate that inhibition of Axl effectively attenuates virus infection of WT DCs.[32] Given the available kinome profiling and other preclinical data on these agents, none of them appear likely to demonstrate pharmacology primarily related to inhibition of Mer.

3.2. Novel Mer Inhibitors with Activity in *In Vivo* Models of Antitumor, Anticoagulation, and Antiviral Indications

Since the first Mer X-ray structure was published in 2009,[51] several novel small molecules designed to target Mer have emerged (Fig. 19.3). Both

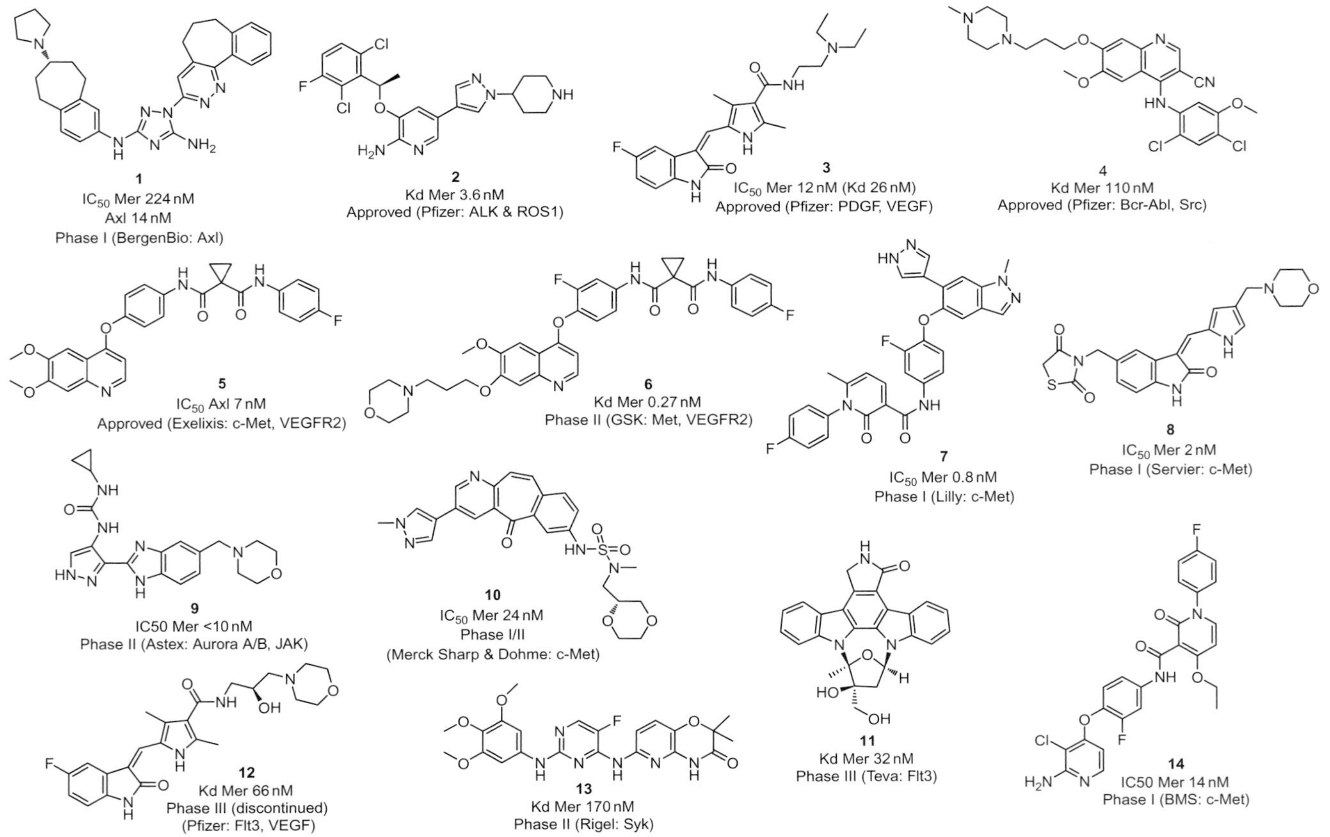

1
IC_{50} Mer 224 nM
Axl 14 nM
Phase I (BergenBio: Axl)

2
Kd Mer 3.6 nM
Approved (Pfizer: ALK & ROS1)

3
IC_{50} Mer 12 nM (Kd 26 nM)
Approved (Pfizer: PDGF, VEGF)

4
Kd Mer 110 nM
Approved (Pfizer: Bcr-Abl, Src)

5
IC_{50} Axl 7 nM
Approved (Exelixis: c-Met, VEGFR2)

6
Kd Mer 0.27 nM
Phase II (GSK: Met, VEGFR2)

7
IC_{50} Mer 0.8 nM
Phase I (Lilly: c-Met)

8
IC_{50} Mer 2 nM
Phase I (Servier: c-Met)

9
IC50 Mer <10 nM
Phase II (Astex: Aurora A/B, JAK)

10
IC_{50} Mer 24 nM
Phase I/II
(Merck Sharp & Dohme: c-Met)

11
Kd Mer 32 nM
Phase III (Teva: Flt3)

12
Kd Mer 66 nM
Phase III (discontinued)
(Pfizer: Flt3, VEGF)

13
Kd Mer 170 nM
Phase II (Rigel: Syk)

14
IC50 Mer 14 nM
Phase I (BMS: c-Met)

Figure 19.2 Current clinical agents with TAM activity and their intended target(s).

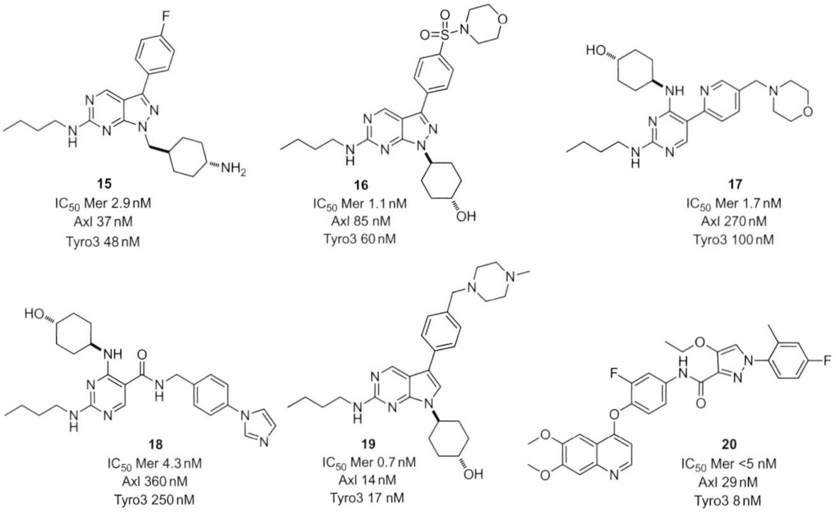

15
IC_{50} Mer 2.9 nM
Axl 37 nM
Tyro3 48 nM

16
IC_{50} Mer 1.1 nM
Axl 85 nM
Tyro3 60 nM

17
IC_{50} Mer 1.7 nM
Axl 270 nM
Tyro3 100 nM

18
IC_{50} Mer 4.3 nM
Axl 360 nM
Tyro3 250 nM

19
IC_{50} Mer 0.7 nM
Axl 14 nM
Tyro3 17 nM

20
IC_{50} Mer <5 nM
Axl 29 nM
Tyro3 8 nM

Figure 19.3 Novel Mer-targeted inhibitors.

UNC569 (**15**)[52] and UNC1062 (**16**)[53] belong to the pyrazolopyrimidine scaffold.[54] Compound **15** is the first small molecule Mer inhibitor discovered through a structure-based drug design approach with good enzymatic activities against the TAM family and promising DMPK properties in mice.[52] Compound **16** is a useful *in vitro* tool compound with improved activity and selectivity, although limited solubility and DMPK properties.[53] Recently, two new Mer selective inhibitors with a pyrimidine scaffold, UNC2250 (**17**)[55] and UNC2881 (**18**)[33], have been discovered using a pseudo-ring replacement strategy.[56] Compound **17** blocks both steady-state phosphorylation of endogenous Mer and ligand-stimulated activation of a chimeric EGFR-Mer protein and decreases colony-forming potential in rhabdoid and NSCLC tumor cells,[55] while **18** inhibits collagen-induced platelet aggregation.[33] The first small molecule Mer inhibitor with extensive *in vivo* activity reported is UNC2025 (**19**).[57] Compound **19** is very potent (Mer $K_i = 0.16$ nM) and can be administered orally (mouse bioavailability = 100%). Treatment with **19** significantly prolongs survival in a mouse model of B-cell ALL with minimal and manageable side effects.[58] In addition, a new pan-TAM inhibitor, LDC1267 (**20**), efficiently reduces metastases dependent on NK cells by enhancing antimetastatic NK cell activity *in vivo*.[13] For example, when **20** is administrated intraperitoneally to mice challenged with B16F10 melanoma, metastatic spread is markedly reduced (Fig. 19.4A).[13] It is also effective in the 4T1 liver metastasis model

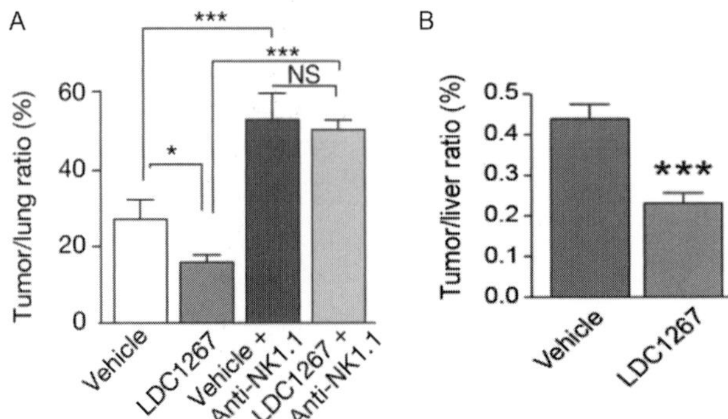

Figure 19.4 Control of metastases by TAM receptor inhibition. (A) Tumor-to-lung ratios in vehicle- and LDC1267-treated wild-type control or NK1.1-depleted mice. $n = 8$ each. $*P < 0.05$, and $***P < 0.001$; NS, not significant. Data are mean values $\pm$ s.e.m. (B) Relative sizes of 4T1 liver micrometastases in syngeneic mice treated with vehicle or LDC1267 (100 mg/kg) via oral gavage. Mean values $\pm$ s.e.m. are shown on day +21 after initiation of LDC1267 therapy (day +27 after orthotopic tumor inoculation into the mammary fat pad). $***P < 0.001$ (Student's *t*-test, $n = 10$ mice each). *Used with permission from Nature* ***2014****, 507, 508–512.* (See the color plate.)

via oral administration (Fig. 19.4B).[13] Taken together, these data indicate small molecule Mer inhibitors can be used in an *in vivo* therapeutic setting as both a direct antitumor agent and as a stimulant to the innate immune system.

4. FUTURE DIRECTIONS AND CONCLUSIONS

Mer is a homeostatic regulator in adult tissues and organ systems. It plays important roles in clearance of apoptotic cells, survival of cancer cells, immune avoidance by enveloped viruses, and aggregation of platelets. Inhibition of Mer triggers the innate immune system to fight viral infections and tumor metastasis, decreases survival of Mer-expressing cancer cells, and blocks thrombosis. Small molecule Mer inhibitors will facilitate further understanding of Mer biology and may eventually yield therapeutic agents for a number of indications.

REFERENCES

1. Lemke, G. *Cold Spring Harb. Perspect. Biol.* **2013**, *5*(11), a009076.
2. (a) Linger, R. M. A.; Keating, A. K.; Earp, H. S.; Graham, D. K. *Adv. Cancer Res.*; *100*, . Academic Press: New York, 200835–83; (b) Lemke, G.; Rothlin, C. V. *Nat. Rev. Immunol.* **2008**, *8*(5), 327–336.

3. Chen, J.; Carey, K.; Godowski, P. J. *Oncogene* **1997**, *14*(17), 2033–2039.
4. Caberoy, N. B.; Zhou, Y.; Li, W. *EMBO J.* **2010**, *29*(23), 3898–3910.
5. Caberoy, N. B.; Alvarado, G.; Bigcas, J.-L.; Li, W. *J. Cell. Physiol.* **2012**, *227*(2), 401–407.
6. (a) Linger, R. M. A.; Keating, A. K.; Earp, H. S.; Graham, D. K. Expert Opin. Ther. Targets. *14*(10), 1073–1090; (b) Brandao, L.; Migdall-Wilson, J.; Eisenman, K.; Graham, D. K. *Crit. Rev. Oncog.* **2011**, *16*(1–2), 47–63.
7. Scott, R. S.; McMahon, E. J.; Pop, S. M.; Reap, E. A.; Caricchio, R.; Cohen, P. L.; Earp, H. S.; Matsushima, G. K. *Nature* **2001**, *411*(6834), 207–211.
8. (a) Graham, D. K.; Dawson, T. L.; Mullaney, D. L.; Snodgrass, H. R.; Earp, H. S. *Cell Growth Differ.* **1994**, *5*(6), 647–657; (b) Lu, Q.; Lemke, G. *Science* **2001**, *293*(5528), 306–311; (c) Neubauer, A.; Fiebeler, A.; Graham, D. K.; O'Bryan, J. P.; Schmidt, C. A.; Barckow, P.; Serke, S.; Siegert, W.; Snodgrass, H. R.; Huhn, D. *Blood* **1994**, *84*(6), 1931–1941.
9. (a) Cohen, P. L.; Caricchio, R.; Abraham, V.; Camenisch, T. D.; Jennette, J. C.; Roubey, R. A.; Earp, H. S.; Matsushima, G.; Reap, E. A. *J. Exp. Med.* **2002**, *196*(1), 135–140; (b) Guttridge, K. L.; Luft, J. C.; Dawson, T. L.; Kozlowska, E.; Mahajan, N. P.; Varnum, B.; Earp, H. S. *J. Biol. Chem.* **2002**, *277*(27), 24057–24066; (c) Hu, B.; Jennings, J. H.; Sonstein, J.; Floros, J.; Todt, J. C.; Polak, T.; Curtis, J. L. *Am. J. Respir. Cell Mol. Biol.* **2004**, *30*(5), 687–693; (d) Todt, J. C.; Hu, B.; Curtis, J. L. *J. Leukoc. Biol.* **2004**, *75*(4), 705–713.
10. Wu, Y.; Singh, S.; Georgescu, M. M.; Birge, R. B. *J. Cell Sci.* **2005**, *118*(Pt 3), 539–553.
11. Carrera Silva, E. A.; Chan, P. Y.; Joannas, L.; Errasti, A. E.; Gagliani, N.; Bosurgi, L.; Jabbour, M.; Perry, A.; Smith-Chakmakova, F.; Mucida, D.; Cheroutre, H.; Burstyn-Cohen, T.; Leighton, J. A.; Lemke, G.; Ghosh, S.; Rothlin, C. V. *Immunity* **2013**, *39*(1), 160–170.
12. Caraux, A.; Lu, Q.; Fernandez, N.; Riou, S.; Di Santo, J. P.; Raulet, D. H.; Lemke, G.; Roth, C. *Nat. Immunol.* **2006**, 7(7), 747–754.
13. Paolino, M.; Choidas, A.; Wallner, S.; Pranjic, B.; Uribesalgo, I.; Loeser, S.; Jamieson, A. M.; Langdon, W. Y.; Ikeda, F.; Fededa, J. P.; Cronin, S. J.; Nitsch, R.; Schultz-Fademrecht, C.; Eickhoff, J.; Menninger, S.; Unger, A.; Torka, R.; Gruber, T.; Hinterleitner, R.; Baier, G.; Wolf, D.; Ullrich, A.; Klebl, B. M.; Penninger, J. M. *Nature* **2014**, *507*, 508–512.
14. Behrens, E. M.; Gadue, P.; Gong, S. Y.; Garrett, S.; Stein, P. L.; Cohen, P. L. *Eur. J. Immunol.* **2003**, *33*(8), 2160–2167.
15. (a) Schlegel, J.; Sambade, M. J.; Sather, S.; Moschos, S. J.; Tan, A. C.; Winges, A.; Deryckere, D.; Carson, C. C.; Trembath, D. G.; Tentler, J. J.; Eckhardt, S. G.; Kuan, P. F.; Hamilton, R. L.; Duncan, L. M.; Miller, C. R.; Nikolaishvili-Feinberg, N.; Midkiff, B. R.; Liu, J.; Zhang, W.; Yang, C.; Wang, X.; Frye, S. V.; Earp, H. S.; Shields, J. M.; Graham, D. K. *J. Clin. Invest.* **2013**, *123*(5), 2257–2267; (b) Haas, M. J. *SciBX: Science-Business eXchange* **2013**, *6*(16).
16. Cook, R. S.; Jacobsen, K. M.; Wofford, A. M.; DeRyckere, D.; Stanford, J.; Prieto, A. L.; Redente, E.; Sandahl, M.; Hunter, D. M.; Strunk, K. E. *J. Clin. Invest.* **2013**, *123*(8), 3231.
17. Carmichael, M. *Nature* **2013**, *498*(7455), S14–S15.
18. Arora, A.; Scholar, E. M. *J. Pharmacol. Exp. Ther.* **2005**, *315*(3), 971–979.
19. de Visser, K. E.; Eichten, A.; Coussens, L. M. *Nat. Rev. Cancer* **2006**, *6*(1), 24–37.
20. Cummings, C. T.; DeRyckere, D.; Earp, H. S.; Graham, D. K. *Clin. Cancer Res.* **2013**, *19*(19), 5275–5280.
21. (a) Graham, D. K.; Salzberg, D. B.; Kurtzberg, J.; Sather, S.; Matsushima, G. K.; Keating, A. K.; Liang, X.; Lovell, M. A.; Williams, S. A.; Dawson, T. L.; Schell, M. J.; Anwar, A. A.; Snodgrass, H. R.; Earp, H. S. *Clin. Cancer Res.* **2006**,

12(9), 2662–2669; (b) Yeoh, E. J.; Ross, M. E.; Shurtleff, S. A.; Williams, W. K.; Patel, D.; Mahfouz, R.; Behm, F. G.; Raimondi, S. C.; Relling, M. V.; Patel, A.; Cheng, C.; Campana, D.; Wilkins, D.; Zhou, X.; Li, J.; Liu, H.; Pui, C. H.; Evans, W. E.; Naeve, C.; Wong, L.; Downing, J. R. *Cancer Cell* **2002**, *1*(2), 133–143; (c) Brandao, L.; Winges, A.; Christoph, S.; Sather, S.; Migdall-Wilson, J.; Schlegel, J.; McGranahan, A.; Gao, D.; Liang, X.; DeRyckere, D. *Blood Cancer J.* **2013**, *3*(1), e101.

22. Lee-Sherick, A.; Eisenman, K.; Sather, S.; McGranahan, A.; Armistead, P.; McGary, C.; Hunsucker, S.; Schlegel, J.; Martinson, H.; Cannon, C.; Keating, A. K.; Earp, H. S.; Liang, X.; DeRyckere, D.; Graham, D. K. *Oncogene* **2013**, *32*, 5359–5368.
23. Linger, R. M.; Cohen, R. A.; Cummings, C. T.; Sather, S.; Migdall-Wilson, J.; Middleton, D. H.; Lu, X.; Baron, A. E.; Franklin, W. A.; Merrick, D. T.; Jedlicka, P.; Deryckere, D.; Heasley, L. E.; Graham, D. K. *Oncogene* **2013**, *32*, 3420–3431.
24. (a) Keating, A. K.; Kim, G. K.; Jones, A. E.; Donson, A. M.; Ware, K.; Mulcahy, J. M.; Salzberg, D. B.; Foreman, N. K.; Liang, X.; Thorburn, A.; Graham, D. K. *Mol. Cancer Ther.* **2010**, *9*(5), 1298–1307; (b) Rogers, A. E.; Le, J. P.; Sather, S.; Pernu, B. M.; Graham, D. K.; Pierce, A. M.; Keating, A. K. *Oncogene* **2012**, *31*(38), 4171–4181; (c) Wang, Y.; Moncayo, G.; Morin, P., Jr.; Xue, G.; Grzmil, M.; Lino, M. M.; Clement-Schatlo, V.; Frank, S.; Merlo, A.; Hemmings, B. A. *Oncogene* **2013**, *32*(7), 872–882.
25. Tworkoski, K.; Singhal, G.; Szpakowski, S.; Zito, C. I.; Bacchiocchi, A.; Muthusamy, V.; Bosenberg, M.; Krauthammer, M.; Halaban, R.; Stern, D. F. *Mol. Cancer Res.* **2011**, *9*(6), 801–812.
26. Keating, A. K.; Salzberg, D. B.; Sather, S.; Liang, X.; Nickoloff, S.; Anwar, A.; Deryckere, D.; Hill, K.; Joung, D.; Sawczyn, K. K.; Park, J.; Curran-Everett, D.; McGavran, L.; Meltesen, L.; Gore, L.; Johnson, G. L.; Graham, D. K. *Oncogene* **2006**, *25*(45), 6092–6100.
27. Thompson, M. R.; Kaminski, J. J.; Kurt-Jones, E. A.; Fitzgerald, K. A. *Viruses* **2011**, *3*(6), 920–940.
28. Yan, N.; Chen, Z. J. *Nat. Immunol.* **2012**, *13*(3), 214–222.
29. (a) Jemielity, S.; Wang, J. J.; Chan, Y. K.; Ahmed, A. A.; Li, W.; Monahan, S.; Bu, X.; Farzan, M.; Freeman, G. J.; Umetsu, D. T.; Dekruyff, R. H.; Choe, R. H. *PLoS Pathog.* **2013**, *9*(3), e1003232; (b) Mercer, J. *Cell Host Microbe* **2011**, *9*(4), 255–257.
30. (a) Mercer, J.; Helenius, A. *Science* **2008**, *320*(5875), 531–535; (b) Mercer, J.; Helenius, A. *Ann. N.Y. Acad. Sci.* **2010**, *1209*, 49–55.
31. Rothlin, C. V.; Ghosh, S.; Zuniga, E. I.; Oldstone, M. B.; Lemke, G. *Cell* **2007**, *131*(6), 1124–1136.
32. Bhattacharyya, S.; Zagórska, A.; Lew, E. D.; Shrestha, B.; Rothlin, C. V.; Naughton, J.; Diamond, M. S.; Lemke, G.; Young, J. A. *Cell Host Microbe* **2013**, *14*(2), 136–147.
33. Zhang, W.; McIver, A. L.; Stashko, M. A.; Deryckere, D.; Branchford, B. R.; Hunter, D.; Kireev, D.; Miley, M. J.; Norris-Drouin, J.; Stewart, W. M.; Lee, M.; Sather, S.; Zhou, Y.; Di Paola, J. A.; Machius, M.; Janzen, W. P.; Earp, H. S.; Graham, D. K.; Frye, S. V.; Wang, X. *J. Med. Chem.* **2013**, *56*(23), 9693–9700.
34. Angelillo-Scherrer, A.; Burnier, L.; Flores, N.; Savi, P.; DeMol, M.; Schaeffer, P.; Herbert, J. M.; Lemke, G.; Goff, S. P.; Matsushima, G. K.; Earp, H. S.; Vesin, C.; Hoylaerts, M. F.; Plaisance, S.; Collen, D.; Conway, E. M.; Wehrle-Haller, B.; Carmeliet, P. *J. Clin. Invest.* **2005**, *115*(2), 237–246.
35. Chen, C.; Li, Q.; Darrow, A. L.; Wang, Y.; Derian, C. K.; Yang, J.; de Garavilla, L.; Andrade-Gordon, P.; Damiano, B. P. *Arterioscler. Thromb. Vasc. Biol.* **2004**, *6*, 1118–1123.
36. Branchford, B. R.; Brzezinski, C.; Sather, S.; Brodsky, G.; DeRyckere, D.; Zhang, W.; Liu, J.; Earp, H. S.; Wang, X.; Frye, S. V. *Blood* **2013**, *122*(21), 2296.

37. Wnuk-Lipinska, K.; Gausdal, G.; Sandal, T.; Frink, R.; Hinz, S.; Hellesøy, M.; Ahmed, L.; Haugen, H.; Xiao, L.; Blø, M. *Clin. Cancer Res.* **2014**, *20*(Suppl. 2), B30.
38. Cui, J. J.; Tran-Dubé, M.; Shen, H.; Nambu, M.; Kung, P.-P.; Pairish, M.; Jia, L.; Meng, J.; Funk, L.; Botrous, I. *J. Med. Chem.* **2011**, *54*(18), 6342–6363.
39. (a) Demetri, G. D.; van Oosterom, A. T.; Garrett, C. R.; Blackstein, M. E.; Shah, M. H.; Verweij, J.; McArthur, G.; Judson, I. R.; Heinrich, M. C.; Morgan, J. A. *Lancet* **2006**, *368*(9544), 1329–1338; (b) Blay, J.-Y.; Reichardt, P. *Expert Rev. Anticancer Ther.* **2009**, *9*(6), 831–838.
40. Cortes, J. E.; Kantarjian, H. M.; Brümmendorf, T. H.; Kim, D.-W.; Turkina, A. G.; Shen, Z.-X.; Pasquini, R.; Khoury, H. J.; Arkin, S.; Volkert, A. *Blood* **2011**, *118*(17), 4567–4576.
41. Yakes, F. M.; Chen, J.; Tan, J.; Yamaguchi, K.; Shi, Y.; Yu, P.; Qian, F.; Chu, F.; Bentzien, F.; Cancilla, B. *Mol. Cancer Ther.* **2011**, *10*(12), 2298–2308.
42. Choueiri, T. K.; Vaishampayan, U.; Rosenberg, J. E.; Logan, T. F.; Harzstark, A. L.; Bukowski, R. M.; Rini, B. I.; Srinivas, S.; Stein, M. N.; Adams, L. M. *J. Clin. Oncol.* **2013**, *31*(2), 181–186.
43. Yan, S. B.; Peek, V. L.; Ajamie, R.; Buchanan, S. G.; Graff, J. R.; Heidler, S. A.; Hui, Y.-H.; Huss, K. L.; Konicek, B. W.; Manro, J. R. *Invest. New Drugs* **2013**, *31*(4), 833–844.
44. Burbridge, M. F.; Bossard, C. J.; Saunier, C.; Fejes, I.; Bruno, A.; Léonce, S.; Ferry, G.; Da Violante, G.; Bouzom, F.; Cattan, V. *Mol. Cancer Ther.* **2013**, *12*(9), 1749–1762.
45. Howard, S.; Berdini, V.; Boulstridge, J. A.; Carr, M. G.; Cross, D. M.; Curry, J.; Devine, L. A.; Early, T. R.; Fazal, L.; Gill, A. L. *J. Med. Chem.* **2008**, *52*(2), 379–388.
46. Katz, J. D.; Jewell, J. P.; Guerin, D. J.; Lim, J.; Dinsmore, C. J.; Deshmukh, S. V.; Pan, B. S.; Marshall, C. G.; Lu, W.; Altman, M. D.; Dahlberg, W. K.; Davis, L.; Falcone, D.; Gabarda, A. E.; Hang, G.; Hatch, H.; Holmes, R.; Kunii, K.; Lumb, K. J.; Lutterbach, B.; Mathvink, R.; Nazef, N.; Patel, S. B.; Qu, X.; Reilly, J. F.; Rickert, K. W.; Rosenstein, C.; Soisson, S. M.; Spencer, K. B.; Szewczak, A. A.; Walker, D.; Wang, W.; Young, J.; Zeng, Q. *J. Med. Chem.* **2011**, *54*(12), 4092–4108.
47. Knapper, S.; Burnett, A. K.; Littlewood, T.; Kell, W. J.; Agrawal, S.; Chopra, R.; Clark, R.; Levis, M. J.; Small, D. *Blood* **2006**, *108*(10), 3262–3270.
48. Kiselyov, A.; Balakin, K. V.; Tkachenko, S. E. *Expert Opin. Investig. Drugs* **2007**, *16*(1), 83–107.
49. Braselmann, S.; Taylor, V.; Zhao, H.; Wang, S.; Sylvain, C.; Baluom, M.; Qu, K.; Herlaar, E.; Lau, A.; Young, C. *J. Pharm. Exp. Ther.* **2006**, *319*(3), 998–1008.
50. Schroeder, G. M.; An, Y.; Cai, Z. W.; Chen, X. T.; Clark, C.; Cornelius, L. A.; Dai, J.; Gullo-Brown, J.; Gupta, A.; Henley, B.; Hunt, J. T.; Jeyaseelan, R.; Kamath, A.; Kim, K.; Lippy, J.; Lombardo, L. J.; Manne, V.; Oppenheimer, S.; Sack, J. S.; Schmidt, R. J.; Shen, G.; Stefanski, K.; Tokarski, J. S.; Trainor, G. L.; Wautlet, B. S.; Wei, D.; Williams, D. K.; Zhang, Y.; Fargnoli, J.; Borzilleri, R. M. *J. Med. Chem.* **2009**, *52*(5), 1251–1254.
51. Huang, X.; Finerty, P., Jr.; Walker, J. R.; Butler-Cole, C.; Vedadi, M.; Schapira, M.; Parker, S. A.; Turk, B. E.; Thompson, D. A.; Dhe-Paganon, S. *J. Struct. Biol.* **2009**, *165*(2), 88–96.
52. Liu, J.; Yang, C.; Simpson, C.; DeRyckere, D.; Van, D. A.; Miley, M. J.; Kireev, D.; Norris-Drouin, J.; Sather, S.; Hunter, D.; Korboukh, V. K.; Patel, H. S.; Janzen, W. P.; Machius, M.; Johnson, G. L.; Earp, H. S.; Graham, D. K.; Frye, S. V.; Wang, X. *ACS Med. Chem. Lett.* **2012**, *3*, 129–134.
53. Liu, J.; Zhang, W.; Stashko, M. A.; DeRyckere, D.; Cummings, C. T.; Hunter, D.; Yang, C.; Jayakody, C. N.; Cheng, N.; Simpson, C.; Norris-Drouin, J.; Sather, S.;

Kireev, D.; Janzen, W. P.; Earp, H. S.; Graham, D. K.; Frye, S. V.; Wang, X. *Eur. J. Med. Chem.* **2013**, *65*, 83–93.
54. Wang, X.; Liu, J.; Yang, C.; Zhang, W.; Frye, S.; Kireev, D. WO2011146313A1, 2011.
55. Zhang, W.; Zhang, D.; Stashko, M. A.; DeRyckere, D.; Hunter, D.; Kireev, D.; Miley, M. J.; Cummings, C.; Lee, M.; Norris-Drouin, J.; Stewart, W. M.; Sather, S.; Zhou, Y.; Kirkpatrick, G.; Machius, M.; Janzen, W. P.; Earp, H. S.; Graham, D. K.; Frye, S. V.; Wang, X. *J. Med. Chem.* **2013**, *56*(23), 9683–9692.
56. Wang, X.; Zhang, W.; Kireev, D.; Zhang, D.; McIver, A. WO2013177168A1, 2013.
57. Zhang, W.; DeRyckere, D.; Hunter, D.; Liu, J.; Stashko, M.; Minson, K. A.; Cummings, C.; Lee, M.; Glaros, T. G.; Newton, D. L.; Sather, S.; Zhang, D.; Kireev, D. B.; Janzen, W. P.; Earp, H. S.; Graham, D. K.; Frye, S. V.; Wang, X. *J. Med. Chem.* **2014**, http://dx.doi.org/10.1021/jm500749d.
58. Graham, D. K.; DeRyckere, D.; Lee-Sherick, A. B.; Winges, A.; Wang, X.; Zhang, W.; Liu, J.; Frye, S. V.; Earp, H. S., III. In: *American Association of Cancer Researchers Pediatric Cancer at the Crossroads Conference, San Diego, CA*, 2013.

SECTION 7

Topics in Drug Design and Discovery

Section Editor: Peter R. Bernstein
PhaRmaB LLC, Rose Valley, Pennsylvania

CHAPTER TWENTY

Disease-Modifying Agents for the Treatment of Cystic Fibrosis

Bradley D. Tait, John P. Miller
Proteostasis Therapeutics Inc., Cambridge, Massachusetts, USA

Contents

Annual Reports in Medicinal Chemistry, Volume 49
ISSN 0065-7743
http://dx.doi.org/10.1016/B978-0-12-800167-7.00020-1

1. INTRODUCTION

1.1. Classes of CFTR Mutations

Cystic fibrosis (CF) is a genetic disease caused by mutations in the CFTR gene that affect biosynthesis or activity of the protein at the levels of transcription, folding, trafficking, degradation, or channel opening. The lack of CFTR chloride and bicarbonate conductance activities reduces the hydration and pH of the apical surfaces of epithelial linings, resulting in thick mucus accumulation in the pancreas, intestines, testes, liver, and lungs.[1] The lungs are particularly at risk of infection and inflammation, which is a major cause of morbidity and mortality in CF.

There are more than 1900 mutations in CFTR that cause CF which fall into five widely recognized classes.[2] Class I mutations (10% of CF patients worldwide) prevent the synthesis of CFTR protein. Class II mutations (70%) inhibit folding and stability of CFTR protein, leading to degradation in the endoplasmic reticulum (ER), and prevention of trafficking to the cell surface. The most common Class II mutation is caused by deletion of Phe508 (F508del CFTR). Class III mutations (2-3%), which are gating mutations, traffic normally to the cell surface but have reduced channel activity. Class IV mutations (<2%) reduce the conductance of ions through the protein channel. Class V mutations (<1%) generally have a splicing defect that reduces the expression of CFTR protein without altering the amino acid sequence.[3] Class IV and V mutations generally confer milder clinical phenotypes than the first three classes.[4]

Rescue of the cellular phenotypes conferred by the different classes are approached by different strategies. Molecules that promote read through of premature termination codons (PTC) treat Class I by enabling the synthesis of functional CFTR from mRNAs that would otherwise be degraded through nonsense-mediated decay or encode truncated, nonfunctional protein. Correctors treat Class II mutants, enhancing the level of protein trafficked to the cell surface either by acting directly on the misfolded CFTR protein (pharmacological chaperones) or by improving the protein folding environment (proteostasis regulators). Potentiators treat Class III mutants by improving the activity of the channels at the cell surface. Compounds are generally classified based on their ability to treat a primary protein defect, but these agents also have the potential to be combined to provide enhanced function.[5] Several excellent reviews have been published on CFTR modulators.[6]

1.2. Cellular Assays

Engineered cell lines have been developed to enable high-throughput screening of compound libraries for the identification of modulators of mutant CFTR function. Such cell lines have been used to successfully identify correctors and potentiators. The cell lines are generally engineered to express a mutant form of CFTR, predominantly F508del. The F508del CFTR-dependent ion conductance is measured by assaying for changes in membrane potential,[7,8] or for enhanced fluorescence quenching of a halide-sensitive fluorescent protein reporter.[9]

Recently, the isolation and development of intestinal cell organoids from patients with F508del and other CF-causing mutations[10] has provided a primary cell system that does not rely on exogenous expression of mutant CFTR. This system may prove to be an advantageous new method for compound library screening.

The pathophysiological consequences of CFTR mutations are most severe in the lung. Hence, primary lung cells are the gold standard for assessing phenotypic rescue. Isolated primary human bronchial epithelial lung cells (HBEs) are differentiated at an air-liquid interface, sufficiently recreating their native environment such that coordinated cilia beating and mucous production can be observed with well-differentiated cells *in vitro*.[11,12] Electrophysiological studies in Ussing chambers using HBEs provide direct measures of endogenous mutant CFTR chloride conductance activity.

The use of functional assays to drive the structure–activity relationship (SAR) to clinical candidates demonstrates the value of using phenotypic screens to discover drugs.

1.3. Other Strategies to Correct Airway Surface Liquid Defects

This chapter focuses on correcting CFTR function by improving protein expression, trafficking, or enhancing its opening by small molecules. Gene therapy has the potential to rescue any CFTR mutation, but overcoming the body's host defense against this approach has proved challenging.[13] The reduced hydration and elevated inflammation in CF airway has been targeted by approaches beyond rescue of CFTR function. A recent review on alternative strategies to correct the airway surface defect is recommended.[14] Some of the potential strategies for ameliorating CF disease include: blocking of Na absorption by epithelial sodium channel (ENaC) inhibition (GS9411,[15] GS5737,[16] picolinamides[17]), stimulation of calcium-activated chloride channels through ionophores, induction of intracellular calcium levels through P2Y2 receptor (denufosol),[18] improvement of hydration by osmolytics (mannitol,[19] hypertonic saline[20]), and anti-inflammatories[21] (BIIL 284 BS,[22] CTX-4430[23]).

2. TREATING CLASS I DEFECTS

2.1. Ataluren (PTC124, 1)

Ataluren is a read-through agent that suppresses PTCs, restoring functional protein production from genes disrupted by nonsense mutations.[24,25] The mechanism of action of Ataluren is still the subject of some debate.[26,27] Ataluren was tested in CF patients with Class I mutations and showed a trend for FEV_1 (forced expiratory volume in 1 s) improvement of about 3%. A subpopulation analysis indicated patients not treated with aminoglycosides showed a 6% improvement in FEV_1.[28] Ataluren is currently in Phase 3 studies in patients who are not on aminoglycosides.[29]

2.2. NB124 (2)

NB124 is a synthetic aminoglycoside designed by eliminating structural components that mediate antibacterial effects while retaining the components that promote read through of PTCs.[30] It was effective in cells with different Class I genotypes, with a 10-fold improvement in therapeutic index over first-generation compounds. NB124 improved CFTR function to ~7% of wt-CFTR activity in HBEs and demonstrated *in vivo* activity. Addition of ivacaftor (**12**, a potentiator) further enhanced activity.

3. TREATING CLASS II DEFECTS (CORRECTORS)

Correctors are compounds that help Class II mutant CFTR overcome folding, stability, and degradation to reach the cell surface. The most common CFTR mutation is F508del. Class II mutations account for 70% of all mutations worldwide. In the Cystic Fibrosis Foundation Patient Registry, 87% of CF patients have at least one F508del, with 47% being homozygous and 40% being heterozygous for F508del combined with another CF-causing mutation.[31] A recent review presents a list of modulators that correct the folding and trafficking of F508del.[32] Strategies include direct correction of F508del CFTR using pharmacological chaperones,[33] enhancing the folding environment using proteostasis regulators,[34] and modulating the cellular stress response to enhance expression and stability of the mutant protein.[35] Compounds that correct CFTR trafficking may also show efficacy in other protein folding/trafficking diseases.[36]

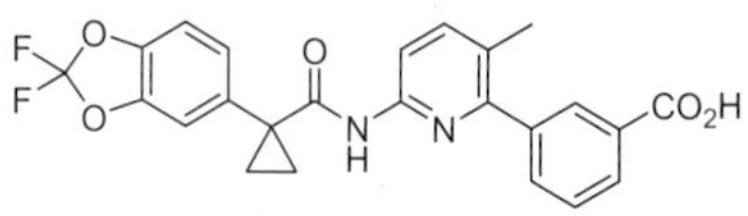

3 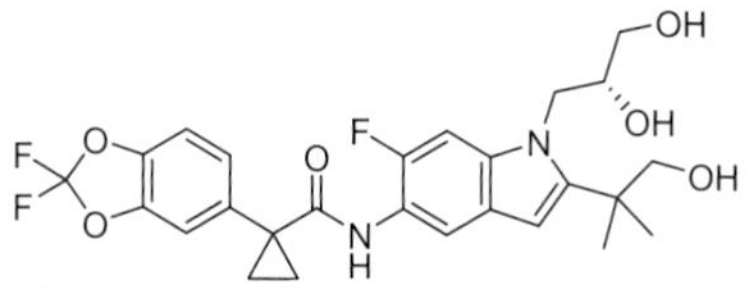 4 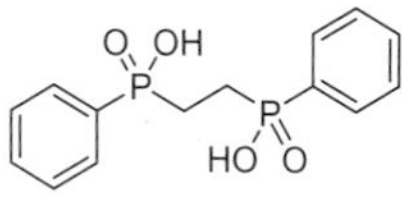 5 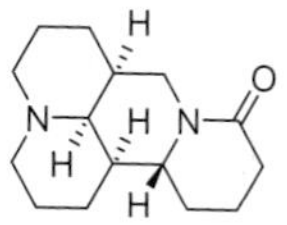 6 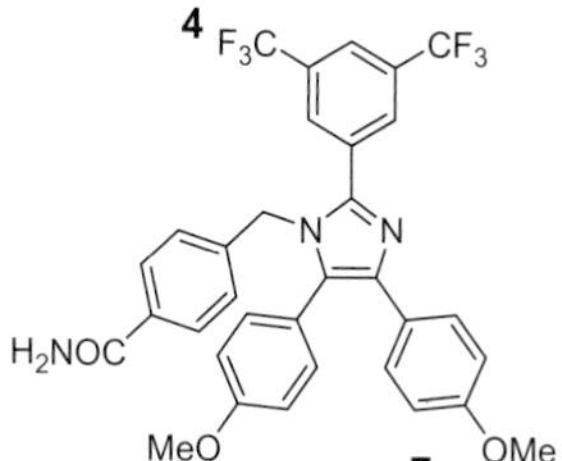 7 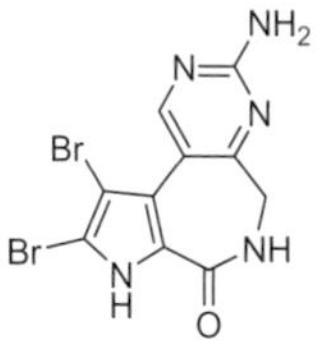 8 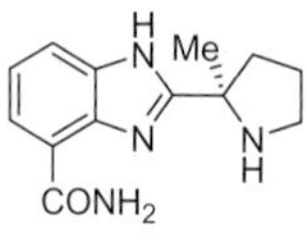 9 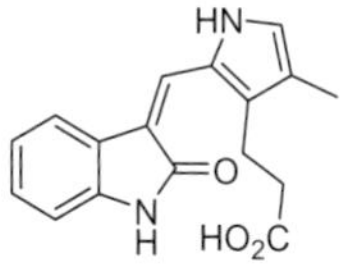 10 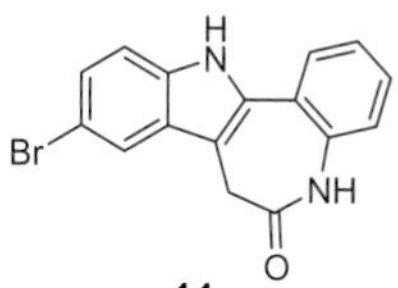 11

3.1. VX-809 (Lumacaftor, 3)

In primary patient F508del/F508del HBEs, the corrector VX-809 improved the chloride channel flux to ~14% of non-CF HBEs. Ivacaftor (**12**, a potentiator) reached ~9%, and the combination of VX-809 with ivacaftor reached ~25% of non-CF HBEs.[37] VX-809 most likely acts as a pharmacological chaperone, stabilizing the interaction of the mutant NBD1 domain with the first membrane spanning domain (MSD1) of the protein.[38–40]

In homozygous F508del patients in the clinic, neither VX-809 nor ivacaftor significantly improved FEV_1 when dosed as a single agent. A Phase 3 study of VX-809 in combination with ivacaftor demonstrated an improvement in FEV_1 of ~2.8–3.3% in patients homozygous for the F508del mutation.[41]

3.2. VX-661 (4)

In primary patient F508del/F508del HBEs, VX-661 improved the chloride channel flux to ~9% of non-CF HBE activity, ivacaftor reached ~8%, and the combination increased the activity to ~26%.[42] The results of a 12-week Phase 2 study of VX-661 in combination with ivacaftor in CF patients homozygous for the F508del mutation are expected in 2014.[43]

3.3. FDL169

A high-throughput screen (HTS) identified FD1027382 (structure not disclosed) as a novel F508del CFTR corrector that demonstrates activity in promoting iodide influx in Cystic Fibrosis Bronchial Epithelial (CFBE) cells transfected with F508del CFTR. Screening in primary cultures of F508del CFTR HBEs confirmed that FD1027382 (20 μM) restored approximately 20% of the chloride ion flux response of VX-809. Hit to lead optimization delivered multiple compounds with maximum efficacy equivalent to positive controls (VX-809) at ≤1 μM.[44] FDL169 is in IND-enabling toxicity studies for an orally available small molecule for the treatment of F508del CF patients.[45]

3.4. 407882 (5)

A structure-based virtual screen of the F508del nucleotide binding domain 1 (NBD1) using the DOCK program identified new correctors of F508del trafficking.[46] The bis-phosphinic acid, 407882, disrupts the NBD1–Keratin 8 interaction, resulting in 15% of non-CF HBE activity in F508del HBEs at 1 μM.

3.5. Matrine (6)

Modulation of the chaperone quality control pathways is a potential strategy for correcting defective CFTR folding.[47] Matrine, a quinolizidine alkaloid, was found to downregulate chaperone Hsc70 and release F508del CFTR from the ER to the cell surface, resulting in greater protein levels of F508del at the cell surface.[48]

3.6. Apoptozole (7)

Apoptozole binds to Hsc70 and Hsp70 with dissociation constants of 0.21 and 0.14 μM, respectively. The compound suppresses ubiquitinylation of F508del CFTR by blocking interaction of the mutant with the Hsc70 chaperone and the ubiquitin ligase CHIP, thus enhancing trafficking and increasing cell surface stability.[49] In F508del-CFTR-expressing Human Embryonic Kidney 293 (HEK) cells, 200 nM **7** increased chloride channel activity to 20% of wild-type (wt-CFTR) expressing cells.

3.7. Latonduine A (8)

Pacific marine sponge extracts were screened for correction of F508del CFTR misfolding, resulting in the identification of latonduines as active compounds. Confocal microscopy indicated trafficking improved up to 45% of wt-CFTR surface expression with Latonduine A.[50] Pull down studies with biotinylated and azido-analogs identified Poly[ADP-ribose] polymerase 3 (PARP)-3 as a primary target of these compounds. Additional work with ABT-888 (**9**) and siRNA knockdown confirmed PARPs as a potential target for improving CFTR trafficking.[51]

3.8. Kinase Inhibitors

A kinase inhibitor library, biased toward FDA-approved drugs and compounds that progressed to the clinic, was screened in a Yellow Fluorescent Protein (YFP) reporter HEK 293 cell assay to identify compounds that rescue F508del CFTR activity.[52] Selected hits were confirmed with western blot staining for CFTR maturation and short-circuit current measurement in Ussing chamber in F508del epithelial Madin–Darby Canine Kidney cells and patient HBEs. Kinase inhibitors targeting FGFR/FEGFR/PDGFR (SU5402, **10**), Ras/Raf/MEK/ERK/p38, and GSK-3β (kenpaullone, **11**) pathways rescue F508del to levels of 10–30% of wt-CFTR.

3.9. Other Correctors

Proteostasis modulators have been identified which have activity in HBEs alone and enhance the activity of other correctors in combination.[53] Cyclic tetrapeptides provide correction by modulating the protein homeostasis environment through both epigenetic and nonepigenetic mechanisms.[54] Ambroxol increases CFTR and ENaC expression as well as chloride transport in a CF cell model.[55]

4. TREATING CLASS III DEFECTS (POTENTIATORS)

12 **13** **14**

4.1. VX-770 (Ivacaftor, 12)

VX-770 (**12**) was the first agent approved to treat the underlying cause of CF.[56] Ivacaftor was approved in January 2012 for the treatment of patients with the G551D mutation found in 4.3% of CF patients.[57] Post-approval data from the PERSIST analysis showed that 144 weeks of continuous treatment with ivacaftor provided sustained effects in lung function (~6% FEV_1 improvement), body weight, and other measures.[58] Genistein (**13**), used in primary HBE assays as a potentiator, gives activity similar to VX-770 in these assays, albeit with a higher EC_{50}.[59]

4.2. RP193

Inhibitors of CDK/GSK-3β (RP193, **14**) are reported to correct F508del. These molecules have also been shown to potentiate wild-type, F508del, and G551D CFTR.[60] SAR on the pyrrolo[2,3-b]pyrazine series identified RP193 with reduced toxicity that retains CFTR potentiation.

4.3. GLPG1837

A halide-sensitive YFP assay was used in F508del CFTR CFBe41o cells to identify active compounds with potencies around 1 μM (structures not revealed).[61] Medicinal chemistry optimization improved potency to <10 nM. The YFP cell line data were predictive for activity in the HBE Ussing assay. Maximal channel opening was greater than 200% of maximum of VX-770 activity. The compound shows better potency in several Class III mutants. Confirmation in Ussing chamber measurements showed strong and dose–responsive activity, both on VX-809-rescued F508del homozygotes and on G551D/F508del heterozygotes.

4.4. Other Potentiators

Additional compounds reported to have potentiation activity include: resveratrol,[62] dehydrocostuslactone,[63] and dihydropyridines.[64–66]

5. COMPOUNDS WITH DUAL ACTIVITY

To fully optimize the functional activity of F508del CFTR, it will likely be necessary to both correct the trafficking defects as well as potentiate the channel opening once at the cell surface.[67] This may be done through combinations of correctors and potentiators (previously mentioned ongoing trial of ivacaftor/lumacaftor), or through the identification of single compounds that possess both activities. Single molecule corrector-potentiators (CoPos) have been reported,[68] and may provide improvement for both functional defects in a single therapeutic. In addition, a number of molecules that provide anti-inflammatory activities also possess F508del CFTR corrector and/or potentiator activity.

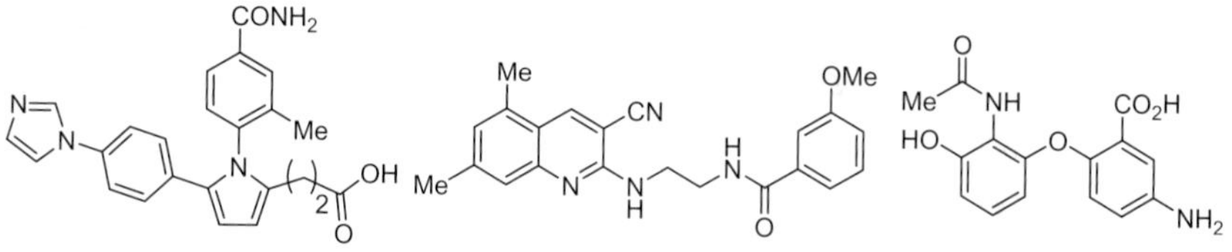

CONNH2 N N Me N OH 2 O **15**

Me CN Me N N H H N O OMe **16**

O Me NH CO2H HO O NH2 **17**

5.1. N6022 (15) and N91115

S-Nitrosoglutathione Reductase (GSNOR) regulates *S*-nitrosothiols and nitric oxide levels through catabolism of *S*-nitrosoglutathione.[69] GSNOR inhibitors demonstrate anti-inflammatory, CFTR corrector,[70,71] and potentiation activities.[72] GSNOR inhibitors may act through the Hsp70/Hsp90 organizing protein (Hop) pathway.[73] N6022 is an inhibitor of GSNOR with an IC_{50} of 8 nM,[74] in development for asthma and F508del CFTR patients as an injectable.[75]

N91115 (structure not reported) is an oral prodrug of N6022.[76] N91115 at 10 μM corrects to about 10% of wt-CFTR in F508del patient cells and corrects CFTR function in F508del CFTR mice.[77] A Phase 1 clinical trial was initiated in March of 2014.[78]

5.2. CoPo-22 (16)

A HTS in Fischer rat thyroid (FRT) cells was performed on 110,000 compounds resulting in the identification of a cyanoquinoline series with dual correction/potentiation activity.[79] Compounds were identified with corrector-only, potentiator-only, and dual activity.[80] Molecular modeling studies indicate that a flexible tether and a relatively short link between the cyanoquinoline and arylamide are important CoPo features.

5.3. Hyalout4 (17)

CFTR exports not only naturally hydrating substances such as chloride, but also hyaluronan.[81] Compounds were prepared to mimic the key structural features of the hyaluronan chain based on diaryl analogs to enhance membrane permeability.[82] Hyalout4 was found to act as a potentiator of F508del CFTR at 100 μM. Preincubation for 24 h increased the activity by 2.6-fold over immediate treatment, indicating this molecule also possesses corrector activity.

5.4. Other Compounds with Dual Activity

Trimethylangelicin was reported to have dual activity, reducing IL-8 expression and potentiating CFTR chloride transport.[83]

6. CONCLUSIONS

Advances in the treatment of CF symptoms have continued to improve lifespan and quality of life for patients with this debilitating disease.[84] Mean survival for CF patients has shown significant improvement from 28 years (1988–1992) to 38 years (2008–2012).[85] However, 23 years passed between the identification of CFTR as the genetic defect that causes CF and the first agent for treating the underlying cause of CF reaching the market.[86] The approval of ivacaftor for CF patients with Class III CFTR-gating mutations demonstrates the realization of the potential to address the underlying cause of CF with small molecules.[87] A large unmet medical need still remains for disease-modifying therapies to treat the majority of CF patients who have one or more copies of the Class II F508del mutation. A variety of approaches to correct the multiple defects imposed by this mutation are being pursued. Combinations of these strategies may ultimately prove to be the successful approach for the treatment of F508del CF.

Hopefully, these efforts will eventually enable the development of treatments for all CF patients.

REFERENCES

1. Berkebile, A. R.; McCray, P. B., Jr. *Int. J. Biochem. Cell Biol.* **2014**, *52*, 124.
2. Boyle, M. P.; De Boeck, K. *Lancet Respir. Med.* **2013**, *1*(2), 158.
3. Rogan, M. P.; Stoltz, D. A.; Hornick, D. B. *Chest* **2011**, *139*(6), 1480.
4. Ferec, C.; Cutting, G. R. *Cold Spring Harb. Perspect. Med.* **2012**, *2*, a009480.
5. Lane, M. A.; Doe, S. J. *Clin. Med.* **2014**, *14*(1), 76.
6. Hadida, S.; Van Goor, F.; Grootenhuis, D. J. *Annu. Rep. Med. Chem.* **2010**, *45*, 157; Rowe, S. M.; Verkman, A. S. *Cold Springs Harb. Perspect. Med.* **2013**, *3*(7), a009761/1.
7. Van Goor, F.; Straley, K. S.; Cao, D.; Gonzalez, J.; Hadida, S.; Hazlewood, A.; Joubran, J.; Knapp, T.; Makings, L. R.; Miller, M.; Neuberger, T.; Olson, E.; Panchenko, V.; Rader, J.; Singh, A.; Stack, J. H.; Tung, R.; Grootenhuis, P. D. J.; Negulescu, P. *Am. J. Physiol. Lung Cell. Mol. Physiol.* **2006**, *290*(6), L1117.
8. Maitra, R.; Sivashanmugam, P.; Warner, K. J. *Biomol. Screen.* **2013**, *18*(9), 1132.
9. Vijftigschild, L. A. W.; van der Ent, C. K.; Beekman, J. M. *Cytometry A* **2013**, *83A*(6), 576.
10. Dekkers, J. F.; Wiegerinck, C. L.; de Jonge, H. R.; Bronsveld, I.; Janssens, H. M.; de Winter-de Groot, K. M.; Brandsma, A. M.; de Jong, N. W. M.; Bijvelds, M. J. C.; Scholte, B. J.; Nieuwenhuis, E. E. S.; van den Brink, S.; Clevers, H.; van der Ent, C. K.; Middendorp, S.; Beekman, J. M. *Nat. Med.* **2013**, *19*(7), 939.
11. Randell, S. H.; Fulcher, M. L.; O'Neal, W. K.; Olsen, J. C. *Methods Mol. Biol.* **2011**, *742*, 285.
12. Neuberger, T.; Burton, B.; Clark, H.; Van Goor, F. *Methods Mol. Biol.* **2011**, *741*, 39 (Cystic Fibrosis Volume 1).
13. Armstrong, D. K.; Cunningham, S.; Davies, J. C.; Alton, E. W. *Arch. Dis. Child.* **2014**, *99*, 465.
14. Donaldson, S. H.; Galietta, L. *Cold Spring Harb. Perspect. Med.* **2013**, *3*(6), a009787.
15. O'Riordan, T. G.; Donn, K. H.; Hodsman, P.; Ansede, J. H.; Newcomb, T.; Lewis, S. A.; Flitter, W. D.; White, V. S.; Johnson, M. R.; Montgomery, A. B.; Warnock, D. G.; Boucher, R. C. *J. Aerosol Med. Pulmonary Drug Delivery* **2013**, *26*, 1.
16. Donn, K. H.; Hirsh, A. J.; Ansede, J. H.; O'Riordan, T. G.; Baker, W. R.; Wright, C. D.; Sabater, J.; Abraham, W. M.; Boucher, R. C.; Johnson, M. R.; Phillips, G. B. In: *27th Annual North American Cystic Fibrosis Conference, Salt Lake City, UT* 2013, Poster 124.
17. Norman, P. *Exp. Opin. Ther. Pat.* **2014**, *24*(7), 1.
18. Ratjen, F.; Durham, T.; Navratil, T.; Schaberg, A.; Accurso, F. J.; Wainwright, C.; Barnes, M.; Moss, R. B. *J. Cyst. Fibros.* **2012**, *11*(6), 539.
19. Bilton, D.; Bellon, G.; Charlton, B.; Cooper, P.; De Boeck, K.; Flume, P. A.; Fox, H. G.; Gallagher, C. G.; Geller, D. E.; Haarman, E. G.; Hebestreit, H. U.; Kolbe, J.; Lapey, A.; Robinson, P.; Wu, J.; Zuckerman, J. B.; Aitken, M. L. *J. Cyst. Fibros.* **2013**, *12*(4), 367.
20. Moore, B. M.; Laguna, T. A.; Liu, M.; McNamara, J. J. *Pediatr. Pulmonol.* **2013**, *48*(8), 747.
21. Prescott, W. A., Jr.; Johnson, C. E. *Pharmacotherapy* **2005**, *25*(4), 555.
22. Konstan, M. W.; Doring, G.; Heltshe, S. L.; Lands, L. C.; Hilliard, K. A.; Koker, P.; Bhattacharya, S.; Staab, A.; Hamilton, A. *J. Cyst. Fibros.* **2014**, *13*, 148.
23. Springman, E. B.; Bhatt, L.; Pugh, M.; Grosswald, R. In: *27th Annual North American Cystic Fibrosis Conference, Salt Lake City, UT* 2013, Poster 268 and 269.

24. Finkel, R. S.; Flanigan, K. M.; Wong, B.; Bonnemann, C.; Sampson, J.; Sweeney, H. L.; Reha, A.; Northcutt, V. J.; Elfring, G.; Barth, J.; Peltz, S. W. *PLoS One* **2013**, *8*(12), e81302.
25. Wilschanski, M.; Miller, L. L.; Shoseyov, D.; Blau, H.; Rivlin, J.; Aviram, M.; Cohen, M.; Armoni, S.; Yaakov, Y.; Pugatsch, T.; Cohen-Cymberknoh, M.; Miller, N. L.; Reha, A.; Northcutt, V. J.; Hirawat, S.; Donnelly, K.; Elfring, G. L.; Ajayi, T.; Kerem, E. *Eur. Respir. J.* **2011**, *38*(1), 59.
26. McElroy, S. P.; Nomura, T.; Torrie, L. S.; Warbrick, E.; Gartner, U.; Wood, G.; McLean, W. H. I. *PLoS Biol.* **2013**, *11*(6), e1001593.
27. Lentini, L.; Melfi, R.; Di Leonardo, A.; Spinello, A.; Barone, G.; Pace, A.; Palumbo Piccionello, A.; Pibiri, I. *Mol. Pharm.* **2014**, *11*, 653.
28. Peltz, S. W.; Morsy, M.; Welch, E. M.; Jacobson, A. *Annu. Rev. Med.* **2013**, *64*, 407.
29. Ataluren listed in Phase 3 on ClinicalTrials.gov.
30. Xue, X.; Mutyam, V.; Tang, L.; Biswas, S.; Du, M.; Jackson, L. A.; Dai, Y.; Belakhov, V.; Shalev, M.; Chen, F.; Schacht, J.; Bridges, R. J.; Baasov, T.; Hong, J.; Bedwell, D. M.; Rowe, S. M. *Am. J. Respir. Cell Mol. Biol.* **2014**, *50*(4), 805.
31. http://www.cff.org/UploadedFiles/research/ClinicalResearch/PatientRegistryReport/2012-CFF-Patient-Registry.pdf.
32. Birault, V.; Solari, R.; Hanrahan, J.; Thomas, D. Y. *Curr. Opin. Chem. Biol.* **2013**, *17*(3), 353.
33. Leidenheimer, N. J.; Ryder, K. G. *Pharmacol. Res.* **2014**, *83*, 10.
34. Balch, W. E.; Roth, D. M.; Hutt, D. M. *Cold Spring Harb. Perspect. Biol.* **2011**, *3*(2), a004499.
35. Nieddu, E.; Pollarolo, B.; Merello, L.; Schenone, S.; Mazzei, M. *Curr. Pharm. Des.* **2013**, *19*(19), 3476.
36. Sampson, H. M.; Lam, H.; Chen, P.-C.; Zhang, D.; Mottillo, C.; Mirza, M.; Qasim, K.; Shrier, A.; Shyng, S.-L.; Hanrahan, J. W.; Thomas, D. Y. *Orphanet J. Rare Dis.* **2013**, *8*, 11.
37. Van Goor, F.; Hadida, S.; Grootenhuis, P. D. J.; Burton, B.; Stack, J. H.; Straley, K. S.; Decker, C. J.; Miller, M.; McCartney, J.; Olson, E. R.; Wine, J. J.; Frizzell, R. A.; Ashlock, M.; Negulescu, P. A. *Proc. Natl. Acad. Sci. U.S.A.* **2011**, *108*(46), 18843.
38. Okiyoneda, T.; Veit, G.; Dekkers, J. F.; Bagdany, M.; Soya, N.; Xu, H.; Roldan, A.; Verkman, A. S.; Kurth, M.; Simon, A.; Heqedus, T.; Beekman, J. M.; Lukacs, G. L. *Nat. Chem. Biol.* **2013**, *9*(7), 444.
39. Loo, T. W.; Bartlett, M. C.; Clarke, D. M. *Biochem. Pharmacol.* **2013**, *86*(5), 612.
40. Ren, H. Y.; Grove, D. E.; De La Rosa, O.; Houck, S. A.; Sopha, P.; Van Goor, F.; Hoffman, B. J.; Cyr, D. M. *Mol. Biol. Cell* **2013**, *24*(19), 3016.
41. http://files.shareholder.com/downloads/VRTX/2745727222x0x764209/03b924cd-8524-4c49-b1dc-1e99037ba491/VRTX_News_2014_6_24_General.pdf.
42. Donaldson, S. In: *36th European CF Conference, Lisbon, Portugal* 2013 https://www.ecfs.eu/conferences/lisbon-2013/data-presentations.
43. Leiden, J. J. P. In: Morgan Healthcare Conference, San Francisco, *CA* 2014.
44. Patron, T.; Valdez, R.; Bhatt, P.; Deshpande, A.; Krouse, M.; Barsukov, G.; Handley, K.; Finnegan, D.; Nugent, R.; Zawistoski, M.; Shang, H.; Kanawade, A.; Li, T.; Sui, J.; Kwok, I.; Mai, V.; Miranda, L.; Lesser, A.; Smith, P.; Layer, E.; Kaczmarek, S.; Kolodziej, A. F.; Fitzpatrick, R.; Cole, B. M. In: 27th Annual North American Cystic Fibrosis Conference, Salt Lake City, *UT* 2013 Poster 50.
45. http://www.flatleydiscoverylab.com/news/2013/10/10/flatley-discovery-lab-llc-nominates-clinical-candidate-for-treatment-of-cystic-fibrosis.
46. Odolczyk, N.; Fritsch, J.; Norez, C.; Servel, N.; FariadaCunha, M.; Bitam, S.; Kupniewska, A.; Wiszniewski, L.; Colas, J.; Tarnowski, K.; Tondelier, D.; Roldan, A.; Saussereau, E. L.; Melin-Heschel, P.; Wieczorek, G.; Lukacs, G. L.;

Dadlez, M.; Faure, G.; Herrmann, H.; Ollero, M.; Becq, F.; Zielenkiewicz, P.; Edelman, A. *EMBO Mol. Med.* **2013**, *5*(10), 1484.
47. Calamini, B.; Silva, M. C.; Madoux, F.; Hutt, D. M.; Khanna, S.; Chalfant, M. A.; Saldanha, S. A.; Hodder, P.; Tait, B. D.; Garza, D.; Balch, W. E.; Morimoto, R. I. *Nat. Chem. Biol.* **2012**, *8*(2), 185.
48. Basile, A.; Pascale, M.; Franceschelli, S.; Nieddu, E.; Mazzei, M. T.; Fossa, P.; Turco, M. C.; Mazzie, M. *J. Cell. Physiol.* **2012**, *227*(9), 3317.
49. Cho, H. J.; Gee, H. Y.; Baek, K.-H.; Ko, S.-K.; Park, J.-M.; Lee, H.; Kim, N.-D.; Lee, M. G.; Shin, I. *J. Am. Chem. Soc.* **2011**, *133*(50), 20267.
50. Carlile, G. W.; Keyzers, R. A.; Teske, K. A.; Robert, R.; Williams, D. E.; Linington, R. G.; Gray, C. A.; Centko, R. M.; Yan, L.; Anjos, S. M.; Sampson, H. M.; Zhang, D.; Liao, J.; Hanrahan, J. W.; Andersen, R. J.; Thomas, D. Y. *Chem. Biol.* **2012**, *19*(10), 1288.
51. Anjos, S. M.; Robert, R.; Waller, D.; Zhang, D. L.; Balghi, H.; Sampson, H. M.; Ciciriello, F.; Lesimple, P.; Carlile, G. W.; Goepp, J.; Liao, J.; Ferraro, P.; Phillipe, R.; Dantzer, F.; Hanrahan, J. W.; Thomas, D. Y. *Front. Pharmacol. Ion Channels Channelopathies* **2012**, *3*, 165.
52. Trzci'nska-Daneluti, A. M.; Nguyen, L.; Jiang, C.; Fladd, C.; Uehling, D.; Prakesch, M.; Al-Awar, R.; Rotin, D. *Mol. Cell. Proteomics* **2012**, *11*(9), 745.
53. Haeberlein, M.; Cullen, M. D.; Miller, J.; Garza, D.; Green, O.; Drew, L.; Guiliano, K.; Dobbs, W.; Longo, K. A.; Wachi, S.; Golji, J.; Smart, J. L.; Bridges, R. J.; Thakerar, A.; Hutt, D. M.; Balch, W. E.; Enyedy, M. J.; Reinhart, P. H.; Tait, B. D. In: *27th Annual North American Cystic Fibrosis Conference, Salt Lake City, UT* 2013, Abstract 260.
54. Hutt, D. M.; Olsen, C. A.; Vickers, C. J.; Herman, D.; Chalfant, M. A.; Montero, A.; Leman, L. J.; Burkle, R.; Maryanoff, B. E.; Balch, W. E.; Ghadiri, M. R. *ACS Med. Chem. Lett.* **2011**, *2*(9), 703.
55. Varelogianni, G.; Hussain, R.; Strid, H.; Oliynyk, I.; Roomans, G. M.; Johannesson, M. *Cell Biol. Int.* **2013**, *37*(11), 1149.
56. Davis, P. B.; Yasothan, U.; Kirkpatrick, P. *Nat. Rev. Drug Discov.* **2012**, *11*(5), 349.
57. Deeks, E. D. *Drugs* **2013**, *73*(14), 1595.
58. McKone, E.; Borowitz, D.; Drevinek, P.; Griese, M.; Konstan, M. W.; Wainwright, C.; Ratgen, F.; Sermet-Gaudelus, I.; Plant, B.; Jiang, Y.; Gilmartin, G.; Davies, J. C. In: *27th Annual North American Cystic Fibrosis Conference, Salt Lake City, UT* 2013, Poster 227.
59. Van Goor, F.; Hadida, S.; Grootenhuis, P. D. J.; Burton, B.; Cao, D.; Neuberger, T.; Turnbull, A.; Singh, A.; Joubran, J.; Hazlewood, A.; Zhou, J.; McCartney, J.; Arumugam, V.; Decker, C.; Yang, J.; Young, C.; Olson, E. R.; Wine, J. J.; Frizzell, R. A.; Ashlock, M.; Negulescu, P. *Proc. Natl. Acad. Sci. U.S.A.* **2009**, *106*(44), 18825.
60. Dannhoffer, L.; Billet, A.; Jollivet, M.; Melin-Heschel, P.; Faveau, C.; Becq, F. *Front. Pharmacol. Ion Channels Channelopathies* **2011**, *2*, 48.
61. Conrath, K.; Andrews, M.; VanderPlas, S.; Sonck, K.; Gees, M.; Jans, M.; Nelles, L.; Gentzsch, M.; Wigerinck, P. In: *27th Annual North American Cystic Fibrosis Conference, Salt Lake City, UT* 2013, Poster 41.
62. Yu, B.; Zhang, Y.; Sui, Y.; Yang, S.; Luan, J.; Wang, X.; Ma, T.; Yang, H. *Pharmazie* **2013**, *68*(11), 877.
63. Wang, X.; Zhang, T. F.; Yu, B.; Yang, S.; luan, J.; Liu, X.; Tang, H. *J. Asian Nat. Prod. Res.* **2013**, *15*(8), 855.
64. Budriesi, R.; Ioan, P.; Leoni, A.; Pedemonte, N.; Locatelli, A.; Micucci, M.; Chiarini, A.; Galietta, L. J. *J. Med. Chem.* **2011**, *54*(11), 3885.
65. Visentin, S.; Ermondi, G.; Medana, C.; Pedemonte, N.; Galietta, L.; Caron, G. *Eur. J. Med. Chem.* **2012**, *55*, 188.

66. Giampieri, M.; Vanthuyne, N.; Nieddu, E.; Mazzei, M. T.; Anzaldi, M.; Pedemonte, N.; Galietta, L. J. V.; Roussel, C.; Mazzei, M. *Chem. Med. Chem.* **2012**, 7(10), 1799.
67. Amaral, M. D.; Farinha, C. M. *Curr. Pharm. Des.* **2013**, *19*(19), 3497.
68. Lukacs, G. L.; Verkman, A. S. *Trends Mol. Med.* **2012**, *18*(2), 81.
69. Sun, X.; Wasley, J. W. F.; Qiu, J.; Blonder, J. P.; Stout, A. M.; Green, L. S.; Strong, S. A.; Colagiovanni, D. B.; Richards, J. P.; Mutka, S. C.; Chun, L.; Rosenthal, G. J. *ACS Med. Chem. Lett.* **2011**, *2*(5), 402.
70. Colagiovanni, D. B.; Drolet, D. W.; Langlois-Forget, E.; Piché, M. P.; Looker, D.; Rosenthal, G. J. *Regul. Toxicol. Pharmacol.* **2012**, *62*(1), 115.
71. Oliynyk, I.; Hussain, R.; Amin, A.; Johannesson, M.; Roomans, G. M. *Exp. Mol. Pathol.* **2013**, *94*(3), 474.
72. http://www.n30pharma.com/october-15-2013/.
73. Marozkina, N. V.; Yemen, S.; Borowitz, M.; Liu, L.; Plapp, M.; Sun, F.; Islam, R.; Erdmann-Gilmore, P.; Townsend, R. R.; Lichti, C. F.; Mantri, S.; Clapp, P. W.; Randell, S. H.; Gaston, B.; Zaman, K. *Proc. Natl. Acad. Sci.* **2010**, *107*(25), 11393.
74. Green, L. S.; Chun, L. E.; Patton, A. K.; Sun, X.; Rosenthal, G. J.; Richards, J. P. *Biochemistry* **2012**, *51*(10), 2157.
75. http://www.n30pharma.com/docs/news/2013-03-12-6022-CF-First-Patient.pdf http://www.n30pharma.com/october-15-2013/.
76. http://www.n30pharma.com/docs/news/2013-10-15-NACFC.pdf.
77. Mutka, S.; Blonder, J.; Quinney, N. L.; Mehra, N.; Patton, A. K.; Looker, D. L.; Gentzsch, M.; Scoggin, C.; Gabriel, S. E. http://www.n30pharma.com/march-21-2013/.
78. http://www.n30pharma.com/march-12-2014/.
79. Phuan, P. W.; Yang, B.; Knapp, J. M.; Wood, A. B.; Lukacs, G. L.; Kurth, M. J.; Verkman, A. S. *Mol. Pharmacol.* **2011**, *80*(4), 683.
80. Knapp, J. M.; Wood, A. B.; Phuan, P.-W.; Lodewyk, M. W.; Tantillo, D. J.; Verkman, A. S.; Kurth, M. J. *J. Med. Chem.* **2012**, *55*(3), 1242.
81. Schulz, T.; Schumacher, U.; Prante, C.; Sextro, W.; Prehm, P. *Pathobiology* **2010**, 77(4), 200.
82. Nowakowska, E.; Schulz, T.; Molenda, N.; Schillers, H.; Prehm, P. *J. Cell. Biochem.* **2012**, *113*(1), 156.
83. Tamanini, A.; Borgatti, M.; Finotti, A.; Piccagli, L.; Bezzerri, V.; Favia, M.; Guerra, L.; Lampronit, I.; Bianchi, N.; Dall'Acqua, F.; Vedaldi, D.; Salvador, A.; Fabbri, E.; Mancini, I.; Nicolis, E.; Casavola, V.; Cabrini, G.; Gambari, R. *Am. J. Physiol. Lung Cell. Mol. Physiol.* **2011**, *300*(3 Pt. 1), L380.
84. Stevens, D. P.; Marshall, B. C. *BMJ Qual. Saf.* **2014**, *23*, i1.
85. Cystic Fibrosis Foundation Therapeutics. *Patient Registry* **2012**, http://www.cff.org/livingwithcf/qualityimprovement/patientregistryreport/.
86. Riordan, J. R.; Rommens, J. M.; Kerem, B. S.; Alon, N.; Rozmahel, R.; Grzelczak, Z.; Zielenski, J.; Lok, S.; Plavsic, N.; Chou, J. L.; Drumm, M. C.; Iannuzzi, M. C.; Collins, F. S.; Tsui, L. C. *Science* **1989**, *245*(4922), 1066.
87. http://investors.vrtx.com/releasedetail.cfm?ReleaseID=827435.

CHAPTER TWENTY-ONE

Advancements in Stapled Peptide Drug Discovery & Development

Vincent Guerlavais, Tomi K. Sawyer
Aileron Therapeutics, Cambridge, Massachusetts, USA

Contents

1. INTRODUCTION

Peptides are an important class of medicines, with more than 60 marketed drugs and several hundred currently in clinical trials. Key properties common to peptides are high potency, excellent selectivity, and minimal toxicity due to off-target effects. Copaxone®, Lupron®, Sandostatin®, Zoladex®, and Victoza® each commanded global sales of more than US$ 1 billion in 2011.[1]

It is noteworthy that about 50% of the FDA-approved peptides are peptide-based macrocycles. Macrocyclic structures leverage their size and complexity when modulating challenging protein targets with extended-binding sites. Second, cyclization confers many critical drug-like properties such as metabolic stability and enhanced pharmacokinetics.[2]

Annual Reports in Medicinal Chemistry, Volume 49
ISSN 0065-7743
http://dx.doi.org/10.1016/B978-0-12-800167-7.00021-3

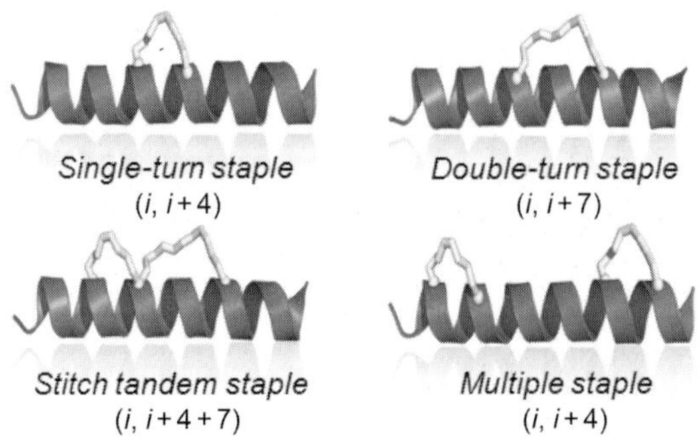

Figure 21.1 Stabilized α-helical peptides a.k.a. SPs. (See the color plate.)

With the exception of cyclosporine A, many peptide-based drugs have been limited to extracellular therapeutic targets due to their poor cell permeability. As a result, there has been a renaissance in constrained peptide-based molecular scaffolds that can efficiently improve potency, selectivity, and achieve cellular uptake.[3,4] Relative to what has been achieved to date for small molecules or biologics, compelling evidence has emerged for a new modality of macrocyclic peptides that can access the cytosol and/or nucleus of cells to modulate intracellular protein–protein interactions.[5,6]

The use of hydrocarbon stapled peptides (SPs; Fig. 21.1) is one of the most impressive recent advances to address the biology of protein complexes mediated by α-helices.[7,8]

SPs leverage a unique set of properties that capitalize on 25 years of genetic research that has unveiled the major drivers of complex diseases, including cancer, endocrine/metabolic disorders, and inflammation. Stapling technology locks peptides into their biologically active shape and imparts unprecedented pharmaceutical stability within the body. SP drugs are derived from natural peptides and are designed to specifically target protein–protein interactions both inside and outside the cell (Fig. 21.2).[10–12]

2. BENEFICIAL EFFECTS ATTRIBUTED TO THE HYDROCARBON STAPLE

SPs contain an all-hydrocarbon cross-link (the staple moiety) between successive turns of the peptide α-helix. The staple moiety is introduced by the regiospecific incorporation of two α-methyl, α-alkenylglycine residues

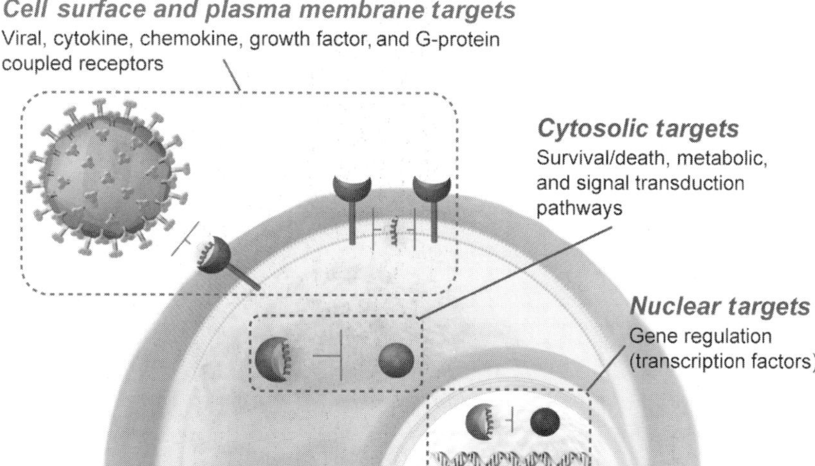

Figure 21.2 Examples of therapeutic targets and pathways for SPs. *From Ref. 9 with permission of publisher.* (See the color plate.)

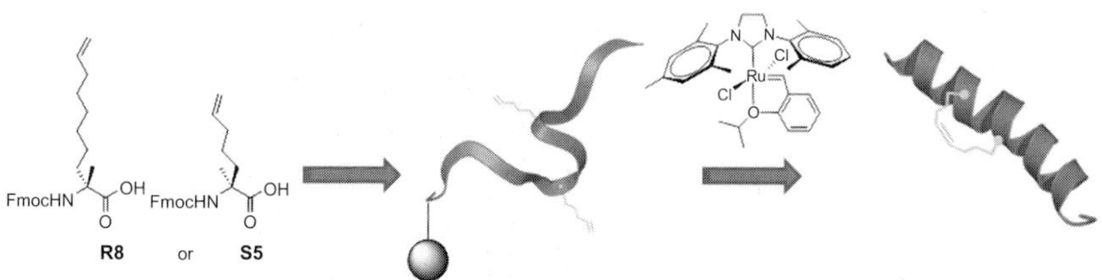

Figure 21.3 Synthetic route for preparing SPs. (See the color plate.)

(R8 or S5), having defined stereochemical configuration and hydrocarbon side-chain length, followed by ruthenium-ring closing olefin metathesis (Fig. 21.3).[13,14]

2.1. Enhancing Pharmacokinetic Properties

Although early efforts in hydrocarbon stapling were focused on the development of cell-permeable, stabilized α-helical peptide (SAHB) analogs for modulation of intracellular protein–protein interactions, this stapling technology can also be easily applied for extending the pharmacological activity of native α-helical peptides that modulate extracellular receptor targets. The beneficial application of hydrocarbon stapling to redress some of the biophysical and pharmacological liabilities of a lengthy α-helical peptide ligand for an extracellular receptor was first performed in the field of anti-human immunodeficiency virus (HIV) therapy.[15] Specifically, a pair of i, $i+4$ staples was incorporated into the sequence of the 37-mer peptide T649v, a potent HIV-1 fusion inhibitor. Circular dischroism (CD) and proteolysis experiments confirmed the enhanced α-helical induction and remarkable protease resistance conferred by the doubly SP named SAH-gp41$_{(626–662)}$ (A–B). Consequently, SAH-gp41$_{(626–662)}$ (A–B) displayed enhanced antiviral activity and longer systemic exposure *in vivo* compared to the corresponding singly stapled analogs and unmodified peptides T649v and Enfuvirtide.

Following this original work, a number of groups independently published similar studies confirming the benefits of hydrocarbon stapling on other native α-helical peptides of therapeutic significance. In each of these studies, stapling technology effectively improved the *in vivo* pharmacokinetic properties of apolipoprotein,[16] conantokin derived from the venoms of marine cone sails,[17] galanin,[18] neuropeptide Y,[18] and peptide-based membrane fusion inhibitor of hepatitis C virus[19] while retaining their pharmacological activities. More recently, the first side-by-side comparison of hydrocarbon staple versus lactam-bridge cyclizations was reported.[20] The two cyclization strategies were applied on the vasoactive intestinal peptide and produced 30-mer analogs with enhanced VPAC2 agonist potency (SP 11, $EC_{50}=0.049$ nM) as measured by glucose-dependent insulin secretion activity in pancreatic rat islets.[20] Interestingly, the helical structure reinforcement produced by both cyclization techniques according the CD analysis did not offer any advantages in terms of resistance of trypsin degradation. In fact, upon short incubation with trypsin, only small traces (<2%) of the parent lactam or olefin SPs were detected. Importantly, this finding implicates the likely requirement of installing multiple staples into lengthy peptides in order to protect the full sequence as demonstrated with the HIV-1 fusion inhibitor, T649v.[15] A recent study has investigated two experimental methods (i.e., CD spectroscopy versus hydrogen/deuterium exchange-mass spectrometry [HDX-MS] analysis) to predict proteolytic stability of stapled borealin-based peptides and showed that resistance to trypsin degradation correlated well with HDX-MS analysis.[21] Furthermore, for several peptides analyzed in this study, a doubly SP showed more resistance to trypsin degradation than its singly stapled counterparts.

2.2. Generating Cell Permeability

The cell permeability of SPs is ultimately derived from the reinforced α-helical structure of the peptides, in a manner consistent with the known structural properties of transmembrane receptors which incorporate single or multiple helical motifs, and the topological and biophysical properties of the peptides, including the contribution of the hydrocarbon staple itself. The cellular uptake of SPs has been repeatedly demonstrated using confocal fluorescence microscopy which requires the utilization of fluorescently labeled analogs,[10] and mechanisms whereby SPs undergo cellular uptake is an active area of investigation. Early published works reported that cellular uptake of SPs based on B-cell lymphoma 2 (BCL-2) domains was probably occurring through endocytic vesicle trafficking.[22] However, the intrinsic cell permeability of another SP called NYAD-1,[23] which inhibits the capsid assembly of HIV-1, was described as preferentially adopting parallel orientation to permeate through the lipid bilayers even at low-solution concentrations. Therefore, the authors concluded that the membrane interaction behavior of NYAD-1 was similar to antimicrobial peptides (AMPs) such as magainin or melittin.[24] This is an interesting observation as it is well known that the largest class of AMPs is represented by positively charged linear peptides whose biological activity is closely linked to their propensity to become amphipathic helices upon interaction with the negatively charged surface of the bacterial cell membrane. Investigations of the impact of inserting hydrocarbon staples into α-helical AMPs on their biological and biophysical properties have been performed, including studies of i, $i+4$ stapled analogs of lasioglossin III, melectin,[25] and esculentin-2EM.[26] Both studies demonstrated that stapling increased helical content and protease resistance, and in the study on lasioglossin III and melectin, it was found that the SPs exhibited a remarkable increase in hemolytic activity,[25] while undesired for developing AMPs as antibiotics, this observation suggests that α-helical stabilization for such stapled and positively charged α-helical AMPs may not match with their mechanism of membrane interaction. Furthermore, such results may challenge the widespread belief that a neutral or positive overall charge is optimal for achieving optimal cellular uptake of peptides.[10,22] Most significantly, the preclinical development of a negatively charged stapled peptide ATSP-7041, which effects reactivation of the p53 pathway via its dual specificity for MDM2 and MDMX, underscores different mechanisms for cell permeability of SPs.[27]

2.3. Improved Target Affinity and Target Specificity

An inherent advantage stapling α-helical peptides confers is that the hydrocarbon moiety enhances pharmacophore binding through decreasing entropic penalty while enhancing enthalpic interactions.[11] For example, a library of SAHBs, modeled after the BH3 domains of human MCL-1 protein, was generated in order to identify potent and selective inhibitors of MCL-1.[28] Among the MCL-1 SAHBs tested, MCL-1 $SAHB_D$ had high α-helical content according to CD analysis as well as the strongest binding activity ($K_D = 10$ nM). Structural analysis revealed that the hydrocarbon staple of MCL-1 $SAHB_D$ exhibited hydrophobic contacts at the perimeter of the core peptide–protein binding interface, and contributed to the enhanced affinity of MCL-1 $SAHB_D$ without compromising specificity. MCL-1 $SAHB_D$ showed disruption of the antiapoptotic MCL-1/BAK protein–protein interaction *in vitro* and in cells as well as sensitized BAK-mediated mitochondrial cytochrome *c* release. Computational studies using an all-atom molecular dynamics (MD) simulations study performed on the complex between MDM2 and a set of SPs including biologically active SAH-p53-8 have been reported.[29,30] The MD study confirmed that increasing the α-helicity of the DNA-binding domain of p53 through stapling results in stronger binding to MDM2. However, the authors pointed out that the location of the staples along the sequence was very critical. Indeed, poor binders were characterized by staples pointing into solution while the good binders had the staples "draped" over the surface of protein target (MDM2), thus enhancing the affinity of the peptides by additional hydrophobic interactions. The surface area buried by this interaction was calculated to be roughly equal to 200 Å^2. The contribution of binding of the fully hydrocarbon linker in SAH-p53-8 was similar to that contributed by the two most important amino acids for binding interaction, Phe19 or Trp23. Other investigations have highlighted the direct binding contribution by a staple moiety, as exemplified by a series of SPs which bind at the coactivator protein-binding site of estrogen receptors (ERs).[31] The most potent binder of the ERs, SP6, displayed K_D of 75 and 155 nM for ERα and ERβ, respectively. Consistent with previous finding for stapled p53 and MCL-1 peptides, the staple moiety of SP6 for ERs enhances the peptide's affinity by preorganizing the unbound state as an α-helix and by effecting key hydrophobic contacts with the target.

3. DRUG DISCOVERY: PRECLINICAL RESEARCH

3.1. BCL-2 Pathway Modulators

The BCL-2 family is composed of pro- and antiapoptotic proteins that control the mitochondrial pathway of programmed cell death. These include the antiapoptotic proteins (e.g., BCL-2, BCL-XL, and MCL-1), proapoptotic proteins (e.g., BAX and BAK), and activating BH3 domain-only proteins (e.g., B-cell lymphoma 2 interacting mediator of cell death (BIM) and BH3-interacting domain death agonist (BID)) and all play critical roles in the induction of apoptosis via BAX and BAK activation. BCL-2 proteins sequester their proapoptotic counterparts and activators through formation of heterodimeric complexes that prevent apoptotic signaling. Drug development efforts have sought to advance molecules that bind to the binding surface of the antiapoptotic proteins BCL-2 and MCL-1, thus competitively inhibiting their ability to sequester BH3-only proteins.

The first all-hydrocarbon-stapled proapoptotic peptide was designed from the α-helix of BCL-2 domain derived from the BH3-only protein, BID and BIM.[32,33] In the early studies, it was shown that BIM and BID single i, $i+4$ stapled 21-mer peptides (referred to as $SAHB_A$ peptides) exhibited high helicity, protease resistance, and cell permeability molecules that had high-affinity binding to the prosurvival family members BCL-2, BCL-XL, and MCL-1.[32,33] Two different research groups independently established the cell-killing activity of the 21-mer BIM-$SAHB_A$ in K562 and U937 leukemia cell lines.[33,34] In additional studies, it was demonstrated that both BIM and BID $SAHB_A$ engaged directly BAX at a newly identified trigger site distinct of the antiapoptotic protein-mediated canonical BH3-binding groove to induce cytochrome *c*-driven apoptosis.[35,36] Furthermore, the direct activation of the other proapoptotic protein, BAK, by the same SPs was reported independently by two research groups.[37,38] More significantly, the treatment of mice bearing a human acute myeloid leukemia graft with the BIM-$SAHB_A$ induced 50% tumor growth inhibition with loss of body weight.[33] Interestingly, combination of BIM-$SAHB_A$ with the BCL-2/BCL-XL inhibitor ABT-263 has shown marked synergy *in vivo* to achieve increased tumor growth suppression.[33] This is an important finding as in certain cancers the antiapoptotic protein MCL-1 is overexpressed and is the major driver of the resistance to both ABT-263 and its optimized analog ABT-199 that are currently the most advanced drug candidates targeting the BCL-2 pathway. Interestingly, an MCL-1 SAHB peptide was also generated by screening a library of stabilized α-helix of BCL-2 domains (SAHBs) and was identified as a selective inhibitor of MCL-1.[28] Finally, it is noteworthy to point out that recent studies also demonstrated a potential therapeutic application of such SAHBs beyond cancer relative to potential treatment of diabetes.[39]

3.2. Wnt Pathway Modulators

The Wnt/β-catenin pathway signal transduction cascade regulates the expression of numerous genes involved in cell differentiation, proliferation, and survival. Inappropriate and chronic activity of the canonical Wnt pathway results in increased β-catenin expression which is involved in the development of benign and malignant tumors. On the other hand, attenuation of this pathway also contributes to a number of human diseases including osteoporosis, neurodegenerative diseases, and diabetes.

A compelling drug discovery strategy for inhibition of the β-catenin pathway has focused on the interaction between β-catenin and T-cell factor (TCF) transcription factor in the nucleus.[40] The fluorescein-labeled SP, fStAx-3, derived from axin (467–481) and possessing an *i*, $i+4$ staple was identified using an fluorescence polarization (FP) assay, as the SP with the highest affinity for β-catenin ($K_D = 60$ nM relative to the parent 15-mer linear beta-catenin binding domain (CBD) axin peptide $K_D = 5$ μM). Lead optimization to further enhance-binding affinity (utilizing phage display libraries) and cell permeability (modifying positive charge) was performed on fStAx-3 and yielded the cellularly active fluorescein-labeled SP fStAx-35R which was found to induce growth inhibition of Wnt-dependent cancer cells without affecting the growth of cancer cells insensitive to deregulated Wnt signaling. After treatment with aStAx-35R at 10 μM for 5 days, inhibition of cellular proliferation ranging from 52% to 74% for the Wnt-dependent cell lines (e.g., DLD1, SW480, and HCT116) was observed.

More recently, an independent investigation of axin (469–482)-based SPs identified SAHPA1 to interfere with the endogenous interaction between axin and β-catenin in the cytosol.[41] Reverse-transcription-polymerase chain reaction studies in HEK293T cells showed SAHPA1 to increase the expression of Wnt/β-catenin target genes such as c-myc, axin2, and cyclin D1. Surprisingly, a comparative analysis of SAHPA1 and fStAx-35R[41] (vide supra) on Wnt/β-catenin signaling cells showed opposed effects by these two SPs. Specifically, fStAx-35R repressed Wnt/β-catenin signaling in both cells lines, whereas SAHPA1 enhanced Wnt/β-catenin signaling in HEK293T and was inactive in SW480 cells. It was suggested that the divergent activities of fStAx-35R and SAHPA1 might be due to their distinct subcellular localizations (i.e., SAHPA1 in the cytoplasm and fStAx-35R in the nucleus).

Importantly, another strategy for inhibiting Wnt signaling pathway has been proposed and has shown promising results.[42] This approach is based on blocking the interaction between β-catenin and BCL9, a coactivator for β-catenin-mediated transcription that is highly expressed in tumors but not in the cells of origin. Hydrocarbon stapling was applied to generate the stabilized peptide SAH-BCL9$_B$ which was effective to dissociate preformed BCL9/β-catenin complexes in a dose-dependent manner (IC_{50} of 135 nM). Colo320 and MM1S cell lines treated with SAH-BCL9$_B$ showed reduced mRNA levels for Wnt/β-catenin target genes such as VEGF, c-MYC, LGR5, LEF1, and axin2. Noteworthy, SAH-BCL9$_B$ was found to effectively suppress tumor growth *in vivo* without apparent toxicity using two established mouse xenograft models of Wnt-driven cancer (i.e., Colo320 and INA-6).

3.3. HIV Inhibitors

Acquired immunodeficiency syndrome (AIDS) is a disease of the human immune system caused by infection with HIV. AIDS remains a major worldwide health problem especially in developing countries where combating the epidemic must overcome societal issues. Therefore, intensive drug discovery efforts are being made to improve upon current therapeutic approaches and, ultimately, find a possible cure to AIDS.

Among new strategies under investigation, SPs may provide enhanced pharmacological activity of previously developed peptide-based HIV-1 fusion inhibitors.[15] Also, other studies have demonstrated proof of concept in cell-based assays that SPs may provide opportunity to inhibit HIV-1 by engaging intracellular targets. Specifically, it has been shown by NMR chemical shift perturbation analysis that different series of SPs can bind to the critical dimerization interface of the C-terminal domain of HIV-1 capsid protein.[23,41] In particular, the SP NYAD-1 binds to C-terminal capsid in the low micromolar range as determined by NMR analysis, and it has been shown to impair the viral assembly in cell-free and cell-based *in vitro* systems. Broad-spectrum antiviral efficacy of NYAD-1 at low micromolar potency has been demonstrated in two different cell lines (i.e., MT-2 and PBMC) using a large panel of HIV-1 isolates with varying subtype, coreceptor use, and drug-resistance status.

Another attractive target for the development of highly specific anti-HIV agents is HIV-1 integrase (IN) that has no mammalian counterpart and is indispensable for the stable infection of host cells because it catalyzes the insertion of viral DNA into the genome. Two independent studies have demonstrated that SPs derived from either Vpr[43] or LEDGF/p75,[44] both exemplifying known viral proteins involved in the pre-integration complex, effectively inhibit the HIV-1 IN catalysis. In the former study,[43] stapled Vpr-based peptide 6S showed similar inhibitory levels as the parent linear compound in both the 3′-end processing ($IC_{50}=2.4$ μM) and strand transfer reactions ($IC_{50}=0.84$ μM) used for assessing *in vitro* IN inhibitory activity.[43] However, in contrast to the corresponding linear peptide, the stapled Vpr-based peptide 6S displayed significant anti-HIV activity at concentrations >2.5 μM in the MT-4 Luc assay. Furthermore, the anti-HIV activity of stapled Vpr-based peptide 6S was confirmed in the p24 assay ($EC_{50}=6.46$ μM) and the MTT assay ($EC_{50}=3.55$ μM). In the latter study,[44] a series of stapled α1 region-derived NL6 peptides were synthesized and NLH6 was identified as a lead molecule with IC_{50} values of 9 ± 1 μM for 3′-processing and 6 ± 1 μM for strand transfer. Another closely related SP analog, NLH16, from this was shown to exhibit the most potent HIV replication inhibition ($IC_{50}=5$ μM) in MT-4 cells and it also displayed a promising *in vitro* therapeutic index ($CC_{50}/EC_{50}=8$).

4. DRUG DEVELOPMENT AND P53 REACTIVATION

4.1. From *In Vitro* to *In Vivo* Proof of Concept

The human transcription factor protein p53 plays a central role in maintaining the integrity of the genome in cells. As a tumor suppressor, p53 induces cell-cycle arrest and apoptosis in response to DNA damage and cellular stress. In about 50% of all human cancers wherein p53 protein is not mutated, its function is primarily inactivated by overexpression of its negative regulators MDM2 and/or MDMX. Consequently, the p53: MDM2/MDMX interactions have revealed MDM2 and MDMX to be very attractive cancer targets.[45]

Recently, it was demonstrated that a SP, SAH-p53-8, derived from the 15-residue α-helical transactivation domain of p53 exhibited high-affinity binding for both MDM2 and MDMX as well as induced dose-dependent inhibition of SJSA-1 cell viability at low micromolar level.[29] MD studies and X-ray crystallographic analysis provided detailed insights into the contribution of the staple for the increased binding affinity.[30,46,47] Further biological investigations using a series of solid tumor cell lines expressing different expression levels of MDM2 and MDMX reveal that SAH-p53-8, in contrast to the MDM2-specific antagonist nutlin-3, was able to reactivate p53 and induce apoptosis irrespective of MDMX levels.[48] Importantly, intravenous administration of SAH-p53-8 to mice bearing an MDMX expressing and nutlin-3-resistant cancer resulted in a p53-response in the tumor cells and suppressed tumor growth.[48] Also, other studies reported that about 65% of human melanoma cell lines expressed high-MDMX protein level and were resistant to treatment with nutlin-3. Importantly, such melanoma cell lines were very responsive to the dual-specific MDM2/MDMX antagonist SAH-p53-8.[49]

Most recently, it has been reported that ATSP-7041 (Fig. 21.4),[27,50] a more specific and potent SP than SAH-8-p53, suppressed tumor growth *in vivo* of multiple human xenograft models.[27] ATSP-7041 exhibited remarkable biological efficacy (>50-fold enhancement of cellular potency to SAH-p53-8) that correlated with its nanomolar-binding affinity to both MDM2 and MDMX, its efficient cell-penetrating properties, its selective p53-dependent submicromolar cellular activity, and its capacity to exhibit a more durable, on-target effects on p53 signaling than the small molecule clinical candidate MDM2-selective inhibitor RG7112.[27] The favorable drug-like and pharmacokinetic properties of ATSP-7041 indicate the potential for convenient clinical dosing regimens.[27] A next-generation dual-specific MDM2/MDMX antagonist SP, ALRN-6924, has been identified as a promising development candidate for phase 1 clinical testing.

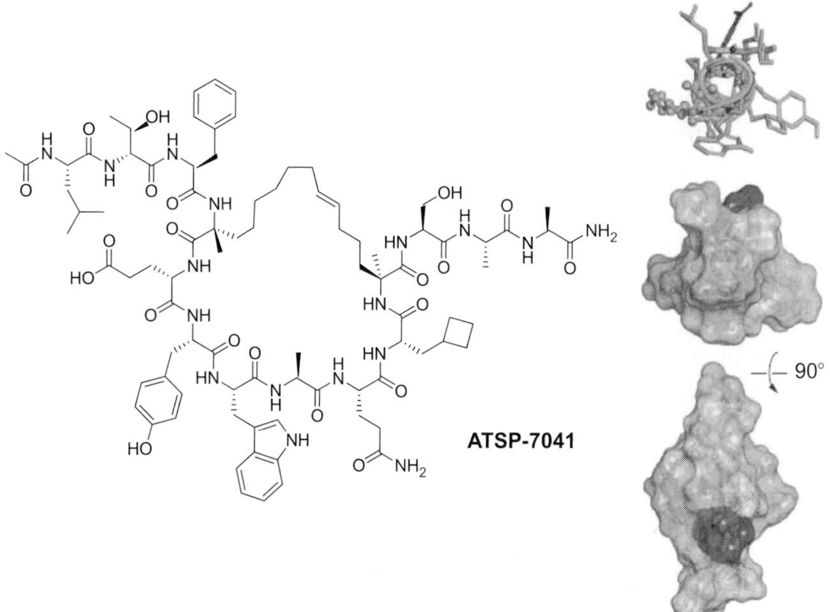

Figure 21.4 2D and 3D structural representations of ATSP-7041. *From Ref. 27 with permission of publisher.* (See the color plate.)

4.2. Advantages with SPs over Small Molecules

Application of the SP strategy to dual inhibitors of MDM2/MDMX has shown the potential to deliver high efficacy for restoring p53 across a broader spectrum of human cancers. In particular, it has been demonstrated that high levels of MDMX in tumors containing wild-type p53 (e.g., melanoma or breast cancer) are responsible for the resistance observed upon treatment with MDM2-specific small molecule antagonist.[27,49] A unique feature of such SPs (versus small molecules) is their ability to efficiently interact with both MDM2 and MDMX. The multifunctionality-based pharmacophore and higher conformational plasticity of SPs may favor their ability for dual specificity for both MDM2 and MDMX.[45] Recently, it was demonstrated that MDM2 mutations conferring resistance to nutlin-3 did not confer resistance to sMTide-02, an SP structurally similar to ATSP-7041.[51,52] Also, binding kinetics from BIAcore experiments demonstrated a significant difference in the on-rate/off-rate for the SP ATSP-7041 and the small molecule RG7112.[27] In addition to its high affinity to MDM2 ($K_D = 0.91$ nM) and MDMX ($K_D = 2.31$ nM), ATSP-7041 was shown to have a slower dissociation half-life from MDM2 (43 min) versus the small molecule RG7112 (6 min). This slow off-rate suggested that ATSP-7041 might exhibit a more durable effect on p53 signaling and this was observed in wash-out experiments in MCF-7 cells in which activation of p53, MDM2, and p21 protein showed prolonged downstream activation of such p53-dependent protein expression levels (up to 48 h) as compared with RG7112 which was essentially ineffective after 4 h.[27]

5. CONCLUDING REMARKS

As illustrated by the growing number of publications, SPs provide novel drug modality with many applications. Because it was not the intent to comprehensively review the known literature relating to SPs, compelling studies illustrating the modulation of other therapeutically relevant protein–protein interactions remain.[53–59] Nevertheless, the studies described here have shown that the incorporation of a staple moiety into α-helical peptides can significantly improve important pharmaceutical properties, including proteolytic resistance, target affinity, cell permeability, and plasma half-lives *in vivo*. There is significant hope that this new drug modality will achieve clinical proof of concept and fulfill the promise of delivering breakthrough medicines on previously intractable targets.

REFERENCES

1. Kaspar, A. A.; Reichert, J. M. *Drug Discov. Today* **2013**, *18*(17/18), 807.
2. Giordanetto, F.; Kihlberg, J. *J. Med. Chem.* **2014**, *57*, 278.
3. Dharanipragada, R. *Future Med. Chem.* **2013**, *5*(7), 831.
4. SciBX Collections, April 2013, http://viewer.zmags.com/publication/38257327#/38257327/1.
5. Ivanov, A. A.; Khuri, F. R.; Fu, H. *Trends Pharmacol. Sci.* **2013**, *34*, 393.
6. Azzarito, V.; Long, K.; Murphy, N. S.; Wilson, A. J. *Nat. Chem.* **2013**, *5*, 161.
7. Higueruelo, A. P.; Jubb, H.; Blundell, T. L. *Curr. Opin. Pharmacol.* **2013**, *13*(5), 791.
8. Milroy, L.-G.; Brunsveld, L. *Future Med. Chem.* **2013**, *5*(18), 2175.
9. Sawyer, T. K. *Chem. Biol. Drug Des.* **2009**, *73*, 3.
10. Verdine, G. L.; Hilinski, G. J. *Methods Enzymol.* **2012**, *503*, 3.
11. Verdine, G. L.; Hilinski, G. J. *Drug Discov. Today Technol.* **2012**, *9*(1), e41.
12. Walensky, L. D.; Bird, G. H. *J. Med. Chem.* **2014**, http://dx.doi.org/10.1021/jm4011675.
13. Schafmeister, C. E.; Po, J.; Verdine, G. L. *J. Am. Chem. Soc.* **2000**, *122*, 5891.
14. Kim, Y.; Grossmann, T. N.; Verdine, G. L. *Nat. Protoc.* **2011**, *6*, 761.
15. Bird, G. H.; Madani, N.; Perry, A. F.; Princiotto, A. M.; Supko, J. G.; He, X.; Gavathiotis, E.; Sodroski, J. G.; Walensky, L. D. *Proc. Natl. Acad. Sci. U.S.A.* **2010**, *107*(32), 14093.
16. Sviridov, D. O.; Ikpot, I. Z.; Stonik, J.; Drake, S. K.; Amar, M.; Osei-Hwedieh, D. O.; Piszczek, G.; Turner, S.; Remaley, A. T. *Biochem. Biophys. Res. Commun.* **2011**, *410*(3), 446.
17. Platt, R. J.; Han, T. S.; Green, B. R.; Smith, M. D.; Skalicky, J.; Gruszczynski, P.; White, H. S.; Olivera, B.; Bula, G.; Gajewiak, J. *J. Biol. Chem.* **2012**, *287*, 20727.
18. Green, B. R.; Klein, B. D.; Lee, H.-K.; Smith, M. D.; White, H. S.; Bulaj, G. *Bioorg. Med. Chem.* **2013**, *21*, 303.
19. Cui, H.-K.; Qing, J.; Guo, Y.; Wang, Y.-J.; Cui, L.-J.; He, T.-H.; Zhang, L.; Liu, L. *Bioorg. Med. Chem.* **2013**, *21*(12), 3547.
20. Giordanetto, F.; Revell, J. D.; Knerr, L.; Hostettler, M.; Paunovic, A.; Priest, C.; Janefeldt, A.; Gill, A. *ACS Med. Chem. Lett.* **2013**, *4*(12), 1163.
21. Shi, X. E.; Wales, T. E.; Elkin, C.; Kawahata, N.; Engen, J. R.; Annis, D. A. *Anal. Chem.* **2013**, *85*(23), 11185.
22. Bird, G. H.; Bernal, F.; Pitter, K.; Walensky, L. D. *Methods Enzymol.* **2008**, *446*, 369.
23. Zhang, H.; Zhao, Q.; Bhattacharya, S.; Waheed, A. A.; Tong, X.; Hong, A.; Heck, S.; Curreli, F.; Goger, M.; Cowburn, D.; Freed, E. O.; Debnath, A. K. *J. Mol. Biol.* **2008**, *378*, 565.
24. Sun, T.-L.; Sun, Y.; Lee, C.-C.; Huang, H. W. *Biophys. J.* **2012**, *104*(9), 1923.
25. Chapuis, H.; Jirina Slaninova, J.; Bednarova, L.; Monincova, L.; Budesınsky, M.; Cerovsky, V. *Amino Acids* **2012**, *43*(5), 2047.
26. Pham, T. K.; Kim, D. H.; Lee, B. J.; Kim, Y. W. *Bioorg. Med. Chem. Lett.* **2013**, *23*(24), 6717.
27. Chang, Y. S.; Graves, B.; Guerlavais, V.; Tovar, C.; Packman, K.; To, K.-H.; Olson, K. A.; Kesavan, K.; Gangurde, P.; Mukherjee, A.; Baker, T.; Darlak, K.; Elkin, C.; Filipovic, Z.; Qureshi, F. Z.; Cai, H.; Berry, P.; Feyfant, E.; Shi, X. E.; Horstick, J.; Annis, D. A.; Manning, A. M.; Fotouhi, N.; Nash, H.; Vassilev, L. T.; Sawyer, T. K. *Proc. Natl. Acad. Sci. U.S.A.* **2013**, *110*(36), E3445.
28. Stewart, M. L.; Fire, E.; Keating, A. E.; Walensky, L. D. *Nat. Chem. Biol.* **2010**, *6*(8), 595.
29. Bernal, F.; Tyler, A. F.; Korsmeyer, S. J.; Walensky, L. D.; Verdine, G. L. *J. Am. Chem. Soc.* **2007**, *129*, 2456.
30. Joseph, T. L.; Lane, D.; Verma, C. *Cell Cycle* **2010**, *9*(22), 4560.
31. Phillips, C.; Roberts, L. R.; Schade, M.; Bazin, R.; Bent, A.; Davies, N. L.; Moore, R.; Pannifer, A. D.; Pickford, A. R.; Prior, S. H.; Read, C. M.; Scott, A.; Brown, D. G.; Xu, B.; Irving, S. L. *J. Am. Chem. Soc.* **2011**, *133*, 9696.

32. Walensky, L. D. *Science* **2004**, *305*(5689), 1466.
33. LaBelle, J. L.; Katz, S. G.; Bird, G. H.; Gavathiotis, E.; Stewart, M. L.; Lawrence, C.; Fisher, J. K.; Godes, M.; Pitter, K.; Kung, A. L.; Walensky, L. D. *J. Clin. Invest.* **2012**, *122*, 2018.
34. Okamoto, T.; Segal, D.; Zobel, K.; Fedorova, A.; Yang, H.; Fairbrother, W. J.; Huang, D. C. S.; Smith, B. J.; Deshayes, K.; Czabotar, P. E. *ACS Chem. Biol.* **2014**, *9*(3), 838.
35. Gavathiotis, E.; Suzuki, M.; Davis, M. L.; Pitter, K.; Bird, G. H.; Katz, S. G.; Tu, H.-C.; Kim, H.; Cheng, E. H.-Y.; Tjandra, N.; Walensky, L. D. *Nature* **2008**, *455*(7216), 1076.
36. Gavathiotis, E.; Reyna, D. E.; Davis, M. L.; Bird, G. H.; Walensky, L. D. *Mol. Cell* **2010**, *40*, 481.
37. Leshchiner, E. S.; Braun, C. R.; Bird, G. H.; Walensky, L. D. *Proc. Natl. Acad. Sci. U.S.A.* **2013**, *110*(11), E986.
38. Moldoveanu, T.; Grace, C. R.; Llambi, F.; Nourse, A.; Fitzgerald, P.; Gehring, K.; Kriwacki, R. W.; Green, D. R. *Nat. Struct. Mol. Biol.* **2013**, *20*(5), 589.
39. Szlyk, B.; Braun, C. R.; Ljubicic, S.; Patton, E.; Bird, G. H.; Osundiji, M. A.; Matschinsky, F. M.; Walensky, L. D.; Danial, N. N. *Nat. Struct. Mol. Biol.* **2013**, *21*(1), 36–42.
40. Grossmann, T. M.; Yeh, J. T.-H.; Bowman, B. R.; Chu, Q.; Moellering, R. E.; Verdine, G. L. *Proc. Natl. Acad. Sci. U.S.A.* **2012**, *109*(44), 17942.
41. Zhang, H.; Curreli, F.; Waheed, A. A.; Mercredi, P. Y.; Mehta, M.; Bhargava, P.; Scacalossi, D.; Tong, X.; Lee, S.; Cooper, A.; Summers, M. F.; Freed, E. O.; Debnath, A. K. *Retrovirology* **2013**, *10*, 136.
42. Takada, K.; Zhu, D.; Bird, G. H.; Sukhdeo, K.; Zhao, J.-J.; Mani, M.; Lemieux, M.; Carrasco, D. E.; Ryan, J.; Horst, D.; Fulciniti, M.; Munshi, N. C.; Xu, W.; Kung, A. L.; Shivdasani, R. A.; Walensky, L. D.; Carrasco, D. R. *Sci. Transl. Med.* **2012**, *4*, 148ra117.
43. Nomura, W.; Aikawa, H.; Ohashi, N.; Urano, E.; Metifiot, M.; Fujino, M.; Maddali, K.; Ozaki, T.; Nozue, A.; Narumi, T.; Hashimoto, C.; Tanaka, T.; Pommier, Y.; Yamamoto, N.; Komano, J. A.; Murakami, T.; Tamamura, H. *ACS Chem. Biol.* **2013**, *8*, 2235.
44. Long, Y.-Q.; Huang, S.-X.; Zawahir, Z.; Xu, Z.-L.; Li, H.; Sanchez, T. W.; Zhi, Y.; De Houwer, S.; Christ, F.; Debyser, Z.; Neamati, N. *J. Med. Chem.* **2013**, *56*, 5601.
45. Hoe, K. K.; Verma, C. S.; Lane, D. P. *Nat. Rev. Drug Discov.* **2014**, *13*, 217.
46. Guo, Z.; Mohanty, U.; Noehre, J.; Sawyer, T. K.; Sherman, W.; Krilov, G. *Chem. Biol. Drug Des.* **2010**, *75*, 348.
47. Baek, S.; Kutchukian, P. S.; Verdine, G. L.; Huber, R.; Holak, T. A.; Won Lee, K.; Popowicz, G. M. *J. Am. Chem. Soc.* **2012**, *134*, 103.
48. Bernal, F.; Wade, M.; Godes, M.; Davis, T. N.; Whitehead, D. G.; Kung, A. L.; Wahl, G. M.; Walensky, L. D. *Cancer Cell* **2010**, *18*, 411.
49. Gembarska, A.; Luciani, F.; Fedele, C.; Russell, E. A.; Dewaele, M.; Villar, S.; Zwolinska, A.; Haupt, Sue; de Lange, J.; Yip, D.; Goydos, J.; Haigh, J. J.; Haupt, Y.; Larue, L.; Jochemsen, A.; Shi, H.; Moriceau, G.; Lo, R. S.; Ghanem, G.; Shackleton, M.; Bernal, F.; Marine, J.-C. *Nat. Med.* **2012**, *18*(8), 1239.
50. Guerlavais, V.; Darlak, K.; Graves, B.; Tovar, C.; Packman, K.; Olson, K.; Kesavan, K.; Gangurde, P.; Horstick, J.; Mukherjee, A.; Baker, T.; Shi, X. E.; Lentini, S.; Sun, K.; Irwin, S.; Feyfant, E.; To, T.; Filipovic, Z.; Elkin, C.; Pero, J.; Santiago, S.; Bruton, T.; Sawyer, T.; Annis, A.; Fotouhi, N.; Manning, T.; Nash, H.; Vassilev, L. T.; Chang, Y. S.; Sawyer, T. K. In: Proceedings of the 23rd American Peptide Symposium, Hawaii, 2013; p 217.
51. Brown, C. J.; Quah, S. T.; Jong, J.; Goh, A. M.; Chiam, P. C.; Khoo, K. H.; Choong, M. L.; Lee, M. A.; Yurlova, L.; Zolghadr, K.; Joseph, T. L.; Verma, C. S.; Lane, D. P. *ACS Chem. Biol.* **2013**, *8*, 506.

52. Jia Wei, S.; Joseph, T.; Chee, S.; Li, L.; Yurlova, L.; Zolghadr, K.; Brown, C.; Lane, D.; Verma, C.; Ghadessy, F. *PLoS One* **2013**, *8*(11), e81068.
53. Hao, Y.; Wang, C.; Cao, B.; Hirsch, B. M.; Song, J.; Markowitz, S. D.; Ewing, R. M.; Sedwick, D.; Liu, L.; Zheng, W.; Wang, Z. *Cancer Cell* **2013**, *23*, 1.
54. Kim, W.; Bird, G. H.; Neff, T.; Guo, G.; AKerenyi, M.; Walensky, L. D.; Orkin, S. H. *Nat. Chem. Biol.* **2013**, *9*(10), 643.
55. Lama, D.; Quah, S. T.; Verma, C. S.; Lakshminarayanan, R.; Beuerman, R. W.; Lane, D. P.; Brown, C. J. *Sci. Rep.* **2013**, *3*, 3451.
56. Spiegel, J.; Cromm, P. M.; Itzen, A.; Goody, R. S.; Grossmann, T. N.; Waldmann, H. *Angew. Chem. Int. Ed. Engl.* **2014**, *53*, 1.
57. Frank, A. O.; Vangamudi, B.; Feldkamp, M. D.; Souza-Fagundes, E. M.; Luzwick, J. W.; Cortez, David; Olejniczak, E. T.; Waterson, A. G.; Rossanese, O. W.; Chazin, W. J.; Fesik, S. W. *J. Med. Chem.* **2014**, *57*(6), 2455.
58. Yang, Y.; Schmitz, R.; Mitala, J.; Whiting, A.; Xiao, W.; Ceribelli, M.; Wright, G. W.; Zhao, H.; Yang, Y.; Xu, W.; Rosenwald, A.; Ott, G.; Gascoyne, R. D.; Connors, J. M.; Rimsza, L. M.; Campo, E.; Jaffe, E. S.; Delabie, J.; Smeland, E. B.; Rita, M.; Braziel, R. M.; Raymond, R.; Tubbs, R. R.; Cook, J. R.; Weisenburger, D. D.; Chan, W. C.; Wiestner, D.; Kruhlak, M. J.; Iwai, K.; Bernal, F.; Staudt, L. M. *Cancer Discov.* **2014**, *4*(4), 480.
59. Cui, H.-K.; Zhao, B.; Li, Y.; Guo, Y.; Hu, H.; Liu, L.; Chen, Y.-G. *Cell Res.* **2013**, *23*, 581.

CHAPTER TWENTY-TWO

Cytochrome P450 Enzyme Metabolites in Lead Discovery and Development

Sylvie E. Kandel*, Larry C. Wienkers†, Jed N. Lampe‡

*XenoTech, LLC, Lenexa, Kansas, USA

†Amgen, Inc., Seattle, Washington, USA

‡Department of Pharmacology, Toxicology, and Therapeutics, The University of Kansas Medical Center, Kansas City, Kansas, USA

Contents

1. INTRODUCTION

The cytochrome P450 (CYP) enzyme superfamily is well known for its role in oxidative drug metabolism[1–4]; however in recent years, there has been a growing interest in the identification of novel therapeutic agents from CYP metabolites. Utilizing their unique oxidative chemistry,[5,6] CYPs are able to catalyze a wide variety of unusual reactions on small-to-medium-sized organic substrates that can lead to new chemical entities with novel

Annual Reports in Medicinal Chemistry, Volume 49
ISSN 0065-7743
http://dx.doi.org/10.1016/B978-0-12-800167-7.00022-5

pharmacological activity. CYP enzymes also play major roles in plant biosynthetic pathways, generating many natural products that have been developed into potent therapeutic agents.[7,8] Additionally, several of the drugs derived from natural products are further transformed by drug metabolizing CYPs *in vivo* into biologically active metabolites, some of which are more potent than the parent compound.[9] More recently, microscale analytical technologies have been employed for the generation and identification of CYP metabolites as lead compounds.[10] In conjunction with the development of bioreactor technology, whereby CYP oxidative transformations may be scaled-up for quantitative production of metabolites, this has ushered in the possibility of utilizing CYPs as a platform for lead discovery and development.[10,11] This review will highlight some of the recent examples of drug leads identified from CYP metabolites and the intriguing possibilities of using CYPs as catalysts for future drug discovery and development.

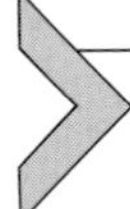

2. IDENTIFICATION OF CYP-MODIFIED NATURAL PRODUCTS AS DRUG LEADS

A large variety of natural products that have been developed into successful drugs contain CYPs in their biosynthetic pathways. These include antibiotic, antimitotic, antineoplastic, antihypertensive, and antiarrhythmic agents.[4] Many of these compounds are secondary metabolites that are involved in plant or microbial defense pathways.[4,12] The unique oxidative chemistry provided by CYPs allows tailoring specific functionalities onto complex carbon skeletons to fine tune their biological activities. In this way, millions of years of chemical warfare between microbes, plants, and animals have produced chemical entities that are exquisitely specific for their targets. Only recently have concerted efforts been made to identify new lead compounds from known CYP biosynthetic pathways involved in the generation of natural products, yet these may prove promising in the years to come. A few examples from a variety of classes are illustrated below.

2.1. Antineoplastic Agents

The potent antimitotic agent Taxol (paclitaxel), originally isolated from endophytic fungi inhabiting the bark of the Pacific yew tree (*Taxus brevifolia*), is thought to have up to 14 CYPs involved in its biosynthetic pathway.[12,13] Initially, there were significant concerns regarding obtaining enough of the compound due to the ecological impact of harvesting the trees. These concerns were alleviated when a commercial scale semisynthetic route, based on extraction from yew pine needles instead of the bark, was identified.[14] Production now relies on large scale plant cell fermentation with the paclitaxel-producing fungus *Penicillium raistrickii*.[15] Despite Taxol's widespread use, problems with off-target effects and the potential for resistance have led to a desire to identify new analogs. One approach was to clone several of the paclitaxel pathway CYPs for expression in the yeast *Saccharomyces cerevisiae*, in an attempt to both increase yield and produce novel analogs through combinatorial biosynthesis.[16] A total of 8 of the 14 CYPs thought to be involved in the pathway were cloned and expressed in *S. cerevisiae*, and this allowed for production of baccatin III, an intermediate in paclitaxel biosynthesis that could function as a precursor for the semisynthesis of novel paclitaxel analogs.[16]

In another case, an alternative retrometabolic approach was used by Guengerich and colleagues to identify novel chemotherapeutic agents based on a previously known pharmacophore.[17] The serendipitous discovery that several human CYPs are able to metabolize indole to indigo and indirubin led to the hypothesis that they might also be able to generate lead compounds for tyrosine kinase inhibition, since indole is a known pharmacophore for many of these enzymes.[18] Guengerich and colleagues added a variety of commercially available substituted indole compounds to bacterial cultures expressing various human CYP2A6 mutants generated by directed evolution.[19] Extracts from these cultures were screened against the kinases CDK1, CDK5, and GSK-3b, and from these initial screenings, they were able to identify several indirubin-based inhibitors that were an order of magnitude more potent than indirubin itself, and to characterize their individual structures using ^{1}H NMR.[19] An approach such as this, employing enzyme mutagenesis and enzymatic coupling to produce novel compound libraries of previously known pharmacophores, may be of particular benefit for scaffolds which are synthetically difficult.

2.2. Antiprotozoal Agents

The most profound advancement in the treatment of malaria in recent decades has been the development of artemisinin, a sesquiterpene lactone endoperoxide isolated from *Artemisia annua* spp. (Chinese wormwood)[8] (Fig. 22.1).

This has generated significant interest in cloning the entire biosynthetic pathway for expression in a compliant heterologous host, such as *S. cerevisiae*. Keasling and colleagues were successful in modifying the yeast mevalonate

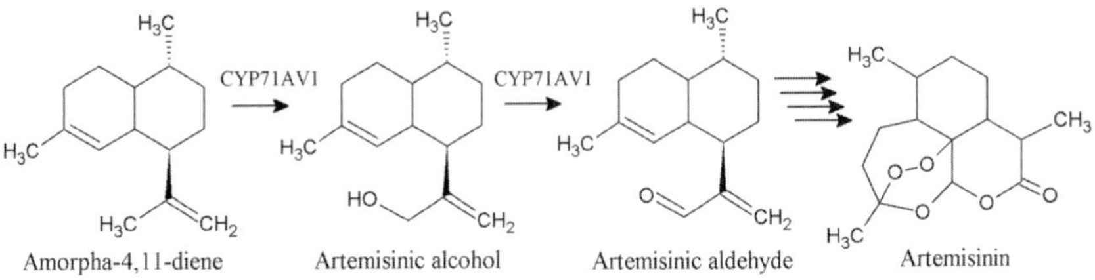

Figure 22.1 Biosynthetic pathway for the antimalarial artemisinin.

pathway and introducing the genes encoding amorphadiene synthase and CYP71AV1 from *A. annua*.[20] In this modified yeast strain, they demonstrated significantly enhanced production and cellular export of artemisinic acid, the direct precursor to artemisinin.[20] In a subsequent study, after engineering a plant CYP reductase and a second CYP in the pathway from *A. annua*, they were able to produce the final artemisinin product in ~20-fold increased yield, reducing the production cost dramatically.[21]

2.3. Antifungal Agents

One of the more efficacious agents used to treat systemic fungal infections is amphotericin B, a polyene secondary metabolite isolated from *Streptomyces nodosus*[22] (Fig. 22.2). Its primary mechanism of action is to sequester ergosterol, an important component of fungal cell membranes. However, toxicities resulting from off-target effects in host cell membranes have limited its use.[22] Previous work demonstrated that diminishing the negative charge on the exocyclic carboxyl group significantly reduced toxicity.[23] Carmody and colleagues thus created a series of *S. nodosus* mutants with targeted deletions in the *amphN* CYP gene locus, a CYP with known tailoring function in the production of amphotericin B, to generate amphotericin analogs where the exocyclic carboxyl groups were substituted by methyl group functionalities[24] (Fig. 22.2). These analogs retained antifungal activity while exhibiting reduced hemolytic toxicity. A future effort in this area might focus on other structural perturbations of the molecule utilizing the principals of combinatorial biosynthesis to generate analogs with reduced toxicity and improved efficacy.

Another successful antifungal agent, griseofulvin, first isolated from the mold *Penicillium griseofulvum*, has seen additional interest lately due to its antiproliferative effects in certain cancer cell model systems.[25] A recently identified CYP in its biosynthetic pathway has been shown to be essential for formation of the grisan scaffold.[26] Using this enzyme, along with certain others involved in the biosynthetic pathway, Cacho and colleagues were, for

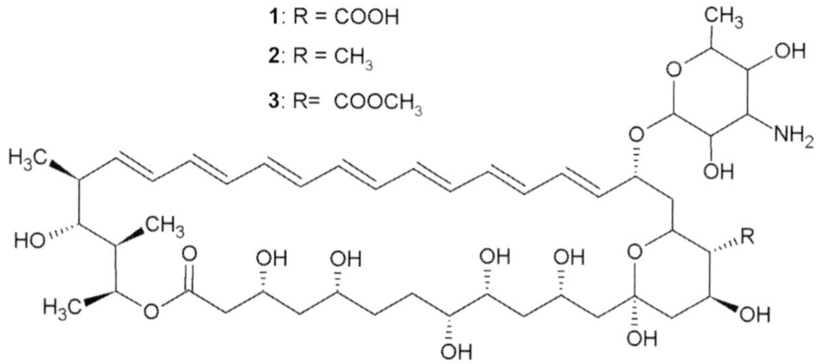

1: R = COOH

2: R = CH_3

3: R= $COOCH_3$

Figure 22.2 Amphotericin B (**1**) and its analogs (**2**) and (**3**)[23].

the first time, able to synthesize griseofulvin entirely *in vitro* from acetyl-CoA and malonyl-CoA feedstocks.[26] Understanding the role of the CYP in the orcinol and phloroglucinol ring coupling reactions opens up the possibility of creating griseofulvin analogs that may have useful applications as antifungal or antineoplastic agents. A number of exciting technologies to combine CYP active metabolite generation with structural elucidation and activity studies have recently surfaced, all of which will no doubt be a boon for novel lead discovery.[10,27]

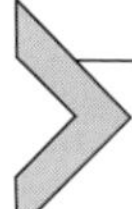

3. IDENTIFICATION OF PHARMACOLOGICALLY ACTIVE METABOLITES OF KNOWN DRUGS

Since the early days of the study of drug metabolism, it has been known that metabolites can be pharmacologically active.[28,29] Probably the most classic, and earliest, example of this phenomenon was the discovery of the antibiotic sulfanilamide as a metabolite of the drug Prontosil.[3] Subsequently, the strategy of developing prodrugs most commonly took advantage of metabolism to release an active form of a drug from a pharmacologically inactive precursor.[3] However, later studies also identified biologically active metabolites from *active* parent drugs.[30–32] The metabolism of minoxidil is an early example of this.[33] Although potent hypotensive effects are seen with the parent compound in animal models, the effects in humans are exclusively due to the sulfonated metabolite.[34,35] More recent studies have suggested that up to 22% of the top 50 drugs that are prescribed in the United States produce active metabolites that are essential to their pharmacological effects.[32]

3.1. Statin Metabolites

Statins are widely prescribed drugs[36] that act to lower circulating cholesterol levels by inhibiting hepatic HMG-CoA reductase.[37] The statins were originally derived from fungal secondary metabolites.[38] Interestingly, several fungal CYPs are known to be involved in the statin biosynthetic pathway.[39] One of the original statins to be developed was compactin (mevastatin); however, it was never brought to market due to adverse effects in a canine animal model.[36] The first statin approved for clinical use was lovastatin (Mevacor) from the fungus *Aspergillus terreus*.[40] Since then, a series of statin analogs has been introduced, including the best-selling atorvastatin (Lipitor) and rosuvastatin (Crestor), both of which have been suggested to have active metabolites.[41–43] Atorvastatin is predominately metabolized by CYP3A4 to form active *ortho*-(2-hydroxy) and *para*-(4-hydroxy) metabolites (Fig. 22.3), as well as several beta-oxidation metabolites.[43–45]

Both 2-hydroxyatorvastatin and 4-hydroxyatorvastatin exhibit plasma concentrations similar to that of atorvastatin, albeit with a somewhat reduced elimination half-life.[46] In the case of rosuvastatin, it undergoes *N*-demethylation at the sulfonamide nitrogen to generate *N*-desmethylrosuvastatin, which also seems to be active, although at reduced potency.[47] Interestingly, both the organic anion transporting polypeptides and CYPs have been demonstrated to be important determinants of the disposition of these drugs, implying that interindividual pharmacogenetic differences may play a role in statin distribution and efficacy *in vivo*.[43,47] As new "omics" technologies become available to analyze the CYP/transporter phenotype of an individual, treatment may be able to be paired to a particular phenotype for improved response.[48] Certainly, the fact that some of the metabolites demonstrate improved pharmacokinetics over the parent compounds suggests that similar metabolites of other statins should be more fully investigated in regards to efficacy. Given that some of the CYPs involved in the fungal production of statins have been cloned and characterized for their activities, it may now be possible to produce these specific metabolites in large quantities for further clinical study and production.[49]

3.2. Antibiotics

Similar to the statins, many of the antibiotics currently in clinical use involve CYPs at critical steps in their biosynthetic pathways.[4,50] These include several different classes of compounds isolated from *Streptomyces* spp., such as erythromycin, tetracycline, amphotericin B, avermectin, and adriamycin.[4]

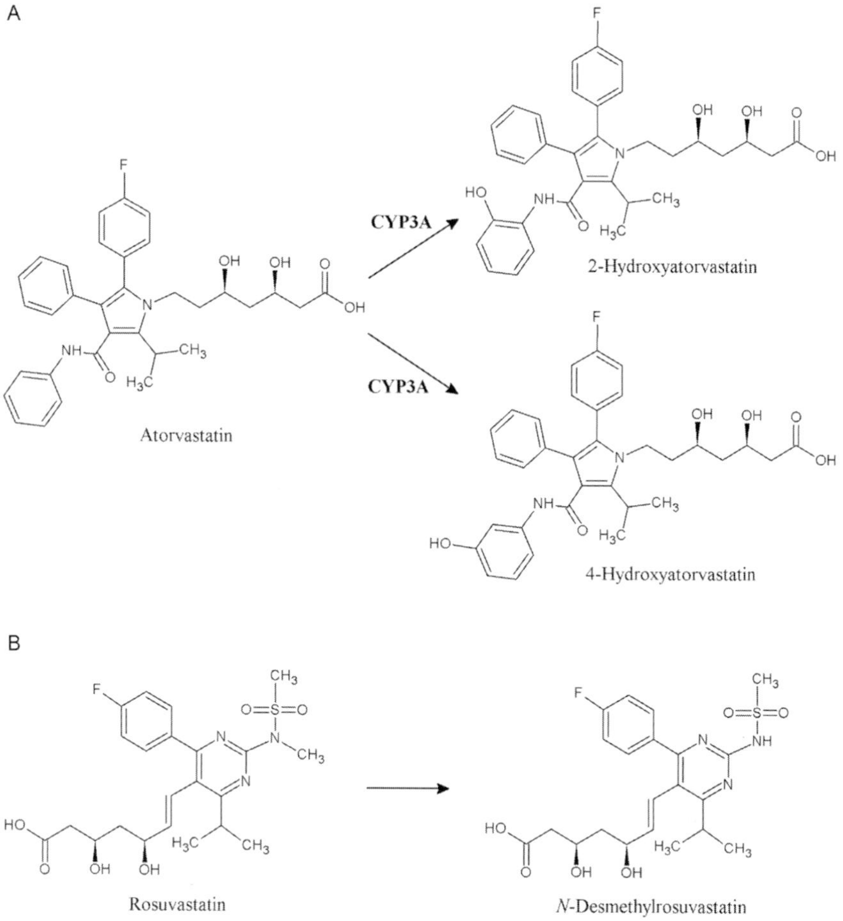

Figure 22.3 (A) Active metabolites of atorvastatin and (B) active metabolite of rosuvastatin.

Many of the CYPs involved in these biosynthetic pathways perform tailoring functions, such as the addition of hydroxyl groups or endoperoxide bridges, near the end of the biosynthetic pathways.

The macrolide antibiotics are an important class of antibiotics in use worldwide.[4,51] While newer analogs such as azithromycin and clarithromycin have replaced erythromycin in many countries, erythromycin is still used substantially for some indications, especially in the Third World.[51] Erythromycin is transformed *in vivo* into 8,9-anhydroerythromycin.[52] Although this derivative possesses no antimicrobial activity itself, it is a potent motilide—a mimic of the peptide motilin that causes duodenal contractions.[52,53] In this case, the daughter compound has therapeutic properties that are distinct from the parent drug. In contrast, the major CYP metabolite of clarithromycin, 14-hydroxyclarithromycin (Fig. 22.4), exhibits potent antimicrobial activity in its own right.[54]

Although this metabolite has improved activity against certain *Legionella* species, its efficacy against penicillin resistant *Streptococcus pneumoniae* is lower than the parent drug, thereby demonstrating an altered spectrum of antimicrobial activity.[54] Several of the bacterial CYPs in the erythromycin biosynthetic pathway have been cloned and expressed in *Escherichia coli*, allowing for further structure–function characterization and the possibility of combinatorial biosynthesis for the production of new antibiotic derivatives, or compounds with entirely new therapeutic properties.[55,56]

3.3. Antidepressants

Thioridazine, an older piperidine antipsychotic still used in some countries, undergoes sequential *S*-oxidation of the thiomethyl group of the thiazine ring to produce the sulfoxide (mesoridazine) and the sulfone (sulforidazine), each of which has been developed independently as a drug[57] (Fig. 22.5).

Both metabolites have higher affinity for D_2, D_3, D_4 dopaminergic, and α_1 noradrenergic receptors than thioridazine itself,[58] and the metabolites are less protein bound than the parent drug, suggesting that the majority of receptor occupancy is due to the metabolites and not the parent compound.[58]

The anxiolytic buspirone, a 5-HT_{1A} agonist, is metabolized to both 6-hydroxybuspirone and an *N*-dealkylated form, both of which are active.[59] The 6-hydroxy metabolite is known to be relatively metabolically stable and contributes substantially to the anxiolytic activity of buspirone,[59] which led GlaxoSmithKline to attempt development of this metabolite under the

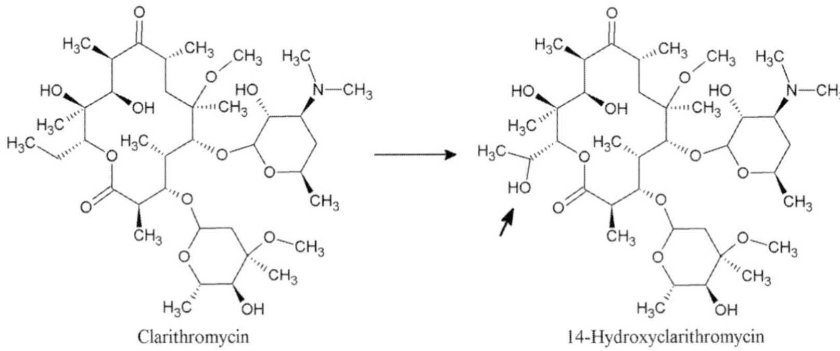

Clarithromycin → 14-Hydroxyclarithromycin

Figure 22.4 Clarithromycin and its active metabolite 14-hydroxyclarithromycin.

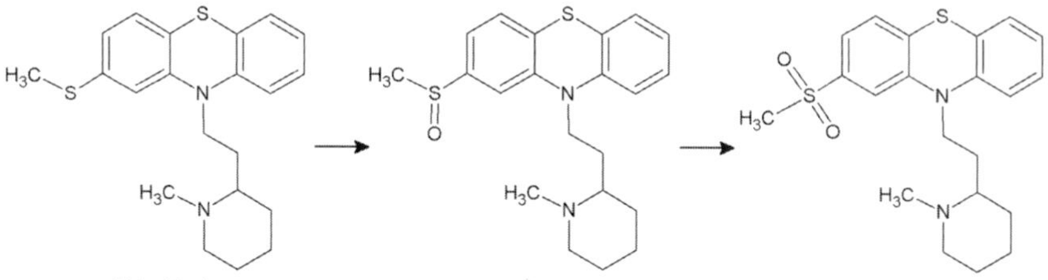

Thioridazine Mesoridazine Sulforidazine

Figure 22.5 Sequential oxidation of thioridazine.

name radafaxine. This suggests that investigation of other hydroxylated versions of the parent compound may be profitable due to their improved pharmacokinetic profiles.

3.4. Antihistamines

Terfenadine, the first nonsedating antihistamine, was initially marketed in the United States in 1985 for the treatment of allergic rhinitis. However, it was withdrawn from clinical use in the late 1990s due to concerns regarding arrhythmia caused by abnormal QT (Q wave to T wave depolarization and repolarization) interval prolongation.[60] This effect was exacerbated by pharmacokinetic drug interactions with azole antifungal agents, such as ketoconazole.[60] Terfenadine is known to undergo extensive first pass metabolism by CYP3A4, primarily producing the carboxylic acid metabolite (fexofenadine).[61] Most of the antihistaminic (H_1 receptor) effect is seen with the metabolite but fexofenadine exhibits reduced cardiac toxicity, which led to its development as a second generation nonsedating antihistamine. In this case, toxicity seen with the parent compound was avoided by an understanding of the pharmacokinetics and activity of the CYP-generated metabolite.

3.5. Muscarinic Antagonists

Tolterodine is a M_2/M_3 muscarinic receptor antagonist used to treat urinary incontinence and overactive bladder.[62] It is predominately metabolized by CYP2D6 to produce 5-hydroxymethyltolerodine (desfesoterodine; Fig. 22.6).[63]

Remarkably, although both the parent and metabolite exhibit similar antimuscarinic activities, the metabolite is 10-fold less protein bound.[64] Therefore, in individuals possessing CYP2D6 extensive metabolizer phenotypes, the desfesoterodine metabolite is predominantly responsible for the

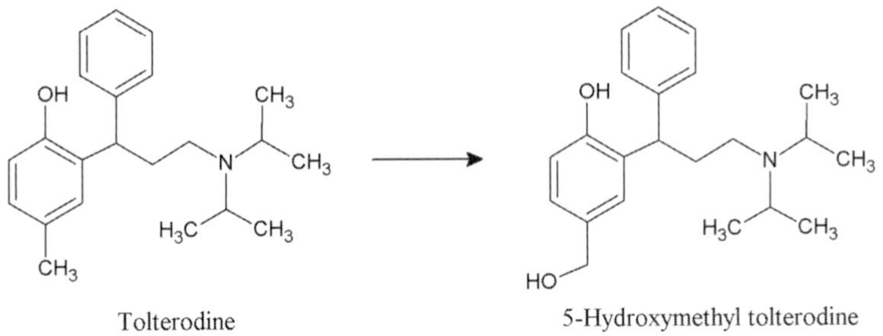

Tolterodine 5-Hydroxymethyl tolterodine

Figure 22.6 Metabolism of tolterodine to its active metabolite, 5-hydroxymethyl tolterodine.

pharmacological effects observed.[63,64] Conversely, in poor metabolizers, the parent compound is the major contributor. Additionally, animal studies have demonstrated that less of the metabolite than the parent drug penetrates the blood–brain barrier, suggesting that the metabolite is less likely to cause off-target central nervous system effects. These data led to the development of fesoterodine, an ester prodrug of the metabolite. This example illustrates how interindividual CYP genetic variability can affect the efficacy through production of an active metabolite.

3.6. Antineoplastic Agents

The tamoxifen CYP2D6 metabolites, norendoxifen, afimoxifene, and endoxifen, were demonstrated to be potent inhibitors of CYP19A1 (aromatase) with a K_i for norendoxifen of 35 nM.[9] Extensive CYP2D6 metabolizers are thus more responsive to therapy with tamoxifen than moderate or poor metabolizers. Additionally, norendoxifen exhibited excellent selectivity for CYP19A1 inhibition when screened against a variety of drug metabolizing CYPs, including CYP2B6, 2C9, 2C19, 2D6, and 3A4.[9] Thus, active metabolites like norendoxifen contribute directly to the anticancer effect seen with this drug and may also serve as potent and selective lead compounds upon which to build novel CYP19A1 inhibitors.

4. FUTURE DIRECTIONS AND CONCLUSIONS

One of the more innovative and exciting approaches to the identification and exploitation of active CYP metabolites has been the development of microscale and microfluidic technologies for the production and analysis of CYP metabolic products.[10] Using these technologies, it is possible to screen a number of CYP-generated metabolites simultaneously. These new approaches will not only accelerate the pace of CYP metabolite lead discovery but also dramatically reduce its cost. Microscale, and other, emerging technologies hold the promise of incorporating CYP metabolite lead identification into part of the drug discovery process.

CYP enzymes are found in every kingdom of life and are essential for many important biological functions, including endogenous and xenobiotic metabolism, secondary metabolite production, and organism homeostasis. In drug discovery and development, they have traditionally been considered a liability. However, the multitalented CYPs can also be a platform for drug discovery and development through the production of active metabolites, natural product lead synthesis, or even as drug targets. As new technologies emerge for metabolite production and characterization, the number of drug candidates produced from CYPs is likely to increase substantially.

ACKNOWLEDGMENTS

This work was supported by the National Center of Research Resources (Grant P20-RR021940), the National Institute of General Medical Sciences [Grant P20- GM103549-07], and the Kansas IDeA Network of Biomedical Research Excellence (Grant QH846868-K-INBRE; J. N. L.) from the National Institutes of Health.

REFERENCES

1. Klingenberg, M. *Arch. Biochem. Biophys.* **1958**, *75*, 376–386.
2. Omura, T.; Sato, R. *J. Biol. Chem.* **1962**, *237*, 1375–1376.
3. Silverman, R. B. *The Organic Chemistry of Drug Design and Drug Action*, 2nd ed.; Elsevier: San Francisco, 2004.
4. Ortiz de Montellano, P. R. *Cytochrome P450: Structure, Mechanism, and Biochemistry*, 3rd ed.; Kluwer Academic/Plenum Publishers: New York, 2005.
5. Ortiz de Montellano, P. R. *Chem. Rev.* **2010**, *110*, 932–948.
6. Guengerich, F. P.; Munro, A. W. *J. Biol. Chem.* **2013**, *288*, 17065–17073.
7. Fuchs, D. A.; Johnson, R. K. *Cancer Treat Rep.* **1978**, *62*, 1219–1222.
8. *Chin. Med. J. (Engl.)* **1979**, *92*, 811–816.
9. Lu, W. J.; Xu, C.; Pei, Z.; Mayhoub, A. S.; Cushman, M.; Flockhart, D. A. *Breast Cancer Res. Treat.* **2012**, *133*, 99–109.
10. Cusack, K. P.; Koolman, H. F.; Lange, U. E.; Peltier, H. M.; Piel, I.; Vasudevan, A. *Bioorg. Med. Chem. Lett.* **2013**, *23*, 5471–5483.
11. Wollenberg, L. A.; Kabulski, J. L.; Powell, M. J.; Chen, J.; Flora, D. R.; Tracy, T. S.; Gannett, P. M. *Appl. Biochem. Biotechnol.* **2014**, *172*, 1293–1306.
12. Morant, M.; Bak, S.; Moller, B. L.; Werck-Reichhart, D. *Curr. Opin. Biotechnol.* **2003**, *14*, 151–162.
13. Engels, B.; Dahm, P.; Jennewein, S. *Metab. Eng.* **2008**, *10*, 201–206.
14. Holton, R. A. *J. Am. Chem. Soc.* **1984**, *106*, 5731–5732.
15. Parekh, S.; Srinivasan, V.; Horn, M. *Adv. Appl. Microbiol.* **2008**, *63*, 105–143.
16. Dejong, J. M.; Liu, Y.; Bollon, A. P.; Long, R. M.; Jennewein, S.; Williams, D.; Croteau, R. B. *Biotechnol. Bioeng.* **2006**, *93*, 212–224.
17. Wu, Z. L.; Aryal, P.; Lozach, O.; Meijer, L.; Guengerich, F. P. *Chem. Biodivers.* **2005**, *2*, 51–65.

18. Gillam, E. M.; Notley, L. M.; Cai, H.; De Voss, J. J.; Guengerich, F. P. *Biochemistry* **2000**, *39*, 13817–13824.
19. Polychronopoulos, P.; Magiatis, P.; Skaltsounis, A. L.; Myrianthopoulos, V.; Mikros, E.; Tarricone, A.; Musacchio, A.; Roe, S. M.; Pearl, L.; Leost, M.; Greengard, P.; Meijer, L. *J. Med. Chem.* **2004**, *47*, 935–946.
20. Ro, D. K.; Paradise, E. M.; Ouellet, M.; Fisher, K. J.; Newman, K. L.; Ndungu, J. M.; Ho, K. A.; Eachus, R. A.; Ham, T. S.; Kirby, J.; Chang, M. C.; Withers, S. T.; Shiba, Y.; Sarpong, R.; Keasling, J. D. *Nature* **2006**, *440*, 940–943.
21. Paddon, C. J.; Westfall, P. J.; Pitera, D. J.; Benjamin, K.; Fisher, K.; McPhee, D.; Leavell, M. D.; Tai, A.; Main, A.; Eng, D.; Polichuk, D. R.; Teoh, K. H.; Reed, D. W.; Treynor, T.; Lenihan, J.; Fleck, M.; Bajad, S.; Dang, G.; Dengrove, D.; Diola, D.; Dorin, G.; Ellens, K. W.; Fickes, S.; Galazzo, J.; Gaucher, S. P.; Geistlinger, T.; Henry, R.; Hepp, M.; Horning, T.; Iqbal, T.; Jiang, H.; Kizer, L.; Lieu, B.; Melis, D.; Moss, N.; Regentin, R.; Secrest, S.; Tsuruta, H.; Vazquez, R.; Westblade, L. F.; Xu, L.; Yu, M.; Zhang, Y.; Zhao, L.; Lievense, J.; Covello, P. S.; Keasling, J. D.; Reiling, K. K.; Renninger, N. S.; Newman, J. D. *Nature* **2013**, *496*, 528–532.
22. Hartsel, S.; Bolard, J. *Trends Pharmacol. Sci.* **1996**, *17*, 445–449.
23. Cheron, M.; Cybulska, B.; Mazerski, J.; Grzybowska, J.; Czerwinski, A.; Borowski, E. *Biochem. Pharmacol.* **1988**, *37*, 827–836.
24. Carmody, M.; Murphy, B.; Byrne, B.; Power, P.; Rai, D.; Rawlings, B.; Caffrey, P. *J. Biol. Chem.* **2005**, *280*, 34420–34426.
25. Finkelstein, E.; Amichai, B.; Grunwald, M. H. *Int. J. Antimicrob. Agents* **1996**, *6*, 189–194.
26. Cacho, R. A.; Chooi, Y. H.; Zhou, H.; Tang, Y. *ACS Chem. Biol.* **2013**, *8*, 2322–2330.
27. Weeks, A. M.; Chang, M. C. *Biochemistry* **2011**, *50*, 5404–5418.
28. Mueller, G. C.; Miller, J. A. *J. Biol. Chem.* **1948**, *176*, 535–544.
29. Brodie, B. B.; Axelrod, J.; Soberman, R.; Levy, B. B. *J. Biol. Chem.* **1949**, *179*, 25–29.
30. Daniel, W.; Netter, K. J. *Naunyn–Schmiedeberg's Arch. Pharmacol.* **1988**, *337*, 105–110.
31. Heikinheimo, O. *Clin. Pharmacokinet.* **1997**, *33*, 7–17.
32. Fura, A. *Drug Discov. Today* **2006**, *11*, 133–142.
33. Messenger, A. G.; Rundegren, J. *Br. J. Dermatol.* **2004**, *150*, 186–194.
34. Pluss, R. G.; Orcutt, J.; Chidsey, C. A. *J. Lab. Clin. Med.* **1972**, *79*, 639–647.
35. Buhl, A. E.; Waldon, D. J.; Baker, C. A.; Johnson, G. A. *J. Invest. Dermatol.* **1990**, *95*, 553–557.
36. Simons, J. *Fortune* **2003**, *147*, 58–62, 66, 68.
37. Endo, A. *J. Lipid Res.* **1992**, *33*, 1569–1582.
38. Brown, A. G.; Smale, T. C.; King, T. J.; Hasenkamp, R.; Thompson, R. H. *J. Chem. Soc. Perkin* **1976**, *1*, 1165–1170.
39. Roche, V. F. *Am. J. Pharm. Educ.* **2005**, *69*, 546–560.
40. Alberts, A. W.; Chen, J.; Kuron, G.; Hunt, V.; Huff, J.; Hoffman, C.; Rothrock, J.; Lopez, M.; Joshua, H.; Harris, E.; Patchett, A.; Monaghan, R.; Currie, S.; Stapley, E.; Albers-Schonberg, G.; Hensens, O.; Hirshfield, J.; Hoogsteen, K.; Liesch, J.; Springer, J. *Proc. Natl. Acad. Sci. U.S.A.* **1980**, *77*, 3957–3961.
41. Pfizer; Pfizer, Ed.; Pfizer: Groton, 2013; Vol. 2013, p Web site describes the cholesterol lowering agent lipitor.
42. AstraZeneca; AstraZeneca, Ed.; AstraZeneca, Inc.: London, UK, 2012; Vol. 2014, p Web site describes the cholesterol lowering agent crestor.
43. Lennernas, H. *Clin. Pharmacokinet.* **2003**, *42*, 1141–1160.
44. Lins, R. L.; Matthys, K. E.; Verpooten, G. A.; Peeters, P. C.; Dratwa, M.; Stolear, J. C.; Lameire, N. H. *Nephrol. Dial. Transplant.* **2003**, *18*, 967–976.

45. Jacobsen, W.; Kuhn, B.; Soldner, A.; Kirchner, G.; Sewing, K. F.; Kollman, P. A.; Benet, L. Z.; Christians, U. *Drug Metab. Dispos.* **2000**, *28*, 1369–1378.
46. Mason, R. P.; Walter, M. F.; Day, C. A.; Jacob, R. F. *J. Biol. Chem.* **2006**, *281*, 9337–9345.
47. Schneck, D. W.; Birmingham, B. K.; Zalikowski, J. A.; Mitchell, P. D.; Wang, Y.; Martin, P. D.; Lasseter, K. C.; Brown, C. D.; Windass, A. S.; Raza, A. *Clin. Pharmacol. Ther.* **2004**, *75*, 455–463.
48. Meyer, U. A.; Zanger, U. M.; Schwab, M. *Annu. Rev. Pharmacol. Toxicol.* **2013**, *53*, 475–502.
49. Khera, S.; Hu, N. *Anal. Bioanal. Chem.* **2013**, *405*, 6009–6018.
50. Chooi, Y. H.; Hong, Y. J.; Cacho, R. A.; Tantillo, D. J.; Tang, Y. *J. Am. Chem. Soc.* **2013**, *135*, 16805–16808.
51. Washington, J. A., 2nd.; Wilson, W. R. *Mayo Clin. Proc.* **1985**, *60*, 189–203.
52. Peeters, T.; Matthijs, G.; Depoortere, I.; Cachet, T.; Hoogmartens, J.; Vantrappen, G. *Am. J. Physiol.* **1989**, *257*, G470–G474.
53. Steinmetz, W. E.; Shapiro, B. L.; Roberts, J. J. *J. Med. Chem.* **2002**, *45*, 4899–4902.
54. Kohno, Y.; Yoshida, H.; Suwa, T.; Suga, T. *Antimicrob. Agents Chemother.* **1989**, *33*, 751–756.
55. Andersen, J. F.; Tatsuta, K.; Gunji, H.; Ishiyama, T.; Hutchinson, C. R. *Biochemistry* **1993**, *32*, 1905–1913.
56. Shinde, P. B.; Han, A. R.; Cho, J.; Lee, S. R.; Ban, Y. H.; Yoo, Y. J.; Kim, E. J.; Kim, E.; Song, M. C.; Park, J. W.; Lee, D. G.; Yoon, Y. J. *J. Biotechnol.* **2013**, *168*, 142–148.
57. Svendsen, C. N.; Bird, E. D. *Psychopharmacology (Berl)* **1986**, *90*, 316–321.
58. Richtand, N. M.; Welge, J. A.; Logue, A. D.; Keck, P. E., Jr.; Strakowski, S. M.; McNamara, R. K. *Neuropsychopharmacology* **2007**, *32*, 1715–1726.
59. Dockens, R. C.; Salazar, D. E.; Fulmor, I. E.; Wehling, M.; Arnold, M. E.; Croop, R. *J. Clin. Pharmacol.* **2006**, *46*, 1308–1312.
60. Thompson, D.; Oster, G. *JAMA* **1996**, *275*, 1339–1341.
61. Lalonde, R. L.; Lessard, D.; Gaudreault, J. *Pharm. Res.* **1996**, *13*, 832–838.
62. Van Kerrebroeck, P.; Kreder, K.; Jonas, U.; Zinner, N.; Wein, A. *Urology* **2001**, *57*, 414–421.
63. Brynne, N.; Dalen, P.; Alvan, G.; Bertilsson, L.; Gabrielsson, J. *Clin. Pharmacol. Ther.* **1998**, *63*, 529–539.
64. Pahlman, I.; Gozzi, P. *Biopharm. Drug Dispos.* **1999**, *20*, 91–99.

SECTION 8

Case Histories and NCEs

Section Editor: Joanne Bronson
Bristol-Myers Squibb, Wallingford, Connecticut

CHAPTER TWENTY-THREE

Case History: Forxiga™ (Dapagliflozin), a Potent Selective SGLT2 Inhibitor for Treatment of Diabetes

William N. Washburn
Bristol-Myers Squibb, Princeton, New Jersey, USA

Contents

1. INTRODUCTION

Type 2 diabetes mellitus (T2DM) is a major health issue that threatens to overwhelm the health resources of nations worldwide. Currently, 382 million patients are estimated to have T2DM; however, this number is projected to rise to 592 million by 2035 as an increasing percentage of the world's population adopts a Western diet and sedentary lifestyle.[1] T2DM is a metabolic disease in which blood glucose levels become elevated as a consequence of a progressive decrease in the body's responsiveness to insulin, a pancreatic hormone that plays a fundamental role in glucose regulation. During the initial stages of the disease, peripheral insulin resistance and improper first-phase insulin release cause the pancreas to secrete ever-increasing amounts of insulin to maintain normal postprandial glucose levels. Eventually, the required

Annual Reports in Medicinal Chemistry, Volume 49
ISSN 0065-7743
http://dx.doi.org/10.1016/B978-0-12-800167-7.00023-7

insulin levels exceed the pancreatic capacity, resulting in the onset of hyperglycemia, the hallmark of T2DM. Over time, the burden from glycosylation of tissue proteins and altered glucose metabolism results in neuropathy, retinopathy, nephropathy, and microvascular and macrovascular complications which give rise to an increased incidence of blindness, gangrene, renal failure, stroke, and cardiovascular (CV) events.

For a normal healthy individual, the liver, pancreas, skeletal muscle, adipose deposits, and kidney are the organs that are primarily responsible for glucose homeostasis. Current widely utilized treatments target these organs, with the exception of the kidney, to (1) increase glucose cellular uptake, (2) increase glucose metabolism, or (3) decrease hepatic glucose output. Since most current antidiabetic agents require insulin in order to be effective, monotherapy drugs will eventually fail to adequately control the patient's blood sugar as insulin levels decline due to progressive exhaustion and death of pancreatic beta cells. Although antidiabetic drug combinations can compensate for declining insulin levels, the need for new modes of treatment that can be safely combined with current agents is underscored by the inability of ~40% of the T2DM patient population to maintain glycated hemoglobin (hemoglobin A1c, HbA1c) levels at the American Diabetes Association recommended goal of <7.0 with existing therapies.[2,3]

Sodium–glucose cotransporter 2 (SGLT2) inhibitors are an attractive class of antidiabetic agents because, unlike current therapies, they offer a means to utilize the kidney to alleviate hyperglycemia.[4–6] Moreover, since the mechanism of action of SGLT2 inhibitors is not dependent on insulin, efficacy should not decline over time. In addition, coadministration with other agents would be complementary and present minimal risk of hyperglycemia since the mechanism of action of SGLT2 inhibitors is orthogonal to that of insulin-dependent antidiabetic agents.[7]

2. RENAL RECOVERY OF GLUCOSE

In the course of removal of metabolic waste materials, the kidneys of a healthy individual filter ~180 g of glucose from ~180 L of blood daily.[8,9] However, as the glomerular filtrate passes through the proximal tubules, virtually all glucose dissolved in the filtrate is recovered by the action of SGLT1 and SGLT2 residing in the lumen-facing apical membranes of the epithelial cell surface. Both transporters couple the transport of glucose with the transport of Na^+ using the free energy gained from Na^+ transport to compensate for that required for glucose transport.[10]

As much as 90% of glucose recovery is carried out by SGLT2, a low-affinity, high-capacity glucose transporter that is expressed only in the S1 and S2 segments of the proximal tubule.[11–13] Any glucose not captured by SGLT2 is recovered by SGLT1, a high-affinity, low-capacity glucose transporter expressed in the S3 segment. SGLT1 is also expressed in the small intestine and the heart.[14] The role of SGLT1 in heart is not known; however, in the small intestine, SGLT1 is responsible for absorption of glucose and galactose. Pursuit of selective SGLT2 inhibitors was supported by the profile of individuals with familial renal glucosuria (excretion of glucose into the urine). Despite lacking functional SGLT2 transporters, these individuals are healthy in all aspects except for being profoundly glucosuric, spilling as much as 140 g of glucose daily in their urine.[15,16] In contrast, SGLT1 inhibition appears undesirable since individuals with nonfunctional SGLT1 transporters have glucose–galactose malabsorption, which is manifested by severe diarrhea.[17]

Figure 23.1 illustrates renal glucose recovery prior to and after SGLT2 inhibition. The rate of glucose filtration increases linearly with glucose concentration in blood. The recovery rate is essentially coincident with the filtration rate; however, the recovery capacity is finite. Once plasma glucose concentrations exceed the transport maximum (T_m), glucosuria increases linearly with blood glucose levels. An SGLT2 inhibitor would reversibly lower the glucose threshold value from ~180 to 200 mg/dL to a much lower value depending on the extent of inhibition, thereby providing a noninsulin-dependent mechanism to reduce blood sugar levels. Reduction

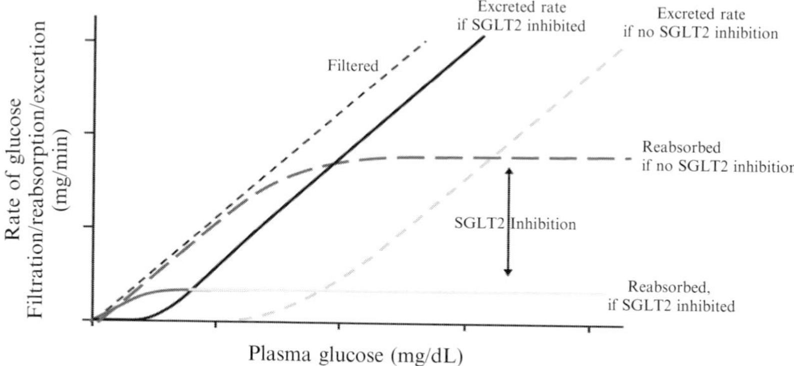

Figure 23.1 Renal processing of glucose: dependence of rates of filtration, recovery, and excretion on plasma glucose concentration in absence and presence of an SGLT2 inhibitor. (See the color plate.)

of glycemic levels by this mechanism is dependent on the mass of glucose excreted daily in urine, which will be proportional to the volume of glomerular filtrate, the glucose concentration in the filtrate, and both the extent and duration of inhibition over the 24-h period. The efficacy of an SGLT2 inhibitor will be decreased for T2DM patients with diminished renal function.

3. O-GLUCOSIDE SGLT2 INHIBITORS

Although the glucosuric response induced by O-glucoside phlorizin **1**, a naturally occurring, competitive inhibitor of SGLT1 and SGLT2, has been well documented, applicability to treatment of diabetes was not widely appreciated prior to its investigation by Rosetti in diabetic rodents and human volunteers, which suggested that phlorizin potentially could provide an effective mechanism for treating T2DM if its poor ADME (absorption, distribution, metabolism and excretion) properties were improved.[18] It was widely recognized that glucosidase cleavage of the labile O-glucoside bond of phlorizin contributed to its unfavorable ADME profile (low bioavailability and rapid clearance).[19] Pharmaceutical industry involvement did not begin until SGLT1 and SGLT2 were identified, cloned, and shown to be nonselectively inhibited by phlorizin. Identification of the SGLT1 and SGLT2 transporters and establishment of the predominant role of the SGLT2 transporter for renal glucose recovery provided a new antidiabetic target for medicinal chemists.

The discovery program at Bristol-Myers Squibb utilized two disclosures as starting points to identify a selective SGLT2 inhibitor amenable to once daily oral administration. The first was a disclosure from Tanabe showing that more lipophilic dihydrochalcone analogs of phlorizin, as exemplified by **2a**, had improved bioavailability when administered as methyl carbonate prodrugs (e.g., **2b**) to suppress glucosidase-mediated degradation prior to absorption.[20] The second was a report from Wyeth showing that oral administration of benzyl pyrazolones such as **3** produced a glucosuric response in mice (Fig. 23.2).[21]

Although **3** had glucosuric activity in mice, it was inactive in rats. A critical development in our program was the realization that a mouse-specific conversion of the benzylpyrazolone **3** to an active glycosylated metabolite, which we hypothesized was the O-glucoside **4**, would account for the glucosuric response observed for mice after oral administration of **3**. Conversely, the absence of the active metabolite in rats would be consistent with the inactivity

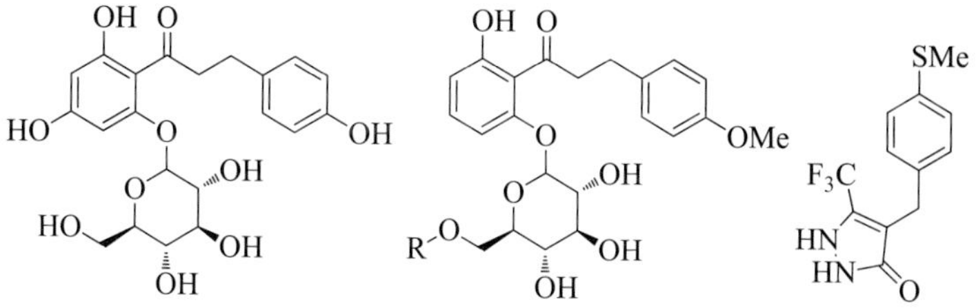

1 Phlorizin

2a R = H
2b R = C(O)OMe

3

Figure 23.2 Glucosuric agents disclosed prior to 1999.

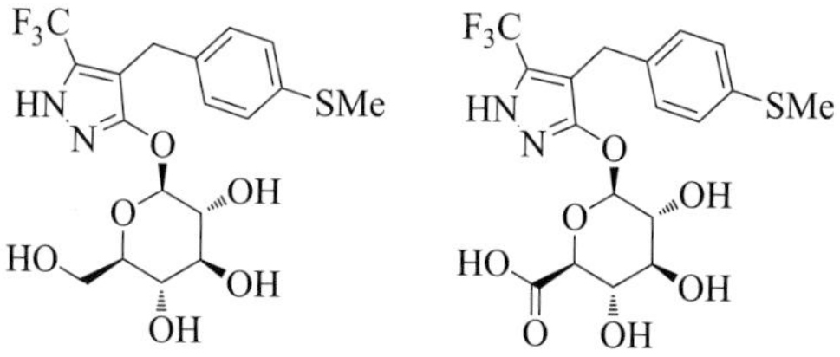

4

5

Figure 23.3 Pyrazolone O-glycosides.

in this species. The requirement for *in vivo* conversion to an active species would also explain the lack of SGLT *in vitro* activity exhibited by **3**. Subsequent characterization confirmed that **4** was a potent SGLT2 inhibitor (22 nM SGLT2 EC_{50} with 15-fold selectivity vs. SGLT1) that was rapidly cleaved by glucosidases. After oral administration of **3** or **4** to mice and rats, analysis of rat urine revealed only glucuronide **5**, which was devoid of SGLT activity, whereas mouse urine contained a mixture of **4** and **5**. This discovery that **3** was glucosylated and glucuronidated in mice but only glucuronidated in rats not only explained the species-dependent glucosuric response but also discouraged further pursuit of the pyrazalone O-glucosides since glucosylation is a rarely utilized metabolic pathway in humans (Fig. 23.3).

The SGLT2 inhibitory and glucosuric properties of the O-glucosides **2a** and **4** provided insight into a pharmacophore model for SGLT2 inhibition (Fig. 23.4). Both **2a** and **4** are O-glucosides of a hydroxylated, vicinally substituted central planar ring linked via a spacer to a distal aryl ring. This proposed pharmacophore requires vicinal substitution of the central aryl ring to achieve proper spatial orientation of the two pendant rings. This hypothesis predicted that O-glucosides of *ortho* benzylphenols **6**, which are generated by merging the two structures, would be potent SGLT2 inhibitors.

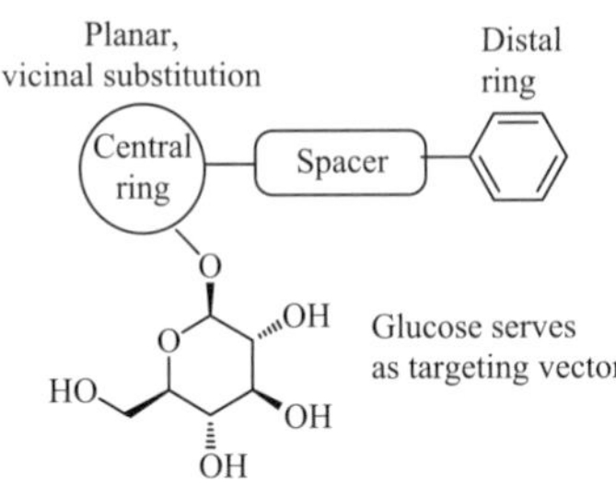

Figure 23.4 Pharmacophore for O-glucoside SGLT2 inhibitor.

The *in vitro* potency of **6a** and related analogs was superior to any previously reported O-glucosides, with several examples having SGLT2 EC_{50}s $\leq$ 10 nM and $\geq$100-fold selectivity versus SGLT1 (Table 23.1). *Meta* or *ortho* substitution on the distal ring decreased SGLT2 inhibitory potency by 3- and 12-fold, respectively, whereas lipophilic *para* substituents enhanced SGLT2 potency as much as 20-fold over that of the parent **6a** (Table 23.1). The finding that large extended *para* substituents such as styrenyl (**6g**) were better accommodated than a branched alkyl group such as *t*-butyl (**6h**) suggested that the distal ring inserted into a deep and narrow binding pocket. Methylation of C3′ or C5′ of the central aryl ring of **6** did not alter SGLT2 potency, but did increase SGLT1 affinity 5- to 10-fold. In contrast, substitution at C4′ or C6′ reduced SGLT2 potency 15- to 50-fold.

Although the lead compound **6e** was not hydrolyzed by gut glucosidases, it was susceptible to glucosidases in rat kidney and monkey liver, raising the possibility that the extent of glucosuria could be species dependent. This concern was heightened upon finding that the glucosuric response produced upon oral administration of 1 mg/kg of **6e** to mice was 10 and 100 times greater than that obtained with rats and monkeys, respectively. Moreover, the predominant species present in blood and urine of mice was **6e**, whereas in rats, it was the corresponding glucuronide. The subsequent finding that oral administration of 2-(4-methylbenzyl)phenol, the aglycone of **6e**, induced dose-dependent glucosuria in mice confirmed our concern that glucosuric responses obtained with mice would overestimate the response in humans. These findings greatly diminished interest in **6e** because a weak glucosuric response in humans was predicted based on the expected similarity in metabolism of **6e** by humans to that of rats and monkeys.

The susceptibility of **6e** and related analogs to glucosidase-mediated degradation prompted pursuit of three possible avenues to improve metabolic stability. First, substitution at C3′ or C6′ to provide steric hindrance and

Table 23.1 SGLT SAR (Structure Activity Relationships) for O-glucosides of *ortho* benzylphenols **6**

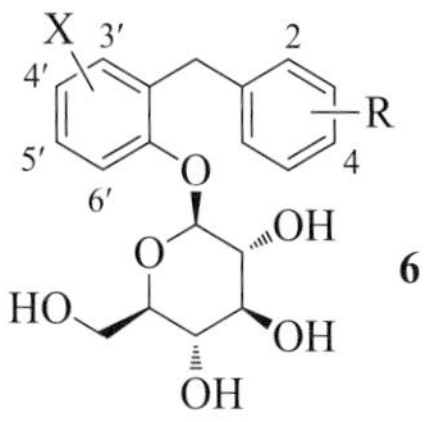

Compound #	X	R	SGLT2 EC_{50} (nM)	SGLT1 EC_{50} (nM)	Selectivity vs. SGLT1
6a	H	H	41	2400	60
6b	H	2-OMe	570	>10,000	
6c	H	3-OMe	136	>10,000	
6d	H	4-OMe	10	>10,000	
6e	H	4-Me	8	2400	300
6f	H	4-Et	2	585	290
6g	H	4-Styrenyl	18	2700	150
6h	H	4-*t*-Butyl	65	>10,000	
6i	3′-Me	4-Me	10	250	25
6j	6′-Me	4-Me	49	>10,000	>200

disfavor glucosidase cleavage offered some promise as **6i** and **6j** were markedly less susceptible to cleavage and exhibited 2- to 3-fold enhanced glycemic reduction. Second, C-glucosides based on replacement of the anomeric oxygen of **6e** with a methylene as in **7** or removal of the anomeric oxygen as in **8** (Fig. 23.5) would be expected to improve metabolic stability, although SGLT2 potency for **7** and **8** was reduced by 60- and 1000-fold, respectively. Third, an unanticipated C-glucoside side product **17** (see below) isolated during exploration of *o*-hydroxybenzamide O-glucosides early in the program suggested an alternative C-glucoside motif with potential for improved potency.

At the start of the program, we investigated whether **10**, the O-glucoside benzamide counterpart of **2**, offered any advantage with respect to potency,

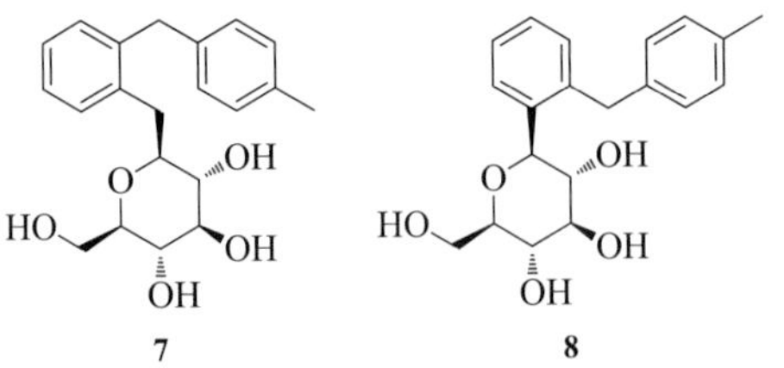

Figure 23.5 C-Glucoside analogs.

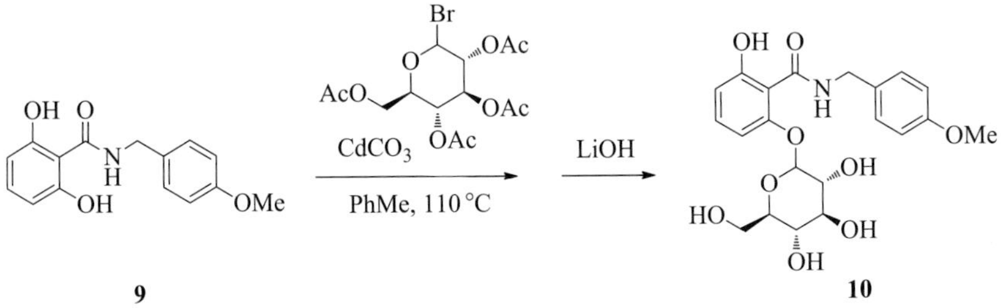

Scheme 23.1 Synthesis of **10**.

selectivity, and stability to glucosidases. Aglycone **9** was converted to **10** by analogy to the literature route reported for **2a** (Scheme 23.1). Despite an 8-fold loss in SGLT2 potency relative to **2a**, the profile of **10** appeared promising as SGLT1 selectivity was improved 3-fold and **10** was not hydrolyzed by glucosidases.

Encouraged by this finding, the SAR of the amide series was briefly explored but found to exhibit little variation in potency. Given the poor prognosis for increasing potency and the failure to observe significant glucosuria after oral administration to Swiss Webster mice, further investigation of the benzamide phenolic O-glucoside series ceased. Prior to termination of this series, we found that subjecting the tertiary phenolic benzamide **11** to the same glucosylation conditions employed for **9** failed to generate any **12** but did form four unexpected products in ~1% yield (Scheme 23.2). All were derived from glucosylation of aglycone **13**, which was shown to have been formed in low yield from **11** due to inadequate pH control during the glycosylation reaction. The two isomeric O-glucosides **14** and **15** as well as the bis-O-glucoside **16** were not pursued since the SGLT2 EC_{50} for each was >5000 nM. In contrast to the discouraging SGLT2 profile of previously disclosed *ortho* C-glucosides,[22] the fourth component was a C-glucoside **17** that exhibited sufficient promise (SGLT2 EC_{50} = 1300 nM; 30-fold selectivity vs. SGLT1) to warrant further consideration. At that time, we did not know if the activity of **17** was due to the

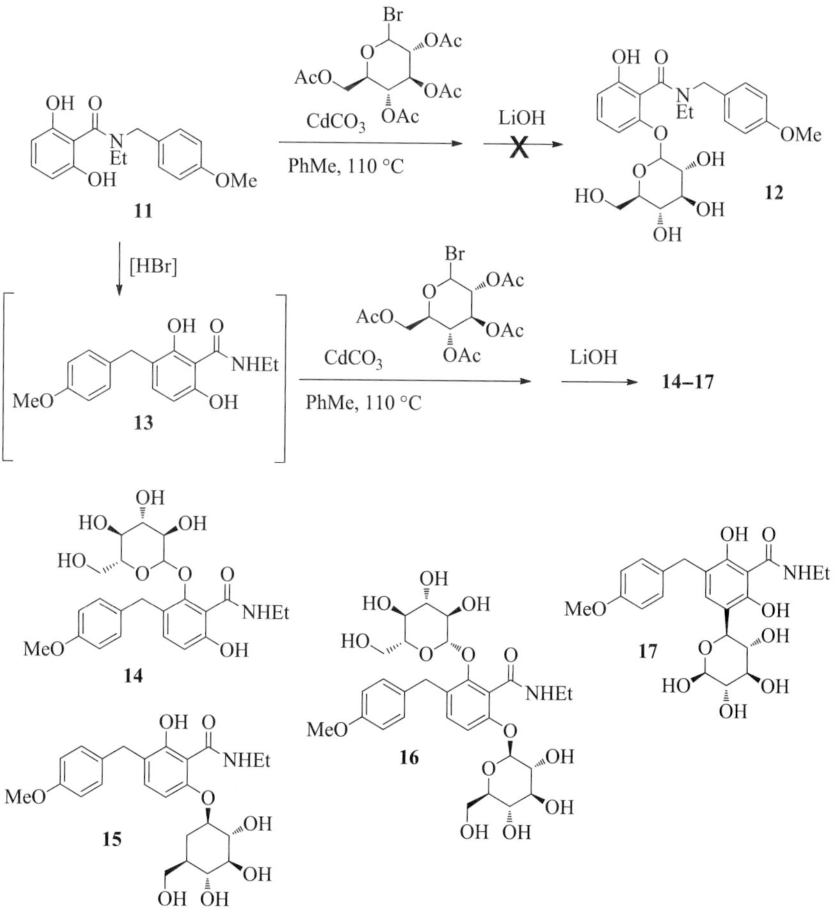

Scheme 23.2 Glycosylated products obtained with 2,6-dihydroxybenzamides.

substitution pattern of the central phenyl ring or due to the atypical polar substituents of the aglycone **13**. Further pursuit of **17** was suspended when synthetic efforts to generate more material failed.

4. C-ARYL GLUCOSIDE SGLT2 INHIBITORS

Ultimately compound **17** proved to be an essential lead for the discovery of C-glucosides such as dapagliflozin and related analogs. The structural similarity of **6** and **17** suggested that the SAR for SGLT2 activity should be similar. This realization that the polar substituents of **17** would strongly disfavor SGLT2 potency redirected the synthetic effort to the preparation of **18a**, the *meta* C-glucoside counterpart of **6e**. The direction of the program changed with the finding that **18a** was a potent SGLT2 inhibitor (EC_{50} = 22 nM) exhibiting >600-fold selectivity versus SGLT1 and no inhibition at 10 μM of GLUT 1 and GLUT 4, two members of the facilitated glucose transporter family that transport glucose across plasma membranes of mammalian cells. Computational studies with **6**, **7**, and **18** revealed that **6** and **18** can access a low energy conformation not accessible to **7**, thereby providing a possible explanation for the inactivity of **7**. Although chemotypes **6** and **18** were pursued concurrently, within a few months all effort became focused on the C-glucoside series **18** analogs due to greater *in vitro* potency and selectivity and in particular the ability to elicit greater plasma glucose reductions for longer duration in rats.

The SGLT2 SAR for distal ring substitution in *meta* C-aryl glucosides **18** was similar to that of *o*-benzylphenolic O-glucosides **6** but with the added benefit of introducing greater selectivity versus SGLT1 (EC_{50} > 10,000 nM). *ortho* substitution was unfavorable. For example, introduction of an *o*-ethyl group as in **18c** decreased potency 20-fold relative to parent **18b** (Table 23.2). *Meta* substituents such as in **18d** provided no advantage but were tolerated. Potency was

Table 23.2 SGLT SAR for C-glucosides of distal substituted diarylmethanes **18**

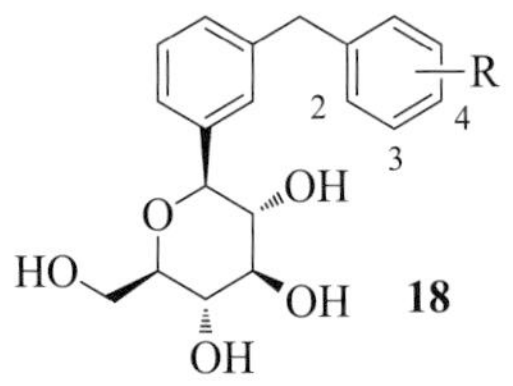

Compound #	R	SGLT2 EC_{50} (nM)	SGLT1 EC_{50} (nM)
18a	4-Me	22	>10,000
18b	H	190	>10,000
18c	2-Et	4000	
18d	3-Me	510	
18e	4-Et	10	>10,000
18f	4-OMe	12	>10,000
18g	4-SMe	6	>10,000
18h	4-CO_2H	340	
18i	4-$NHSO_2Ph$	390	

strongly enhanced by *para* substitution: small lipophilic groups such as Et (**18e**), MeO (**18f**), or MeS (**18g**) increased SGLT2 potency by 20- to 25-fold. Incorporation of polar groups such as CO_2H (**18h**) or $NHSO_2Ph$ (**18i**) did not reduce potency more than 2-fold. The α-anomer of **18e** had a 20-fold decrease in affinity compared to the β-anomer **18e**, underscoring the marked preference for β-linked aglycone.

Replacement of the methylene linker between the phenyl rings with alternative spacers reduced SGLT2 potency, suggesting that spatial orientation of the distal ring was important.[23,24] If the distal ring was unsubstituted, replacement with a longer spacer such as an ethano or propano group decreased potency 3- to 5-fold compared to **18b**. However, if the distal ring was substituted with a *para* methyl, the potency decrease with the longer spacer was 10- to 30-fold versus **18a**. This differential response suggested that the binding pocket for the aglycone will accommodate an unsubstituted distal ring in many orientations with little change in binding energy; however, there is only one conformation that will orient the *para* substituent in a pocket to maximize binding energy. Incorporation of alternative one atom spacers for the methylene was also not beneficial. Compared to **18a**, affinity decreased 25-fold for the corresponding diphenyl ether, 3-fold for the diphenyl sulfide, and 13-fold if the methylene was replaced with CHMe.

Substitution of the open positions of the central aryl ring of **18e** with a methyl group was systematically investigated (Table 23.3). Methyl groups at C5′ (**19a**) or C6′ (**19b**) produced minor SGLT2 potency increases of 2- and 5-fold, respectively, without altering SGLT1 affinity. In contrast, C2′ methylation (**19c**) decreased SGLT2 potency 10-fold with no apparent change in SGLT1 affinity. A methyl group attached at C4′ as in **19d** increased both SGLT1 and SGLT2 affinities ~15-fold. Since substituents at C4 and C4′ produced the largest synergistic increase in SGLT2 affinity, the program focused on delineation of the *in vitro* and *in vivo* SAR for >40 combinations of lipophilic substituents at those two positions.

Most of these C4- and C4′-substituted compounds (e.g., **19e, 19f, 19g**) exhibited EC_{50} values of 1–3 nM for inhibition of α-methylglucoside uptake by CHO cells transfected with human SGLT2, suggesting that the capacity of the *in vitro* assay to distinguish among very potent inhibitors had been exceeded. The compounds were rank ordered for potency *in vivo* using a rat model that employed Sprague–Dawley rats made diabetic by prior administration of streptozotocin (STZ), a pancreatic beta-cell selective toxin. After oral administration of 0.1 mg/kg of the SGLT2 inhibitor, hourly blood glucose levels were measured over 5 h during which time the

Table 23.3 SGLT SAR for methylation of central ring of C-aryl glucoside

19

Compound	R	X	SGLT2 EC_{50} (nM)	SGLT1 EC_{50} (nM)	% Decrease in blood glucose[a]
19a	5′-Me	4-Et	5.6	>4000	17
19b	6′-Me	4-Et	2.7	>8000	28
19c	2′-Me	4-Et	98	>8000	
19d	4′-Me	4-Et	0.6	600	41
19e	4′-Me	4-SMe	1.4	770	47
19f (dapaglifozin)	4′-Cl	4-OEt	1.1	1390	47
19g	4′-Cl	4-Et	0.9	810	47

[a]In STZ rats at 5 h post 0.1 mg/kg p.o. dose after vehicle correction.

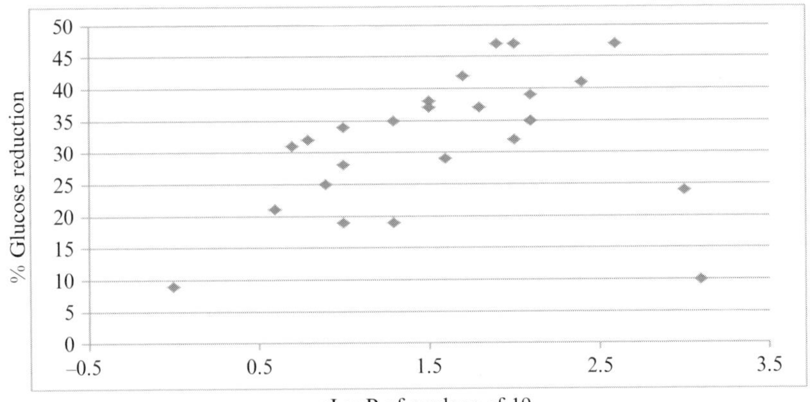

Figure 23.6 Correlation of cLogP with % reduction of blood glucose after correction for the vehicle response at 5 h after oral administration of 0.1 mg/kg dose of SGLT2 inhibitors **19** to food-restricted diabetic STZ rats. (See the color plate.)

rats were deprived of food (Fig. 23.6). Depending on the compound, the percent decrease in blood glucose from the t_0 value of ~580 mg/dL ranged from 8% to 47% after correction for the glycemic decrease of the control cohort.

In vivo efficacy appeared to exhibit a U-shaped relationship with cLogP, with the maximum response being obtained when cLogP was 2–2.5 (Fig. 23.6). Moreover, free fraction exhibited a linear dependence on cLogP with the result that free fractions of 5–10% were obtained when cLogP values were ~2–2.5 for this chemotype (data not shown). A possible explanation for the efficacy–lipophilicity relationship is that insufficient levels of drug were filtered by the kidney to achieve a maximum glucosuric response if the free fraction was too low, whereas clearance was too fast to maintain a maximum glucosuric response throughout the 5-h study if the free fraction was too high. Once this pattern was perceived, cLogP values heavily influenced which combinations of substituents were considered.

Substituent selection was subject to several other constraints. SGLT2 *in vitro* potency was greatest when the C4 substituent for the distal ring was methyl, chloro, fluoro, ethoxy, ethyl, methoxy, or thiomethyl. In addition, SGLT2 potency was enhanced if fluoro, methyl, or chloro was attached at C4′, whereas incorporation of groups larger than methoxy or ethyl at this position greatly reduced SGLT2 activity. Although distal ring C4 substituents did not impact SGLT1 potency, the central ring C4′ substituent markedly modulated SGLT1 potency. If C4′ were hydrogen or fluorine, SGLT1 binding was least favored, resulting in ~2000-fold selectivity versus SGLT1; if C4′ were methyl or chlorine, the increase in SGLT1 potency caused the selectivity to drop to ~700–1000; if C4′ were methoxy or ethyl, the selectivity decreased to ~20-fold due to the increase in SGLT1 affinity.[23]

Profiles of three compounds **19e**, **19f**, and **19g** in the STZ rat assay were particularly encouraging. A 47% reduction in blood glucose levels was seen with these compounds, which was greater than that produced by any other compound when administered at 0.1 mg/kg. In addition, dose-dependent efficacy was obtained with doses as low as 0.01 mg/kg. Upon evaluation in 24-h acute studies using ZDF (Zucker diabetic fatty) rats, which are genetically predisposed to become diabetic, **19e, 19f**, and **19g** produced comparable and robust glycemic reductions that were maintained throughout a subsequent 2-week subchronic study. Compound **19f** ultimately replaced the original lead **19e** after **19e** was found to exhibit low but unacceptable oxidative instability when formulated with typical pharmaceutical excipients. Compound **19f** began clinical development as BMS-512148; however, it later became known as dapagliflozin.[25] Compound **19g** (BMS-655956) entered clinical trials as a back-up for dapagliflozin, but its development was halted after completion of a SAD study due to the unimpeded progression of dapagliflozin through Phase II and Phase III studies.

5. SYNTHESIS OF DAPAGLIFLOZIN

The current synthetic route for dapagliflozin portrayed in Scheme 23.3 incorporates many process improvements over the initial route used to prepare *m*-diarylmethane C-glucosides.[26,27] In the original route, benzyl ether-protecting groups were employed throughout the synthesis until removal by hydrogenolysis in the final step. Et_3SiH reduction of the intermediary *O*-methyl glucoside generated a 5:1 mixture of β/α protected aryl glucosides. As a consequence of all intermediates being syrups, purification was challenging for the desired β-C-aryl glucosides, which exist as amorphous solids. In the optimized route, trimethylsilyl ethers were employed during the organolithium addition to the protected gluconolactone. Prior to reduction of the crude *O*-methyl glucoside, the material was peracetylated. Under the specified conditions requiring addition of one equivalent of water, Et_3SiH reduction of the *O*-methyl glucoside yielded almost exclusively the crystalline tetraacetoxy β-C-glucoside, thereby enabling purification prior to alkaline hydrolysis to liberate the final product. To date, all attempts to induce crystallization of these amorphous C-aryl glucosides have failed. However, after considerable effort, the process group found that these C-glucosides will cocrystallize in the presence of L-phenylalanine or L-proline, or in some cases with diols to form stable 1:1 complexes which can be purified to homogeneity by repeated crystallizations.

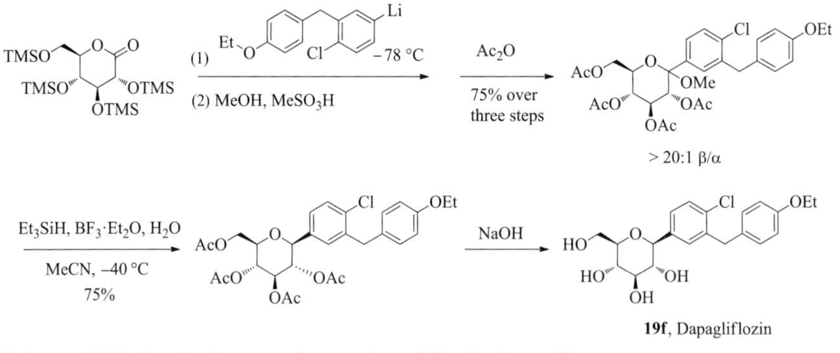

TMSO, TMSO, OTMS, OTMS — (1) Et–O–C₆H₄–CH₂–C₆H₃(Cl)–Li, −78 °C; (2) MeOH, $MeSO_3H$ — Ac_2O, 75% over three steps — AcO, AcO, OAc, OAc, OMe, Cl, OEt; > 20:1 β/α

Et_3SiH, $BF_3 \cdot Et_2O$, H_2O; MeCN, −40 °C; 75% — AcO, AcO, OAc, OAc, Cl, OEt — NaOH — HO, HO, OH, OH, Cl, OEt

19f, Dapagliflozin

Scheme 23.3 Synthetic route for employed for C-glucosides.

6. PRECLINICAL PROFILING STUDIES WITH DAPAGLIFLOZIN

Further *in vitro* characterization of dapagliflozin revealed it to be a reversible, potent SGLT2 inhibitor ($K_i = 0.55$ nM) with high selectivity versus all other SGLT transporters.[28] Dose-dependent glucosuria accompanied by as much as a 10-fold increase in urine volume was observed during a 6-h study with ZDF rats that were food-restricted following oral administration of dapagliflozin at 0.01, 0.1, 1, and 10 mg/kg. The amount of glucose spilt in urine per 400 g of body weight increased from 0.24 g for vehicle treated control animals to 2.2, 4.1, 4.2, and 4.2 g, respectively, at the increasing doses of dapagliflozin.[29] As a consequence of the drug-induced glucosuria, plasma glycemic levels of the ZDF rats receiving 1 and 10 mg/kg of dapagliflozin were effectively normalized at 6-h postdose; in addition, the 0.1 mg/kg dose achieved near-normalized glucose levels. At 24-h postdose, all cohorts except the group receiving the 0.01 mg/kg dose exhibited 20–60% glycemic reductions. A subsequent 2-week study with dapagliflozin administered once daily to ZDF rats for 2 weeks at 0.01, 0.1, and 1 mg/kg confirmed that robust dose-dependent plasma glucose reductions were maintained throughout the study under fed or fasting conditions (Fig. 23.7). No evidence for hypoglycemia was detected even after the animals experienced a severe overnight fast. After completion of this 2-week study, the profound insulin resistance due to glucotoxicity prior to treatment appeared to have been ameliorated based on hyperinsulinemic–euglycemic clamp studies showing that the glucose disposal rate had increased and that hepatic glucose output had decreased. Pancreatic staining revealed an increase in both insulin content and beta-cell mass suggestive of beta-cell

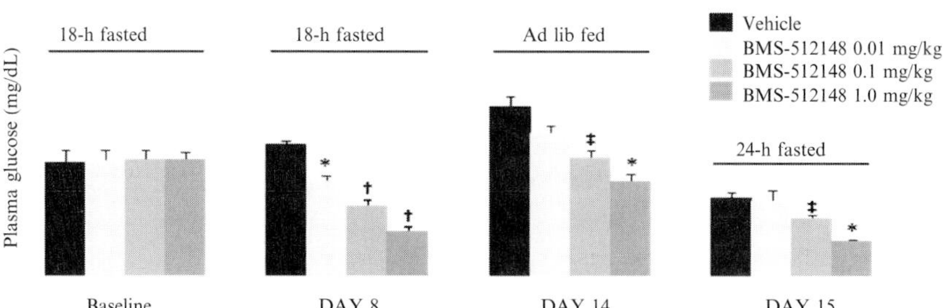

Figure 23.7 Dose-dependent reductions in plasma glucose of ZDF rats during a 2-week study with dapagliflozin (**19f**) administered daily. $^{*}p < 0.0001$; $^{+}p < 0.05$.

preservation. In summary, these pharmacological findings obtained using modestly glucosuric ZDF diabetic rats supported advancement of dapagliflozin.

A subsequent 4-week study assessed weight loss for *ad lib* fed diet-induced obese Sprague–Dawley rats when dapagliflozin was administered p.o. at 0.5, 1, and 5 mg/kg once daily.[30] In addition, a second cohort administered 5 mg/kg daily of dapagliflozin was included which was pair fed (daily rations were the same as that consumed by the vehicle control on the prior day). All *ad lib* fed-treated cohorts increased their water and food intake; however, the caloric increase did not fully compensate for the calories lost as glucosuria (Table 23.4). The decrease in respiratory quotient, which indicates a metabolic shift requiring increased reliance on fatty acids as an alternative energy source, was consistent with the weight loss being primarily due to loss of fat mass. It is noteworthy that comparison of the two cohorts receiving 5 mg/kg of dapagliflozin revealed the *ad lib* cohort increased food intake to compensate for only a third of the weight loss of the pair fed cohort. This finding bolstered expectations that dapagliflozin also would induce T2DM patients to lose weight.

The ADME profile determined for dapagliflozin was highly supportive of clinical development.[31] Plasma protein binding of dapagliflozin ranged from 91% in human to 95% in rat. All evidence indicated that dapagliflozin did not inhibit or upregulate Cyp450 enzymes. Upon incubation with mouse, rat, dog, monkey, and human hepatocytes, dapagliflozin was qualitatively converted to a similar mixture of Phase I (O-deethylation and

Table 23.4 Four-week weight loss study with dapagliflozin in diet-induced obese Sprague–Dawley rats

		Dapagliflozin-treated cohorts (mg/kg)			
Properties	**Vehicle**	**0.5**	**1.0**	**5**	**5 (pair fed to veh.)**
Decrease in body weight from Day 1	0.5%	4.3%	4.1%	6.1%	9.9%
Increase in total H_2O intake vs. vehicle		50%	77%	94%	82%
Increase in total kcal consumed vs. vehicle		7%	14%	14%	0
Decrease in adipose mass from Day 1	2.5%	4.1%	8.8%	7.3%	12.6%
Respiratory quotient (Day 15)	0.96	0.8	0.8	0.81	0.78

hydroxylation of the diarylmethane moiety to a diarylcarbinol) and Phase II (glucuronidation of the C3 sugar hydroxyl) metabolites. Despite being a P-glycoprotein 1 substrate, oral bioavailability was high due to high permeability. The pharmacokinetic parameters measured in rats, dogs, and monkeys predicted that oral exposure, clearance, and elimination half-life in humans would be conducive to a single daily administration.

Clinical studies confirmed these predictions. Dapagliflozin was a low-clearance compound for which the terminal half-life was 13.8 h. At least 75% of a 50-mg oral dose was absorbed within 24 h; T_{max} was achieved in <1 h. Urinary excretion was the major elimination route comprising >75% of the dose primarily as the C3 sugar hydroxyl glucuronide that was generated by UGT1A9.[32]

7. CLINICAL STUDIES WITH DAPAGLIFLOZIN

Dapagliflozin was found to be safe and well tolerated by healthy volunteers in double-blinded, placebo-controlled, single ascending and a 2-week multiple ascending dose studies for which the highest daily doses were 500 and 100 mg, respectively.[33,34] Dose-dependent glucosuria was obtained with no evidence of hypoglycemia. Administration of ≥20 mg of dapagliflozin produced near maximum urinary glucose excretion (UGE) of ~60 g per day. Dose-dependent UGE, which reached a plateau at ~75 g, was obtained in a subsequent Phase II double-blind, 2-week, multiple ascending dose study in which 47 T2DM patients with unimpaired renal function were administered 5, 25, and 100 mg of dapagliflozin or placebo.[34] Both fasting serum glucose levels and the area under the curve for glucose excursion following an oral glucose tolerance test exhibited significant improvement on day 14, thereby establishing dapagliflozin to be efficacious in diabetic patients.

More than 19 Phase II/III studies comprising more than 8500 T2DM patients have shown dapagliflozin to be safe and efficacious when administered at 5 or 10 mg daily as monotherapy or as add-on studies with insulin, sitagliptin, metformin, pioglitazone, or sulfonyureas.[35] Comparator studies with metformin and glipizide revealed dapagliflozin not to be inferior after 52 weeks. No significant increase in CV events was observed during additional studies with patients at higher CV risk. Studies in T2DM patients with various stages of impaired renal function supported administration to patients with modest renal impairment; however, dapagliflozin is contraindicated for patients with glomerular filtration rates <60 mL/min.

When administered once daily as a 5- or 10-mg tablet, T2DM patients achieved improved glycemic control manifested by HbA1c reductions of ≥0.5% with minimal increased risk of hypoglycemia. Additional benefits were a 2–3% reduction in body weight and a trend for a modest 3–5 mm reduction in systolic blood pressure. Although urinary tract infections were only slightly increased over placebo, the incidence of genital tract infections was approximately double the incidence rate observed for the placebo-treated group. However, genital tract infections were self-reported by patients; no clinical confirmation occurred. Virtually all were readily resolved by topical treatments necessitating few if any withdrawals from drug treatment.

8. CONCLUSION

The evolution from phlorizin to dapagliflozin illustrates the power of opportunistically combining systematic pharmacophore evaluation with detailed attention to unexpected findings. Publication of the C-glucoside patents from Bristol-Myers Squibb transformed the approach that the pharmaceutical industry was pursuing for SGLT2 inhibitors from O-glucoside- to C-glucoside-based structures and stimulated more than 20 groups to initiate research efforts to identify alternative novel C-glucoside containing SGLT2 inhibitors.[36,37] These efforts identified eight other SGLT2 inhibitors which progressed into advanced clinical trials, all of which incorporate the same meta C-glycosylated diarylmethane pharmacophore that was first employed with dapagliflozin. As a result, the phlorizin-induced glucosuria first noted as a medical curiosity over 150 years ago is now being utilized to provide a paradigm-breaking novel insulin-independent means to treat T2DM.

Dapagliflozin was well tolerated in preclinical studies and human clinical trials. Moreover, the PK profile, low dose, and clean safety profile enabled combination with other antidiabetic agents in a single formulation. Due to the safety and efficacy obtained in multiple Phase II/III clinical trials, the FDA, the European Commission, and health authorities of 40 other countries have approved use of dapagliflozin as monotherapy or add-on therapy with other antidiabetic agents including insulin. Dapagliflozin is marketed as Farxiga™ in the United States and as Forxiga™ worldwide. European Health Authority has recently approved Xigduo™, a dapagliflozin/metformin combination. This paradigm shift for treatment of T2DM with dapagliflozin will provide major benefits for patients with T2DM.

REFERENCES

1. International Diabetes Federation *IDF Diabetes Atlas*, 6th ed.; International Diabetes Federation: Brussels, Belgium, 2013. http://www.idf.org/diabetesatlas.
2. de Pablos-Velasco, P.; Bradley, C.; Eschwège, E. E.; Gönder-Frederick, L. A.; Parhofer, K. G.; Vandenberghe, H.; Simon, D. *Diabetologia* **2010**, *53*(Suppl. 1), 1012.
3. Wong, K.; Glovaci, D.; Malik, S.; Franklin, S. S.; Wygant, G.; Iloeje, U.; Kan, H.; Wong, N. D. *J. Diabetes Complicat.* **2012**, *26*, 169.
4. Marsenic, O. *Am. J. Kidney Dis.* **2009**, *53*, 875.
5. Chao, E. C.; Henry, R. R. *Nat. Rev. Drug Discov.* **2010**, *9*, 551.
6. Washburn, W. N. *J. Med. Chem.* **2009**, *52*, 1785.
7. Whaley, J. M.; Tirmenstein, M.; Reilly, T. P.; Poucher, S. M.; Saye, J.; Parikh, S.; List, J. F. *Diabetes Metab. Syndr. Obes.* **2012**, *9*, 551.
8. Gerich, J. E. *Diabet. Med.* **2010**, *27*, 136.
9. Mather, A.; Pollock, C. *Kidney Int.* **2011**, *79* (Suppl. 120), S1.
10. Wright, E. M.; Loo, D. D. F.; Hirayama, B. A. *Physiol. Rev.* **2011**, *91*, 733.
11. Vallon, V.; Platt, K. A.; Cunard, R.; Schroth, J.; Whaley, J.; Thomson, S. C.; Koepsell, H.; Rieg, T. *J. Am. Soc. Nephrol.* **2011**, *22*, 104.
12. Sabolic, I.; Vrhovac, I.; Eror, D. B.; Gerasimova, M.; Rose, M.; Breljak, D.; Ljubojevic, M.; Brzica, H.; Sebastiani, A.; Thal, S. C.; Sauvant, C.; Kipp, H.; Vallon, V.; Koepsell, H. *Am. J. Physiol. Cell Physiol.* **2012**, *302*, C1174.
13. Chen, J.; Williams, S.; Ho, S.; Loraine, H.; Hagan, D.; Whaley, J. M.; Feder, J. N. *Diabetes Ther.* **2010**, *1*, 57.
14. Sabolic, I.; Skarica, M.; Gorboulev, V.; Ljubojević, M.; Balen, D.; Herak-Kramberger, C. M.; Koepsell, H. *Am. J. Physiol. Renal Physiol.* **2006**, *290*, F913.
15. Calado, J.; Santer, R.; Rueff, J. *Kidney Int.* **2011**, *79*, S7.
16. Santer, R.; Calado, J. *Clin. J. Am. Soc. Nephrol.* **2010**, *5*, 133.
17. Wright, E. M.; Hirayama, B. A.; Loo, D. F. *J. Intern. Med.* **2007**, *261*, 32.
18. Rossetti, L.; Shulman, G. I.; Zawalich, W.; DeFronzo, R. A. *J. Clin. Invest.* **1987**, *80*, 1037.
19. Ehrenkranz, J. R.; Lewis, N. G.; Kahn, C. R.; Roth, J. *Diabetes Metab. Res. Rev.* **2005**, *21*, 31.
20. Tsujihara, K.; Hongu, M.; Saito, K.; Inamasu, M.; Arakawa, K.; Oku, A.; Matsumoto, M. *Chem. Pharm. Bull. (Tokyo)* **1996**, *44*, 1174.
21. Kees, K. L.; Fitzgerald, J. J.; Steiner, K. E.; Mattes, J. F.; Mihan, B.; Tosi, T.; Mondoro, D.; McCaleb, M. L. *J. Med. Chem.* **1996**, *39*, 3920.
22. Link, J. T.; Sorensen, B. K. *Tetrahedron Lett.* **2000**, *41*, 9213.
23. Washburn, W. N. SGLT2 inhibitors in development. In: *New Therapeutic Strategies for Type 2 Diabetes: Small Molecule Approaches*, Jones, R. M. Ed.; 1st ed.; Royal Chemical Society: Cambridge, UK, 2012; p 29.
24. Ellsworth, B. A.; Meng, W.; Patel, M.; Girotra, R. N.; Wu, G.; Sher, P.; Hagan, D.; Obermeier, M.; Humphreys, W. G.; Robertson, J. G.; Wang, A.; Han, S.; Waldron, T.; Morgan, N. N.; Whaley, J. M.; Washburn, W. N. *Bioorg. Med. Chem. Lett.* **2008**, *18*, 4770.
25. Meng, W.; Ellsworth, B. A.; Nirschl, A. A.; McCann, P. J.; Patel, M.; Girotra, R. N.; Wu, G.; Sher, P. M.; Morrison, E. P.; Biller, S. A.; Zahler, R.; Deshpande, P. P.; Pullockaran, A.; Hagan, D. L.; Morgan, N.; Taylor, J. R.; Obermeier, M. T.; Humphreys, W. G.; Khanna, A.; Discenza, L.; Robertson, J. G.; Wang, A.; Han, S.; Wetterau, J. R.; Janovitz, E. B.; Flint, O. P.; Whaley, J. M.; Washburn, W. N. *J. Med. Chem.* **2008**, *51*, 1145.
26. Deshpande, P. P.; Ellsworth, B. A.; Buono, F. G.; Pullockaran, A.; Singh, J.; Kissick, T. P.; Huang, M.-H.; Lobinger, H.; Denzel, T.; Mueller, R. H. *J. Org. Chem.* **2007**, *72*, 9746.

27. Deshpande, P. P.; Singh, J.; Pullockaran, A.; Kissick, T.; Ellsworth, B. A.; Gougoutas, J. Z.; Dimarco, J.; Fakes, M.; Reyes, M.; Lai, C.; Lobinger, H.; Denzel, T.; Ermann, P.; Crispino, G.; Randazzo, M.; Gao, Z.; Randazzo, R.; Lindrud, M.; Rosso, V.; Buono, F.; Doubleday, W. W.; Leung, S.; Richberg, P.; Hughes, D.; Washburn, W. N.; Meng, W.; Volk, K. J.; Mueller, R. H. *Org. Process Res. Dev.* **2012**, *16*, 577.
28. Uveges, A.; Hagan, D.; Onorato, J.; Whaley, J. M. In: *71st ADA Scientific Sessions, San Diego, CA*, **2011**, Poster P-987.
29. Han, S.; Hagan, D. L.; Taylor, J. R.; Xin, L.; Meng, W.; Biller, S. A.; Wetterau, J. R.; Washburn, W. N.; Whaley, J. M. *Diabetes* **2008**, *57*, 1723.
30. Devenny, J. J.; Godonis, H. E.; Harvey, S. J.; Rooney, S.; Cullen, M. J.; Pelleymounter, M. A. *Obesity* **2012**, *20*, 1645.
31. Obermeier, M. T.; Yao, M.; Khanna, A.; Koplowitz, B.; Zhu, M.; Li, W.; Komoroski, B.; Kasichayanula, S.; Discenza, L.; Washburn, W.; Meng, W.; Ellsworth, B. A.; Whaley, J. M.; Humphreys, W. G. *Drug Metab. Dispos.* **2010**, *38*, 405.
32. Kasichayanula, S.; Liu, X.; Lacreta, F.; Griffen, S. C.; Boulton, D. W. *Clin. Pharmacokinet.* **2014**, *53*, 17.
33. Komoroski, B.; Vachharajani, N.; Boulton, D.; Kornhauser, D.; Geraldes, M.; Li, L.; Pfister, M. *Clin. Pharmacol. Ther.* **2009**, *85*, 520.
34. Komoroski, B.; Vachharajani, N.; Feng, Y.; Li, L.; Kornhauser, D.; Pfister, M. *Clin. Pharmacol. Ther.* **2009**, *85*, 513.
35. Washburn, W. N.; Poucher, S. M. *Expert Opin. Invest. Drugs* **2013**, *22*, 463.
36. Ellsworth, B.; Washburn, W. N.; Sher, P. M.; Wu, G.; Meng, W. US Patent 6,414,126, 2002.
37. Ellsworth, B.; Washburn, W. N.; Sher, P. M.; Wu, G.; Meng, W. US Patent 6,515,117, 2003.

CHAPTER TWENTY-FOUR

Case History: Kalydeco® (VX-770, Ivacaftor), a CFTR Potentiator for the Treatment of Patients with Cystic Fibrosis and the *G551D-CFTR* Mutation

Sabine Hadida*, Frederick Van Goor*, Kirk Dinehart†, Adam R. Looker†, Peter Mueller†, Peter D.J. Grootenhuis*
*Vertex Pharmaceuticals Incorporated, San Diego, California, USA
†Vertex Pharmaceuticals Incorporated, Boston, Massachusetts, USA

Contents

1. INTRODUCTION

Cystic fibrosis (CF) is the most common life-threatening autosomal recessive disorder in Caucasian populations.[1] The disease affects approximately 70,000 patients worldwide[2] and is most prevalent in North America and Europe. CF is caused by mutations in the cystic fibrosis transmembrane conductance regulator (*CFTR*) gene. The *CFTR* gene encodes a protein

Annual Reports in Medicinal Chemistry, Volume 49
ISSN 0065-7743
http://dx.doi.org/10.1016/B978-0-12-800167-7.00024-9

highly expressed in epithelial tissues that functions as a chloride- and bicarbonate-selective ion channel and is involved in the regulation of salt as well as water transport and pH balance[3–5] in multiple organs, such as the lung, pancreas, liver, and intestinal tract.[6] CF-causing gene mutations impair the function of CFTR, leading to clinical manifestations[7] such as abnormal sweat electrolytes, chronic and progressive respiratory disease, exocrine pancreatic dysfunction, digestive problems, and infertility.[8] However, the primary cause of morbidity and mortality in CF is lung disease. In the lung, mutations in the *CFTR* gene lead to reduced quantity or function of CFTR protein at the cell surface, resulting in reduced CFTR-mediated Cl^- and fluid secretion and increased epithelial Na^+ channel-mediated Na^+ and fluid absorption.[6,9,10] This imbalance between Cl^- secretion and Na^+ absorption, and the resulting dehydration of the airway surface, likely contribute to the damaging cascade of mucus accumulation blocking the airway, as well as to infection, inflammation, and tissue destruction that characterizes CF lung disease.[11] The loss of CFTR-mediated HCO_3^- secretion may also contribute to production of the thick, sticky mucus that plugs the passageways in the lung.[12]

Initial therapies to treat CF targeted the downstream disease consequences secondary to the loss of CFTR function[13] and included airway clearance therapies, mucolytic treatments, anti-infective, and anti-inflammatory drugs. Although these treatments have helped to extend the median predicted age of survival,[14] there remains a large medical need for more efficacious treatments that address the underlying defect of CF. Toward this goal, it was hypothesized that small molecule therapies that increase CFTR function could address the downstream consequences of *CFTR* gene dysfunction and slow the progression of the disease. Kalydeco® (VX-770, ivacaftor, **1**) is the first small molecule therapy that targets CFTR to treat the underlying cause of CF, opening a new era in the treatment of CF (Fig. 24.1).

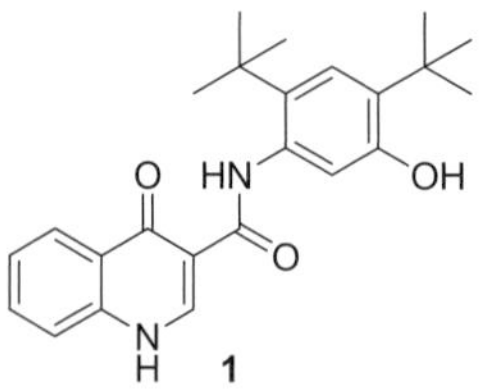

Figure 24.1 Kalydeco® (VX-770, ivacaftor, **1**).

2. CFTR AS A DRUG DISCOVERY TARGET

CFTR is a member of the ATP-binding cassette superfamily and is localized in the apical membrane of epithelial cells.[15] CFTR is a 1480 amino acid protein composed of two sets of six helical membrane-spanning domains (MSD1 and MSD2) that form the channel pore, two cytoplasmic nucleotide-binding domains (NBD1 and NBD2) that bind and hydrolyze ATP, and a regulatory "R" domain with multiple protein kinase A and protein kinase C phosphorylation sites.[16,17] Over 1900 mutations in the CFTR protein have been reported.[18] Approximately 90% of the CF patients carry a loss-of-function *CFTR* mutation on at least one allele that results in deletion of phenylalanine 508 (F508del) in NBD1.[7] The F508del mutation prevents the proper domain folding and assembly of the multidomain CFTR protein during its biogenesis in the endoplasmic reticulum and also interferes with trafficking, membrane stability, and channel gating.[8,19–21] As a result, the number of channels present in the membrane is far less than observed in cells expressing wild-type CFTR.[22–25] Another relatively common mutation that affects 4–5% of CF patients[26] is the *G551D* mutation. This mutation results in CFTR protein which is delivered to the cell surface but with severely impaired channel gating activity; as a result, the channel opens less frequently than normal.

There has been growing interest in small molecule therapies that increase mutant CFTR function and thereby slow down the progression of the disease. These therapies, broadly classified as CFTR modulators, have been extensively reviewed and include, among others, CFTR potentiators and CFTR correctors.[27] CFTR potentiators act in the presence of endogenous or pharmacological CFTR activators to increase the channel gating activity of the CFTR protein located at the cell surface, leading to enhanced chloride transport. CFTR correctors act by increasing the delivery and amount of functional CFTR protein to the cell surface, resulting in enhanced ion transport. CFTR potentiators and correctors may be coadministered to maximize the clinical efficacy and/or therapeutic window. With the goal to identify development candidates with each mechanism, we carried out two drug discovery programs in parallel, one targeting the identification of CFTR potentiators and the other pursuing CFTR correctors. This chapter describes our approach toward the discovery of CFTR potentiators.

3. THE DISCOVERY OF CFTR POTENTIATORS

Functional cell-based high-throughput screening (HTS) assays were developed to screen a variety of chemically distinct scaffolds for their ability to potentiate CFTR. Hits were followed up with extensive medicinal chemistry and iterative structure–activity relationship (SAR) analyses using a combination of approaches such as parallel synthesis and traditional medicinal chemistry. The activity of the most interesting compounds was confirmed *in vitro* in human bronchial epithelial (HBE) cells isolated from bronchi of CF subjects carrying the *F508del/F508del* and *G551D/F508del* mutations, as validated animal models for CF lung disease were not available. A complete data package, including *in vitro* selectivity, physicochemical properties, *in vitro* and *in vivo* ADME profile, and preclinical drug safety assessment, was then generated for these compounds and the information was integrated into the design of new analogs.

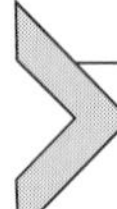

4. MEDICINAL CHEMISTRY EFFORTS CULMINATING IN IVACAFTOR

Approximately 228,000 compounds from the Vertex screening collection were tested in a cell-based fluorescence membrane potential assay in NIH 3T3 cells expressing the F508del CFTR mutation to help identify chemically distinct and novel scaffolds for further medicinal chemistry evaluation.[28] Following hit confirmation in additional assays (patch clamp experiments, selectivity, cell viability, and Ussing chambers studies in F508del HBEs), the four potentiator scaffolds shown in Fig. 24.2 were identified.

2 **3** **4** **VRT-484 (5)**

MW = 368; cLogP = 2.9
EC_{50} = 2.1 μM (F508del 3T3)
EC_{50} = 1.5 μM (F508del HBE)
EC_{50} = 12.2 μM (G551D/F508del HBE)

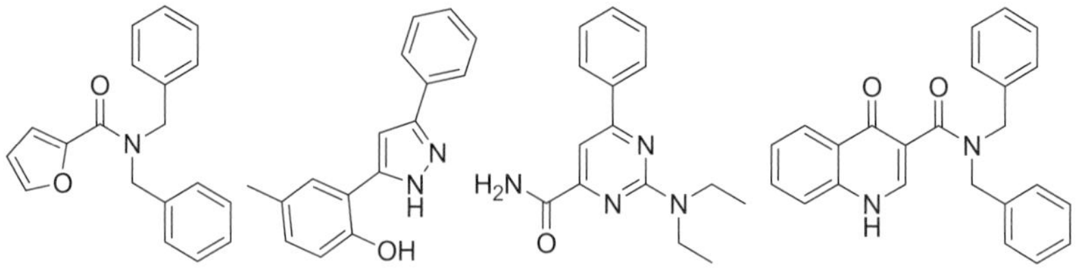

Figure 24.2 CFTR potentiator hits from HTS.

4.1. Hit-to-Lead Efforts

Hit-to-lead medicinal chemistry efforts were initiated around all 4 validated hits. SAR trends emerged from analog synthesis around hit **5**, whereas flat SAR was observed around hits **2**, **3**, and **4**. Continued, extensive experimentation therefore concentrated on quinolinone scaffold **5** and is described below. Compound **5** had a potency of 2.1 μM in the primary medicinal chemistry assay in F508del NIH 3T3 cells and a similar potency in the F508del HBE cells. Compound **5** was also active in HBEs expressing the G551D mutation. Initial work around chemotype **5** concentrated on variations of the amine portion of the molecule while maintaining the quinolinone amide. A diverse set of analogs prepared from primary and secondary amines with aliphatic, aromatic, and heterocyclic substitutions were synthesized and tested in the F508del NIH 3T3 potentiator assay. Unless noted otherwise, all the data reported in this chapter were generated in the F508del NIH 3T3-based potentiator assay. Data for many of these compounds, plotted as a function of their molecular weight (MW) and cLogP and grouped by potency ranges, are represented in Fig. 24.3.

The analysis of the data revealed interesting and unexpected trends. We discovered that potency could be improved with the incorporation of large lipophilic substituents *ortho* to the aniline ring (i.e., isopropyl aniline). Of particular interest was 6-indolyl derivative **6**, the first compound with an EC_{50} of 100 nM identified by the project team. Interestingly, 5- and 7-indolyl regioisomers showed reduced activity relative to **6**. We

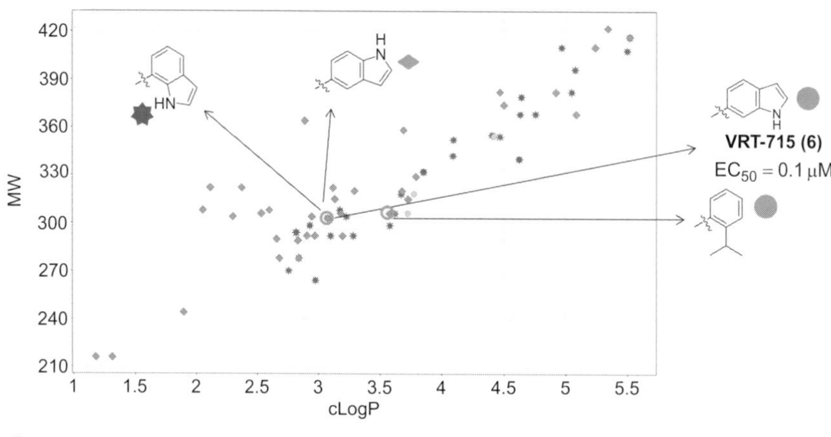

Figure 24.3 Amide exploration around **5**. (See the color plate.)

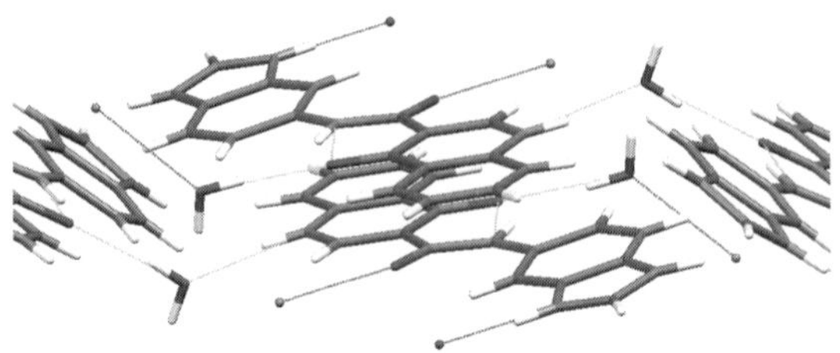

Figure 24.4 X-ray structure of **6**. (See the color plate.)

hypothesized that the indole NH was involved in a hydrogen-bonding interaction leading to this significant potency improvement. The activity of **6** was determined in F508del/F508del HBEs and the EC_{50} was confirmed at 50 nM. Unfortunately, **6** was insoluble in both aqueous and organic solvents due to its high aromatic character. The X-ray crystal structure of a hydrate of **6** (Fig. 24.4) further highlighted the presence of an intramolecular hydrogen bond between the amide hydrogen and the lone pair of electrons on the carbonyl oxygen of the quinolinone that imposed a planar conformation to the structure. In addition, a dense network of intermolecular hydrogen bonds and extensive π–π stacking between the molecules in the crystal structure further contributed to the poor solubility of the molecule.

4.2. Reducing the Planarity of VRT-715

In an effort to improve the solubility of compound **6**, the chemistry strategy, as shown in Fig. 24.5, then centered around distorting the planarity of **6** by modifying the amide linker or by incorporating substitutions on the aryl amine moiety to increase its three dimensionality. As illustrated by compounds **7**, **8**, **9**, and **10**, linker modifications (approach A) led to a dramatic reduction in CFTR activity although these compounds displayed improved solubilities.

Modification of the amide moiety (approach B) was more successful. Alkyl substituents were initially placed around the indole ring of **6** (Fig. 24.6). These groups had a detrimental effect on potency at positions 1 and 7, whereas they were tolerated at positions 2 and 4. Potency improved with addition of lipophilic groups at positions 3 and 5, with *tert*-butyl analogs **13** and **16** being the most potent indole derivatives prepared. Reduction of the indole ring of **6** to indoline **17** led to a slight loss in potency. Ring expansion of **17** to tetrahydroquinoline **18** led to a further loss in potency. However, the potency was regained with insertion of a "*tert*-butyl-like" group at the 4-position as shown in 4,4-dimethyltetrahydroquinoline **19**.

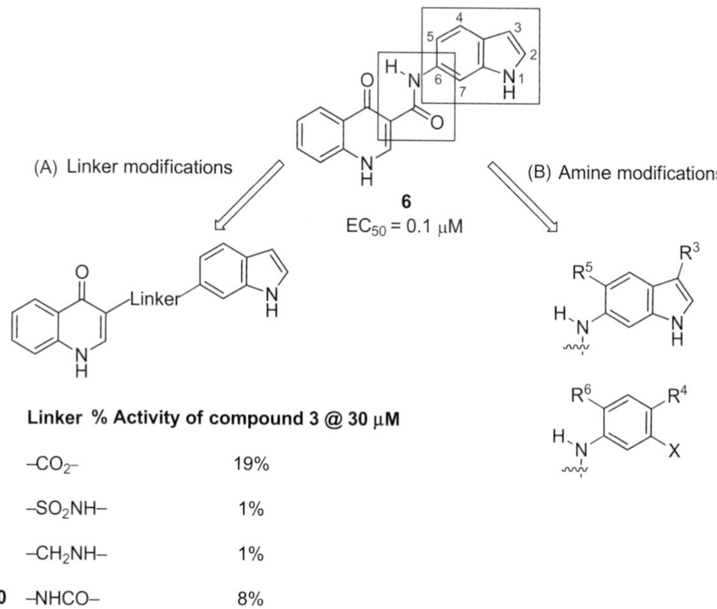

	Linker	% Activity of compound 3 @ 30 μM
7	$-CO_2-$	19%
8	$-SO_2NH-$	1%
9	$-CH_2NH-$	1%
10	–NHCO–	8%

Figure 24.5 Chemistry strategy to disrupt the planarity of **6**.

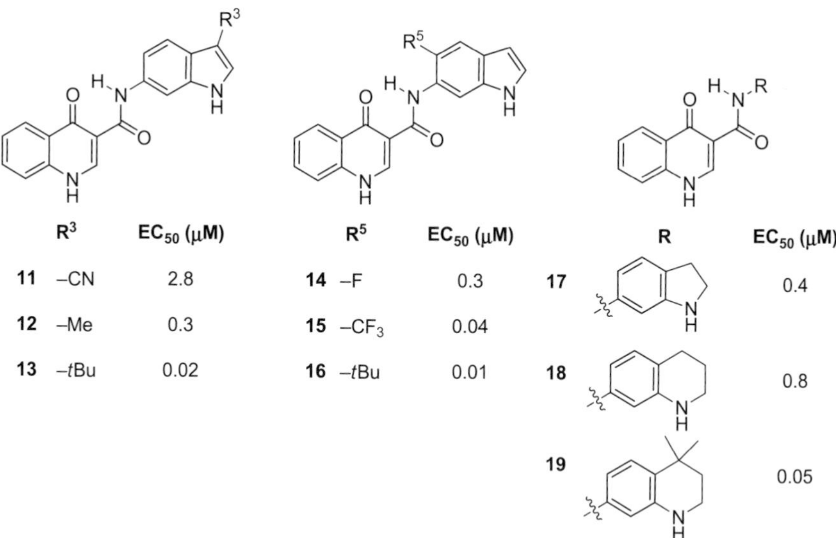

	R^3	EC_{50} (μM)
11	–CN	2.8
12	–Me	0.3
13	–*t*Bu	0.02

	R^5	EC_{50} (μM)
14	–F	0.3
15	$-CF_3$	0.04
16	–*t*Bu	0.01

	R	EC_{50} (μM)
17	(indolin-6-yl)	0.4
18	(tetrahydroquinolin-7-yl)	0.8
19	(4,4-dimethyltetrahydroquinolin-7-yl)	0.05

Figure 24.6 Amide SAR in the bicyclic analogs.

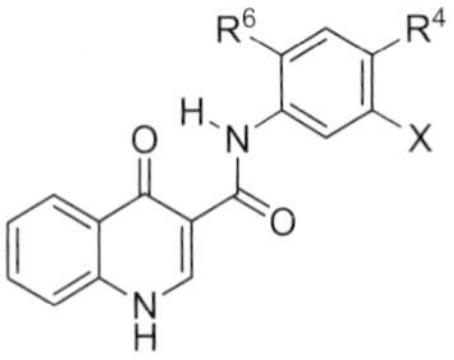

	R^4	R^6	X	EC_{50} (μM)
20	–Et	–H	$-NH_2$	1.7
21	–*t*Bu	–H	$-NH_2$	0.1
22	–H	–*t*Bu	$-NH_2$	0.5
23	–*t*Bu	–H	–H	0.1
24	–*t*Bu	–H	–F	0.1
25	–*t*Bu	–H	$-NHCOCH_3$	3.5
26	–*t*Bu	–H	$-SO_2NH_2$	5.1
27	–*t*Bu	–H	–OH	0.003
28	–*t*Bu	–F	–OH	0.002
29	–*t*Bu	$-CF_3$	–OH	0.003
1	–*t*Bu	–*t*Bu	–OH	0.003

Figure 24.7 Amide SAR in the monocyclic analogs.

Ring opening of **19** provided *tert*-butyl aniline **21** (Fig. 24.7) with a twofold loss in potency. Reducing the size of the group at the 4-position in compound **20** or moving the *tert*-butyl group to the 2-position in compound **22** led to 17-fold and 5-fold potency reductions, respectively. Removal of the aniline moiety in compound **23** or replacement with a fluorine atom in **24** led to compounds equipotent with **21**. Substitution of the aniline moiety with other hydrogen bond donors such as carbamate **25** or sulfonamide **26** resulted in significant potency loss, whereas replacement of the aniline group in **21** with phenol in **27** led to 33-fold potency improvements. Addition of substituents of varying size and electronic character at the 6-position of **27** as shown with compounds **28**, **29**, and **1** retained activity.

4.3. Final Compound Selection

CFTR potentiators with EC_{50}s below 100 nM were tested for their ability to increase CFTR-mediated chloride transport in Ussing chamber studies using F508del/F508del and F508del/G551D HBE primary cell cultures. Profiling was carried out to determine *in vitro* ADME properties such as metabolic stability in rat, dog, monkey, and human liver microsomes;

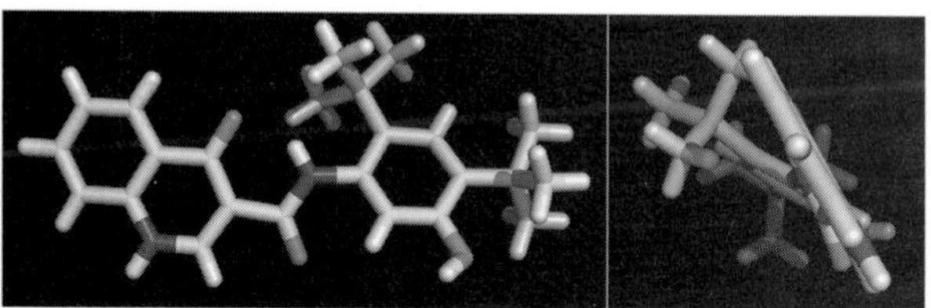

Figure 24.8 Views of X-ray structure of **1**. (See the color plate.)

cytochrome P450 inhibition; *in vitro* selectivity (hERG, MDS lead profiling screen); and *in vivo* rat and dog pharmacokinetic properties (iv, po). These efforts led to the discovery of **1** as a lead compound. As shown from two poses of an X-ray structure of **1** (Fig. 24.8), the addition of *tert*-butyl groups at R^4 and R^6 was key to breaking the planarity of the compound.

5. PRECLINICAL PROPERTIES OF IVACAFTOR

The detailed *in vitro* preclinical characterization of **1** has been published.[29] Compound **1** is a potent CFTR potentiator with an EC_{50} of 3 nM in F508del NIH 3T3 cells and 22 nM on residual CFTR in F508del HBE cells. When tested in our assays, **1** was 10–100-fold more potent than other known CFTR potentiators such as Genistein,[30] NS1619,[31] SF-03,[32] $\Delta F508_{act}$-02,[33] or compound **3**.[28] Further profiling of **1** revealed its attributes and challenges. Compound **1** is a broad-acting CFTR potentiator that acts at multiple CFTR-gating mutations beyond G551D, as well as CFTR mutations associated with defects in protein processing and/or function.[34,35] Of particular interest is its activity in HBEs expressing the G551D/F508del mutation where an EC_{50} of 236 nM was determined. Compound **1** displayed an excellent selectivity profile when profiled at Ricerca against 160 receptors and enzymes. Compound **1** also exhibited an outstanding *in vivo* pharmacokinetic profile with low clearance and good oral bioavailability in both rat and dog. The plasma protein binding of **1** was high with >99% bound in all species tested. To assess the relevance of these data, the activity of **1** was determined in F508del HBE in the presence of increasing amounts of human serum. A shift in the potency of **1** of less than threefold was observed when 20% human serum was added to the Ussing chambers experiments, leading us to believe that the high level of plasma protein binding was not a "show-stopper" for the compound. Since there is some protein in the HBE cell media, even without the external

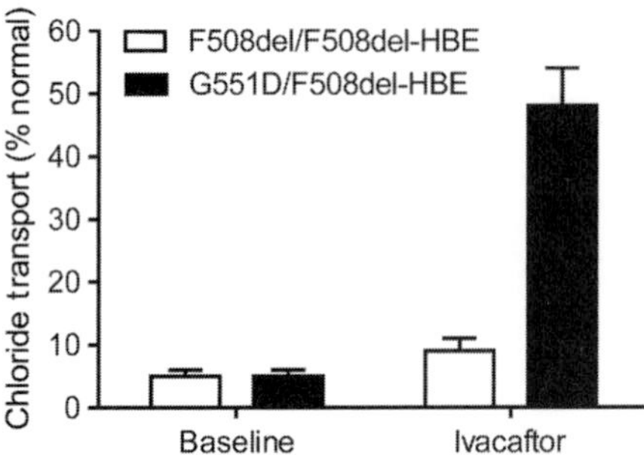

Figure 24.9 Compound **1** restored CFTR function *in vitro*.

addition of human serum, it is possible that the intrinsic potency of **1** was underestimated due its high level of protein binding.

The *in vitro* pharmacological characterization of **1** is shown in Fig. 24.9. In Ussing chambers studies using G551D/F508del HBEs, **1** restored CFTR function to ~50% of the normal function. With F508del/F508del HBEs, in which only minimal amounts of CFTR are delivered to the cell surface, CFTR function was restored to ~10% of the normal function.

Compound **1** was then used *in vitro* to test the hypothesis that restoring CFTR function through an increase in chloride flux would result in a decrease in sodium reabsorption and, as a result, would lead to an increase in the hydration of the airway and the increase in cilia beat frequency. To test this hypothesis, G551D/F508del HBEs were treated for 5 days in parallel with either 10 μM of **1** or DMSO. Fluorescent beads were added to the surface of the cells to assess the fluidity of the mucus through the movement of these beads. In the absence of **1**, the cell surface in the G551D/F508del HBE cultures was dehydrated and the ciliary beat frequency was low compared to HBE cultures derived from individuals without CFTR mutations. This was consistent with the loss of chloride and fluid transport. Visual inspection revealed a faster flow over the cells treated with **1** than over the cells incubated with DMSO. Quantitation of the cilia beat frequency, an indicator of the fluidity of the mucus secreted by the HBEs, revealed a fourfold increase in cells treated with **1** in comparison with DMSO-treated cells. This exciting *in vitro* result confirmed our initial hypothesis that restoring chloride secretion through CFTR modulation would result in rehydration of the airway surface in the lung.

6. FORMULATION DEVELOPMENT

We evaluated **1** *in silico* and found that it met the Lipinski "rule of 5"[36] as it displayed 3 hydrogen bond donors, 5 hydrogen bonds acceptors, a

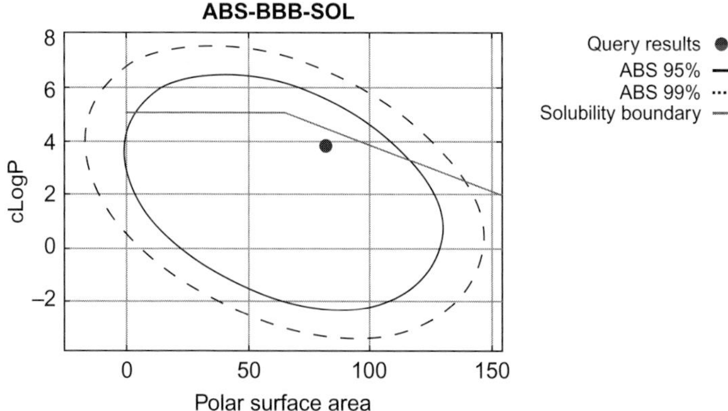

Figure 24.10 *In silico* prediction of oral absorption for **1**.

relatively low-molecular weight of 392 Da and a calculated cLogP of 3.82. These properties suggest that **1** has the right profile to be orally absorbed. *In silico* absorption models, as shown in Fig. 24.10,[37] also predicted **1** to be orally available. However, measured properties revealed that **1** had poor aqueous solubility (<0.05 μg/ml), a measured LogP of 5.68 and a relatively high melting point of 292 °C. These data suggested that a traditional drug development paradigm could be challenging for **1** and that solubility enhancements would be important to enable its development.

To enhance the solubility of **1**, several approaches were extensively evaluated in parallel and included the preparation of salts, the formation of cocrystals and solvates, and the development of amorphous dispersions. Solubility enhancements of varying degrees were achieved with these approaches. However, the unexpectedly high glass transition temperature of **1**, which was measured to be 175 °C, is an indicator of the stability of the amorphous form of the compound. Based upon the surprising stability of the amorphous form of **1**, we targeted the use of spray-dried dispersion (SDD) formulations for development. Figure 24.11 shows the formulation effect in the oral exposure of **1** administered to rats. Oral bioavailabilities around 5% were observed when a crystalline suspension of **1** was administered to rats at doses of 50 and 200 mg/kg. Significantly better exposures, with oral bioavailabilities ranging from 80% to 100%, were observed when **1** was administered as an SDD formulation. With this formulation, a desirable linear exposure over a dose range from 50 to 400 mg/kg was observed. Overall, the stabilized amorphous form of **1** was ~32-fold more bioavailable than crystalline forms. Similar results were obtained when the SDD formulation was tested in higher species.

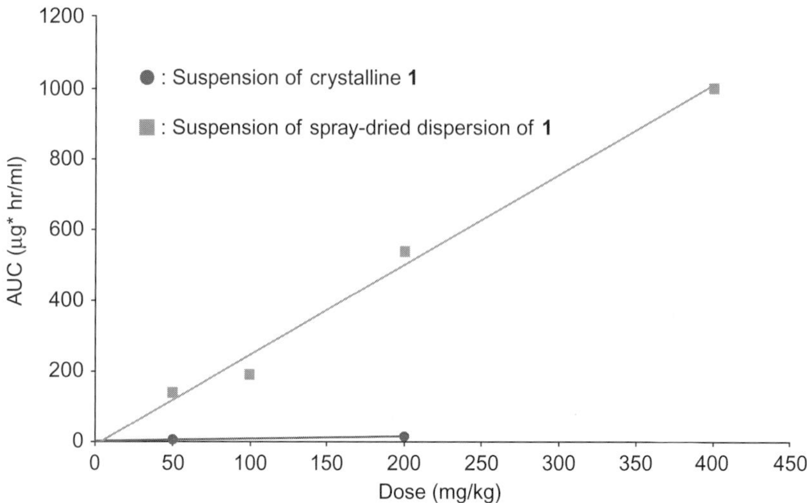

Figure 24.11 Formulation effect on oral exposure of **1** in rats. (See the color plate.)

In light of this data, considerable effort was spent trying to maximize both the solubility and the stability of the SDD formulation. We evaluated a number of combinations of **1** and hydroxypropylmethylcellulose acetate succinate (HPMCAS). Three distinct combinations of **1** and HPMCAS with 90% drug load, 80% drug load, and 50% drug load were compared to the neat amorphous form of **1**. Solubility being the major driver for the oral bioavailability of **1**, the kinetic solubility of these formulations was determined in fed simulated intestinal fluid (FeSSIF) and the data are summarized in Fig. 24.12. The analysis of the data revealed that solid dispersions of **1** and polymer stabilize the amorphous form of **1** in aqueous media. Unexpectedly, the most favorable combination was determined to be the 50% drug load formulation as it maintained a solubility of 0.6 mg/ml for a period of 24 h.

We also evaluated the stability of these SDD formulations using open dish stress studies over an 8-week period. Accelerated stress conditions included a range of temperatures from 40 to 80 °C and relative humidity of 75% and 100%. At the end of these studies, traces of crystalline material were only observed under the most extreme conditions. Using this information, a model was built to estimate the time it would take to see 10% crystallization of **1**. The model predicted that it would take more than 10 years at 25 °C and 100% relative humidity. We now had identified a solid form of **1** that could be progressed into development.

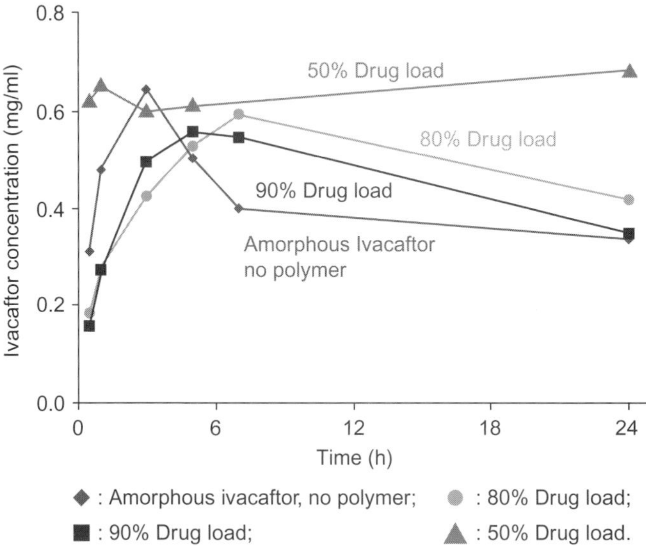

Figure 24.12 Kinetic solubility of **1** in FeSSIF. (See the color plate.)

During the chemical manufacturing process, **1** is prepared as a crystalline active pharmaceutical ingredient (API) with a high overall molar yield (>80%). This API is subsequently taken through a spray drying process to lead to an SDD formulation with a kinetic aqueous solubility of 0.6 mg/ml, a surprising 12,000-fold enhancement over the crystalline API. These properties are compatible with a formulation that can withstand the process of tabletting, packaging, and storage required for a product to be developed.

7. CLINICAL STUDIES

In 2006, Phase I testing of **1** in normal subjects and CF volunteers was initiated. Following a single dose in the fed state, **1** displayed a terminal half-life of approximately 12 h.[38] During the same year, the FDA granted Fast Track designation to **1** given the serious and life-threatening nature of CF. A two-part multicenter, randomized, double-blind, placebo-controlled, Phase II study in adults with CF who carried a least one copy of the *G551D* CFTR-gating mutation started in the second quarter of 2007. A total of 39 patients received oral doses (25, 75, or 150 mg or placebo) of **1** every 12 h for 14 days (part 1) or 150 or 250 mg or placebo for 28 days (part 2). The very promising results of an interim analysis were announced on March 27, 2008. Patients in part 1 of the trial showed significant improvements in key indicators of CF such as lung function, as measured by an increase in FEV_1 (percent predicted forced expiratory volume in 1 s), nasal potential difference measurements of chloride transport, and sweat chloride levels. It was the first time a drug had lowered the sweat chloride levels in CF patients, in some cases to below the diagnostic threshold for CF. Full analysis of the Phase II data confirmed the earlier analysis where the median relative change in the primary outcome FEV_1 was found to be 8.7%.[39] Few severe side effects were reported and all were resolved without discontinuation.

The Phase II results prompted us to initiate the Phase III registration studies STRIVE ($n=167$ patients)[40] and ENVISION ($n=52$ patients).[41] In the STRIVE trial, the patients (12 years or older) were treated for 48 weeks and had at least one copy of the *G551D* mutation. A dose of 150 mg q12h improved the lung function as measured by statistically significant absolute change in FEV_1 by 10% points compared with placebo. Similar results were obtained in the ENVISION trial in patients aged 6–11 years. In STRIVE, it was found that **1** significantly reduced the risk of pulmonary exacerbations. Interestingly, patient's body weight also improved suggesting that orally administered **1** restores chloride transport in multiple organs. The PERSIST study was an open-label rollover study of patients who received either **1** or placebo in STRIVE or ENVISION. Compound **1** was generally well tolerated, with a low discontinuation rate and >90% compliance. Given the relatively low number of *G551D* patients, **1** was further evaluated in people who were homozygous for the much more common *F508del* mutation. The trial ($n=112$ patients) was primarily powered to assess the safety of **1**. **1** was found to be well tolerated; however, improved lung function was not demonstrated. On January 31 2012, the FDA approved **1** for the treatment of CF in patients 6 years of age and older carrying the *G551D* mutation. Given that *in vitro* studies demonstrated that **1** also potentiates multiple gating mutations, several other trials are on-going post-approval.

8. CONCLUSION

Compound **1** was discovered after significant efforts utilizing a combination of HTS followed by extensive medicinal chemistry optimization and iterative SAR analysis on multiple scaffolds. Critical aspects of the identification and development of **1** include availability of functional membrane potential assays in recombinant NIH 3T3 cells and in primary HBE cells derived from people with CF and the use of a stabilized amorphous formulation to achieve good oral bioavailability. Compound **1** received approval by the FDA, the EMA, Health Canada and TGA Australia for the treatment of people with CF age 6 years and older who have a *G551D* mutation in the *CFTR* gene. Additional clinical studies using **1**, alone or in combination with a CFTR corrector, are underway.

REFERENCES

1. Kazarian, H. *Hum. Mutat.* **1994**, *4*, 167.
2. Adapted from The molecular genetic epidemiology of CF. Report of a joint meeting of WHO/ECFTN/ICF(M)A/ECFS. Accessed 07/22/2009. UK CF Trust Patient Registry Report **2008**.
3. Knowles, M. R.; Stutts, M. J.; Spock, A.; Fischer, N.; Gatzy, J. T.; Boucher, R. C. *Science* **1983**, *221*, 1067.
4. Quinton, P. M. *Nature* **1983**, *301*, 421.
5. Riordan, J. R.; Rommens, J. M.; Kerem, B.; Alon, N.; Rozmahel, R.; Grzelczak, Z.; Zielenski, J.; Lok, S.; Plavsic, N.; Chou, J. L.; Drumm, M. L.; Iannuzzi, M. C.; Collins, F. C.; Tsui, L. *Science* **1989**, *245*, 1066.
6. Boucher, R. C. *Trends Mol. Med.* **2007**, *13*, 231.
7. Castellani, C.; Cuppens, H.; Macek, M.; Cassiman, J. J.; Kerem, E.; Durie, P.; Tullis, E.; Assael, B. M.; Bombieri, C.; Brown, A.; Casals, T.; Claustres, M.; Cutting, G. R.; Dequeker, E.; Dodge, J.; Doull, I.; Farrell, P.; Ferec, C.; Girodon, E.; Johannesson, M.; Kerem, B.; Knowles, M.; Munck, A.; Pignatti, P. F.; Radojkovic, D.; Rizzotti, P.; Schwarz, M.; Stuhrmann, M.; Tzetis, M.; Zielenski, J.; Elborn, J. S. *J. Cyst. Fibros.* **2008**, *7*, 179.
8. Farrell, P. M.; Rosenstein, B. J.; White, T. B.; Accurso, F. J.; Castellani, C.; Cutting, G. R.; Durie, P. R.; Legrys, V. A.; Massie, J.; Parad, R. B.; Rock, M. J.; Campbell, P. W., III *J. Pediatr.* **2008**, *153*, S4.
9. Joo, N. S.; Irokawa, T.; Robbins, R. C.; Wine, J. J. *J. Biol. Chem.* **2006**, *281*, 7392.
10. Matsui, H.; Grubb, B. R.; Tarran, R.; Randell, S. H.; Gatzy, J. T.; Davis, C. W.; Boucher, R. C. *Cell* **1998**, *95*, 1005.
11. Gibson, R. L.; Burns, J. L.; Ramsey, B. W. *Am. J. Respir. Crit. Care Med.* **2003**, *168*, 918.
12. Quinton, P. M. *Lancet* **2008**, *372*, 415.
13. Frerichs, C.; Smyth, A. *Expert. Opin. Pharmacother.* **2009**, *10*, 1191.
14. O'Sullivan, B. P.; Freedman, S. D. *Lancet* **2009**, *373*, 1891.
15. Pedemonte, N.; Lukacs, G. L.; Du, K.; Caci, E.; Zegarra-Moran, O.; Galietta, L. J. V.; Verkman, A. S. *J. Clin. Invest.* **2005**, *115*, 2564.
16. Van Goor, F.; Hadida, S.; Grootenhuis, P. D. J. *Top. Med. Chem.* **2008**, *3*, 29.
17. Sheppard, D. N.; Welsh, M. J. *Physiol. Rev.* **1999**, *79*, S23.
18. http://genet.sickkids.on.ca/StatisticsPage.html.
19. Noone, P. G.; Knowles, M. R. *Respir. Res.* **2001**, *2*, 328.
20. Cantin, A. M.; Bilodeau, G.; Ouellet, C.; Liao, J.; Hanrahan, J. W. *Am. J. Physiol. Cell Physiol.* **2006**, *290*, C262.
21. Cantin, A. M.; Hanrahan, J. W.; Bilodeau, G.; Ellis, L.; Dupuis, A.; Liao, J.; Zielenski, J.; Durie, P. *Am. J. Respir. Crit. Care Med.* **2006**, *173*, 1139.
22. Cheng, S. H.; Gregory, R. J.; Marshall, J.; Paul, S.; Souza, D. W.; White, G. A.; O'Riordan, C. R.; Smith, A. E. *Cell* **1990**, *63*, 827.
23. Denning, G. M.; Ostedgaard, L. S.; Welsh, M. J. *J. Cell Biol.* **1992**, *118*, 551.
24. Lukacs, G. L.; Mohamed, A.; Kartner, N.; Chang, X. B.; Riordan, J. R.; Grinstein, S. *EMBO J.* **1994**, *13*, 6076.
25. Ward, C. L.; Kopito, R. R. *J. Biol. Chem.* **1994**, *269*, 25710.

26. http://www.cff.org/UploadedFiles/research/ClinicalResearch/2011-Patient-Registry.pdf.
27. Hadida, S.; Van Goor, F.; Grootenhuis, P. D. *Annu. Rep. Med. Chem.* **2010**, *45*, 157.
28. Van Goor, F.; Strayley, K. S.; Cao, D.; Gonzalez, J.; Hadida, S.; Hazlewood, A.; Joubran, J.; Knapp, T.; Lewis, L. R.; Miller, M.; Neuberger, T.; Olson, E.; Pachenko, V.; Rader, J.; Singh, A.; Stack, J. H.; Tung, R.; Grootenhuis, P. D. J.; Negulescu, P. *Am. J. Physiol. Lung Cell. Mol. Physiol.* **2006**, *290*, L1117.
29. Van Goor, F.; Hadida, S.; Grootenhuis, P. D. J.; Burton, B.; Cao, D.; Neuberger, T.; Turnbull, A.; Singh, A.; Joubran, J.; Hazlewood, A.; Zhou, J.; McCartney, J.; Arumugam, V.; Decker, C.; Yang, J.; Young, C.; Olson, E. R.; Wine, J. J.; Frizzell, R. A.; Ashlock, M.; Negulescu, P. *Proc. Natl. Acad. Sci. U.S.A.* **2009**, *106*, 18825.
30. Morana, O.; Zegarra-Moran, O. *FEBS Lett.* **2005**, *579*, 3979.
31. Al-Nakkash, L.; Hu, S.; Li, M.; Hwang, T. *J. Pharmacol. Exp. Ther.* **2001**, *296*, 464.
32. Suen, Y. F.; Robins, L.; Yang, B.; Verkman, A. S.; Nantz, M. H.; Kurth, M. J. *Bioorg. Med. Chem. Lett.* **2006**, *16*, 537.
33. Yang, H.; Shelat, A. A.; Guy, R. K.; Gopinath, V. S.; Ma, T.; Du, K.; Lukacs, G. L.; Taddei, A.; Folli, C.; Pedemonte, N.; Galietta, L. J. V.; Verkman, A. S. *J. Biol. Chem.* **2003**, *278*, 35079.
34. Yu, H.; Burton, B.; Huang, C.; Worley, J.; Cao, D.; Johnson, J. P., Jr.; Urrutia, A.; Joubran, J.; Seepersaud, S.; Sussky, K.; Hoffman, B. J.; Van Goor, F. *J. Cyst. Fibros.* **2012**, *11*, 237.
35. Van Goor, F.; Yu, H.; Burton, B.; Hoffman, B. J. *J. Cyst. Fibros.* **2013**, *13*, 29.
36. Lipinski, C. A.; Lombardo, F.; Dominy, B. W.; Feeney, P. J. *Adv. Drug Deliv. Rev.* **1997**, *23*, 3.
37. Egan, W. J.; Merz, K. M., Jr.; Baldwin, J. J. *J. Med. Chem.* **2000**, *43*, 3867.
38. Deeks, E. D. *Drugs* **2013**, *73*, 1595.
39. Accurso, F. J.; Rowe, S. M.; Clancy, J. P.; Boyle, M. P.; Dunitz, J. M.; Durie, P. R.; Sagel, S. D.; Hornick, D. B.; Konstan, M. W.; Donaldson, S. H.; Moss, R. B.; Pilewski, J. M.; Rubenstein, R. C.; Uluer, A. Z.; Aitken, M. L.; Freedman, S. D.; Rose, L. M.; Mayer-Hamblett, N.; Dong, Q.; Zha, J.; Stone, A. J.; Olson, E. R.; Ordoñez, C. L.; Campbell, P. W.; Ashlock, M. A.; Ramsey, B. W. *New Engl. J. Med.* **2010**, *363*, 1991.
40. Ramsey, B. W.; Davies, J.; McElvaney, N. G.; Tullis, E.; Bell, S. C.; Dřevínek, P.; Griese, M.; McKone, E. F.; Wainwright, C. E.; Konstan, M. W.; Moss, R.; Ratjen, F.; Sermet-Gaudelus, I.; Rowe, S. M.; Dong, Q.; Rodriguez, S.; Yen, K.; Ordoñez, C.; Elborn, J. S.; VX08-770-102 Study Group. *New Engl. J. Med* **2011**, *365*, 1663.
41. Davies, J. C.; Wainwright, C. E.; Canny, G. J.; Chilvers, M. A.; Howenstine, M. S.; Munck, A.; Mainz, J. G.; Rodriguez, S.; Li, H.; Yen, K.; Ordoñez, C. L.; Ahrens, R. *Am. J. Respir. Crit. Care Med.* **2013**, *187*, 1219.

CHAPTER TWENTY-FIVE

Case History: Xeljanz™ (Tofacitinib Citrate), a First-in-Class Janus Kinase Inhibitor for the Treatment of Rheumatoid Arthritis

Mark E. Flanagan, Matthew F. Brown, Chakrapani Subramanyam, Michael J. Munchhof[1]

Pfizer Worldwide Medicinal Chemistry, Groton, Connecticut, USA

[1]Current address: 266 West Road, Salem, CT 06420, USA.

Contents

1. INTRODUCTION

Xeljanz™ (tofacitinib citrate) represents the first in a new class of Janus kinase (JAK) inhibitors for the treatment of rheumatoid arthritis (RA). It operates by inhibiting the JAK pathways involved in the signaling of a number of cytokines that have been implicated in the pathogenesis of

Annual Reports in Medicinal Chemistry, Volume 49
ISSN 0065-7743
http://dx.doi.org/10.1016/B978-0-12-800167-7.00025-0

the disease. RA is a chronic and debilitating condition that affects up to 1% of the adult population in the United States, disproportionately affecting women in the 40–70 age range.[1] RA is an autoimmune disease, and while its causes are not completely understood, a consistent theme in the pathology of the disease begins with a break in T- and B-cell tolerance against self-antigen, resulting in undesirable immune activity and tissue destruction primarily affecting the synovial joints. For much of the previous century, clinical treatment for RA was restricted to so-called conventional synthetic disease-modifying antirheumatic drugs (or csDMARDs), some of which are still used today (e.g., methotrexate and sulfasalazine).[2] However, efficacy and tolerability can be an issue with these drugs. Furthermore, the mechanisms of action of these csDMARDs are not fully understood.[3] Research during the past 25 years has brought to the forefront the role of immune cells and cytokines in driving the pathology of RA, which has led to more targeted and effective therapies that are available today.[4] However, until the first global approval of Xeljanz™ on November 6, 2012, these targeted therapies were limited to biologics, primarily monoclonal antibodies, designed to bind to specific cytokines or their extracellular receptor domains, thus blocking downstream activity. While biologics represent a significant step forward in the treatment of RA, they do have some limitations in terms of how they are administered, requiring subcutaneous injection or intravenous infusion. Furthermore, patients may not respond, exhibit partial response or lose response over time, due in part to neutralizing antibodies.[5] Consequently, a need for new treatment options remained.

For this case history, we will provide details around how we initially became interested in the JAK enzymes and what makes them compelling drug targets. Select elements of the medicinal chemistry program leading to the discovery of tofacitinib (**1**, Fig. 25.1) will be discussed, along with an analysis of the features of this molecule that contribute to its safety, efficacy, and favorable properties as an oral drug.

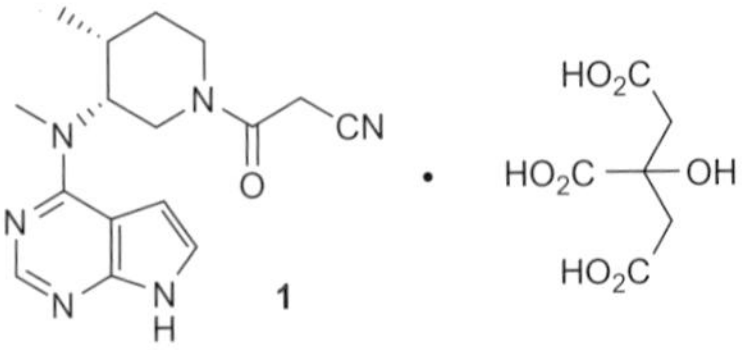

Figure 25.1 Xeljanz™ (tofacitinib citrate—CP-690,550).

2. RATIONALE FOR TARGETING THE JAK ENZYMES

Biologic disease-modifying antirheumatic drugs have clearly demonstrated the utility of blocking specific cytokines in the treatment of RA. However, because of their size (~150 kDa), their activity is restricted to extracellular events in cytokine messaging. Another way to modulate the activity of cytokines is to target the intracellular signaling pathways that they utilize in transferring information from the extracellular environment to the nucleus of the cell.[6] The JAK/STAT pathways represent one such set of signaling networks and were the focus of our efforts.

Cytokines signal through a complex network involving several different receptor superfamilies.[7] The JAKs are involved in the signal transduction for the class I and II receptors and operate as illustrated in Fig. 25.2.[8,9] These enzymes are constitutively expressed and associate with the different subunits of the receptors, operating in pairs.[10] Cytokine binding results in activation of the JAKs, ultimately leading to the phosphorylation of signal transducers and activators of transcription (STAT) proteins, allowing them to dimerize and translocate to the nucleus of the cell. Once in the nucleus,

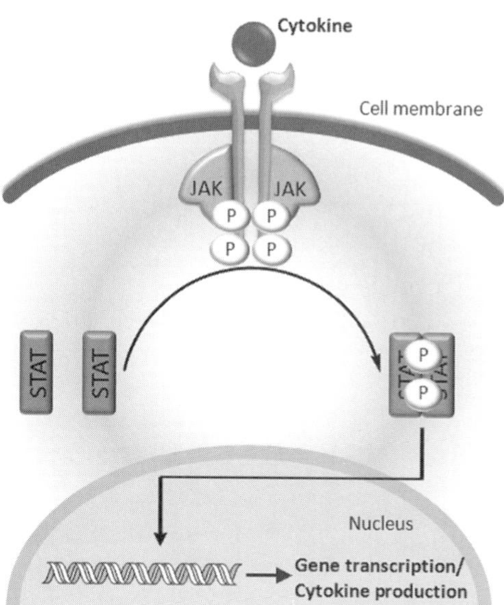

Figure 25.2 JAK/STAT signaling. *Adapted with permission from Ref. 10. Copyright 2003 Macmillan Publishers Ltd.* (See the color plate.)

the STAT dimers engage target genes, thereby affecting their expression and leading to changes in cellular function.

The JAKs are compelling as drug targets because within the complex network of signal transduction, the JAK/STAT pathways appear to be non-redundant. Furthermore, numerous cytokines implicated in the pathogenesis of a variety of inflammatory diseases (including RA) signal through the JAK pathways.[10] Small-molecule blockade of the JAKs has been shown to attenuate the signal (or message) coming from a given cytokine or cytokines.[11] Consequently, if that message is associated with some process or event in the inflammatory response, a JAK inhibitor offers the opportunity of partial and reversible modulation of that response. Figure 25.3A lists the cytokines that have been implicated in the pathogenesis of RA and indicates those (boxed) known to utilize the JAK pathways.[12,13] These cytokines, however, do not all signal through a single pathway (or receptor). This is a consequence of there being four JAK enzymes (JAK1, JAK2, JAK3, and TYK2) that pair in different combinations as illustrated in Fig. 25.3B. Cytokine specificity for a specific receptor provides a way to differentiate the messages associated with the various cytokines. For example, JAK3 always pairs with JAK1, forming the receptors that the six known common gamma-chain cytokines (IL-2, -4, -7, -9, -15, -21) signal through that are primarily associated with adaptive immunity.[14] JAK1 can also pair with JAK2 and TYK2, forming receptors that other cytokines signal through that are primarily associated with the innate immune response, including a number of proinflammatory cytokines thought to be important in RA (e.g., IL-6). JAK2 is the only JAK family member known

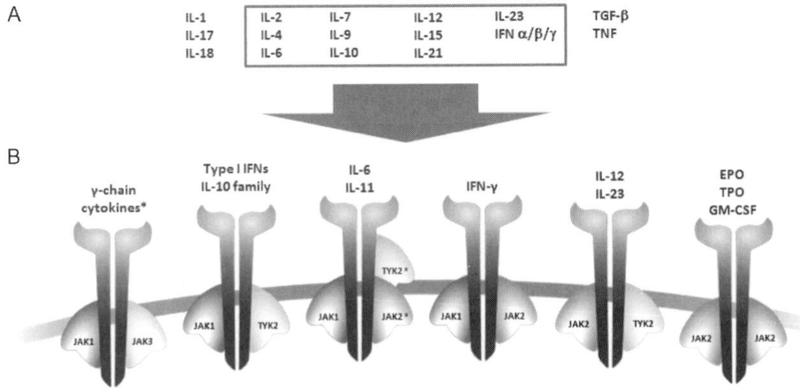

Figure 25.3 (A) Cytokines implicated in the pathogenesis of RA; those boxed utilize the JAK pathways. (B) JAK pathways resulting from combinations of four JAK enzymes. *IL-2, IL-4, IL-7, IL-9, IL-15, and IL-21. (See the color plate.)

to pair with itself, controlling downstream signaling for certain cytokines (e.g., IL-3 and IL-5) as well as growth factors such as erythropoietin (EPO), thrombopoietin (TPO), and granulocyte-macrophage colony-stimulating factor (GMCSF). Consequently, in addition to playing a role in the signaling of proinflammatory cytokines, JAK2 is also involved in other processes such as red blood cell (RBC) regulation.

In considering the appropriate balance of JAK inhibition for RA, certain of these pathways (and therefore JAKs) appear more desirable to modulate than others. Therefore, the challenges in targeting the JAK family are similar to those for many drug discovery approaches. Finding the appropriate balance of inhibition to modulate disease-causing pathways, while avoiding inhibition of pathways important for normal homeostasis, is critical to a successful drug discovery program.

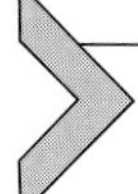

3. MEDICINAL CHEMISTRY EFFORTS CULMINATING IN THE IDENTIFICATION OF TOFACITINIB[15]

Initial reports and public disclosures occurring in the early 1990s drew our attention to JAK3 as a potential target for the prevention of solid organ transplant rejection.[16,17] At that time, JAK3 was known to be predominantly expressed in lymphoid cells (B-cells, T-cells, and NK cells); less was known about the expression of the other JAKs.[13,18] There was also human genetic data linking mutations in the common γ-chain of the IL-2 receptor family and in the JAK3 enzyme with a severe-combined-immune-deficient phenotype in humans.[19] This provided confidence that partial inhibition of JAK3 with a small molecule would result in selective immune modulation restricted to the hematopoietic system. As will be discussed in the subsequent sections, finding a selective JAK3 inhibitor presented a significant challenge, especially since structural information for the JAKs was not available in the mid-1990s when this medicinal chemistry program was first undertaken. However, empirical observations during the course of this 3-year medicinal chemistry program ultimately led to tofacitinib, which inhibits not only JAK3 but also JAK1 and the associated downstream signaling of a number of proinflammatory cytokines. The excellent efficacy observed with tofacitinib in preclinical models of RA[20] and transplant rejection[21,22] coupled with acceptable preclinical safety data led to development efforts in both indications, which over time focused primarily on RA. The following sections will describe in more detail the various steps leading to the discovery of tofacitinib.

3.1. Lead Identification: High-Throughput Screening

While today's compound libraries across the pharmaceutical industry contain a wide variety of kinase inhibitor series, this was not the case at the time of our JAK3 high-throughput screen in 1995–1996. In fact, relatively few kinase discovery programs had been prosecuted prior to this time due, in part, to a healthy level of skepticism regarding the ability to identify selective inhibitors that bind in the highly conserved enzyme ATP site. While pursuit of kinase inhibitors as cancer therapeutics was gaining support, less enthusiasm existed for programs targeting a chronic therapy for a non-oncology indication as it was assumed that adequate safety would be difficult to achieve. However, the human genetic data provided a compelling case for JAK3 inhibitors as immune modulators, so the decision was made to move forward with lead identification efforts.

Screening the corporate compound collection of approximately 400,000 compounds with an ELISA-based JAK3 enzyme assay led to the identification of a number of hits, the most interesting of which was pyrrolopyrimidine **2** (Fig. 25.4). This analog had been prepared in the Pfizer epidermal growth factor receptor (EGFR) oncology program. It demonstrated moderate JAK3 enzyme inhibitory and cell-based activity (enzyme $IC_{50} \sim 250$ nM, IL-2 blast $IC_{50} \sim 2400$ nM) and some degree of selectivity versus EGFR, insulin receptor kinase, and Lck (in-house data). Initially, JAK family selectivity was only assessed versus JAK2 as this was known to be involved in EPO signaling, and inhibitory activity could result in untoward effects on RBCs. While no structural data were available throughout the tofacitinib discovery program, the pyrrolopyrimidine moiety found in compound **2** was eventually shown to be a "hinge-binding" motif. Over

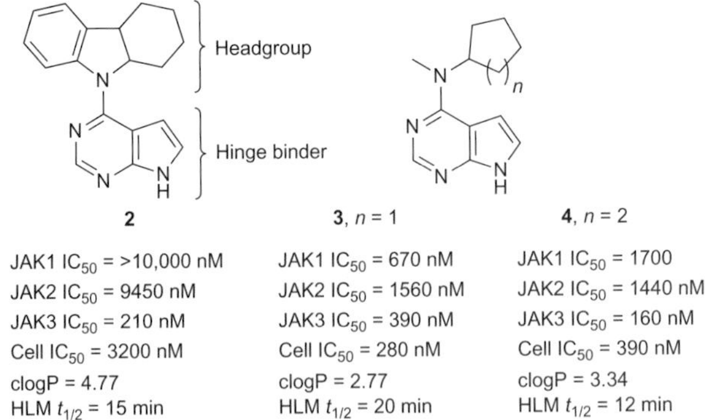

Figure 25.4 Program lead (**2**) and early analogs exhibiting improved cell potency.

the past two decades, this has become a "privileged structure," successfully utilized across the industry in the design of various ATP-competitive JAK inhibitors.

3.2. Early Structure–Activity Relationships: Developing a Pharmacophore Model

With our lead molecule in hand, structural modifications were implemented to address issues of JAK3 potency, selectivity, and drug-like properties. Compounds were tested in cell-free JAK kinase assays using a solid-phase ELISA format. Most were also evaluated in a T-cell blast cellular proliferation assay and in human liver microsome (HLM) incubations to assess *in vitro* metabolism. Promising compounds were then tested in a number of cell specificity assays to gain a better understanding of pathway selectivity and effects outside the hematopoietic compartment. Additional details of this screening strategy and assay conditions have been reported.[15]

We suspected that the pyrrolopyrimidine of **2** would be an important part of the pharmacophore since it bore resemblance to the purine of ATP. This was in fact the case; attempts to replace or modify this heteroaryl group generally led to significantly reduced JAK enzyme and/or cell potency. As a consequence, subsequent efforts focused on modifications of the amino headgroup where early structure–activity relationships (SARs) suggested more flexibility in terms of affecting kinase potency and physical property space.

Early modifications of the headgroup demonstrated that the aromatic ring of **2** could be eliminated while maintaining JAK3 potency (e.g., **3** and **4**). These fully saturated headgroups were unique at the time as published kinase inhibitors being prosecuted across the industry for other targets typically incorporated an anilide headgroup. This opened up a variety of structural and stereochemical options to pursue as we began to further modify this part of the molecule.

A breakthrough came as a result of empirical observations, relating amino headgroup structure with an improvement in cell potency. Compounds possessing an *N*-methylcycloalkyl headgroup motif (e.g., **3** and **4**) exhibited an increase in JAK1 inhibition and corresponding improvement in cell potency, suggesting that inhibition of both JAK3 and JAK1 (or JAK1 alone) may be important for functional activity. While these analogs exhibited only modest JAK1 potency in the ELISA assay, subsequent studies have shown that in other kinase assay configurations, such as caliper assays, these compounds and tofacitinib more potently inhibit JAK1. This latter point will be discussed in more detail in a subsequent section. Importantly,

4 represented a minimum pharmacophore of interest, containing a pyrrolopyrimidine hinge-binding heterocycle and amino headgroup scaffold ready for further elaboration.

3.3. Informing Headgroup Structure: High-Speed Analoging and Natural Products

As previously mentioned, at the time of this medicinal chemistry program, there were no X-ray crystal structures available for any of the JAK enzymes. In the absence of this structural information, we used high-speed analoging (HSA) as a means to explore questions of potency and selectivity around this pharmacophore (**4**). To expand the SAR in this chemotype, an HSA library was pursued in a two-step process as depicted in Fig. 25.5. A diverse set of cyclic ketones were reacted with methylamine in the presence of a polymer-supported borohydride reagent. The resulting amine products were then reacted (S_NAr) with 4-chloropyrrolo[2,3-*d*]pyrimidine affording ~100 new test compounds following purification, taking advantage of novel catch and release methodologies to isolate final products. A number of SAR findings were observed from this endeavor, key of which was a potency advantage for compounds containing a small alkyl substituent at either C-2′ or C-5′ of a cyclohexyl ring (e.g., Fig. 25.6, **5**–**7**). The C-2′,C-5′-dimethyl compound **7** was the most potent analog prepared in this study, exhibiting 20 nM potency against the JAK3 kinase and 340 nM in the IL-2 blast cellular assay. In addition to providing a potency improvement, the corresponding C-2′-methyl group, which also appears on the piperidine of tofacitinib, would later be associated with a specific binding interaction with the JAKs, likely contributing to an improvement in kinome selectivity.

The discovery of **7** represented a significant step forward in amino headgroup evolution, particularly as it relates to JAK3 potency. However, some fundamental questions about this structure remained. First, **7** existed as a mixture of isomers, creating questions as to the optimal stereochemistry

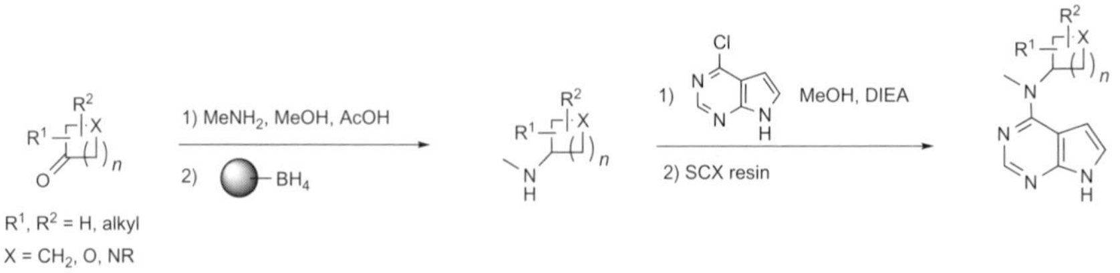

Figure 25.5 HSA protocol. DIEA, diisopropylethylamine; SCX, sulfonic acid cation exchange.

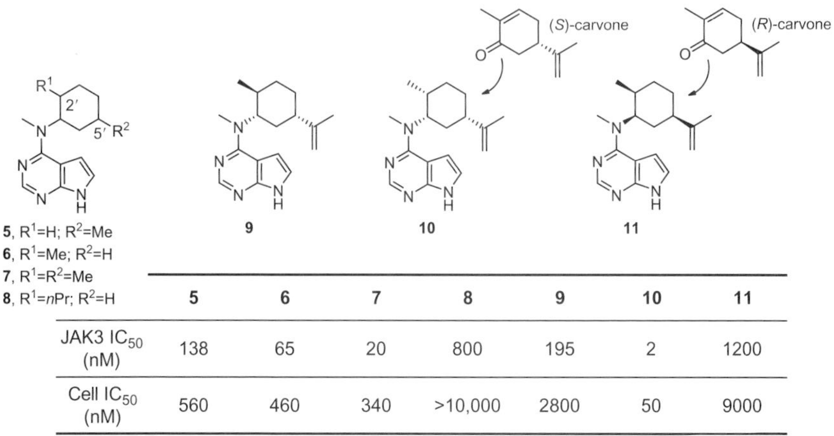

	5	6	7	8	9	10	11
JAK3 IC_{50} (nM)	138	65	20	800	195	2	1200
Cell IC_{50} (nM)	560	460	340	>10,000	2800	50	9000

Figure 25.6 HSA and carvone-derived analogs.

associated with its three chiral centers. Also, while several HSA compounds provided some limited SAR regarding which substitutions were accommodated at C-2′ and C-5′ (Fig. 25.6), questions remained regarding what other groups might be tolerated at these positions on the cyclohexane ring.

To explore this, we turned to natural products with similar substitution patterns and used them as synthetic building blocks. Data from these analogs taught us that groups larger than methyl at C-2′ were not well tolerated, consistent with HSA results (e.g., **8**). Larger groups, however, were accommodated at C-5′. Importantly, analogs prepared from the natural terpenoid, carvone indicated that the optimal relative and absolute stereochemistry around the cyclohexane ring was the all *cis*-configuration derived from (*S*)-carvone (**9–11**). Compound **10** exhibited promising kinase and cell potency; however, solubility and metabolic stability were poor as indicated by its short HLM half-life and lack of bioavailability in rodents (rat %F < 7). In an attempt to improve the physicochemical properties of this compound, the isopropenyl group of **10** was used as a synthetic handle to explore this vector. Several analogs in this "carvone series" (**12**) were eventually prepared and evaluated, confirming that a variety of substitutions were tolerated at this position in terms of JAK potency. However, in the end, it was determined that the carvone scaffold was intrinsically too lipophilic to provide a proper balance between metabolic stability and enzyme and cell potency. Additionally, the synthetic complexity of the subseries limited the scope of chemical space that could be explored at the C-5′ position. As a consequence, the team began to consider more polar and synthetically tractable headgroup scaffolds such as the piperidine analogues **13** to reduce the overall lipophilicity of the compounds and expand our understanding of the SAR in the C-5′ region of the pharmacophore.

12 13

X = –C(O)–, –C(O)O–, –C(O)NR′–, $-SO_2-$

3.4. Optimizing Property Space and ADME

In considering carvone replacements, piperidine **13**, represented an attractive alternative for further exploration. The initial piperidine analogs were prepared as an approximately 10:1 mixture of diastereomers favoring *cis*, which were eventually separated and the optimal 3*R*,4*R*-configuration established as the series progressed.[15] This template had the potential to address several of the liabilities of the cycloalkyl-derived compounds. First, introduction of a heteroatom in the ring decreased the clog*P* by at least one unit as compared to similar cyclohexyl compounds. Second, the piperidine nitrogen provided a synthetic handle to rapidly explore SAR in the C-5′ region of the carvone lead. Finally, the introduction of nitrogen in the ring simplified the structures by eliminating a stereocenter present in the carvone series. In pursuing the piperidine series, a concerted effort was made to balance polarity with potency, while attempting to limit the introduction of unnecessary molecular weight. While parameters such as ligand efficiency and lipophilic efficiency (LipE) were not widely known or used at the time of this program, the team's strategy to move to the piperidine series was consistent with the design elements behind LipE. As such, a retrospective analysis of LipE reveals the progress that was made in improving this parameter during this stage of the program; these data are included in Table 25.1.

Initial results with the piperidine series were encouraging, demonstrating that the ring nitrogen was tolerated, as was a broad range of functionality appended to this position. Following an initial wide-ranging survey of chemical space, efforts were focused on four subseries: amides, sulfonamides, ureas, and amines derived from reductive aminations (e.g., **13b**). Less than 100 nM *in vitro* kinase potency was achieved in all four series, which translated to varying degrees of cellular potency. However, in all four series, balancing potency, selectivity, and ADME properties turned out to be challenging. This was due, in part, to the fact that there was not a second site in

Table 25.1 Data for *cis*-racemic piperidine analogs and **10**

Cpd	-X-R	JAK1 IC_{50} (nM)	JAK2 IC_{50} (nM)	JAK3 IC_{50} (nM)	Cell IC_{50} (nM)	c$\log P$	HLM $t_{1/2}$ (min)	LipE[a]
10	NA	370	30	2	50	4.83	14	3.9
13a	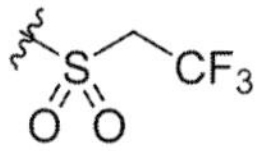	270	790	14	1260	2.83	29	5.0
13b	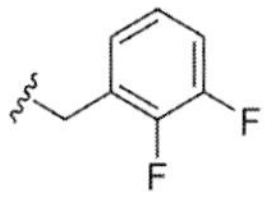	>10,000	1440	42	42	4.61	19	2.8
13c	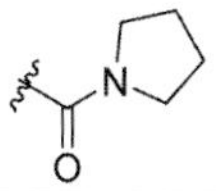	ND	142	6	156	3.30	4.5	4.9
13d	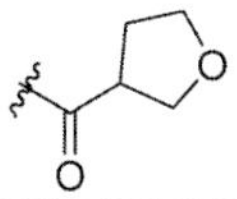	>10,000	1040	40	2500	2.01	91	5.4
13e	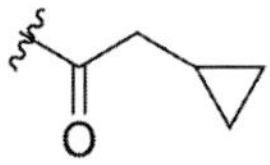	1910	460	20	230	2.55	14	5.1
13f	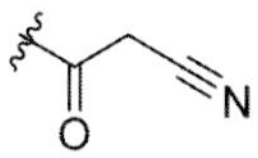	110	66	3	40	1.52	>100	7.0[b]

[a]LipE = $pIC_{50} - c\log P$ (based on JAK3 IC_{50}).
[b]LipE for 3*R*,4*R*-analog (**1**) = 7.5.

the pharmacophore that could be modified to "tune" or improve properties or potency, requiring all optimization to occur via the piperidine nitrogen substituent.

As the piperidine series evolved and we began to build a larger data set of *in vitro* HLM data, a clear trend emerged demonstrating that a lower $c\log P$ (<2.5) was necessary to achieve good microsomal stability, which correlated well with observed *in vivo* clearance in rat and dog. Attempts to achieve potency in the required lipophilicity range in several of the subseries turned out to be difficult. However, smaller, less lipophilic amide substituents did provide excellent potency, which led to a focus on this subseries. Exploration of this chemical space ultimately led to the cyanoacetamide (**13f**; *cis*-racemic form of **1**).[23] This small polar side chain struck the right balance, providing potency and microsomal stability. Chiral separation and kinetic resolution techniques eventually provided access to the more potent, 3*R*,4*R*-enantiomer of **13f** (**1**) which, in addition to providing excellent potency, also exhibited an acceptable *in vivo* pharmacokinetic profile and other desirable properties. Furthermore, the polar, low-molecular weight (MW = 312 Da, $c \log P = 1.52$) nature of **1** resulted in good aqueous solubility (>4 mg/mL). Extensive crystallization efforts identified a highly crystalline and nonhygroscopic salt form of **1** with citric acid, which had favorable properties for tablet formulation and oral administration.[24] The latter was first observed in preclinical studies where tofacitinib citrate exhibited good oral bioavailability in rat, dog, and monkey ($F = 27\%$, 78%, and 48%, respectively).[15] These data nicely predicted the 74% bioavailability that would eventually be determined for human. Taken together, these attributes contributed to the decision to advance tofacitinib to clinical trials and ultimately to the market.

4. SELECTIVITY AND PHARMACOLOGY OF TOFACITINIB

At the outset of the program, we considered what would be an appropriate selectivity profile within the JAK family for a small-molecule inhibitor. Given the information available at the time regarding protein expression and our interest in transplant rejection as an indication, the decision was made to target JAK3. However, even beyond JAK family selectivity, a more fundamental selectivity question arose regarding how to target the JAKs within the human kinome. Finding an ATP-competitive and selective kinase inhibitor safe enough for use in a chronic indication represented a significant challenge. Fortunately, our lead provided a good baseline starting point in this regard; therefore, as the medicinal chemistry program progressed and our analogs began to make additional stereospecific interactions with the JAKs, kinome selectivity improved dramatically.

Internal kinase selectivity assessments using a panel of approximately 100 enzymes indicated tofacitinib as being >1000-fold selective at 1 μM drug concentrations relative to kinases outside the JAK family.[21] Similar results were reported by external groups studying tofacitinib.[25] While empirical observations ultimately led us to a compound with this selectivity profile (**1**), high-resolution cocrystal structures available today with all four JAK enzymes have helped provide insights into the high degree of kinome selectivity.[26,27] As indicated in Fig. 25.7, tofacitinib binds deep in the ATP-binding site of JAK3 making interactions with several features of the enzyme. As anticipated, the pyrrolopyrimidine of tofacitinib binds to the hinge region. In this binding mode, the piperidine headgroup orients such that the cyanoacetamide projects toward the P-loop, filling a pocket and

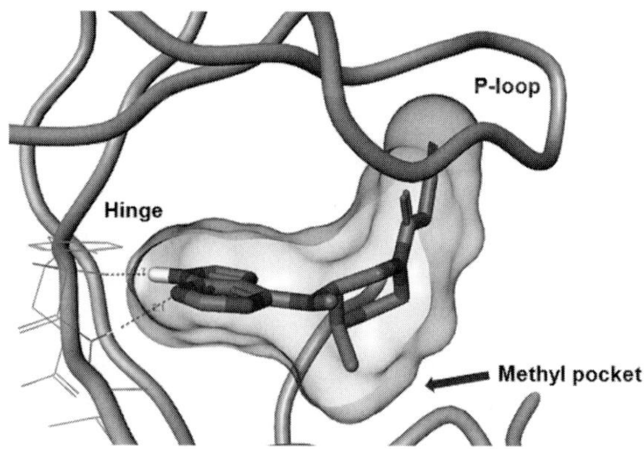

Figure 25.7 Tofacitinib cocrystal structure with JAK3. (See the color plate.)

likely making favorable interactions in this region of the kinase. This also places the piperidine ring methyl in a lipophilic pocket toward the C-terminal lobe at the base of the active site. Tofacitinib's ability to compete with ATP while still exhibiting a high degree of selectivity for the JAK family is thought to result from this combination of common structural features shared with ATP, along with the unique elements of tofacitinib binding that nicely compliment features of the JAK family ATP-binding sites. As a consequence, the kinome selectivity observed for tofacitinib is a likely contributor to its overall safety profile and lack of off-target pharmacology.

While exploitable differences exist for gaining kinome selectivity, selectivity within the JAK family is complicated by the similarity in ATP-binding sites where tofacitinib resides. This complicates IC_{50} measurements, particularly with regard to cell-free assays using enzyme catalytic domains. Original solid-phase ELISA assay IC_{50}s indicated tofacitinib to be primarily a JAK3 inhibitor, with 20- and 100-fold selectivity for JAK2 and JAK1, respectively (Table 25.2).[28] However, in other reports using caliper assays,

Table 25.2 Historic and contemporary enzymatic and cellular data for tofacitinib

	Enzymatic data; IC_{50} (nM)				**Cellular data; IC_{50} (nM)**		
Format	JAK1	JAK2	JAK3	TYK2	*IL-15 P-stat5*	*IL-6 P-stat1*	*IL-12 P-stat4*
ELISA[21]	112	20	1	ND	55.8	75.4	409
Caliper[29] [ATP] = K_M	3.2	4.1	1.6	34	*IFNα P-stat3*	*IL-23 P-stat3*	*EPO P-stat5*
Caliper[30] [ATP] = 1 mM	15.1	77.4	55.0	489	35	229	302

tofacitinib is reported to be a pan-JAK inhibitor, although somewhat less active against TYK2.[29] It is important to point out that all of these values were collected in the presence of K_M concentrations of ATP. When caliper assays are run at more physiologically relevant levels of ATP (1 mM), tofacitinib more potently inhibits JAK1 as compared to JAK3 (15 and 55 nM, respectively), reflecting the ~10-fold difference in K_M's for ATP between the two enzymes (JAK1, $K_M \sim 40$ μM; JAK3, $K_M \sim 4$ μM).[30]

Given these discrepancies, a potentially more reliable way to assess potency and pathway selectivity is through the use of appropriate cellular assays. In this regard, the original tissue culture assays wherein tofacitinib tested 11 nM (IL-2 blast T-cell proliferation assay) are in reasonably good agreement with recently reported IC_{50}s determined in primary cells using various cytokine stimuli.[30] In both cases, selectivity has been observed for the JAK1/3 and JAK1/2 pathways relative to the JAK2/2 pathway. As mentioned earlier, EPO and TPO signal through JAK2/2 pathways and are involved in hematopoiesis. Importantly, the level of functional selectivity observed for tofacitinib has proven to be manageable clinically in RA trials in terms of identifying efficacious doses that minimize hematologic effects.

5. PRECLINICAL PROPERTIES OF TOFACITINIB

Since the treatment of organ transplant rejection was the initial plan for tofacitinib, it was first tested in *in vivo* efficacy models of allograft rejection. The first efficacy study was run in the murine heterotopic cardiac transplant model. In this study, prolonged graft survival was observed in a dose-dependent manner. Improved graft survival was also observed in other rodent models as well as in nonhuman primates.[21,22]

In preclinical studies of destructive inflammatory arthritis, tofacitinib demonstrated efficacy in both mouse and rat models, exhibiting dose-dependent improvements in endpoints. Tofacitinib was shown to be efficacious in the collagen-induced arthritis model, exhibiting improvements in arthritis scores and inflammatory biomarkers. Treatment with efficacious doses of tofacitinib resulted in the rapid reduction of these endpoints within 4 h. Continued treatment with tofacitinib significantly improved arthritis scores within 48 h. Furthermore, inflammatory cell infiltrates, including T-cells and macrophages, were reduced within 7 days of treatment.[31]

Tofacitinib was also evaluated in the rat adjuvant-induced arthritis (AIA) model.[29,32] Treatment during development of disease resulted in a dose-dependent decrease in paw swelling and normalization of inflammatory mediators (IL-6 and IL-17). Importantly, osteoclast-mediated bone resorption, common in AIA, was reduced with tofacitinib treatment. Additionally, tofacitinib treatment resulted in a decrease in the level of receptor activator of nuclear factor kappa-B ligand (RANKL), resulting in a reduction in osteoclasts and associated bone resorption.

6. CLINICAL PROPERTIES OF TOFACITINIB

Tofacitinib originally entered the clinic to be evaluated as a treatment for renal transplant rejection.[33] While preclinical and Phase 2 data reported for transplant rejection were promising, the encouraging efficacy in preclinical RA disease models, along with a variety of other factors led to the re-focusing of development efforts toward RA as an initial indication. Overall, the RA development program consisted of 22 Phase 1 studies, 8 Phase 2 studies, 6 Phase 3 studies, and 2 open-label, long-term extension studies. The Phase 3 study designs are presented in Table 25.3. Safety and tolerability were assessed in a demographically diverse RA patient population. The development program demonstrated that tofacitinib consistently improved the signs and symptoms of RA, physical functioning, and other patient-reported outcomes, including health-related quality of life, across multiple lines of therapy. Furthermore, radiographic data showed a reduction of progression of structural joint damage. Safety findings observed in the overall tofacitinib RA program are consistent for a drug that modulates immune function, including serious and other important infections (e.g., tuberculosis and other opportunistic infections) and malignancies, including lymphoma.

Table. 25.3 Phase 3 RA study design for tofacitinib

	ORAL Start A3921069	**ORAL Standard A3921064**	**ORAL Scan A3921044**	**ORAL Sync A3921046**	**ORAL Solo A3921045**	**ORAL Step A3921032**
Population	MTX Naive	MTX-IR	MTX-IR	DMARD-IR	DMARD-IR	TNFi-IR
Background treatment	None	MTX	MTX	Nonbiologic DMARD(s)	None	MTX
Study duration	2 years	1 year	2 years	1 year	6 months	6 months
Number of patients	958	717	797	792	610	399

DMARD, disease-modifying antirheumatic drug; IR, inadequate responder; MTX, methotrexate; TNFi, tumor necrosis factor inhibitor.

Other adverse events observed clinically with tofacitinib include herpes zoster infections, gastrointestinal perforations, decreased neutrophil and lymphocyte counts, liver enzyme elevations, and lipid elevations.[34] Anemia was mostly mild to moderate in severity in Phase 3 studies and occurred at similar frequencies to the placebo groups over 3 months of placebo treatment, confirming efforts to minimize JAK2 inhibition.[35]

In addition to RA, tofacitinib is being developed for a number of other inflammatory indications. These include oral and topical treatments for psoriasis (Phases 3 and 2, respectively),[36,37] ulcerative colitis (Phase 3),[38] Crohn's disease (Phase 2), psoriatic arthritis (Phase 3), and ankylosing spondylitis (Phase 2).

7. CONCLUSIONS

The JAK enzymes have proven to be an important group of drug targets, with potential applications across a range of inflammatory and other indications.[30] The approval of Xeljanz™ (tofacitinib citrate) represents a number of firsts in this regard. For example, tofacitinib is the first approved kinase inhibitor for a non-oncology indication, a notion that was given a low probability of success when this program was initiated. The program benefited from a quality lead series, providing a good starting point with respect to a number of drug parameters. We prosecuted the program in the absence of guiding crystal structures, instead making progress through judicious use of HSA to maximize discoveries through empiricism, combined with the targeted use of natural products to optimize stereospecific interactions with the enzymes. Finally, by applying good medicinal chemistry principles, we eventually identified tofacitinib, a low-dose, oral drug and first-in-class treatment for RA.

ACKNOWLEDGMENTS

The authors would like to thank the extended JAK team for their contributions to the discovery work described herein. This work was funded by Pfizer Inc.

REFERENCES

1. Sacks, J. J.; Luo, Y.-H.; Helmick, C. G. *Arthritis Care Res. (Hoboken)* **2010**, *62*, 460.
2. van Vollenhoven, R. F. *Nat. Rev. Rheumatol.* **2009**, *5*, 531.
3. Feely, M. G.; Erickson, A.; O'Dell, J. R. *Expert. Opin. Pharmacother.* **2009**, *10*, 2095.
4. Feldmann, M. *Nat. Rev. Immunol.* **2002**, *2*, 364.
5. Harrold, L. R.; Reed, G. W.; Kremer, J. M.; Curtis, J. R.; Solomon, D. H.; Hochberg, M. C.; Greenberg, J. D. *Ann. Rheum. Dis.* **2014**, E-ISSN: 1468–2060.

6. Mavers, M.; Ruderman, E. M.; Perlman, H. *Curr. Rheumatol. Rep.* **2009**, *11*, 378.
7. Baker, S. J.; Rane, S. G.; Reddy, E. P. *Oncogene* **2007**, *26*, 6724.
8. Leonard, W. J.; O'Shea, J. J. *Annu. Rev. Immunol.* **1998**, *16*, 293.
9. Gadina, M.; Hilton, D.; Johnston, J. A.; Morinobu, A.; Lighvani, A.; Zhou, Y.-J.; Visconti, R.; O'Shea, J. J. *Curr. Opin. Immunol.* **2001**, *13*, 363.
10. Shuai, K.; Liu, B. *Nat. Rev. Immunol.* **2003**, *3*, 900.
11. Pesu, M.; Laurence, A.; Kishore, N.; Zwillich, S. H.; Chan, G.; O'Shea, J. J. *Immunol. Rev.* **2008**, *223*, 132.
12. McInnes, I. B.; Schett, G. *Nat. Rev. Immunol.* **2007**, *7*, 429.
13. O'Sullivan, L. A.; Liongue, C.; Lewis, R. S.; Stephenson, S. E. M.; Ward, A. C. *Mol. Immunol.* **2007**, *44*, 2497.
14. Hofmann, S. R.; Ettinger, R.; Zhou, Y.-J.; Gadina, M.; Lipsky, P.; Siegel, R.; Candotti, F.; O'Shea, J. J. *Curr. Opin. Allergy Clin. Immunol.* **2002**, *2*, 495.
15. Flanagan, M. E.; Blumenkopf, T. A.; Brissette, W. H.; Brown, M. F.; Casavant, J. M.; Chang, S.-P.; Doty, J. L.; Elliott, E. A.; Fisher, M. B.; Hines, M.; Kent, C.; Kudlacz, E. M.; Lillie, B. M.; Magnuson, K. S.; McCurdy, S. P.; Munchhof, M. J.; Perry, B. D.; Sawyer, P. S.; Strelevitz, T. J.; Subramanyam, C.; Sun, J.; Whipple, D. A.; Changelian, P. S. *J. Med. Chem.* **2010**, *53*, 8468.
16. Johnston, J. A.; Bacon, C. M.; Riedy, M. C.; O'Shea, J. J. *J. Leukoc. Biol.* **1996**, *60*, 441.
17. Witthuhn, B. A.; Silvennoinen, O.; Miura, O.; Lai, K. S.; Cwik, C.; Liu, E. T.; Ihle, J. N. *Nature (London)* **1994**, *370*, 153.
18. O'Shea, J. J.; Plenge, R. *Immunity* **2012**, *36*, 542.
19. Russell, S. M.; Tayebi, N.; Nakajima, H.; Riedy, M. C.; Roberts, J. L.; Aman, M. J.; Migone, T. S.; Noguchi, M.; Markert, M. L.; Buckley, R. H.; O'Shea, J. J.; Leonard, W. J. *Science* **1995**, *270*, 797.
20. Milici, A. J.; Kudlacz, E. M.; Audoly, L.; Zwillich, S.; Changelian, P. *Arthritis Res. Ther.* **2008**, *10*(1), R14.
21. Changelian, P. S.; Flanagan, M. E.; Ball, D. J.; Kent, C. R.; Magnuson, K. S.; Martin, W. H.; Rizzuti, B. J.; Sawyer, P. S.; Perry, B. D.; Brissette, W. H.; McCurdy, S. P.; Kudlacz, E. M.; Conklyn, M. J.; Elliott, E. A.; Koslov, E. R.; Fisher, M. B.; Strelevitz, T. J.; Yoon, K.; Whipple, D. A.; Sun, J.; Munchhof, M. J.; Doty, J. L.; Casavant, J. M.; Blumenkopf, T. A.; Hines, M.; Brown, M. F.; Lillie, B. M.; Subramanyam, C.; Chang, S.-P.; Milici, A. J.; Beckius, G. E.; Moyer, J. D.; Su, C.; Woodworth, T. G.; Gaweco, A. S.; Beals, C. R.; Littman, B. H.; Fisher, D. A.; Smith, J. F.; Zagouras, P.; Magna, H. A.; Saltarelli, M. J.; Johnson, K. S.; Nelms, L. F.; Des, E. S. G.; Hayes, L. S.; Kawabata, T. T.; Finco-Kent, D.; Baker, D. L.; Larson, M.; Si, M.-S.; Paniagua, R.; Higgins, J.; Holm, B.; Reitz, B.; Zhou, Y.-J.; Morris, R. E.; O'Shea, J. J.; Borie, D. C. *Science* **2003**, *302*, 875.
22. Kudlacz, E.; Perry, B.; Sawyer, P.; Conklyn, M.; McCurdy, S.; Brissette, W.; Flanagan, M.; Changelian, P. *Am. J. Transplant.* **2004**, *4*, 51.
23. Blumenkopf, T. A.; Flanagan, M. E.; Munchhof, M. J. US Patent 0,053,782 A1, 2001.
24. Flanagan, M. E.; Li, Z. J. US Patent 7,803,805 B2, 2010.
25. Karaman, M. W.; Herrgard, S.; Treiber, D. K.; Gallant, P.; Atteridge, C. E.; Campbell, B. T.; Chan, K. W.; Ciceri, P.; Davis, M. I.; Edeen, P. T.; Faraoni, R.; Floyd, M.; Hunt, J. P.; Lockhart, D. J.; Milanov, Z. V.; Morrison, M. J.; Pallares, G.; Patel, H. K.; Pritchard, S.; Wodicka, L. M.; Zarrinkar, P. P. *Nat. Biotechnol.* **2008**, *26*, 127.
26. Williams, N. K.; Bamert, R. S.; Patel, O.; Wang, C.; Walden, P. M.; Wilks, A. F.; Fantino, E.; Rossjohn, J.; Lucet, I. S. *J. Mol. Biol.* **2009**, *387*, 219.
27. Chrencik, J. E.; Patny, A.; Leung, I. K.; Korniski, B.; Emmons, T. L.; Hall, T.; Weinberg, R. A.; Gormley, J. A.; Williams, J. M.; Day, J. E.; Hirsch, J. L.;

Kiefer, J. R.; Leone, J. W.; Fischer, H. D.; Sommers, C. D.; Huang, H.-C.; Jacobsen, E. J.; Tenbrink, R. E.; Tomasselli, A. G.; Benson, T. E. *J. Mol. Biol.* **2010**, *400*, 413.

28. Changelian, P. S.; Moshinsky, D.; Kuhn, C. F.; Flanagan, M. E.; Munchhof, M. J.; Harris, T. M.; Doty, J. L.; Sun, J.; Kent, C. R.; Magnuson, K. S.; Perregaux, D. G.; Sawyer, P. S.; Kudlacz, E. M. *Blood* **2008**, *111*, 2155.
29. Meyer, D. M.; Jesson, M. I.; Li, X.; Elrick, M. M.; Funckes-Shippy, C. L.; Warner, J. D.; Gross, C. J.; Dowty, M. E.; Ramaiah, S. K.; Hirsch, J. L.; Saabye, M. J.; Barks, J. L.; Kishore, N.; Morris, D. L. *J. Inflamm. (Lond.)* **2010**, **7**, 41.
30. (a) Clark, J. D.; Flanagan, M. E.; Telliez, J.-B. *J. Med. Chem.* **2014**, *57*(12), 5023–5038; (b) Thorarensen, A.; Banker, M. E.; Fensome, A.; Telliez, J.-B.; Juba, B.; Vincent, F.; Czerwinski, R. M.; Casimiro-Garcia, A. *ACS Chem. Biol.* **2014**, *9*(7), 1552–1558.
31. Lin, T. H.; Hegen, M.; Quadros, E.; Nickerson-Nutter, C. L.; Appell, K. C.; Cole, A. G.; Shao, Y.; Tam, S.; Ohlmeyer, M.; Wang, B.; Goodwin, D. G.; Kimble, E. F.; Quintero, J.; Gao, M.; Symanowicz, P.; Wrocklage, C.; Lussier, J.; Schelling, S. H.; Hewet, A. G.; Xuan, D.; Krykbaev, R.; Togias, J.; Xu, X.; Harrison, R.; Mansour, T.; Collins, M.; Clark, J. D.; Webb, M. L.; Seidi, K. J. *Arthritis Rheum.* **2010**, *62*, 2283.
32. Ghoreschi, K.; Jesson, M. I.; Li, X.; Lee, J. L.; Ghosh, S.; Alsup, J. W.; Warner, J. D.; Tanaka, M.; Steward-Tharp, S. M.; Gadina, M.; Thomas, C. J.; Minnerly, J. C.; Storer, C. E.; LaBranche, T. P.; Radi, Z. A.; Dowty, M. E.; Head, R. D.; Meyer, D. M.; Kishore, N.; O'Shea, J. J. *J. Immunol.* **2011**, *186*, 4234.
33. Busque, S.; Leventhal, J.; Brennan, D. C.; Steinberg, S.; Klintmalm, G.; Shah, T.; Mulgaonkar, S.; Bromberg, J. S.; Vincenti, F.; Hariharan, S.; Slakey, D.; Peddi, V. R.; Fisher, R. A.; Lawendy, N.; Wang, C.; Chan, G. *Am. J. Transplant.* **2009**, *9*, 1936.
34. Fleischmann, R. *Curr. Opin. Rheumatol.* **2012**, *24*, 335.
35. Xeljanz, US prescribing information. Available on-line at www.Pfizer.com.
36. Boy, M. G.; Wang, C.; Wilkinson, B. E.; Chow, V. F.-S.; Clucas, A. T.; Krueger, J. G.; Gaweco, A. S.; Zwillich, S. H.; Changelian, P. S.; Chan, G. *J. Invest. Dermatol.* **2009**, *129*, 2299.
37. Papp, K. A.; Menter, A.; Strober, B.; Langley, R. G.; Buonanno, M.; Wolk, R.; Gupta, P.; Krishnaswami, S.; Tan, H.; Harness, J. A. *Br. J. Dermatol.* **2012**, *167*, 668.
38. Sandborn, W. J.; Ghosh, S.; Panes, J.; Vranic, I.; Su, C.; Rousell, S.; Niezychowski, W. *N. Engl. J. Med.* **2012**, *367*, 616.

CHAPTER TWENTY-SIX

New Chemical Entities Entering Phase III Trials in 2013

Gregory T. Notte
Gilead Sciences Inc., San Mateo, California, USA

Content

Selection Criteria

- The Phase III clinical trial must have been registered with ClinicalTrials.gov.[a]
- The chemical structure must be available. References describing the medicinal chemistry discovery effort are included if available.
- It must be the first time that this compound has reached Phase III for any indication as a single agent or in combination.
- The compound must be synthetic in origin. The following classes of drugs are not included: biologics; inorganic or organometallic compounds; ssRNA; dendrimers; endogenous substances; radiopharmaceuticals; natural polypeptides; herbal extracts.
- New formulations or single enantiomers of a previously approved drug are not included; novel prodrugs are included.
- Compounds meeting the criteria are shown as the free base or free acid except those containing quaternary nitrogens.
- This list was compiled using publically available information.[b] It is intended to give an overview of small molecule chemical matter entering Phase III and may not be all-inclusive.

[a] The selection criteria this year included compounds whose Phase III trial began in 2012, but the trial was not registered with Clinicaltrials.gov or the structure was not disclosed until 2013.

[b] The following two websites were used extensively in compiling this information: http://www.clinicaltrials.gov/ http://www.ama-assn.org/ama/pub/physician-resources/medical-science/united-states-adopted-names-council.page.Any additional information could be obtained via a web-search or from the Sponsor's website.

Annual Reports in Medicinal Chemistry, Volume 49
ISSN 0065-7743
http://dx.doi.org/10.1016/B978-0-12-800167-7.00026-2

Facts and Figures

- In 2013, there were 1390 Phase III trials registered at ClinicalTrials.gov that were classified as having a "drug intervention."
- Of the registered trials, 43 molecules (3.1%) met the selection criteria.
- For the molecules contained herein[c]:
  - Average molecular weight = 540 (range = 151.7–1893.0)
  - Average c Log D = 1.98 (range = −8.26–7.87)
- The top five indications from all trials were:
  - Type 1 and 2 diabetes (70 trials)
  - Breast cancer (35 trials)
  - Hepatitis C (31 trials)
  - Chronic obstructive pulmonary disorder (28 trials)
  - Rheumatoid arthritis (28 trials)

Overview of results from last 3 years

	Phase III trials registered	NCEs (%)	Avg MW	Avg *c* Log *D*
2011	1537	42 (2.7%)	542	3.6[d]
2012	1426	39 (2.7%)	505	1.97
2013	1390	43 (3.1%)	540	1.98

	NCEs	Approved	NDA under review	Discontinued[e]
2011	42	4 (10%)	7 (17%)	12 (28%)
2012	39	3 (8%)	6 (15%)	1 (3%)
2013	43	–	–	–

[c] Calculated using ACD labs software. MIM-D3 and Velcalcetide were excluded from the *c* Log *D* averages due to the large negative values that were obtained.

[d] These values were reported as *c* Log *P* and calculated using Chemdraw. Starting in 2012, ACD Labs was used to calculate *c* Log *D*.

[e] The designation of "discontinued" was applied to molecules for which publically available information indicated they would not be approved in the indication which was referenced in the original NCE chapter. In some cases, additional clinical development may still be ongoing.

1. Aclerastide (DSC-127)[1]

Sponsor: Derma
MW/c Log *D*: 913.03/−1.81
CAS#: 227803-63-6
Start/End Date: Feb 2013–Jul 2015
Indication: Diabetic foot ulcers
Route of Admin: Topical
MOA: Angiotensin receptor agonist
ClinicalTrials.gov Identifier NCT01830348

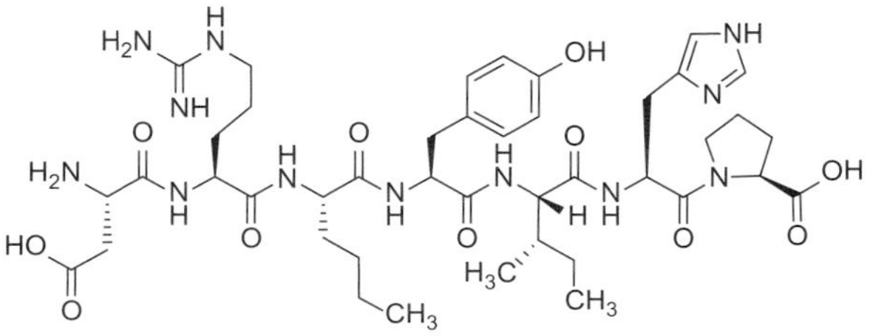

2. ARN-509 (JNJ-56021927)[2]

Sponsor: Aragon
MW/c Log *D*: 477.43/1.30
CAS#: 956104-40-8
Start/End Date: Sep 2013–Dec 2016
Indication: Castration-resistant prostate cancer
Route of Admin: Oral, 240 mg, qd
MOA: Androgen receptor antagonist
ClinicalTrials.gov Identifier: NCT01946204

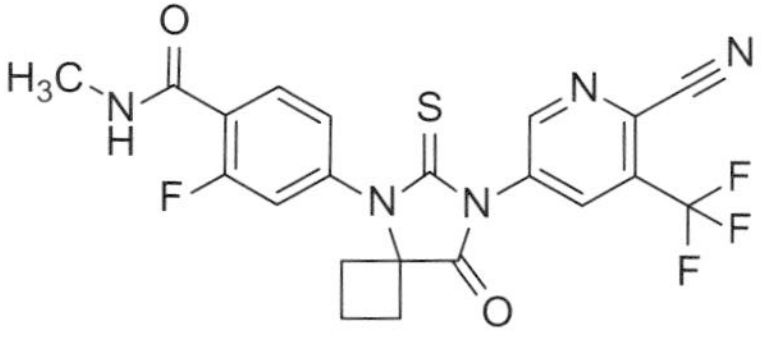

3. ASP-2151 (M-5520)[3]

Sponsor: Maruho
MW/*c* Log *D*: 482.55/2.10
CAS#: 841301-32-4
Start/End Date: Aug 2013–Dec 2014
Indication: Herpes simplex
Route of Admin: Oral, 200 mg, qd
MOA: Helicase-primase inhibitor
ClinicalTrials.gov Identifier: NCT01959295

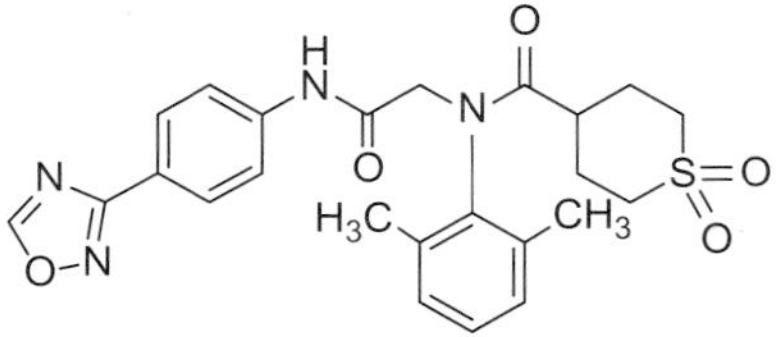

Continued

4. Beclabuvir (BMS-791325)[4]
Sponsor: Bristol-Myers Squibb
MW/c Log *D*: 659.84/1.99
CAS#: 958002-33-0
Start/End Date: Dec 2013–Aug 2014
Indication: Hepatitis C virus
Route of Admin: Oral, 75 mg, bid
MOA: NS5B polymerase inhibitor
ClinicalTrials.gov Identifier: NCT01973049

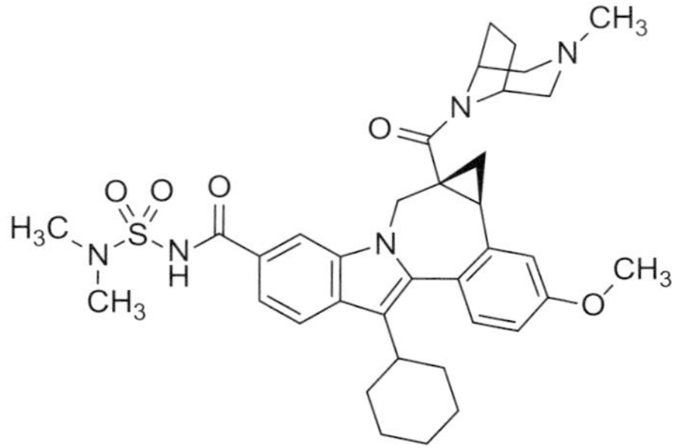

5. Binimetinib (MEK-162)[5]
Sponsor: Novartis
MW/*c* Log *D*: 441.23/5.42
CAS#: 606143-89-9
Start/End Date: Jul 2013–Dec 2014
Indication: Metastatic or unresectable cutaneous melanoma
Route of Admin: Oral, 45 mg, bid
MOA: MEK inhibitor
ClinicalTrials.gov Identifier: NCT01763164

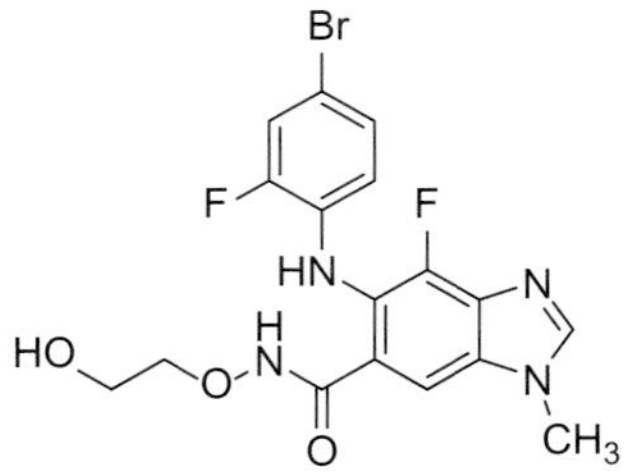

6. Cadazolid (ACT-179811)[6]
Sponsor: Actelion
MW/*c* Log *D*: 585.55/0.78
CAS#: 1025097-10-2
Start/End Date: Nov 2013–Nov 2015
Indication: Clostridium difficile-associated diarrhea
Route of Admin: Oral, 250 mg, bid
MOA: Quinolone antibiotic
ClinicalTrials.gov Identifier: NCT01987895

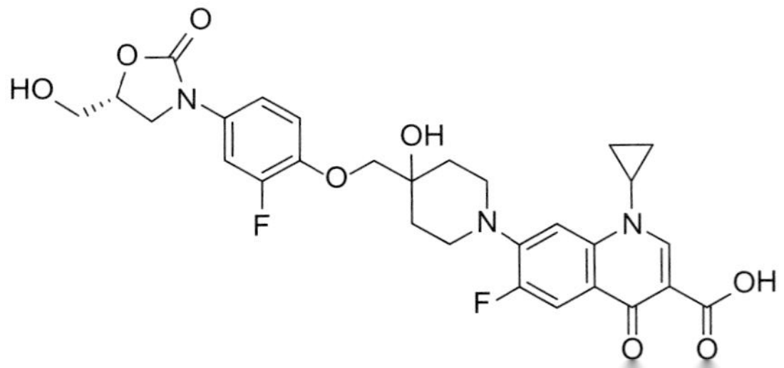

7. Cebranopadol (GRT-6005)[7]
Sponsor: Grünenthal
MW/*c* Log *D*: 378.48/4.15
CAS#: 863513-91-1
Start/End Date: Nov 2013–Mar 2016
Indication: Cancer related pain
Route of Admin: Oral, 0.2–1.0 mg, qd
MOA: Opiate receptor-like 1 agonist
ClinicalTrials.gov Identifier: NCT01964378

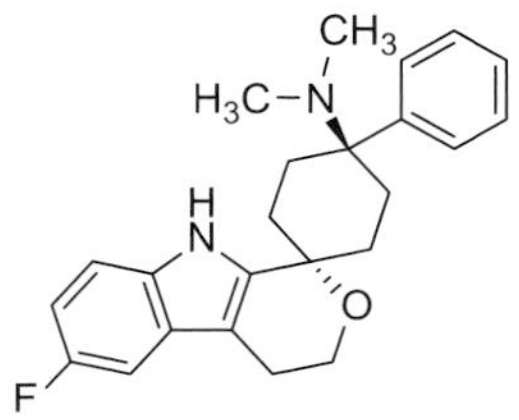

8. Ceritinib (LDK-378)[8]
Sponsor: Novartis
MW/*c* Log *D*: 558.14/2.24
CAS#: 1032900-25-6
Start/End Date: Jun 2013–Jul 2017
Indication: Non-small cell lung cancer
Route of Admin: Oral, 750 mg, qd
MOA: ALK inhibitor
ClinicalTrials.gov Identifier: NCT01828112

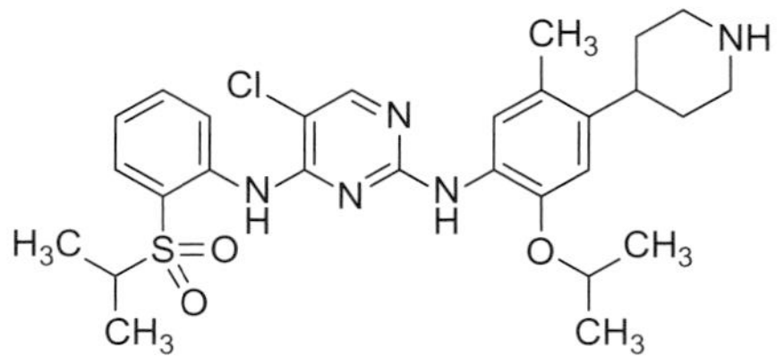

9. Decernotinib (VX-509)[9]
Sponsor: Vertex
MW/*c* Log *D*: 392.38/2.26
CAS#: 944842-54-0
Start/End Date: Apr 2013–Mar 2016
Indication: Rheumatoid arthritis
Route of Admin: Oral
MOA: JAK3 Inhibitor
ClinicalTrials.gov Identifier: NCT01830985

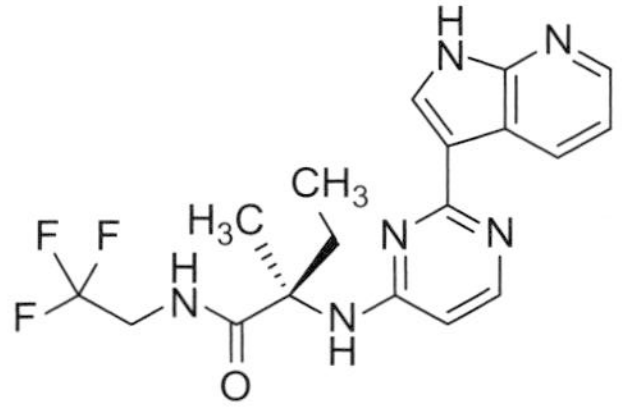

Continued

10. Delafloxacin (ABT-492)[10]
Sponsor: Melinta Therapeutics
MW/*c* Log *D*: 440.76/−1.10
CAS#: 352458-37-8
Start/End Date: Apr 2013–Jun 2014
Indication: Acute bacterial skin infections
Route of Admin: iv
MOA: DNA topoisomerase inhibitor
ClinicalTrials.gov Identifier: NCT01811732

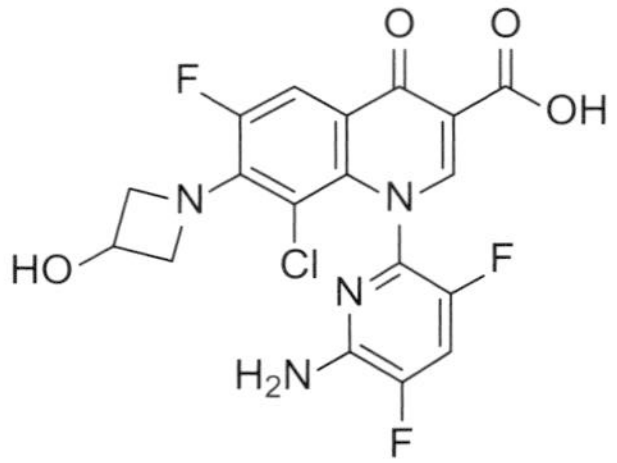

11. Duvelisib (IPI-145)[11]
Sponsor: Infinity
MW/*c* Log *D*: 416.86/4.60
CAS#: 1201438-56-3
Start/End Date: Nov 2013–Aug 2015
Indication: Chronic lymphocytic leukemia
Route of Admin: Oral, bid
MOA: PI3Kδ/γ inhibitor
ClinicalTrials.gov Identifier: NCT02004522

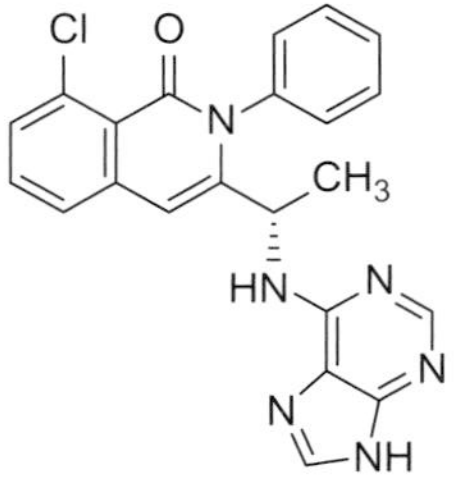

12. Elobixibat (AZD-7806)[12]
Sponsor: Ferring
MW/*c* Log *D*: 695.89/2.71
CAS#: 439087-18-0
Start/End Date: Apr 2013–Aug 2014
Indication: Chronic idiopathic constipation
Route of Admin: Oral, 5–10 mg
MOA: Ileal bile acid transporter inhibitor
ClinicalTrials.gov Identifier: NCT01827592

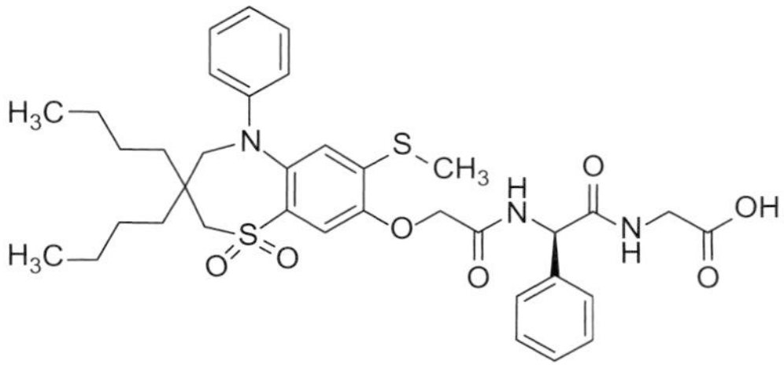

13. Emixustat (ACU-4429)[13]
Sponsor: Acucela
MW/c Log D: 263.38/0.88
CAS#: 1141777-14-1
Start/End Date: Feb 2013–Jul 2016
Indication: Dry age-related macular degeneration
Route of Admin: Oral, 2.5–10 mg, qd
MOA: Retinoid isomerohydrolase inhibitor
ClinicalTrials.gov Identifier: NCT01802866

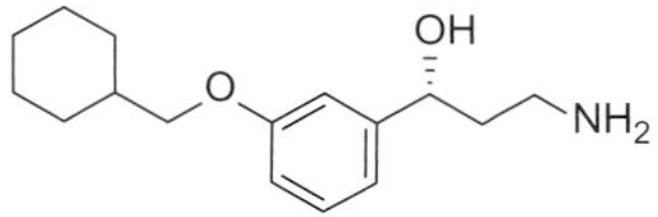

14. Encorafenib (LGX-818)[14]
Sponsor: Novartis
MW/c Log D: 540.01/1.09
CAS#: 1269440-17-6
Start/End Date: Sep 2013–Jun 2017
Indication: Metastatic melanoma with BRAF V600 mutation
Route of Admin: Oral, 300–450 mg, qd (alone or in combination with MEK-162)
MOA: B-Raf inhibitor
ClinicalTrials.gov Identifier: NCT01909453

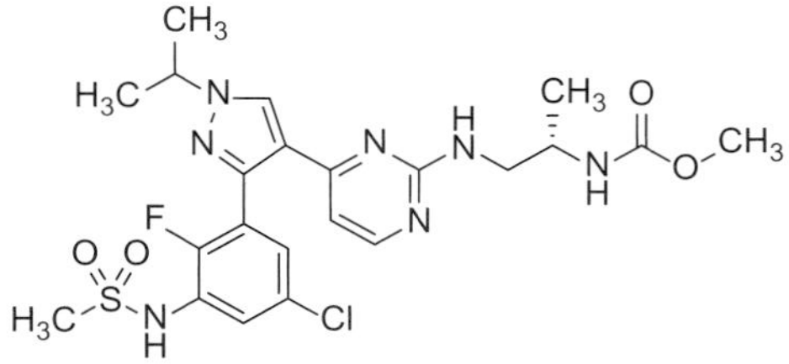

15. Eravacycline (TP-434)[15]
Sponsor: Tetraphase
MW/c Log D: 558.56/−2.48
CAS#: 1207283-85-9
Start/End Date: Aug 2013–Jan 2015
Indication: Complicated intra-abdominal infection
Route of Admin: iv
MOA: Tetracycline antibiotic
ClinicalTrials.gov Identifier: NCT01844856

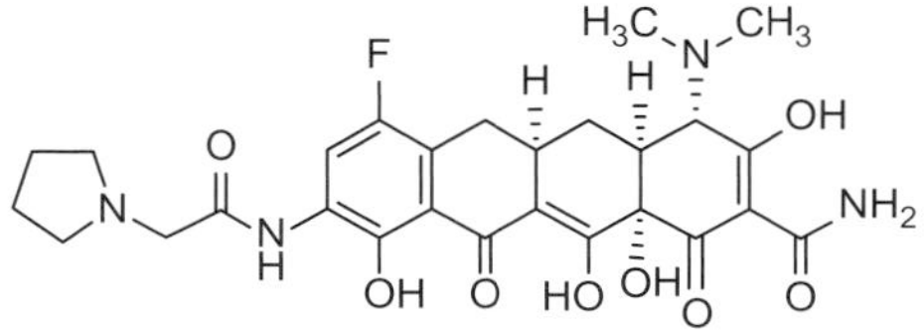

Continued

16. Erismodegib (LDE-225)[16]
Sponsor: Novartis
MW/c Log D: 485.5/5.42
CAS#: 956697-53-3
Start/End Date: May 2013–Aug 2016
Indication: Medulloblastoma
Route of Admin: Oral
MOA: Smoothened antagonist
ClinicalTrials.gov Identifier: NCT01708174

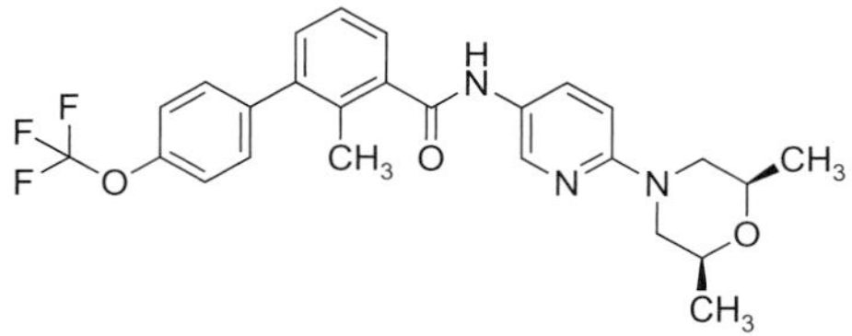

17. Ertugliflozin (PF-4971729)[17]
Sponsor: Merck Sharp & Dohme
MW/c Log D: 436.88/6.49
CAS#: 1210344-57-2
Start/End Date: Oct 2013–Aug 2015
Indication: Type 2 diabetes mellitus
Route of Admin: Oral, 5–15 mg, qd
MOA: SGLT2 inhibitor
ClinicalTrials.gov Identifier: NCT01958671

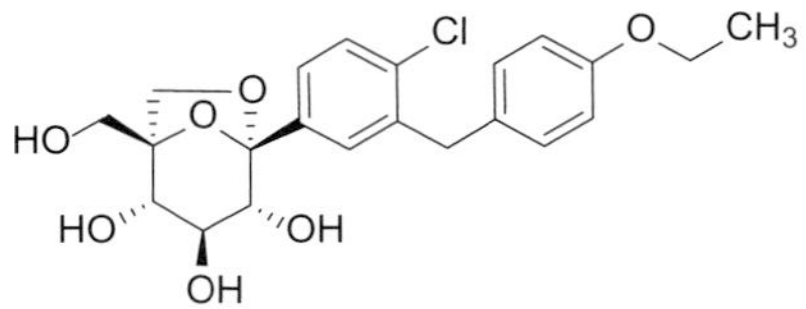

18. Favipiravir (T-705)[18]
Sponsor: MediVector
MW/c Log D: 157.1/−1.72
CAS#: 259793-96-9
Start/End Date: Dec 2013–Dec 2014
Indication: Influenza
Route of Admin: Oral, bid
MOA: DNA directed RNA polymerase inhibitor
ClinicalTrials.gov Identifier: NCT02008344

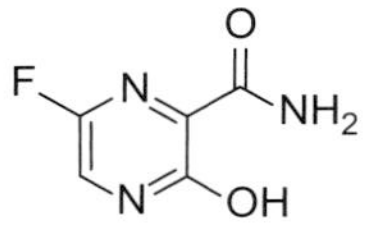

19. Latanoprostene bunod (NCX-116)[19] Sponsor: Bausch & Lomb MW/*c* Log *D*: 507.62/3.97 CAS#: 860005-21-6 Start/End Date: Jan 2013–Jun 2014 Indication: Open-angle glaucoma Route of Admin: ophthalmic solution, QD MOA: Nitric oxide donor ClinicalTrials.gov Identifier: NCT01749930	HO, HO, OH, O, O, O, NO_2
20. LEE-011[20] Sponsor: Novartis MW/*c* Log *D*: 434.54/−1.96 CAS#: 1211441-98-3 Start/End Date: Dec 2013–Jan 2017 Indication: Advanced breast cancer Route of Admin: Oral, 600 mg, qd MOA: CDK4/6 inhibitor ClinicalTrials.gov Identifier: NCT01958021	HN, N, N, H, N, N, N, N, O, CH_3, CH_3
21. Lu AE58054[21] Sponsor: Lundbeck MW/*c* Log *D*: 398.37/2.59 CAS#: 467459-31-0 Start/End Date: Oct 2013–Sep 2015 Indication: Alzheimer's disease Route of Admin: Oral, 30–60 mg, qd MOA: 5 HT_6R antagonist ClinicalTrials.gov Identifier: NCT01955161	F, HN, H, N, O, F, F, F, F

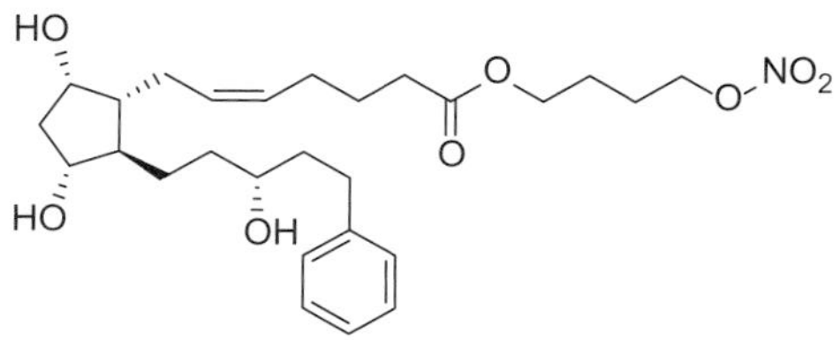

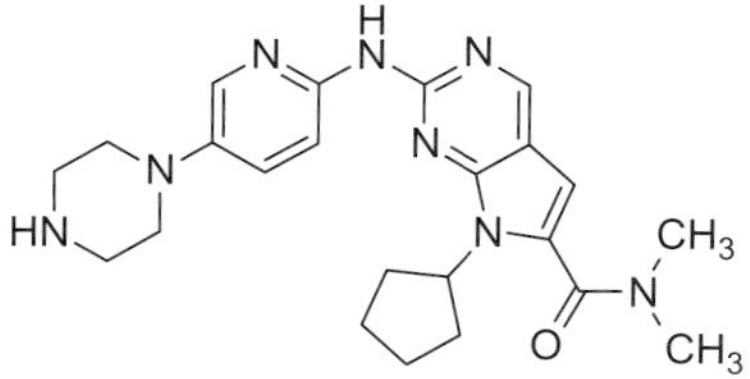

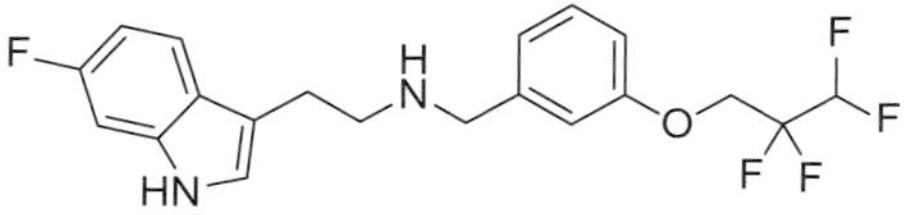

Continued

22. Lumacaftor (VX-809)[22]
Sponsor: Vertex
MW/*c* Log *D*: 452.41/2.29
CAS#: 936727-05-8
Start/End Date: Mar 2013–May 2014
Indication: Cystic fibrosis, homozygous for the F508del CFTR Mutation
Route of Admin: Oral, 600 mg, qd/400 mg, bid
MOA: ΔF508-CFTR corrector
ClinicalTrials.gov Identifier: NCT01807923

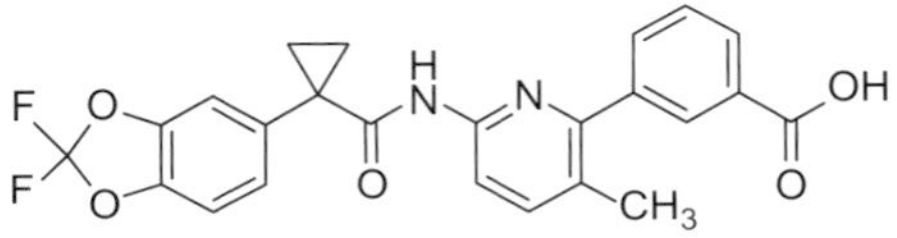

23. MIM-D3[23]
Sponsor: Mimetogen
MW/*c* Log *D*: 580.54/−8.26
CAS#: 263251-78-1
Start/End Date: Oct 2013–Jun 2014
Indication: Dry eye
Route of Admin: Ophthalmic solution
MOA: TrkA receptor agonist
ClinicalTrials.gov Identifier: NCT01960010

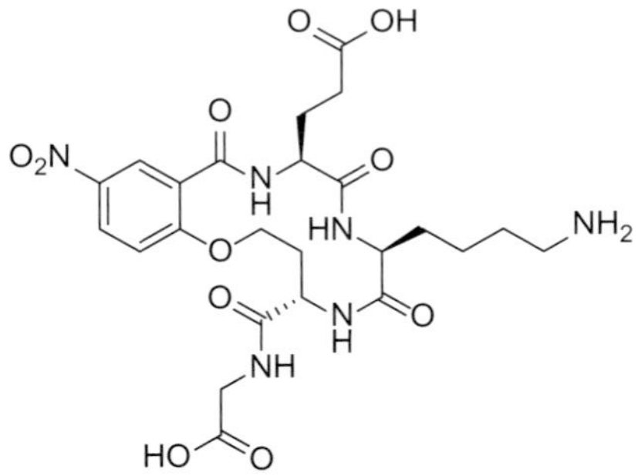

24. Momelotinib (GS-0387)[24]
Sponsor: Gilead
MW/*c* Log *D*: 414.46/1.21
CAS# 1056634-68-4
Start/End Date: Oct 2013–Jun 2016
Indication: Primary myelofibrosis
Route of Admin: Oral, qd
MOA: JAK1/2 inhibitor
ClinicalTrials.gov Identifier: NCT01969838

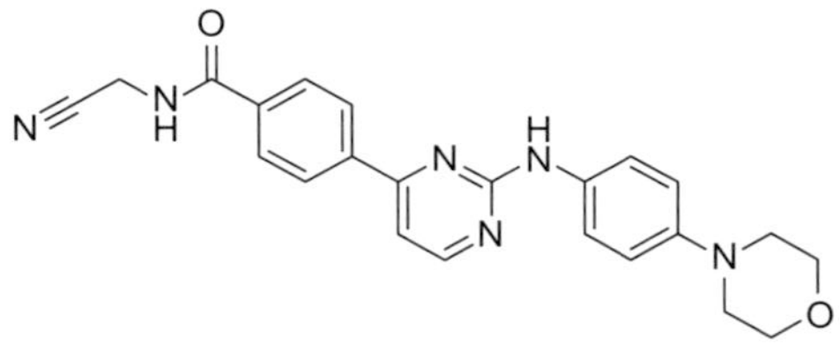

25. Naldemedine (S-297995)[25]

Sponsor: Shionogi
MW/*c* Log *D*: 570.64/1.57
CAS#: 916072-89-4
Start/End Date: Aug 2013–Jul 2015
Indication: Opioid-induced constipation
Route of Admin: Oral, qd
MOA: Opioid receptor antagonist
ClinicalTrials.gov Identifier: NCT01965158

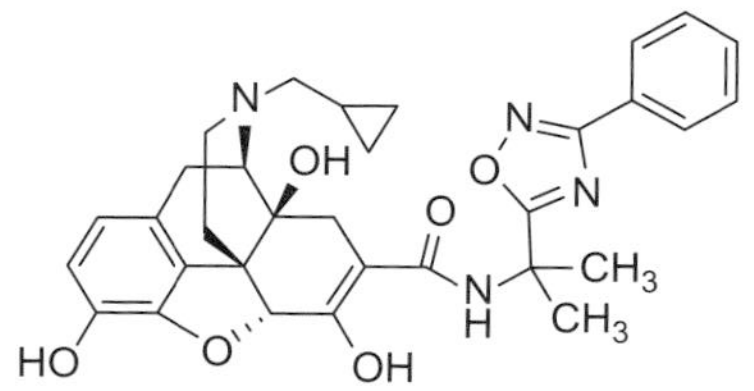

26. Niraparib (MK-4827)[26]

Sponsor: Tesaro
MW/*c* Log *D*: 320.39/0.49
CAS#: 1038915-60-4
Start/End Date: Jun 2013–Mar 2016
Indication: Platinum-sensitive ovarian cancer
Route of Admin: Oral, qd
MOA: PARP-1/2 inhibitor
ClinicalTrials.gov Identifier: NCT01847274

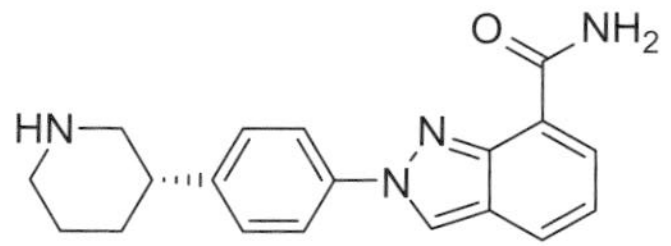

27. Olaparib (AZD-2281)[27]

Sponsor: AstraZeneca
MW/*c* Log *D*: 434.46/0.0
CAS#: 763113-22-0
Start/End Date: Aug 2013–Jul 2016
Indication: BRCA mutated ovarian cancer
Route of Admin: Oral, 300 mg, bid
MOA: PARP-1/2/3 inhibitor
ClinicalTrials.gov Identifier: NCT01844986

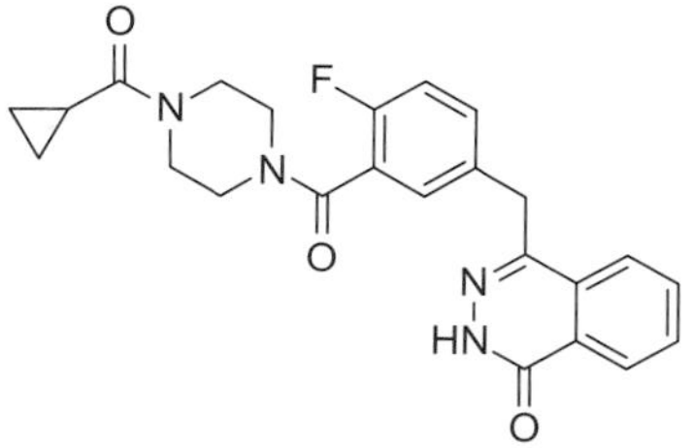

Continued

28. Ombitasvir (ABT-267)[28]
Sponsor: AbbVie
MW/*c* Log *D*: 894.11/6.29
CAS#: 1258226-87-7
Start/End Date: Nov 2012–Oct 2013
Indication: HCV infection
Route of Admin: Oral, qd
MOA: NS5A inhibitor
ClinicalTrials.gov Identifier: NCT01716585

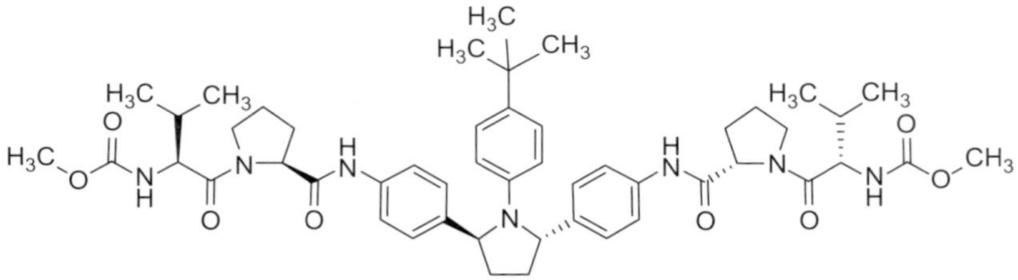

29. Pacritinib (SB-1518)[29]
Sponsor: Cell therapeutics
MW/*c* Log *D*: 472.58/2.04
CAS#: 937272-79-2
Start/End Date: Dec 2012–Aug 2014
Indication: Primary myelofibrosis
Route of Admin: Oral, 400 mg, qd
MOA: Jak2 inhibitor
ClinicalTrials.gov Identifier: NCT01773187

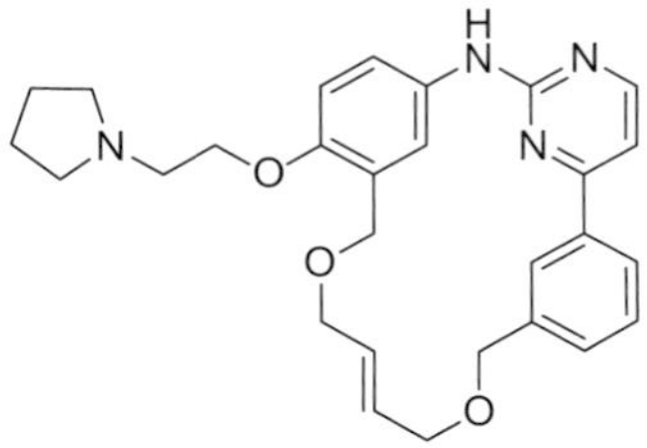

30. Palbociclib (PD-332991)[30]
Sponsor: Pfizer
MW/*c* Log *D*: 447.53/−0.10
CAS#: 571190-30-2
Start/End Date: Feb 2013–Mar 2015
Indication: Breast cancer
Route of Admin: Oral, 125 mg, qd
MOA: CDK 4/6 inhibitor
ClinicalTrials.gov Identifier: NCT01740427

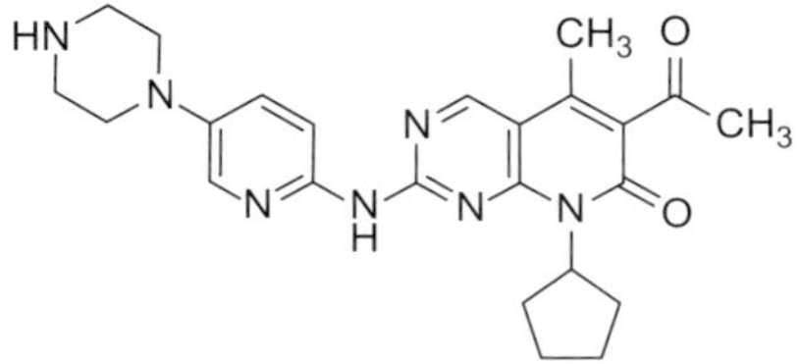

31. Pefcalcitol (M-518101)[31]
Sponsor: Maruho
MW/*c* Log *D*: 519.54/5.42
CAS#: 381212-03-9
Start/End Date: Oct 2013–Sep 2014
Indication: Plaque psoriasis
Route of Admin: Topical
MOA: Vitamin D analog
ClinicalTrials.gov Identifier: NCT01989429

32. Polmacoxib (CG100649)[32]
Sponsor: Crystal
Genomics
MW/*c* Log *D*: 361.39/2.35
CAS#: 301692-76-2
Start/End Date: Mar 2013–Dec 2013
Indication: Osteoarthritis
Route of Admin: Oral, 2 mg
MOA: Cyclooxygenase 2 inhibitor
ClinicalTrials.gov Identifier: NCT01765296

33. Reparixin (DF-1681)[33]
Sponsor: Dompé
MW/*c* Log *D*: 283.39/−0.47
CAS#: 266359-83-5
Start/End Date: Jul 2012–May 2014
Indication: Islet transplantation in diabetes mellitus type 1
Route of Admin: iv
MOA: CXCR1/CXCR2/IL-8 antagonist
ClinicalTrials.gov Identifier: NCT01817959

Continued

34. Retagliptin (SP-2086)[34]

Sponsor: Jiangsu HengRui Medicine
MW/*c* Log *D*: 464.36/1.73
CAS#: 1174122-54-3
Start/End Date: Apr 2013–Jan 2014
Indication: Type 2 diabetes mellitus
Route of Admin: Oral, 50 mg, qd or bid
MOA: Dipeptidyl peptidase IV inhibitor
ClinicalTrials.gov Identifier: NCT01970046

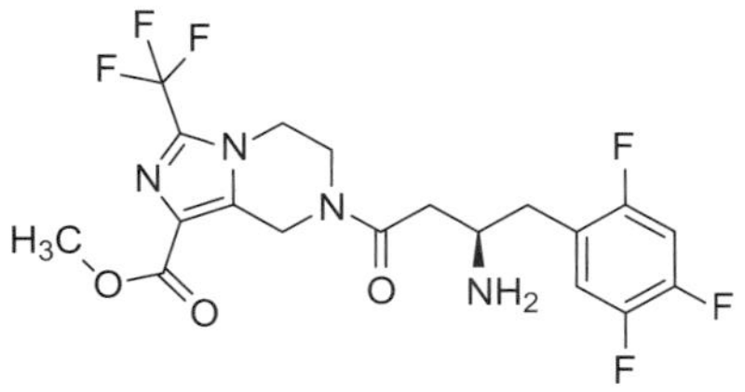

35. Selumetinib (ARRY-886)[35]

Sponsor: AstraZeneca
MW/*c* Log *D*: 457.68/5.55
CAS#: 606143-52-6
Start/End Date: Sep 2013–Jul 2016
Indication: Non-small cell lung cancer with KRAS mutation
Route of Admin: Oral, 75 mg, bid
MOA: MEK inhibitor
ClinicalTrials.gov Identifier: NCT01933932

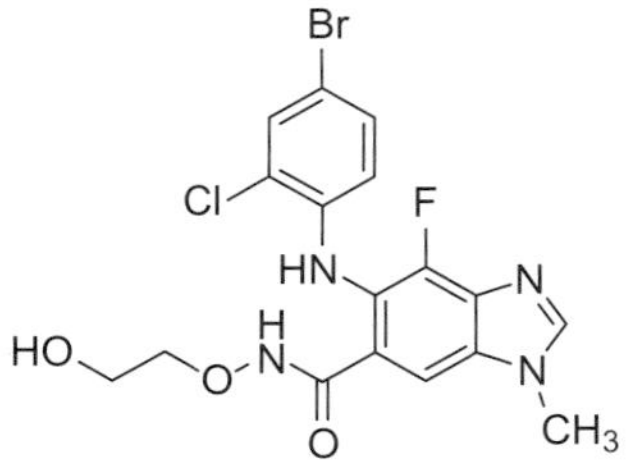

36. Talazoparib (BMN-673)[36]

Sponsor: BioMarin
MW/*c* Log *D*: 380.35/1.91
CAS#: 1207456-01-6
Start/End Date: Oct 2013–Jun 2016
Indication: BRCA mutated breast cancer
Route of Admin: Oral, 1.0 mg, qd
MOA: PARP-1/2 inhibitor
ClinicalTrials.gov Identifier: NCT01945775

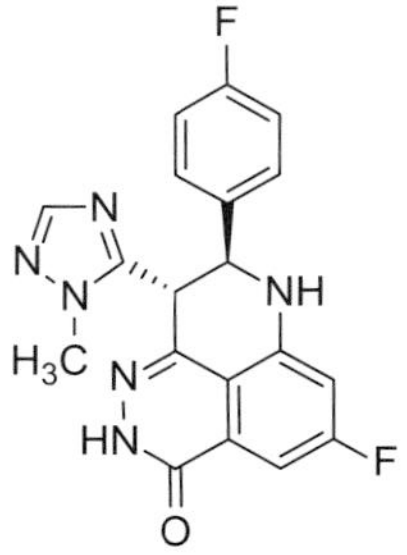

37. Tenofovir alafenamide (GS-7340)[37]
Sponsor: Gilead
MW/*c* Log *D*: 476.47/2.20
CAS#: 379270-37-8
Start/End Date: Dec 2012–Sep 2014
Indication: HIV infection
Route of Admin: Oral, 10 mg, qd
MOA: Reverse transcriptase inhibitor
ClinicalTrials.gov Identifier: NCT01780506

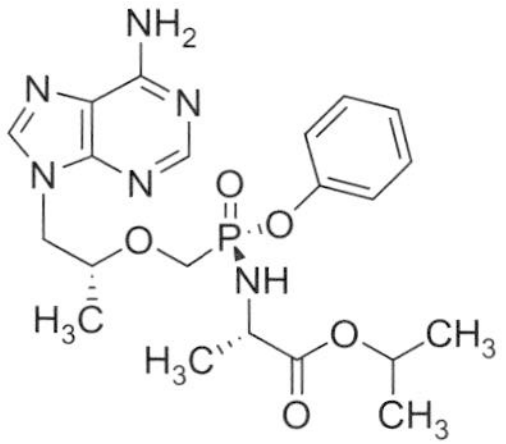

38. Tetrabenazine-d6 (SD-809)[38]
Sponsor: Auspex
MW/*c* Log *D*: 323.46/3.43
CAS#: 1221885-59-1
Start/End Date: Jun 2013–Aug 2014
Indication: Chorea
Route of Admin: Oral, 6–12 mg
MOA: Vesicular monoamine transporter 2 inhibitor
ClinicalTrials.gov Identifier: NCT01795859

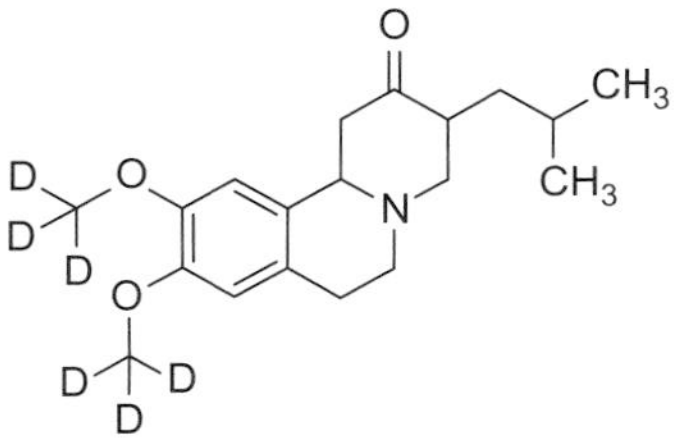

39. Velcalcetide (AMG-416)[39]
Sponsor: Amgen
MW/*c* Log *D*: 1016.19/–
CAS#: 1262780-97-1
Start/End Date: Jul 2013–May 2015
Indication: Secondary hyperparathyroidism
Route of Admin: iv
MOA: Calcium-sensing receptor agonist
ClinicalTrials.gov Identifier: NCT01785875

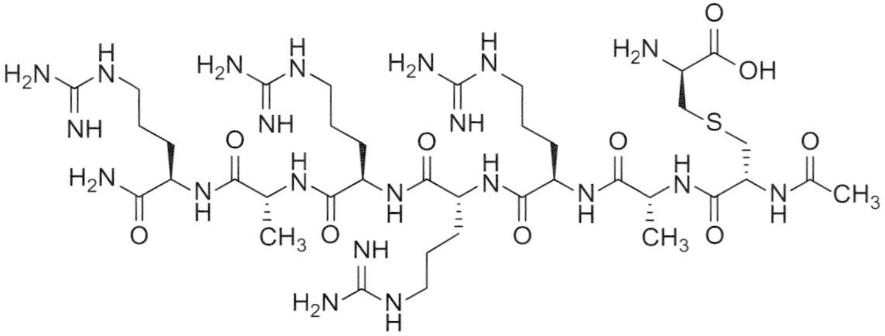

Continued

40. Venetoclax (ABT-199)[40]

Sponsor: AbbVie
MW/*c* Log *D*: 868.44/7.87
CAS#: 1257044-40-8
Start/End Date: Dec 2013–Aug 2018
Indication: Chronic lymphocytic leukemia
Route of Admin: Oral, 400 mg, qd
MOA: Bcl-2 inhibitor
ClinicalTrials.gov Identifier: NCT02005471

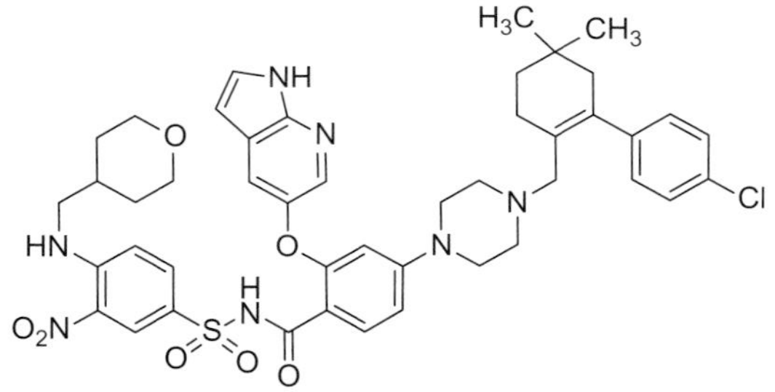

41. Veruprevir (ABT-450)[41]

Sponsor: AbbVie
MW/*c* Log *D*: 765.88/−0.87
CAS#: 1216941-48-8
Start/End Date: Aug 2012–Jan 2014
Indication: HCV infection
Route of Admin: Oral, qd, with ritonavir
MOA: NS3 protease inhibitor
ClinicalTrials.gov Identifier: NCT01674725

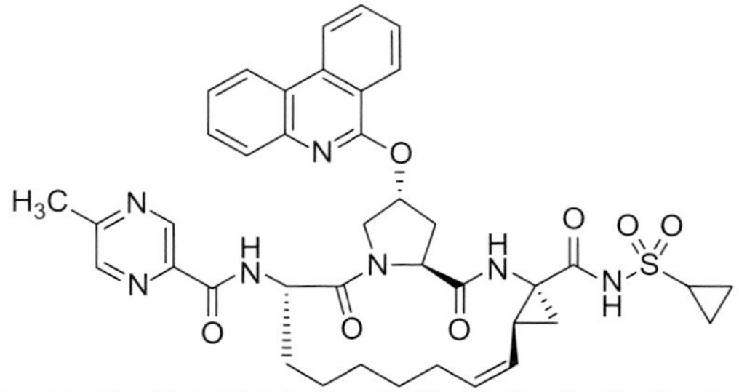

42. Volasertib (BI-6727)[42]

Sponsor: Boehringer Ingelheim
MW/*c* Log *D*: 618.81/1.72
CAS#: 755038-65-4
Start/End Date: Jan 2013–Jan 2016
Indication: Acute myeloid leukemia
Route of Admin: iv
MOA: Polo-like kinase-1 inhibitor
ClinicalTrials.gov Identifier: NCT01721876

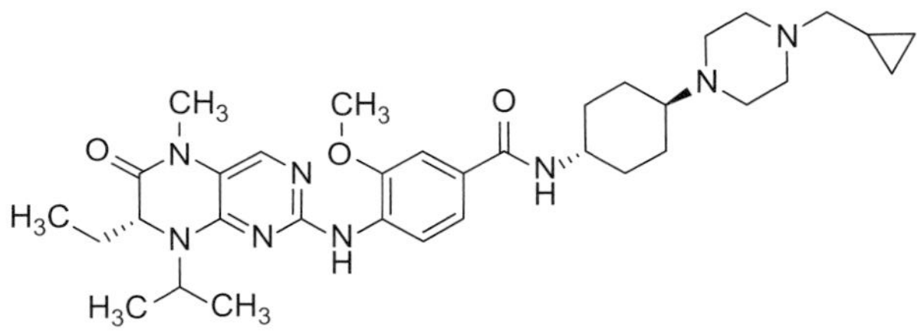

43. Zoptarelin Doxorubicin (AN-152)[43]

Sponsor: AEterna Zentaris
MW/*c* Log *D*: 1893.01/−2.44
CAS#: 139570-93-7
Start/End Date: Apr 2013–Dec 2015
Indication: Endometrial cancer
Route of Admin: iv
MOA: DNA topoisomerase II inhibitor
ClinicalTrials.gov Identifier: NCT01767155

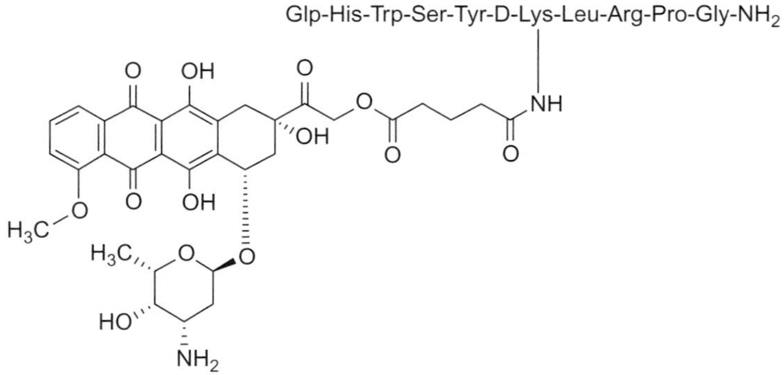

Glp-His-Trp-Ser-Tyr-D-Lys-Leu-Arg-Pro-Gly-NH_2

REFERENCES

1. Rodgers, K.; Verco, S.; Bolton, L.; Dizerega, G. *Expert Opin. Investig. Drugs* **2011**, *20*, 1575.
2. Clegg, N. J.; Wongvipat, J.; Joseph, J. D.; Tran, C.; Ouk, S.; Dilhas, A.; Chen, Y.; Grillot, K.; Bischoff, E. D.; Cai, L.; Aparicio, A.; Dorow, S.; Arora, V.; Shao, G.; Qian, J.; Zhao, H.; Yang, G.; Cao, C.; Sensintaffar, J.; Wasielewska, T.; Herbert, M. R.; Bonnefous, C.; Darimont, B.; Scher, H. I.; Smith-Jones, P.; Klang, M.; Smith, N. D.; De Stanchina, E.; Wu, N.; Ouerfelli, O.; Rix, P. J.; Heyman, R. A.; Jung, M. E.; Sawyers, C. L.; Hager, J. H. *Cancer Res.* **2012**, *72*, 1494.
3. Chono, K.; Katsumata, K.; Kontani, T.; Kobayashi, M.; Sudo, K.; Yokota, T.; Konno, K.; Shimizu, Y.; Suzuki, H. *J. Antimicrob. Chemother.* **2010**, *65*, 1733.
4. Gao, M. Gardiner, D. F.; Lemm, J. A.; McPhee, F.; Voss, S. A. Patent Application US 2012/196794, 2012.
5. Fritsch, C.; Huang, X.; Boehm, M.; Di Tomaso, E.; Cosaert J. G. C. E. Patent Application WO 2013/066483, 2013.
6. Locher, H. H.; Seiler, P.; Chen, X.; Schroeder, S.; Pfaff, P.; Enderlin, M.; Klenk, A.; Fournier, E.; Hubschwerlen, C.; Ritz, D.; Kelly, C. P.; Keck, W. *Antimicrob. Agents Chemother.* **2014**, *58*, 892.
7. Frosch, S.; Linz, K.; Bloms-Funke, P. Patent Application WO 2013/113857, 2013. http://www.ama-assn.org/resources/doc/usan/cebranopadol.pdf.
8. Marsilje, T. H.; Pei, W.; Chen, B.; Lu, W.; Uno, T.; Jin, Y.; Jiang, T.; Kim, S.; Li, N.; Warmuth, M.; Sarkisova, Y.; Sun, F.; Steffy, A.; Pferdekamper, A. C.; Li, A. G.; Joseph, S. B.; Kim, Y.; Liu, B.; Tuntland, T.; Cui, X.; Gray, N. S.; Steensma, R.; Wan, Y.; Jiang, J.; Chopiuk, G.; Li, J.; Gordon, W. P.; Richmond, W.; Johnson, K.; Chang, J.; Groessl, T.; He, Y. Q.; Phimister, A.; Aycinena, A.; Lee, C. C.; Bursulaya, B.; Karanewsky, D. S.; Seidel, H. M.; Harris, J. L.; Michellys, P. Y. *J. Med. Chem.* **2013**, *56*, 5675.
9. Tanoury, G. J.; Jung, Y. C.; Magdziak, D.; Looker, A. Patent Application WO 2013/006634, 2013.
10. Haight, R. A.; Arizman, Z. S.; Barnes, M. D.; Benz, J. N.; Gueffier, X. F.; Henry, F. R.; Hsu, C. M.; Lee, C. E.; Morin, L.; Pearl, K. B.; Peterson, M. J.; Plata, D. J.; Willcox, D. R. *Org. Process Res. Dev.* **2006**, *10*, 751.
11. Boyle, D. L.; Kim, H.; Topolewski, K.; Bartok, B.; Firestein, G. S. *J. Pharmacol. Exp. Ther.* **2014**, *348*, 271.
12. Tack, J.; Corsetti, M. *Drugs Fut.* **2012**, *37*, 475.
13. Kubota, R.; Boman, N. L.; David, R.; Mallikaarjun, S.; Patil, S.; Birch, D. *Retina* **2012**, *32*, 183, http://www.ama-assn.org/resources/doc/usan/emixusat.pdf.
14. Verma, D.; Krishnamachari, Y.; Shen, X.; Lee, H.; Li, P.; Singh, R.; Tan, L. Patent Application WO/2013078264, 2013.
15. Xiao, X. Y.; Hunt, D. K.; Zhou, J.; Clark, R. B.; Dunwoody, N.; Fyfe, C.; Grossman, T. H.; O'Brien, W. J.; Plamondon, L.; Rönn, M.; Sun, C.; Zhang, W. J.; Sutcliffe, J. A. *J. Med. Chem.* **2012**, *55*, 597.
16. Pan, S.; Wu, X.; Jiang, J.; Gao, W.; Wa, Y.; Cheng, D.; Han, D.; Liu, J.; Englund, N. P.; Wang, Y.; Peukert, S.; Miller-Moslin, K.; Yuan, J.; Guo, R.; Matsumoto, M.; Vattay, A.; Jiang, Y.; Tsao, J.; Sun, F.; Pferdekamper, A. C.; Dodd, S.; Tuntland, T.; Maniara, W.; Kelleher, J. F., III; Yao, Y.; Warmuth, M.; Williams, J.; Dorsch, M. *ACS Med. Chem. Lett.* **2010**, *1*, 130.
17. Mascitti, V.; Maurer, T. S.; Robinson, R. P.; Bian, J.; Boustany-Kari, C. M.; Brandt, T.; Collman, B. M.; Kalgutkar, A. S.; Klenotic, M. K.; Leininger, M. T.; Lowe, A.; Maguire, R. J.; Masterson, V. M.; Miao, Z.; Mukaiyama, E.; Patel, J. D.; Pettersen, J. C.; Préville, C.; Samas, B.; She, L.; Sobol, Z.; Steppan, C. M.;

Stevens, B. D.; Thuma, B. A.; Tugnait, M.; Zeng, D.; Zhu, T. *J. Med. Chem.* **2011**, *54*, 2952.

18. Kiso, M.; Takahashi, K.; Sakai-Tagawa, Y.; Shinya, K.; Sakabe, S.; Le, Q. M. *Proc. Natl. Acad. Sci. U.S.A.* **2010**, *107*, 882.
19. Krauss, A. H.; Impagnatiello, F.; Toris, C. B.; Gale, D. C.; Prasanna, G.; Borghi, V.; Chiroli, V.; Chong, W. K.; Carreiro, S. T.; Ongini, E. *Exp. Eye Res.* **2011**, *93*, 250, http://www.ama-assn.org/resources/doc/usan/latanoprostene-bunod.pdf.
20. Calienni, J. V.; Chen, G. P.; Gong, B.; Kapa, P. K.; Saxena, V. Patent Application US 2012/0115878.
21. Arnt, J.; Bang-Andersen, B.; Grayson, B.; Bymaster, F. P.; Cohen, M. P.; DeLapp, N. W.; Giethlen, B.; Kreilgaard, M.; McKinzie, D. L.; Neill, J. C.; Nelson, D. L.; Nielsen, S. M.; Poulsen, M. N.; Schaus, J. M.; Witten, L. M. *Int. J. Neuropsychopharmacol.* **2010**, *13*, 1021.
22. Okiyoneda, T.; Veit, G.; Dekkers, J. F.; Bagdany, M.; Soya, N.; Xu, H.; Roldan, A.; Verkman, A. S.; Kurth, M.; Simon, A.; Hegedus, T.; Beekman, J. M.; Lukacs, G. L. *Nat. Chem. Biol.* **2013**, *9*, 444.
23. Maliartchouk, S.; Feng, Y.; Ivanisevic, L.; Debeir, T.; Cuello, A. C.; Burgess, K.; Saragovi, H. U. *Mol. Pharmacol.* **2000**, *57*, 385.
24. Burns, C. J.; Bourke, D. G.; Andrau, L.; Bu, X.; Charman, S. A.; Donohue, A. C.; Fantino, E.; Farrugia, M.; Feutrill, J. T.; Joffe, M.; Kling, M. R.; Kurek, M.; Nero, T. L.; Nguyen, T.; Palmer, J. T.; Phillips, I.; Shackleford, D. M.; Sikanyika, H.; Styles, M.; Su., S.; Treutlein, H.; Zeng, J.; Wilks, A. F. *Bioorg. Med. Chem. Lett.* **2009**, *19*, 5887.
25. Mashimo, A.; Ichio, S. Patent Application WO 2013/172297, 2013. http://www.ama-assn.org/resources/doc/usan/naldemedine.pdf.
26. Jones, P.; Altamura, S.; Boueres, J.; Ferrigno, F.; Fonsi, M.; Giomini, C.; Lamartina, S.; Monteagudo, E.; Ontoria, J. M.; Orsale, M. V.; Palumbi, M. C.; Pesci, S.; Roscilli, G.; Scarpelli, R.; Schultz-Fademrecht, C.; Toniatti, C.; Rowley, M. *J. Med. Chem.* **2009**, *52*, 7170.
27. Menear, K. A.; Adcock, C.; Boulter, R.; Cockcroft, X. L.; Copsey, L.; Cranston, A.; Dillon, K. J.; Drzewiecki, J.; Garman, S.; Gomez, S.; Javaid, H.; Kerrigan, F.; Knights, C.; Lau, A.; Loh, V. M.; Matthews, I. T.; Moore, S.; O'Connor, M. J.; Smith, G. C.; Martin, N. M. *J. Med. Chem.* **2008**, *51*, 6581.
28. Kapoor, M. J. Patent Application WO 2013/101552, 2013. Note: Trial began in 2012, but structure was not available until 2013.
29. William, A. D.; Lee, A. C.; Blanchard, S.; Poulsen, A.; Teo, E. L.; Nagaraj, H.; Tan, E.; Chen, D.; Williams, M.; Sun, E. T.; Goh, K. C.; Ong, W. C.; Goh, S. K.; Hart, S.; Jayaraman, R.; Pasha, M. K.; Ethirajulu, K.; Wood, J. M.; Dymock, B. W. *J. Med. Chem.* **2011**, *54*, 4638.
30. Toogood, P. L.; Harvey, P. J.; Repine, J. T.; Sheehan, D. J.; VanderWel, S. N.; Zhou, H.; Keller, P. R.; McNamara, D. J.; Sherry, D.; Zhu, T.; Brodfuehrer, J.; Choi, C.; Barvian, M. R.; Fry, D. W. *J. Med. Chem.* **2005**, *48*, 2388.
31. Shimizu, K.; Kawase, A.; Haneishi, T.; Kato, Y.; Kinoshita, K.; Ohmori, M.; Furuta, Y.; T, Emura.; Kato, N.; Mitsui, T.; Yamaguchi, K.; Morita, K.; Sekiguchi, N.; Yamamoto, T.; Matsushita, T.; Shimaoka, S.; Sugita, A.; Morikawa, K. *Bioorg. Med. Chem. Lett.* **2006**, *16*, 3323.
32. Shin, S. S.; Byun, Y.; Lim, K. M.; Choi, J. K.; Lee, K.; Moh, J. H.; Kim, J. K.; Jeong, Y. S.; Kim, J. Y.; Choi, Y. H.; Koh, H.; Park, Y.; Oh, Y. I.; Noh, M.; Chung, S. *J. Med. Chem.* **2004**, *47*, 792.
33. Allegretti, M.; Bertini, R.; Cesta, M. C.; Bizzarri, C.; Di Bitondo, R.; Di Cioccio, V.; Galliera, E.; Berdini, V.; Topai, A.; Zampella, G.; Russo, V.; Di Bello, N.; Nano, G.;

Nicolini, L.; Locati, M.; Fantucci, P.; Florio, S.; Colotta, F. *J. Med. Chem.* **2005**, *48*, 4312, Note: trial was begun in 2012, but not registered until 2013.
34. Yuan, K.; Sun, P. Patent Application WO 2010/111905, 2010.
35. Revill, P.; Serradell, N.; Bolos, J.; Bozzo, J. *Drugs Fut.* **2006**, *31*, 85.
36. Shen, Y.; Rehman, F. L.; Feng, Y.; Boshuizen, J.; Bajrami, I.; Elliott, R.; Wang, B.; Lord, C. J.; Post, L. E.; Ashworth, A. *Clin. Cancer Res.* **2013**, *19*, 5003.
37. Lee, W. A.; He, G.; Eisenberg, E.; Cihlar, T.; Swaminathan, S.; Mulato, A.; Cundy, K. C. *Antimicrob. Agents Chemother.* **2005**, *49*, 1898, Note: trial was begun in 2012, but not registered until 2013.
38. Gant, T. G.; Zhang, C.; Shahbaz, M. Patent Application WO 2011/153157, 2011. http://download.ama-assn.org/resources/doc/usan/x-pub/dutetrabenazine.pdf.
39. Walter, S.; Baruch, A.; Dong, J.; Tomlinson, J. E.; Alexander, S. T.; Janes, J.; Hunter, T.; Yin, Q.; Maclean, D.; Bell, G.; Mendel, D. B.; Johnson, R. M.; Karim, F. *J. Pharmacol. Exp. Ther.* **2013**, *346*, 229.
40. Souers, A. J.; Leverson, J. D.; Boghaert, E. R.; Ackler, S. L.; Catron, N. D.; Chen, J.; Dayton, B. D.; Ding, H.; Enschede, S. H.; Fairbrother, W. J.; Huang, D. C.; Hymowitz, S. G.; Jin, S.; Khaw, S. L.; Kovar, P. J.; Lam, L. T.; Lee, J.; Maecker, H. L.; Marsh, K. C.; Mason, K. D.; Mitten, M. J.; Nimmer, P. M.; Oleksijew, A.; Park, C. H.; Park, C. M.; Phillips, D. C.; Roberts, A. W.; Sampath, D.; Seymour, J. F.; Smith, M. L.; Sullivan, G. M.; Tahir, S. K.; Tse, C.; Wendt, M. D.; Xiao, Y.; Xue, J. C.; Zhang, H.; Humerickhouse, R. A.; Rosenberg, S. H.; Elmore, S. W. *Nat. Med.* **2013**, *19*, 202.
41. Sheikh, A. Y.; Moiz, D.; Pal, A. E.; Gong, Y.; Brackemeyer, P. J.; Zhang, G. G.; Wagaw, S. Patent Application WO2014/011840, 2014.
42. Rudolph, D.; Steegmaier, M.; Hoffmann, M.; Grauert, M.; Baum, A.; Quant, J. C.; Haslinger, C.; Garin-Chesa, P.; Adolf, G. R. *Clin. Cancer Res.* **2009**, *15*, 3094.
43. Engel, J.; Emons, G.; Pinski, A.; Schally, V. *Expert Opin. Investig. Drugs* **2012**, *21*, 891.

CHAPTER TWENTY-SEVEN

To Market, To Market—2013

Joanne Bronson*, Amelia Black[†], Murali Dhar[‡], Bruce Ellsworth[§], J. Robert Merritt[¶]

*Bristol-Myers Squibb Company, Wallingford, Connecticut, USA
[†]Bristol-Myers Squibb Company, Redwood City, California, USA
[‡]Bristol-Myers Squibb Company, Princeton, New Jersey, USA
[§]Bristol-Myers Squibb Company, Pennington, New Jersey, USA
[¶]Kean University, Union, New Jersey, USA

Contents

Annual Reports in Medicinal Chemistry, Volume 49
ISSN 0065-7743
http://dx.doi.org/10.1016/B978-0-12-800167-7.00027-4

OVERVIEW

This year's To Market, To Market chapter provides summaries for 27 compounds (25 small molecules and two monoclonal antibodies) that received approval for the first time in any country in 2013.[1] Fifteen of the first-time approvals were from the United States, including all seven of the new anticancer agents. The total US Food and Drug Administration (FDA) new drug approvals were lower in 2013 (27) than in 2012 (39), but were on par with long-term averages (these totals include drugs previously approved outside the United States and agents outside the scope of this chapter).[2] Three of the US first-time approvals were from the FDA's new breakthrough therapy designation: ibrutininb for mantle cell lymphoma (MCL), obinutuzumab for chronic lymphocytic leukemia (CLL), and sofosbuvir for hepatitis C. Five first-time approvals were from Japan, three from the European Union, two from Canada, and one each from India and Russia. The approval of saroglitazar from India is noteworthy in that it is the first drug to be discovered, developed, and approved wholly in India.[3] In terms of disease areas, anticancer drugs once again dominated, with seven first-time approvals: five small molecules, one antibody drug conjugate (ADC), and one monoclonal antibody. Anti-infective drugs followed closely in number with six in total: two approvals for hepatitis C, two for human immunodeficiency virus (HIV) infection, one pharmacokinetic (PK) enhancer for codosing with HIV protease inhibitors (PIs), and one antifungal agent. The following overview is organized by therapeutic area, with approvals covered in this chapter briefly summarized first, followed by the mention of additional approvals of interest that are outside of the scope of the chapter (e.g., combination therapies and imaging agents).

Kinase inhibitors continue to have a major impact in the oncology area, with four of the five small molecule anticancer drugs coming from this class. **Afatinib (Gilotrif™)** was approved by the US FDA for first-line treatment of patients with metastatic non-small cell lung cancer (NSCLC) and with tumors that have epidermal growth factor receptor (EGFR) exon 19 deletions or exon 21 (L858R) substitution. Unlike the first-generation reversible EGFR-tyrosine kinase inhibitors (EGFR-TKI) erlotinib and gefitinib, afatinib functions as an irreversible inhibitor by covalently binding directly to the ATP-binding site in the kinase domains of both EGFR and HER2. Progression-free survival (PFS) with afatinib was 4.2 months longer than with traditional chemotherapy. The recommended dose of afatinib is 40 mg administered orally once daily. **Dabrafenib (Tafinlar®)** was approved by the US FDA for the treatment of patients with unresectable or metastatic melanoma with $BRAF^{V600E}$ mutation as detected by a FDA-approved test. It is the second approved BRAF inhibitor for this indication, the first being vemurafenib in 2011. The median PFS was 5.1 months in dabrafenib recipients compared with 2.7 months in dacarbazine recipients. The recommended dose of dabrafenib is 150 mg administered orally twice daily. **Trametinib (Mekinist™)** is the third drug approved by the US FDA for the treatment of patients with unresectable or metastatic melanoma with $BRAF^{V600E}$ or $BRAF^{V600K}$ mutations as detected by an FDA-approved test. Unlike vemurafenib and dabrafenib, which are TKIs with specificity for $BRAF^{V600E}$, trametinib is a MEK1 and MEK2 inhibitor that targets the mitogen-activated protein kinase (MAPK)/extracellular signal-regulated kinase (ERK) kinase (MEK) pathway. Median PFS duration was 4.8 months with trametinib treatment compared to 1.5 months in the chemotherapy arm. The recommended dose of trametinib is 2 mg administered orally once daily. **Ibrutinib (Imbruvica™)**, which received breakthrough designation status, was approved by the US FDA for the treatment of patients with MCL who have received at least one prior therapy. Ibrutinib is a first-in-class inhibitor of Bruton's tyrosine kinase (Btk) and works by irreversibly binding to cysteine-481 in the active site of Btk thereby inhibiting phosphorylation of tyrosine-223 and affecting downstream B-cell signaling pathways. Overall response rate at a dose of 560 mg/day was 69% and the median PFS was 13.9 months for all MCL patients. The recommended dose of ibrutinib is 560 mg administered orally once daily. In February 2014, the US FDA approved ibrutinib for the treatment of CLL at a dose of 420 mg administered orally once daily. The fifth small molecule anticancer drug is **pomalidomide (Pomalyst®)**, a 4-amino analog of thalidomide that was approved by the US FDA for the treatment of multiple myeloma (MM) in patients with disease progression after receiving other cancer therapeutics. Pomalidomide has enhanced potency and an improved toxicity profile relative to thalidomide, and acts by a similar immunomodulatory mechanism. In patients who were refractory to bortezomib and lenalidomide, PFS was nearly twice as long for patients receiving pomalidomide and low-dose dexamethasone than for patients receiving high-dose dexamethasone alone. Pomalidomide is given orally, once daily with a recommended dose of 4 mg on days 1–21 of repeated 28-day cycles.

Two biologic agents were approved as anticancer drugs in 2013. **Ado-trastuzumab emtansine (Kadcyla®)** is the first US FDA-approved ADC for treating human EGFR (HER2)-positive metastatic breast cancer (mBC). Ado-trastuzumab emtansine (T-DM1) is composed of the HER2-targeted antibody trastuzumab (Herceptin®) linked via a stable uncleavable thioether to a potent cytotoxic maytansine derivative (microtubule polymerization inhibitor), with an average of 3.5 cytotoxic molecules linked to each antibody. Median PFS was 9.6 months for T-DM1 versus 6.4 months for lapatinib plus capecitabine. Ado-trastuzumab ematasine was approved at a recommended dose of 3.6 mg/kg administered as an intravenous infusion every 3 weeks (q3w) until disease progression or unacceptable toxicity. The US FDA approved the glycoengineered, type II anti-CD20 antibody **obinutuzumab (Gazyva®)** in combination with chlorambucil chemotherapy for the treatment of people with previously untreated CLL. Obinutuzumab was the first drug to be approved under the FDA's breakthrough therapy designation. Obinutuzumab was designed to have improved therapeutic efficacy compared with the previously developed type I anti-CD20 antibodies, rituximab, and ofatumumab. Obinutuzumab plus chlorambucil gave a significantly longer PFS compared with chlorambucil alone (median 23 months vs. 10.9 months) and longer duration of PFS compared with rituximab plus chlorambucil (26.7 months vs. 15.2 months). Obinutuzumab was approved with a recommended dose of 1000 mg (with the exception of 100 mg on day 1 and 900 mg on days on the first cycle), administered by intravenous infusion on days 1, 2, 8, and 15 of six treatment cycles each of 28 days duration. Chlorambucil is given orally at 0.5 mg/kg on days 1 and 15 of each treatment cycle.

In the anti-infectives arena, **simeprevir** is a second generation hepatitis C virus (HCV) PI that was approved in 2013 in Japan (**Sovriad™**), Canada (**Galexos™**), and the United States (**Olysio™**) for the treatment of genotype 1 HCV infection in combination with pegylated interferon and ribavirin (PR). Simeprevir has high protease affinity due to induced fit binding in an extended S2 subsite, and unlike the first-generation PIs boceprevir and telaprevir, does not act through formation of a covalent reversible intermediate. Simeprevir in combination with PR showed sustained virologic response (SVR) rates that were superior to comparator drugs in patients who were treatment naïve as well as in patients who did not respond or who relapsed after previous interferon-based therapy. Simeprevir is approved as a 100 or 150 mg dose administered orally once daily for 12 weeks in combination with PR. **Sofosbuvir (Sovaldi™)** was approved in the United States for the treatment of HCV infection as a component of a combination antiviral treatment regimen. It was the third drug approved by the US FDA in 2013 under breakthrough therapy designation. Sofosbuvir is a novel anti-HCV agent that acts as a prodrug for 2′-F-2′-*C*-methyluridine monophosphate, which is converted to the corresponding triphosphate analog in cells, where it acts as an inhibitor of the HCV NS5B RNA-dependent RNA polymerase. In clinical trials, sofosbuvir was effective in treating multiple genotypes of the HCV and was effective in patients who could not tolerate or take an interferon-based treatment regimen. Sofosbuvir was approved as a 400 mg tablet, taken once daily with or without food. The recommended duration of treatment and comedication depend on the HCV genotype: 12 weeks in combination with PR for genotype 1, 12 weeks in combination with ribavirin for genotype 2, and in combination with ribavirin for 24 weeks for genotype 3.

Two HIV integrase strand transfer inhibitors (INSTIs), **dolutegravir (Tivicay®)** and **elvitegravir (Vitekta™)**, were approved for the treatment of HIV infection in combination with other antiretroviral drugs. Raltegravir, which was approved in 2007, is the other drug in this class. Dolutegravir was noninferior to raltegravir for decreasing viral load in treatment-naïve patients and was superior to raltegravir in patients who had previously received nucleoside/nucleotide reverse transcriptase inhibitors (NRTIs). Unlike elvitegravir, dolutegravir does not require PK boosting. Dolutegravir is given orally, once daily and is available as a 50 mg tablet to be taken in combination with other anti-HIV drugs. Elvitegravir, which had been approved in the United States in 2012 as part of the fixed-dose combination (FDC) drug Stribild®, was approved in the EU as a single agent to be used as part of combination therapy. The standalone formulation of elvitegravir is given orally, once daily, as an 85 or 150 mg tablet and must be coadministered with a ritonavir-boosted HIV PI. **Cobicistat (Tybost®)** was approved in the EU as a PK enhancer of atazanavir or darunavir as part of antiretroviral combination therapy in HIV-1-infected adults. Cobicistat was previously approved as one of the components of Stribild®, where it serves as a PK enhancer of elvitegravir. Cobicistat is similar to ritonavir in its Cyp3A inhibition properties, but by design, is inactive itself as an anti-HIV agent. In clinical studies, cobicistat was comparable to ritonavir in enhancing the PK of atazanavir and darunavir and was noninferior to ritonavir in a Phase III efficacy and safety trial in combination with atazanavir and emtricitabine/tenofovir disopoxil fumarate. Cobicistat was approved at 150 mg once daily as a PK enhancer for the HIV PIs atazanavir (300 mg once daily) and darunavir (800 mg once daily) as part of antiretroviral combination therapy in adults with HIV-1 infection.

Rounding out the anti-infective area, **efinaconazole (Jublia®)** was approved in Canada as a 10% topical solution for the treatment of onychomycosis, a fungal disease of the nails. Like other azole antifungal agents, efinaconazole acts by disrupting fungal cell membranes through inhibition of sterol 14α-demethylase. Efinaconazole has potent antifungal activity against clinical isolates of dermatophytes and retains activity in the presence of keratin, indicating that more unbound drug is available at the site of action. Efinaconazole gave higher rates of complete cure of onychomycosis (15.2–17.8%) versus vehicle-treatment (3.3–5.5%). These results are two- to threefold better than in typical studies with other topical antifungal agents, and are comparable to treatment with oral therapies. Efinaconazole was approved as a 10% nonlacquer solution.

There were three small molecules and two new combination therapies approved for the treatment of type 2 diabetes mellitus (T2DM) in the endocrine disease area. **Canagliflozin (Invokana®)** is the first US-approved sodium-glucose cotransporter (SGLT) inhibitor for the treatment of T2DM. Inhibition of renal SGLT suppresses glucose reabsorption, which permits glucose excretion into urine and reduction of hyperglycemia. Diabetes patients given 100 or 300 mg of canagliflozin once per day orally showed significant reductions in hemoglobin A1C (HbA_{1C}) versus placebo after 26 weeks. Both doses also reduced plasma glucose levels, blood pressure (BP), and body weight (BW) while increasing high-density lipoprotein (HDL)cholesterol levels. Canagliflozin has been studied as a standalone therapy and in combination with other T2DM therapies. Canagliflozin is given orally, once daily and is available as 100 or 300 mg tablets. The glucagon-like peptide-1 (GLP-1) analog **Lixisenatide (Lumyxia®)** received approval in the EU for the treatment of T2DM. GLP-1 mimetics, such as exenatide, liraglutide, and lixisenatide, act upon the pancreas to promote glucose-dependent insulin secretion and to decrease glucagon secretion, and also act on the gastrointestinal tract to delay gastric emptying thereby reducing glucose load. Lixisenatide produced a significant reduction in HbA_{1C} versus placebo in multiple studies at a 20 μg daily dose. Lixisenatide demonstrated slightly fewer adverse events (AEs) as compared to liraglutide (20 μg lixisenatide vs. 1.8 mg liraglutide maintenance doses). Lixisenatide is available as a solution for injection in a prefilled pen that provides either 10 or 20 μg of lixisenatide in each dose and is administered once daily. **Saroglitazar (Lipaglyn™)** is a peroxisome proliferator-activated receptor (PPAR) antagonist that was approved for use by the Drug Controller General of India for the treatment of diabetic dyslipidemia and hypertriglyceridemia that is not controlled by statin therapy. In diabetic patients with elevated triglycerides (TGs), saroglitazar at a dose of 4 mg reduced plasma TGs by −45.7% and total cholesterol (TC) by −6.9%, with very low-density lipoprotein (VLDL) being the predominantly lowered species (−46.1% vs. placebo). Saroglitazar is the first drug discovered, developed, and approved wholly in India, representing a milestone for the pharmaceutical industry in India. The drug is available in tablet form of 4 mg dose for oral administration. **Alogliptin (Nesina®)**, which was previously approved in Japan in 2010, was approved in the United States as a single agent and as part of two combination products, **Kazano**[4] (alogliptin and metformin hydrochloride) and **Oseni**[5] (alogliptin and pioglitazine). Alogliptin alone gives reductions in HbA_{1C} of 0.4–0.6% compared with placebo after 26 weeks of treatment. Kazano resulted in additional reductions in HbA_{1C} of 1.1% over alogliptin alone and 0.5% over metformin alone. Oseni resulted in additional reductions in HbA_{1C} of 0.4–0.9% over alogliptin alone and 0.4–0.6% over pioglitazone alone.

In the metabolic disease area, there was one approval for hypercholesteremia, one for lipodystrophy, and one for obesity. **Mipomersen (Kynamro®)** was approved with orphan drug status by the US FDA for patients with homozygous familial hypercholesterolemia (HoFH) as an adjunct to lipid-lowering medications and diet. Mipomersen is a 20-mer phosphorothioate antisense oligonucleotide (ASO) designed to inhibit apolipoprotein B100 (Apo B100) transcription by Watson–Crick pairing to the mRNA sequence to target the complex for silencing and elimination. Mipomersen has broad tissue distribution and a 31-day elimination half-life. Treatment of HoFH patients with 200 mg of mipomersen resulted in a 24.7% reduction in low-density lipoprotein-cholesterol (LDL-C) versus 3.3% in the placebo group. Mipomersen was approved for use as 1 mL of a 200 mg/mL solution in saline for subcutaneous injection administered once weekly for the reduction of plasma lipids in HoFH patients. **Metreleptin (Myalept™)** is a leptin analog that was approved in Japan for the treatment of lipodystrophy and in the United States in early 2014 for the treatment of generalized lipodystrophy. Lipodystrophy is a heterogeneous group of very rare disorders characterized by generalized or partial loss of adipose tissue and leptin deficiency. Metreleptin is a recombinant *N*-methionyl analog of the leptin protein. In an open-label registrational trial in patients with generalized lipodystrophy and low-leptin levels, metreleptin treatment for a mean of 2.7 years (range of 3.6 months to 10.9 years), with dose adjustments on an individual basis, gave significant reductions in HbA_{1C}, fasting glucose, and TGs after 1 year of treatment. Metreleptin is administered as a subcutaneous injection at doses ranging from 0.02 to 0.08 mg/kg daily depending on gender, BW, and clinical response. **Cetilistat (Oblean®)** was approved in Japan for the treatment of obesity, limited to patients with both T2DM and dyslipidemia, and with a body mass index (BMI) ≥ 25 kg/m^2 in spite of dietary treatment and/or exercise therapy. As with orlistat, cetilistat acts via inhibition of pancreatic lipases in the gut to inhibit fat absorption and thereby reduce caloric uptake from diet. Cetilistat works presystemically with no detectable drug levels in plasma. In patients who maintained a hypocaloric diet, treatment with cetilistat three times a day for 12 weeks with doses of 60, 120, or 240 mg led to ~2 kg BW loss at all doses and dose proportional reductions of low-density lipoprotein (LDL) cholesterol. Cetilistat is approved as an oral dose of 120 mg taken three times a day immediately after each meal.

Two first-time approvals were achieved in the cardiovascular disease area for the treatment of pulmonary hypertension: **macitentan (Opsumit®)** and **riociguat (Adempas®)**. Macitentan, which received US FDA approval for the treatment of pulmonary arterial hypertension (PAH) to delay disease progression, is an endothelin receptor A and B antagonist that acts to relax pulmonary arteries. Treatment of PAH patients with macitentan for at least 2 years delayed disease progression and increased exercise capacity relative to treatment with placebo. Macitentan was approved as a 10 mg oral daily dose. Riociguat is a guanylate cyclase stimulator that was approved in Canada and the United States for the treatment of patients with chronic thromboembolic pulmonary hypertension (CTEPH) after surgical treatment, inoperable CTEPH, and for the treatment of adults with PAH. Riociguat is the first approved chemotherapeutic option for the treatment of CTEPH. In patients with CTEPH and PAH, riociguat given three times daily was significantly more effective than placebo in improving the 6-min walking distance (6MWD). The recommended initiation dose of riociguat is 1 mg administered orally three times daily, or 0.5 mg daily for patients who may not tolerate the hypotensive effect of riociguat, followed by an increase of 0.5 mg as tolerated to a maximum of 2.5 mg three times daily.

The two approvals in the central nervous system disease area were for the Parkinson's disease (PD) treatment **istradefylline (Nouriast®)** and the antidepressant **vortioxetine (Brintellix)**. Istradefylline was approved in Japan for adjunctive treatment of PD. Istradefylline acts by antagonism of the adenosine A_{2A} receptor, which is colocalized with dopamine D_2 receptors in the striatum, to enhance dopamine D_2-dependent signaling. In clinical trials with istradefylline given in combination with levo-dopa, significant decreases in undesired immobility (OFF time) and improvements in desired mobility without dyskinesia (ON time) were achieved compared with levo-dopa alone. Istradefylline was approved as adjunctive treatment for PD at an oral dose of 20 or 40 mg, once daily. Vortioxetine was approved in the United States for the treatment of major depressive disorder (MDD). Vortioxetine was discovered from a designed multiple ligand approach and acts as a serotonin transport (SERT) inhibitor, a 5-HT1A receptor agonist, and a 5-HT3A receptor antagonist, in addition to having affinity at other receptors. Vortioxetine showed statistically significant improvements in MDD compared with placebo in several clinical trials as measured on standard depression rating scales. The recommended starting dose of vortioxetine is 10 mg administered orally once daily, followed by an increase to 20 mg/day, as tolerated. **Vizamyl (18F-flutemetamol)** was approved in the United States as a radioactive diagnostic drug for detecting β-amyloid in the brain using positron emission tomography (PET) imaging.[6] It is used as part of the evaluation for Alzheimer's disease and is the second diagnostic drug for imaging β-amyloid after Amyvid, which was approved in 2012.

In the respiratory disease area, one single agent and three new combination therapies were approved for the treatment of chronic obstructive pulmonary disease (COPD). **Olodaterol (Striverdi®)**, which was approved in Russia, Canada, and the European Union, is a long-acting β_2-adrenergic receptor agonist that acts as a bronchodilator via relaxation of airway smooth muscles. In clinical trials with patients with moderate to severe COPD, olodaterol gave significant improvements in lung function as measured by forced expiratory volume in 1 s (FEV_1). Olodaterol is approved as a 5 μg dose given once daily the using the Respimat® Soft Mist™ inhaler for the long-term maintenance bronchodilator treatment of airflow obstruction in patients with COPD. The three approved combination drugs are inhaler products based on coadministration of a β_2-adrenoceptor agonist with either an anticholinergic agent or an inhaled corticosteroid. The US FDA approved **Anoro Ellipta (umeclidinium/vilanterol)** as a combination inhalation powder for the once daily, long-term maintenance treatment of airflow obstruction in patients with COPD.[7] Umeclidinium is an anticholinergic agent that prevents large airway muscles from tightening; vilanterol is the long-acting β_2-adrenergic agonist component that relaxes airway muscles. **Breo Ellipta (fluticasone furoate/vilanterol)** was also approved in the United States as an inhalation treatment for COPD.[8] In this combination, vilanterol is coadministered with the inhaled corticosteroid fluticasone furoate, which acts to decrease inflammation in the lungs. Breo Ellipta is indicated for long term, once daily, maintenance treatment of airflow obstruction and for reducing exacerbations in patients with COPD. **Ultibro Breezhaler (indacaterol/glycopyrronium bromide)** was approved in Germany and the Netherlands as a fixed-dose combination for maintenance treatment to relieve symptoms in adult patients with COPD.[9] Indacaterol is β_2-adrenoceptor agonist and glycopyrronium bromide is an anticholinergic agent.

There were single first-time approvals in several disease areas. In the immunology area, **dimethyl fumarate (Tecfidera®)** was approved by the US FDA for the treatment of relapsing forms of multiple sclerosis (MS). While its mechanism is not completely understood, dimethyl fumarate increases anti-inflammatory cytokines (IL-10, IL-4, and IL-6), decreases proinflammatory cytokines (IL-1β, IL-6, and TNF-α), and activates the nuclear factor E2-related factor 2 (Nrf2) pathway to protect neuronal cells. Dimethyl fumarate has been used for the treatment of psoriasis in Europe since 1994 and has a favorable long-term safety profile. In a 2-year trial, 28% of patients receiving dimethyl fumarate twice daily were "disease-free" at the end of the study versus only 15% for placebo. In another 2-year trial, dimethyl fumarate gave a 44% reduction in annualized relapse rate versus placebo, while a conventional MS therapeutic, glatiramer acetate, afforded only a 29% reduction in relapse rate. Tecfidera® (dimethyl fumarate) is given orally, twice daily with or without food as a hard gelatin delayed-release capsule at a starting dose of 120 mg. After 7 days, the dose is increased to a maintenance level of 240 mg twice daily for the duration of therapy. **Acotiamide (Acofide®)** was approved in Japan in the gastrointestinal disease area for the treatment of postprandial fullness, upper abdominal bloating, and early satiation due to functional dyspepsia (FD). Acotiamide enhances acetylcholine release from enteric neurons through inhibition of acetyl cholinesterase and selective muscarinic acetylcholine M_1 and M_2 receptor antagonism, thereby enhancing gastric emptying and gastric accommodation. Acotiamide was efficacious in clinical trials in 52.2% of treated patients compared with 34.8% in the placebo group. The recommended dose of acotiamide is 100 mg taken orally three times daily. In the area of women's health, the US FDA approved **ospemifene (Osphena™)** for the treatment of moderate to severe dyspareunia, a symptom of vulvar and vaginal atrophy (VVA), due to menopause. Ospemifene is a selective estrogen receptor (ER) modulator (SERM) and the first nonhormonal, nonestrogen for the treatment of moderate to severe dyspareunia in women with menopausal VVA. It binds to ERα and ERβ with tissue-specific estrogenic agonist/antagonist effects. The efficacy and safety of ospemifene was established in three placebo-controlled clinical trials (two 12-week efficacy trials and one 52-week long-term safety trial). The recommended dose of ospemifene is 60 mg administered orally once daily with food. A final approval of note is the combination therapy Simbrinza™, which was approved in the United States for reduction of elevated intraocular pressure in patients with primary open-angle glaucoma or ocular hypertension.[10] **Simbrinza™** is a fixed-dose combination of brinzolamide, a carbonic anhydrase inhibitor, and brimonidine, an α_2-adrenergic receptor agonist. It is the only fixed-dose combination therapy for glaucoma that does not include a beta blocker.

1. ACOTIAMIDE (DYSPEPSIA)[11–17]

Class:	Acetylcholinesterase (AChE) inhibitor
Country of origin:	Japan
Originator:	Zeria Pharmaceutical
First introduction:	Japan
Introduced by:	Zeria/Astellas
Trade name:	Acofide®
CAS registry no:	773092-05-0 (hydrochloride-trihydrate) 185106-16-5 (free base)
Molecular weight:	541.06 (hydrochloride-trihydrate) 450.5 (free base)

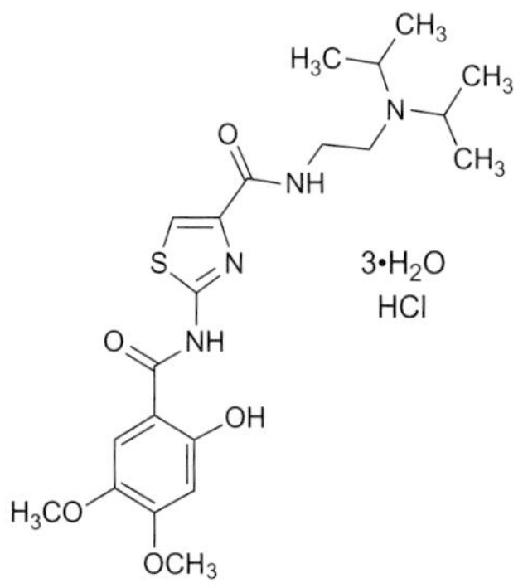

In June 2013, the Ministry of Health, Labor and Welfare in Japan approved acotiamide (also referred to as Z-338 and YM443), for the treatment of postprandial fullness, upper abdominal bloating, and early satiation due to functional dyspepsia (FD). It is the first drug approved worldwide for the treatment of FD diagnosed by Rome III criteria (*vide infra*). The annual incidence rate of FD worldwide is estimated to be between 1% and 6%.[11] Dyspepsia is broadly defined as pain or discomfort in the upper abdomen, particularly in the epigastrium. The Rome III criteria define FD as the presence of chronic dyspeptic symptoms in the absence of underlying structural or metabolic disease that readily explains the symptoms.[12] Chemotherapeutic options include treatment with a prokinetic agent (e.g., cisapride, tegaserod, and domperidone) and a proton pump inhibitor (e.g., omeprazole), although their efficacy is limited. Other management strategies include dietary and lifestyle modifications in those with mild, intermittent symptoms, and eradication of *Helicobacter pylori* in those with severe symptoms. Acetylcholine is a key excitatory neurotransmitter in the enteric nervous system as well as at the neuronal synapse. Inhibition of acetylcholinesterase (AChE) has the potential to stimulate contractility in the gastrointestinal tract. Acotiamide enhances acetylcholine release from enteric neurons through selective muscarinic acetylcholine receptor M_1, M_2 antagonism, and inhibition of AChE, thereby enhancing gastric emptying and gastric accommodation. Acotiamide inhibits the acetylcholine-induced Ca^{2+}-activated Cl^- current with an IC_{50} of 1.8 μM in oocytes expressing muscarinic M_1 receptors and in oocytes expressing muscarinic M_2 receptors, acotiamide inhibited the acetylcholine-induced K^+ currents with an IC_{50} of 10.1 μM. However, studies in conscious dogs, guinea-pig gastric muscle strips and assays with human cholinesterase indicate that acotiamide inhibits AChE with a K_i of 610 nM suggesting that acotiamide stimulates gastric motility mainly by inhibiting AChE activation.[13] In a restraint stress-induced rat model, acotiamide significantly improved delayed gastric emptying and feeding inhibition but did not affect normal gastric emptying or feeding in intact rats.[14] A synthetic route to acotiamide that employs pyridine hydrochloride to selectively cleave a 2-substituted methyl ether from a 2,4,5-trimethoxy benzamide intermediate, as a key step, has been reported.[15]

In a placebo-controlled double-blind study conducted at a single center in Japan in healthy patients, the mean values for C_{max}, T_{max}, AUC, clearance, and $T_{1/2}$ for acotiamide dosed at 100 mg were approximately 30.82 ng/mL, 2.42 h, 171 ng h/mL, 698 L/h, and 13.31 h, respectively.[16] *In vitro* protein binding for acotiamide is ~85% in human plasma. The efficacy of acotiamide was established in a number of Phase II and III clinical trials.[16] In a Phase III randomized, multicenter, placebo-controlled clinical study[17] patients received acotiamide 100 mg orally three times daily ($n=452$) or placebo ($n=445$). Patients who classified as responders according to a global assessment of overall treatment efficacy (primary endpoint) were 52.2% of those receiving acotiamide and 34.8% in the placebo group ($p < 0.001$). In addition, the elimination rate for all three meal-related symptoms was 15.3% among patients receiving acotiamide compared with 9.0% in the placebo group ($p = 0.004$). Increased TG, serum c-glutamyltransferase, or prolactin levels, and nasopharyngitis were the most common adverse events observed during acotiamide clinical trials. The recommended dose of acotiamide is 100 mg taken orally three times daily.

2. ADO-TRASTUZUMAB EMTANSINE (ANTICANCER)[18–22]

Class:	Recombinant monoclonal ADC
Type:	Humanized, IgG1κ, anti-HER2, linked to cytotoxic agent DM1
Country of origin:	United States
Originator:	Genentech, a member of the Roche Group
First introduction:	United States
Introduced by:	Genentech
Trade name:	Kadcyla®
CAS registry no:	1018448-65-1
Expression System:	Recombinant Chinese Hamster Ovary CHO-cell line. Small molecule components are produced by chemical synthesis
Molecular weight:	148.5 kDa

T-DM1 is a human epidermal growth factor receptor (HER2)-targeted antibody drug conjugate (ADC) that was approved in February 2013 by the US FDA for use as a single agent for the treatment of patients with HER2-positive, metastatic breast cancer (mBC) who previously received treatment with trastuzumab and a taxane, separately, or in combination. Patients either have been treated for metastatic disease or have developed disease recurrence during or within 6 months of completing adjuvant therapy. Breast cancer is the most prevalent form of cancer in women and the second leading cause of cancer death in the United States.[18] According to the American Cancer Society, approximately 229,000 people in the United States will be diagnosed with breast cancer with 40,000 deaths in 2012. Approximately, 20–25% of breast cancers have overexpression of the HER2 gene, which is associated with the pathogenesis of more aggressive tumors and with poorer prognosis for patients. The single agent trastuzumab (Herceptin®) has modest activity in mBC treatment, while trastuzumab plus taxane-based chemotherapy has significant increases in overall survival (OS) and progression-free survival (PFS) compared to chemotherapy alone. Nevertheless, mBC eventually progresses in most patients and the chemotherapy-associated toxicity is a significant source of patient morbidity. A clear unmet need for more effective and better-tolerated therapies for HER2-positive mBC was recognized.[19] ADCs are designed to specifically deliver cytotoxic agents to tumor cells while minimizing normal tissue toxicity, thereby improving the therapeutic index of the cytotoxic agent. T-DM1 is composed of trastuzumab linked to the potent cytotoxic microtubule polymerization inhibitor DM1 (derivative of maytansine) via a stable uncleavable thioether linker. T-DM1 is produced by chemically crosslinking the cytotoxic maytansinoid derivative to the lysine residues of trastuzumab such that there is an average of 3.5 cytotoxic molecules linked to each antibody. In addition to delivering DM1 to tumor cells, T-DM1 retains the effector functions of trastuzumab, including inhibition of HER2-mediated signal transduction and activation of antibody-dependent cell-mediated cytotoxicity. In preclinical efficacy studies, T-DM1 induced cell-cycle arrest and apoptosis in HER2-positive cancer cells and demonstrated activity in trastuzumab and lapatinib-resistant HER2-positive cancer models.[20] T-DM1 is the first FDA-approved ADC for treating HER2-positive mBC.

The Phase I trial was a dose-escalation study (0.3–4.8 mg/kg) of T-DM1 given intravenously q3w or weekly to 24 patients with advanced HER2-positive breast cancer who had previously used a trastuzumab-based regime.[21] A trend toward faster clearance was observed at T-DM1 doses $\leq$1.2 mg/kg; however, at 2.4–4.8 mg/kg doses q3w, T-DM1-exhibited linear PKs with a C_{max} of 76.2 $\pm$ 19.1 μg/mL, AUC_{inf} of 300 $\pm$ 65.8 mg mL, terminal half-life of 3.1 $\pm$ 0.7 days and clearance of 12.7 $\pm$ 3.56 mL/day/kg for the 3.6 mg/kg cohort. After T-DM1 dosing, total trastuzumab versus T-DM1 concentrations declined more slowly and mean concentrations of DM1 were consistently low (4.57 $\pm$ 1.33 ng/mL mean) with undetectable levels within a few days after dosing. The EMILLA Phase III trial assessed the efficacy and safety of T-DM1, as compared with lapatinib plus capecitabine, in HER2-positive advanced breast cancer patients previously treated with trastuzumab and a taxane.[22] The primary end points were PFS and OS. In the trial, 991 patients were randomized in a 1:1 ratio to either T-DM1 or lapatinib plus capecitabine treatments. Patients received T-DM1 (3.6 mg/kg intravenous q3w) or lapatinib (1250 mg oral daily) plus capecitabine (1000 mg/m^2 oral every 12 h on days 1–14 of each 21-day treatment cycle) until progressive disease or unmanageable toxicity. Median PFS was 9.6 months for T-DM1 versus 6.4 months for lapatinib plus capecitabine ($P < 0.001$) and median OS was 30.9 months for T-DM1 versus 25.1 months for lapatinib plus capecitabine ($P < 0.001$). Results for all secondary endpoints also favored T-DM1. The most common grade 3–4 adverse events for T-DM1 were thrombocytopenia (12.9%), elevated aspartate aminotransferase (4.3%), and elevated alanine aminotransferase (2.9%). Ado-trastuzumab emtansine was approved at a recommended dose of 3.6 mg/kg administered as an intravenous infusion q3w until disease progression or unacceptable toxicity.

3. AFATINIB (ANTICANCER)[23–29]

Class:	Tyrosine Kinase Inhibitor
Country of origin:	United States
Originator:	Boehringer-Ingelheim
First introduction:	United States
Introduced by:	Boehringer-Ingelheim
Trade name:	Gilotrif™
CAS registry no:	850140-72-6 (dimaleate salt) 439081-18-2 (free base)
Molecular weight:	718.1 (dimaleate salt) 485.9 (free base)

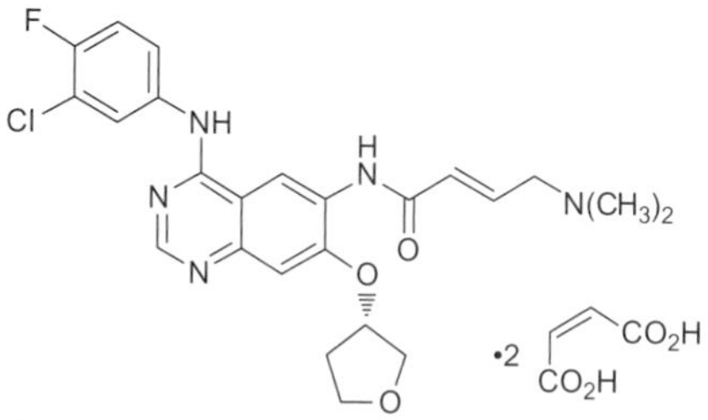

In July 2013, the US FDA approved afatinib (also referred to as BIBW-2992), for the first-line treatment of patients with metastatic non-small cell lung cancer (NSCLC) and with tumors that have epidermal growth factor receptor (EGFR) exon 19 deletions or exon 21 (L858R) substitution. Lung cancer is the leading cause of cancer death in men and women worldwide. An estimated 226,160 new cases of lung cancer were expected to be diagnosed in the United States in 2012.[23] Non-small cell lung carcinoma is the most common type comprising approximately 85% of all lung cancers of which 75% are metastatic.[24] Chemotherapeutic options include treatment with platinum-based regimens. However, survival outcomes are poor with a median overall survival (OS) duration of 8–11 months. About 10–20% of patients with NSCLC have mutations in the EGFR, the most common being deletions in exon 19 (the Leucine-Arginine-Glutamic acid-Alanine, or LREA, deletion) and a single amino acid substitution in exon 21 (L858R). These activating mutations occur in the catalytic domain, therefore targeting EGFR has become an important strategy in the treatment of NSCLC. Erlotinib and gefitinib are first-generation EGFR-Tyrosine kinase inhibitors (EGFR-TKIs)

that bind reversibly to the kinase domain and effectively inhibit both wild-type and mutated EGFR.[25] However, disease progression typically occurs after a median of 10–14 months of treatment with a reversible EGFR-TKI. In addition, a T790M missense mutation (gatekeeper mutation) in exon 20 accounts for 50–60% of patients with disease progression while on a first-generation EGFR-TKI. Afatinib functions as an irreversible inhibitor by covalently binding directly to the ATP-binding site in the kinase domains of both EGFR (Cys 773) and HER2 (Cys 805; HER-2 is the preferred dimerization partner of EGFR) resulting in downregulation of EGFR signaling. Afatinib is a potent inhibitor of wild-type and mutant forms (L858R) of EGFR (IC_{50s} of 0.5 and 0.4 nM, respectively), and HER2 ($IC_{50}=14$ nM), but about 100-fold more active against the gefitinib resistant L858R–T790M EGFR double mutant, with an IC_{50} of 10 nM.[26] Consistent with its *in vitro* activity, afatinib induces tumor regression in xenograft and transgenic lung cancer models, with superior activity over erlotinib.[26] A synthetic route to afatinib that employs the displacement of a phenylsulfonyl group to install the (*S*)-3-hydoxytetrahydrofuran ring and a modified Horner–Wadsworth–Emmons reaction with {[4-(3-chloro-4-phenylamino)-7-((*S*)-tetrahydrofuran-3-yloxy)-quinazolin-6-ylcarbamoyl]-methyl}-phosphonate and dimethylaminoacetaldehyde-hydrogen sulfite adduct to install the eneamide moiety, has been reported.[27]

At dose ranges of 20–50 mg, AUC and C_{max} increases were slightly more than dose proportional, the median T_{max} ranges from 2 to 5 h, the mean absolute bioavailability is 92% (at 20 mg dose) compared to solution and the elimination half-life is 37 h after repeat dosing in cancer patients. Afatinib is highly protein bound (~95%) and has an apparent volume of distribution during the terminal phase of 4500 L.[28] The efficacy of afatinib was established in a randomized, multicenter, open-label Phase III clinical study of 345 participants with metastatic NSCLC whose tumors harbored EGFR mutations.[29] Patients were randomized (2:1) to receive afatinib 40 mg orally once daily ($n=230$) or up to six cycles of pemetrexed/cisplatin ($n=115$) with progression-free survival (PFS) as the primary endpoint and overall response rate (ORR) and OS as secondary endpoints. Statistically significant improvements in PFS were observed in patients randomly assigned to afatinib (median PFS increase of 4.2 months) compared to those receiving pemetrexed/cisplatin. There were no statistically significant differences between the arms in OS. Diarrhea, rash/dermatitis, acneiform, stomatitis, paronychia, dry skin, decreased appetite, and pruritus were the most common adverse events (≥20%) observed during afatinib clinical trials. The recommended dose of afatinib is 40 mg administered orally once daily.

4. CANAGLIFLOZIN (ANTIDIABETIC)[30–42]

Class:	SGLT 2 inhibitor
Country of origin:	Japan
Originator:	Mitsubishi Tanabe Pharma
First introduction:	United States
Introduced by:	Janssen Inc.
Trade name:	Invokana®
CAS registry no:	842133-18-0
Molecular weight:	444.52

In March 2013, the US FDA approved canagliflozin (JNJ-28431754; TA-7284) for the treatment of type 2 diabetes in adults.[30] In type 2 diabetes, ineffective production and use of insulin results in elevated blood sugar levels which, if left untreated, may cause heart disease, stroke, retinopathy, kidney failure, and limb amputation due to poor blood flow. The World Health Organization (WHO) reports that type 2 diabetes accounts for 90% of the nearly 350 million cases of diabetes worldwide and projects that diabetes will be the 7th leading cause of death by 2030.[31] The front-line treatment for type 2 diabetes is metformin, either as mono-therapy or in combination with other medications. Second-line treatments include sulfonyl ureas, meglitinides, and thiazolidine diones. Insulin therapy may be required for patients with unresolved hyperglycemia. Many of these medications cause significant adverse effects and are often contraindicated in patients with impaired kidney function or cardiovascular disease.[32] Canagliflozin is the first US-approved sodium-glucose co-transporter (SGLT) inhibitor for the treatment of type 2 diabetes. Inhibition of renal SGLT suppresses glucose reabsorption, which permits glucose excretion into urine and reduction of hyperglycemia.[33] Canagliflozin was discovered through structural modifications of phlorizin, a known inhibitor of renal glucose reabsorption.[33] Early modifications of OH groups on the glucose moiety were insufficient to adequately impair hydrolysis by intestinal β-glucosidase.[33,34] Introduction of the C-glucoside moiety, as in the clinical candidate dapagliflozin,[35,36] afforded sufficient resistance to hydrolysis. Finally, incorporation of the thiophene moiety in canagliflozin provided improved potency for hSGLT2 (exclusive to kidney), $IC_{50} = 2.2$ nM, while offering significant selectivity over hSGLT1 (in kidney and heart), $IC_{50} = 910$ nM.[34] Hyperglycemic, high-fat (HF) diet fed mice (KK strain) that received a single 3 mg/kg oral dose of canagliflozin had a 48% reduction in blood glucose levels after 6 h versus vehicle-treated mice.[34] Noteworthy in the multistep synthesis of canagliflozin is the stereoselective formation of the β-*C*-glucoside which is accomplished by coupling of the aryllithium aglycone with 2,3,4,6-tetra-*O*-trimethylsilyl-β-D-gluconolactone followed by desilylation and stereoselective reduction with triethylsilane and boron trifluoride etherate.[34]

In type 2 diabetes patients, canagliflozin has a 10–12 h half-life, is highly protein bound (99%), and is excreted primarily as glucuronide metabolites in bile and urine.[37,38] In a Phase III trial, patients given 100 or 300 mg of canagliflozin orally, once daily showed significant reductions in hemoglobin A1C (HbA_{1C}) versus placebo after 26 weeks.[39] Both doses also reduced plasma glucose levels, blood pressure, and body weight while increasing HDL cholesterol levels.[38] A dose-ranging trial evaluating canagliflozin as an add-on to metformin in patients whose type 2 diabetes was inadequately controlled, showed significantly improved glycemic control at all doses studied (50, 100, 200, and 300 mg).[40] Evaluation of canagliflozin add-on versus glimepiride add-on in patients whose type 2 diabetes was inadequately controlled with metformin alone, showed that 300 mg canagliflozin once daily lowered HbA_{1C} more effectively than 300 mg glimepiride once daily.[40] All trials reported increased incidences of genital infections with canagliflozin.[39–41] Commonly reported adverse events are female genital mycotic infections and urinary tract infections.[42] Canagliflozin is contraindicated in patients with severe renal impairment or hypersensitivity to the drug. Invokana® (canagliflozin) is given orally, once daily and is available as 100 or 300 mg tablets.

5. CETILISTAT (ANTIOBESITY)[43–52]

Class:	Pancreatic lipase inhibitor
Country of origin:	United Kingdom
Originator:	Alizyme PLC
First introduction:	Japan
Introduced by:	Takeda and Norgine
Trade name:	Oblean®
CAS registry no:	282526-98-1
Molecular weight:	401.58

Cetilistat (also known as ATL-962) was approved in September 2013 by the Japanese Ministry of Health, Labor and Welfare for the treatment of obesity, limited to patients with both type 2 diabetes mellitus (T2DM) and dyslipidemia, and with a body mass index (BMI) $\geq 25\ kg/m^2$ in spite of dietary treatment and/or exercise therapy.[43] The World Health Organization estimates that in 2008, 12% of adults aged 20 or over were obese, while in Japan the incidence is 4.5% of the population.[44] Obesity is a major risk factor for type 2 diabetes, hyperlipidemia, cardiovascular disease, and death.[45] Currently approved therapies for the treatment of obesity include phentermine/topiramate combination, lorcacerin, orlistat, and a pancreatic lipase inhibitor.[46] As with orlistat, cetilistat works via inhibition of pancreatic lipases in the gut to inhibit fat absorption and thereby reduce caloric uptake from diet. The medicinal chemistry program has not been described in the scientific literature, but the patent describing cetilistat also describes the synthesis of analogs with varied aryl substituents and lipophilic tails.[47] The synthesis of cetilistat involves condensation of a hexadecylcarbonochloridate with 2-amino-5-methylbenzoic acid; other analogs were synthesized by varying the carbonochloridate and 2-aminobenzoic acid components. Cetilistat is a potent inhibitor of human and rat pancreatic lipase with IC_{50}s of 15 and 136 nM, respectively,[48] with little inhibition of trypsin or chymotrypsin.[49] Cetilistat given at 3–100 mg/kg dose-dependently reduced absorption after an oral fat (Intrafat 20%) challenge in Sprague–Dawley rats as measured by a 45–90% reduction in circulating triglyceride AUC_{0-6h}. The effect of cetilistat on body weight (BW) was assessed in F344 rats acclimatized to a high fat (HF) (45% fat content) diet and then fed a HF diet formulated with cetilistat. Cetilistat doses correlating with doses of 4.9, 14.9, and 50.7 mg/kg/day produced approximately a 15–25 g reduction in BW relative to control.

In Phase I clinical trials, cetilistat was found to increase dietary fat excretion relative to placebo in all dose groups (doses from 50 to 300 mg, tid) and to a comparable level to orlistat 120 mg tid.[50] Cetilistat works presystemically with no detectable drug levels in plasma. In 12-week studies, 371 patients who maintained a hypocaloric diet were randomized to placebo or cetilistat 60, 120, and 240 mg, tid. Drug treatment led to ~2 kg BW loss (placebo subtracted) that was equivalent across dose groups. Significant dose response for reductions of LDL-C were observed +0.1 (placebo), −0.19, −0.295, and −0.366 mmol/L for 60, 120, and 240 mg, tid dose groups, respectively, with generally mild and GI-related adverse events.[51] In a separate trial, small (−0.14% placebo subtracted) reductions in glycated hemoglobin (HbA_{1C}) and changes in BW were observed, similar to those observed for orlistat; however, the proportion of patients experiencing GI side effects was significantly lower for cetilistat versus orlistat (83.3% vs. 86.6%, $p < 0.001$) with lower rates of AEs that led to withdrawal (2.5% vs. 11.6%). The authors claim that differential physicochemical characteristics could effect binding to fat micelles in the gut compartment, thereby differentiating the compounds.[52] Based on the results of these clinical trials for efficacy and safety of a 120 mg dose, administered thrice daily, the Japanese Ministry of Health, Labor and Welfare approved cetilistat for use in Japan.[43]

6. COBICISTAT (ANTIVIRAL, PHARMACOKINETIC ENHANCER)[53–59]

Class:	Cytochrome P450 3A inhibitor
Country of origin:	United States
Originator:	Gilead
First introduction:	European Union
Introduced by:	Gilead
Trade name:	Tybost®
CAS registry no:	1004316-88-4
Molecular weight:	776.02

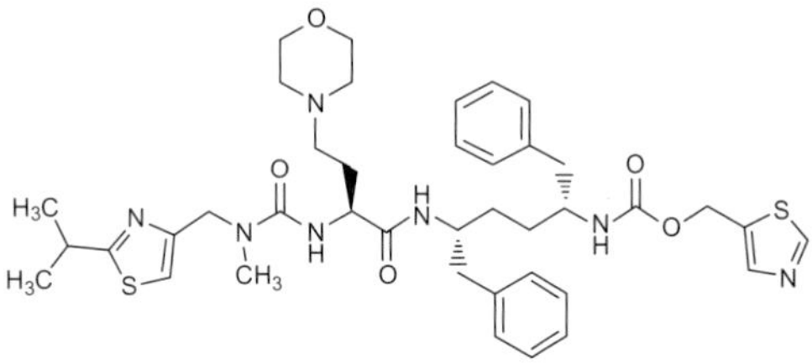

In September 2013, cobicistat (also known as GS-9350) was approved in the European Union as a pharmacokinetic (PK) enhancer of atazanavir 300 mg once daily or darunavir 800 mg once daily as part of antiretroviral combination therapy in human immunodeficiency virus-1 (HIV-1) infected adults.[53,54] Cobicistat was previously approved as one of the components of Stribild® (a fixed-dose combination of emtriciabine, tenofovir, elvitegravir, and cobicistat) in the United States in August 2012 and in the European Union in May 2013. In this single-tablet regimen, cobicistat serves as a PK enhancer of elvitegravir. The WHO estimated that 35.3 million people were living with HIV infection at the end of 2012.[55] Advanced stages of HIV infection lead to acquired immune deficiency syndrome (AIDS), a condition with high rates of mortality in which affected individuals are susceptible to opportunistic infections. The standard of care for the treatment of HIV infection is highly active antiretroviral therapy (HAART), consisting of a combination of multiple drugs with different mechanisms of action in the viral life cycle. HAART provides durable and full suppression of HIV-1 replication, and is highly effective in delaying disease progression and reducing mortality. Protease inhibitors (PIs) are often used in HAART due to their excellent antiviral activity and relatively low rate of resistance development. Since, most PIs are metabolized rapidly by Cyp3A enzymes, these agents are typically codosed with the Cyp3A inhibitor ritonavir, which was originally approved as an HIV PI, but is now used at subtherapeutic doses as a PK enhancer. Cobicistat was discovered from efforts to find a PK enhancer that retained the Cyp3A inhibitory properties of ritonavir but addressed disadvantages of ritonavir such as potential to induce HIV resistance with subtherapeutic doses, induction of drug metabolizing enzymes (Cyp, p-glycoprotein, and uridine diphosphate glucuronyltransferase), and association with lipid disorders.[56] Cobicistat is an analog of ritonavir in which the hydroxyl group that forms key interactions with the HIV protease active site has been removed. In addition, a morpholine group has been incorporated to improve solubility and physicochemical properties. Cobicistat is similar to ritonavir in its Cyp3A inhibition properties, including inactivation kinetics ($k_{inact}=0.44\ min^{-1}$ and $K_i=939$ nM vs. $k_{inact}=0.23\ min^{-1}$ and $K_i=256$ nM for ritonavir) and potency (for midazolam as substrate, Cyp3A $IC_{50}=0.154$ μM vs. 0.107 μM for ritonavir), but is inactive against HIV protease and HIV replication in a cell-based assay. Cobicistat was weaker than ritonavir as an activator of the PXR receptor, was inactive in a lipid accumulation assay in human adipocytes, showed lower inhibition of glucose uptake, and had greatly improved aqueous solubility. Cobicistat is prepared via the symmetric intermediate (*2R*,*5R*)-1,6-diphenylhexane-2,5-diamine by coupling with an activated carbonate to afford the thiazolylmethyl carbamate, followed by installation of the morpholine-containing amide via coupling with a fully functionalized acid precursor.[56]

The PK profile of cobicistat was assessed in healthy volunteers.[57] Exposures increased in a nonlinear fashion as expected for a compound that inhibits its own clearance pathway (164-fold increase for a single dose of 50 versus 400 mg; 47-fold increase for multiple-day dosing of 50–300 mg). Cobicistat has a short-plasma half-life (single dose, 1.41–5.2 h; multiple doses, 2.19–8.07 h). The maximal level of Cyp3A4 inhibition, as assessed through coadministration of the Cyp3A4 substrate midazolam, was seen at the 200 mg dose. Cobicistat was well tolerated at all doses. The PK enhancing effects of cobicistat on atazanavir and darunavir were assessed in Phase I studies.[58] Atazanavir (300 mg)/ritonavir (100 mg) and atazanavir (300 mg)/cobicistat (150 mg) gave bioequivalent exposures of atazanavir. Similar results were seen with darunavir. A Phase III randomized, double-blind, double-dummy, active-controlled trial (GS-US-216-0114) was conducted to evaluate the efficacy and safety of cobicistat (344 patients) versus ritonavir (348 patients) as a PK enhancer of atazanavir in combination with emtricitabine/tenofovir disoproxil fumarate in treatment-naïve patients.[59] The primary endpoint (HIV-1 RNA load of <50 copies/mL at week 48) was seen in 85% of patients in the cobicistat arm compared with 87% of patients receiving ritonavir. Safety and tolerability were also similar. The most common adverse events were related to elevated bilirubin levels, and occurred in 40.7% of cobicistat patients compared with 36.2% of ritonavir patients. This trial showed cobicistat to have high efficacy and noninferiority to ritonavir. The use of darunavir in combination with cobicistat is based on PK studies; no efficacy data for this combination is available to date. Cobicistat was approved at 150 mg once daily as a PK enhancer for the HIV PIs atazanavir (300 mg once daily) and darunavir (800 mg once daily) as part of antiretroviral combination therapy in adults with HIV-1 infection.

7. DABRAFENIB (ANTICANCER)[60–65]

Class:	B-Raf^{V600E} kinase inhibitor
Country of origin:	United States
Originator:	GlaxoSmithKline
First introduction:	United States
Introduced by:	GlaxoSmithKline
Trade name:	Tafinlar®
CAS registry no:	1195768-06-9 (methanesulfonic acid salt) 1195765-45-7 (free base)
Molecular weight:	615.6 (methanesulfonic acid salt); 519.5 (free base)

In May 2013, the US FDA approved dabrafenib (also referred to as GSK 2118436) for the treatment of patients with unresectable or metastatic melanoma with the BRAF^{V600E} mutation as detected by a FDA-approved test. The National Cancer Institute estimated that more than 76,000 new cases of melanoma and 9000 deaths due to melanoma would occur in the United States in 2013. Although immunotherapeutic (IL-2) and chemotherapeutic options (dacarbazine) exist, overall survival rates are not high with these treatments. Oncogenic B-raf signaling is implicated in approximately 50% of melanomas and dabrafenib, like its predecessor vemurafenib (the first marketed BRAF inhibitor), has been developed as a targeted therapy for the BRAF gene. Substitution of valine to glutamic acid in the kinase domain due to mutation at nucleotide 1799 of the BRAF gene leads to the constitutive activation of B-raf kinase, which in turn results in excessive cell proliferation due to dysregulated downstream signaling and gene expression. Dabrafenib was identified from a screen of an oncology-directed kinase collection,[60] followed by extensive structure–activity relationships (SAR) on an initial thiazole lead.[61] Dabrafenib is a potent inhibitor of B-Raf^{V600E} kinase ($IC_{50}=0.65$ nM) compared to its potency against wild-type B-raf ($IC_{50}=3.2$ nM). It also inhibits other kinases (e.g., CRAF) and other mutant B-raf kinases (BRAF^{V600K} and BRAF^{V600D}) with enzyme IC_{50}s of <5 nM and is fairly selective versus a panel of 270 kinases.[62] Consistent with its *in vitro* activity, oral administration of dabrafenib inhibits the growth of B-Raf^{V600E} mutant melanoma (A375P) and colon cancer (Colo205) human tumor xenografts growing subcutaneously in immunocompromised mice.[61,62] Key steps in the synthesis of dabrafenib[61,63] are condensation of an aryl sulfonamide ester with the lithium anion of 2-chloro-4-methylpyrimidine to generate a ketone intermediate and bromination of the ketone intermediate with *N*-bromosuccinamide followed by cyclization with *tert*-butyl thioamide to afford the desired thiazole core.

At single oral doses of 12–300 mg, dabrafenib demonstrates dose proportional PKs, although upon twice daily dosing the exposure was less than dose proportional. The median time to maximum plasma concentration (C_{max}) was 2 h and the mean bioavailability was 95%. The mean terminal half-life of dabrafenib was 8 h after oral administration and the apparent volume of distribution was 70.3 L. Dabrafenib is highly protein bound (~99.7%).[64] The efficacy of dabrafenib was studied in an international, multicenter, randomized (3:1), open-label, active-controlled trial conducted in 250 patients with previously untreated BRAF^{V600E} mutation-positive, unresectable, or metastatic melanoma. Patients were randomized to receive dabrafenib 150 mg (po, twice daily, $n=187$) or dacarbazine (1000 mg/m^2) intravenously q3w ($n=63$). The estimated median progression-free survival time (main efficacy outcome) was 5.1 months in dabrafenib recipients and 2.7 months in dacarbazine recipients.[65] Hyperkeratosis, headache, pyrexia, arthralgia, papilloma, alopecia, and palmar–plantar erythrodysesthesia syndrome were the most common adverse events (≥20%) observed during dabrafenib clinical trials. The recommended dose of dabrafenib is 150 mg administered orally twice daily.

8. DIMETHYL FUMARATE (MULTIPLE SCLEROSIS)[66–80]

Class:	Nrf2 pathway activator
Country of origin:	United States
Originator:	Biogen Idec
First introduction:	United States
Introduced by:	Biogen Idec
Trade name:	Tecfidera®
CAS registry no:	624-49-7
Molecular weight:	144.12

In March 2013, the US FDA approved dimethyl fumarate for the treatment of relapsing forms of multiple sclerosis (MS).[66] The National MS Society reports incidences of 4.2 million individuals with MS worldwide with 400,000 cases in the United States based on an extrapolation of US census data from 2000.[67] Standard intravenous treatments for MS include: interferon beta, glatiramer acetate, natalizumab, and mitoxantrone. Recently approved oral therapeutics are fingolimod (Gilenya®) and teriflunomide (Aubagio®). Dimethyl fumarate is the newest oral therapeutic for MS. While its mechanism is not completely understood, dimethyl fumarate increases anti-inflammatory cytokines (IL-10, IL-4, and IL-6), decreases proinflammatory cytokines (IL-1β, IL-6, and TNF-α), and activates the Nrf2 pathway to protect neuronal cells.[68] Nrf2 is activated by covalent bond-forming electrophiles such as dimethyl fumarate, a Michael acceptor.[69] Dimethyl fumarate has been used for the treatment of psoriasis in Europe since 1994 and has a favorable long-term safety profile.[70] An exploratory study in patients with relapsing remitting MS showed significant reductions in MS lesions after 18 weeks of treatment with 720 mg/day of dimethyl fumarate.[71] Evaluation of dimethyl fumarate in a mouse experimental autoimmune encephalomyelitis model of MS resulted in reduced spinal cord macrophage inflammation.[72] Dimethyl fumarate is obtained in high purity by esterification of fumaric acid with methanol and catalytic sulfuric acid.[73]

A study of healthy subjects receiving a combination of dimethyl fumarate (120 mg) and monomethyl fumarate (95 mg) showed that dimethyl fumarate is rapidly hydrolyzed to monomethyl fumarate with no detection of dimethyl fumarate or fumaric acid in the serum.[74]

Monomethyl fumarate is metabolized through the citric acid cycle resulting in excretion during respiration as CO_2.[75] Extended T_{lag} (absorption lag time) and T_{max} were observed in subjects taking the drug with breakfast.[74] The package insert states that taking dimethyl fumarate with a high-fat, high-calorie meal decreases C_{max} by 40% with no effect on overall AUC, which may decrease the incidence of flushing.[76] A Phase II study in 257 patients with relapsing remitting MS showed that patients receiving 240 mg of dimethyl fumarate three times daily for 24 weeks had a 69% reduction in new gadolinium-enhanced lesions and a 32% annualized reduction in relapse rates versus placebo.[77] Phase III DEFINE and CONTROL trials were conducted with a combined total of over 2600 relapsing remitting MS patients. The 2-year DEFINE trial utilized combined clinical and radiological measures to demonstrate that 28% of patients receiving 240 mg of dimethyl fumarate twice daily were "disease-free" at the end of the study versus only 15% for placebo.[78] The 2-year CONFIRM trial showed that patients receiving 240 mg of dimethyl fumarate twice daily had a 44% reduction in annualized relapse rate versus placebo.[79] A comparator arm in the CONFIRM trial showed that a conventional MS therapeutic, 20 mg/day s.c. of glatiramer acetate, afforded only a 29% reduction in relapse rate.[79] Flushing and nausea were the most commonly reported adverse events.[80] Tecfidera® (dimethyl fumarate) is given orally, twice daily with or without food as a hard gelatin delayed-release capsule at a starting dose of 120 mg. After 7 days, the dose is increased to a maintenance level of 240 mg twice daily for the duration of therapy.

9. DOLUTEGRAVIR (ANTIVIRAL)[81–91]

Class:	INSTI
Country of origin:	United States
Originator:	Shionogi & GlaxoSmithKline
First introduction:	United States
Introduced by:	ViiV Healthcare & GlaxoSmithKline
Trade name:	Tivicay®
CAS registry no:	1051375-16-6 (neutral form) 1051375-19-9 (sodium salt)
Molecular weight:	419.4 (neutral form) 441.4 (sodium salt)

In August 2013, the US FDA approved dolutegravir (also referred to as S/GSK1349572) for the treatment of HIV-1 infection in adults and children ages 12 years and older in combination with other antiretroviral drugs.[81] Dolutegravir was approved in Canada in November 2013. HIV/AIDS remains a global epidemic with 35 million people infected, including 2.3 million new infections as of 2012.[82] AIDS deaths are in decline largely due to the availability of antiretroviral treatments. These treatments typically target three virus-encoded enzymes: reverse transcriptase, protease, and, most recently, integrase. Inhibition of HIV integrase prevents the fusion of viral DNA 3′-hydroxyl groups to host DNA phosphate groups, thus preventing viral DNA insertion and ultimately viral replication.[83] Dolutegravir joins raltegravir and elvitegravir (this chapter of ARMC) as the latest of three FDA-approved HIV integrase strand transfer inhibitors (INSTIs). Dolutegravir was discovered by rational design from a literature diketo acid HIV integrase inhibitor utilizing X-ray coordinates to predict ideal bond angles between the diketone and distal benzyl group.[84] In dolutegravir, the monocyclic component of the reported inhibitor was replaced with the tricyclic carbamoyl pyridone moiety. The researchers postulated that the appropriate arrangement of three oxygens would permit chelation with two magnesium ions in the binding site thus affording improved potency. Ultimately, this arrangement along with further modifications afforded dolutegravir, a potent inhibitor of HIV integrase ($IC_{50}=1.7$ nM).[85] Noteworthy in the multistep synthesis of dolutegravir is formation of the fused tricycle with correct stereochemistry. This is accomplished by condensation of a pyridinone aldehyde ester with *R*-3-aminobutanol in dichloromethane with acetic acid as catalyst and microwave heating to afford a 20:1 mixture favoring the desired diastereomer.[85]

A Phase III SPRING-2 study evaluated dolutegravir (50 mg once daily) versus raltegravir (400 mg bid) in antiretroviral-naïve adults with HIV-1. Both drugs were coformulated with standard nucleoside/nucleotide reverse transcriptase inhibitors (NRTIs). After 48 weeks, 88% of dolutegravir-treated patients and 85% of raltegravir-treated patients had less than 50 copies/mL of HIV-1 RNA, demonstrating that dolutegravir was noninferior to raltegravir for the treatment-naïve patients. Furthermore, a lower incidence of emergent resistance was observed in the patients receiving dolutegravir.[86] A similar Phase III SAILING study in patients who had previously received NRTIs but not INSTIs demonstrated superiority of dolutegravir over raltegravir for decreasing viral load in treatment experienced patients, especially since dolutegravir can be administered once daily.[87] At a dose of 50 mg once daily in healthy subjects, dolutegravir is well tolerated, has a half-life of 15 h and does not inhibit Cyp3A.[88] It is eliminated, primarily unchanged, in the feces and as a glucuronide or *N*-dealkylated metabolite in urine.[89] Unlike elvitegravir, dolutegravir does not require pharmacokinetic boosting[90] and can be administered without formulation as a fixed-dose combination (FDC). Commonly reported adverse events are: hypersensitivity reactions; serum liver biochemistry changes in patients with hepatitis B or C; fat redistribution; and immune reconstitution syndrome.[91] Dolutegravir (Tivicay®) is given orally, once daily and is available as a 50 mg tablet. It can be administered twice daily for patients who have INSTI resistance or are taking UGT1A/Cyp3A inducers.

10. EFINACONAZOLE (ANTIFUNGAL)[92–98]

Class:	14α-Demethylase inhibitor
Country of origin:	Japan
Originator:	Kaken Pharmaceuticals
First introduction:	Canada
Introduced by:	Valeant
Trade name:	Jublia®
CAS registry no:	164650-44-6
Molecular weight:	348.39

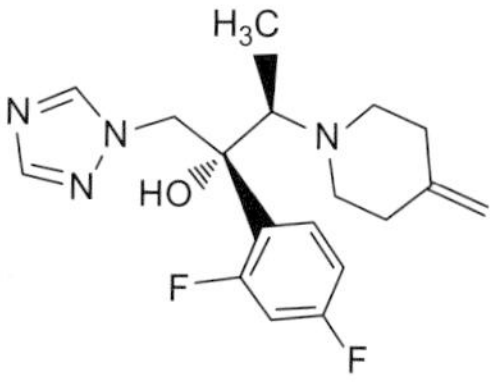

In October 2013, efinaconazole (also known as KP-103) was approved in Canada as a 10% topical solution for the treatment of onychomycosis.[92] Onychomycosis is a fungal disease of the nails that is caused by infection of the nail bed, nail plate, and surrounding tissue with dermatophytes, yeasts, or molds. Onychomycosis occurs in 10% of the general population, with the incidence increasing to 50% in people 70 years of age or older.[93] More than 30% of diabetic individuals suffer with onychomycosis. Treatment of onychomycosis is challenging due to slow growth of nails and poor penetration of drugs to the site of infection. Oral antifungal agents are effective, but can suffer from safety and drug–drug interaction issues, particularly in the elderly. Topical antifungal agents have the advantage of being applied directly at the site of infection, however, poor penetration into infected tissue often limits efficacy by the topical route. Like other azole antifungal agents, efinaconazole acts by disrupting fungal cell membranes through inhibition of sterol 14α-demethylase, an enzyme involved in the biosynthesis of ergosterol, which is a key component of the fungal cell membrane.[94]

Efinaconazole has potent antifungal activity against clinical isolates of dermatophytes, including *Trichophyton mentagrophyes* ($MIC_{80}=0.125$ μg/mL) and *Trichophyton rubrum* ($MIC_{80}=0.25$ μg/mL), as well as against *Candida* and *Malassezia* species.[95] Unlike other antifungal agents, efinaconazole retains activity in the presence of keratin, indicating that more unbound drug is available at the site of action. Efinaconazole is efficacious in guinea-pig models of fungal infection.[95] Efinaconazole is prepared by reaction of an epoxide intermediate with 4-methylenepiperidine.[96]

Studies in healthy volunteers and patients with onychomycosis indicated that application of a 10% solution efinaconazole resulted in low-systemic exposure.[92] In one study, after application of 10% efinaconazole for 7 days to all 10 toenails in healthy volunteers, plasma levels of efinaconazole were 0.54 ng/mL and in patients with severe onychomycosis, plasma levels were 0.67 ng/mL after 28 days of treatment.[97] The efficacy of a 10% topical solution of efinaconazole was established in two Phase III randomized, double-blind studies in patients with mild to moderate onychomycosis, defined as having 20–50% clinical involvement of the target toenail.[98] Patients were randomized (3:1) to efinaconazole or vehicle and treated once daily for 48 weeks, with follow-up assessment at week 52. The primary efficacy end point was the proportion of patients showing complete cure, defined as having both no clinical involvement of the target toenail and mycological cure (negative potassium hydroxide examination and negative fungal culture). In the first study (870 patients), 17.8% of patients showed complete cure with 10% efinaconazole treatment compared with 3.3% of the vehicle-treated group. Similar results were seen in the second study (785 patients), where 15.2% of efinaconazole-treated patients showed complete cure versus 5.5% for the vehicle arm. These results are two to threefold better than in typical studies with other topical antifungal agents, and are comparable to treatment with oral therapies. A secondary measure of treatment success was defined as reduction of the affected area of the toenail to $\leq$10%. In the first study, treatment success ranged from 21.3% to 44.8% for efinaconazole versus 5.6–16.8% with vehicle. In the second study, treatment success was 17.9–40.2% for efinaconazole compared with 7.0–15.4% for vehicle. Efinaconazole 10% solution was generally well tolerated, with adverse events being similar in the drug- and vehicle-treated arms. The most common adverse events were related to irritation at the application site, as well as nasopharyngitis and upper respiratory tract infection. Efinazocnazole was approved as a 10% nonlacquer solution for the treatment of onychomycosis.

11. ELVITEGRAVIR (ANTIVIRAL)[99–108]

Class:	INSTI
Country of origin:	Japan
Originator:	Torii Pharmaceuticals (subsidiary of Japan Tobacco)
First introduction:	European Union, United States
Introduced by:	Gilead Sciences
Trade name:	Vitekta™ (Europe), Stribild® (as combo in the United States and Japan)
CAS registry no:	697761-98-1
Molecular weight:	447.9

In November 2013, the European Medicines Agency (EMA) approved elvitegravir (also known as GS 9137 and JTK 303) as a single agent to be used as part of an antiviral regimen that includes a ritonavir-boosted protease inhibitor for the treatment of HIV-1 in adults without mutations indicative of elvitegravir resistance.[99] In August 2012, the US FDA had previously approved Stribild®, a fixed-dose, combined formulation of elvitegravir, cobicistat, emtricitabine, and tenofovir disoproxil fumarate, for patients with HIV-1 who have not received prior treatment.[100] Elvitegravir is the second of three marketed HIV integrase strand transfer inhibitors (INSTIs) including raltegravir and dolutegravir (this volume of ARMC). Elvitegravir was discovered by modification of a literature naphthyridine HIV integrase inhibitor in which the naphthyridine core served as a bioisostere for the diketo acid moiety in an original series.[101] Serendipitously, a 4-quinolone-3-carboxylic acid precursor en route to the desired bioisosteric glyoxylic acid demonstrated modest integrase inhibition ($IC_{50} = 1600$ nM).[102] Further derivatization led to elvitegravir with enhanced inhibition of integrase strand transfer ($IC_{50} = 7.2$ nM) and significant antiviral activity ($EC_{50} = 0.9$ nM).[102] Elvitegravir was prepared in seven synthetic steps from 2,4-difluoro-5-iodobenzoic acid.[102] The corresponding acid chloride was coupled to ethyl 3-(dimethylamino) acrylate and further substituted with *S*-valinol. Base promoted cyclization afforded the quinolone which was protected as silyl ether. Negishi coupling installed the 2-fluoro-3-chlorobenzyl moiety. Subsequent hydrolysis and methoxylation afforded elvitegravir.

Elvitegravir was evaluated in 40 HIV-1-infected patients who were not receiving antiretroviral therapy. When coadministered with ritonavir (50 mg once daily), elvitegravir (400 mg and 800 mg twice daily) significantly reduced viral load and levels of HIV-1 RNA after 10 days.[103] Elvitegravir is extensively metabolized by Cyp3A4 and subsequently glucuronidated. Coadministration with a Cyp3A4 inhibitor, such as ritonavir, is required to extend the half-life of elvitegravir.[104] A Phase III study in over 1000 treatment-naïve patients demonstrated that a combined formulation of elvitegravir, cobicistat (a Cyp3A4 inhibitor), emtricitabine, and tenofovir was noninferior to a combination of raltegravir, emtricitabine, and tenofovir. Furthermore, patients receiving the elvitegravir combination had fewer abnormal results in liver function tests.[105] Stribild®, the combined formulation of elvitegravir (150 mg), cobicistat (150 mg), emtricitabine (300 mg), and tenofovir disoproxil (245 mg) is given orally, once daily with food as a tablet.[106] Stribild® has a boxed warning for lactic acidosis, severe hepatomegaly with steatosis, and possible exacerbation of hepatitis B due to previously reported cases involving the nucleoside analogs present in this combination.[106] Another Phase III trial demonstrated that elvitegravir (150 mg once daily) was noninferior to raltegravir (400 mg twice daily) in HIV-1 patients who were failing standard antiretroviral treatment.[107] The researchers surmised that once daily dosing could improve patient compliance for elvitegravir versus raltegravir.[107] Vitekta™, the standalone formulation of elvitegravir, is given orally, once daily, with food as an 85 or 150 mg tablet and must be coadministered with a ritonavir-boosted protease inhibitor.[108]

12. IBRUTINIB (ANTICANCER)[109–114]

Class:	Tyrosine Kinase Inhibitor
Country of origin:	United States
Originator:	Celera/Pharmacyclics
First introduction:	United States
Introduced by:	Pharmacyclics and Johnson & Johnson
Trade name:	Imbruvica™
CAS registry no:	936563-96-1
Molecular weight:	440.5

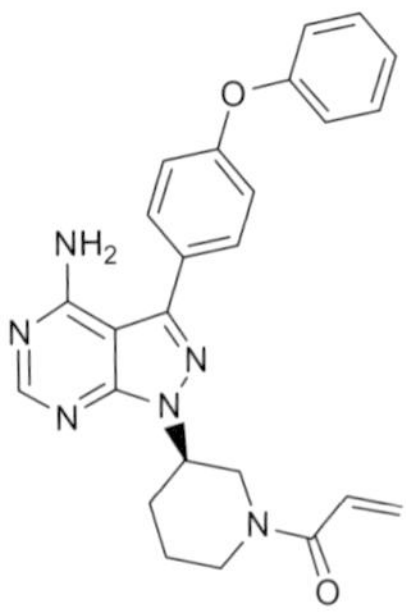

O
NH_2
N
N
N
N
N
O

In November 2013, the US FDA approved ibrutinib (also referred to as PCI-32765), for the treatment of patients with mantle cell lymphoma (MCL) who had received at least one prior therapy. MCL is a rare and aggressive type of blood cancer that accounts for ~6% of all B-cell non-Hodgkin lymphoma cases in the United States.[109] Treatment with the monoclonal antibody rituximab plus hyperCVAD (cyclophosphamide, vincristine, doxorubicin, dexamethasone alternating with methotrexate, and cytarabine) is recommended as aggressive induction therapy in newly diagnosed patients. Other chemotherapeutic options include treatment with bortezomib for patients who have received at least one prior therapy or with lenalidomide, the first oral agent for the treatment MCL. Ibrutinib is the second oral agent approved for the treatment of MCL. It works by irreversibly inhibiting Bruton's tyrosine kinase (Btk) leading to the inhibition of B-cell receptor signaling and resulting in the reduction of malignant B-cell proliferation and induction of cell death. Btk plays an important role in the differentiation, development, proliferation, and survival of B cells via activation of cell-cycle regulators and regulating the expression of pro- and anti-apoptotic proteins. Aberrant Btk activity results in a variety of B-cell malignancies including MCL. Ibrutinib inhibits Btk by irreversibly binding to cysteine-481 in the active site thereby inhibiting phosphorylation of tyrosine-223 and affecting downstream B-cell signaling pathways.[110] Ibrutinib is a potent inhibitor of Btk (IC_{50}=0.5 nM) and is efficacious in canine models of B-cell lymphoma. At dose ranges of 2.5–20 mg/kg, there was full occupancy of Btk in peripheral blood and tumor tissue for 24 h.[110] A synthetic route to ibrutinib that employs Suzuki coupling of 3-iodo-1*H*-pyrazolo[3,4-*d*]pyrimidin-4-amine with (4-phenoxyphenyl)boronic acid followed by Mitsunobu reaction with *N*-Boc-3-hydroxypiperidine as key steps has been reported.[111a,b]

Ibrutinib is rapidly absorbed after oral administration with a median T_{max} of 1–2 h. Dose proportional increase in exposures were noted from 420 to 840 mg/day and the steady state AUC in patients at 560 mg is 953 ± 705 ng h/mL. The half-life of ibrutinib is 4–6 h and the apparent clearance is ~1000 L/h. Ibrutinib exhibits reversible binding to human plasma proteins (97.3%) and has an apparent volume of distribution at steady state of ~10,000 L.[112a,b] The efficacy and safety of ibrutinib in MCL was established in a multicenter, open-label, single-arm Phase II clinical study of 111 patients who had three prior anticancer therapies.[113] Ibrutinib administered orally at 560 mg once daily demonstrated an overall response rate (ORR, primary endpoint) of 69%. The median response time for the 75 patients who responded was 1.9 months (5.5 months to complete response) and the median progression free survival (PFS) for all patients was 13.9 months. Thrombocytopenia, diarrhea, neutropenia, anemia, fatigue, musculoskeletal pain, peripheral edema, upper respiratory tract infection, nausea, bruising, dyspnea, constipation, rash, abdominal pain, vomiting, and decreased appetite were the most common adverse events (≥20%) observed during ibrutinib MCL clinical trials. The recommended dose of ibrutinib for MCL is 560 mg administered orally once daily. In February 2014, the US FDA also approved ibrutinib for CLL.[114] The recommended dose of ibrutinib for CLL is 420 mg administered orally once daily.

13. ISTRADEFYLLINE (PARKINSON'S DISEASE)[115–122]

Class:	Adenosine A_{2A} receptor antagonist
Country of origin:	Japan
Originator:	Kyowa Hakko Kirin
First introduction:	Japan
Introduced by:	Kyowa Hakko Kirin
Trade name:	Nouriast®
CAS registry no:	155270-99-8
Molecular weight:	384.43

In March 2013, istradefylline (also known as KW-6002) was approved in Japan for adjunctive treatment of Parkinson's disease (PD).[115] PD is a chronic, progressive neurological disorder that is characterized by impairment of motor function (bradykinesia, tremors, rigidity, and postural disturbances) and accompanied by additional symptoms including mood disorders and cognitive decline. The prevalence of PD is estimated at 1–2% of the population worldwide, with emergence of the disease typically occurring at age 50 and older.[116] The pathophysiology of PD involves the progressive loss of dopamine-containing neurons in the substantia nigra, causing loss of dopamine input to motor regions of the brain. The resulting imbalance in striatal output pathways leads to the classic motor deficit symptoms. Current treatment options are focused on increasing or replacing dopamine, such as administration of levo-dopa (a dopamine precursor), dopamine metabolism inhibitors, and dopamine receptor agonists. Over time, these agents become less effective (wearing-off), with less predictable switching between desired mobility (the ON state) and undesired immobility (the OFF state). Istradefylline acts by antagonism of the adenosine A_{2A} receptor, which is colocalized with dopamine D_2 receptors in the striatum, to enhance dopamine D_2-dependent signaling. Istradefylline is a light-sensitive compound and has been evaluated *in vitro* under low-light conditions to prevent isomerization of the (*E*)-styryl group and decomposition.[117,118] Istradefylline has a K_i of 2.2 nM for the rat adenosine A_{2A} receptor and an ED_{50} of 0.03 mg/kg, po, in reversal of haloperidol-induced catalepsy in mice.[117] Further characterization showed istradefylline to have a K_i of 12 nM for the human adenosine A2A receptor, to be highly selective, and to be a functional competitive antagonist.[118] Istradefylline has activity alone and in combination with levo-dopa in preclinical animal models of PD.[119] The synthesis of istradefylline was accomplished by 1-ethyl-3-(3-dimethylaminopropyl)carbodiimide-mediated coupling of 5,6-diamino-1,3-diethyluracil with 3,4-dimethoxycinnamic acid, followed by cyclization upon treatment with aqueous NaOH, and selective methylation (MeI, K_2CO_3, and DMF).[117]

In clinical studies, istradefylline had a long half-life (>60 h) and showed dose proportional increases in AUC and C_{max} in single and multiple ascending dose studies.[115] Istradefylline is an inhibitor of Cyp3A and p-glycoprotein. A population analysis of data combined from several Phase I and II/III studies in patients and normal, healthy volunteers showed that the istradefylline PK profile fits a 2-compartment model with first-order absorption.[120] At steady state, istradefylline exposure increased by 35% in the presence of Cyp3A4 inhibitors and decreased by 38% in smokers. A PET study in healthy male volunteers with [^{11}C]KW-6002 showed >90% receptor occupancy with daily oral doses of greater than 5 mg of istradefylline.[121] Several Phase II and III trials have been conducted with istradefylline as adjunct therapy to levo-dopa for PD patients.[115] All but one of these studies showed a significant decrease in OFF time. As an example, a Phase III double-blind study was conducted in Japan with 373 patients randomized to receive placebo ($n=126$), 20 mg/day of istradefylline ($n=123$), or 40 mg/day of istradefylline ($n=124$) for 12 weeks in combination with a stable regimen of levo-dopa and other anti-PD medications.[122] The primary end point was an improvement in total awake time spent in the OFF state. The change in daily OFF time was significantly reduced in both drug treatment groups, with similar effects in the 20 mg/day (−0.99 h) and 40 mg/day (−96 h) compared with placebo (−0.23 h). The lack of dose response may be due to saturation of receptor binding at doses above 5 mg. Daily ON time was assessed as a secondary measure of efficacy, with improvements in ON time without troublesome dyskinesia being seen with both istradefylline treatment groups compared with placebo. In addition, the percentage of patients reported to be "much" or "very much improved" was 28.7% in the 40 mg/day group, 20.8% in the 20 mg/day group, and 10.7% with placebo. The most frequently reported treatment-emergent AE was mild to moderate dyskinesia (involuntary movement). Istradefylline was approved as adjunctive treatment for PD at an oral dose of 20 or 40 mg, once daily.

14. LIXISENATIDE (ANTIDIABETIC)[123–131]

Class:	GLP-1 analog
Country of origin:	Denmark
Originator:	Zealand Pharma A/S
First introduction:	European Union
Introduced by:	Sanofi
Trade name:	Lumyxia®
CAS registry no:	827033-10-3
Molecular weight:	4858.49

H-HGEGTFTSDLSKQMEEEAVRLFIEWLKNGGPSSGAPPSKKKKKK-NH_2

Lixisenatide (also known as AVE0010) is a glucagon-like peptide 1 (GLP-1) analog that received marketing approval in February 2013 from the European Commission for the treatment of type 2 diabetes mellitus (T2DM). The incidence of T2DM was estimated at 371 million worldwide in 2012 and is expected to rise to 552 million by 2030.[123] In addition to lifestyle modifications, there are a number of pharmaceutical treatments to manage diabetes, including insulin, metformin, sulfonylureas, GLP-1 mimetics, dipeptidyl-peptidase IV (DPP-IV) inhibitors, sodium glucose transporter 2 inhibitors, thiazolidinediones, and alpha glucosidase inhibitors;[124] however, in a recent review, ~40% of patients do not achieve desired glycated hemoglobin (HbA_{1C}) levels.[125] Diabetes mellitus results in increased incidence of microvascular (neuropathy, retinopathy, and nephropathy) and macrovascular (peripheral artery disease, myocardial infarction, and stroke) complications. GLP-1 mimetics, such as exenatide, liraglutide, and lixisenatide act upon the pancreas to promote glucose-dependent insulin secretion and to decrease glucagon secretion. They also act on the gastrointestinal tract to delay gastric emptying thereby reducing glucose load postprandially.[126]

GLP-1 is a native peptide that is secreted in the gut in reaction to nutrients, but it has a short (2–3 min) half-life in plasma due to the action of DPP-IV. Several modified GLP-1 analogs confer DPP-IV resistance and prolonged plasma half-life, as demonstrated by exenatide (half-life = 2–3 h), which was first isolated and identified from the saliva of Helodermatidae lizards (a.k.a. Gila monster).[127] In an attempt to further improve plasma half-life, lixisenatide was designed via C-terminal deletion of a proline residue and attachment of six lysine residues, resulting in a fourfold improvement in binding affinity relative to the native GLP-$1_{7\text{–}36}$.[128] Synthesis was achieved via standard Merrifield coupling procedures to produce the 44 amino acid peptide. A single dose (1 nmol/kg) administered intraperitoneally to *db/db* mice (a diabetic mice strain that lacks the leptin receptor) produced a maximally efficacious effect in reduction of plasma glucose after an acute oral glucose challenge. After 42 days of treatment of *db/db* mice (starting at 6 weeks age), similar reductions in HbA_{1C}, a recognized marker of glycemic control, were observed for the three treatment groups (1, 10, and 100 nmol/kg HbA_{1C} values 6.5–7%) versus vehicle-treated animals (HbA_{1C} ~ 8.4%).[129]

Subcutaneous injection of 20 μg of lixisenatide in humans results in 84 pg/mL peak plasma levels at 2 h with a volume of distribution of ~100 L. Renal elimination represents the predominant clearance route with a plasma half-life of 3 h. Mass balance has not been evaluated, since lixisenatide is composed of natural amino acids, and its degradation products are expected to follow normal catabolic pathways.[130] The lixisenatide clinical program was recently reviewed.[126] Lixisenatide produced 0.3–0.88% reduction in HbA_{1C} versus placebo across multiple studies at 20 μg daily dose. Lixisenatide demonstrates similar efficacy to liraglutide once daily and exenatide twice daily but, in direct comparisons, lixisenatide demonstrated slightly fewer adverse events (AEs) as compared to liraglutide (20 μg lixisenatide vs. 1.8 mg liraglutide maintenance doses). Commonly observed AEs include nausea and vomiting, allergic reactions (0.4%), and hypoglycemia that was often associated with insulin and sulfonylurea comedication, leading to a precaution for its use in combination with these products. In the recommendation letter, "The CHMP, on the basis of quality, safety and efficacy data submitted, considers there to be a favorable benefit-to-risk balance . . ."[131] Lixisenatide is available as a solution for injection in a prefilled pen that provides either 10 or 20 μg of lixisenatide in each dose.

15. MACITENTAN (ANTIHYPERTENSIVE)[132–138]

Class:	Endothelin antagonist
Country of origin:	Switzerland
Originator:	Actelion Pharmaceuticals Ltd.
First introduction:	United States
Introduced by:	Actelion Pharmaceuticals Ltd.
Trade name:	Opsumit®
CAS registry no:	441798-33-0
Molecular weight:	588.273

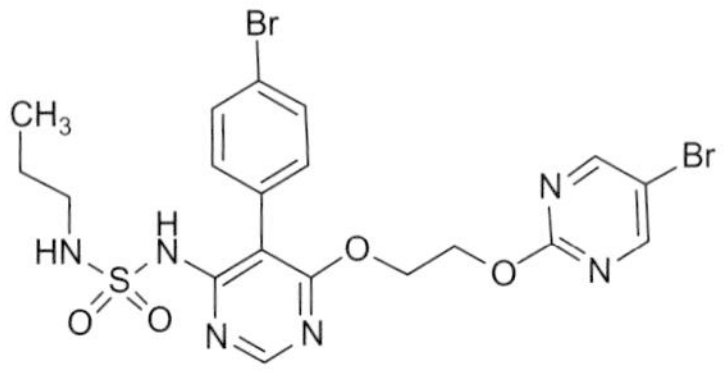

Macitentan (also known as ACT-064992) received US FDA approval in October 2013 for the treatment of pulmonary arterial hypertension (PAH) (WHO group I) to delay disease progression.[132] PAH is a disease that is characterized by elevated pulmonary vascular resistance that often leads to heart failure and death.[133] The incidence of PAH is estimated to be between 15 and 50 million.[134] There is a particularly high risk of developing PAH in individuals with connective tissue diseases such as scleroderma (7–27%), polymyositis/dermatomyositis (25%), systemic lupus erythematosus (0.5–43%), and mixed connective tissue disease (50–60%).[135] Treatment options include phosphodiesterase type 5 inhibitors, prostacyclins, and the endothelin receptor antagonists bosentan and ambrisentan. Macitentan was discovered through SAR studies starting with the bosentan structure with three main goals: (1) to increase potency for both endothelin receptor A and B (ET_A and ET_B) subtypes; (2) to improve tissue distribution to reach the target receptors; and (3) to avoid bile salt transport inhibition.[136] Starting with the bosentan sulfonamido-pyrimidinyl central core, potency was increased 10-fold via incorporation of a bromopyrimidinyl ethylene glycol ether, as found in the clinical endothelin antagonist, T-0201. An aryl ether in bosentan was replaced with the bromophenyl group in macitentan, and a substituent on the 2-position of the central pyrimidine was replaced with hydrogen. Several sulfonamides and alkyl sulfamates were explored, with the propylsulfamate providing the best combination of *in vitro* potency, especially for ET_B antagonism, and *in vivo* efficacy. Whereas bosentan clinically demonstrates increases in ALT (>3 times the upper limit of normal (ULN)), macitentan was predicted to have low risk of ALT increase due to its decreased bile salt transport inhibition *in vitro*.[136] Macitentan is synthesized from a symmetrical 5-(4′-bromophenyl)-3,6-dichloropyrimidine that undergoes sequential nucleophilic aromatic substitution chloride displacements with potassium propylsulfamide and ethylene glycol. Deprotonation of the remaining hydroxyl group of the ethylene glycol fragment and reaction with 5-bromo-2-chloropyrimidine gives macitentan in a three-step linear sequence.[136] In conscious hypertensive Dahl salt-sensitive rats, administration of 3 mg/kg of macitentan produced a maximal blood pressure (BP) lowering effect of −25 mm Hg in the absence of any heart rate changes, whereas the maximal BP effect for bosentan in this model was −7 mm Hg.

Upon administration of a single 10 mg oral dose of macitentan in humans, plasma C_{max} (288 nM) occurs at 6 h with an elimination half-life of 15 h.[137] The major metabolite results from *N*-dealkylation to give an active primary sulfamide with a T_{max} of 44 h and a long elimination half-life of 44 h. In drug disposition studies, 73.6% of the radioactivity was recovered in urine and feces, with urine as the predominant route for elimination. Clinical efficacy and safety was evaluated in a multicenter, double-blind placebo-controlled study (SERAPHIN) including 742 PAH patients characterized with WHO class II, III, or IV disease (more severe form of PAH than class I).[138] The patients were randomized into three groups: placebo (n=250), 3 mg (n=250), and 10 mg (n=242) of macitentan for 3 years of intended treatment. Treatment with 3 and 10 mg of macitentan reduced the incidence of the composite primary endpoint, which was the first event related to PAH or death from any cause versus placebo (hazard ratio for 3 mg: 0.70, and for 10 mg: 0.55, both in favor of macitentan treatment). Patients were also assessed for exercise capacity in a 6-min walk distance test. The placebo group experienced a mean decreased distance of 9.4 m whereas treatment groups experienced increased distance of 7.4 and 12.5 m for 3 and 10 mg groups, respectively. The primary adverse events included upper respiratory tract infection (20% and 15% for 3 and 10 mg, respectively, vs. 13.3% for placebo) and nasopharyngitis (14.8% and 14.0% for 3 and 10 mg, respectively, vs. 10.4% for placebo). The incidence of ALT increases of >3-fold above the upper limit of normal were lower than placebo (3.6% and 3.4% for 3 and 10 mg, respectively, vs. 4.5% in placebo), confirming the design feature of minimizing bile salt transport inhibition. The US FDA approved macitentan in 10 mg oral daily dose for patients suffering from PAH (WHO group I) with the requirement to implement a Risk Evaluation and Mitigation Strategy (REMS) program to monitor for terategenicity.[132]

16. METRELEPTIN (LIPODYSTROPHY)[139–150]

Class:	Leptin analog
Country of origin:	United States
Originator:	Amgen
First introduction:	Japan
Introduced by:	Amylin Pharmaceuticals (subsidiary Astra Zeneca) and Shionogi & Co. Ltd.
Trade name:	Myalept™
CAS registry no:	186018-45-1
Molecular weight:	16,156 Da

MVPIQKVQDD TKTLIKTIVT RINDISHTQS VSSKQKVTGL DFIPGLHPIL 50
TLSKMDQTLA VYQQILTSMP SRNVIQISND LENLRDLLHV LAFSKSCHLP 100
WASGLETLDS LGGVLEASGY STEVVALSRL QGSLQDMLWQ LDLSPGC 147
Disulfide bridge location 97–147

Metreleptin was approved by the Japanese Ministry of Health, Labor and Welfare in March 2013 for the treatment of lipodystrophy and by the US FDA in February 2014 for the treatment of generalized lipodystrophy. Lipodystrophy is a heterogeneous group of very rare disorders characterized by generalized or partial loss of adipose tissue and leptin deficiency. Although robust epidemiology data on the condition are not available, lipodystrophy is considered an extremely rare disorder with an estimated prevalence for generalized lipodystrophy of less than 1 per million.[139] Patients with generalized lipodystrophy have near complete lack of adipose tissue, resulting in a deficiency of hormones produced by adipocytes and ectopic lipid deposition in tissues and organs that do not normally store lipid such as liver and skeletal muscle. This ectopic lipid deposition predisposes patients to severe insulin resistance, diabetes mellitus, and/or hypertriglyceridemia that can result in life-threatening comorbidities such as acute pancreatitis, steatohepatitis, and accelerated atherosclerotic disease.[140] Currently, there are no approved therapies for the treatment of lipodystrophy, and standard therapies for diabetes and hypertriglyceridemia are often insufficient due to the underlying pathophysiology and severity of the metabolic abnormalities. Metreleptin is a recombinant *N*-methionyl analog of the leptin protein[141]; leptin itself was identified via positional cloning in studies on the cause of obesity in the *ob/ob* mouse strain, which lacks the leptin gene.[142] While there have been many studies of leptin in obesity, this first approval of a leptin analog is directed toward the treatment of lipodystrophy. Toward that end, the effects of leptin administration were studied in a transgenic mouse model of lipodystrophy (with constitutive activation of sterol-regulatory-element-binding-protein-1c) that results in lack of white adipose tissue.[143] Since, adipocytes serve as a storage depot for triglycerides (TG), the lack of adipocytes results in ectopic lipid deposition in organs such as liver causing hepatomegaly and severe insulin resistance. Treatment with recombinant leptin infusion for 12 days reversed insulin resistance and diabetes and corrected liver TG content to near-normal. The improvement of metabolic abnormalities with leptin administration in the setting of persistent lack of adipose tissue in this mouse model indicates that leptin replacement is more important than adipose tissue *per se* for the maintenance of metabolic homeostasis. Based on this preclinical evidence, metreleptin was advanced to the clinic to explore its use for the treatment of lipodystrophy in humans.

When metreleptin was administered via subcutaneous injection to healthy adult subjects without lipodystrophy, peak leptin concentrations occurred at ~4 h. The half-life of metreleptin is about 4–5 h, and nonclinical studies suggest renal clearance of metreleptin.[144] Metreleptin has been explored for several indications[145]; however, the initial regulatory approvals were focused on the treatment of rare forms of lipodystrophy. In a small open-label trial,[146] nine patients with lipodystrophy (eight with generalized lipodystrophy and one with partial lipodystrophy) were treated with metreleptin. Metreleptin treatment resulted in dramatic improvement in metabolic parameters for all nine patients: mean glycated hemoglobin (HbA_{1C}) was reduced from a baseline of 9.1% to 7.2%, nearly reaching the American Diabetes Association goal of $<7\%$ for diabetes therapies;[147] mean TG levels decreased by 60%; and mean liver volume decreased by 28%. These improvements in metabolic abnormalities occurred in the setting of discontinuation or large reductions in concomitant antidiabetes therapy. Controlled withdrawal of metreleptin in one patient led to increased TG and insulin levels that were corrected by resumption of metreleptin therapy. In an open-label registrational trial in 48 patients with generalized lipodystrophy and low-leptin levels (median 0.7 ng/mL for males, 1.0 ng/mL for females), metreleptin treatment for a median of 2.7 years (range 3.6 months to 10.9 years) was studied for effects on metabolic parameters, including diabetes and hypertriglyceridemia.[144] Administration of metreleptin was weight-based with subsequent adjustments based on individual response. Over the entire study period, the mean weighted average dose for males was 2.93 mg/day and for females was 5.29 mg/day. Mean reductions in HbA_{1C} (−2% from baseline mean of 8.7%), fasting glucose (−49 mg/dL from baseline mean of 174 mg/dL), and TGs (−55% from baseline median of 348 mg/dL) were observed after 1 year of treatment, indicating significant metabolic improvements with metreleptin treatment in this patient population. The most common adverse events (occurring in ≥10%) included headache, hypoglycemia, decreased weight, and abdominal pain. In a published subgroup analysis of 27 patients with generalized or partial lipodystrophy, 23 (86%) met criteria for definite or borderline steatohepatitis at baseline based on liver biopsy, whereas only nine (33%) had definite or borderline steatohepatitis after metreleptin treatment (mean duration 26 months).[148] Metreleptin for subcutaneous injection was approved in Japan as replacement therapy to treat patients with lipodystrophy.[149] Metreleptin was also approved by the US FDA in February 2014 as replacement therapy to treat the complications of leptin deficiency, in addition to diet, in patients with congenital or acquired generalized lipodystrophy. Metreleptin is available only under a risk evaluation and mitigation strategy (REMS) program due to the potential risks of lymphoma (based on three cases of T-cell lymphoma in patients with acquired generalized lipodystrophy in the clinical studies) and the potential risks associated with the development of neutralizing antibodies (reported in two [6%] lipodystrophy patients and three obese subjects without lipodystrophy in clinical studies).[144,150]

17. MIPOMERSEN (ANTIHYPERCHOLESTEREMIC)[151–158]

Class:	Anti-sense oligonucleotide
Country of origin:	United States
Originator:	ISIS Pharmaceuticals
First introduction:	United States
Introduced by:	Genzyme Corporation
Trade name:	Kynamro®
CAS registry no:	629167-92-6
Molecular weight:	7594.80

19 Na^+ 5′-<u>G-^{Me}C-^{Me}C-^{Me}U-^{Me}C</u>-A-G-T-^{Me}C-T-G^{Me}-C-T-T-^{Me}C-<u>G-^{Me}C-A-^{Me}C-^{Me}C</u>-3′

Mipomersen is a 20-mer phosphorothioate oligonuleotide with underlined residues containing 2′-*O*-(2-methoxyethyl) nucleosides; all other residues are 2′-deoxynucleosides. Substitution at the 5-position of cytosine (C) and uracil (U) bases with a methyl group is indicated by ^{Me}C or ^{Me}U.

Mipomersen (also known as ISIS 301012) was approved with orphan drug status by the US FDA in January 2013 as an adjunct to lipid-lowering medications and diet to reduce low density lipoprotein-cholesterol (LDL-C), apolipoprotein B (Apo B), total cholesterol (TC), and non-high density lipoprotein-cholesterol (non-HDL-C) in patients with homozygous familial hypercholesterolemia (HoFH).[151] The global incidence of HoFH is estimated to be 1 in 1 million. Patients with HoFH are characterized with untreated LDL-C concentrations of >500 mg/dL or treated levels of ≥300 mg/dL.[152] Due to the congenital nature of the disease,[153] hypercholesterolemia often presents itself early in life for HoFH patients. Left untreated, many HoFH patients die in their 20s and 30s. The HoFH population has a high incidence of coronary atherosclerosis and obstructive coronary artery disease as a result of hypercholesterolemia. Current treatments for HoFH involve statin therapy (e.g., simvastatin) in combination with other lipid-lowering agents such as ezetimibe, niacin, bile acid sequestrants, fibrates, omega-3 fatty acids, and the microsomal TG transfer protein inhibitor lomitapide. In patients who have improperly controlled cholesterol levels or are intolerant to statin treatment, LDL apheresis may be initiated to reduce acute hypercholesterolemia; however, this treatment is not widely available, it is invasive, and high-LDL levels tend to be reestablished rapidly (1–2 weeks), requiring frequent treatment.[152]

Mipomersen is an anti-sense oligonucleotide (ASO) designed to inhibit Apo B100 transcription by Watson–Crick pairing to the mRNA sequence to target the complex for silencing and elimination, often via RNase H1 degradation.[154,155] The modified oligonucleotide has phosphorothioate linkages and two 2′-methoxyethyl (MOE) regions that provide the RNA specificity, affinity, and nuclease resistance, as well as a DNA gap region to recruit RNase H1 for destruction of the complex. Mipomersen is synthesized via solid-phase nucleotide synthesis involving sequential addition of nucleotide phosphoramidates, with subsequent phosphorothiolation via phenylacetyldisulfide.[156] Initial preclinical characterization employed the corresponding mouse Apo B1-specific ASO, ISIS-147764, in high-fat fed LDL receptor deficient mice. When administered at 25–100 mg/kg, ISIS-147764 produced dose-dependent reductions in hepatic Apo B1 and LDL-C by 60–90%, reducing plasma lipids below those of LDLR−/− chow-fed animals.[157] Corresponding 50–90% reductions in aortic atherosclerosis were observed. At 75 mg/kg, increases in ALTs were observed after 12 weeks of administration of ISIS-147764. Mipomersen itself was evaluated in transgenic mice overexpressing human Apo B100 via intraperitoneal injection of 25 mg/kg twice weekly, whereupon it demonstrated near complete knockdown of circulating hAPO B100 levels after 3 weeks with a sustained effect for the duration of the 11-week treatment, returning to baseline after a 10-week washout.[158]

In humans, a 200 mg subcutaneous dose of mipomersen demonstrates good bioavailability (54–78% vs. intravenous dosing) achieving a C_{max} of 2.73 μg/mL at ~4 h with broad tissue distribution and a long (31 days) elimination half-life.[154] The predominant metabolites are the result of endonuclease cleavage with elimination of parent and metabolites in urine. In preclinical species, mipomersen demonstrates significant accumulation in liver, which is the primary site of action; however, such disposition studies have not been completed in humans. In a randomized, placebo-controlled Phase III clinical trial in 51 patients with HoFH and average baseline LDL-C of 420 mg/dL, 34 patients were treated with 200 mg of mipomersen for 24 weeks, resulting in a 24.7% reduction in LDL-C versus 3.3% reduction in the placebo group ($n=17$). In aggregated human clinical trials, 62% of treated patients demonstrated increases in hepatic fat content (>5%) as assessed by magnetic resonance imaging and, of those patients, 16% of patients had one value of alanine aminotransferase above the upper limit of normal. Other adverse events involved injection site reactions (84%) and flu-like symptoms (30%) that occur within 2 days of administration.[151] Mipomersen was approved for use as 1 mL of a 200 mg/mL solution in saline for subcutaneous injection administered once weekly for the reduction of plasma lipids in HoFH patients, with a recommended Risk Evaluation and Mitigation Strategy (REMS) plan to monitor for elevated liver enzymes in clinical practice.

18. OBINUTUZUMAB (ANTICANCER)[159–164]

Class:	Recombinant monoclonal antibody
Type:	Glycoengineered, humanized IgG1κ, and anti-CD20
Country of origin:	United States
Originator:	GlycArt Biotechnology AG
First introduction:	United States
Introduced by:	Genentech, a member of the Roche Group
Trade name:	Gazyva®
CAS registry no:	949142-50-1
Expression System:	Recombinant using GlycArt's Chinese Hamster Ovary Cells (CHO) engineered to overexpress GnTIII[159]
Molecular weight:	146.1 kDa

In November 2013, the US FDA approved the glycoengineered, type II anti-CD20 antibody obinutuzumab (also known as GA101) in combination with chlorambucil chemotherapy for the treatment of people with previously untreated chronic lymphocytic leukemia (CLL). Obinutuzumab is the first drug approved under the FDA's breakthrough therapy designation, created in 2012 to quicken the pace of development and review of drugs for serious conditions. CLL is a slowly progressing B-cell blood and bone marrow cancer. According to the National Cancer Institute, 15,680 Americans will be diagnosed and 4580 will die from CLL in 2013. Rituximab, a marketed anti-CD20 antibody, has been widely used either alone or in combination with multiagent chemotherapy for a variety of lymphoproliferative disorders. However, indolent lymphomas, such as CLL, remain incurable, with patients relapsing after treatment, highlighting the need for more effective therapies.[160] Anti-CD20 antibodies are categorized as type I or type II depending on their mode of binding and their cytolytic activities. Obinutuzumab is a unique type II anti-CD20 antibody designed to have improved therapeutic efficacy compared with the previously developed type I anti-CD20 antibodies, rituximab and ofatumumab. The Fc portion of obinutuzumab was glycoengineered to reduce fucosylation of the Fc carbohydrate, resulting in increased FcγRIIIa affinity and antibody-dependent cellular cytotoxicity (ADCC) potency. As expected for a type II antibody, obinutuzumab demonstrated lower complement-mediated cytotoxicity, more potent mediation of cell death via the nonclassical apoptosis pathway and increased ADCC as compared with rituximab in preclinical studies.[161] In addition, obinutuzumab induced a stronger antitumor effect in mouse xenograft models of human lymphoma than rituximab and ofatumumab,[162] supporting clinical investigation of this third generation anti-CD20 antibody.

The pharmacokinetic profile of obinutuzumab was studied in a Phase I trial of 22 previously-treated patients with relapsed or refractory CD20-positive non-Hodgkin lymphoma or CLL.[160] Obinutuzumab was administered intravenously weekly for up to four treatments during the induction phase with doses ranging from 200 to 2000 mg. Eight patients proceeded to maintenance therapy and received 1–8 additional infusions. During the 4-week induction phase, higher concentrations were achieved with higher doses administered, and the C_{max} and C_{trough} values increased. Obinutuzumab serum concentrations were consistently higher in responding versus nonresponding patients. After the last dose of induction, a rapid decline in obinutuzumab concentration was observed. During maintenance therapy, peak serum levels were proportional to the administered dose, but trough levels were low in all cohorts. In addition, in population PK studies, the steady state mean volume of distribution of obinutuzumab was ~3.8 L, the mean terminal clearance was 0.09 L/day, and the elimination half-life was 28 days.[163] A two-stage, three-arm Phase III trial was conducted with 781 treatment-naïve CLL patients.[164] Patients received chlorambucil alone, obinutuzumab plus chlorambucil, or rituximab plus chlorambucil in six 28-day cycles. Patients who received obinutuzumab plus chlorambucil had a significantly reduced risk of disease progression or death and a significantly longer progression-free survival (PFS) than those who received chlorambucil alone (median 23 vs. 10.9 months). Obinutuzumab plus chlorambucil demonstrated superiority over rituximab plus chlorambucil with significantly longer duration of PFS (26.7 vs. 15.2 months). The most common grade 3–4 adverse events (AEs) with obinutuzumab plus chlorambucil versus chlorambucil alone were neutropenia (34% vs. 16%), infusion-related reactions (21% vs. 0%), thrombocytopenia (11% vs. 3%), leucopenia (5% vs. 0%), and anemia (4% vs. 5%). Grade 3 and 4 infusion-related AEs occurred in 20% of the obinutuzumab plus chlorambucil group versus 4% of the rituximab plus chlorambucil group. Obinutuzumab was approved with a recommended dose of 1000 mg, administered intravenously, with the exception of the first infusions in cycle 1, which are administered on day 1 (100 mg) and day 2 (900 mg). Doses are on days 1, 2, 8, and 15 of six treatment cycles each of 28 days duration. Chlorambucil is given orally at 0.5 mg/kg on days 1 and 15 of each treatment cycle.

19. OLODATEROL (CHRONIC OBSTRUCTIVE PULMONARY DISEASE)[165–174]

Class:	β2-Adrenergic receptor agonist
Country of origin:	European Union
Originator:	Boehringer-Ingelheim
First introduction:	Russia
Introduced by:	Boehringer-Ingelheim
Trade name:	Striverdi®
CAS registry no:	868049-49-4 (free base) 869477-96-3 (hydrochloride)
Molecular weight:	386.18 (free base) 422.90 (hydrochloride salt)

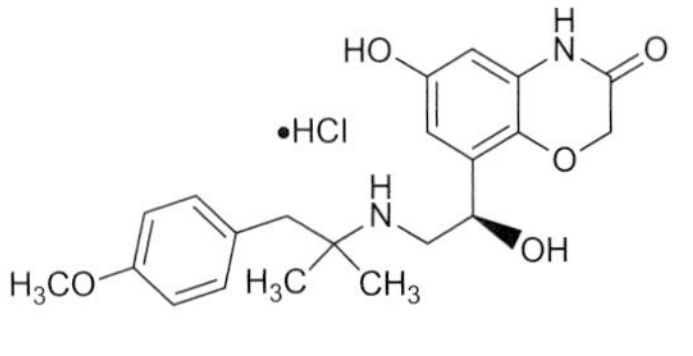

In 2013, olodaterol (also known BI-1744 CL) was approved for the treatment of chronic obstructive pulmonary disease (COPD) in Russia (March), Canada (June), and the European Union (October).[165] Olodaterol is delivered via the Respimat® Soft Mist™ inhaler. COPD is a progressive, irreversible lung disease that is characterized by chronic airflow limitation and is associated with an abnormal inflammatory response of the lungs to noxious particles or gasses.[166] COPD is a major cause of chronic morbidity and mortality worldwide, with an estimated 64 million people having COPD in 2004, and more than 3 million deaths attributed to COPD in 2005.[167] The primary risk factor for COPD is exposure to tobacco smoke (direct or second-hand). Other risk factors include air pollution and exposure to occupational dusts and chemicals. Treatment options for COPD include bronchodilators and anti-inflammatory agents. Olodaterol is a long-acting β_2-adrenergic receptor agonist that acts as a bronchodilator via relaxation of airway smooth muscles. Olodaterol was discovered from an effort to identify a β_2-adrenergic receptor agonist that could be given once daily and with a superior safety profile over known β_2-adrenergic receptor agonists.[168] Olodaterol is a potent β_2 agonist ($EC_{50}=0.1$ nM, intrinsic activity = 88%) and is highly selective for the β_2 receptor over β_1 and β_3 receptors.[168,169] In preclinical models in guinea pigs and dogs, olodaterol had a rapid onset of action and provided bronchoprotection over 24 h. Mechanistic studies indicated that olodaterol forms a highly stable complex with the β_2-adrenergic receptor, with a dissociation half-life of 17.8 h.[170] The synthesis of olodaterol proceeds through an (*R*)-styrene epoxide that is prepared via enantioselective reduction of an α-chloroketone intermediate.[168] Ring opening of the epoxide with 1-(4-methoxyphenyl)-2-methylpropan-2-amine provides olodaterol. The crystalline hydrochloride salt of olodaterol has appropriate properties for formulation in a dry powder inhalation device.

The pharmacokinetic profile of olodaterol was determined as part of a trial to assess the dose response and time response of bronchodilator efficacy following single doses of 2–40 μg of olodaterol administered the Respimat® Soft Mist™ inhaler.[171] Maximum concentrations were reached within 10 min after administration and exposures increased in a dose proportional manner. All treatments were safe and well tolerated. The efficacy of olodaterol was assessed in two pairs of confirmatory 48-week Phase III trials in which olodaterol was given in conjunction with usual maintenance therapy. All four studies were randomized, double blinded, and placebo controlled. The primary end point was improvement in lung function as measured by forced expiratory volume in 1 sec (FEV_1). In one pair of studies (624 and 642 patients with moderate to severe COPD), olodaterol was given at 5 or 10 μg once daily using the Respimat® Soft Mist™ inhaler and compared with placebo.[165,172] There were significant improvements in FEV_1 at week 12 versus placebo in both the 5 and 10 μg dose groups; this effect was maintained at week 48. In the second pair of studies (904 and 934 patients with moderate to severe COPD), olodaterol was given at 5 or 10 μg once daily using the Respimat® Soft Mist™ inhaler and compared with formoterol given at 12 μg twice daily via Aerolizer and placebo.[165,173] There was significant improvement in FEV_1 at weeks 12 and 48 versus placebo in both the 5 and 10 μg dose groups. There were no statistically significant differences between the olodaterol and formoterol treatment arms. In both sets of trials, the 5 and 10 μg doses show comparable efficacy. In a pooled analysis of the Phase III trials, the most common serious adverse events (AEs) were COPD exacerbation (4.7%) and pneumonia (1.6%).[174] The number of deaths reported during treatment was similar in olodaterol- and placebo-treated patients (1.5%). AEs that occurred at a rate of ≥2% with olodaterol compared with placebo-included nasopharyngitis, upper respiratory tract infection, and bronchitis. Olodaterol is indicated for the management of COPD and is not indicated for the treatment of asthma. Olodaterol is approved as a 5 μg dose given once daily using the Respimat® Soft Mist™ inhaler for the long-term maintenance bronchodilator treatment of airflow obstruction in patients with COPD.

20. OSPEMIFENE (DYSPAREUNIA)[175–181]

Class:	Estrogen Receptor agonist-antagonist
Country of origin:	Finland
Originator:	Tess Diagnostics and Pharmaceuticals/Hormos Medical/QuatRx
First introduction:	United States
Introduced by:	Shionogi
Trade name:	Osphena™
CAS registry no:	128607-22-7
Molecular weight:	378.9

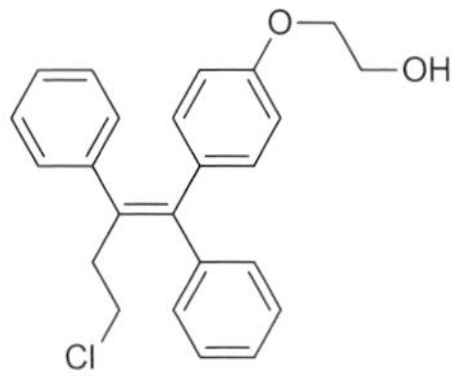

In February 2013, the US FDA approved ospemifene (also referred to as FC1271a), for the treatment of moderate to severe dyspareunia, a symptom of vulvar and vaginal atrophy (VVA), due to menopause. It is estimated that there are 150 million postmenopausal women worldwide with 40–70% suffering from VVA. In a recent Menopause-Specific Quality of Life Questionnaire, 58% ($n = 1561$) of postmenopausal women reported less frequent but severe symptoms that interfered with work, sexual activity, mood, concentration, and overall quality of life.[175] First-line treatment options for dyspareunia include topical vaginal estrogen therapy, lubricants, and moisturizers. Although currently available estrogen replacement therapies alleviate symptoms associated with VVA, nonhormonal therapies are needed since estrogen therapy is associated with a risk of adverse events such as endometrial hyperplasia and venous thromboembolism. Ospemifene is a selective estrogen receptor (ER) modulator (SERM) and the first nonhormonal, nonestrogen for the treatment of moderate to severe dyspareunia in women with menopausal VVA. It binds to ERα ($IC_{50} \sim 800$ nM) and ERβ ($IC_{50} \sim 1600$ nM)[176] with tissue-specific estrogenic agonist/antagonist effects. Treatment with ospemifene increases the thickness of the vaginal tissue thereby decreasing fragility of the tissue and reducing potential for pain during sexual intercourse. However, the exact cellular mechanisms of induction of various tissue-specific responses of ospemifene are not well understood. In a preclinical model of menopause in ovariectomized rats,[177] ospemifene increased vaginal epithelial height to a greater extent compared with another SERM, raloxifene, at a 10 mg/kg/day dose. A synthetic route to ospemifene that employs the McMurry coupling of (4-(2-hydroxyethoxy)phenyl)-(phenyl)methanone and 3-chloropropiophenone in the presence of titanium tertrachloride and zinc powder to yield predominantly the desired *Z*-isomer has been reported in the patent literature.[178]

The pharmacokinetic profile of ospemifene was evaluated in postmenopausal women under fasted conditions at a dose of 60 mg.[179] Mean C_{max} and AUC were 533 ng/mL and 4165 ng h/mL, respectively. The median T_{max} was ~2 h and the elimination half-life was 26 h. Ospemifene is highly protein bound (>99%) and has an apparent volume of distribution of 448 L. The efficacy and safety of ospemifene were established in three placebo-controlled clinical trials (two 12-week efficacy trials and one 52-week long-term safety trial).[180,181] In one of these double-blind trials,[180] 826 postmenopausal women were randomized 1:1:1 to receive ospemifene 30 or 60 mg/day or placebo orally for 12 weeks. Change from baseline to 12 weeks in the percentage of superficial and parabasal cells on the vaginal smear, change in vaginal pH, and change in severity of most bothersome symptoms (vaginal dryness or dyspareunia) were the four primary end points. Statistical significance (vs. placebo) was achieved for all four end points in this clinical trial at the 60 mg dose. Hot flush, vaginal discharge, muscle spasms, genital discharge, and hyperhidrosis were the most common adverse events (≥1%) observed during ospemifene clinical trials. Ospemifene comes with a black box warning for endometrial cancer and cardiovascular disorders. The recommended dose of ospemifene is 60 mg administered orally once daily with food.

21. POMALIDOMIDE (ANTICANCER)[182–195]

Class:	Immunomodulator
Country of origin:	United States
Originator:	Celgene Corporation
First introduction:	United States
Introduced by:	Celgene Corporation
Trade name:	Pomalyst®
CAS registry no:	19171-19-8
Molecular weight:	273.24

In February 2013, the US FDA approved pomalidomide (also known as CC4047) for the treatment of multiple myeloma (MM) in patients with disease progression after receiving other cancer therapeutics.[182] The National Cancer Institute reports 77,617 cases of MM with an estimated 22,000 new cases diagnosed each year in the United States as of 2013.[183] There were approximately 11,000 MM-related deaths in the United States in 2013, but 5-year survival rates have improved from 25.6% in 1989 to an average of 43.2% in 2003–2009. Common first-line treatments for MM include autologous stem cell transplant; chemotherapeutic agents such as melphalan, bisphosphonates, bortezomib, steroids such as prednisone and dexamethasone, thalidomide, and lenalidomide; and combinations of these treatments.[184] In 2006, the FDA approved thalidomide, typically in combination with dexamethasone for the treatment of newly diagnosed MM patients.[185] Thalidomide-treated MM patients typically receive 200 mg daily and can expect to experience adverse reactions that may cause up to 30% of patients to discontinue therapy.[186] Pomalidomide is a 4-amino analog of thalidomide with enhanced potency and an improved toxicity profile. Pomalidomide and thalidomide exert their effects by modulation of immunity, inhibition of angiogenesis, interference with the bone/tumor microenvironment, and inhibition of the cereblon protein.[187] Pomalidomide potently inhibited *in vitro* proliferation in a variety of human MM cell lines, $IC_{50} \sim 10$ nM, while thalidomide showed almost no inhibition up to 100 μM.[188] In mouse MM tumor models, 50 mg/kg daily doses of pomalidomide resulted in marked inhibition of tumor growth after 15 days of treatment and complete regression in 3–6 weeks versus thalidomide-treated controls at the same dose.[188] Pomalidomide is prepared by condensation of 4-nitrophthalic anhydride with 3-aminopiperidine-2,6-dione followed by catalytic hydrogenation of the nitro group.[189,190]

A small Phase I trial conducted with 24 relapsed and refractory MM patients given 1–10 mg of pomalidomide daily determined that 2 mg/day was the maximum tolerated dose and that the drug had a median half-life of 7 h. Adverse events included dose-limiting neutropenia and thrombosis.[191] A study in healthy subjects given 2 mg of ^{14}C-labeled pomalidomide showed that while the parent compound predominated in circulation, extensive metabolism, primarily by cytochrome P450-mediated hydroxylation followed by glucuronidation, lead to elimination of the resulting metabolites mostly in urine.[192] Pomalidomide's approval was based on the Phase II MM-002 study of 221 patients refractory to previous treatments with lenalidomide and bortezomib. This study demonstrated that pomalidomide (4 mg once daily on days 1–21 of a 28-day cycle) with or without LoDex (40 mg once weekly) was well tolerated and effective with an overall response rate of 30% for the combined therapy.[193] In additional Phase II studies, 35 patients with MM refractory to bortezomib and lenalidomide who received 2 mg of pomalidomide daily and 40 mg of dexamethasone weekly demonstrated a 49% response rate, with a 6-month survival rate of 78%. Thirty-five patients receiving a higher 4 mg dose showed no further improvement in response rate. A Phase III study in 435 similar refractory patients explored pomalidomide (4 mg daily) with low-dose dexamethasone (40 mg weekly) versus high-dose dexamethasone alone (40 mg twice weekly). Progression-free survival was nearly twice as long for patients receiving pomalidomide and low-dose dexamethasone than for patients receiving high-dose dexamethasone alone.[194] Commonly reported adverse events are: fatigue, asthenia, neutropenia, anemia, constipation, nausea, diarrhea, dyspnea, upper respiratory tract infections, back pain, and pyrexia.[195] Pomalidomide has a box warning for embryo-fetal toxicity and venous thromboembolism. Pomalyst® (pomalidomide) is given orally, once daily and is available as 1–4 mg capsules. The recommended dose is 4 mg taken orally on days 1–21 of repeated 28-day cycles.[195]

22. RIOCIGUAT (PULMONARY HYPERTENSION)[196–203]

Class:	Guanylate cyclase stimulator
Country of origin:	Germany
Originator:	Bayer
First introduction:	Canada/United States
Introduced by:	Bayer
Trade name:	Adempas®
CAS registry no:	625115-55-1
Molecular weight:	422.4

In September 2013, Health Canada approved riociguat (also referred to as BAY 63-2521), for the treatment of patients with chronic thromboembolic pulmonary hypertension (CTEPH) after surgical treatment or inoperable CTEPH and for the treatment of adults with pulmonary arterial hypertension (PAH). It was approved by the US FDA for the same indication in October 2013. CTEPH is defined as mean pulmonary artery pressure >25 mm Hg persisting for at least 6 months after a pulmonary embolism is diagnosed.[196] Although the actual incidence of CTEPH is unknown, it appears to occur in 2–4% of patients after acute pulmonary embolism. Pulmonary endarterectomy (removal of blood clots from pulmonary arteries) is a surgical option for the treatment of CTEPH, but has a number of limitations. Riociguat is the first approved chemotherapeutic option for the treatment of CTEPH. In healthy individuals, endothelial cells lining the blood vessel wall generate the vasodilator nitric oxide (NO). NO increases production of the signaling molecule cyclic guanosine monophosphate (cGMP) by activating the enzyme soluble guanylate cyclase (sGC). In patients suffering from PAH and CTEPH, there is endothelial dysfunction because of insufficient stimulation of the NO–sGC–cGMP pathway. The resulting change of vascular tone (thickening of arterial walls), increased cellular proliferation, fibrosis, and inflammation eventually lead to increased BP, heart failure, and death. Riociguat has a dual mode of action and works by (a) sensitizing sGC to the body's NO by stabilizing NO–sGC binding and (b) an NO-independent, direct stimulation of sGC via a different binding site. This process restores the NO–sGC–cGMP pathway and leads to increased generation of cGMP with subsequent vasodilation.[197] In preclinical *in vivo* studies, treatment of hypoxic mice that have fully established PAH with riociguat at 10 mg/kg once daily improved pulmonary hemodynamics, right heart hypertrophy, and structural remodeling of the lung vasculature.[198] Riociguat was identified through optimization of a benzyl indazole compound YC-1 that was initially discovered by screening for compounds that inhibited platelet aggregation by stimulation of cGMP synthesis.[199] A synthetic route to riociguat that employs the condensation of an amidine-azaindazole with a phenyldiazo-substituted malonic dinitrile to form the diaminopyrimidine ring as a key step has been reported.[200]

The pharmacokinetic profile of riociguat was evaluated in healthy adult volunteers at a dose range of 0.25–5 mg.[201] At the 2.5 mg dose, mean C_{max} and AUC were ~63.5 μg/L and ~348.5 μg h/L, respectively, although there was significant intra- and interindividual variability particularly for AUC. The median T_{max} was ~1.5 h and the elimination half-life was 12 h in patients and 7 h in healthy volunteers. Riociguat has an absolute bioavailability of 94%, is highly protein bound (~95%), and has an apparent volume of distribution of 30 L. The efficacy of riociguat in CTEPH was established in a 16-week, randomized, double-blind, multicenter, placebo-controlled Phase III clinical study (CHEST-1).[202] In patients ($n=173$), riociguat ≤2.5 mg three times daily was significantly more effective ($p<0.001$) than placebo ($n=88$) in improving the 6-min walking distance (6MWD) at 16 weeks (primary end point). The efficacy of riociguat in PAH was established in a 12-week, double-blind, multicenter, placebo-controlled Phase III trial (PATENT-1).[203] In patients ($n=254$), riociguat ≤2.5 mg three times daily was significantly more effective ($p<0.001$) than placebo ($n=126$) in improving the mean 6MWD (primary end point). The 6MWD increased from baseline (361 m) by 30 m in the riociguat group, whereas it decreased from baseline (368 m) by 6 m in the placebo group. Headache, dizziness, dyspepsia/gastritis, nausea, diarrhea, hypotension, vomiting, anemia, gastroesophageal reflux, and constipation were the most common adverse events (≥3%) observed during riociguat clinical trials. Riociguat comes with a black box warning for embryo-fetal toxicity. The recommended initiation dose of riociguat is 1 mg administered orally three times daily. However, for patients who may not tolerate the hypotensive effect of riociguat, the starting dose is 0.5 mg three times daily. It is recommended to increase the dosing by 0.5 mg no sooner than 2 weeks after initial dosing as tolerated to a maximum of 2.5 mg three times daily.

23. SAROGLITAZAR (ANTIDIABETIC)[204–212]

Class:	Peroxisome proliferator-activated receptor agonist
Country of origin:	India
Originator:	Cadila Healthcare Ltd.
First introduction:	India
Introduced by:	Zydus Discovery
Trade name:	Lipaglyn™
CAS registry no:	1334214-20-8 (1/2 Mg salt), 495399-09-2 (free acid)
Molecular weight:	901.42 (1/2 Mg salt), 439.57 (free acid)

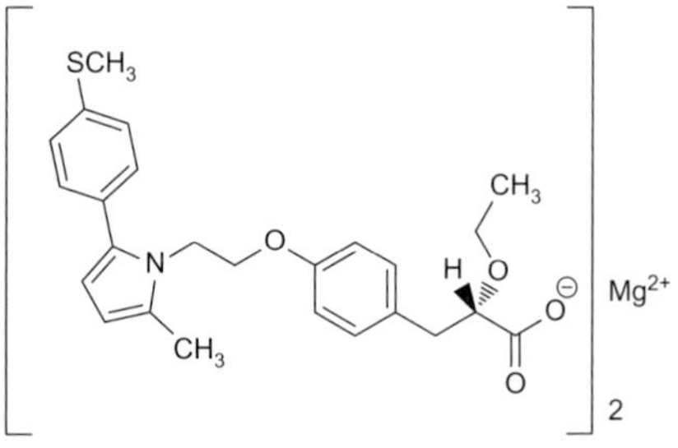

Saroglitazar (also known as ZYH1) was approved for use by the Drug Controller General of India for the treatment of diabetic dyslipidemia and hypertriglyceridemia that is not controlled by statin therapy. Diabetes prevalence is estimated at 371 million worldwide. The odds ratio for acute myocardial infarction (male: 2.67, female: 4.26) and dyslipidemia (male: 3.76, female: 4.42) demonstrate the heightened cardiovascular risk for patients with diabetes.[204,205] In South Asia, the population-attributable coronary risk due to dyslipidemia is nearly double the rate in the European Union. Dyslipidemia is typically treated with statin therapy to reduce LDL cholesterol, but there are several populations where evidence of benefit for cardiovascular outcomes is lacking.[206] Selective peroxisome proliferator-activated receptor (PPAR) agents are employed in clinical practice, such as fenofibrate (PPARα agonist) for hyperlipidemia and rosiglitazone (PPARγ agonist) for diabetes.[3] Saroglitazar is the first approved agent with dual PPAR agonist activity, with EC_{50}s of 0.00065 and 3 nM, respectively, for the α- and γ-PPAR isoforms in HepG2 cells.[207] The medicinal chemistry program has not been described in the scientific literature, but numerous substituents on the pyrrole ring are described in the saroglitazar patent.[208] These substituted pyrroles were synthesized via ethanolamine condensation with substituted diketones and subsequent activation of the hydroxyethylpyrrole for phenolic etherification with (*S*)-2-ethoxy-3-(4-hydroxyphenyl)propanoic acid.[213] The efficacy of saroglitazar was evaluated in *db/db* mice (genetically diabetic animals that lack the leptin receptor), wherein it produced a 17–55% reduction in serum triglycerides (TGs) after 12 days of administration of saroglitazar (0.01–3 mg/kg).[207]

In a Phase I human clinical trial, saroglitazar was demonstrated to have good tolerability up to a 128 mg dose. When the marketed 4 mg dose of saroglitazar was administered, plasma C_{max} of 0.77 μM was achieved at 0.71 h with an elimination half-life of 2.9 h and AUC_{last} of 1.83 μM h.[209] Elimination was not observed in urine and preclinical results indicate that hepatobiliary excretion of drug is the main clearance mechanism. The compound is extensively protein bound (~96%) with a high volume of distribution (20.14 L).[210] In a 12-week clinical trial in diabetic patients with elevated TGs (>200 mg/dL), saroglitazar 4 mg reduced plasma TGs by −45.7% and total cholesterol by −6.9%, with very low density lipoprotein (VLDL) being the predominantly lowered species (−46.1% vs. placebo).[211] Saroglitazar also produced mild antidiabetic effects, reducing placebo subtracted fasting plasma glucose by 4.7%. The most frequently reported adverse event was gastritis (4/99 vs. placebo 1/102). Commonly observed PPAR effects on hemodilution and edema were not observed.[212] Saroglitazar is the first drug discovered, developed, and approved wholly in India, representing a milestone for the pharmaceutical industry in India.

24. SIMEPREVIR (ANTIVIRAL)[214–223]

Class:	Hepatitis C Virus NS3/4A Protease Inhibitor
Country of origin:	Ireland and Sweden
Originator:	Tibotec and Medivir
First introduction:	Japan
Introduced by:	Janssen Pharmaceutical
Trade name:	Sovriad™
CAS registry no:	923604-59-5
Molecular weight:	749.94

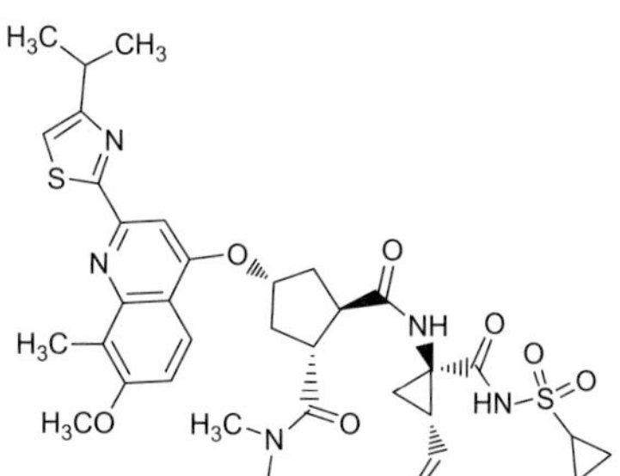

In September 2013, simeprevir (also known as TMC435) was approved in Japan (trade name Sovriad™) for the treatment of genotype 1 hepatitis

C virus (HCV) infection in combination with pegylated interferon and ribavirin (PR).[214] Simeprevir was approved for the same indication in November 2013 in the United States (trade name Olysio™)[215a] and Canada (trade name Galexos™).[215b] Worldwide, more than 150 million people are estimated to be infected with HCV.[216] Chronic infection with HCV occurs in >75% of infected individuals, and can lead to liver cirrhosis and hepatocellular carcinoma. It is estimated that 350,000 people die each year as a result of chronic HCV infection. There are six genotypes of HCV, with genotype 1 having the broadest distribution and being the most common genotype in North and South America, Europe, Asia, and Australia. The standard of care for the treatment of chronic hepatitis C has been based on PR combination therapy, but there are limitations due to variable response rates and adverse events. In 2011, the HCV NS3/4A protease inhibitors (PIs) boceprevir and telaprevir were the first new direct-acting antiviral agents approved for the treatment of HCV infection. Both of these agents are covalent, reversible serine trap PIs that incorporate a reactive α-ketoamide electrophilic center. Simeprevir is the third HCV PI to receive approval and was discovered from an effort to optimize a novel series of cyclopentane-core macrocyclic HCV PIs.[217,218] Unlike the earlier PIs, simeprevir does not rely on formation of a covalent intermediate to inhibit the enzyme, but instead gains binding affinity through a large P2 quinoline substituent that occupies an extended S2 subsite of HCV protease by induced fit. This pocket is not occupied by inhibitors such as telaprevir and boceprevir. Other key features of simeprevir are truncation of the P3 capping group (the *N*-methyl amide), use of an acylsulfonamide as an acid isostere, and incorporation of an isopropylthiazole group to give improved permeability. Simeprevir is a potent NS3/4A PI (K_i = 0.36 nM), with antiviral activity in the HCV replicon assay (genotype 1b EC_{50} = 7.8 nM; genotype 1a EC_{50} = 28.4 nM).[218,219] It is 25-fold less potent against HCV genotype 2, >1000 less potent for HCV genotype 3, but has 3-fold better potency for HCV genotype 4. Although simeprevir is highly protein bound (>99%), there is only a 2.4-fold shift in replicon EC_{50} values in the presence of 50% human serum albumin. Simeprevir has a favorable preclinical pharmacokinetic (PK) profile, with systemic exposure supportive of once daily dosing based on coverage of effective antiviral concentrations in the liver and a favorable liver-to-plasma ratio. The process route to simeprevir starts with *trans*cyclopentanone-3,4-dicarboxylic acid, which is reduced to an alcohol, cyclized to a lactone acid, and then treated with cinchonidine to obtain a crystalline salt in high-enantiomeric purity. Further elaboration involves Mitsunobu reaction to install the P2 group and ring closing metathesis to form the macrocyclic ring.[218]

The PK profile of simeprevir was determined in Phase I studies.[220] In healthy volunteers, once daily dosing of 100, 200, or 400 mg showed prolonged absorption ($T_{max}=4$ h) and greater than dose proportional increases in exposures. Trough plasma concentrations at all doses were well above replicon EC_{50} values. HCV patients received a 200 mg dose once daily for 5 days and showed higher exposure than in healthy volunteers, likely due impaired liver function and slower elimination of simeprevir. Plasma levels of HCV RNA dropped rapidly in all patients, with a median 3.5-$\log_{10}$ IU/mL drop by day 3. Simeprevir has been evaluated in more than 20 Phase II and III clinical trials.[214] The open-label CONCERTO-2 and CONCERTO-3 Phase III trials in Japan with 100 mg simeprevir once daily in combination with PR in patients with genotype 1 HCV infection showed sustained virologic response (SVR; absence of detectable HCV RNA) at 12 weeks of 52.8% for simeprevir in prior nonresponders and 95.9% in prior relapsers.[221] The multinational PROMISE Phase III trial ($n=393$) was a placebo-controlled, randomized trial in patients who had relapsed after previous interferon-based therapy.[222] Simeprevir at 150 mg once daily with PR for 12 weeks gave an SVR of 79.2% compared with 36.1% for placebo plus PR. On-treatment failure and relapse rates were significantly lower in the simeprevir arm. The QUEST-1 and QUEST-2 Phase III trials in treatment-naïve patients given 150 mg simeprevir once daily with PR for 12 weeks gave SVR rates of ~80%.[223] Most patients receiving simeprevir were able to shorten therapy to 24 weeks. Simeprevir was generally well tolerated, with the most common adverse events being rash, pruritis, nausea, and headache. In Japan, simeprevir is indicated for the treatment of chronic hepatitis C genotype 1 infection in patients who are treatment naïve with high-HCV RNA levels or who failed previous interferon-based therapy as a 100 mg capsule administered orally once daily for 12 weeks in combination with PR. In the United States, the recommended dose is 150 mg orally once daily in combination with PR, followed by either 12 or 36 additional weeks of PR depending on response status.

25. SOFOSBUVIR (ANTIVIRAL)[224–235]

Class:	Hepatitis C Virus NS5B polymerase inhibitor
Country of origin:	United States
Originator:	Pharmasset
First introduction:	United States
Introduced by:	Gilead
Trade name:	Sovaldi™
CAS registry no:	1190307-88-0
Molecular weight:	529.45

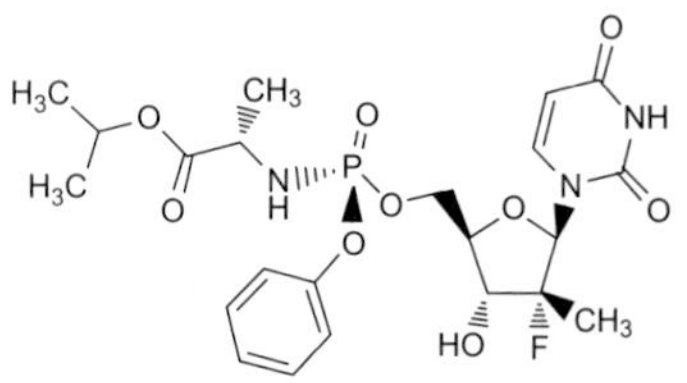

In December 2013, sofosbuvir (also known as GS-7977 and PSI-7977) was approved in the United States for the treatment of hepatitis C virus (HCV) infection as a component of a combination antiviral treatment regimen. The Center for Disease Control estimates that there are 2.7–3.9 million cases of HCV infection in the United States, with more than 16,000 new cases occurring each year.[224a] Worldwide, more than 130–150 million individuals are estimated to be infected with HCV.[224b] Approximately 80% of infected individuals develop chronic infection, which can lead to liver cirrhosis and hepatocellular carcinoma. Until recently, the primary treatment option for hepatitis C was a combination of pegylated interferon and ribavirin (PR).[225] However, sustained virologic response (SVR) to these agents is achieved in only 40–60% of patients, and adverse side effects lead to high rates of discontinuation. Direct-acting antiviral agents (DAAs) have been the focus of the search for novel therapies, with HCV NS3/4A PIs being the first class of approved DAAs. Sofosbuvir is a novel anti-HCV agent that acts as a prodrug for 2′-F-2′-*C*-methyluridine monophosphate, which is converted to the corresponding triphosphate analog in cells, where it acts as an inhibitor of the HCV NS5B RNA-dependent RNA polymerase. Sofosbuvir was discovered from an effort to enhance the activity of the parent nucleoside by bypassing rate-limiting monophosphorylation with a prodrug that would liberate the intact monophosphate in the liver, where it would then be converted by cellular kinases to the active triphosphate species.[226,227] In addition, the prodrug was designed to be amenable to oral delivery. In the initial synthesis of sofosbuvir, the *iso*-propyl ester of (L)-alanine was coupled with phenyl dichlorophosphate to provide a diastereomeric intermediate that was coupled with the uridine nucleoside. The diastereomeric mixture (GS-9851) was shown to produce high levels of triphosphate *in vitro* in primary hepatocytes and in the livers of rats, dogs, and monkeys after oral dosing. The individual diastereomers were obtained by chromatography or by crystallization. The diastereomer with the S_p configuration at the phosphorous center (sofosbuvir) was found to be >10-fold more potent in an HCV replicon assay than the corresponding R_p diastereomer (EC_{90}s of 0.42 and 7.5 μM, respectively). Sofosbuvir is not cytotoxic at concentrations up to 100 μM. Sofosbuvir demonstrated broad genotype coverage using replicon cells from genotypes 1a, 1b, 2a, 2b, and 3a.[228] Crossresistance and selection studies showed that the S282T mutation in NS5B polymerase confers resistance to sofosbuvir. A multigram scale synthesis of sofosbuvir was reported that involves coupling of a diastereomerically pure pentafluorophenol phosphoramidate intermediate with 2′-F-2′-*C*-methyluridine.[229]

The major systemic drug exposure following oral administration of sofosbuvir is from the inactive metabolite 2′-F-2′-*C*-methyluridine (GS-331007). Pharmacokinetic studies in humans were initially performed with GS-9851, a mixture of sofosbuvir and its diastereomer, which is metabolized through the same pathway as sofosbuvir. In healthy subjects receiving single oral doses of 25–800 mg GS-9851, the C_{max} for GS-331007 was 7-fold higher than for GS-9851, and the AUC was 41-fold higher for the metabolite than for GS-9851.[230] In treatment-naïve patients infected with HCV genotype 1, GS09851 was given orally at 50, 100, 200, or 400 mg once daily for 3 days.[231] Increases in C_{max} and AUC for GS-9851 and its metabolites were less than dose proportional, particularly at the highest doses. GS-331007 was the primary drug-related moiety in the plasma and urine. In a randomized, 28-day dose-ranging trial in treatment-naïve patients with HCV genotype 1, sofosbuvir was given once daily at oral doses of 100, 200, or 400 mg in combination with PR.[232] Sofosbuvir showed a median T_{max} of 1 h and a half-life of 0.48–0.75 h. The metabolite GS331007 had a T_{max} of 4 h and a half-life of 7.27–11.8 h. After a single 400 mg oral dose of ^{14}C-sofosbuvir, sofosbuvir, and GS-331007 accounted for 4% and >90% of systemic exposure, respectively.[233] The mean total recovery of the dose was 80% in urine and 14% in feces. Sofosbuvir is 61–65% bound to human plasma protein.[233]

The safety and efficacy of sofosbuvir in combination with ribavirin with or without peginterferon were evaluated in five Phase III trials in HCV patients infected with genotypes 1–6 and one Phase III study in HCV/HIV-coinfected patients.[233] The POSITRON trial (blinded, placebo-controlled, and genotype 2 or 3) assessed a 12-week sofosbuvir–ribavirin regimen ($n=207$) versus placebo ($n=71$) in patients for whom treatment with peginterferon was not an option.[234] The SVR (absence of detectable

HCV RNA) was 78% in the sofosbuvir–ribavirin arm compared with 0% for placebo. The FUSION trial (genotype 2 or 3) evaluated patients who had not responded to prior interferon therapy.[234] The SVR was 50% for patients receiving 12 weeks of sofosbuvir–ribavirin treatment versus 73% with 16 weeks of treatment. In the NEUTRINO Phase III trial (open-label, $n=327$, HCV genotypes 1, 4, 5, or 6), an SVR of 90% was reported after 12 weeks of treatment with 400 mg once daily of sofosbuvir in combination with PR.[235] In the FISSION Phase III trial (noninferiority, $n=499$, HCV genotype 2 or 3), an SVR of 67% was achieved in both the sofosbuvir–ribavirin and the PR groups, with higher response rates for sofosbuvir–ribavirin treatment of genotype 2-infected patients (97%) compared with genotype 3-infected patients (57%).[235] The VALENCE trial evaluated sofosbuvir in combination with ribavirin for 24 weeks in patients with genotype 3 HCV and found an SVR of 84% ($n=250$).[233] For patients in this study with genotype 2 virus ($n=73$), the SVR was 93%. Sofosbuvir was generally safe and well tolerated, with the most common adverse events being fatigue, headache, nausea, and insomnia. In the PHOTON open-label trial in HCV/HIV-coinfected patients treated with sofosbuvir–ribavirin, SVRs were 76% for HCV genotype 1 ($n=114$, 24 weeks), 88% for HCV genotype 2 ($n=26$, 12 weeks), and 92% for HCV genotype 3 ($n=13$, 24 weeks). Overall rates of discontinuation in these trials were low. Sofosbuvir was approved as a 400 mg tablet, taken once daily with or without food. The recommended duration of treatment and comedication depend on HCV genotype: 12 weeks in combination with PR for genotype 1, 12 weeks in combination with ribavirin for genotype 2, and in combination with ribavirin for 24 weeks for genotype 3.

26. TRAMETINIB (ANTICANCER)[236–242]

Class:	MEK1/MEK2 inhibitor
Country of origin:	Japan
Originator:	Japan Tobacco
First introduction:	United States
Introduced by:	GlaxoSmithKline
Trade name:	Mekinist™
CAS registry no:	1187431-43-1 (dimethylsulfoxide solvate) 871700-17-3 (free base)
Molecular weight:	693.5 (dimethylsulfoxide solvate) 615.3 (parent)

In May 2013, the US FDA approved trametinib (also referred to as GSK1120212 and JTP-74057), for the treatment of patients with unresectable or metastatic melanoma with $BRAF^{V600E}$ or $BRAF^{V600K}$ mutations as detected by an FDA-approved test. Melanoma is the fifth leading cause of cancer death in men and seventh in women[236] and an estimated 76,000 new cases are expected to be diagnosed in the United States in 2013. Prior to 2011, dacarbazine and IL-2 were the only FDA-approved treatments for melanoma, however, overall survival rates are not high with these treatments. Vemurafenib and dabrafenib have been developed as targeted therapies for the BRAF gene since oncogenic B-raf signaling is seen in approximately 50% of melanomas. Activation of the MAPK/ERK kinase (MEK) pathway due to mutation of upstream factors such as BRAF leads to increased growth and proliferation of cancer cells.[237] Two forms of MEK (MEK1 and MEK2) are known to exist and are the only known substrates for BRAF. Preclinical data suggests that MEK inhibition suppresses tumor growth in melanoma $BRAF^{V600E}$ xenograft models.[238] An initial high-throughput cellular screen directed at inducers of cyclin-dependent kinase 4/6 inhibitor $p15^{INK4b}$ led to the identification of a hit that was later confirmed to have MEK1/2 inhibitory activity. Extensive lead optimization led to the identification of trametinib[239] which is a potent ATP non-competitive inhibitor of MEK1 and MEK2 ($IC_{50} = 0.7$ and 0.9 nM, respectively, with initially unphosphorylated MEK).[240] It also showed inhibitory activity in ACHN and HT-29 cancer cell lines (IC_{50}s of 9.8 and 0.57 nM, respectively). Consistent with its *in vitro* activity, trametinib showed significant antitumor activity in a $KRAS^{G12S}$ A549 tumor xenograft model where near to complete tumor growth inhibition (TGI) was observed at 5.0 and 2.5 mg/kg (92% and 87% TGI, respectively).[239] Broad antitumor activity was seen in other xenograft models as well.[240] A synthetic route to trametinib that employs a base catalyzed rearrangement of a pyrido[2,3-*d*]pyrimidine core to pyrido[4,3-*d*]pyrimidine, as a key step has been reported.[239] The dimethyl sulfoxide solvate was selected as the final form, since it gave consistent exposures and showed desirable solid state properties.[239]

The pharmacokinetic profile of trametinib was studied in a multicenter Phase I clinical trial in patients with advanced solid tumors. At a dose of 2 mg, median time to maximum concentration (C_{max}) was 1.5 h, half-life was ~4 days, and the mean absolute bioavailability was 72%. Trametinib is highly protein bound (~97%) and has an apparent volume of distribution of 214 L.[241] The safety and efficacy of trametinib was established in an international, multicenter, open-label, active-controlled Phase III trial in 322 patients with $BRAF^{V600E}$ or $BRAF^{V600K}$ mutation-positive, unresectable, or metastatic melanoma.[242] Patients were randomized (2:1) to receive trametinib 2 mg orally once daily ($n = 214$) or chemotherapy ($n = 108$) consisting of either dacarbazine or paclitaxel with progression-free survival (PFS) as the primary endpoint and overall survival (OS) as secondary endpoint. Median PFS duration was 4.8 months in the trametinib arm and 1.5 months in the chemotherapy arm. The OS at 6 months was significantly higher for patients in the trametinib arm versus chemotherapy (81% vs. 67%). Rash, diarrhea, and lymphedema were the most common adverse events (≥20%) observed during trametinib clinical trials. The recommended dose of trametinib is 2 mg administered orally once daily.

27. VORTIOXETINE (ANTIDEPRESSANT)[243–252]

Class:	Selective serotonin reuptake inhibitor
Country of origin:	Denmark
Originator:	Lundbeck
First introduction:	United States
Introduced by:	Takeda
Trade name:	Brintellix™
CAS registry no:	508233-74-7 (free base) 960203-27-4 (HBr salt)
Molecular weight:	298.45 (free base) 379.36 (HBr salt)

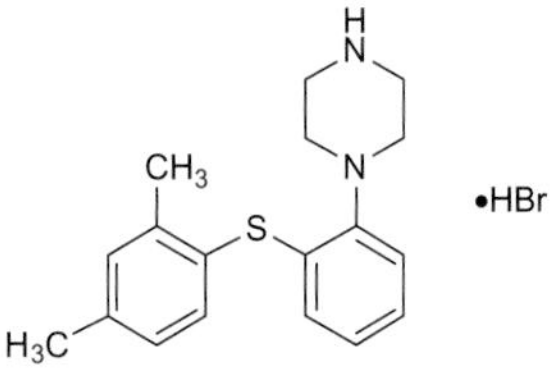

In September 2013, vortioxetine (also known as Lu AA21004) was approved in the United States for the treatment of major depressive disorder (MDD).[243] MDD affects ~7% of adults in the United States each year and is one of the most common mental disorders in the United States.[244] A study in 2010 concluded that MDD ranks second worldwide as a cause of disability.[245] MDD is characterized by severely depressed mood with symptoms that include persistent sadness, feelings of hopelessness, low-energy levels, and loss of interest in pleasurable activities. Treatments for MDD include antidepressant medications, psychotherapy, and electroconvulsive therapy. Among antidepressant treatments, serotonin transport (SERT) inhibitors such as fluoxetine, paroxetine, and citalopram, are the most commonly prescribed medications. Although these selective serotonin reuptake inhibitors (SSRIs) are generally safe and effective, there are limitations including variable rates of response and side effects such as sexual dysfunction and weight gain. In addition, onset of antidepressant action is typically 1–2 weeks after initiation of treatment. SSRIs act by blocking the serotonin (5-hydroytryptamine, 5-HT) transporter, which leads to increased levels of extracellular 5-HT. The resulting acute stimulation of presynaptic 5-HT1A autoreceptors reduces 5-HT cell firing and 5-HT release, but repeated dosing leads to desensitization of 5-HT1A autoreceptors, thereby increasing 5-HT levels and augmenting serotonergic neurotransmission. Vortioxetine was discovered from a designed multiple ligand approach to identifying an antidepressant agent that combined SERT inhibition with 5-HT1A agonism to more rapidly desensitize 5-HT1A receptors and 5-HT3A antagonism to improve mood and cognitive function.[246,247] Vortioxetine has a human SERT $IC_{50}=5.4$ nM, an $EC_{50}=200$ nM as a human 5-HT1A receptor agonist (efficacy = 96%; $K_i=39$ nM), and an $IC_{50}=12$ nM as a human 5-HT3A receptor antagonist ($K_i=3.7$ nM). It has weak inhibition of the dopamine and norepinephrine transporters, but high affinity for the human β1-noradrenergic receptor ($K_i=46$ nM), human 5-HT1B receptor ($K_i=33$ nM, partial agonist), and the human 5-HT-7 receptor ($K_i=19$ nM, antagonist). In rats, vortioxetine significantly increased 5-HT levels in the brain with moderate SERT occupancy, suggesting the benefit of interacting with multiple targets. Vortioxetine also showed antidepressant activity in preclinical models.[246,247] Vortioxetine is prepared by palladium-mediated coupling of piperazine with 1-iodo-2,4-dimethylbenzene and 2-bromobenzenethiol or with 2,4-dimethylbenzenethiol and 1-bromo-2-idodbenzene.[246,248]

Pharmacokinetic studies with vortioxetine in healthy volunteers showed an absolute bioavailability of 75% (intravenous) and a half-life of 69 h.[249] After oral dosing (normalized to a 10 mg single dose), maximum concentrations were achieved at 8 h and the mean elimination half-life was 57 h. There was a linear dose–exposure response up to 75 mg in single dosing and up to 60 mg in multiple dosing studies. Vortioxetine is highly protein bound (98%) and undergoes extensive metabolism via oxidation and glucuronidation, with Cyp2D6-mediated oxidation producing the major metabolite, 3-methyl-4-(2-piperazine-1-yl-phenylsulfanyl)-benzoic acid.[250] Dose adjustment is required if vortioxetine is given with a Cyp2D6 inhibitor (reduced dose) or a Cyp2D6 inducer (increase dose). Vortioxetine is a weak-inactive Cyp inhibitor.[246] The efficacy of vortioxetine was demonstrated in several Phase III studies.[250] In one randomized, double-blind, placebo-controlled 8-week trial in adults with MDD ($n=560$), there was a statistically significant reduction in the total score on the Hamilton Depression Rating Scale after 8 weeks of oral treatment with 10 mg vortioxetine, once daily, compared with placebo (−16.2 vs. −11.3).[251] In another 8-week study (n = 608), vortioxetine at 15 or 20 mg per day was superior to placebo as measured by the primary efficacy end point of a change from baseline in the Montgomery–Asberg Depression Rating Scale (−17.2 for 15 mg of vortioxetine, −18.8 for 20 mg of vortioxetine, and −11.7 for placebo).[252] Vortioxetine was well tolerated, with common adverse events of nausea, headache, diarrhea, dry mouth, and dizziness. The recommended starting dose of vortioxetine is 10 mg administered orally once daily, followed by an increase to 20 mg/day, as tolerated. The prescribing information carries a warning for increased risk of suicidal thought and behavior in children, adolescents, and young adults taking antidepressants.[250]

REFERENCES

1. The collection of new therapeutic entities first launched in 2012 originated from the following sources: Prous Integrity Database; Thomson-Reuters Pipeline Database; The Pink Sheet; Drugs@FDA Website; FDA News Releases.
2. Mullard, A. *Nature Rev. Drug Discov.* **2013**, *13*, 85.
3. Agrawal, R. *Curr. Drug Targets* **2014**, *15*, 151.
4. http://www.accessdata.fda.gov/drugsatfda_docs/label/2013/203414s000lbl.pdf.
5. http://www.accessdata.fda.gov/drugsatfda_docs/label/2013/022426s000lbl.pdf.
6. http://www.accessdata.fda.gov/drugsatfda_docs/label/2013/203137s000lbl.pdf.
7. http://www.accessdata.fda.gov/drugsatfda_docs/label/2013/203975s000lbl.pdf.
8. http://www.accessdata.fda.gov/drugsatfda_docs/label/2013/204275s000lbl.pdf.
9. http://www.ema.europa.eu/docs/en_GB/document_library/EPAR_-_Product_Information/human/002679/WC500151255.pdf.
10. http://www.accessdata.fda.gov/drugsatfda_docs/label/2013/204251s000lbl.pdf.link.
11. Tack, J.; Talley, N. J. *Nat. Rev. Gastroenterol. Hepatol.* **2013**, *10*, 134.
12. Tack, J.; Talley, N. J.; Camilleri, M.; Holtmann, G.; Hu, P.; Malagelada, J.-R.; Stanghellini, V. *Gastroenterology* **2006**, *130*, 1466.
13. Matsunaga, Y.; Tanaka, T.; Yoshinaga, K.; Ueki, S.; Hori, Y.; Eta, R.; Kawabata, Y.; Yoshii, K.; Yoshida, K.; Matsumura, T.; Furuta, S.; Takei, M.; Tack, J.; Itoh, Z. *J. Pharmacol. Exp. Ther.* **2011**, *336*, 791.
14. Seto, K.; Sasaki, T.; Katsunuma, K.; Kobayashi, N.; Tanaka, K.; Tack, J. *Neurogastroenterol. Motil.* **2008**, *20*, 1051.
15. Nagasawa, M.; Murata, M.; Nishioka, H.; Kurimoto, T.; Ueki, S.; Kitagawa, O. US Patent 5,981,557, **1999**.
16. Nolan, M. L.; Scott, L. J. *Drugs* **2013**, *73*, 1377.
17. Matsueda, K.; Hongo, M.; Tack, J.; Saito, Y.; Kato, H. *Gut* **2012**, *61*, 821.
18. Boyraz, B.; Sendur, M. A. N.; Aksoy, S.; Babacan, T.; Roach, E. C.; Kizilarslanoglu, M. C.; Petekkaya, I.; Altundag, K. *Curr. Med. Res. Opin.* **2013**, *29*, 405.
19. Hurvitz, S. A.; Dirix, L.; Kocsis, J.; Bianchi, G. V.; Lu, J.; Vinholes, J.; Guardino, E.; Song, C.; Tong, B.; Ng, V.; Chu, Y.; Perez, E. A. *J. Clin. Oncol.* **2013**, *31*, 1157.
20. Krop, I.; Winer, E. P. *Clin. Cancer Res.* **2014**, *20*, 15.

21. Girish, S.; Gupta, M.; Wang, B.; Lu, D.; Krop, I. E.; Vogel, C. L.; Burris, H. A., III; LoRusso, P. M.; Yi, J.; Saad, O.; Tong, B.; Chu, Y.; Holden, S.; Joshi, A. *Cancer Chemother. Pharmacol.* **2012**, *69*, 1229.
22. Verma, S.; Miles, D.; Gianni, L.; Krop, I. E.; Welslau, M.; Baselga, J.; Pegram, M.; Oh, D.; Diéras, V.; Guardino, E.; Fang, L.; Lu, M. W.; Olsen, S.; Blackwell, K. *N. Engl. J. Med.* **2012**, *367*, 1783.
23. Siegel, R.; Naishadham, D.; Jemal, A. *Cancer J. Clin.* **2012**, *62*, 10.
24. National Comprehensive Cancer Network. *NCCN Clinical Practice Guidelines in Oncology: Non-Small Cell Lung Cancer*; 2012. Available from: http://www.nccn.org/professionals/physician_gls/f_guidelines.asp.
25. Riely, G. J.; Politi, K. A.; Miller, V. A.; Pao, W. *Clin. Cancer Res.* **2006**, *12*, 7232.
26. Li, D.; Ambrogio, L.; Shimamura, T.; Kubo, S.; Takahashi, M.; Chirieac, L. R.; Padera, R. F.; Shapiro, G. I.; Baum, A.; Himmelsbach, F.; Rettig, W. J.; Meyerson, M.; Solca, F.; Greulich, H.; Wong, K. K. *Oncogene* **2008**, *27*, 4702.
27. Schroeder, J.; Dziewas, G.; Fachinger, T.; Jaeger, B.; Reichel, C.; Renner, S. WO Patent Application 2007/085638, **2007**.
28. Dungo, R. T.; Keating, G. M. *Drugs* **2013**, *73*, 1503.
29. Sequist, L. V.; Yang, J. C.-H.; Yamamoto, N.; O'Byrne, K.; Hirsh, V.; Mok, T.; Geater, S. L.; Orlov, S.; Tsai, C.-H.; Boyer, M.; Su, W.-C.; Bennouna, J.; Kato, T.; Gorbunova, V.; Lee, K. H.; Shah, R.; Massey, D.; Zazulina, V.; Shahidi, M.; Schuler, M. *J. Clin. Oncol.* **2013**, *31*, 3327.
30. http://www.fda.gov/NewsEvents/Newsroom/PressAnnouncements/ucm345848.htm (last accessed 19.02.2014).
31. http://www.who.int/mediacentre/factsheets/fs312/en/ (last accessed 19.02.2014).
32. Devitt, M. *Am. Fam. Physician* **2013**, *87*, 140.
33. Tsujihara, T.; Hongu, M.; Saito, K.; Kawanishi, H.; Kuriyama, K.; Matsumoto, M.; Oku, A.; Ueta, K.; Tsuda, M.; Saito, A. *J. Med. Chem.* **1999**, *42*, 5311.
34. Nomura, S.; Sakamaki, S.; Hongu, M.; Kawanishi, E.; Koga, Y.; Sakamoto, T.; Yamamoto, Y.; Ueta, K.; Kimata, H.; Nakayama, K.; Tsuda-Tsukimoto, M. *J. Med. Chem.* **2010**, *53*, 6355.
35. Ellsworth, B.; Washburn, W.; Sher, P.; Wu, G.; Meng, W. US Patent 6,414,126, **2002**.
36. Meng, W.; Ellsworth, B.; Nirschl, A.; McCann, P.; Patel, M.; Girotra, R.; Wu, G.; Sher, P.; Morrison, E.; Biller, S.; Zahler, R.; Deshpande, P.; Pullockaran, A.; Hagan, D.; Morgan, N.; Taylor, J.; Obermeier, M.; Humphreys, W.; Khanna, A.; Discenza, L.; Robertson, J.; Wang, A.; Han, S.; Wetterau, J.; Janovitz, E.; Flint, O.; Whaley, J.; Washburn, W. *J. Med. Chem.* **2008**, *51*, 1145.
37. Scheen, A. *Clin. Pharmacokinet.* **2014**, *53*, 213.
38. Devineni, D.; Curtin, C.; Polidori, D.; Gutierrez, M.; Murphy, J.; Rusch, S.; Rothenberg, P. *J. Clin. Pharmacol.* **2013**, *53*, 601.
39. Stenlo, K.; Cefalu, W.; Kim, K.; Alba, M.; Usiskin, K.; Tong, C.; Canovatchel, W.; Meininger, G. *Diabetes Obes. Metab.* **2012**, *15*, 372.
40. Rosenstock, J.; Aggarwal, N.; Polidori, D.; Zhao, Y.; Arbit, D.; Usiskin, K.; Capuano, G.; Canovatchel, W. *Diabetes Care* **2012**, *35*, 1232.
41. Cefalu, W.; Leiter, L.; Yoon, K.; Arias, P.; Niskanen, L.; Xie, J.; Balis, D.; Canovatchel, W.; Meininger, G. *Lancet* **2013**, *382*, 941.
42. http://www.invokanahcp.com/prescribing-information.pdf (last accessed 28.02.2014).
43. Takeda Pharmaceuticals Press Release, www.takeda.com/news/2013/20130920_5991.html (20.09.2013).
44. WHO Global Health Observatory, http://www.who.int/gho/ncd/risk_factors/overweight/en/ (last accessed 11.03.2014).
45. Bray, G. A. *J. Med. Chem.* **2006**, *49*, 4001.
46. Yanovski, S. Z.; Yanovski, J. A. *J. Am. Med. Assoc.* **2013**, *311*, 74.

47. Hodson, H.; Downham, R.; Mitchell, T.J.; Carr, B.J.; Dunk, C.R.; Palmer, R.M.J. US Patent 6,656,934, **2003**.
48. Yamada, Y.; Kato, T.; Ogino, H.; Ashina, S.; Kato, K. *Horm. Metab. Res.* **2008**, *40*, 539.
49. Padwal, R. *Curr. Opin. Investig. Drugs* **2008**, *9*, 414.
50. Bryson, A.; de la Motte, S.; Dunk, C. *Br. J. Clin. Pharmcol.* **2009**, *67*, 309.
51. Kopelman, P.; Bryson, A.; Hickling, R.; Rissanen, A.; Rossner, S.; Toubro, S.; Valensi, P. *Int. J. Obes.* **2007**, *31*, 494.
52. Kopelman, P.; Groot, G.; de, H.; Rissanen, A.; Rossner, S.; Toubro, S.; Palmer, R.; Hallam, R.; Bryson, A.; Hickling, R. I. *Obesity* **2009**, *18*, 108.
53. http://www.gilead.com/news/press-releases/2013/9/european-commission-approves-gilead-sciences-tybost-a-new-boosting-agent-for-hiv-therapy.
54. http://www.ema.europa.eu/docs/en_GB/document_library/Summary_of_opinion_-_Initial_authorisation/human/002572/WC500146622.pdf.
55. http://www.who.int/gho/hiv/en/.
56. Xu, L.; Liu, H.; Murray, B. P.; Callebaut, C.; Lee, M. S.; Hong, A.; Strickley, R. G.; Tsai, L. K.; Stray, K. M.; Wang, Y.; Rhodes, G. R.; Desai, M. C. *ACS Med. Chem. Lett.* **2010**, *1*, 209.
57. Mathias, A. A.; German, P.; Murray, B. P.; Wei, L.; Jain, A.; West, S.; Warren, D.; Hui, J.; Kearney, B. P. *Nature* **2010**, *87*, 322.
58. Shah, B. M.; Schafer, J. J.; Priano, J.; Squires, K. E. *Pharmacotherapy* **2013**, *33*, 1107.
59. Gallant, J. E.; Koenig, E.; Andrade-Villanueva, J.; Chetchotisakd, P.; DeJesus, E.; Antunes, F.; Arastéh, K.; Moyle, G.; Rizzardini, G.; Fehr, J.; Liu, Y.; Zhong, L.; Callebaut, C.; Szwarcberg, J.; Rhee, M. S.; Cheng, A. K. *J. Infect. Dis.* **2013**, *208*, 32.
60. Stellwagen, J.; Adjabeng, G.; Arnone, M.; Dickerson, S.; Han, C.; Hornberger, K.; King, A.; Mook, R.; Petrov, K.; Rheault, T.; Rominger, C.; Rossanese, O.; Smitheman, K.; Waterson, A.; Uehling, D. *Bioorg. Med. Chem. Lett.* **2011**, *21*, 4436.
61. Rheault, T. R.; Stellwagen, J. C.; Adjabeng, G. M.; Hornberger, K. R.; Petrov, K. G.; Waterson, A. G.; Dickerson, S. H.; Mook, R. A., Jr.; Laquerre, S. G.; King, A. J.; Rossanese, O. W.; Arnone, M. R.; Smitheman, K. N.; Kane-Carson, L. S.; Han, C.; Moorthy, G. S.; Moss, K. G.; Uehling, D. E. *ACS Med. Chem. Lett.* **2013**, *4*, 358.
62. Laquerre, S.; Arnone, M.; Moss, K.; Yang, J.; Fisher, K.; Kane-Carson, L. S.; Smitheman, K.; Ward, J.; Heidrich, B.; Rheault, T.; Adjabeng, G.; Hornberger, K.; Stellwagen, J.; Waterson, A.; Han, C.; Mook, R. A., Jr.; Uehling, D.; King, A. J. In: *21st AACR-NCI-EORTC International Conference on Molecular Targets and Cancer Therapeutics, Boston, MA* 2009, Abstract B88.
63. Adams, J. L.; Dickerson, S. H.; Johnson, N. W.; Kuntz, K.; Petrov, K.; Ralph, J. M.; Rheault, T. R.; Schaaf, G.; Stellwagen, J.; Tian, X.; Uehling, D. E.; Waterson, A. G.; Wilson, B. WO Patent Application 2009/137391, **2009**.
64. GlaxoSmithKline. *Tafinlar® (Dabrafenib Capsules): US Prescribing Information*, **2013**. http://us.gsk.com/products/assets/us_tafinlar.pdf.
65. Hauschild, A.; Grob, J.-J.; Demidov, L. V.; Jouary, T.; Gutzmer, R.; Millward, M.; Rutkowski, P.; Blank, C. U.; Miller, W. H.; Kaempgen, E.; Algarra, S.-M.; Karaszewska, B.; Mauch, C.; Chiarion-Sileni, V.; Martin, A.-M.; Swann, S.; Haney, P.; Mirakhur, B.; Guckert, M. E.; Goodman, V.; Chapman, P. B. *Lancet* **2012**, *380*, 358.
66. http://www.accessdata.fda.gov/drugsatfda_docs/appletter/2013/204063Orig1s000ltr.pdf (last accessed 29.01.2014).
67. http://www.nationalmssociety.org/about-the-society/ms-prevalence/index.aspx (last accessed 29.01.2014).
68. Sobieraj, D.; Coleman, C. *Formulary* **2012**, *47*, 386.

69. Wilson, A.; Kerns, J.; Callahan, J.; Moody, C. *J. Med. Chem.* **2013**, *56*, 7463.
70. Mrowietz, U.; Asadullah, K. *Trends Mol. Med.* **2005**, *11*, 43.
71. Schimrigh, S.; Brune, N.; Hellwig, K.; Lukas, C.; Bellenberg, B.; Rieks; Hoffmann, V.; Pohlau, D.; Przuntek, H. *Eur. J. Neurol.* **2006**, *13*, 604.
72. Schilling, S.; Goelz, S.; Linker, R.; Luehder, F.; Gold, R. *Clin. Exp. Immunol.* **2006**, *145*, 101.
73. Guzowski, J.; Kiesman, W.; Irdam, E. Patent Application WO2012/170923, **2012**.
74. Litjens, N.; Burggraaf, J.; van Strijen, E.; van Gulpen, C.; Mattie, H.; Schoemaker, R.; van Dissel, J.; Thio, H.; Nibbering, P. *Br. J. Clin. Pharmacol.* **2004**, *58*, 429.
75. Werdenberg, D.; Joshi, R.; Wolffram, S.; Merkle, H.; Langguth, P. *Biopharm. Drug Dispos.* **2003**, *24*, 259.
76. http://www.tecfidera.com/pdfs/full-prescribing-information.pdf (last accessed 29.01.2014).
77. Kappos, L.; Gold, R.; Miller, D.; MacManus, D.; Havrdova, E.; Limmroth, V.; Polman, C.; Schmierer, K.; Yousry, T.; Yang, M.; Eraksoy, M.; Meluzinova, E.; Rektor, I.; Dawson, K.; Sandrock, A.; O'Neill, G. *Lancet* **2008**, *372*, 1463.
78. Giovannoni, G.; Gold, R.; Kappos, L.; Arnold, D.; Bar-Or, A.; Selmaj, K.; Zhang, A.; Sheikh, S.; Dawson, K. *Neurology* **2012**, *78,* PD5.005. (AAN Meeting Abstracts).
79. Fox, R.; Miller, D.; Phillips, T.; Kita, M.; Hutchinson, M.; Havrdova, E.; Yang, M.; Zhang, R.; Viglietta, V.; Dawson, K. *Neurology* **2012**, *78,* S01.003 (AAN Meeting Abstracts).
80. Atal, S.; Atal, S. *Int. J. Basic Clin. Pharmacol.* **2013**, *2*, 849.
81. http://www.fda.gov/NewsEvents/Newsroom/PressAnnouncements/ucm364744.htm (last accessed 28.02.2014).
82. http://www.unaids.org/en/media/unaids/contentassets/documents/epidemiology/2013/gr2013/UNAIDS_Global_Report_2013_en.pdf (last accessed 28.02.2014).
83. Hare, S.; Smith, S.; Métifiot, M.; Jaxa-Chamiec, A.; Pommier, Y.; Hughes, S.; Cherepanov, P. *Mol. Pharmacol.* **2011**, *80*, 565.
84. Wai, J.; Egbertson, M.; Payne, L.; Fisher, T.; Embrey, M.; Tran, L.; Melamed, J.; Langford, H.; Guare, J.; Zhuang, L.; Grey, V.; Vacca, J.; Holloway, M.; Naylor-Olsen, A.; Hazuda, D.; Felock, P.; Wolfe, A.; Stillmock, K.; Schleif, W.; Gabryelski, L.; Young, S. *J. Med. Chem.* **2000**, *43*, 4923.
85. Johns, B.; Kawasuji, T.; Weatherhead, J.; Taishi, T.; Temelkoff, D.; Yoshida, H.; Akiyama, T.; Taoda, Y.; Murai, H.; Kiyama, R.; Fuji, M.; Tanimoto, N.; Jeffrey, J.; Foster, S.; Yoshinaga, T.; Seki, T.; Kobayashi, M.; Sato, A.; Johnson, M.; Garvey, E.; Fujiwara, T. *J. Med. Chem.* **2013**, *56*, 5901.
86. Raffi, F.; Rachlis, A.; Stellbrink, H.-J.; Hardy, W.; Torti, C.; Orkin, C.; Bloch, M.; Podzamczer, D.; Pokrovsky, V.; Pulido, F.; Almond, S.; Margolis, D.; Brennan, C.; Min, S. *Lancet* **2013**, *381*, 735.
87. Cahn, P.; Pozniak, A.; Mingrone, H.; Shuldyakov, A.; Brites, C.; Andrade-Villanueva, J.; Richmond, G.; Buendia, C.; Fourie, J.; Ramgopal, M.; Hagins, D.; Felizart, F.; Madruga, J.; Reuter, T.; Newman, T.; Small, C.; Lombaard, J.; Grinsztejn, B.; Dorey, D.; Underwood, M.; Griffith, S.; Min, S. *Lancet* **2013**, *382*, 700.
88. Min, S.; Song, I.; Borland, J.; Chen, S.; Lou, Y.; Fujiwara, T.; Piscitelli, S. *Antimicrob. Agents Chemother.* **2010**, *54*, 254.
89. Reese, M.; Savina, P.; Generaux, G.; Tracey, H.; Humphreys, J.; Kanaoka, E.; Webster, L.; Harmon, K.; Clarke, J.; Polli, J. *Drug Metab. Dispos.* **2013**, *41*, 353.
90. Rathbun, R.; Lockhart, S.; Miller, M.; Liedtke, M. *Ann. Pharmacother.* **2013**, *20*, 1.
91. http://www.accessdata.fda.gov/drugsatfda_docs/label/2013/204790lbl.pdf (last accessed 21.03.2014).
92. Patel, T.; Dhillon, S. *Drugs* **2013**, *73*, 1977.
93. Thomas, J.; Jacobson, G. A.; Narkowicz, C. K.; Peterson, G. M.; Burnet, H.; Sharpe, C. *J. Clin. Pharm. Ther.* **2010**, *35*, 497.

94. Tatsumi, Y.; Nagashima, M.; Shibanushi, T.; Iwata, A.; Kangawa, Y.; Inui, F.; Jo Siu, W. J.; Pillai, R.; Nishiyama, Y. *Antimicrob. Agents Chemother.* **2013**, *57*, 2405.
95. Tatsumi, Y.; Yokoo, M.; Arika, T. *Antimicrob. Agents Chemother.* **2001**, *45*, 1493.
96. (a) Ogura, H.; Kobayashi, H.; Nagai, K.; Nishida, T.; Naito, T.; Tatsumi, Y.; Yokoo, M.; Arika, T. *Chem. Pharm. Bull* **1999**, *47*, 1417; (b) Mimura, M.; Watanabe, M.; Ishiyama, N.; Yamada, T. WO2012029836, **2012**.
97. Jarratt, M.; Jo Siu, W.; Yamakawa, E.; Kodera, N.; Pillai, R.; Smith, K. *J. Drugs Dermatol.* **2013**, *12*, 1010.
98. Elewski, B. E.; Rich, P.; Pollak, R.; Pariser, D. M.; Watanabe, S.; Senda, H.; Ieda, C.; Smith, K.; Pillasi, R.; Ramakrishna, T.; Olin, J. T. *J. Am. Acad. Dermatol.* **2013**, *68*, 600.
99. http://www.gilead.com/news/press-releases/2013/11/european-commission-approves-gileads-vitekta-an-integrase-inhibitor-for-the-treatment-of-hiv1-infection (last accessed 26.03.2014).
100. http://www.fda.gov/NewsEvents/Newsroom/PressAnnouncements/ucm317004.htm (last accessed 26.03.2014).
101. Hazuda, D.; Young, S.; Guare, J.; Anthony, N.; Gomez, R.; Wai, J.; Vacca, J.; Handt, L.; Motzel, S.; Klein, H.; Dornadula, G.; Danovich, R.; Witmer, M.; Wilson, K.; Tussey, L.; Schleif, W.; Gabryelski, L.; Jin, L.; Miller, M.; Casimiro, D.; Emini, E.; Shiver, J. *Science* **2004**, *305*, 528.
102. Sato, M.; Motomura, T.; Aramaki, H.; Matsuda, T.; Yamashita, M.; Ito, Y.; Kawakami, H.; Matsuzaki, Y.; Watanabe, W.; Yamataka, K.; Ikeda, S.; Kodama, E.; Matsuoka, M.; Shinkai, H. *J. Med. Chem.* **2006**, *49*, 1506.
103. DeJesus, E.; Berger, D.; Markowitz, M.; Cohen, C.; Hawkins, T.; Ruane, P.; Elion, R.; Farthing, C.; Zhong, L.; Cheng, A.; McColl, D.; Kearney, B. *J. Acquir. Immune Defic. Syndr.* **2006**, *43*, 1.
104. Ramanathan, S.; Shen, G.; Cheng, A.; Kearney, B. *J. Acquir. Immune Defic. Syndr.* **2007**, *44*, 274.
105. DeJesus, E.; Rockstroh, J.; Henry, K.; Molina, J.; Gathe, J.; Ramanathan, S.; Wei, X.; Yale, K.; Szwarcberg, J.; White, K.; Cheng, A.; Kearney, B. *Lancet* **2012**, *379*, 2429.
106. http://www.accessdata.fda.gov/drugsatfda_docs/label/2012/203100s000lbl.pdf (last accessed 28.03.2014).
107. Molina, J.; LaMarca, A.; Andrade-Villanueva, J.; Clotet, B.; Clumeck, N.; Liu, Y.; Zhong, L.; Margot, N.; Cheng, A.; Chuck, S. *Lancet* **2012**, *12*, 27.
108. http://www.ema.europa.eu/docs/en_GB/document_library/EPAR_-_Product_Information/human/002577/WC500155576.pdf (last accessed 28.03.2014).
109. Morton, L. M.; Wang, S. S.; Devesa, S. S.; Hartge, P.; Weisenburger, D. D.; Linet, M. S. *Blood* **2006**, *107*, 265.
110. Honigberg, L. E.; Smith, A. M.; Sirisawad, M.; Verner, E.; Loury, D.; Chang, B.; Li, S.; Pan, Z.; Thamm, D. H.; Miller, R. A.; Buggy, J. J. *Proc. Natl. Acad. Sci.* **2010**, *107*, 13075.
111. (a) Pan, Z.; Scheerens, H.; Li, S.; Schultz, B. E.; Sprengeler, P. A.; Burrill, L. C.; Mendonca, R. V.; Sweeney, M. D.; Scott, K. C. K.; Grothaus, P. G.; Jeffery, D. A.; Spoerke, J. M.; Honigberg, L. A.; Young, P. R.; Dalrymple, S. A.; Palmer, J. T. *ChemMedChem* **2007**, *2*, 58; (b) Honigberg, L.; Verner, E.; Pan, Z. US Patent Application 0076921, **2008**.
112. (a) http://dailymed.nlm.nih.gov/dailymed/lookup.cfm?setid=0dfd0279-ff17-4ea9-89be-9803c71bab44 (last accessed 22.02.2014). (b) Sukbunthemg, J.; Jejukar, P.; Chan, S.; Tran, A. L.; Moussa, D.; James, D. F.; Loury, D. In: *ASCO Annual Meeting*; Chicago, IL. **2013**. Abstract 7056.
113. Wang, M. L.; Rule, S.; Martin, P.; Goy, A.; Auer, R.; Kahl, B. S.; Jurczak, W.; Advani, R. H.; Romaguera, J. E.; Williams, M. E.; Barrientos, J. C.; Chmielowska, E.; Radford, J.; Stilgenbauer, S.; Dreyling, M.; Jedrzejczak, W. W.;

Johnson, P.; Spurgeon, S. E.; Li, L.; Zhang, L.; Newberry, K.; Ou, Z.; Cheng, N.; Fang, B.; McGrivy, J.; Clow, F.; Buggy, J. J.; Chang, B. Y.; Beaupre, D. M.; Kunkel, L. A.; Blum, K. A. *N. Engl. J. Med.* **2013**, *369*, 507.
114. http://www.fda.gov/NewsEvents/Newsroom/PressAnnouncements/ucm385764.html.
115. Dungo, R.; Deeks, E. D. *Drugs* **2013**, *73*, 875.
116. World Health Organization Report on Neurological Disorders: Public Health Challenges, **2006**, 140. http://www.who.int/mental_health/neurology/neurological_disorders_report_web.pdf.
117. Shimada, J.; Koike, N.; Nonaka, H.; Shiozaki, S.; Yanagawa, K.; Kanda, T.; Kobayashi, H.; Ishimura, M.; Nakamura, J.; Kase, H.; Suzuki, F. *Bioorg. Med. Chem. Lett.* **1997**, *18*, 2349.
118. Saki, M.; Yamada, K.; Koshimura, E.; Sasaki, K.; Kanda, T. *Naunyn-Schmiedebergs Arch. Pharmacol.* **2013**, *386*, 963.
119. Jenner, P. *Expert Opin. Investig. Drugs* **2005**, *14*, 729.
120. Knebel, W.; Rao, N.; Uchimura, T.; Mori, A.; Fisher, J.; Gastonguay, M. R.; Chaikin, P. *J. Clin. Pharmacol.* **2011**, *51*, 40.
121. Brooks, D. J.; Doder, M.; Osman, S.; Luthra, S. K.; Hirani, E.; Hume, S.; Kase, H.; Kilborn, J.; Martindill, S.; Mori, A. *Synapse* **2008**, *62*, 671.
122. Mizuno, Y.; Kondo, T.; The Japanese Istradefylline Study Group. *Movement Disord.* **2013**, *28*, 1138.
123. International Diabetes Federation. *IDF Diabetes Atlas*, 5th ed.; International Diabetes Federation: Brussels, Belgium, 2012. http://www.idf.org/diabetesatlas.
124. American Diabetes Association website: http://www.diabetes.org/living-with-diabetes/treatment-and-care/medication/oral-medications/what-are-my-options.html.
125. Rotenstein, L. S.; Kozak, B. M.; Shivers, J. P.; Yarchoan, M.; Close, J.; Close, K. L. *Clin. Diabetes* **2012**, *30*, 44.
126. Petersen, A. B.; Christensen, M. *Diabetes Metab. Syndr. Obes.* **2013**, *6*, 217.
127. Parkes, D. G.; Mace, K. F.; Trautmann, M. E. *Expert Opin. Drug Discov.* **2013**, *8*, 219.
128. Christensen, M.; Knop, F. K.; Vilsboll, T.; Holst, J. J. *Expert Opin. Investig. Drugs* **2011**, *20*, 549.
129. Thorkildsen, C.; Neve, S.; Larsen, B. D.; Meier, E.; Petersen, J. S. *J. Pharmacol. Exp. Ther.* **2003**, *307*, 490.
130. Committee for Medicinal Products for Human Use, European Medicines Agency, Assessment Report EMA/CHMP/703852/2012.
131. Committee for Medicinal Products for Human Use, European Medicines Agency, Summary of Opinion, EMA/CHMP/706420/2012.
132. Temple, R. *US FDA Approval Letter*; CDER: Silver Spring, MD, 2013.
133. Farber, H. W.; Loscalzo, J. *N. Engl. J. Med.* **2004**, *351*, 1655.
134. Sutendra, G.; Michelakis, E. D. *Sci. Transl. Med.* **2013**, *5*, 1.
135. Ahmed, S.; Palevsky, H. I. *Rheum. Dis. Clin. N. Am.* **2014**, *40*, 103.
136. Bolli, M. H.; Boss, C.; Binkert, C.; Buchmann, S.; Bur, D.; Hess, P.; Iglarz, M.; Meyer, S.; Rein, J.; Rey, M.; Treiber, A.; Clozel, M.; Fishli, W.; Weller, T. *J. Med. Chem.* **2012**, *55*, 7849.
137. Bruderer, S.; Hopfgartner, G.; Seiberling, M.; Wank, J.; Sidharta, P. N.; Treiber, A.; Dingemanse, J. *Xenobiotica* **2012**, *42*, 901.
138. Pulido, T.; Adzerikho, I.; Channick, R. N.; Delcroix, M.; Galie, N.; Ghofrani, H.-A.; Jansa, P.; Jing, Z.-C.; Le Brun, F.-O.; Mehta, S.; Mettelholzer, C. M.; Perchenet, L.; Sastry, B. K. S.; Sitbon, O.; Souza, R.; Torbicki, A.; Zeng, X.; Rubin, L. J.; Simonneau, G. *N. Engl. J. Med.* **2013**, *369*, 809.
139. Garg, A. *N. Engl. J. Med.* **2004**, *350*, 1220.
140. Chan, J. L.; Oral, E. A. *Endocrine Pract.* **2010**, *16*, 310.

141. Pelleymounter, M. A.; Cullen, M. J.; Baker, M. B.; Hecht, R.; Winters, D.; Boone, T.; Collins, F. *Science* **1995**, *269*, 540.
142. Zhang, Y.; Proenca, R.; Maffei, M.; Barone, M.; Leopold, L.; Friedman, J. M. *Nature* **1994**, *372*, 425.
143. Shimomura, I.; Hammer, R. E.; Ikemoto, S.; Brown, M. S.; Goldstein, J. L. *Nature* **1999**, *401*, 73.
144. Myalept label, downloaded from: http://www.myalept.com/index.aspx.
145. Chou, K.; Perry, C. M. *Drugs* **2013**, *73*, 989.
146. Oral, E. A.; Simha, V.; Ruiz, E.; Andewelt, A.; Premkumar, A.; Snell, P.; Wagner, A. J.; DePaoli, A. M.; Reitman, M. L.; Taylor, S. I.; Gorden, P.; Garg, A. *N. Engl. J. Med.* **2002**, *346*, 570.
147. American Diabetes Association. *Diabetes Care* **2014**, *37*, S14.
148. Zadeh, E. S.; Lungu, A. O.; Cochran, E. K.; Brown, R. J.; Ghany, M. G.; Heller, T.; Kleiner, D. E.; Gorden, P. *J. Hepatol.* **2013**, *59*, 131.
149. http://www.pmda.go.jp/english/service/pdf/drugs/metreleptin_mar2013_e.pdf.
150. Liscinsky, M. US FDA Press Release, February 25, **2014**, www.fda.gov/newsevents/newsroom/pressannouncements/ucm387060.htm.
151. Kynamro Prescribing Label, www.fda.gov/downloads/drugs/drugsafety/ucm337730.pdf.
152. Raal, F. J.; Santos, R. D. *Atherosclerosis* **2012**, *223*, 262–268.
153. Cuchel, M.; Meagher, E. A.; Theron, H. d. T.; Blom, D. J.; Marais, A. D.; Hegele, R. A.; Averna, M. R.; Sirtori, C. R.; Shah, P. K.; Gaudet, D.; Stefanutti, C.; Vigna, G. B.; Du Plessis, A. M. E.; Propert, K. J.; Sasiela, W. J.; Bloedon, L. T.; Rader, D. J. *Lancet* **2013**, *381*, 40.
154. Crooke, S. T.; Geary, R. S. *Br. J. Clin. Pharmacol.* **2012**, *76*, 269.
155. Seth, P. P.; Siwkowski, A.; Allerson, C. R.; Vasquez, G.; Lee, S.; Prakash, T. P.; Wancewicz, E. V.; Witchell, D.; Swayze, E. E. *J. Med. Chem.* **2009**, *52*, 10.
156. Krotz, A.; Gorman, D.; Mataruse, P.; Foster, C.; Godbout, J. D.; Coffin, C. C.; Scozzari, A. N. *Org. Process Res. Dev.* **2004**, *8*, 852.
157. Mullick, A. E.; Fu, W.; Graham, M. J.; Lee, R. G.; Witchell, D.; Bell, T. A.; Whipple, C. P.; Crooke, R. M. *J. Lipid Res.* **2011**, *52*, 885.
158. Merki, E.; Graham, M. J.; Mullick, A. E.; Miller, E. R.; Crooke, R. M.; Pitas, R. E.; Witztum, J. L.; Tsimikas, S. *Circulation* **2008**, *118*, 743.
159. Beck, A.; Reichert, J. M. *MAbs* **2012**, *4*, 419.
160. Sehn, L. H.; Assouline, S. E.; Stewart, D. A.; Mangel, J.; Gayscoyne, R. D.; Fine, G.; Frances-Lasserre, S.; Carlile, D. J.; Crump, M. *Blood* **2012**, *119*, 5118.
161. Dalle, S.; Reslan, L.; Besseyre de Horts, T.; Herveau, S.; Herting, F.; Plesa, A.; Friess, T.; Umana, P.; Klein, C.; Dumontet, C. *Mol. Cancer Ther.* **2011**, *10*, 178.
162. Herter, S.; Herting, F.; Mundigi, O.; Waldhauer, I.; Weinzieri, T.; Fauti, T.; Muth, G.; Ziegler-Landesberger, D.; Van Puijenbroek, E.; Lang, S.; Duong, M. N.; Reslan, L.; Gerdes, C. A.; Friess, T.; Baer, U.; Burtscher, H.; Weidner, M.; Dumonte, C.; Umana, P.; Niederfellner, G.; Bacac, M.; Klein, C. *Mol. Cancer Ther.* **2013**, *12*, 2031.
163. Cameron, F.; McCormack, P. L. *Drugs* **2014**, *74*, 147.
164. Goede, V.; Fischer, K.; Busch, R.; Engelke, A.; Eichhorst, B.; Wendtner, C. M.; Chagorova, T.; de la Serna, J.; Dilhuydy, M.; Illmer, T.; Opat, S.; Owen, C. J.; Samoylova, O.; Kreuzer, K.; Stilgenbauer, S.; Döhner, H.; Langerak, A. W.; Ritgen, M.; Kneba, M.; Asikanius, E.; Humphrey, K.; Wenger, M.; Hallek, M. *N. Engl. J. Med.* **2014**, *370*, 1101.
165. Gibb, A.; Yang, L. P. H. *Drugs* **2013**, *73*, 1841.
166. (a) Mannino, D. M.; Buist, A. S. *Lancet* **2007**, *370*, 765; (b) http://www.goldcopd.org/uploads/users/files/GOLD_Report_2013_Feb20.pdf.
167. http://www.who.int/mediacentre/factsheets/fs315/en/.

168. Bouyssou, T.; Hoenke, C.; Rudolf, K.; Lustenberger, P.; Pestel, S.; Sieger, P.; Lotz, R.; Heine, C.; Büttner, F. H.; Schnapp, A.; Konetzki, I. *Bioorg. Med. Chem. Lett.* **2010**, *20*, 1410.
169. Bouyssou, T.; Casarosa, P.; Naline, E.; Pestel, S.; Konetzki, I.; Devillier, P.; Schnapp, A. *J. Pharmacol. Exp. Ther.* **2010**, *334*, 53.
170. Casarosa, P.; Kollak, I.; Kiechle, T.; Ostermann, A.; Schnapp, A.; Kiesling, R.; Pieper, M.; Sieger, P.; Gantner, F. *J. Pharmacol. Exp. Ther.* **2011**, *337*, 600.
171. Van Noord, J. A.; Smeets, J. J.; Drenth, B. M.; Rascher, J.; Pivovarova, A.; Hamilton, A. L.; Cornelissen, P. J. G. *Pulmon. Pharmacol. Ther.* **2011**, *24*, 666.
172. Ferguson, G.; Feldman, G; Hofbauer, P.; Hamilton, A.; Allen, L.; Korducki, L.; Sachs, P. *Annual Congress of the European Respiratory Society*; Barcelona, Spain, **2013**. Abstract no. 187.
173. Koch, A.; Pizzichini, E.; Hamilton, A.; Hart, L.; Korducki, L.; De Salvo, M. C.; Paggiaro, P. *Annual Congress of the European Respiratory Society*; Barcelona, Spain, **2013**, Abstract no. 764.
174. http://www.hc-sc.gc.ca/dhp-mps/prodpharma/sbd-smd/drug-med/sbd_smd_2013_striverdi_respimat_155649-eng.php.
175. Williams, R. E.; Levine, K. B.; Kalilani, L.; Lewis, J.; Clark, R. V. *Maturitas* **2009**, *62*, 153.
176. Qu, Q.; Zheng, H.; Dahllund, J.; Laine, A.; Cockcroft, N.; Peng, Z.; Koskinen, M.; Hemminki, K.; Kangas, L.; Vaananen, K.; Harkonen, P. *Endocrinology* **2000**, *141*, 809.
177. Unkila, M.; Kari, S.; Yatkin, E.; Lammintausta, R. *J. Steroid Biochem. Mol. Biol.* **2013**, *138*, 107.
178. Eklund, L.; Nilsson, J. WO Patent Application 2011/089385, **2011**.
179. http://dailymed.nlm.nih.gov/dailymed/lookup.cfm?setid=8462d6ab-e3cd-4efa-a360-75bf8f917287 (last accessed 22.02.2014).
180. Bachmann, G. A.; Komi, J. O.; Ospemifene Study Group. *Menopause* **2010**, *17*, 480.
181. Portman, D. J.; Bachmann, G. A.; Simon, J. A.; Ospemifene Study Group. *Menopause* **2013**, *20*, 623.
182. http://www.fda.gov/newsevents/newsroom/pressannouncements/ucm338895.htm (last accessed 4.02.2014).
183. http://seer.cancer.gov/statfacts/html/mulmy.html (last accessed 4.02.2014).
184. Rajkumar, S. *Nat. Rev. Clin. Oncol.* **2011**, *8*, 479.
185. http://www.cancer.gov/cancertopics/druginfo/fda-thalidomide (last accessed 11.02.2014).
186. http://www.thalomid.com/pdf/thalomid_pi.pdf (last accessed 11.02.2014).
187. McCurdy, A.; Lacy, M. *Ther. Adv. Hematol.* **2013**, *4*, 211.
188. Lentzsch, S.; Rogers, M.; LeBlanc, R.; Birsner, A.; Shah, J.; Treston, A.; Anderson, K.; D'Amato, R. *Cancer Res.* **2002**, *62*, 2300.
189. Muller, G.; Stirling, D.; Chen, R. US Patent 5,635,517, **1997**.
190. Muller, G.; Chen, R.; Huang, S.; Corral, L.; Wong, L.; Patterson, R.; Chen, Y.; Kaplan, G.; Stirling, D. *Bioorg. Med. Chem. Lett.* **1999**, *9*, 1625.
191. Schey, S.; Fields, P.; Bartlett, J. B.; Clarke, I.; Ashan, G.; Knight, R.; Streetly, M.; Dalgleish, A. *J. Clin. Oncol.* **2004**, *22*, 3269.
192. Hoffmann, M.; Kasserra, C.; Reyes, J.; Schafer, P.; Kosek, J.; Capone, L.; Parton, A.; Kim-Kang, H.; Surapaneni, S.; Kumar, G. *Cancer Chemother. Pharmacol.* **2013**, *71*, 489.
193. Richardson, P.; Siegel, D.; Vij, R.; Hofmeister, C.; Baz, R.; Jaganneth, S.; Chen, C.; Lonial, S.; Jakubowiak, A.; Bahlis, N.; Song, K.; Belch, A.; Raje, N.; Shustik, C.; Lentzsch, S.; Lacy, M.; Mikhael, J.; Matous, J.; Vesole, D.; Chen, M.; Zaki, M.; Jacques, C.; Yu, Z.; Anderson, K. *Blood* **2014**, *123*, 1826.
194. Terpos, E.; Kanellias, N.; Christoulas, D.; Kastritis, E.; Dimopoulos, M. *Oncol. Targets Ther.* **2013**, *6*, 531.
195. http://www.pomalyst.com/wp-content/uploads/2013/08/prescribing_information.pdf (last accessed 15.02.2014).

196. Piazza, G.; Goldhaber, S. Z. *N. Engl. J. Med.* **2011**, *364*, 351.
197. Schermuly, R. T.; Janssen, W.; Weissmann, N.; Stasch, J.-P.; Grimminger, F.; Ghofrani, H. A. *Expert Opin. Investig. Drugs* **2011**, *20*, 567.
198. Schermuly, R. T.; Stasch, J. P.; Pullamsetti, S. S.; Middendorff, R.; Muller, D.; Schluter, K. D.; Dingendorf, A.; Hackemack, S.; Kolosionek, E.; Kaulen, C.; Dumitrascu, R.; Weissmann, N.; Mittendorf, J.; Klepetko, W.; Seeger, W.; Ghofrani, H. A.; Grimminger, F. *Eur. Respir. J.* **2008**, *32*, 881.
199. Yoshina, S.; Tanaka, A.; Kuo, S.-C. *Yukugaku Zasshi* **1978**, *98*, 204.
200. Mittendorf, J.; Weigand, S.; Alonso-Alija, C.; Bischoff, E.; Feurer, A.; Gerisch, M.; Kern, A.; Knorr, A.; Lang, D.; Muenter, K.; Radtke, M.; Schirok, H.; Schlemmer, K.-H.; Stahl, E.; Straub, A.; Wunder, F.; Stasch, J.-P. *ChemMedChem* **2009**, *4*, 853.
201. Frey, R.; Muck, W.; Unger, S.; Artmeier-Brandt, U.; Weimann, G.; Wensing, G. *J. Clin. Pharmacol.* **2008**, *48*, 926.
202. Ghofrani, H.-A.; D'Armini, A. M.; Grimminger, F.; Hoeper, M. M.; Jansa, P.; Kim, N. H.; Mayer, E.; Simonneau, G.; Wilkins, M. R.; Fritsch, A.; Neuser, D.; Weimann, G.; Wang, C. *N. Engl. J. Med.* **2013**, *369*, 319.
203. Ghofrani, H.-A.; Galie, N.; Grimminger, F.; Grunig, E.; Humbert, M.; Jing, Z.-C.; Keogh, A. M.; Langleben, D.; Kilama, M. O.; Fritsch, A.; Neuser, D.; Rubin, L. *N. Engl. J. Med.* **2013**, *369*, 330.
204. Yusuf, S.; Hawken, S.; Ounpuu, S.; Dans, T.; Avezum, A.; Lanas, F.; McQueen, M.; Budaj, A.; Pais, P.; Varigos, J.; Lisheng, L. *Lancet* **2004**, *364*, 937.
205. Fruchart, J.-C.; Davignon, J.; Hermans, M. P.; Al-Rubeaan, K.; Amarenco, P.; Assmann, G.; Barter, P.; Betteridge, J.; Bruckert, E.; Cuevas, A.; Farnier, M.; Ferrannini, E.; Fioretto, P.; Gensest, J.; Ginsberg, H. N.; Gotto, A. M.; Hu, D.; Kadowaki, T.; Kodama, T.; Kremp, M.; Matsuzawa, Y.; Millán Núñez-Cortés, J.; Monfil, C. C.; Ogawa, H.; Plutzky, J.; Rader, D. J.; Sadikot, S.; Santos, R. D.; Shlyakhto, E.; Sritara, P.; Sy, R.; Tall, A.; Tan, C. E.; Tokgozoglu, L.; Toth, P. P.; Valensi, P.; Wanner, C.; Zambon, A.; Zhu, J.; Zimmet, P. *Cardiovasc. Diabetol.* **2014**, *13*, 1.
206. Stone, N. J.; Robinson, J. G.; Lichtenstein, A. H.; Bairey Merz, C. N.; Blum, C. B.; Eckel, R. H.; Goldberg, A. C.; Gordon, D.; Levy, D.; Lloyd-Jones, D. M.; McBride, P.; Schwartz, J. S.; Shero, S. T.; Smith, S. C., Jr; Watson, K.; Wilson, P. W. F. *J. Am. Coll. Cardiol.* **2014**, *63*, 2889.
207. http://www.lipaglyn.com/downloads/Lipaglyn_Preclinical_Studies.ppsx.
208. Lohray, B. B.; Lohray, V. B.; Barot, V. K. G. USPTO 6,987,123 B2, **2006**.
209. Jani, R. H.; Kansagra, K.; Jain, M. R.; Patel, H. *Clin. Drug Investig.* **2013**, *33*, 809.
210. Lipaglyn http://lipaglyn.com/doctor.html (accessed 25.03.2014), **2014**.
211. Jani, R. H.; Pai, V.; Jha, P.; Jariwala, G.; Mukhopadhyay, S.; Bhansali, A.; Joshi, S. *Diabetes Technol. Ther.* **2014**, *16*, 63.
212. www.fda.gov/downloads/drugs/guidancecomplianceregulatoryinformation/guidances/ucm071624.pdf.
213. http://www.accessdata.fda.gov/drugsatfda_docs/label/2013/022271s000lbl.pdf.
214. Vaidya, A.; Perry, C. M. *Drugs* **2013**, *73*, 2093.
215. (a) http://www.fda.gov/newsevents/newsroom/pressannouncements/ucm376449.htm. (b) http://hepcbc.ca/2013/11/galexos-simeprevir-approved-by-health-canada/.
216. Hajarizadeh, B.; Grebely, J.; Dore, G. J. *Nature Rev. Gastroenterol. Hepatol.* **2013**, *10*, 553.
217. Raboisson, P.; de Kock, H.; Rosenquist, A.; Nilsson, M.; Salvador-Oden, L.; Lin, T.-I.; Roue, N.; Ivanov, V.; Wahling, H.; Wickstrom, K.; Hamelink, E.; Edlund, M.; Vrang, L.; Vendeville, S.; Van de Vreken, W.; McGowan, D.; Tahri, A.; Hu, L.; Bouton, C.; Lenz, O.; Delouvroy, F.; Pille, G.; Surleraux, D.; Wigerinck, P.; Samuelsson, B.; Simmen, K. *Bioorg. Med. Chem. Lett.* **2008**, *18*, 4853.

218. Rosenquist, A.; Samuelsson, B.; Johansson, P.-O.; Cummings, M. D.; Lenz, O.; Raboisson, P.; Simmen, K.; Vendeville, S.; de Kock, H.; Nilsson, M.; Horvath, A.; Kalmeijer, R.; de la Rosa, G.; Beumont-Mauviel, M. *J. Med. Chem.* **2014**, *57*, 1673.
219. Lin, T.-I.; Lenz, O.; Fanning, G.; Verbinnen, T.; Delouvroy, F.; Scholliers, A.; Vermeiren, K.; Rosenquist, A.; Edlund, M.; Samuelsson, B.; Vrang, L.; de Kock, H.; Wigerinck, P.; Raboisson, P.; Simmen, K. *Antimicrob. Agents Chemother.* **2009**, *53*, 1377.
220. Reesink, H. W.; Fanning, G. C.; Abou Farha, K.; Weegink, C.; Van Vliet, A.; Van't Klooster, G.; Lenz, O.; Aharchi, F.; Marien, K.; Van Remoortere, P.; de Kock, H.; Broeckaert, F.; Meyvisch, P.; van Beirendonck, E.; Simmen, K.; Verloes, R. *Gastroenterology* **2010**, *138*, 913.
221. Izumi, N.; Hayashi, N.; Kumada, H.; Okanoue, T.; Tsubouchi, H.; Yatsuhashi, H.; Kato, M.; Ki, R.; Komada, Y.; Seto, C.; Goto, S. *J. Gastroenterol.* **2014**, *49*, 941.
222. Forns, X.; Lawitz, E.; Zeuzem, S.; Gane, E.; Bronowicki, J. P.; Andreone, P.; Horban, A.; Brown, A.; Peeters, M.; Lenz, O.; Ouwerkerk-Mahadevan, S.; Scott, J.; De La Rosa, G.; Kalmeijer, R.; Sinha, R.; Beumont-Mauviel, M. *Gastroenterology* **2014**, *146*, 1669.
223. (a) Jacobson, I.; Dore, G. J.; Foster, G. R.; Fried, M. W.; Radu, M.; Rafalskiy, V. V.; Moroz, L.; Craxì, A.; Peeters, M.; Lenz, O.; Ouwerkerk-Mahadevan, S.; Kalmeijer, R.; Beumont-Mauviel, M. *J. Hepatol.* **2013**, *58*, S574, Abstract 1425; (b) Manns, M.; Marcellin, P.; Poordad, F.; de Araujo, E. S. A.; Buti, M.; Horsmans, Y.; Ewa Janczewska, E. J.; Villamil, F.; Peeters, M.; Lenz, O.; Ouwerkerk-Mahadevan, S.; Kalmeijer, R.; Beumont-Mauviel, M. *J. Hepatol.* **2013**, *58*, S568, Abstract 1413.
224. (a) http://www.cdc.gov/HEPATITIS/Statistics/index.htm. (b) http://www.who.int/mediacentre/factsheets/fs164/en/.
225. Ahn, J.; Flamm, S. L. *Gastroenterol. Hepatol.* **2014**, *10*, 90.
226. Sofia, M. J.; Bao, D.; Chang, W.; Du, J.; Nagarathnam, D.; Rachakonda, S.; Ganapati Reddy, P.; Ross, B. S.; Wang, P.; Zhang, H.-R.; Bansal, S.; Espirity, C.; Keilman, M.; Lam, A. M.; Micolochick Steuer, H. M.; Niu, C.; Otto, M. J.; Furman, P. A. *J. Med. Chem.* **2010**, *53*, 7202.
227. Murakami, E.; Tolstykh, T.; Bao, H.; Niu, C.; Micolochick Steuer, H. M.; Bao, D.; Chang, W.; Espiritu, C.; Bansal, S.; Lam, A. M.; Otto, M. J.; Sofia, M. J.; Furman, P. A. *J. Biol. Chem.* **2010**, *285*, 34337.
228. Lam, A. M.; Espiritu, C.; Bansal, S.; Micolochick Steuer, H. M.; Niu, C.; Zennou, V.; Keilman, M.; Zhu, Y.; Lan, S.; Otto, M. J.; Furman, P. A. *Antimicrob. Agents Chemother.* **2012**, *56*, 3359.
229. Ross, B. S.; Ganapati Reddy, P.; Zhang, H.-R.; Rachakonda, S.; Sofia, M. J. *J. Org. Chem.* **2011**, *76*, 8311.
230. Rodriguez-Torres, M.; Lawitz, E.; Kowdley, K. V.; Nelson, D. R.; DeJesus, E.; McHutchison, J. G.; Cornpropst, M. T.; Mader, M.; Albanis, E.; Jiang, D.; Hebner, C. M.; Symonds, W. T.; Berrey, M. M.; Lalezari, J. *J. Hepatol.* **2013**, *58*, 663.
231. Denning, J.; Cornpropst, M.; Flach, S. D.; Berrey, M. M.; Symonds, W. T. *Antimicrob. Agents Chemother.* **2013**, *57*, 1201.
232. Lawitz, E.; Rodriguez-Torres, M.; Denning, J. M.; Albanis, E.; Cornpropst, M.; Berrey, M. M.; Symonds, W. T. *Antimicrob. Agents Chemother.* **2013**, *57*, 1209.
233. http://www.accessdata.fda.gov/drugsatfda_docs/label/2013/204671s000lbl.pdf.
234. Jacobson, I. M.; Gordon, S. C.; Kowdley, K. V.; Yoshida, E. M.; Rodriguez-Torres, M.; Sulkowski, M. S.; Shiffman, M. L.; Lawitz, E.; Everson, G.; Bennett, M.; Schiff, E.; Tarek Al-Assi, M.; Subramanian, G. M.; Lin, M.; McNally, J.; Brainard, D.; Symonds, W. T.; McHutchison, J. G.; Patel, K.; Feld, J.; Pianko, S.; Nelson, D. R. *N. Engl. J. Med.* **2013**, *368*, 1867.

235. Lawitz, E.; Mangia, A.; Wyles, D.; Rodriguez-Torres, M.; Hassanein, T.; Gordon, S. C.; Schultz, M.; Davis, M. N.; Kayali, Z.; Reddy, K. R.; Jacobson, I. M.; Kowdley, K. V.; Nyberg, L.; Subramanian, G. M.; Hyland, R. H.; Arterburn, S.; Jiang, D.; McNally, J.; Brainard, D.; Symonds, W. T.; McHutchison, J. G.; Sheikh, A. M.; Younossi, Z.; Gane, E. J. *N. Engl. J. Med.* **2013**, *368*, 1878.
236. Culos, K. A.; Cuellar, S. *Ann. Pharmacother.* **2013**, *47*, 519.
237. McCubrey, J. A.; Milella, M.; Tafuri, A.; Martelli, A. M.; Lunghi, P.; Bonati, A.; Cervello, M.; Lee, J. T.; Steelman, L. S. *Curr. Opin. Investig. Drugs* **2008**, *9*, 614.
238. Solit, D. B.; Garraway, L. A.; Pratilas, C. A.; Sawai, A.; Getz, G.; Basso, A.; Ye, Q.; Lobo, J. M.; She, Y.; Osman, I.; Golub, T. R.; Sebolt-Leopols, J.; Sellers, W. R.; Rosen, N. *Nature* **2006**, *439*, 358.
239. Abe, H.; Kikuchi, S.; Hayakawa, K.; Iida, T.; Nagahashi, N.; Maeda, K.; Sakamoto, J.; Matsumoto, N.; Miura, T.; Matsumura, K.; Seki, N.; Inaba, T.; Kawasaki, H.; Yamaguchi, T.; Kakefuda, R.; Nanayama, T.; Kurachi, H.; Hori, Y.; Yoshida, T.; Kakegawa, J.; Watanabe, Y.; Gilmartin, A. G.; Richter, M. C.; Moss, K. G.; Laquerre, S. G. *ACS Med. Chem. Lett.* **2011**, *2*, 320.
240. Gilmartin, A. G.; Bleam, M. R.; Groy, A.; Moss, K. G.; Minthorn, E. A.; Kulkarni, S. G.; Rominger, C. M.; Erskine, S.; Fisher, K. E.; Yang, J.; Zappacosta, F.; Annan, R.; Sutton, D.; Laquerre, S. G. *Clin. Cancer Res.* **2011**, *17*, 989.
241. Infante, J. R.; Fecher, L. A.; Falchook, G. S.; Nallapareddy, S.; Gordon, M. S.; Becerra, C.; DeMarini, D. J.; Cox, D. S.; Xu, Y.; Morris, S. R.; Peddareddigari, V. G. R.; Le, N. T.; Hart, L.; Bendell, J. C.; Eckhardt, G.; Kurzrock, R.; Flaherty, K.; Burris, H. A.; Messersmith, W. A. *Lancet Oncol.* **2012**, *13*, 773.
242. Flaherty, K. T.; Robert, C.; Hersey, P.; Nathan, P.; Garbe, C.; Milhem, M.; Demidov, L. V.; Hassel, J. C.; Rutkowski, P.; Mohr, P.; Dummer, R.; Trefzer, U.; Larkin, J. M. G.; Utikal, J.; Dreno, B.; Nyakas, M.; Middleton, M. R.; Becker, J. C.; Casey, M.; Sherman, L. J.; Wu, F. S.; Ouellet, D.; Martin, A.-M.; Patel, K.; Schadendorf, D. *N. Engl. J. Med.* **2012**, *367*, 107.
243. Gibb, A.; Deeks, E. D. *Drugs* **2014**, *74*, 135.
244. http://www.nimh.nih.gov/statistics/1mdd_adult.shtml.
245. Ferrari, A. J.; Charlson, F. J.; Norman, R. E.; Patten, S. B.; Freedman, G.; Murray, C. J. L.; Vos, T.; Whiteford, H. A. *PLoS Med.* **2013**. http://dx.doi.org/10.1371/journal.pmed.1001547.
246. Bang-Andersen, B.; Ruhland, T.; Jørgensen, M.; Smith, G.; Frederiksen, K.; Jensen, K. G.; Zhong, H.; Nielsen, S. M.; Hogg, S.; Mørk, A.; Stensbøl, T. B. *J. Med. Chem.* **2011**, *54*, 3206.
247. Mørk, A.; Pehrson, A.; Brennum, L. T.; Møller Nielsen, S.; Zhong, H.; Lassen, A. B.; Miller, S.; Westrich, L.; Boyle, N. J.; Sanchez, C.; Fischer, C. W.; Liebenberg, N.; Wegener, G.; Bundgaard, C.; Hogg, S.; Bang-Andersen, B.; Stensbøl, T. B. *J. Pharmacol. Exp. Ther.* **2012**, *340*, 666.
248. Christensen, K. Patent Application WO2013/102573, **2013**.
249. Areberg, J.; Søgaard, B.; Høger, A.-M. *Basic Clin. Pharmacol. Toxicol.* **2012**, *111*, 198.
250. http://www.accessdata.fda.gov/drugsatfda_docs/label/2013/204447s000lbl.pdf.
251. Henigsberg, N.; Mahableshwarkar, A. R.; Jacobsen, P.; Chen, Y.; Thase, M. E. *J. Clin. Psychiatry* **2012**, *73*, 953.
252. Boulenger, J.-P.; Loft, H.; Olsen, C. K. *Int. Clin. Physchopharmacol.* **2014**, *29*, 138.

KEYWORD INDEX, VOLUME 49

Note: Page numbers followed by "*f*" indicate figures, "*t*" indicate tables, and "*np*" indicate footnotes.

Q

R

S

T

CUMULATIVE CHAPTER TITLES KEYWORD INDEX, VOLUME 1 – 49

CUMULATIVE NCE INTRODUCTION INDEX, 1983–2013

GENERIC NAME	INDICATION	YEAR INTRODUCED	ARMC VOL., (PAGE)
abacavir sulfate	antiviral	1999	35 (333)
abarelix	anticancer	2004	40 (446)
abatacept	antiarthritic	2006	42 (509)
abiraterone acetate	anticancer	2011	47 (505)
acarbose	antidiabetic	1990	26 (297)
aceclofenac	antiinflammatory	1992	28 (325)
acemannan	wound healing agent	2001	37 (259)
acetohydroxamic acid	urinary tract/bladder disorders	1983	19 (313)
acetorphan	antidiarrheal	1993	29 (332)
acipimox	antihypercholesterolemic	1985	21 (323)
acitretin	antipsoriasis	1989	25 (309)
aclidinium bromide	chronic obstructive pulmonary disorder	2012	48 (481)
acotiamide	dyspepsia	2013	49 (447)
acrivastine	antiallergy	1988	24 (295)
actarit	antiinflammatory	1994	30 (296)
adalimumab	antiarthritic	2003	39 (267)
adamantanium bromide	antibacterial	1984	20 (315)
adefovir dipivoxil	antiviral	2002	38 (348)
ado-trastuzumab emtansine	anticancer	2013	49 (449)
adrafinil	sleep disorders	1986	22 (315)
AF-2259	antiinflammatory	1987	23 (325)
afatinib	anticancer	2013	49 (451)
aflibercept	ophthalmologic, macular degeneration	2011	47 (507)
afloqualone	muscle relaxant	1983	19 (313)
agalsidase alfa	Fabry's disease	2001	37 (259)
alacepril	antihypertensive	1988	24 (296)
alcaftadine	ophthalmologic (allergic conjunctivitis)	2010	46 (444)
alclometasone dipropionate	antiinflammatory	1985	21 (323)
alefacept	antipsoriasis	2003	39 (267)
alemtuzumab	anticancer	2001	37 (260)
alendronate sodium	osteoporosis	1993	29 (332)
alfentanil hydrochloride	analgesic	1983	19 (314)
alfuzosin hydrochloride	antihypertensive	1988	24 (296)
alglucerase	Gaucher's disease	1991	27 (321)
alglucosidase alfa	Pompe disease	2006	42 (511)
aliskiren	antihypertensive	2007	43 (461)
alitretinoin	anticancer	1999	35 (333)

GENERIC NAME	INDICATION	YEAR INTRODUCED	ARMC VOL., (PAGE)
alminoprofen	analgesic	1983	19 (314)
almotriptan	antimigraine	2000	36 (295)
alogliptin	antidiabetic	2010	46 (446)
alosetron hydrochloride	irritable bowel syndrome	2000	36 (295)
alpha-1 antitrypsin	emphysema	1988	24 (297)
alpidem	anxiolytic	1991	27 (322)
alpiropride	antimigraine	1988	24 (296)
alteplase	antithrombotic	1987	23 (326)
alvimopan	post-operative ileus	2008	44 (584)
ambrisentan	pulmonary hypertension	2007	43 (463)
amfenac sodium	antiinflammatory	1986	22 (315)
amifostine	cytoprotective	1995	31 (338)
aminoprofen	antiinflammatory	1990	26 (298)
amisulpride	antipsychotic	1986	22 (316)
amlexanox	antiasthma	1987	23 (327)
amlodipine besylate	antihypertensive	1990	26 (298)
amorolfine hydrochloride	antifungal	1991	27 (322)
amosulalol	antihypertensive	1988	24 (297)
ampiroxicam	antiinflammatory	1994	30 (296)
amprenavir	antiviral	1999	35 (334)
amrinone	congestive heart failure	1983	19 (314)
amrubicin hydrochloride	anticancer	2002	38 (349)
amsacrine	anticancer	1987	23 (327)
amtolmetin guacil	antiinflammatory	1993	29 (332)
anagliptin	antidiabetic	2012	48 (483)
anagrelide hydrochloride	antithrombotic	1997	33 (328)
anakinra	antiarthritic	2001	37 (261)
anastrozole	anticancer	1995	31 (338)
angiotensin II	anticancer adjuvant	1994	30 (296)
anidulafungin	antifungal	2006	42 (512)
aniracetam	cognition enhancer	1993	29 (333)
anti-digoxin polyclonal antibody	antidote, digoxin poisoning	2002	38 (350)
APD	osteoporosis	1987	23 (326)
apixaban	antithrombotic	2011	47 (509)
apraclonidine hydrochloride	antiglaucoma	1988	24 (297)
aprepitant	antiemetic	2003	39 (268)
APSAC	antithrombotic	1987	23 (326)
aranidipine	antihypertensive	1996	32 (306)
arbekacin	antibacterial	1990	26 (298)
arformoterol	antiasthma	2007	43 (465)
argatroban	antithrombotic	1990	26 (299)
arglabin	anticancer	1999	35 (335)
aripiprazole	antipsychotic	2002	38 (350)

GENERIC NAME	INDICATION	YEAR INTRODUCED	ARMC VOL., (PAGE)
armodafinil	sleep disorders	2009	45 (478)
arotinolol hydrochloride	antihypertensive	1986	22 (316)
arteether	antimalarial	2000	36 (296)
artemisinin	antimalarial	1987	23 (327)
asenapine	antipsychotic	2009	45 (479)
aspoxicillin	antibacterial	1987	23 (328)
astemizole	antiallergy	1983	19 (314)
astromycin sulfate	antibacterial	1985	21 (324)
atazanavir	antiviral	2003	39 (269)
atomoxetine	attention deficit hyperactivity disorder	2003	39 (270)
atorvastatin calcium	antihypercholesterolemic	1997	33 (328)
atosiban	premature labor	2000	36 (297)
atovaquone	antiparasitic	1992	28 (326)
auranofin	antiarthritic	1983	19 (314)
avanafil	male sexual dysfunction	2011	47 (512)
axitinib	anticancer	2012	48 (485)
azacitidine	anticancer	2004	40 (447)
azelaic acid	acne	1989	25 (310)
azelastine hydrochloride	antiallergy	1986	22 (316)
azelnidipine	antihypertensive	2003	39 (270)
azilsartan	antihypertensive	2011	47 (514)
azithromycin	antibacterial	1988	24 (298)
azosemide	diuretic	1986	22 (316)
aztreonam	antibacterial	1984	20 (315)
balofloxacin	antibacterial	2002	38 (351)
balsalazide disodium	ulcerative colitis	1997	33 (329)
bambuterol	antiasthma	1990	26 (299)
barnidipine hydrochloride	antihypertensive	1992	28 (326)
beclobrate	antihypercholesterolemic	1986	22 (317)
bedaquiline	antibacterial	2012	48 (487)
befunolol hydrochloride	antiglaucoma	1983	19 (315)
belatacept	immunosuppressant	2011	47 (516)
belimumab	lupus	2011	47 (519)
belotecan	anticancer	2004	40 (449)
benazepril hydrochloride	antihypertensive	1990	26 (299)
benexate hydrochloride	antiulcer	1987	23 (328)
benidipine hydrochloride	antihypertensive	1991	27 (322)
beraprost sodium	antiplatelet	1992	28 (326)
besifloxacin	antibacterial	2009	45 (482)
betamethasone butyrate propionate	antiinflammatory	1994	30 (297)
betaxolol hydrochloride	antihypertensive	1983	19 (315
betotastine besilate	antiallergy	2000	36 (297)
bevacizumab	anticancer	2004	40 (450)

GENERIC NAME	INDICATION	YEAR INTRODUCED	ARMC VOL., (PAGE)
bevantolol hydrochloride	antihypertensive	1987	23 (328)
bexarotene	anticancer	2000	36 (298)
biapenem	antibacterial	2002	38 (351)
bicalutamide	anticancer	1995	31 (338)
bifemelane hydrochloride	nootropic	1987	23 (329)
bilastine	antiallergy	2010	46 (449)
bimatoprost	antiglaucoma	2001	37 (261)
binfonazole	sleep disorders	1983	19 (315)
binifibrate	antihypercholesterolemic	1986	22 (317)
biolimus drug-eluting stent	coronary artery disease, antirestenotic	2008	44 (586)
bisantrene hydrochloride	anticancer	1990	26 (300)
bisoprolol fumarate	antihypertensive	1986	22 (317)
bivalirudin	antithrombotic	2000	36 (298)
blonanserin	antipsychotic	2008	44 (587)
boceprevir	antiviral	2011	47 (521)
bopindolol	antihypertensive	1985	21 (324)
bortezomib	anticancer	2003	39 (271)
bosentan	antihypertensive	2001	37 (262)
bosutinib	anticancer	2012	48 (489)
brentuximab	anticancer	2011	47 (523)
brimonidine	antiglaucoma	1996	32 (306)
brinzolamide	antiglaucoma	1998	34 (318)
brodimoprin	antibacterial	1993	29 (333)
bromfenac sodium	antiinflammatory	1997	33 (329)
brotizolam	sleep disorders	1983	19 (315)
brovincamine fumarate	cerebral vasodilator	1986	22 (317)
bucillamine	immunomodulator	1987	23 (329)
bucladesine sodium	congestive heart failure	1984	20 (316)
budipine	Parkinson's disease	1997	33 (330)
budralazine	antihypertensive	1983	19 (315
bulaquine	antimalarial	2000	36 (299)
bunazosin hydrochloride	antihypertensive	1985	21 (324)
bupropion hydrochloride	antidepressant	1989	25 (310)
buserelin acetate	hormone therapy	1984	20 (316)
buspirone hydrochloride	anxiolytic	1985	21 (324)
butenafine hydrochloride	antifungal	1992	28 (327)
butibufen	antiinflammatory	1992	28 (327)
butoconazole	antifungal	1986	22 (318)
butoctamide	sleep disorders	1984	20 (316)
butyl flufenamate	antiinflammatory	1983	19 (316)
cabazitaxel	anticancer	2010	46 (451)
cabergoline	antiprolactin	1993	29 (334)
cabozantinib	anticancer	2012	48 (491)
cadexomer iodine	wound healing agent	1983	19 (316)

GENERIC NAME	INDICATION	YEAR INTRODUCED	ARMC VOL., (PAGE)
cefuzonam sodium	antibacterial	1987	23 (331)
celecoxib	antiarthritic	1999	35 (335)
celiprolol hydrochloride	antihypertensive	1983	19 (317
centchroman	contraception	1991	27 (324)
centoxin	immunomodulator	1991	27 (325)
cerivastatin	antihypercholesterolemic	1997	33 (331)
certolizumab pegol	irritable bowel syndrome	2008	44 (592)
cetilistat	antiobesity	2013	49 (454)
cetirizine hydrochloride	antiallergy	1987	23 (331)
cetrorelix	infertility	1999	35 (336)
cetuximab	anticancer	2003	39 (272)
cevimeline hydrochloride	antixerostomia	2000	36 (299)
chenodiol	gallstones	1983	19 (317)
CHF-1301	Parkinson's disease	1999	35 (336)
choline alfoscerate	cognition enhancer	1990	26 (300)
choline fenofibrate	antihypercholesterolemic	2008	44 (594)
cibenzoline	antiarrhythmic	1985	21 (325)
ciclesonide	antiasthma	2005	41 (443)
cicletanine	antihypertensive	1988	24 (299)
cidofovir	antiviral	1996	32 (306)
cilazapril	antihypertensive	1990	26 (301)
cilostazol	antithrombotic	1988	24 (299)
cimetropium bromide	antispasmodic	1985	21 (326)
cinacalcet	hyperparathyroidism	2004	40 (451)
cinildipine	antihypertensive	1995	31 (339)
cinitapride	gastroprokinetic	1990	26 (301)
cinolazepam	anxiolytic	1993	29 (334)
ciprofibrate	antihypercholesterolemic	1985	21 (326)
ciprofloxacin	antibacterial	1986	22 (318)
cisapride	gastroprokinetic	1988	24 (299)
cisatracurium besilate	muscle relaxant	1995	31 (340)
citalopram	antidepressant	1989	25 (311)
cladribine	anticancer	1993	29 (335)
clarithromycin	antibacterial	1990	26 (302)
clevidipine	antihypertensive	2008	44 (596)
clevudine	antiviral	2007	43 (466)
clobenoside	antiinflammatory	1988	24 (300)
cloconazole hydrochloride	antifungal	1986	22 (318)
clodronate disodium	calcium regulation	1986	22 (319)
clofarabine	anticancer	2005	41 (444)
clopidogrel hydrogensulfate	antithrombotic	1998	34 (320)
cloricromen	antithrombotic	1991	27 (325)
clospipramine hydrochloride	antipsychotic	1991	27 (325)

GENERIC NAME	INDICATION	YEAR INTRODUCED	ARMC VOL., (PAGE)
cobicistat	antiviral pharmacokinetic enhancer)	2013	49 (456)
colesevelam hydrochloride	antihypercholesterolemic	2000	36 (300)
colestimide	antihypercholesterolemic	1999	35 (337)
colforsin daropate hydrochloride	congestive heart failure	1999	35 (337)
conivaptan	hyponatremia	2006	42 (514)
corifollitropin alfa	infertility	2010	46 (455)
crizotinib	anticancer	2011	47 (525)
crofelemer	antidiarrheal	2012	48 (494)
crotelidae polyvalent immune fab	antidote, snake venom poisoning	2001	37 (263)
cyclosporine	immunosuppressant	1983	19 (317)
cytarabine ocfosfate	anticancer	1993	29 (335)
dabigatran etexilate	anticoagulant	2008	44 (598)
dabrafenib	anticancer	2013	49 (458)
dalfampridine	multiple sclerosis	2010	46 (458)
dalfopristin	antibacterial	1999	35 (338)
dapagliflozin	antidiabetic	2012	48 (495)
dapiprazole hydrochloride	antiglaucoma	1987	23 (332)
dapoxetine	premature ejaculation	2009	45 (488)
daptomycin	antibacterial	2003	39 (272)
darifenacin	urinary tract/bladder disorders	2005	41 (445)
darunavir	antiviral	2006	42 (515)
dasatinib	anticancer	2006	42 (517)
decitabine	myelodysplastic syndromes	2006	42 (519)
defeiprone	iron chelation therapy	1995	31 (340)
deferasirox	iron chelation therapy	2005	41 (446)
defibrotide	antithrombotic	1986	22 (319)
deflazacort	antiinflammatory	1986	22 (319)
degarelix acetate	anticancer	2009	45 (490)
delapril	antihypertensive	1989	25 (311)
delavirdine mesylate	antiviral	1997	33 (331)
denileukin diftitox	anticancer	1999	35 (338)
denopamine	congestive heart failure	1988	24 (300)
denosumab	osteoporosis	2010	46 (459)
deprodone propionate	antiinflammatory	1992	28 (329)
desflurane	anesthetic	1992	28 (329)
desloratadine	antiallergy	2001	37 (264)
desvenlafaxine	antidepressant	2008	44 (600)
dexfenfluramine	antiobesity	1997	33 (332)
dexibuprofen	antiinflammatory	1994	30 (298)
dexlansoprazole	antiulcer	2009	45 (492)
dexmedetomidine hydrochloride	sleep disorders	2000	36 (301)

GENERIC NAME	INDICATION	YEAR INTRODUCED	ARMC VOL., (PAGE)
dexmethylphenidate hydrochloride	attention deficit hyperactivity disorder	2002	38 (352)
dexrazoxane	cardioprotective	1992	28 (330)
dezocine	analgesic	1991	27 (326)
diacerein	antiinflammatory	1985	21 (326)
didanosine	antiviral	1991	27 (326)
dilevalol	antihypertensive	1989	25 (311)
dimethyl fumarate	multiple sclerosis	2013	49 (460)
diquafosol tetrasodium	ophthalmologic (dry eye)	2010	46 (462)
dirithromycin	antibacterial	1993	29 (336)
disodium pamidronate	osteoporosis	1989	25 (312)
divistyramine	antihypercholesterolemic	1984	20 (317)
docarpamine	congestive heart failure	1994	30 (298)
docetaxel	anticancer	1995	31 (341)
dofetilide	antiarrhythmic	2000	36 (301)
dolasetron mesylate	antiemetic	1998	34 (321)
dolutegravir	antiviral	2013	49 (461)
donepezil hydrochloride	Alzheimer's disease	1997	33 (332)
dopexamine	congestive heart failure	1989	25 (312)
doripenem	antibacterial	2005	41 (448)
dornase alfa	cystic fibrosis	1994	30 (298)
dorzolamide hydrochloride	antiglaucoma	1995	31 (341)
dosmalfate	antiulcer	2000	36 (302)
doxacurium chloride	muscle relaxant	1991	27 (326)
doxazosin mesylate	antihypertensive	1988	24 (300)
doxefazepam	anxiolytic	1985	21 (326)
doxercalciferol	hyperparathyroidism	1999	35 (339)
doxifluridine	anticancer	1987	23 (332)
doxofylline	antiasthma	1985	21 (327)
dronabinol	antiemetic	1986	22 (319)
dronedarone	antiarrhythmic	2009	45 (495)
drospirenone	contraception	2000	36 (302)
drotrecogin alfa	antisepsis	2001	37 (265)
droxicam	antiinflammatory	1990	26 (302)
droxidopa	Parkinson's disease	1989	25 (312)
duloxetine	antidepressant	2004	40 (452)
dutasteride	benign prostatic hyperplasia	2002	38 (353)
duteplase	anticoagulant	1995	31 (342)
ebastine	antiallergy	1990	26 (302)
eberconazole	antifungal	2005	41 (449)
ebrotidine	antiulcer	1997	33 (333)
ecabet sodium	antiulcer	1993	29 (336)
ecallantide	angioedema, hereditary	2009	46 (464)
eculizumab	hemoglobinuria	2007	43 (468)

GENERIC NAME	INDICATION	YEAR INTRODUCED	ARMC VOL., (PAGE)
edaravone	neuroprotective	2001	37 (265)
edoxaban	antithrombotic	2011	47 (527)
efalizumab	antipsoriasis	2003	39 (274)
efavirenz	antiviral	1998	34 (321)
efinaconazole	antifungal	2013	49 (463)
efonidipine	antihypertensive	1994	30 (299)
egualen sodium	antiulcer	2000	36 (303)
eldecalcitol	osteoporosis	2011	47 (529)
eletriptan	antimigraine	2001	37 (266)
eltrombopag	antithrombocytopenic	2009	45 (497)
elvitegravir	antiviral	2013	49 (465)
emedastine difumarate	antiallergy	1993	29 (336)
emorfazone	analgesic	1984	20 (317)
emtricitabine	antiviral	2003	39 (274)
enalapril maleate	antihypertensive	1984	20 (317)
enalaprilat	antihypertensive	1987	23 (332)
encainide hydrochloride	antiarrhythmic	1987	23 (333)
enfuvirtide	antiviral	2003	39 (275)
enocitabine	anticancer	1983	19 (318)
enoxacin	antibacterial	1986	22 (320)
enoxaparin	anticoagulant	1987	23 (333)
enoximone	congestive heart failure	1988	24 (301)
enprostil	antiulcer	1985	21 (327)
entacapone	Parkinson's disease	1998	34 (322)
entecavir	antiviral	2005	41 (450)
enzalutamide	anticancer	2012	48 (497)
epalrestat	antidiabetic	1992	28 (330)
eperisone hydrochloride	muscle relaxant	1983	19 (318)
epidermal growth factor	wound healing agent	1987	23 (333)
epinastine	antiallergy	1994	30 (299)
epirubicin hydrochloride	anticancer	1984	20 (318)
eplerenone	antihypertensive	2003	39 (276)
epoprostenol sodium	antiplatelet	1983	19 (318)
eprosartan	antihypertensive	1997	33 (333)
eptazocine hydrobromide	analgesic	1987	23 (334)
eptilfibatide	antithrombotic	1999	35 (340)
erdosteine	expectorant	1995	31 (342)
eribulin mesylate	anticancer	2010	46 (465)
erlotinib	anticancer	2004	40 (454)
ertapenem sodium	antibacterial	2002	38 (353)
erythromycin acistrate	antibacterial	1988	24 (301)
erythropoietin	hematopoietic	1988	24 (301)
escitalopram oxolate	antidepressant	2002	38 (354)
eslicarbazepine acetate	anticonvulsant	2009	45 (498)
esmolol hydrochloride	antiarrhythmic	1987	23 (334)

GENERIC NAME	INDICATION	YEAR INTRODUCED	ARMC VOL., (PAGE)
esomeprazole magnesium	antiulcer	2000	36 (303)
eszopiclone	sleep disorders	2005	41 (451)
ethyl icosapentate	antithrombotic	1990	26 (303)
etizolam	anxiolytic	1984	20 (318)
etodolac	antiinflammatory	1985	21 (327)
etoricoxibe	antiarthritic	2002	38 (355)
etravirine	antiviral	2008	44 (602)
everolimus	immunosuppressant	2004	40 (455)
exemestane	anticancer	2000	36 (304)
exenatide	antidiabetic	2005	41 (452)
exifone	cognition enhancer	1988	24 (302)
ezetimibe	antihypercholesterolemic	2002	38 (355)
factor VIIa	haemophilia	1996	32 (307)
factor VIII	hemostatic	1992	28 (330)
fadrozole hydrochloride	anticancer	1995	31 (342)
falecalcitriol	hyperparathyroidism	2001	37 (266)
famciclovir	antiviral	1994	30 (300)
famotidine	antiulcer	1985	21 (327)
fasudil hydrochloride	amyotrophic lateral sclerosis	1995	31 (343)
febuxostat	gout	2009	45 (501)
felbamate	anticonvulsant	1993	29 (337)
felbinac	antiinflammatory	1986	22 (320)
felodipine	antihypertensive	1988	24 (302)
fenbuprol	biliary tract dysfunction	1983	19 (318)
fenoldopam mesylate	antihypertensive	1998	34 (322)
fenticonazole nitrate	antifungal	1987	23 (334)
fesoterodine	urinary tract/bladder disorders	2008	44 (604)
fexofenadine	antiallergy	1996	32 (307)
fidaxomicin	antibacterial	2011	47 (531)
filgrastim	immunostimulant	1991	27 (327)
finasteride	benign prostatic hyperplasia	1992	28 (331)
fingolimod	multiple sclerosis	2010	46 (468)
fisalamine	antiinflammatory	1984	20 (318)
fleroxacin	antibacterial	1992	28 (331)
flomoxef sodium	antibacterial	1988	24 (302)
flosequinan	congestive heart failure	1992	28 (331)
fluconazole	antifungal	1988	24 (303)
fludarabine phosphate	anticancer	1991	27 (327)
flumazenil	antidote, benzodiazepine overdose	1987	23 (335)
flunoxaprofen	antiinflammatory	1987	23 (335)
fluoxetine hydrochloride	antidepressant	1986	22 (320)
flupirtine maleate	analgesic	1985	21 (328)
flurithromycin ethylsuccinate	antibacterial	1997	33 (333)

GENERIC NAME	INDICATION	YEAR INTRODUCED	ARMC VOL., (PAGE)
flutamide	anticancer	1983	19 (318)
flutazolam	anxiolytic	1984	20 (318)
fluticasone furoate	antiallergy	2007	43 (469)
fluticasone propionate	antiinflammatory	1990	26 (303)
flutoprazepam	anxiolytic	1986	22 (320)
flutrimazole	antifungal	1995	31 (343)
flutropium bromide	antiasthma	1988	24 (303)
fluvastatin	antihypercholesterolemic	1994	30 (300)
fluvoxamine maleate	antidepressant	1983	19 (319)
follitropin alfa	infertility	1996	32 (307)
follitropin beta	infertility	1996	32 (308)
fomepizole	antidote, ethylene glycol poisoning	1998	34 (323)
fomivirsen sodium	antiviral	1998	34 (323)
fondaparinux sodium	antithrombotic	2002	38 (356)
formestane	anticancer	1993	29 (337)
formoterol fumarate	chronic obstructive pulmonary disorder	1986	22 (321)
fosamprenavir	antiviral	2003	39 (277)
fosaprepitant dimeglumine	antiemetic	2008	44 (606)
foscarnet sodium	antiviral	1989	25 (313)
fosfluconazole	antifungal	2004	40 (457)
fosfosal	analgesic	1984	20 (319)
fosinopril sodium	antihypertensive	1991	27 (328)
fosphenytoin sodium	anticonvulsant	1996	32 (308)
fotemustine	anticancer	1989	25 (313)
fropenam	antibacterial	1997	33 (334)
frovatriptan	antimigraine	2002	38 (357)
fudosteine	expectorant	2001	37 (267)
fulveristrant	anticancer	2002	38 (357)
gabapentin	anticonvulsant	1993	29 (338)
gabapentin Enacarbil	restless leg syndrome	2011	47 (533)
gadoversetamide	diagnostic	2000	36 (304)
gallium nitrate	calcium regulation	1991	27 (328)
gallopamil hydrochloride	antianginal	1983	19 (3190)
galsulfase	mucopolysaccharidosis VI	2005	41 (453)
ganciclovir	antiviral	1988	24 (303)
ganirelix acetate	infertility	2000	36 (305)
garenoxacin	antibacterial	2007	43 (471)
gatilfloxacin	antibacterial	1999	35 (340)
gefitinib	anticancer	2002	38 (358)
gemcitabine hydrochloride	anticancer	1995	31 (344)
gemeprost	abortifacient	1983	19 (319)
gemifloxacin	antibacterial	2004	40 (458)
gemtuzumab ozogamicin	anticancer	2000	36 (306)

GENERIC NAME	INDICATION	YEAR INTRODUCED	ARMC VOL., (PAGE)
gestodene	contraception	1987	23 (335)
gestrinone	contraception	1986	22 (321)
glatiramer acetate	multiple sclerosis	1997	33 (334)
glimepiride	antidiabetic	1995	31 (344)
glucagon, rDNA	antidiabetic	1993	29 (338)
GMDP	immunostimulant	1996	32 (308)
golimumab	antiinflammatory	2009	45 (503)
goserelin	hormone therapy	1987	23 (336)
granisetron hydrochloride	antiemetic	1991	27 (329)
guanadrel sulfate	antihypertensive	1983	19 (319)
gusperimus	immunosuppressant	1994	30 (300)
halobetasol propionate	antiinflammatory	1991	27 (329)
halofantrine	antimalarial	1988	24 (304)
halometasone	antiinflammatory	1983	19 (320)
histrelin	precocious puberty	1993	29 (338)
hydrocortisone aceponate	antiinflammatory	1988	24 (304)
hydrocortisone butyrate	antiinflammatory	1983	19 (320)
ibandronic acid	osteoporosis	1996	32 (309)
ibopamine hydrochloride	congestive heart failure	1984	20 (319)
ibritunomab tiuxetan	anticancer	2002	38 (359)
ibrutinib	anticancer	2013	49 (466)
ibudilast	antiasthma	1989	25 (313)
ibutilide fumarate	antiarrhythmic	1996	32 (309)
icatibant	angioedema, hereditary	2008	44 (608)
idarubicin hydrochloride	anticancer	1990	26 (303)
idebenone	nootropic	1986	22 (321)
idursulfase	mucopolysaccharidosis II (Hunter syndrome)	2006	42 (520)
iguratimod	antiarthritic	2011	47 (535)
iloprost	antiplatelet	1992	28 (332)
imatinib mesylate	anticancer	2001	37 (267)
imidafenacin	urinary tract/bladder disorders	2007	43 (472)
imidapril hydrochloride	antihypertensive	1993	29 (339)
imiglucerase	Gaucher's disease	1994	30 (301)
imipenem/cilastatin	antibacterial	1985	21 (328)
imiquimod	antiviral	1997	33 (335)
incadronic acid	osteoporosis	1997	33 (335)
indacaterol	chronic obstructive pulmonary disease	2009	45 (505)
indalpine	antidepressant	1983	19 (320)
indeloxazine hydrochloride	nootropic	1988	24 (304)
indinavir sulfate	antiviral	1996	32 (310)
indisetron	antiemetic	2004	40 (459)
indobufen	antithrombotic	1984	20 (319)

GENERIC NAME	INDICATION	YEAR INTRODUCED	ARMC VOL., (PAGE)
influenza virus (live)	antiviral	2003	39 (277)
ingenol mebutate	anticancer	2012	48 (499)
insulin lispro	antidiabetic	1996	32 (310)
interferon alfacon-1	antiviral	1997	33 (336)
interferon gamma-1b	immunostimulant	1991	27 (329)
interferon, b-1a	multiple sclerosis	1996	32 (311)
interferon, b-1b	multiple sclerosis	1993	29 (339)
interferon, gamma	antiinflammatory	1989	25 (314)
interferon, gamma-1	anticancer	1992	28 (332)
interleukin-2	anticancer	1989	25 (314)
ioflupane	diagnostic	2000	36 (306)
ipilimumab	anticancer	2011	47 (537)
ipriflavone	osteoporosis	1989	25 (314)
irbesartan	antihypertensive	1997	33 (336)
irinotecan	anticancer	1994	30 (301)
irsogladine	antiulcer	1989	25 (315)
isepamicin	antibacterial	1988	24 (305)
isofezolac	antiinflammatory	1984	20 (319)
isoxicam	antiinflammatory	1983	19 (320)
isradipine	antihypertensive	1989	25 (315)
istradefylline	Parkinson's disease	2013	49 (468)
itopride hydrochloride	gastroprokinetic	1995	31 (344)
itraconazole	antifungal	1988	24 (305)
ivabradine	antianginal	2006	42 (522)
ivacaftor	cystic fibrosis	2012	48 (501)
ivermectin	antiparasitic	1987	23 (336)
ixabepilone	anticancer	2007	43 (473)
ketanserin	antihypertensive	1985	21 (328)
ketorolac tromethamine	analgesic	1990	26 (304)
kinetin	dermatologic, skin photodamage	1999	35 (341)
lacidipine	antihypertensive	1991	27 (330)
lacosamide	anticonvulsant	2008	44 (610)
lafutidine	antiulcer	2000	36 (307)
lamivudine	antiviral	1995	31 (345)
lamotrigine	anticonvulsant	1990	26 (304)
landiolol	antiarrhythmic	2002	38 (360)
laninamivir octanoate	antiviral	2010	46 (470)
lanoconazole	antifungal	1994	30 (302)
lanreotide acetate	growth disorders	1995	31 (345)
lansoprazole	antiulcer	1992	28 (332)
lapatinib	anticancer	2007	43 (475)
laronidase	mucopolysaccharidosis I	2003	39 (278)
latanoprost	antiglaucoma	1996	32 (311)
lefunomide	antiarthritic	1998	34 (324)

GENERIC NAME	INDICATION	YEAR INTRODUCED	ARMC VOL., (PAGE)
lenalidomide	myelodysplastic syndromes, multiple myeloma	2006	42 (523)
lenampicillin hydrochloride	antibacterial	1987	23 (336)
lentinan	immunostimulant	1986	22 (322)
lepirudin	anticoagulant	1997	33 (336)
lercanidipine	antihypertensive	1997	33 (337)
letrazole	anticancer	1996	32 (311)
leuprolide acetate	hormone therapy	1984	20 (319)
levacecarnine hydrochloride	cognition enhancer	1986	22 (322)
levalbuterol hydrochloride	antiasthma	1999	35 (341)
levetiracetam	anticonvulsant	2000	36 (307)
levobunolol hydrochloride	antiglaucoma	1985	21 (328)
levobupivacaine hydrochloride	anesthetic	2000	36 (308)
levocabastine hydrochloride	antiallergy	1991	27 (330)
levocetirizine	antiallergy	2001	37 (268)
levodropropizine	antitussive	1988	24 (305)
levofloxacin	antibacterial	1993	29 (340)
levosimendan	congestive heart failure	2000	36 (308)
lidamidine hydrochloride	antidiarrheal	1984	20 (320)
limaprost	antithrombotic	1988	24 (306)
linaclotide	irritable bowel syndrome	2012	48 (502)
linagliptin	antidiabetic	2011	47 (540)
linezolid	antibacterial	2000	36 (309)
liraglutide	antidiabetic	2009	45 (507)
liranaftate	antifungal	2000	36 (309)
lisdexamfetamine	attention deficit hyperactivity disorder	2007	43 (477)
lisinopril	antihypertensive	1987	23 (337)
lixisenatide	antidiabetic	2013	49 (470)
lobenzarit sodium	antiinflammatory	1986	22 (322)
lodoxamide tromethamine	antiallergy	1992	28 (333)
lomefloxacin	antibacterial	1989	25 (315)
lomerizine hydrochloride	antimigraine	1999	35 (342)
lomitapide	antihypercholersteremic	2012	48 (504)
lonidamine	anticancer	1987	23 (337)
lopinavir	antiviral	2000	36 (310)
loprazolam mesylate	sleep disorders	1983	19 (321)
loprinone hydrochloride	congestive heart failure	1996	32 (312)
loracarbef	antibacterial	1992	28 (333)
loratadine	antiallergy	1988	24 (306)
lorcaserin hydrochloride	antiobesity	2012	48 (506)

GENERIC NAME	INDICATION	YEAR INTRODUCED	ARMC VOL., (PAGE)
lornoxicam	antiinflammatory	1997	33 (337)
losartan	antihypertensive	1994	30 (302)
loteprednol etabonate	antiallergy	1998	34 (324)
lovastatin	antihypercholesterolemic	1987	23 (337)
loxoprofen sodium	antiinflammatory	1986	22 (322)
lulbiprostone	constipation	2006	42 (525)
luliconazole	antifungal	2005	41 (454)
lumiracoxib	antiinflammatory	2005	41 (455)
lurasidone hydrochloride	antipsychotic	2010	46 (473)
Lyme disease vaccine	Lyme disease	1999	35 (342)
mabuterol hydrochloride	antiasthma	1986	22 (323)
macitentan	pulmonary hypertension	2013	49 (472)
malotilate	hepatoprotective	1985	21 (329)
manidipine hydrochloride	antihypertensive	1990	26 (304)
maraviroc	antiviral	2007	43 (478)
masoprocol	anticancer	1992	28 (333)
maxacalcitol	hyperparathyroidism	2000	36 (310)
mebefradil hydrochloride	antihypertensive	1997	33 (338)
medifoxamine fumarate	antidepressant	1986	22 (323)
mefloquine hydrochloride	antimalarial	1985	21 (329)
meglutol	antihypercholesterolemic	1983	19 (321)
melinamide	antihypercholesterolemic	1984	20 (320)
meloxicam	antiarthritic	1996	32 (312)
mepixanox	respiratory stimulant	1984	20 (320)
meptazinol hydrochloride	analgesic	1983	19 (321)
meropenem	antibacterial	1994	30 (303)
metaclazepam	anxiolytic	1987	23 (338)
metapramine	antidepressant	1984	20 (320)
methylnaltrexone bromide	constipation	2008	44 (612)
metreleptin	lipodystrophy	2013	49 (474)
mexazolam	anxiolytic	1984	20 (321)
micafungin	antifungal	2002	38 (360)
mifamurtide	anticancer	2009	46 (476)
mifepristone	abortifacient	1988	24 (306)
miglitol	antidiabetic	1998	34 (325)
miglustat	Gaucher's disease	2003	39 (279)
milnacipran	antidepressant	1997	33 (338)
milrinone	congestive heart failure	1989	25 (316)
miltefosine	anticancer	1993	29 (340)
minodronic acid	osteoporosis	2009	45 (509)
miokamycin	antibacterial	1985	21 (329)
mipomersen	antihypercholesterolemic	2013	49 (476)
mirabegron	urinary tract/bladder disorders	2011	47 (542)
mirtazapine	antidepressant	1994	30 (303)
misoprostol	antiulcer	1985	21 (329)

GENERIC NAME	INDICATION	YEAR INTRODUCED	ARMC VOL., (PAGE)
mitiglinide	antidiabetic	2004	40 (460)
mitoxantrone hydrochloride	anticancer	1984	20 (321)
mivacurium chloride	muscle relaxant	1992	28 (334)
mivotilate	hepatoprotective	1999	35 (343)
mizolastine	antiallergy	1998	34 (325)
mizoribine	immunosuppressant	1984	20 (321)
moclobemide	antidepressant	1990	26 (305)
modafinil	sleep disorders	1994	30 (303)
moexipril hydrochloride	antihypertensive	1995	31 (346)
mofezolac	analgesic	1994	30 (304)
mogamulizumab	anticancer	2012	48 (507)
mometasone furoate	antiinflammatory	1987	23 (338)
montelukast sodium	antiasthma	1998	34 (326)
moricizine hydrochloride	antiarrhythmic	1990	26 (305)
mosapride citrate	gastroprokinetic	1998	34 (326)
moxifloxacin hydrochloride	antibacterial	1999	35 (343)
moxonidine	antihypertensive	1991	27 (330)
mozavaptan	hyponatremia	2006	42 (527)
mupirocin	antibacterial	1985	21 (330)
muromonab-CD3	immunosuppressant	1986	22 (323)
muzolimine	diuretic	1983	19 (321)
mycophenolate mofetil	immunosuppressant	1995	31 (346)
mycophenolate sodium	immunosuppressant	2003	39 (279)
nabumetone	antiinflammatory	1985	21 (330)
nadifloxacin	antibacterial	1993	29 (340)
nafamostat mesylate	pancreatitis	1986	22 (323)
nafarelin acetate	hormone therapy	1990	26 (306)
naftifine hydrochloride	antifungal	1984	20 (321)
naftopidil	urinary tract/bladder disorders	1999	35 (344)
nalfurafine hydrochloride	pruritus	2009	45 (510)
nalmefene hydrochloride	addiction, opioids	1995	31 (347)
naltrexone hydrochloride	addiction, opioids	1984	20 (322)
naratriptan hydrochloride	antimigraine	1997	33 (339)
nartograstim	leukopenia	1994	30 (304)
natalizumab	multiple sclerosis	2004	40 (462)
nateglinide	antidiabetic	1999	35 (344)
nazasetron	antiemetic	1994	30 (305)
nebivolol	antihypertensive	1997	33 (339)
nedaplatin	anticancer	1995	31 (347)
nedocromil sodium	antiallergy	1986	22 (324)
nefazodone	antidepressant	1994	30 (305)
nelarabine	anticancer	2006	42 (528)
nelfinavir mesylate	antiviral	1997	33 (340)

GENERIC NAME	INDICATION	YEAR INTRODUCED	ARMC VOL., (PAGE)
neltenexine	cystic fibrosis	1993	29 (341)
nemonapride	antipsychotic	1991	27 (331)
nepafenac	antiinflammatory	2005	41 (456)
neridronic acide	calcium regulation	2002	38 (361)
nesiritide	congestive heart failure	2001	37 (269)
neticonazole hydrochloride	antifungal	1993	29 (341)
nevirapine	antiviral	1996	32 (313)
nicorandil	antianginal	1984	20 (322)
nif ekalant hydrochloride	antiarrhythmic	1999	35 (344)
nilotinib	anticancer	2007	43 (480)
nilutamide	anticancer	1987	23 (338)
nilvadipine	antihypertensive	1989	25 (316)
nimesulide	antiinflammatory	1985	21 (330)
nimodipine	cerebral vasodilator	1985	21 (330)
nimotuzumab	anticancer	2006	42 (529)
nipradilol	antihypertensive	1988	24 (307)
nisoldipine	antihypertensive	1990	26 (306)
nitisinone	antityrosinaemia	2002	38 (361)
nitrefazole	addiction, alcohol	1983	19 (322)
nitrendipine	antihypertensive	1985	21 (331)
nizatidine	antiulcer	1987	23 (339)
nizofenzone	nootropic	1988	24 (307)
nomegestrol acetate	contraception	1986	22 (324)
norelgestromin	contraception	2002	38 (362)
norfloxacin	antibacterial	1983	19 (322)
norgestimate	contraception	1986	22 (324)
obinutuzumab	anticancer	2013	49 (478)
OCT-43	anticancer	1999	35 (345)
octreotide	growth disorders	1988	24 (307)
ofatumumab	anticancer	2009	45 (512)
ofloxacin	antibacterial	1985	21 (331)
olanzapine	antipsychotic	1996	32 (313)
olimesartan Medoxomil	antihypertensive	2002	38 (363)
olodaterol	chronic obstructive pulmonary disorder	2013	49 (480)
olopatadine hydrochloride	antiallergy	1997	33 (340)
omacetaxine mepesuccinate	anticancer	2012	48 (510)
omalizumab allergic	antiasthma	2003	39 (280)
omeprazole	antiulcer	1988	24 (308)
ondansetron hydrochloride	antiemetic	1990	26 (306)
OP-1	osteoinductor	2001	37 (269)
orlistat	antiobesity	1998	34 (327)

GENERIC NAME	INDICATION	YEAR INTRODUCED	ARMC VOL., (PAGE)
ornoprostil	antiulcer	1987	23 (339)
osalazine sodium	antiinflammatory	1986	22 (324)
oseltamivir phosphate	antiviral	1999	35 (346)
ospemifene	dyspareunia	2013	49 (482)
oxaliplatin	anticancer	1996	32 (313)
oxaprozin	antiinflammatory	1983	19 (322)
oxcarbazepine	anticonvulsant	1990	26 (307)
oxiconazole nitrate	antifungal	1983	19 (322)
oxiracetam	cognition enhancer	1987	23 (339)
oxitropium bromide	antiasthma	1983	19 (323)
ozagrel sodium	antithrombotic	1988	24 (308)
paclitaxal	anticancer	1993	29 (342)
palifermin	mucositis	2005	41 (461)
paliperidone	antipsychotic	2007	43 (482)
palonosetron	antiemetic	2003	39 (281)
panipenem/ betamipron carbapenem	antibacterial	1994	30 (305)
panitumumab	anticancer	2006	42 (531)
pantoprazole sodium	antiulcer	1995	30 (306)
parecoxib sodium	analgesic	2002	38 (364)
paricalcitol	hyperparathyroidism	1998	34 (327)
parnaparin sodium	anticoagulant	1993	29 (342)
paroxetine	antidepressant	1991	27 (331)
pasireotide	Cushing's Disease	2012	48 (512)
pazopanib	anticancer	2009	45 (514)
pazufloxacin	antibacterial	2002	38 (364)
pefloxacin mesylate	antibacterial	1985	21 (331)
pegademase bovine	immunostimulant	1990	26 (307)
pegaptanib	ophthalmologic (macular degeneration)	2005	41 (458)
pegaspargase	anticancer	1994	30 (306)
peginesatide acetate	hematopoietic	2012	48 (514)
pegvisomant	growth disorders	2003	39 (281)
pemetrexed	anticancer	2004	40 (463)
pemirolast potassium	antiasthma	1991	27 (331)
penciclovir	antiviral	1996	32 (314)
pentostatin	anticancer	1992	28 (334)
peramivir	antiviral	2010	46 (477)
perampanel	anticonvulsant	2012	48 (516)
pergolide mesylate	Parkinson's disease	1988	24 (308)
perindopril	antihypertensive	1988	24 (309)
perospirone hydrochloride	antipsychotic	2001	37 (270)
pertuzumab	anticancer	2012	48 (517)
picotamide	antithrombotic	1987	23 (340)
pidotimod	immunostimulant	1993	29 (343)

GENERIC NAME	INDICATION	YEAR INTRODUCED	ARMC VOL., (PAGE)
piketoprofen	antiinflammatory	1984	20 (322)
pilsicainide hydrochloride	antiarrhythmic	1991	27 (332)
pimaprofen	antiinflammatory	1984	20 (322)
pimecrolimus	immunosuppressant	2002	38 (365)
pimobendan	congestive heart failure	1994	30 (307)
pinacidil	antihypertensive	1987	23 (340)
pioglitazone hydrochloride	antidiabetic	1999	35 (346)
pirarubicin	anticancer	1988	24 (309)
pirfenidone	pulmonary fibrosis, idiopathic	2008	44 (614)
pirmenol	antiarrhythmic	1994	30 (307)
piroxicam cinnamate	antiinflammatory	1988	24 (309)
pitavastatin	antihypercholesterolemic	2003	39 (282)
pivagabine	antidepressant	1997	33 (341)
pixantrone dimaleate	anticancer	2012	48 (519)
plaunotol	antiulcer	1987	23 (340)
plerixafor hydrochloride	stem cell mobilizer	2009	45 (515)
polaprezinc	antiulcer	1994	30 (307)
pomalidomide	anticancer	2013	49 (484)
ponatinib	anticancer	2012	48 (521)
porfimer sodium	anticancer	1993	29 (343)
posaconazole	antifungal	2006	42 (532)
pralatrexate	anticancer	2009	45 (517)
pramipexole hydrochloride	Parkinson's disease	1997	33 (341)
pramiracetam sulfate	cognition enhancer	1993	29 (343)
pramlintide	antidiabetic	2005	41 (460)
pranlukast	antiasthma	1995	31 (347)
prasugrel	antiplatelet	2009	45 (519)
pravastatin	antihypercholesterolemic	1989	25 (316)
prednicarbate	antiinflammatory	1986	22 (325)
pregabalin	anticonvulsant	2004	40 (464)
prezatide copper acetate	wound healing agent	1996	32 (314)
progabide	anticonvulsant	1985	21 (331)
promegestrone	contraception	1983	19 (323)
propacetamol hydrochloride	analgesic	1986	22 (325)
propagermanium	antiviral	1994	30 (308)
propentofylline propionate	cerebral vasodilator	1988	24 (310)
propiverine hydrochloride	urinary tract/bladder disorders	1992	28 (335)
propofol	anesthetic	1986	22 (325)
prulifloxacin	antibacterial	2002	38 (366)
pumactant	respiratory distress syndrome	1994	30 (308)
quazepam	sleep disorders	1985	21 (332)
quetiapine fumarate	antipsychotic	1997	33 (341)
quinagolide	hyperprolactinemia	1994	30 (309)

GENERIC NAME	INDICATION	YEAR INTRODUCED	ARMC VOL., (PAGE)
quinapril	antihypertensive	1989	25 (317)
quinf amideamebicide	antiparasitic	1984	20 (322)
quinupristin	antibacterial	1999	35 (338)
rabeprazole sodium	antiulcer	1998	34 (328)
radotinib	anticancer	2012	48 (523)
raloxifene hydrochloride	osteoporosis	1998	34 (328)
raltegravir	antiviral	2007	43 (484)
raltitrexed	anticancer	1996	32 (315)
ramatroban	antiallergy	2000	36 (311)
ramelteon	sleep disorders	2005	41 (462)
ramipril	antihypertensive	1989	25 (317)
ramosetron	antiemetic	1996	32 (315)
ranibizumab	ophthalmologic (macular degeneration)	2006	42 (534)
ranimustine	anticancer	1987	23 (341)
ranitidine bismuth citrate	antiulcer	1995	31 (348)
ranolazine	antianginal	2006	42 (535)
rapacuronium bromide	muscle relaxant	1999	35 (347)
rasagiline	Parkinson's disease	2005	41 (464)
rebamipide	antiulcer	1990	26 (308)
reboxetine	antidepressant	1997	33 (342)
regorafenib	anticancer	2012	48 (524)
remifentanil hydrochloride	analgesic	1996	32 (316)
remoxipride hydrochloride	antipsychotic	1990	26 (308)
repaglinide	antidiabetic	1998	34 (329)
repirinast	antiallergy	1987	23 (341)
retapamulin	antibacterial	2007	43 (486)
reteplase	antithrombotic	1996	32 (316)
retigabine	anticonvulsant	2011	47 (544)
reviparin sodium	anticoagulant	1993	29 (344)
rifabutin	antibacterial	1992	28 (335)
rifapentine	antibacterial	1988	24 (310)
rifaximin	antibacterial	1987	23 (341)
rifaximin	antibacterial	1985	21 (332)
rilmazafone	sleep disorders	1989	25 (317)
rilmenidine	antihypertensive	1988	24 (310)
rilonacept	genetic autoinflammatory syndromes	2008	44 (615)
rilpivirine	antiviral	2011	47 (546)
riluzole	amyotrophic lateral sclerosis	1996	32 (316)
rimantadine hydrochloride	antiviral	1987	23 (342)
rimexolone	antiinflammatory	1995	31 (348)
rimonabant	antiobesity	2006	42 (537)
riociguat	pulmonary hypertension	2013	49 (486)

GENERIC NAME	INDICATION	YEAR INTRODUCED	ARMC VOL., (PAGE)
risedronate sodium	osteoporosis	1998	34 (330)
risperidone	antipsychotic	1993	29 (344)
ritonavir	antiviral	1996	32 (317)
rivaroxaban	anticoagulant	2008	44 (617)
rivastigmin	Alzheimer's disease	1997	33 (342)
rizatriptan benzoate	antimigraine	1998	34 (330)
rocuronium bromide	muscle relaxant	1994	30 (309)
rofecoxib	antiarthritic	1999	35 (347)
roflumilast	chronic obstructive pulmonary disorder	2010	46 (480)
rokitamycin	antibacterial	1986	22 (325)
romidepsin	anticancer	2009	46 (482)
romiplostim	antithrombocytopenic	2008	44 (619)
romurtide	immunostimulant	1991	27 (332)
ronafibrate	antihypercholesterolemic	1986	22 (326)
ropinirole hydrochloride	Parkinson's disease	1996	32 (317)
ropivacaine	anesthetic	1996	32 (318)
rosaprostol	antiulcer	1985	21 (332)
rosiglitazone maleate	antidiabetic	1999	35 (348)
rosuvastatin	antihypercholesterolemic	2003	39 (283)
rotigotine	Parkinson's disease	2006	42 (538)
roxatidine acetate hydrochloride	antiulcer	1986	22 (326)
roxithromycin	antiulcer	1987	23 (342)
rufinamide	anticonvulsant	2007	43 (488)
rufloxacin hydrochloride	antibacterial	1992	28 (335)
rupatadine fumarate	antiallergy	2003	39 (284)
ruxolitinib	anticancer	2011	47 (548)
RV-11	antibacterial	1989	25 (318)
salmeterol hydroxynaphthoate	antiasthma	1990	26 (308)
sapropterin hydrochloride	phenylketouria	1992	28 (336)
saquinavir mesvlate	antiviral	1995	31 (349)
sargramostim	immunostimulant	1991	27 (332)
saroglitazar	antidiabetic	2013	49 (488)
sarpogrelate hydrochloride	antithrombotic	1993	29 (344)
saxagliptin	antidiabetic	2009	45 (521)
schizophyllan	immunostimulant	1985	22 (326)
seratrodast	antiasthma	1995	31 (349)
sertaconazole nitrate	antifungal	1992	28 (336)
sertindole	antipsychotic	1996	32 (318)
setastine hydrochloride	antiallergy	1987	23 (342)
setiptiline	antidepressant	1989	25 (318)
setraline hydrochloride	antidepressant	1990	26 (309)
sevoflurane	anesthetic	1990	26 (309)

GENERIC NAME	INDICATION	YEAR INTRODUCED	ARMC VOL., (PAGE)
sibutramine	antiobesity	1998	34 (331)
sildenafil citrate	male sexual dysfunction	1998	34 (331)
silodosin	urinary tract/bladder disorders	2006	42 (540)
simeprevir	antiviral	2013	49 (489)
simvastatin	antihypercholesterolemic	1988	24 (311)
sipuleucel-t	anticancer	2010	46 (484)
sitafloxacin hydrate	antibacterial	2008	44 (621)
sitagliptin	antidiabetic	2006	42 (541)
sitaxsentan	pulmonary hypertension	2006	42 (543)
sivelestat	antiinflammatory	2002	38 (366)
SKI-2053R	anticancer	1999	35 (348)
sobuzoxane	anticancer	1994	30 (310)
sodium cellulose phosphate	urinary tract/bladder disorders	1983	19 (323)
sofalcone	antiulcer	1984	20 (323)
sofosbuvir	antiviral	2013	49 (492)
solifenacin	urinary tract/bladder disorders	2004	40 (466)
somatomedin-1	growth disorders	1994	30 (310)
somatotropin	growth disorders	1994	30 (310)
somatropin	growth disorders	1987	23 (343)
sorafenib	anticancer	2005	41 (466)
sorivudine	antiviral	1993	29 (345)
sparfloxacin	antibacterial	1993	29 (345)
spirapril hydrochloride	antihypertensive	1995	31 (349)
spizofurone	antiulcer	1987	23 (343)
stavudine	antiviral	1994	30 (311)
strontium ranelate	osteoporosis	2004	40 (466)
succimer	antidote, lead poisoning	1991	27 (333)
sufentanil	analgesic	1983	19 (323)
sugammadex	neuromuscular blockade, reversal	2008	44 (623)
sulbactam sodium	antibacterial	1986	22 (326)
sulconizole nitrate	antifungal	1985	21 (332)
sultamycillin tosylate	antibacterial	1987	23 (343)
sumatriptan succinate	antimigraine	1991	27 (333)
sunitinib	anticancer	2006	42 (544)
suplatast tosilate	antiallergy	1995	31 (350)
suprofen	analgesic	1983	19 (324)
surfactant TA	respiratory surfactant	1987	23 (344)
tacalcitol	antipsoriasis	1993	29 (346)
tacrine hydrochloride	Alzheimer's disease	1993	29 (346)
tacrolimus	immunosuppressant	1993	29 (347)
tadalafil	male sexual dysfunction	2003	39 (284)
tafamidis	neurodegeneration	2011	47 (550)
tafluprost	antiglaucoma	2008	44 (625)
talaporfin sodium	anticancer	2004	40 (469)

GENERIC NAME	INDICATION	YEAR INTRODUCED	ARMC VOL., (PAGE)
talipexole	Parkinson's disease	1996	32 (318)
taltirelin	neurodegeneration	2000	36 (311)
tamibarotene	anticancer	2005	41 (467)
tamsulosin hydrochloride	benign prostatic hyperplasia	1993	29 (347)
tandospirone	anxiolytic	1996	32 (319)
tapentadol hydrochloride	analgesic	2009	45 (523)
tasonermin	anticancer	1999	35 (349)
tazanolast	antiallergy	1990	26 (309)
tazarotene	antipsoriasis	1997	33 (343)
tazobactam sodium	antibacterial	1992	28 (336)
teduglutide	short bowel syndrome	2012	48 (526)
tegaserod maleate	irritable bowel syndrome	2001	37 (270)
teicoplanin	antibacterial	1988	24 (311)
telaprevir	antiviral	2011	47 (552)
telavancin	antibacterial	2009	45 (525)
telbivudine	antiviral	2006	42 (546)
telithromycin	antibacterial	2001	37 (271)
telmesteine	expectorant	1992	28 (337)
telmisartan	antihypertensive	1999	35 (349)
temafloxacin hydrochloride	antibacterial	1991	27 (334)
temocapril	antihypertensive	1994	30 (311)
temocillin disodium	antibacterial	1984	20 (323)
temoporphin	anticancer	2002	38 (367)
temozolomide	anticancer	1999	35 (349)
temsirolimus	anticancer	2007	43 (490)
teneligliptin	antidiabetic	2012	48 (528)
tenofovir disoproxil fumarate	antiviral	2001	37 (271)
tenoxicam	antiinflammatory	1987	23 (344)
teprenone	antiulcer	1984	20 (323)
terazosin hydrochloride	antihypertensive	1984	20 (323)
terbinafine hydrochloride	antifungal	1991	27 (334)
terconazole	antifungal	1983	19 (324)
teriflunomide	multiple sclerosis	2012	48 (530)
tertatolol hydrochloride	antihypertensive	1987	23 (344)
tesamorelin acetate	lipodystrophy	2010	46 (486)
thrombin alfa	hemostatic	2008	44 (627)
thrombomodulin (recombinant)	anticoagulant	2008	44 (628)
thymopentin	immunomodulator	1985	21 (333)
tiagabine	anticonvulsant	1996	32 (319)
tiamenidine hydrochloride	antihypertensive	1988	24 (311)
tianeptine sodium	antidepressant	1983	19 (324)
tibolone	hormone therapy	1988	24 (312)

GENERIC NAME	INDICATION	YEAR INTRODUCED	ARMC VOL., (PAGE)
ticagrelor	antithrombotic	2010	46 (488)
tigecycline	antibacterial	2005	41 (468)
tilisolol hydrochloride	antihypertensive	1992	28 (337)
tiludronate disodium	Paget's disease	1995	31 (350)
timiperone	antipsychotic	1984	20 (323)
tinazoline	nasal decongestant	1988	24 (312)
tioconazole	antifungal	1983	19 (324)
tiopronin	urolithiasis	1989	25 (318)
tiotropium bromide	chronic obstructive pulmonary disorder	2002	38 (368)
tipranavir	antiviral	2005	41 (470)
tiquizium bromide	antispasmodic	1984	20 (324)
tiracizine hydrochloride	antiarrhythmic	1990	26 (310)
tirilazad mesylate	subarachnoid hemorrhage	1995	31 (351)
tirofiban hydrochloride	antithrombotic	1998	34 (332)
tiropramide hydrochloride	muscle relaxant	1983	19 (324)
tizanidine	muscle relaxant	1984	20 (324)
tofacitinib	antiarthritic	2012	48 (532)
tolcapone	Parkinson's disease	1997	33 (343)
toloxatone	antidepressant	1984	20 (324)
tolrestat	antidiabetic	1989	25 (319)
tolvaptan	hyponatremia	2009	45 (528)
topiramate	anticonvulsant	1995	31 (351)
topotecan hydrochloride	anticancer	1996	32 (320)
torasemide	diuretic	1993	29 (348)
toremifene	anticancer	1989	25 (319)
tositumomab	anticancer	2003	39 (285)
tosufloxacin tosylate	antibacterial	1990	26 (310)
trabectedin	anticancer	2007	43 (492)
trametinib	anticancer	2013	49 (494)
trandolapril	antihypertensive	1993	29 (348)
travoprost	antiglaucoma	2001	37 (272)
treprostinil sodium	antihypertensive	2002	38 (368)
tretinoin tocoferil	antiulcer	1993	29 (348)
trientine hydrochloride	antidote, copper poisoning	1986	22 (327)
trimazosin hydrochloride	antihypertensive	1985	21 (333)
trimegestone	contraception	2001	37 (273)
trimetrexate glucuronate	antifungal	1994	30 (312)
troglitazone	antidiabetic	1997	33 (344)
tropisetron	antiemetic	1992	28 (337)
trovafloxacin mesylate	antibacterial	1998	34 (332)
troxipide	antiulcer	1986	22 (327)
ubenimex	immunostimulant	1987	23 (345)

GENERIC NAME	INDICATION	YEAR INTRODUCED	ARMC VOL., (PAGE)
udenafil	male sexual dysfunction	2005	41 (472)
ulipristal acetate	contraception	2009	45 (530)
unoprostone isopropyl ester	antiglaucoma	1994	30 (312)
ustekinumab	antipsoriasis	2009	45 (532)
vadecoxib	antiarthritic	2002	38 (369)
vaglancirclovir hydrochloride	antiviral	2001	37 (273)
valaciclovir hydrochloride	antiviral	1995	31 (352)
valrubicin	anticancer	1999	35 (350)
valsartan	antihypertensive	1996	32 (320)
vandetanib	anticancer	2011	47 (555)
vardenafil	male sexual dysfunction	2003	39 (286)
varenicline	addiction, nicotine	2006	42 (547)
vemurafenib	anticancer	2011	47 (556)
venlafaxine	antidepressant	1994	30 (312)
vernakalant	antiarrhythmic	2010	46 (491)
verteporfin	ophthalmologic (macular degeneration)	2000	36 (312)
vesnarinone	congestive heart failure	1990	26 (310)
vigabatrin	anticonvulsant	1989	25 (319)
vilazodone	antidepressant	2011	47 (558)
vildagliptin	antidiabetic	2007	43 (494)
vinflunine	anticancer	2009	46 (493)
vinorelbine	anticancer	1989	25 (320)
vismodegib	anticancer	2012	48 (534)
voglibose	antidiabetic	1994	30 (313)
voriconazole	antifungal	2002	38 (370)
vorinostat	anticancer	2006	42 (549)
vortioxetine	antidepressant	2013	49 (496)
xamoterol fumarate	congestive heart failure	1988	24 (312)
ximelagatran	anticoagulant	2004	40 (470)
zafirlukast	antiasthma	1996	32 (321)
zalcitabine	antiviral	1992	28 (338)
zaleplon	sleep disorders	1999	35 (351)
zaltoprofen	antiinflammatory	1993	29 (349)
zanamivir	antiviral	1999	35 (352)
ziconotide	analgesic	2005	41 (473)
zidovudine	antiviral	1987	23 (345)
zileuton	antiasthma	1997	33 (344)
zinostatin stimalamer	anticancer	1994	30 (313)
ziprasidone hydrochloride	antipsychotic	2000	36 (312)
zofenopril calcium	antihypertensive	2000	36 (313)
zoledronate disodium	osteoporosis	2000	36 (314)

GENERIC NAME	INDICATION	YEAR INTRODUCED	ARMC VOL., (PAGE)
zolpidem hemitartrate	sleep disorders	1988	24 (313)
zomitriptan	antimigraine	1997	33 (345)
zonisamide	anticonvulsant	1989	25 (320)
zopiclone	sleep disorders	1986	22 (327)
zucapsaicin	analgesic	2010	46 (495)
zuclopenthixol acetate	antipsychotic	1987	23 (345)

CUMULATIVE NCE INTRODUCTION INDEX, 1983–2013 (BY INDICATION)

GENERIC NAME	INDICATION	YEAR INTRODUCED	ARMC VOL., (PAGE)
gemeprost	abortifacient	1983	19 (319)
mifepristone	abortifacient	1988	24 (306)
azelaic acid	acne	1989	25 (310)
nitrefazole	addiction, alcohol	1983	19 (322)
varenicline	addiction, nicotine	2006	42 (547)
nalmefene hydrochloride	addiction, opioids	1995	31 (347)
naltrexone hydrochloride	addiction, opioids	1984	20 (322)
donepezil hydrochloride	Alzheimer's disease	1997	33 (332)
rivastigmin	Alzheimer's disease	1997	33 (342)
tacrine hydrochloride	Alzheimer's disease	1993	29 (346)
fasudil hydrochloride	amyotrophic lateral sclerosis	1995	31 (343)
riluzole	amyotrophic lateral sclerosis	1996	32 (316)
alfentanil hydrochloride	analgesic	1983	19 (314)
alminoprofen	analgesic	1983	19 (314)
dezocine	analgesic	1991	27 (326)
emorfazone	analgesic	1984	20 (317)
eptazocine hydrobromide	analgesic	1987	23 (334)
flupirtine maleate	analgesic	1985	21 (328)
fosfosal	analgesic	1984	20 (319)
ketorolac tromethamine	analgesic	1990	26 (304)
meptazinol hydrochloride	analgesic	1983	19 (321)
mofezolac	analgesic	1994	30 (304)
parecoxib sodium	analgesic	2002	38 (364)
propacetamol hydrochloride	analgesic	1986	22 (325)
remifentanil hydrochloride	analgesic	1996	32 (316)
sufentanil	analgesic	1983	19 (323)
suprofen	analgesic	1983	19 (324)
tapentadol hydrochloride	analgesic	2009	45 (523)
ziconotide	analgesic	2005	41 (473)
zucapsaicin	analgesic	2010	46 (495)
desflurane	anesthetic	1992	28 (329)
levobupivacaine hydrochloride	anesthetic	2000	36 (308)
propofol	anesthetic	1986	22 (325)
ropivacaine	anesthetic	1996	32 (318)
sevoflurane	anesthetic	1990	26 (309)
ecallantide	angioedema, hereditary	2009	46 (464)
icatibant	angioedema, hereditary	2008	44 (608)
acrivastine	antiallergy	1988	24 (295)
astemizole	antiallergy	1983	19 (314)
azelastine hydrochloride	antiallergy	1986	22 (316)

GENERIC NAME	INDICATION	YEAR INTRODUCED	ARMC VOL., (PAGE)
betotastine besilate	antiallergy	2000	36 (297)
bilastine	antiallergy	2010	46 (449)
cetirizine hydrochloride	antiallergy	1987	23 (331)
desloratadine	antiallergy	2001	37 (264)
ebastine	antiallergy	1990	26 (302)
emedastine difumarate	antiallergy	1993	29 (336)
epinastine	antiallergy	1994	30 (299)
fexofenadine	antiallergy	1996	32 (307)
fluticasone furoate	antiallergy	2007	43 (469)
levocabastine hydrochloride	antiallergy	1991	27 (330)
levocetirizine	antiallergy	2001	37 (268)
lodoxamide tromethamine	antiallergy	1992	28 (333)
loratadine	antiallergy	1988	24 (306)
loteprednol etabonate	antiallergy	1998	34 (324)
mizolastine	antiallergy	1998	34 (325)
nedocromil sodium	antiallergy	1986	22 (324)
olopatadine hydrochloride	antiallergy	1997	33 (340)
ramatroban	antiallergy	2000	36 (311)
repirinast	antiallergy	1987	23 (341)
rupatadine fumarate	antiallergy	2003	39 (284)
setastine hydrochloride	antiallergy	1987	23 (342)
suplatast tosilate	antiallergy	1995	31 (350)
tazanolast	antiallergy	1990	26 (309)
gallopamil hydrochloride	antianginal	1983	19 (319)
ivabradine	antianginal	2006	42 (522)
nicorandil	antianginal	1984	20 (322)
ranolazine	antianginal	2006	42 (535)
cibenzoline	antiarrhythmic	1985	21 (325)
dofetilide	antiarrhythmic	2000	36 (301)
dronedarone	antiarrhythmic	2009	45 (495)
encainide hydrochloride	antiarrhythmic	1987	23 (333)
esmolol hydrochloride	antiarrhythmic	1987	23 (334)
ibutilide fumarate	antiarrhythmic	1996	32 (309)
landiolol	antiarrhythmic	2002	38 (360)
moricizine hydrochloride	antiarrhythmic	1990	26 (305)
nif ekalant hydrochloride	antiarrhythmic	1999	35 (344)
pilsicainide hydrochloride	antiarrhythmic	1991	27 (332)
pirmenol	antiarrhythmic	1994	30 (307)
tiracizine hydrochloride	antiarrhythmic	1990	26 (310)
vernakalant	antiarrhythmic	2010	46 (491)
abatacept	antiarthritic	2006	42 (509)
adalimumab	antiarthritic	2003	39 (267)
anakinra	antiarthritic	2001	37 (261)
auranofin	antiarthritic	1983	19 (314)

GENERIC NAME	INDICATION	YEAR INTRODUCED	ARMC VOL., (PAGE)
celecoxib	antiarthritic	1999	35 (335)
etoricoxibe	antiarthritic	2002	38 (355)
iguratimod	antiarthritic	2011	47 (535)
lefunomide	antiarthritic	1998	34 (324)
meloxicam	antiarthritic	1996	32 (312)
rofecoxib	antiarthritic	1999	35 (347)
tofacitinib	antiarthritic	2012	48 (532)
vadecoxib	antiarthritic	2002	38 (369)
amlexanox	antiasthma	1987	23 (327)
arformoterol	antiasthma	2007	43 (465)
bambuterol	antiasthma	1990	26 (299)
ciclesonide	antiasthma	2005	41 (443)
doxofylline	antiasthma	1985	21 (327)
flutropium bromide	antiasthma	1988	24 (303)
ibudilast	antiasthma	1989	25 (313)
levalbuterol hydrochloride	antiasthma	1999	35 (341)
mabuterol hydrochloride	antiasthma	1986	22 (323)
montelukast sodium	antiasthma	1998	34 (326)
omalizumab allergic	antiasthma	2003	39 (280)
oxitropium bromide	antiasthma	1983	19 (323)
pemirolast potassium	antiasthma	1991	27 (331)
pranlukast	antiasthma	1995	31 (347)
salmeterol hydroxynaphthoate	antiasthma	1990	26 (308)
seratrodast	antiasthma	1995	31 (349)
zafirlukast	antiasthma	1996	32 (321)
zileuton	antiasthma	1997	33 (344)
adamantanium bromide	antibacterial	1984	20 (315)
arbekacin	antibacterial	1990	26 (298)
aspoxicillin	antibacterial	1987	23 (328)
astromycin sulfate	antibacterial	1985	21 (324)
azithromycin	antibacterial	1988	24 (298)
aztreonam	antibacterial	1984	20 (315)
balofloxacin	antibacterial	2002	38 (351)
bedaquiline	antibacterial	2012	48 (487)
besifloxacin	antibacterial	2009	45 (482)
biapenem	antibacterial	2002	38 (351)
brodimoprin	antibacterial	1993	29 (333)
carboplatin	antibacterial	1986	22 (318)
carumonam	antibacterial	1988	24 (298)
cefbuperazone sodium	antibacterial	1985	21 (325)
cefcapene pivoxil	antibacterial	1997	33 (330)
cefdinir	antibacterial	1991	27 (323)
cefditoren pivoxil	antibacterial	1994	30 (297)
cefepime	antibacterial	1993	29 (334)

GENERIC NAME	INDICATION	YEAR INTRODUCED	ARMC VOL., (PAGE)
cefetamet pivoxil hydrochloride	antibacterial	1992	28 (327)
cefixime	antibacterial	1987	23 (329)
cefmenoxime hydrochloride	antibacterial	1983	19 (316)
cefminox sodium	antibacterial	1987	23 (330)
cefodizime sodium	antibacterial	1990	26 (300)
cefonicid sodium	antibacterial	1984	20 (316)
ceforanide	antibacterial	1984	20 (317)
cefoselis	antibacterial	1998	34 (319)
cefotetan disodium	antibacterial	1984	20 (317)
cefotiam hexetil hydrochloride	antibacterial	1991	27 (324)
cefozopran hydrochloride	antibacterial	1995	31 (339)
cefpimizole	antibacterial	1987	23 (330)
cefpiramide sodium	antibacterial	1985	21 (325)
cefpirome sulfate	antibacterial	1992	28 (328)
cefpodoxime proxetil	antibacterial	1989	25 (310)
cefprozil	antibacterial	1992	28 (328)
ceftaroline fosamil	antibacterial	2010	46 (453)
ceftazidime	antibacterial	1983	19 (316)
cefteram pivoxil	antibacterial	1987	23 (330)
ceftibuten	antibacterial	1992	28 (329)
ceftobiprole medocaril	antibacterial	2008	44 (589)
cefuroxime axetil	antibacterial	1987	23 (331)
cefuzonam sodium	antibacterial	1987	23 (331)
ciprofloxacin	antibacterial	1986	22 (318)
clarithromycin	antibacterial	1990	26 (302)
dalfopristin	antibacterial	1999	35 (338)
daptomycin	antibacterial	2003	39 (272)
dirithromycin	antibacterial	1993	29 (336)
doripenem	antibacterial	2005	41 (448)
enoxacin	antibacterial	1986	22 (320)
ertapenem sodium	antibacterial	2002	38 (353)
erythromycin acistrate	antibacterial	1988	24 (301)
fidaxomicin	antibacterial	2011	47 (531)
fleroxacin	antibacterial	1992	28 (331)
flomoxef sodium	antibacterial	1988	24 (302)
flurithromycin ethylsuccinate	antibacterial	1997	33 (333)
fropenam	antibacterial	1997	33 (334)
garenoxacin	antibacterial	2007	43 (471)
gatilfloxacin	antibacterial	1999	35 (340)
gemifloxacin	antibacterial	2004	40 (458)
imipenem/cilastatin	antibacterial	1985	21 (328)

GENERIC NAME	INDICATION	YEAR INTRODUCED	ARMC VOL., (PAGE)
isepamicin	antibacterial	1988	24 (305)
lenampicillin hydrochloride	antibacterial	1987	23 (336)
levofloxacin	antibacterial	1993	29 (340)
linezolid	antibacterial	2000	36 (309)
lomefloxacin	antibacterial	1989	25 (315)
loracarbef	antibacterial	1992	28 (333)
meropenem	antibacterial	1994	30 (303)
miokamycin	antibacterial	1985	21 (329)
moxifloxacin hydrochloride	antibacterial	1999	35 (343)
mupirocin	antibacterial	1985	21 (330)
nadifloxacin	antibacterial	1993	29 (340)
norfloxacin	antibacterial	1983	19 (322)
ofloxacin	antibacterial	1985	21 (331)
panipenem/ betamipron carbapenem	antibacterial	1994	30 (305)
pazufloxacin	antibacterial	2002	38 (364)
pefloxacin mesylate	antibacterial	1985	21 (331)
prulifloxacin	antibacterial	2002	38 (366)
quinupristin	antibacterial	1999	35 (338)
retapamulin	antibacterial	2007	43 (486)
rifabutin	antibacterial	1992	28 (335)
rifapentine	antibacterial	1988	24 (310)
rifaximin	antibacterial	1987	23 (341)
rifaximin	antibacterial	1985	21 (332)
rokitamycin	antibacterial	1986	22 (325)
rufloxacin hydrochloride	antibacterial	1992	28 (335)
RV-11	antibacterial	1989	25 (318)
sitafloxacin hydrate	antibacterial	2008	44 (621)
sparfloxacin	antibacterial	1993	29 (345)
sulbactam sodium	antibacterial	1986	22 (326)
sultamycillin tosylate	antibacterial	1987	23 (343)
tazobactam sodium	antibacterial	1992	28 (336)
teicoplanin	antibacterial	1988	24 (311)
telavancin	antibacterial	2009	45 (525)
telithromycin	antibacterial	2001	37 (271)
temafloxacin hydrochloride	antibacterial	1991	27 (334)
temocillin disodium	antibacterial	1984	20 (323)
tigecycline	antibacterial	2005	41 (468)
tosufloxacin tosylate	antibacterial	1990	26 (310)
trovafloxacin mesylate	antibacterial	1998	34 (332)
abarelix	anticancer	2004	40 (446)
abiraterone acetate	anticancer	2011	47 (505)

GENERIC NAME	INDICATION	YEAR INTRODUCED	ARMC VOL., (PAGE)
ado-trastuzumab emtansine	anticancer	2013	49 (449)
afatinib	anticancer	2013	49 (451)
alemtuzumab	anticancer	2001	37 (260)
alitretinoin	anticancer	1999	35 (333)
amrubicin hydrochloride	anticancer	2002	38 (349)
amsacrine	anticancer	1987	23 (327)
anastrozole	anticancer	1995	31 (338)
arglabin	anticancer	1999	35 (335)
axitinib	anticancer	2012	48 (485)
azacitidine	anticancer	2004	40 (447)
belotecan	anticancer	2004	40 (449)
bevacizumab	anticancer	2004	40 (450)
bexarotene	anticancer	2000	36 (298)
bicalutamide	anticancer	1995	31 (338)
bisantrene hydrochloride	anticancer	1990	26 (300)
bortezomib	anticancer	2003	39 (271)
bosutinib	anticancer	2012	48 (489)
brentuximab	anticancer	2011	47 (523)
cabozantinib	anticancer	2012	48 (491)
camostat mesylate	anticancer	1985	21 (325)
capecitabine	anticancer	1998	34 (319)
carfilzomib	anticancer	2012	48 (492)
catumaxomab	anticancer	2009	45 (486)
cetuximab	anticancer	2003	39 (272)
cladribine	anticancer	1993	29 (335)
clofarabine	anticancer	2005	41 (444)
crizotinib	anticancer	2011	47 (525)
cytarabine ocfosfate	anticancer	1993	29 (335)
dabrafenib	anticancer	2013	49 (458)
dasatinib	anticancer	2006	42 (517)
degarelix acetate	anticancer	2009	45 (490)
denileukin diftitox	anticancer	1999	35 (338)
docetaxel	anticancer	1995	31 (341)
doxifluridine	anticancer	1987	23 (332)
enocitabine	anticancer	1983	19 (318)
enzalutamide	anticancer	2012	48 (497)
epirubicin hydrochloride	anticancer	1984	20 (318)
erlotinib	anticancer	2004	40 (454)
exemestane	anticancer	2000	36 (304)
fadrozole hydrochloride	anticancer	1995	31 (342)
fludarabine phosphate	anticancer	1991	27 (327)
flutamide	anticancer	1983	19 (318)
formestane	anticancer	1993	29 (337)
fotemustine	anticancer	1989	25 (313)

GENERIC NAME	INDICATION	YEAR INTRODUCED	ARMC VOL., (PAGE)
fulveristrant	anticancer	2002	38 (357)
gefitinib	anticancer	2002	38 (358)
gemcitabine hydrochloride	anticancer	1995	31 (344)
gemtuzumab ozogamicin	anticancer	2000	36 (306)
ibritunomab tiuxetan	anticancer	2002	38 (359)
ibrutinib	anticancer	2013	49 (466)
idarubicin hydrochloride	anticancer	1990	26 (303)
imatinib mesylate	anticancer	2001	37 (267)
ingenol mebutate	anticancer	2012	48 (499)
interferon, gamma-1	anticancer	1992	28 (332)
interleukin-2	anticancer	1989	25 (314)
ipilimumab	anticancer	2011	47 (537)
irinotecan	anticancer	1994	30 (301)
ixabepilone	anticancer	2007	43 (473)
lapatinib	anticancer	2007	43 (475)
letrazole	anticancer	1996	32 (311)
lonidamine	anticancer	1987	23 (337)
masoprocol	anticancer	1992	28 (333)
miltefosine	anticancer	1993	29 (340)
mitoxantrone hydrochloride	anticancer	1984	20 (321)
mogamulizumab	anticancer	2012	48 (507)
nedaplatin	anticancer	1995	31 (347)
nelarabine	anticancer	2006	42 (528)
nilutamide	anticancer	1987	23 (338)
nimotuzumab	anticancer	2006	42 (529)
obinutuzumab	anticancer	2013	49 (478)
OCT-43	anticancer	1999	35 (345)
ofatumumab	anticancer	2009	45 (512)
omacetaxine mepesuccinate	anticancer	2012	48 (510)
oxaliplatin	anticancer	1996	32 (313)
paclitaxal	anticancer	1993	29 (342)
panitumumab	anticancer	2006	42 (531)
pazopanib	anticancer	2009	45 (514)
pegaspargase	anticancer	1994	30 (306)
pemetrexed	anticancer	2004	40 (463)
pentostatin	anticancer	1992	28 (334)
pertuzumab	anticancer	2012	48 (517)
pirarubicin	anticancer	1988	24 (309)
pixantrone dimaleate	anticancer	2012	48 (519)
pomalidomide	anticancer	2013	49 (484)
ponatinib	anticancer	2012	48 (521)
porfimer sodium	anticancer	1993	29 (343)
pralatrexate	anticancer	2009	45 (517)

GENERIC NAME	INDICATION	YEAR INTRODUCED	ARMC VOL., (PAGE)
radotinib	anticancer	2012	48 (523)
raltitrexed	anticancer	1996	32 (315)
ranimustine	anticancer	1987	23 (341)
regorafenib	anticancer	2012	48 (524)
ruxolitinib	anticancer	2011	47 (548)
SKI-2053R	anticancer	1999	35 (348)
sobuzoxane	anticancer	1994	30 (310)
sorafenib	anticancer	2005	41 (466)
sunitinib	anticancer	2006	42 (544)
talaporfin sodium	anticancer	2004	40 (469)
tamibarotene	anticancer	2005	41 (467)
tasonermin	anticancer	1999	35 (349)
temoporphin	anticancer	2002	38 (367)
temozolomide	anticancer	1999	35 (349)
temsirolimus	anticancer	2007	43 (490)
topotecan hydrochloride	anticancer	1996	32 (320)
toremifene	anticancer	1989	25 (319)
tositumomab	anticancer	2003	39 (285)
trabectedin	anticancer	2007	43 (492)
trametinib	anticancer	2013	49 (494)
valrubicin	anticancer	1999	35 (350)
vandetanib	anticancer	2011	47 (555)
vemurafenib	anticancer	2011	47 (556)
vinorelbine	anticancer	1989	25 (320)
vismodegib	anticancer	2012	48 (534)
vorinostat	anticancer	2006	42 (549)
zinostatin stimalamer	anticancer	1994	30 (313)
cabazitaxel	anticancer	2010	46 (451)
eribulin mesylate	anticancer	2010	46 (465)
mifamurtide	anticancer	2009	46 (476)
nilotinib	anticancer	2007	43 (480)
romidepsin	anticancer	2009	46 (482)
sipuleucel-t	anticancer	2010	46 (484)
vinflunine	anticancer	2009	46 (493)
angiotensin II	anticancer adjuvant	1994	30 (296)
dabigatran etexilate	anticoagulant	2008	44 (598)
duteplase	anticoagulant	1995	31 (342)
enoxaparin	anticoagulant	1987	23 (333)
lepirudin	anticoagulant	1997	33 (336)
parnaparin sodium	anticoagulant	1993	29 (342)
reviparin sodium	anticoagulant	1993	29 (344)
rivaroxaban	anticoagulant	2008	44 (617)
thrombomodulin (recombinant)	anticoagulant	2008	44 (628)
ximelagatran	anticoagulant	2004	40 (470)

GENERIC NAME	INDICATION	YEAR INTRODUCED	ARMC VOL., (PAGE)
eslicarbazepine acetate	anticonvulsant	2009	45 (498)
felbamate	anticonvulsant	1993	29 (337)
fosphenytoin sodium	anticonvulsant	1996	32 (308)
gabapentin	anticonvulsant	1993	29 (338)
lacosamide	anticonvulsant	2008	44 (610)
lamotrigine	anticonvulsant	1990	26 (304)
levetiracetam	anticonvulsant	2000	36 (307)
oxcarbazepine	anticonvulsant	1990	26 (307)
perampanel	anticonvulsant	2012	48 (516)
pregabalin	anticonvulsant	2004	40 (464)
progabide	anticonvulsant	1985	21 (331)
retigabine	anticonvulsant	2011	47 (544)
rufinamide	anticonvulsant	2007	43 (488)
tiagabine	anticonvulsant	1996	32 (319)
topiramate	anticonvulsant	1995	31 (351)
vigabatrin	anticonvulsant	1989	25 (319)
zonisamide	anticonvulsant	1989	25 (320)
bupropion hydrochloride	antidepressant	1989	25 (310)
citalopram	antidepressant	1989	25 (311)
desvenlafaxine	antidepressant	2008	44 (600)
duloxetine	antidepressant	2004	40 (452)
escitalopram oxolate	antidepressant	2002	38 (354)
fluoxetine hydrochloride	antidepressant	1986	22 (320)
fluvoxamine maleate	antidepressant	1983	19 (319)
indalpine	antidepressant	1983	19 (320)
medifoxamine fumarate	antidepressant	1986	22 (323)
metapramine	antidepressant	1984	20 (320)
milnacipran	antidepressant	1997	33 (338)
mirtazapine	antidepressant	1994	30 (303)
moclobemide	antidepressant	1990	26 (305)
nefazodone	antidepressant	1994	30 (305)
paroxetine	antidepressant	1991	27 (331)
pivagabine	antidepressant	1997	33 (341)
reboxetine	antidepressant	1997	33 (342)
setiptiline	antidepressant	1989	25 (318)
setraline hydrochloride	antidepressant	1990	26 (309)
tianeptine sodium	antidepressant	1983	19 (324)
toloxatone	antidepressant	1984	20 (324)
venlafaxine	antidepressant	1994	30 (312)
vilazodone	antidepressant	2011	47 (558)
vortioxetine	antidepressant	2013	49 (496)
acarbose	antidiabetic	1990	26 (297)
alogliptin	antidiabetic	2010	46 (446)
anagliptin	antidiabetic	2012	48 (483)
canagliflozin	antidiabetic	2013	49 (453)

GENERIC NAME	INDICATION	YEAR INTRODUCED	ARMC VOL., (PAGE)
dapagliflozin	antidiabetic	2012	48 (495)
epalrestat	antidiabetic	1992	28 (330)
exenatide	antidiabetic	2005	41 (452)
glimepiride	antidiabetic	1995	31 (344)
glucagon, rDNA	antidiabetic	1993	29 (338)
insulin lispro	antidiabetic	1996	32 (310)
linagliptin	antidiabetic	2011	47 (540)
liraglutide	antidiabetic	2009	45 (507)
lixisenatide	antidiabetic	2013	49 (470)
miglitol	antidiabetic	1998	34 (325)
mitiglinide	antidiabetic	2004	40 (460)
nateglinide	antidiabetic	1999	35 (344)
pioglitazone hydrochloride	antidiabetic	1999	35 (346)
pramlintide	antidiabetic	2005	41 (460)
repaglinide	antidiabetic	1998	34 (329)
rosiglitazone maleate	antidiabetic	1999	35 (348)
saroglitazar	antidiabetic	2013	49 (488)
saxagliptin	antidiabetic	2009	45 (521)
sitagliptin	antidiabetic	2006	42 (541)
teneligliptin	antidiabetic	2012	48 (528)
tolrestat	antidiabetic	1989	25 (319)
troglitazone	antidiabetic	1997	33 (344)
vildagliptin	antidiabetic	2007	43 (494)
voglibose	antidiabetic	1994	30 (313)
acetorphan	antidiarrheal	1993	29 (332)
crofelemer	antidiarrheal	2012	48 (494)
lidamidine hydrochloride	antidiarrheal	1984	20 (320)
flumazenil	antidote, benzodiazepine overdose	1987	23 (335)
trientine hydrochloride	antidote, copper poisoning	1986	22 (327)
anti-digoxin polyclonal antibody	antidote, digoxin poisoning	2002	38 (350)
fomepizole	antidote, ethylene glycol poisoning	1998	34 (323)
succimer	antidote, lead poisoning	1991	27 (333)
crotelidae polyvalent immune fab	antidote, snake venom poisoning	2001	37 (263)
aprepitant	antiemetic	2003	39 (268)
dolasetron mesylate	antiemetic	1998	34 (321)
dronabinol	antiemetic	1986	22 (319)
fosaprepitant dimeglumine	antiemetic	2008	44 (606)
granisetron hydrochloride	antiemetic	1991	27 (329)
indisetron	antiemetic	2004	40 (459)
nazasetron	antiemetic	1994	30 (305)

GENERIC NAME	INDICATION	YEAR INTRODUCED	ARMC VOL., (PAGE)
ondansetron hydrochloride	antiemetic	1990	26 (306)
palonosetron	antiemetic	2003	39 (281)
ramosetron	antiemetic	1996	32 (315)
tropisetron	antiemetic	1992	28 (337)
amorolfine hydrochloride	antifungal	1991	27 (322)
anidulafungin	antifungal	2006	42 (512)
butenafine hydrochloride	antifungal	1992	28 (327)
butoconazole	antifungal	1986	22 (318)
caspofungin acetate	antifungal	2001	37 (263)
cloconazole hydrochloride	antifungal	1986	22 (318)
eberconazole	antifungal	2005	41 (449)
efinaconazole	antifungal	2013	49 (463)
fenticonazole nitrate	antifungal	1987	23 (334)
fluconazole	antifungal	1988	24 (303)
flutrimazole	antifungal	1995	31 (343)
fosfluconazole	antifungal	2004	40 (457)
itraconazole	antifungal	1988	24 (305)
lanoconazole	antifungal	1994	30 (302)
liranaftate	antifungal	2000	36 (309)
luliconazole	antifungal	2005	41 (454)
micafungin	antifungal	2002	38 (360)
naftifine hydrochloride	antifungal	1984	20 (321)
neticonazole hydrochloride	antifungal	1993	29 (341)
oxiconazole nitrate	antifungal	1983	19 (322)
posaconazole	antifungal	2006	42 (532)
sertaconazole nitrate	antifungal	1992	28 (336)
sulconizole nitrate	antifungal	1985	21 (332)
terbinafine hydrochloride	antifungal	1991	27 (334)
terconazole	antifungal	1983	19 (324)
tioconazole	antifungal	1983	19 (324)
trimetrexate glucuronate	antifungal	1994	30 (312)
voriconazole	antifungal	2002	38 (370)
apraclonidine hydrochloride	antiglaucoma	1988	24 (297)
befunolol hydrochloride	antiglaucoma	1983	19 (315)
bimatoprost	antiglaucoma	2001	37 (261)
brimonidine	antiglaucoma	1996	32 (306)
brinzolamide	antiglaucoma	1998	34 (318)
dapiprazole hydrochloride	antiglaucoma	1987	23 (332)
dorzolamide hydrochloride	antiglaucoma	1995	31 (341)
latanoprost	antiglaucoma	1996	32 (311)
levobunolol hydrochloride	antiglaucoma	1985	21 (328)

GENERIC NAME	INDICATION	YEAR INTRODUCED	ARMC VOL., (PAGE)
tafluprost	antiglaucoma	2008	44 (625)
travoprost	antiglaucoma	2001	37 (272)
unoprostone isopropyl ester	antiglaucoma	1994	30 (312)
lomitapide	antihypercholersteremic	2012	48 (504)
acipimox	antihypercholesterolemic	1985	21 (323)
atorvastatin calcium	antihypercholesterolemic	1997	33 (328)
beclobrate	antihypercholesterolemic	1986	22 (317)
binifibrate	antihypercholesterolemic	1986	22 (317)
cerivastatin	antihypercholesterolemic	1997	33 (331)
choline fenofibrate	antihypercholesterolemic	2008	44 (594)
ciprofibrate	antihypercholesterolemic	1985	21 (326)
colesevelam hydrochloride	antihypercholesterolemic	2000	36 (300)
colestimide	antihypercholesterolemic	1999	35 (337)
divistyramine	antihypercholesterolemic	1984	20 (317)
ezetimibe	antihypercholesterolemic	2002	38 (355)
fluvastatin	antihypercholesterolemic	1994	30 (300)
lovastatin	antihypercholesterolemic	1987	23 (337)
meglutol	antihypercholesterolemic	1983	19 (321)
melinamide	antihypercholesterolemic	1984	20 (320)
mipomersen	antihypercholesterolemic	2013	49 (476)
pitavastatin	antihypercholesterolemic	2003	39 (282)
pravastatin	antihypercholesterolemic	1989	25 (316)
ronafibrate	antihypercholesterolemic	1986	22 (326)
rosuvastatin	antihypercholesterolemic	2003	39 (283)
simvastatin	antihypercholesterolemic	1988	24 (311)
alacepril	antihypertensive	1988	24 (296)
alfuzosin hydrochloride	antihypertensive	1988	24 (296)
aliskiren	antihypertensive	2007	43 (461)
amlodipine besylate	antihypertensive	1990	26 (298)
amosulalol	antihypertensive	1988	24 (297)
aranidipine	antihypertensive	1996	32 (306)
arotinolol hydrochloride	antihypertensive	1986	22 (316)
azelnidipine	antihypertensive	2003	39 (270)
azilsartan	antihypertensive	2011	47 (514)
barnidipine hydrochloride	antihypertensive	1992	28 (326)
benazepril hydrochloride	antihypertensive	1990	26 (299)
benidipine hydrochloride	antihypertensive	1991	27 (322)
betaxolol hydrochloride	antihypertensive	1983	19 (315)
bevantolol hydrochloride	antihypertensive	1987	23 (328)
bisoprolol fumarate	antihypertensive	1986	22 (317)
bopindolol	antihypertensive	1985	21 (324)
bosentan	antihypertensive	2001	37 (262)
budralazine	antihypertensive	1983	19 (315)
bunazosin hydrochloride	antihypertensive	1985	21 (324)

GENERIC NAME	INDICATION	YEAR INTRODUCED	ARMC VOL., (PAGE)
cadralazine	antihypertensive	1988	24 (298)
candesartan cilexetil	antihypertensive	1997	33 (330)
captopril	antihypertensive	1982	13 (086)
carvedilol	antihypertensive	1991	27 (323)
celiprolol hydrochloride	antihypertensive	1983	19 (317)
cicletanine	antihypertensive	1988	24 (299)
cilazapril	antihypertensive	1990	26 (301)
cinildipine	antihypertensive	1995	31 (339)
clevidipine	antihypertensive	2008	44 (596)
delapril	antihypertensive	1989	25 (311)
dilevalol	antihypertensive	1989	25 (311)
doxazosin mesylate	antihypertensive	1988	24 (300)
efonidipine	antihypertensive	1994	30 (299)
enalapril maleate	antihypertensive	1984	20 (317)
enalaprilat	antihypertensive	1987	23 (332)
eplerenone	antihypertensive	2003	39 (276)
eprosartan	antihypertensive	1997	33 (333)
felodipine	antihypertensive	1988	24 (302)
fenoldopam mesylate	antihypertensive	1998	34 (322)
fosinopril sodium	antihypertensive	1991	27 (328)
guanadrel sulfate	antihypertensive	1983	19 (319)
imidapril hydrochloride	antihypertensive	1993	29 (339)
irbesartan	antihypertensive	1997	33 (336)
isradipine	antihypertensive	1989	25 (315)
ketanserin	antihypertensive	1985	21 (328)
lacidipine	antihypertensive	1991	27 (330)
lercanidipine	antihypertensive	1997	33 (337)
lisinopril	antihypertensive	1987	23 (337)
losartan	antihypertensive	1994	30 (302)
manidipine hydrochloride	antihypertensive	1990	26 (304)
mebefradil hydrochloride	antihypertensive	1997	33 (338)
moexipril hydrochloride	antihypertensive	1995	31 (346)
moxonidine	antihypertensive	1991	27 (330)
nebivolol	antihypertensive	1997	33 (339)
nilvadipine	antihypertensive	1989	25 (316)
nipradilol	antihypertensive	1988	24 (307)
nisoldipine	antihypertensive	1990	26 (306)
nitrendipine	antihypertensive	1985	21 (331)
olimesartan Medoxomil	antihypertensive	2002	38 (363)
perindopril	antihypertensive	1988	24 (309)
pinacidil	antihypertensive	1987	23 (340)
quinapril	antihypertensive	1989	25 (317)
ramipril	antihypertensive	1989	25 (317)
rilmenidine	antihypertensive	1988	24 (310)
spirapril hydrochloride	antihypertensive	1995	31 (349)

GENERIC NAME	INDICATION	YEAR INTRODUCED	ARMC VOL., (PAGE)
telmisartan	antihypertensive	1999	35 (349)
temocapril	antihypertensive	1994	30 (311)
terazosin hydrochloride	antihypertensive	1984	20 (323)
tertatolol hydrochloride	antihypertensive	1987	23 (344)
tiamenidine hydrochloride	antihypertensive	1988	24 (311)
tilisolol hydrochloride	antihypertensive	1992	28 (337)
trandolapril	antihypertensive	1993	29 (348)
treprostinil sodium	antihypertensive	2002	38 (368)
trimazosin hydrochloride	antihypertensive	1985	21 (333)
valsartan	antihypertensive	1996	32 (320)
zofenopril calcium	antihypertensive	2000	36 (313)
aceclofenac	antiinflammatory	1992	28 (325)
actarit	antiinflammatory	1994	30 (296)
AF-2259	antiinflammatory	1987	23 (325)
alclometasone dipropionate	antiinflammatory	1985	21 (323)
amfenac sodium	antiinflammatory	1986	22 (315)
aminoprofen	antiinflammatory	1990	26 (298)
ampiroxicam	antiinflammatory	1994	30 (296)
amtolmetin guacil	antiinflammatory	1993	29 (332)
betamethasone butyrate propionate	antiinflammatory	1994	30 (297)
bromfenac sodium	antiinflammatory	1997	33 (329)
butibufen	antiinflammatory	1992	28 (327)
butyl flufenamate	antiinflammatory	1983	19 (316)
canakinumab	antiinflammatory	2009	45 (484)
clobenoside	antiinflammatory	1988	24 (300)
deflazacort	antiinflammatory	1986	22 (319)
deprodone propionate	antiinflammatory	1992	28 (329)
dexibuprofen	antiinflammatory	1994	30 (298)
diacerein	antiinflammatory	1985	21 (326)
droxicam	antiinflammatory	1990	26 (302)
etodolac	antiinflammatory	1985	21 (327)
felbinac	antiinflammatory	1986	22 (320)
fisalamine	antiinflammatory	1984	20 (318)
flunoxaprofen	antiinflammatory	1987	23 (335)
fluticasone propionate	antiinflammatory	1990	26 (303)
golimumab	antiinflammatory	2009	45 (503)
halobetasol propionate	antiinflammatory	1991	27 (329)
halometasone	antiinflammatory	1983	19 (320)
hydrocortisone aceponate	antiinflammatory	1988	24 (304)
hydrocortisone butyrate	antiinflammatory	1983	19 (320)
interferon, gamma	antiinflammatory	1989	25 (314)
isofezolac	antiinflammatory	1984	20 (319)
isoxicam	antiinflammatory	1983	19 (320)

GENERIC NAME	INDICATION	YEAR INTRODUCED	ARMC VOL., (PAGE)
lobenzarit sodium	antiinflammatory	1986	22 (322)
lornoxicam	antiinflammatory	1997	33 (337)
loxoprofen sodium	antiinflammatory	1986	22 (322)
lumiracoxib	antiinflammatory	2005	41 (455)
mometasone furoate	antiinflammatory	1987	23 (338)
nabumetone	antiinflammatory	1985	21 (330)
nepafenac	antiinflammatory	2005	41 (456)
nimesulide	antiinflammatory	1985	21 (330)
osalazine sodium	antiinflammatory	1986	22 (324)
oxaprozin	antiinflammatory	1983	19 (322)
piketoprofen	antiinflammatory	1984	20 (322)
pimaprofen	antiinflammatory	1984	20 (322)
piroxicam cinnamate	antiinflammatory	1988	24 (309)
prednicarbate	antiinflammatory	1986	22 (325)
rimexolone	antiinflammatory	1995	31 (348)
sivelestat	antiinflammatory	2002	38 (366)
tenoxicam	antiinflammatory	1987	23 (344)
zaltoprofen	antiinflammatory	1993	29 (349)
arteether	antimalarial	2000	36 (296)
artemisinin	antimalarial	1987	23 (327)
bulaquine	antimalarial	2000	36 (299)
halofantrine	antimalarial	1988	24 (304)
mefloquine hydrochloride	antimalarial	1985	21 (329)
almotriptan	antimigraine	2000	36 (295)
alpiropride	antimigraine	1988	24 (296)
eletriptan	antimigraine	2001	37 (266)
frovatriptan	antimigraine	2002	38 (357)
lomerizine hydrochloride	antimigraine	1999	35 (342)
naratriptan hydrochloride	antimigraine	1997	33 (339)
rizatriptan benzoate	antimigraine	1998	34 (330)
sumatriptan succinate	antimigraine	1991	27 (333)
zomitriptan	antimigraine	1997	33 (345)
cetilistat	antiobesity	2013	49 (454)
dexfenfluramine	antiobesity	1997	33 (332)
lorcaserin hydrochloride	antiobesity	2012	48 (506)
orlistat	antiobesity	1998	34 (327)
rimonabant	antiobesity	2006	42 (537)
sibutramine	antiobesity	1998	34 (331)
atovaquone	antiparasitic	1992	28 (326)
ivermectin	antiparasitic	1987	23 (336)
quinf amideamebicide	antiparasitic	1984	20 (322)
beraprost sodium	antiplatelet	1992	28 (326)
epoprostenol sodium	antiplatelet	1983	19 (318)
iloprost	antiplatelet	1992	28 (332)
prasugrel	antiplatelet	2009	45 (519)

GENERIC NAME	INDICATION	YEAR INTRODUCED	ARMC VOL., (PAGE)
cabergoline	antiprolactin	1993	29 (334)
acitretin	antipsoriasis	1989	25 (309)
alefacept	antipsoriasis	2003	39 (267)
calcipotriol	antipsoriasis	1991	27 (323)
efalizumab	antipsoriasis	2003	39 (274)
tacalcitol	antipsoriasis	1993	29 (346)
tazarotene	antipsoriasis	1997	33 (343)
ustekinumab	antipsoriasis	2009	45 (532)
amisulpride	antipsychotic	1986	22 (316)
aripiprazole	antipsychotic	2002	38 (350)
asenapine	antipsychotic	2009	45 (479)
blonanserin	antipsychotic	2008	44 (587)
clospipramine hydrochloride	antipsychotic	1991	27 (325)
lurasidone hydrochloride	antipsychotic	2010	46 (473)
nemonapride	antipsychotic	1991	27 (331)
olanzapine	antipsychotic	1996	32 (313)
paliperidone	antipsychotic	2007	43 (482)
perospirone hydrochloride	antipsychotic	2001	37 (270)
quetiapine fumarate	antipsychotic	1997	33 (341)
remoxipride hydrochloride	antipsychotic	1990	26 (308)
risperidone	antipsychotic	1993	29 (344)
sertindole	antipsychotic	1996	32 (318)
timiperone	antipsychotic	1984	20 (323)
ziprasidone hydrochloride	antipsychotic	2000	36 (312)
zuclopenthixol acetate	antipsychotic	1987	23 (345)
drotrecogin alfa	antisepsis	2001	37 (265)
cimetropium bromide	antispasmodic	1985	21 (326)
tiquizium bromide	antispasmodic	1984	20 (324)
eltrombopag	antithrombocytopenic	2009	45 (497)
romiplostim	antithrombocytopenic	2008	44 (619)
alteplase	antithrombotic	1987	23 (326)
anagrelide hydrochloride	antithrombotic	1997	33 (328)
apixaban	antithrombotic	2011	47 (509)
APSAC	antithrombotic	1987	23 (326)
argatroban	antithrombotic	1990	26 (299)
bivalirudin	antithrombotic	2000	36 (298)
cilostazol	antithrombotic	1988	24 (299)
clopidogrel hydrogensulfate	antithrombotic	1998	34 (320)
cloricromen	antithrombotic	1991	27 (325)
defibrotide	antithrombotic	1986	22 (319)
edoxaban	antithrombotic	2011	47 (527)
eptilfibatide	antithrombotic	1999	35 (340)

GENERIC NAME	INDICATION	YEAR INTRODUCED	ARMC VOL., (PAGE)
ethyl icosapentate	antithrombotic	1990	26 (303)
fondaparinux sodium	antithrombotic	2002	38 (356)
indobufen	antithrombotic	1984	20 (319)
limaprost	antithrombotic	1988	24 (306)
ozagrel sodium	antithrombotic	1988	24 (308)
picotamide	antithrombotic	1987	23 (340)
reteplase	antithrombotic	1996	32 (316)
sarpogrelate hydrochloride	antithrombotic	1993	29 (344)
ticagrelor	antithrombotic	2010	46 (488)
tirofiban hydrochloride	antithrombotic	1998	34 (332)
levodropropizine	antitussive	1988	24 (305)
nitisinone	antityrosinaemia	2002	38 (361)
benexate hydrochloride	antiulcer	1987	23 (328)
dexlansoprazole	antiulcer	2009	45 (492)
dosmalfate	antiulcer	2000	36 (302)
ebrotidine	antiulcer	1997	33 (333)
ecabet sodium	antiulcer	1993	29 (336)
egualen sodium	antiulcer	2000	36 (303)
enprostil	antiulcer	1985	21 (327)
esomeprazole magnesium	antiulcer	2000	36 (303)
famotidine	antiulcer	1985	21 (327)
irsogladine	antiulcer	1989	25 (315)
lafutidine	antiulcer	2000	36 (307)
lansoprazole	antiulcer	1992	28 (332)
misoprostol	antiulcer	1985	21 (329)
nizatidine	antiulcer	1987	23 (339)
omeprazole	antiulcer	1988	24 (308)
ornoprostil	antiulcer	1987	23 (339)
pantoprazole sodium	antiulcer	1995	30 (306)
plaunotol	antiulcer	1987	23 (340)
polaprezinc	antiulcer	1994	30 (307)
rabeprazole sodium	antiulcer	1998	34 (328)
ranitidine bismuth citrate	antiulcer	1995	31 (348)
rebamipide	antiulcer	1990	26 (308)
rosaprostol	antiulcer	1985	21 (332)
roxatidine acetate hydrochloride	antiulcer	1986	22 (326)
roxithromycin	antiulcer	1987	23 (342)
sofalcone	antiulcer	1984	20 (323)
spizofurone	antiulcer	1987	23 (343)
teprenone	antiulcer	1984	20 (323)
tretinoin tocoferil	antiulcer	1993	29 (348)
troxipide	antiulcer	1986	22 (327)
abacavir sulfate	antiviral	1999	35 (333)
adefovir dipivoxil	antiviral	2002	38 (348)

GENERIC NAME	INDICATION	YEAR INTRODUCED	ARMC VOL., (PAGE)
amprenavir	antiviral	1999	35 (334)
atazanavir	antiviral	2003	39 (269)
boceprevir	antiviral	2011	47 (521)
cidofovir	antiviral	1996	32 (306)
clevudine	antiviral	2007	43 (466)
darunavir	antiviral	2006	42 (515)
delavirdine mesylate	antiviral	1997	33 (331)
didanosine	antiviral	1991	27 (326)
dolutegravir	antiviral	2013	49 (461)
efavirenz	antiviral	1998	34 (321)
elvitegravir	antiviral	2013	49 (465)
emtricitabine	antiviral	2003	39 (274)
enfuvirtide	antiviral	2003	39 (275)
entecavir	antiviral	2005	41 (450)
etravirine	antiviral	2008	44 (602)
famciclovir	antiviral	1994	30 (300)
fomivirsen sodium	antiviral	1998	34 (323)
fosamprenavir	antiviral	2003	39 (277)
foscarnet sodium	antiviral	1989	25 (313)
ganciclovir	antiviral	1988	24 (303)
imiquimod	antiviral	1997	33 (335)
indinavir sulfate	antiviral	1996	32 (310)
influenza virus (live)	antiviral	2003	39 (277)
interferon alfacon-1	antiviral	1997	33 (336)
lamivudine	antiviral	1995	31 (345)
laninamivir octanoate	antiviral	2010	46 (470)
lopinavir	antiviral	2000	36 (310)
maraviroc	antiviral	2007	43 (478)
nelfinavir mesylate	antiviral	1997	33 (340)
nevirapine	antiviral	1996	32 (313)
oseltamivir phosphate	antiviral	1999	35 (346)
penciclovir	antiviral	1996	32 (314)
peramivir	antiviral	2010	46 (477)
propagermanium	antiviral	1994	30 (308)
raltegravir	antiviral	2007	43 (484)
rilpivirine	antiviral	2011	47 (546)
rimantadine hydrochloride	antiviral	1987	23 (342)
ritonavir	antiviral	1996	32 (317)
saquinavir mesvlate	antiviral	1995	31 (349)
simeprevir	antiviral	2013	49 (489)
sofosbuvir	antiviral	2013	49 (492)
sorivudine	antiviral	1993	29 (345)
stavudine	antiviral	1994	30 (311)
telaprevir	antiviral	2011	47 (552)
telbivudine	antiviral	2006	42 (546)

GENERIC NAME	INDICATION	YEAR INTRODUCED	ARMC VOL., (PAGE)
tenofovir disoproxil fumarate	antiviral	2001	37 (271)
tipranavir	antiviral	2005	41 (470)
vaglancirclovir hydrochloride	antiviral	2001	37 (273)
valaciclovir hydrochloride	antiviral	1995	31 (352)
zalcitabine	antiviral	1992	28 (338)
zanamivir	antiviral	1999	35 (352)
zidovudine	antiviral	1987	23 (345)
cobicistat	antiviral pharmacokinetic enhancer)	2013	49 (456)
cevimeline hydrochloride	antixerostomia	2000	36 (299)
alpidem	anxiolytic	1991	27 (322)
buspirone hydrochloride	anxiolytic	1985	21 (324)
cinolazepam	anxiolytic	1993	29 (334)
doxefazepam	anxiolytic	1985	21 (326)
etizolam	anxiolytic	1984	20 (318)
flutazolam	anxiolytic	1984	20 (318)
flutoprazepam	anxiolytic	1986	22 (320)
metaclazepam	anxiolytic	1987	23 (338)
mexazolam	anxiolytic	1984	20 (321)
tandospirone	anxiolytic	1996	32 (319)
atomoxetine	attention deficit hyperactivity disorder	2003	39 (270)
dexmethylphenidate hydrochloride	attention deficit hyperactivity disorder	2002	38 (352)
lisdexamfetamine	attention deficit hyperactivity disorder	2007	43 (477)
dutasteride	benign prostatic hyperplasia	2002	38 (353)
finasteride	benign prostatic hyperplasia	1992	28 (331)
tamsulosin hydrochloride	benign prostatic hyperplasia	1993	29 (347)
fenbuprol	biliary tract dysfunction	1983	19 (318)
clodronate disodium	calcium regulation	1986	22 (319)
gallium nitrate	calcium regulation	1991	27 (328)
neridronic acide	calcium regulation	2002	38 (361)
dexrazoxane	cardioprotective	1992	28 (330)
brovincamine fumarate	cerebral vasodilator	1986	22 (317)
nimodipine	cerebral vasodilator	1985	21 (330)
propentofylline propionate	cerebral vasodilator	1988	24 (310)
indacaterol	chronic obstructive pulmonary disease	2009	45 (505)
aclidinium bromide	chronic obstructive pulmonary disorder	2012	48 (481)
formoterol fumarate	chronic obstructive pulmonary disorder	1986	22 (321)

GENERIC NAME	INDICATION	YEAR INTRODUCED	ARMC VOL., (PAGE)
olodaterol	chronic obstructive pulmonary disorder	2013	49 (480)
roflumilast	chronic obstructive pulmonary disorder	2010	46 (480)
tiotropium bromide	chronic obstructive pulmonary disorder	2002	38 (368)
aniracetam	cognition enhancer	1993	29 (333)
choline alfoscerate	cognition enhancer	1990	26 (300)
exifone	cognition enhancer	1988	24 (302)
levacecarnine hydrochloride	cognition enhancer	1986	22 (322)
oxiracetam	cognition enhancer	1987	23 (339)
pramiracetam sulfate	cognition enhancer	1993	29 (343)
amrinone	congestive heart failure	1983	19 (314)
bucladesine sodium	congestive heart failure	1984	20 (316)
carperitide	congestive heart failure	1995	31 (339)
colforsin daropate hydrochloride	congestive heart failure	1999	35 (337)
denopamine	congestive heart failure	1988	24 (300)
docarpamine	congestive heart failure	1994	30 (298)
dopexamine	congestive heart failure	1989	25 (312)
enoximone	congestive heart failure	1988	24 (301)
flosequinan	congestive heart failure	1992	28 (331)
ibopamine hydrochloride	congestive heart failure	1984	20 (319)
levosimendan	congestive heart failure	2000	36 (308)
loprinone hydrochloride	congestive heart failure	1996	32 (312)
milrinone	congestive heart failure	1989	25 (316)
nesiritide	congestive heart failure	2001	37 (269)
pimobendan	congestive heart failure	1994	30 (307)
vesnarinone	congestive heart failure	1990	26 (310)
xamoterol fumarate	congestive heart failure	1988	24 (312)
lulbiprostone	constipation	2006	42 (525)
methylnaltrexone bromide	constipation	2008	44 (612)
centchroman	contraception	1991	27 (324)
drospirenone	contraception	2000	36 (302)
gestodene	contraception	1987	23 (335)
gestrinone	contraception	1986	22 (321)
nomegestrol acetate	contraception	1986	22 (324)
norelgestromin	contraception	2002	38 (362)
norgestimate	contraception	1986	22 (324)
promegestrone	contraception	1983	19 (323)
trimegestone	contraception	2001	37 (273)
ulipristal acetate	contraception	2009	45 (530)
biolimus drug-eluting stent	coronary artery disease, antirestenotic	2008	44 (586)

GENERIC NAME	INDICATION	YEAR INTRODUCED	ARMC VOL., (PAGE)
pasireotide	Cushing's Disease	2012	48 (512)
dornase alfa	cystic fibrosis	1994	30 (298)
ivacaftor	cystic fibrosis	2012	48 (501)
neltenexine	cystic fibrosis	1993	29 (341)
amifostine	cytoprotective	1995	31 (338)
kinetin	dermatologic, skin photodamage	1999	35 (341)
gadoversetamide	diagnostic	2000	36 (304)
ioflupane	diagnostic	2000	36 (306)
azosemide	diuretic	1986	22 (316)
muzolimine	diuretic	1983	19 (321)
torasemide	diuretic	1993	29 (348)
ospemifene	dyspareunia	2013	49 (482)
acotiamide	dyspepsia	2013	49 (447)
alpha-1 antitrypsin	emphysema	1988	24 (297)
erdosteine	expectorant	1995	31 (342)
fudosteine	expectorant	2001	37 (267)
telmesteine	expectorant	1992	28 (337)
agalsidase alfa	Fabry's disease	2001	37 (259)
chenodiol	gallstones	1983	19 (317)
cinitapride	gastroprokinetic	1990	26 (301)
cisapride	gastroprokinetic	1988	24 (299)
itopride hydrochloride	gastroprokinetic	1995	31 (344)
mosapride citrate	gastroprokinetic	1998	34 (326)
alglucerase	Gaucher's disease	1991	27 (321)
imiglucerase	Gaucher's disease	1994	30 (301)
miglustat	Gaucher's disease	2003	39 (279)
rilonacept	genetic autoinflammatory syndromes	2008	44 (615)
febuxostat	gout	2009	45 (501)
lanreotide acetate	growth disorders	1995	31 (345)
octreotide	growth disorders	1988	24 (307)
pegvisomant	growth disorders	2003	39 (281)
somatomedin-1	growth disorders	1994	30 (310)
somatotropin	growth disorders	1994	30 (310)
somatropin	growth disorders	1987	23 (343)
factor VIIa	haemophilia	1996	32 (307)
erythropoietin	hematopoietic	1988	24 (301)
peginesatide acetate	hematopoietic	2012	48 (514)
eculizumab	hemoglobinuria	2007	43 (468)
factor VIII	hemostatic	1992	28 (330)
thrombin alfa	hemostatic	2008	44 (627)
malotilate	hepatoprotective	1985	21 (329)
mivotilate	hepatoprotective	1999	35 (343)
buserelin acetate	hormone therapy	1984	20 (316)
goserelin	hormone therapy	1987	23 (336)

GENERIC NAME	INDICATION	YEAR INTRODUCED	ARMC VOL., (PAGE)
leuprolide acetate	hormone therapy	1984	20 (319)
nafarelin acetate	hormone therapy	1990	26 (306)
tibolone	hormone therapy	1988	24 (312)
cinacalcet	hyperparathyroidism	2004	40 (451)
doxercalciferol	hyperparathyroidism	1999	35 (339)
falecalcitriol	hyperparathyroidism	2001	37 (266)
maxacalcitol	hyperparathyroidism	2000	36 (310)
paricalcitol	hyperparathyroidism	1998	34 (327)
quinagolide	hyperprolactinemia	1994	30 (309)
conivaptan	hyponatremia	2006	42 (514)
mozavaptan	hyponatremia	2006	42 (527)
tolvaptan	hyponatremia	2009	45 (528)
bucillamine	immunomodulator	1987	23 (329)
centoxin	immunomodulator	1991	27 (325)
thymopentin	immunomodulator	1985	21 (333)
filgrastim	immunostimulant	1991	27 (327)
GMDP	immunostimulant	1996	32 (308)
interferon gamma-1b	immunostimulant	1991	27 (329)
lentinan	immunostimulant	1986	22 (322)
pegademase bovine	immunostimulant	1990	26 (307)
pidotimod	immunostimulant	1993	29 (343)
romurtide	immunostimulant	1991	27 (332)
sargramostim	immunostimulant	1991	27 (332)
schizophyllan	immunostimulant	1985	22 (326)
ubenimex	immunostimulant	1987	23 (345)
belatacept	immunosuppressant	2011	47 (516)
cyclosporine	immunosuppressant	1983	19 (317)
everolimus	immunosuppressant	2004	40 (455)
gusperimus	immunosuppressant	1994	30 (300)
mizoribine	immunosuppressant	1984	20 (321)
muromonab-CD3	immunosuppressant	1986	22 (323)
mycophenolate mofetil	immunosuppressant	1995	31 (346)
mycophenolate sodium	immunosuppressant	2003	39 (279)
pimecrolimus	immunosuppressant	2002	38 (365)
tacrolimus	immunosuppressant	1993	29 (347)
cetrorelix	infertility	1999	35 (336)
corifollitropin alfa	infertility	2010	46 (455)
follitropin alfa	infertility	1996	32 (307)
follitropin beta	infertility	1996	32 (308)
ganirelix acetate	infertility	2000	36 (305)
defeiprone	iron chelation therapy	1995	31 (340)
deferasirox	iron chelation therapy	2005	41 (446)
alosetron hydrochloride	irritable bowel syndrome	2000	36 (295)
certolizumab pegol	irritable bowel syndrome	2008	44 (592)
linaclotide	irritable bowel syndrome	2012	48 (502)

GENERIC NAME	INDICATION	YEAR INTRODUCED	ARMC VOL., (PAGE)
tegaserod maleate	irritable bowel syndrome	2001	37 (270)
nartograstim	leukopenia	1994	30 (304)
metreleptin	lipodystrophy	2013	49 (474)
tesamorelin acetate	lipodystrophy	2010	46 (486)
belimumab	lupus	2011	47 (519)
Lyme disease vaccine	Lyme disease	1999	35 (342)
avanafil	male sexual dysfunction	2011	47 (512)
sildenafil citrate	male sexual dysfunction	1998	34 (331)
tadalafil	male sexual dysfunction	2003	39 (284)
udenafil	male sexual dysfunction	2005	41 (472)
vardenafil	male sexual dysfunction	2003	39 (286)
laronidase	mucopolysaccharidosis I	2003	39 (278)
idursulfase	mucopolysaccharidosis II (Hunter syndrome)	2006	42 (520)
galsulfase	mucopolysaccharidosis VI	2005	41 (453)
palifermin	mucositis	2005	41 (461)
dalfampridine	multiple sclerosis	2010	46 (458)
dimethyl fumarate	multiple sclerosis	2013	49 (460)
fingolimod	multiple sclerosis	2010	46 (468)
glatiramer acetate	multiple sclerosis	1997	33 (334)
interferon, b-1a	multiple sclerosis	1996	32 (311)
interferon, b-1b	multiple sclerosis	1993	29 (339)
natalizumab	multiple sclerosis	2004	40 (462)
teriflunomide	multiple sclerosis	2012	48 (530)
afloqualone	muscle relaxant	1983	19 (313)
cisatracurium besilate	muscle relaxant	1995	31 (340)
doxacurium chloride	muscle relaxant	1991	27 (326)
eperisone hydrochloride	muscle relaxant	1983	19 (318)
mivacurium chloride	muscle relaxant	1992	28 (334)
rapacuronium bromide	muscle relaxant	1999	35 (347)
rocuronium bromide	muscle relaxant	1994	30 (309)
tiropramide hydrochloride	muscle relaxant	1983	19 (324)
tizanidine	muscle relaxant	1984	20 (324)
decitabine	myelodysplastic syndromes	2006	42 (519)
lenalidomide	myelodysplastic syndromes, multiple myeloma	2006	42 (523)
tinazoline	nasal decongestant	1988	24 (312)
tafamidis	neurodegeneration	2011	47 (550)
taltirelin	neurodegeneration	2000	36 (311)
sugammadex	neuromuscular blockade, reversal	2008	44 (623)
edaravone	neuroprotective	2001	37 (265)
bifemelane hydrochloride	nootropic	1987	23 (329)
idebenone	nootropic	1986	22 (321)

GENERIC NAME	INDICATION	YEAR INTRODUCED	ARMC VOL., (PAGE)
indeloxazine hydrochloride	nootropic	1988	24 (304)
nizofenzone	nootropic	1988	24 (307)
alcaftadine	ophthalmologic (allergic conjunctivitis)	2010	46 (444)
diquafosol tetrasodium	ophthalmologic (dry eye)	2010	46 (462)
pegaptanib	ophthalmologic (macular degeneration)	2005	41 (458)
ranibizumab	ophthalmologic (macular degeneration)	2006	42 (534)
verteporfin	ophthalmologic (macular degeneration)	2000	36 (312)
aflibercept	ophthalmologic, macular degeneration	2011	47 (507)
OP-1	osteoinductor	2001	37 (269)
alendronate sodium	osteoporosis	1993	29 (332)
APD	osteoporosis	1987	23 (326)
denosumab	osteoporosis	2010	46 (459)
disodium pamidronate	osteoporosis	1989	25 (312)
eldecalcitol	osteoporosis	2011	47 (529)
ibandronic acid	osteoporosis	1996	32 (309)
incadronic acid	osteoporosis	1997	33 (335)
ipriflavone	osteoporosis	1989	25 (314)
minodronic acid	osteoporosis	2009	45 (509)
raloxifene hydrochloride	osteoporosis	1998	34 (328)
risedronate sodium	osteoporosis	1998	34 (330)
strontium ranelate	osteoporosis	2004	40 (466)
zoledronate disodium	osteoporosis	2000	36 (314)
tiludronate disodium	Paget's disease	1995	31 (350)
nafamostat mesylate	pancreatitis	1986	22 (323)
budipine	Parkinson's disease	1997	33 (330)
CHF-1301	Parkinson's disease	1999	35 (336)
droxidopa	Parkinson's disease	1989	25 (312)
entacapone	Parkinson's disease	1998	34 (322)
istradefylline	Parkinson's disease	2013	49 (468)
pergolide mesylate	Parkinson's disease	1988	24 (308)
pramipexole hydrochloride	Parkinson's disease	1997	33 (341)
rasagiline	Parkinson's disease	2005	41 (464)
ropinirole hydrochloride	Parkinson's disease	1996	32 (317)
rotigotine	Parkinson's disease	2006	42 (538)
talipexole	Parkinson's disease	1996	32 (318)
tolcapone	Parkinson's disease	1997	33 (343)
sapropterin hydrochloride	phenylketouria	1992	28 (336)
alglucosidase alfa	Pompe disease	2006	42 (511)

GENERIC NAME	INDICATION	YEAR INTRODUCED	ARMC VOL., (PAGE)
alvimopan	post-operative ileus	2008	44 (584)
histrelin	precocious puberty	1993	29 (338)
dapoxetine	premature ejaculation	2009	45 (488)
atosiban	premature labor	2000	36 (297)
nalfurafine hydrochloride	pruritus	2009	45 (510)
pirfenidone	pulmonary fibrosis, idiopathic	2008	44 (614)
ambrisentan	pulmonary hypertension	2007	43 (463)
macitentan	pulmonary hypertension	2013	49 (472)
riociguat	pulmonary hypertension	2013	49 (486)
sitaxsentan	pulmonary hypertension	2006	42 (543)
pumactant	respiratory distress syndrome	1994	30 (308)
mepixanox	respiratory stimulant	1984	20 (320)
surfactant TA	respiratory surfactant	1987	23 (344)
gabapentin Enacarbil	restless leg syndrome	2011	47 (533)
teduglutide	short bowel syndrome	2012	48 (526)
adrafinil	sleep disorders	1986	22 (315)
armodafinil	sleep disorders	2009	45 (478)
binfonazole	sleep disorders	1983	19 (315)
brotizolam	sleep disorders	1983	19 (315)
butoctamide	sleep disorders	1984	20 (316)
dexmedetomidine hydrochloride	sleep disorders	2000	36 (301)
eszopiclone	sleep disorders	2005	41 (451)
loprazolam mesylate	sleep disorders	1983	19 (321)
modafinil	sleep disorders	1994	30 (303)
quazepam	sleep disorders	1985	21 (332)
ramelteon	sleep disorders	2005	41 (462)
rilmazafone	sleep disorders	1989	25 (317)
zaleplon	sleep disorders	1999	35 (351)
zolpidem hemitartrate	sleep disorders	1988	24 (313)
zopiclone	sleep disorders	1986	22 (327)
plerixafor hydrochloride	stem cell mobilizer	2009	45 (515)
tirilazad mesylate	subarachnoid hemorrhage	1995	31 (351)
balsalazide disodium	ulcerative colitis	1997	33 (329)
acetohydroxamic acid	urinary tract/bladder disorders	1983	19 (313)
darifenacin	urinary tract/bladder disorders	2005	41 (445)
fesoterodine	urinary tract/bladder disorders	2008	44 (604)
imidafenacin	urinary tract/bladder disorders	2007	43 (472)
mirabegron	urinary tract/bladder disorders	2011	47 (542)
naftopidil	urinary tract/bladder disorders	1999	35 (344)
propiverine hydrochloride	urinary tract/bladder disorders	1992	28 (335)
silodosin	urinary tract/bladder disorders	2006	42 (540)
sodium cellulose phosphate	urinary tract/bladder disorders	1983	19 (323)
solifenacin	urinary tract/bladder disorders	2004	40 (466)

GENERIC NAME	INDICATION	YEAR INTRODUCED	ARMC VOL., (PAGE)
tiopronin	urolithiasis	1989	25 (318)
acemannan	wound healing agent	2001	37 (259)
cadexomer iodine	wound healing agent	1983	19 (316)
epidermal growth factor	wound healing agent	1987	23 (333)
prezatide copper acetate	wound healing agent	1996	32 (314)

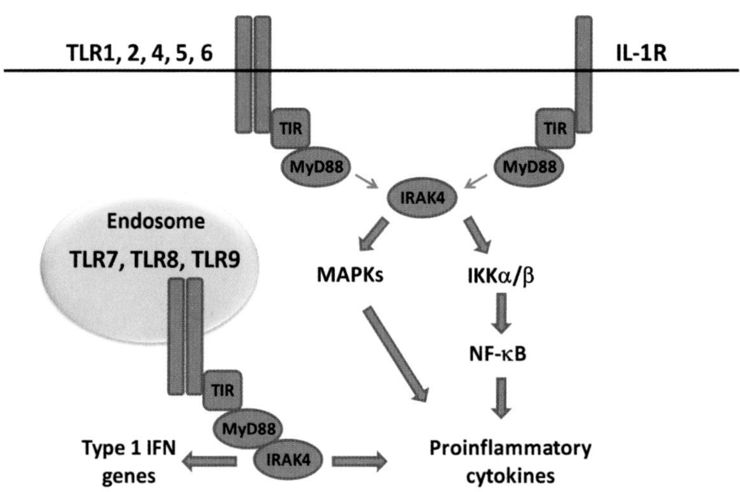

Plate 9.1 Schematic of IRAK4 signaling.

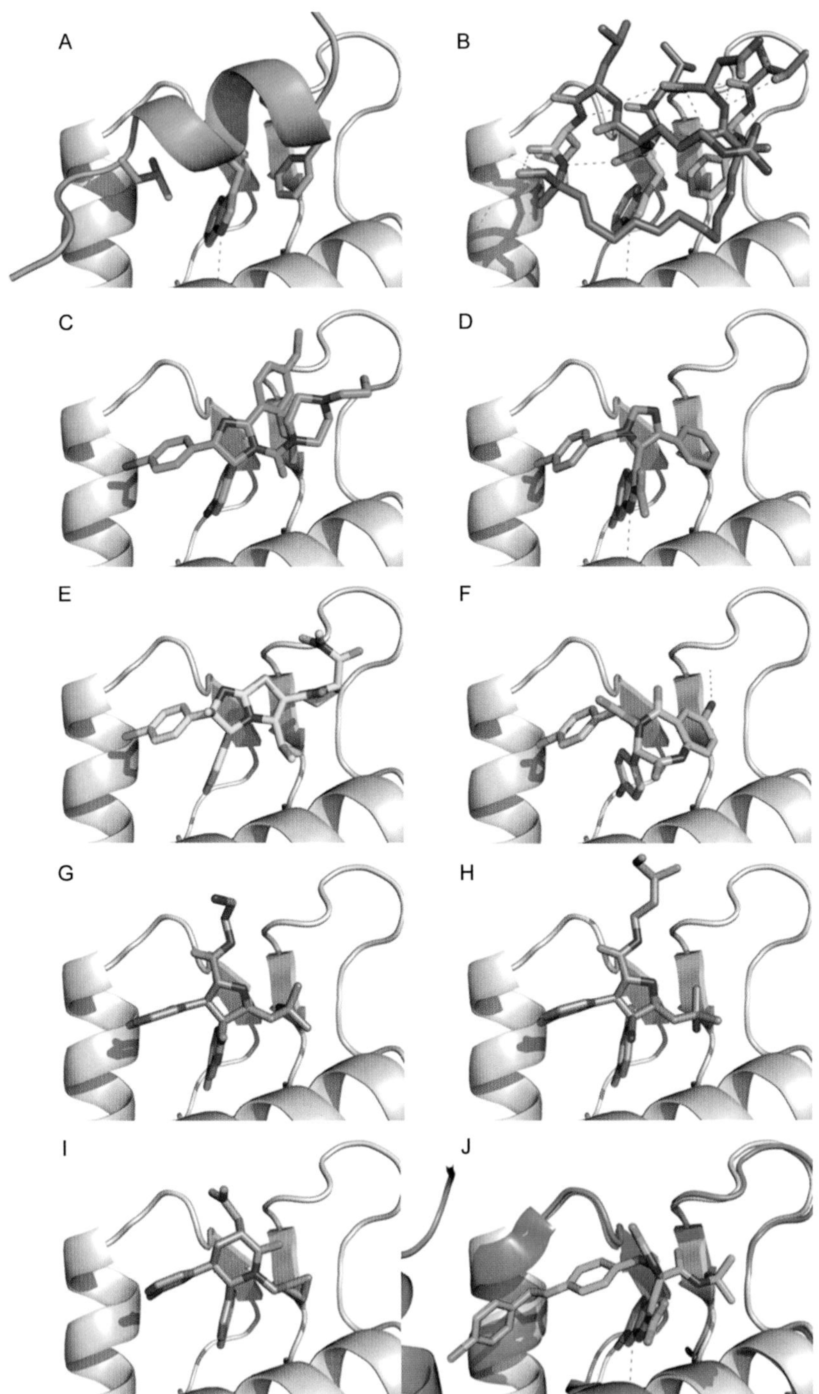

Plate 12.15 See legend on the opposite page.

Plate 12.15 Alignment of different chemotypes cocrystallized with MDM2 superimposed onto PDB: 1YCR. The majority of X-ray structures show only side-chain rotations and small movements of the receptor α-helices. (A) p53 peptide (cyan α-helix with anchoring amino acids $F^{19}W^{23}L^{26}$ rendered as sticks) in the MDM2 binding pocket (gray secondary structure, PDB: 1YCR). W^{23} of p53 forms a hydrogen bond to MDM2's L^{54}. (B) Stapled peptide **42** closely mimics the p53 peptide's anchoring amino acids (pink sticks) and backbone and has an additional hydrocarbon staple (blue stick, PDB: 3V3B) which exhibits multiple contacts to the MDM2 receptor. (C) Nutlin derivative **1** (green sticks) with two hydrophobic *p*-halophenyl substituents occupying the W^{23} and L^{26} pockets and an *o*-ethoxy phenyl moiety targeting the F^{19} pocket (PDB: 1RV1). (D) Tri-substituted indoloimidazole **4** exhibiting an indole substituent closely mimicking p53's W^{23} and two phenyl groups occupying the L^{26} and F^{19} pockets. (PDB: 3LBK). (E) Imidazothiazole **9** exhibiting two hydrophobic *p*-halophenyl substituents occupying the W^{23} and L^{26} pockets and an *iso*-butyric side chain targeting the F^{196} pocket (PDB: 3VZV). (F) Benzodiazepindione **11** with two hydrophobic *p*-halophenyl substituents occupying the W^{23} and L^{26} pockets and a 7-I benzene forming a halogen bond to the backbone carbonyl of MDM2's Q^{72} (PDB: 1T4E). (G) Spirocyclic oxindole **21** featuring an indole mimicking p53's W^{23}, a hydrophobic *p*-halobenzene substituent sitting in the L^{26} pocket, and a bulky alkyl chain introduced F^{19} pocket (PDB: 4JVR). (H) Cyanopyrrolidine **31** featuring two hydrophobic phenyl groups in the W^{23} and L^{26} pockets, a bulky alkyl chain in the F19 pocket, and a solubilizing diol-side chain (PDB: 4JSC). (I) Piperidinone **35** featuring two hydrophobic phenyl groups and a cyclopropyl methyl alkyl chain targeting the $L^{26}W^{23}F^{19}$ pockets, respectively (PDB: 1RV1). (J) Indole-2-carboxylic acid **27** showing an unusual binding mode inducing a never before seen deep binding pocket by ordering the otherwise disordered N-terminus of MDM2 (blue α-helix at the left side). The phenyl group of the extended benzylphenol can be regarded as a fourth pharmacophore element. The indole moiety mimics the W^{23} and the *tert*-butyl group occupies the F^{19} pocket (PDB: 4MDN).

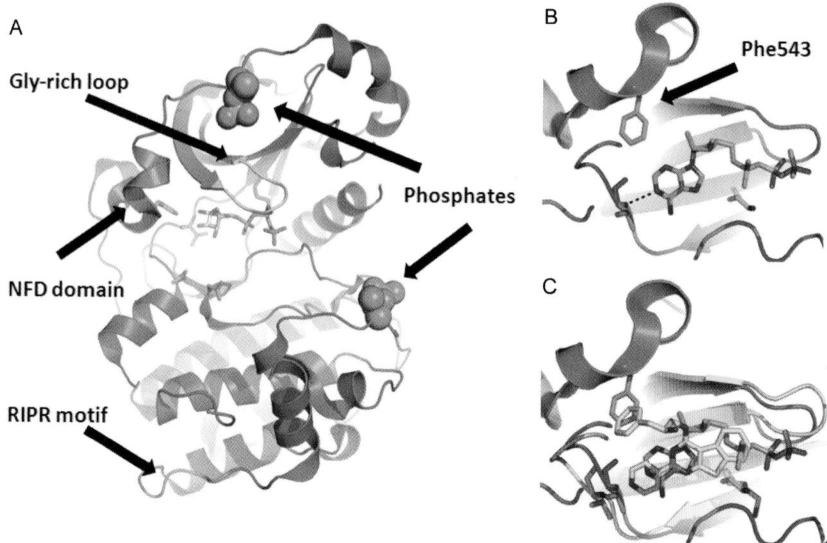

Plate 13.1 Structures of kinase domain (KD)-PKCι. (A) Structure of ATP bound to kinase domain PKCι, highlighting the gly-rich loop (yellow), NFD domain (green), RIPR motif (cyan) and the priming phosphorylation sites (space fill). (B) Structure of ATP bound to the nucleotide-binding cleft, key hydrogen bonding interactions represented by dashed lines along with Phe543 in the NFD motif. (C) Superposition of CRT0066854 (yellow) on the ATP-bound enzyme structure (magenta). Upon binding of CRT0066854 to the nucleotide-binding cleft, F543 is ejected from the active site caused by a steric clash with the ligand phenyl ring moiety (yellow). *Figures adapted from research originally published in* The Biochemical Journal, *see ref. 21. © The Biochemical Society.*

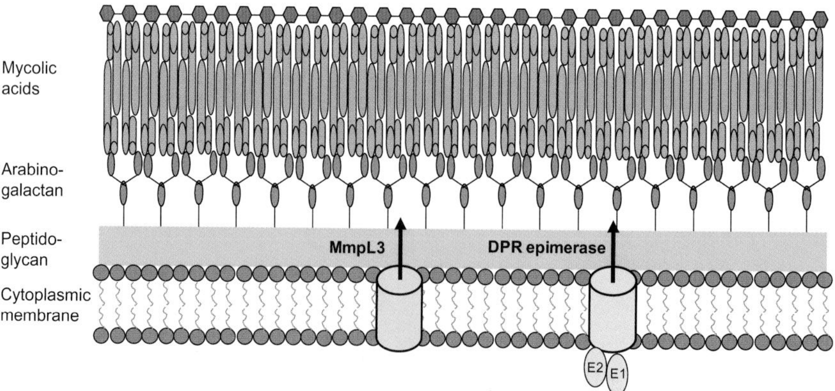

Plate 14.1 Schematic showing the cell wall of *Mycobacterium tuberculosis.*[8]

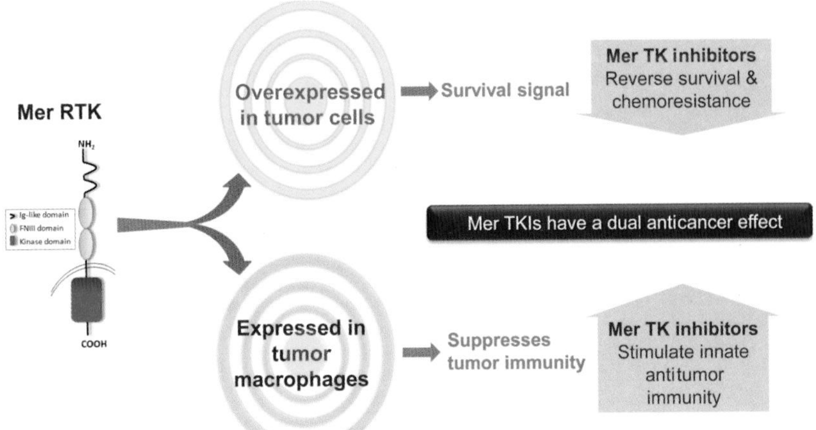

Plate 19.1 Mer RTK is a dual target in cancer.

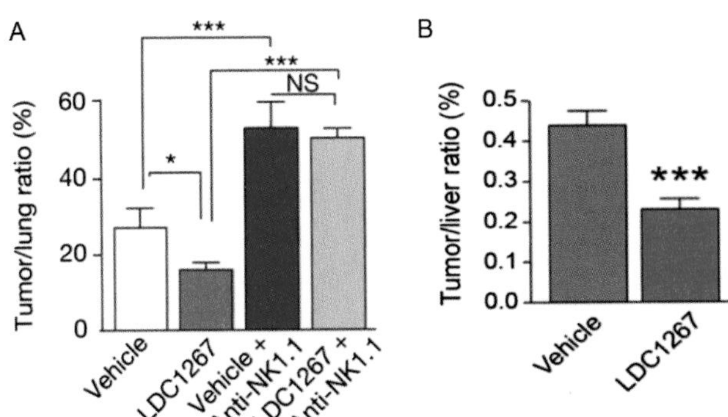

Plate 19.4 Control of metastases by TAM receptor inhibition. (A) Tumor-to-lung ratios in vehicle- and LDC1267-treated wild-type control or NK1.1-depleted mice. $n = 8$ each. $*P < 0.05$, and $***P < 0.001$; NS, not significant. Data are mean values ± s.e.m. (B) Relative sizes of 4T1 liver micrometastases in syngeneic mice treated with vehicle or LDC1267 (100 mg/kg) via oral gavage. Mean values ± s.e.m. are shown on day +21 after initiation of LDC1267 therapy (day +27 after orthotopic tumor inoculation into the mammary fat pad). $***P < 0.001$ (Student's *t*-test, $n = 10$ mice each). *Used with permission from Nature* ***2014****, 507, 508–512.*

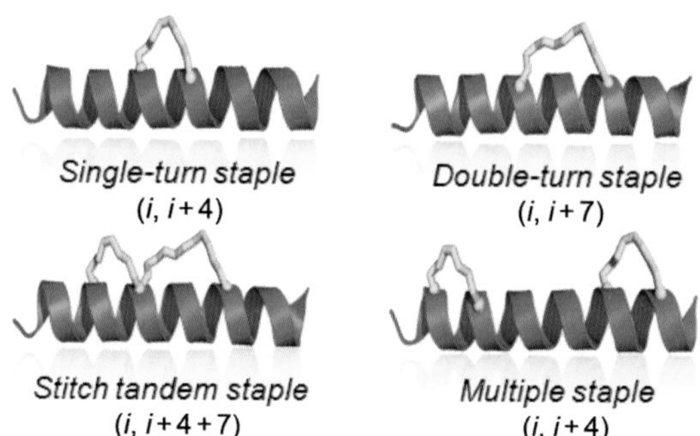

Plate 21.1 Stabilized α-helical peptides a.k.a. SPs.

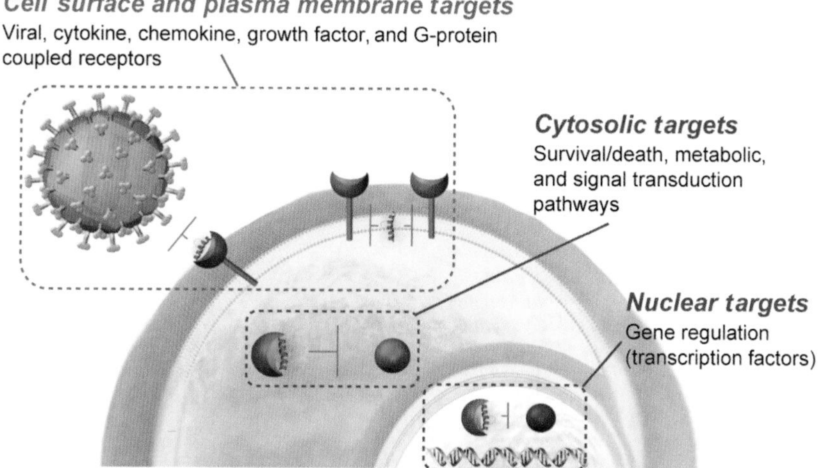

Plate 21.2 Examples of therapeutic targets and pathways for SPs. *From Ref. 9 with permission of publisher.*

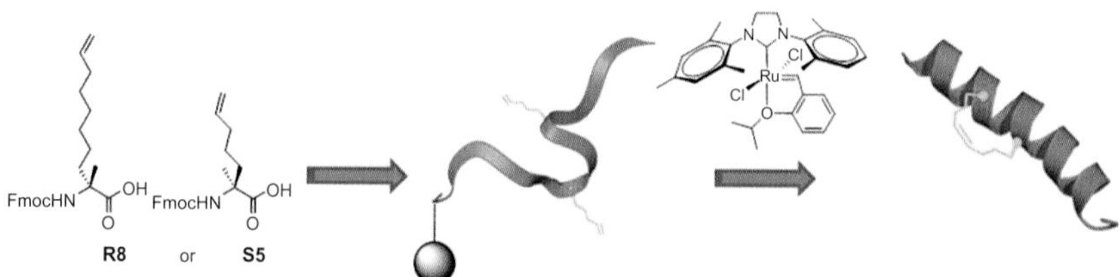

Plate 21.3 Synthetic route for preparing SPs.

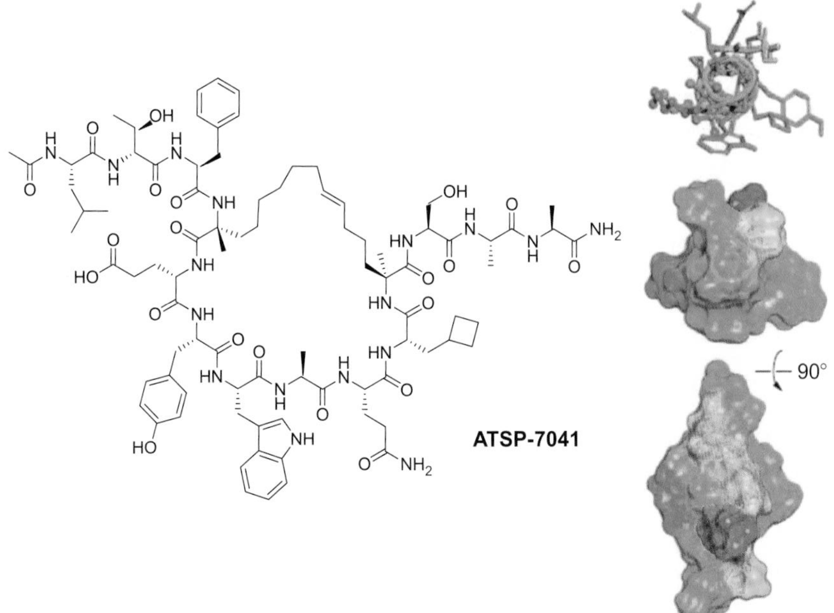

Plate 21.4 2D and 3D structural representations of ATSP-7041. *From Ref. 27 with permission of publisher.*

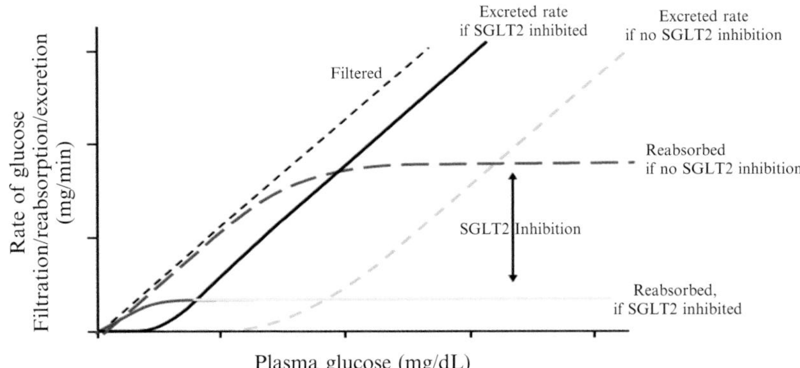

Plate 23.1 Renal processing of glucose: dependence of rates of filtration, recovery, and excretion on plasma glucose concentration in absence and presence of an SGLT2 inhibitor.

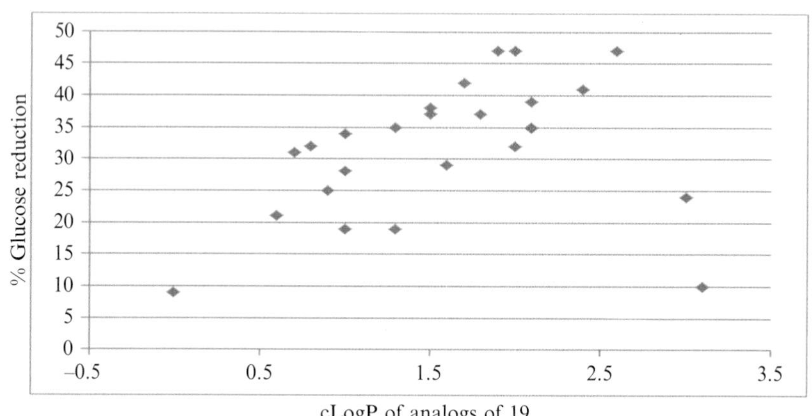

Plate 23.6 Correlation of cLogP with % reduction of blood glucose after correction for the vehicle response at 5 h after oral administration of 0.1 mg/kg dose of SGLT2 inhibitors **19** to food-restricted diabetic STZ rats.

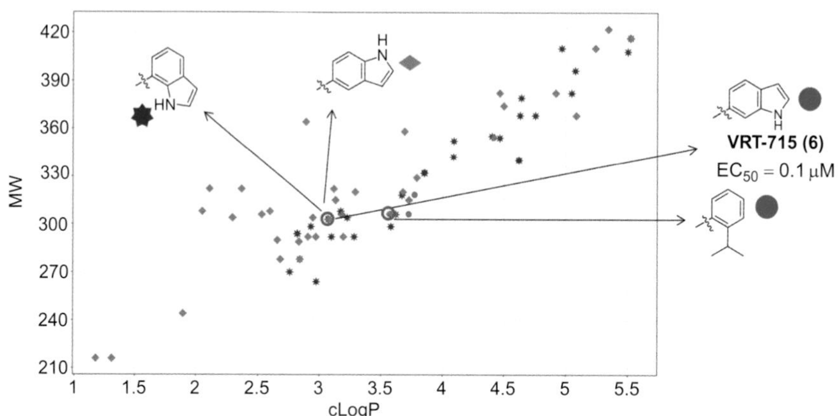

Plate 24.3 Amide exploration around **5**.

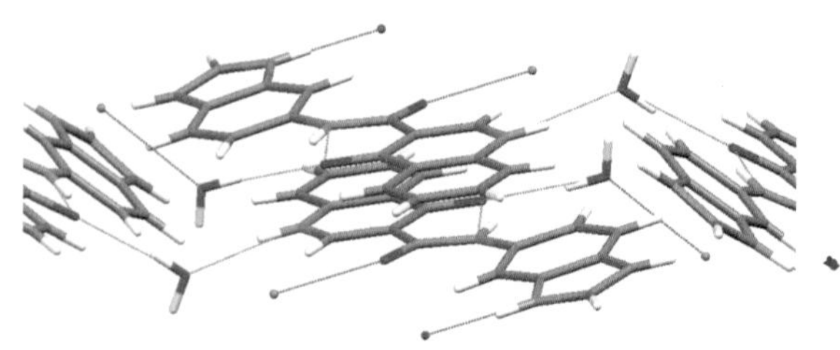

Plate 24.4 X-ray structure of **6**.